LIST OF THE ELEMENTS

Name	Symbol	Atomic Number	Atomic Weight	Name	Symbol	Atomic Number	Atomic Weight
Actinium	Ac	89	227.028	Mercury	Hg	80	200.59
Aluminum	Al	13	26.982	Molybdenum	Mo	42	95.94
Americium	Am	95	243	Neodymium	Nd	60	144.24
Antimony	Sb	51	121.76	Neon	Ne	10	20.180
Argon	Ar	18	39.948	Neptunium	Np	93	237.048
Arsenic	As	33	74.922	Nickel	Ni	28	58.69
Astatine	At	85	210	Niobium	Nb	41	92.906
Barium	Ba	56	137.327	Nitrogen	N	7	14.007
Berkelium	Bk	97	247	Nobelium	No	102	259
Beryllium	Be	4	9.012	Osmium	Os	76	190.23
Bismuth	Bi	83	208.980	Oxygen	O	8	15.999
Bohrium	Bh	107	262	Palladium	Pd	46	106.42
Boron	B	5	10.811	Phosphorus	P	15	30.974
Bromine	Br	35	79.904	Platinum	Pt	78	195.08
Cadmium	Cd	48	112.411	Plutonium	Pu	94	244
Calcium	Ca	20	40.078	Polonium	Po	84	209
Californium	Cf	98	251	Potassium	K	19	39.098
Carbon	C	6	12.011	Praseodymium	Pr	59	140.908
Cerium	Ce	58	140.115	Promethium	Pm	61	145
Cesium	Cs	55	132.905	Protactinium	Pa	91	231.036
Chlorine	Cl	17	35.453	Radium	Ra	88	226.025
Chromium	Cr	24	51.996	Radon	Rn	86	222
Cobalt	Co	27	58.933	Rhenium	Re	75	186.207
Copper	Cu	29	63.546	Rhodium	Rh	45	102.906
Curium	Cm	96	247	Rubidium	Rb	37	85.468
Dubnium	Db	105	262	Ruthenium	Ru	44	101.07
Dysprosium	Dy	66	162.5	Rutherfordium	Rf	104	261
Einsteinium	Es	99	252	Samarium	Sm	62	150.36
Erbium	Er	68	167.26	Scandium	Sc	21	44.956
Europium	Eu	63	151.964	Seaborgium	Sg	106	263
Fermium	Fm	100	257	Selenium	Se	34	78.96
Fluorine	F	9	18.998	Silicon	Si	14	28.086
Francium	Fr	87	223	Silver	Ag	47	107.868
Gadolinium	Gd	64	157.25	Sodium	Na	11	22.990
Gallium	Ga	31	69.723	Strontium	Sr	38	87.62
Germanium	Ge	32	72.61	Sulfur	S	16	32.066
Gold	Au	79	196.967	Tantalum	Ta	73	180.948
Hafnium	Hf	72	178.49	Technetium	Tc	43	98
Hassium	Hs	108	265	Tellurium	Te	52	127.60
Helium	He	2	4.003	Terbium	Tb	65	158.925
Holmium	Ho	67	164.93	Thallium	Tl	81	204.383
Hydrogen	H	1	1.0079	Thorium	Th	90	232.038
Indium	In	49	114.82	Thulium	Tm	69	168.934
Iodine	I	53	126.905	Tin	Sn	50	118.71
Iridium	Ir	77	192.22	Titanium	Ti	22	47.88
Iron	Fe	26	55.845	Tungsten	W	74	183.84
Krypton	Kr	36	83.8	Uranium	U	92	238.029
Lanthanum	La	57	138.906	Vanadium	V	23	50.942
Lawrencium	Lr	103	262	Xenon	Xe	54	131.29
Lead	Pb	82	207.2	Ytterbium	Yb	70	173.04
Lithium	Li	3	6.941	Yttrium	Y	39	88.906
Lutetium	Lu	71	174.967	Zinc	Zn	30	65.39
Magnesium	Mg	12	24.305	Zirconium	Zr	40	91.224
Manganese	Mn	25	54.938	—	Uun	110	269
Meitnerium	Mt	109	266	—	Uuu	111	272
Mendelevium	Md	101	258	—	Uub	112	277

Introductory Chemistry

A Conceptual Focus

INTRODUCTORY CHEMISTRY

A Conceptual Focus

STEVE RUSSO
Cornell University

MIKE SILVER
Hope College

An imprint of Addison Wesley Longman, Inc.

San Francisco • Reading, Massachusetts • New York • Harlow, England
Don Mills, Ontario • Sydney • Mexico City • Madrid • Amsterdam

Editor:	Ben Roberts
Senior Developmental Editor:	Margot Otway
Developmental Editors:	Becky Strehlow, Susan Weisberg
Marketing Manager:	Jennifer Schmidt
Director of Marketing:	Stacy Treco
Associate Project Editor:	Claudia Herman
Producer:	Claire Masson
Publishing Assistants:	Tony Asaro, Rani Lynds
Production Coordination:	Joan Marsh
Production Managment:	Phyllis Niklas
Composition:	Dovetail Publishing Services
Art Director:	Emiko-Rose Koike
Cover Image & Chapter Openers:	Quade Paul, fiVth.com
Text and Cover Designer:	Mark Ong, Side-by-Side Studios
Artists:	Bert Dodson, Blakeley Kim, Ken Probst
Photographer & Photo Research:	Cary Groner
Manufacturing:	Holly Bisso, Virginia Pierce
Prepress House:	H&S Graphics
Cover Printer:	Coral Graphics
Printer and Binder:	Von Hoffmann Press

About the Artists:

Bert Dodson has illustrated over sixty books. He is the author of *Keys to Drawing*, published in eight languages. His distinctive, whimsical cartoons can also be found in *The Way Life Works*, of which he is co-author. Bert lives in Bradford, Vermont.

Quade Paul is a scientific and medical illustrator. He combined digital illustration and 3D modeling techniques to create the stunning cover image and chapter openers in this book. All of the molecular imagery in his artwork is based on actual data. Quade's work also includes editorial illustration, scientific animation, and multimedia projects. He can be contacted at fiVth.com.

Photos courtesy of:

Corel: pp. 85, 286; Houston Astrodome: p. 71; Photodisc: pp. 2, 5, 7, 150, 288, 315, 316, 490, 533, 546; Photo Researcher's Inc.: pp. 95, 97, 117, 164, 211, 212, 356, 383, 546, 602; Prentice-Hall: pp. 5, 6, 78, 79, 210, 315; NASA: pp. 3, 23, 94, 510; National Library of Medicine: p. 1; Tony Stone: pp. 1, 85; Models: p. 182 Coffee cup courtesy Direct Imagination™, p. 380 Lysozyme based on file 1lmq.pdb, p. 588 Hemoglobin based on file 1hbb.pdb.

Library of Congress Catalog Card Number: 99-76686

ISBN: 0-321-01525-8
1 2 3 4 5 6 7 8 9 10—VHP—03 02 01 00

Contents

Preface to the Student

This book was inspired by our desire to give you a permanent and confident grasp of the fundamental concepts of chemistry. Many years after this course, we hope you will still remember the fundamental concepts and be able to apply them to new situations as they arise. To accomplish this goal we have tried to elevate the basic concepts above the flood of information and problem-solving that are required in an introductory chemistry course.

We have not attempted to accomplish our goal by lowering the level of the book, by reducing the number of topics covered, or by shying away from mathematical problem-solving. Quite the opposite. A quick look at the table of contents will reveal that we cover the full range of topics. Indeed, we cover some topics to a greater extent than is traditional. For example, there is a detailed look at the reversibility of reactions in the equilibrium chapter; the kinetics chapter highlights the usefulness of kinetics studies in revealing reaction mechanisms; and the chapter on the mathematical side of chemistry covers precision, accuracy, and uncertainty to great depth. We set the bar high for conceptual understanding. To quote one of our mentors, the late Professor Michell J. Sienko, we believe that this "does not make the subject more difficult but actually makes it clearer."

How, then, do we attempt to accomplish our goal of leaving you with a permanent and confident grasp of the fundamental concepts of chemistry?

1. We make a special effort to break you out of the passive reading mode. This is very important. Students commonly make the mistake of treating a chemistry textbook like a novel, assuming they can simply read through it once and get it. Quite frankly, this is impossible. A chemistry textbook is best read with pencil in hand and a pad of paper nearby—for taking notes, for drawing pictures to represent models and concepts, and for self-testing. We have clusters of practice problems scattered throughout the chapters. But, unfortunately, experience tells us that most students wait until problems are assigned before attempting them, short-circuiting the authors' attempt to get students to exercise their new knowledge while reading. As for worked-out problems, most students simply nod their way through them. No pain, no gain! You have to struggle with a problem to really learn what you do and don't know. We have therefore added something new to our textbook—WorkPatches. WorkPatches are conceptual practice problems, interwoven into the fabric of the text itself and highlighted by a stop sign. As you read, the text leads naturally into the WorkPatch, where a question is asked. The text following the WorkPatch encourages you to try the question because it often discusses the answer without necessarily revealing it (the answer itself is given at the end of the chapter). Your answer to the WorkPatch will often lead to the development of the next concept. Skipping a Work-Patch will be like coming into the middle of a conversation and trying to

understand what is being said. The WorkPatches will help you break passive reading, build understanding, and visualize concepts.

2. Absolutely nothing in each chapter invites you to skip it. There is no convoluted trek through a broken path of text and misplaced illustrations. There are no boxes of special topics, historical perspectives, or applications; no margins crammed with flow diagrams, plots, or tables. There are no long figure legends. Everything of interest has been intricately interwoven into the text itself and used to illustrate the concepts. Figures are always right there, where the text mentions them. The many illustrations help to put concepts into pictorial terms.

3. We have written the text in a conversational tone, inviting you to continue reading. We occasionally describe concepts using everyday language and situations because of their power to evoke unforgettable images and create lasting connections. Also, we are honest, sometimes telling you that some things are more important to remember than others.

4. The book uses humor, often in the form of cartoons, to illustrate concepts and situations.

5. Important concepts are attached to stories and applications in an attempt to make them memorable.

We hope that this approach will help you to get a firm grip on the fundamental concepts of chemistry.

ACKNOWLEDGMENTS

We very much want to acknowledge the people who were so important and valuable to us in the creation of this book. The following reviewers of the various stages of the manuscript are gratefully acknowledged:

Edward Alexander , San Diego Mesa College
Joe Asire, Cuesta College
Barbara Balko, Lewis and Clark College
David Ball, Cleveland State University
Dan Bedgood, Arizona State University
Jack Benefield, Valencia Community College
Kenneth Bennett, Kalamazoo Valley
    Community College
Bill Bornhorst, Grossmont College
Joe Burnett, University of Iowa
Connie Churchill, Oakton Community College
Ana Ciereszko, Miami-Dade Community College
John Cullen, Ricks College
Walter Dean, Lawrence Technological University
Ron Distefano, Northampton Community College
David Dollimore, University of Toledo
Jerry Driscoll, University of Utah
Mary Ann Durick, Bismarck State College
Gary Fisher, De Anza College

Perry Forman, University of Alabama at
    Birmingham
Roger Frampton, Tidewater Community College
Marc Franco, South Seattle Community College
Elizabeth Gaillard, Northern Illinois University
Angela Glisan-King, Wake Forest University
Wendy Gloffke, Cedar Crest College
John Goodwin, Coastal Carolina University
George Goth, Skyline College
Stanley Grenda, University of Nevada, Las Vegas
Elizabeth Griffith, University of South Carolina
James Hardcastle, Texas Woman's University
Claudia Hein, Diablo Valley College
R.A. Hoots, Davis, CA
Jeffrey Hurlbut, Metropolitan State College
James Jacob, University of Rhode Island
Fred Johnson, Brevard Community College
Ray Johnson, Hillsdale College
Stanley Johnson, Orange Coast College

Curtis Keedy, Lewis and Clark College

Keith Kennedy, St. Cloud State University

Roy Kennedy, Massachusetts Bay Community College

Christine Kerr, Montgomery College

Leslie Kinsland, University of Southwestern Louisiana

John Konitzer, McHenry County College

Dave Kort, Mississippi State University

Glenn Kuehn, New Mexico State University

Alfred Lee, City College of San Francisco

Nancy Levinger, Colorado State University

Jimmy Li, College of San Mateo

Ann Loeb, College of Lake County

David Macaulay, William Rainey Harper College

Christopher Makaroff, Miami University

Joe March, University of Alabama at Birmingham

Stanley Marcus, Cornell University

Don Marshall, Sonoma State University

Carol Martinez, Albuquerque Technical/Vocational Institute

Saundra Yancy McGuire, Louisiana State University

Valerie Meehan, City College of San Francisco

Sara Melford, DePaul University

Richard Mitchell, Arkansas State University

Wendell Morgan, Hutchinson Community College

Karl Mueller, Pennsylvania State University

Milica Nedelson, Oakton Community College

Tom Neils, Grand Rapids Community College

Barbara Rainard, Community College of Allegheny County

Don Roach, Miami-Dade Community College

Mike Rodgers, Southeast Missouri State University

Patricia Rogers, University of California, Irvine

Rolland Rue, South Dakota State University

Martha Sanner, Middlesex Community College

Wesley Smith, Ricks College

Dennis Stevens, University of Nevada, Las Vegas

Christine Sullivan, Skyline College

Dave Tanis, Grand Valley State University

Vernon Theilmann, Southwest Missouri State University

Eric Trump, Emporia State University

Kenward Vaughn, Bakersfield College

Trudie Jo Slapar Wagner, Vincennes University

John Weyh, Western Washington University

Thomas Willard, Florida Southern College

Don Williams, Hope College

Joseph Wilson, University of Kentucky

Linda Wilson, Middle Tennessee State University

Our thanks go to Rebecca Strehlow and Susan Weisberg, two extraordinarily talented and dedicated development editors; to Margot Otway, whose editing, invaluable advice, and dedication were exceeded only by her passion for making this the best book possible; to Emi Koike (and her team), the most talented and creative textbook artists in the world; to Pattie Silver-Thompson and Bert Dodson, our cartoonists; to John Cullen and Stan Marcus, for their unbelievably detailed (and picky) job of accuracy checking; to Saundra McGuire, for helping to improve the in-chapter sample and end-of-chapter problems while preparing the study guide; to Joan Keyes and Jonathan Peck at Dovetail Publishing Services, for taking on and excelling at the challenge of laying out such an unconventional book; to Phyllis Niklas, for her remarkable copy editing and facility for holding the entire project together along its (often bumpy) path from typewritten manuscript to printed copy; to Jennifer Schmidt, for her tireless marketing efforts and for being there at so many of the Strategies for Science Teaching Success Workshops to give us moral support and a lift in her car; to Claudia Herman, who handled all the supplements that accompany this book and worked closely with their creators; to David Kort, who worked on the testing program; and to Wendy Gloffke, who authored the accompanying laboratory manual. Last, but certainly not least, thanks to our senior editors, Joan Marsh and Ben Roberts. This book could not have happened without the Herculean efforts of Joan and Ben on our behalf and their fervent and unwavering belief both in this project and in us. We are forever in their debt.

FOR THE STUDENT

The Chemistry Place,™ Special Edition for *Introductory Chemistry:*
A Conceptual Focus
www.chemplace.com/intro/russo
Directed by Dr. Joe March of the University of Alabama, Birmingham

This study tool offers interactive tutorials, practice quizzes, and collaborative group activities—all written specifically to accompany this textbook. For the instructor, the Web site provides course management capabilities including: an online syllabus builder, an online gradebook, and an online quiz generator in which instructors can create quizzes from pre-existing questions or add their own questions.

The Chemistry of Life CD-ROM for Introductory Chemistry (0-8053-3109-3)
By Robert M. Thornton of the University of California, Davis

Through high-quality animations and interactive simulations, this tutorial helps students master crucial concepts such as Atoms and Molecules, Water, Acids and Bases, Gases, Organic Molecules, Carbohydrates, Lipids, Proteins, Nucleic Acids, and Reactions and Enzymes. It also includes diagnostic quizzes and an illustrated glossary.

Student Study Guide and Solutions Manual (0-321-03763-4)
By Saundra Yancy McGuire of Lousiana State University

The Student Study Guide and Solutions Manual provides comprehensive answers and explanations to all of the practice problems in the text and to selected end-of-chapter problems. Learning Objectives highlight the concepts students should learn in each chapter in a clear and easy-to-follow style. Exercises for Self-Testing enable students to test themselves with additional fill-ins and short-answer questions.

Introductory Chemistry Laboratory Manual (0-8053-3119-0)
By Wendy Gloffke of Cedar Crest College

Helps students develop data acquisition, organization, and analysis skills while teaching basic techniques. Students construct their own data tables, answer conceptual questions, and make predictions before performing experiments. They also have the opportunity to visualize and describe molecular-level activity and explain the results. Contains a full suite of laboratory exercises geared to the text.

FOR THE INSTRUCTOR

Instructor's Teaching Guide (0-8053-3104-2)
By Saundra Yancy McGuire of Lousiana State University

Includes chapter summaries, chapter outlines, complete descriptions of appropriate chemical demonstrations for lecture, suggestions for addressing common student misconceptions, and examples of everyday applications of selected topics for lecture use.

B/C Science Digital Library for *Introductory Chemistry: A Conceptual Focus*
(0-8053-3099-2)

An easily searchable resource containing all the visuals from the text. Images
can be downloaded in a variety of formats and inserted into PowerPoint slides
or other presentation tools. Instructors can also download images from this
CD-ROM to their own Web site to assist them in teaching online.

Testing Program
Printed Test Bank (0-321-40557-9); TestGenEQ Win/Mac (0-321-40559-5)
By David A. Kort of Mississippi State University

A complete testing package with more than 700 questions that correspond to
major topics in the text. Available in electronic version or print.

Benjamin/Cummings Custom Laboratory Program

Create a custom laboratory manual by selecting labs and reorganizing experiments from our existing library, or by adding your own.

Instructor's Manual for the Laboratory Manual (0-8053-3108-5)
By Wendy Gloffke of Cedar Crest College

Color Acetates (0-321-40560-9)
Includes 125 full-color acetates.

Chapter 1

What Is Chemistry?

It is the latter part of the nineteenth century, 1896 to be exact. I am overcome with grief. My youngest child is burning with fever. The sickness has spread from her ear to her entire body. Her skin has a scarlet look and she is in great pain. The doctor has applied some tincture of iodine but he does not know the cause of what ails her. He has told us to make arrangements. My beloved child will not see her fourth birthday.

It is the latter part of the twentieth century, 1996 to be exact. My daughter was ill yesterday with an earache. Our pediatrician diagnosed a streptococcus infection and administered the antibiotic amoxicillin. My daughter thought it tasted good, and she is back in preschool today, completely free of fever and pain.

It is the early part of the twenty-first century, 2026 to be exact. We have chosen to have a daughter. Unlike most of today's parents, we will not preselect her IQ. However, we do agree with our genetic counselor that her system should be genetically engineered so that she will be immune to all known bacterial and viral infections.

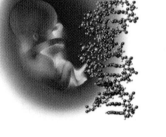

SCIENCE AND TECHNOLOGY 1.1

The span from 1896 to 1996 was 100 years. The year 2026 is less than 30 years away. The pace at which things are changing is accelerating at an unbelievable rate. One hundred years from now will be as different from today as today is from 500 years ago. Within 10 years the entire genetic code of the human genome will be unraveled. One hundred years ago we had never even heard of DNA; today we are beginning to modify it. One hundred years ago we burned coal; in less than half that time from now we'll generate energy in fusion reactors, powered by hydrogen taken from the ocean's water, duplicating the process that occurs within the cores of stars.

Change this rapid is new for humanity. Until recently, each new generation could expect to live pretty much like the one before it. Not any more. We scarcely have time to get used to one change before ten more are upon us. We've barely had time to think about the ethics of birth control, a development of the 1970s, and now fetal cell research, life prolongation, and cloning are knocking at our door. Your personal computer and its software are almost outdated the day you buy them. What is feeding all of this change? The answer is *science*. The dictionary defines **science** as "the experimental investigation and explanation of natural phenomena," or "knowledge from experience." Science starts with a simple question, like "how?" or "why?" How are atoms put together? Why do bats fly at night? Science is the pursuit of knowledge for its own sake, because we are curious. But science itself can't cause change unless something is done with the knowledge it uncovers. For that we need **technology**, the *application* of scientific knowledge. So science is also the pipeline for technology—feeding it and supplying it with ideas. For example, scientists asked, "How are atoms put together?" and their experiments led them to an answer. Today, technologists can use that knowledge to change our lives by developing nuclear medical technologies to treat cancer and by building nuclear bombs. As is so often the case with scientific knowledge, it can be used to achieve very different ends.

Science: The discovery of the structure of the atom

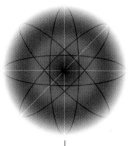

Technology: Applications of that knowledge

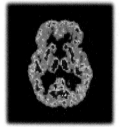

Medical application of atomic energy

Military application of atomic energy

Since science feeds technology, it's not a bad idea to ask the following question: "Is science always right?" Just consider these headlines:

1948 Totally Safe Freons Revolutionize Refrigeration

1975 Freons Destroying Earth's Ozone Layer

1985 Unsaturated Fats Good for Health

1994 Unsaturated Fats Not So Healthy After All

1989 Cold Fusion: Limitless Energy

1991 Cold Fusion: A Delusion?

What is going on? Isn't science only about absolute, fundamental, provable truths? Unfortunately, no. Science, like literature, art, and music, is a human endeavor. And because it is a human endeavor, carried out by human scientists, it would not be wise to bet on science's infallibility. Ego, mistakes, and stubbornness can all get in the way of finding the truth. And what about technology? Is the technological application of scientific knowledge always good? Consider some of the forces that drive technology: the desire to benefit humankind, the desire to make a profit, the desire to be stronger than our enemies. Which motive do you think drives the cigarette industry in the application of their knowledge of the effects of nicotine?

This brings us to some very important questions: Who should decide which areas of science are explored? Who should decide if a scientific result is real or a hoax? Who should decide what is done with scientific knowledge? In a free society the answer is supposed to be the collective "you." After all, what scientists discover and technologists put into practice will dramatically affect how you live. In addition, much of the scientific research in this country is paid for by your tax dollars. Does this mean you have to earn a Ph.D. in chemistry, biology, and physics so that you can make informed decisions? For most people that is not a practical or even a desired solution. But total ignorance of science is not the answer either, since science and technology affect us directly in our everday living.

This book is about the branch of science known as *chemistry*, often considered the "central science" since it forms a bridge between the principles of physics and the practice of biology. **Chemistry** is the study of matter and the transformations it undergoes. To understand this definition we need to focus on two key words: *matter* and *transformation*.

Hydrogen and oxygen *transform* into water and heat in the space shuttle's liquid-fueled engine, helping to propel it into space.

Water produced in rocket engine

MATTER 1.2

Matter can be simply defined as "stuff"—anything that has mass and occupies space. Some matter you can feel and see, like this book. Other matter, like the air that surrounds you, is difficult to detect, but it still exists. It occupies space (the volume of the room), and it has mass. Defining matter as "stuff" is adequate for most everyday situations. Since science often asks very specific questions, however, it almost always requires much more precise and detailed definitions. Chemistry begins defining matter by dividing it into two broad types, *pure substances* and *mixtures*. In **pure substances**, only a single type of matter is present. **Mixtures** occur when two or more pure substances are intermingled with each other. For example, table salt (chemical name, sodium chloride) is a pure substance. So is water. And so is table sugar (chemical name, sucrose). But if you put salt and sugar in a jar together and shake, you have a mixture. Dissolve sugar in water and you have another mixture. Some things that you might not think of as mixtures actually do fit the definition—a rock, for example. In most rocks, you'll see a mixture of different minerals, each a different pure substance.

Chemists further subdivide mixtures into two types, *homogeneous* and *heterogeneous*. **Homogeneous mixtures** (*homo* meaning same) are ones in which the composition of the mixture is identical throughout. A cup of tea with some sugar dissolved in it is an example of a homogeneous mixture. Once well stirred, such a mixture is exactly the same no matter where you sample it.

Homogeneous
mixture
(sugar dissolved in tea)

Heterogeneous
mixture
(granite)

Another name for a homogeneous mixture is a **solution**. Our well-stirred cup of tea with sugar is a solution. The air you are breathing, consisting of a mixture made mostly of nitrogen gas and a smaller amount of oxygen gas, is also a homogeneous mixture or solution. Under normal conditions, no matter where you sample the air in the room, its composition (percent nitrogen and percent oxygen) is the same.

The rock pictured above is an example of a **heterogeneous mixture** (*hetero* meaning different), because its composition is not the same throughout. Sometimes, heterogeneous mixtures can appear to be homogeneous although they aren't. The mixture of table salt and table sugar is a heterogeneous mixture, even if the two substances are ground into a fine powder. This is because a tiny sample taken at one place in the mixture might contain a different ratio of salt to sugar than a sample taken at some other place. It would probably even be possible to get a tiny sample from this mixture that was pure sugar or pure salt, no matter how well you mechanically ground the two together. Only when such a tiny sample has the same composition wherever you sample it (as in the tea with sugar) can you call the mixture a solution.

1.1 WORKPATCH

Classify the following examples of matter as pure substances, heterogeneous mixtures, or solutions.

(a) A piece of wood
(b) An iron nail
(c) A rusty iron nail
(d) A well-stirred mixture of food dye in water
(e) Bee's wax and candle wax ground together
(f) Bee's wax and candle wax melted together, then allowed to solidify

Check your answers to WorkPatch 1.1 against the solutions given at the end of the chapter. Did you get the right answers for the last two items? These items should make you think. Only one of the mixtures of bee's wax and candle wax represents a solution, because only one of the processes (melting or grinding) is capable of producing a homogeneous mixture.

1.1 Which of the following are solutions?
(a) A sample of our atmosphere
(b) Iron and copper powder ground together
(c) Iron and copper powder melted together
(d) Detergent dissolved in water
(e) Tiny oil droplets suspended in water

Answer: (a), (c), and (d) are solutions (homogeneous mixtures); (b) and (e) are heterogeneous mixtures and are thus not solutions.

1.2 Examination of "homogenized" milk under a microscope reveals suspended globules of fat. Is milk a heterogeneous mixture or a homogeneous solution? Explain.

1.3 Fog is a suspension of tiny droplets of water in air. Is fog a heterogeneous mixture or a homogeneous solution? Explain.

Before we leave the definition of matter, we want to say something about what matter is made of. All the matter that exists on our planet—from the air, to the dust in the air, to the ground we walk on, to the water that covers most of the planet, to the life forms that live on it—is made from *elements*. **Elements** are the basic building blocks of matter. At the time of writing, there are 115 known elements, 90 that occur naturally and 25 that can be synthetically prepared. All the known elements have been organized in a tabular form known as the *periodic table*.

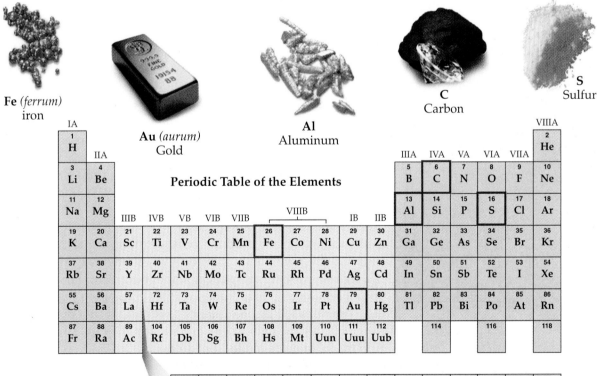

Fe *(ferrum)* iron

Au *(aurum)* Gold

Al Aluminum

C Carbon

S Sulfur

Periodic Table of the Elements

IA																	VIIIA
1 H	IIA											IIIA	IVA	VA	VIA	VIIA	2 He
3 Li	4 Be											5 B	6 C	7 N	8 O	9 F	10 Ne
11 Na	12 Mg	IIIB	IVB	VB	VIB	VIIB	VIIIB			IB	IIB	13 Al	14 Si	15 P	16 S	17 Cl	18 Ar
19 K	20 Ca	21 Sc	22 Ti	23 V	24 Cr	25 Mn	26 Fe	27 Co	28 Ni	29 Cu	30 Zn	31 Ga	32 Ge	33 As	34 Se	35 Br	36 Kr
37 Rb	38 Sr	39 Y	40 Zr	41 Nb	42 Mo	43 Tc	44 Ru	45 Rh	46 Pd	47 Ag	48 Cd	49 In	50 Sn	51 Sb	52 Te	53 I	54 Xe
55 Cs	56 Ba	57 La	72 Hf	73 Ta	74 W	75 Re	76 Os	77 Ir	78 Pt	79 Au	80 Hg	81 Tl	82 Pb	83 Bi	84 Po	85 At	86 Rn
87 Fr	88 Ra	89 Ac	104 Rf	105 Db	106 Sg	107 Bh	108 Hs	109 Mt	110 Uun	111 Uuu	112 Uub		114		116		118

58 Ce	59 Pr	60 Nd	61 Pm	62 Sm	63 Eu	64 Gd	65 Tb	66 Dy	67 Ho	68 Er	69 Tm	70 Yb	71 Lu
90 Th	91 Pa	92 U	93 Np	94 Pu	95 Am	96 Cm	97 Bk	98 Cf	99 Es	100 Fm	101 Md	102 No	103 Lr

The periodic table is so important to chemistry that we will devote most of Chapter 3 to it. (A more complete periodic table, including the full names of all the elements, appears inside the front cover of this book.)

Many elements have names that you are probably quite familiar with—carbon, silver, gold, iron, aluminum, uranium, hydrogen, helium, oxygen, and nitrogen are a few. Chemists represent the elements with one- or two-letter symbols (as shown in the periodic table), some taken from the English names (like C for carbon, H for hydrogen, Al for aluminum) and others taken from the Latin names (like Fe for iron from the Latin word *ferrum*; Au for gold from the Latin word *aurum*).

When we say that elements are the basic building blocks of matter, how basic is basic? For example, suppose you had a piece of pure elemental gold. What would happen if you cut the piece of gold in half, and then in half again, and then again, and then again? How much could you divide the piece and still have elemental gold? People have been trying to answer this question for centuries. Around 400 BC the Greek philosopher Democritus suggested an atomic theory of the universe. Simply put, he said that all things are made up from minute, indivisible, indestructible particles called *atoms*. That this might be true was by no means self-evident. A piece of gold does not appear to be made of individual particles. Nevertheless, Democritus would have said that you can keep dividing up your piece of gold until you get down to one atom of gold. In fact, Democritus was right, but his theory was not an easy sell. Aristotle, another Greek philosopher living around the same time, placed more trust in his senses and said that matter was continuous and not made up of atoms. According to Aristotle, you could keep dividing your gold into smaller and smaller pieces forever; at no point would you reach an indivisible particle. Since Aristotle's proposition seemed more obviously correct, his theory carried the day for more than 2000 years.

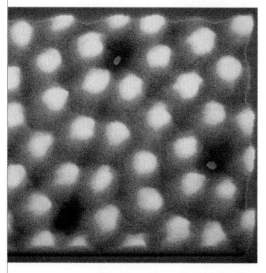

As you will see in Chapter 3, scientists did eventually return to the atomic theory. We now know that all the elements exist as atoms, and an **atom** is the smallest piece possible of an element. Atoms are so tiny that, until recently, scientists thought we would never be able to see them. They spoke too soon. Though no one has seen an atom through an ordinary microscope, in the early 1980s the scanning tunneling microscope produced actual images of individual atoms—like the silicon atoms that appear as bumps on the surface of the silicon crystal shown in the photo at left.

With the knowledge that matter is made up of atoms of the elements, we can now divide pure substances into two types, *elemental substances* and *compounds*. An **elemental substance** is one that is made from atoms of just one element. For example, our piece of pure gold is made from just gold atoms and nothing else. The same is true of a piece of pure iron (or any pure metal), the oxygen you breathe, or the helium gas in a balloon. **Compounds**, on the other hand, are pure substances made from atoms of two or more different elements. Water, for example, is made from atoms of hydrogen and oxygen. A **chemical formula** for a compound indicates the number of atoms of each element that make up the smallest possible piece of that pure substance.

The formula for water, H₂O, tells us that the smallest possible piece of water is made from 2 hydrogen atoms and 1 oxygen atom. Because it is made from two different elements, pure water is classified as a compound. So is table sugar, whose formula is $C_{12}H_{22}O_{11}$. The oxygen in the air you breathe is a different case. Its formula, O_2, tells us that the smallest piece of life-sustaining oxygen gas has 2 oxygen atoms in it. However, though it is made from 2 atoms, both are atoms of the same element, so oxygen is not considered to be a compound; it is an elemental substance. To summarize:

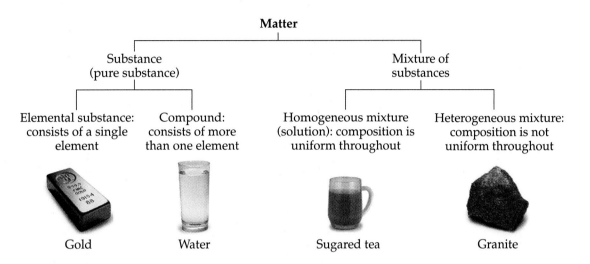

Classifying matter according to this scheme sometimes requires a careful examination. Consider, for example, a glass of water, a glass of lemonade made from a powdered mix, and a glass of lemonade made with fresh-squeezed lemons:

Pure water   Lemonade from   Fresh lemonade
              powdered mix

The water is an example of a pure substance and a compound. The lemonade made from the mix is an example of a homogeneous mixture (a solution). The lemonade made from lemons is an example of a heterogeneous mixture due to the presence of lemon pulp bits floating in it.

PRACTICE PROBLEMS

1.4 Which of the following are compounds?
 (a) Iron oxide (Fe_2O_3)
 (b) Ozone (O_3)
 (c) Iron (Fe)
 (d) Carbon monoxide (CO)
 (e) Propane (C_3H_8)

Answer: Iron oxide (Fe_2O_3), carbon monoxide (CO), and propane (C_3H_8), are all compounds, since all are made from more than one kind of element. Ozone (O_3) and iron (Fe) are examples of elemental substances, not compounds, since each is made of only one type of element.

1.5 Which of the following are compounds?
 (a) Sulfur (S_8)
 (b) A mixture of iron and aluminum powder
 (c) A mixture of O_2 and N_2 gases
 (d) Sulfur dioxide (SO_2)
 (e) Ammonia (NH_3)

1.6 True or false? A compound is a pure substance, but a pure substance need not be a compound. Give examples to prove your answer.

1.3 MATTER AND ITS PHYSICAL TRANSFORMATIONS

Let's get back to the definition of chemistry, which is why we started looking into matter in the first place. We said that chemistry is the study of matter and the transformations it undergoes. Since *transformation* means change, chemistry is the study of changes in matter. And just as the word "matter" had to be subdivided into pure substances and mixtures, we must be very careful with the word "change." In the science of chemistry there are two kinds of changes matter can undergo: *physical* and *chemical*.

A **physical transformation** to a pure substance is one that leaves it as the same substance but in a different physical state. This leads us to another question: What do we mean by a *state of matter*? You are already familiar with the three most common **states of matter**: solid, liquid, and gas (or vapor). We will define these states in a detailed way later in this book, but for now your common knowledge of them will do just fine. We will use water to demonstrate physical changes to matter. If you take a glass of liquid water and put it in a freezer, it will change to the solid state in a process we call **freezing**. This is an example of a physical change because only the state of the substance has changed. It's still water after the change. The reverse of freezing, called **melting**, is another example of a physical change. Naturally, the temperature below which liquid water freezes is also the temperature above which solid water melts, 32°F or 0°C at normal atmospheric pressure. (The Fahrenheit and Celsius temperature scales are described in Chapter 2.) Thus, this temperature is called the freezing point or the melting point for water, depending on which way you are going. When it is heated to the boiling point (212°F or 100°C at normal atmospheric pressure), water will boil and undergo **evaporation**, the change from the liquid to the vapor state. In the reverse of evaporation, **condensation**, the vapor returns to the liquid state. After either change, we still have water, so evaporation and condensation are two more examples of physical changes.

The melting point and boiling point of water are among its physical properties. The **physical properties** of a pure substance characterize its physical state and physical behavior. Other physical properties of water include its color (pure water is colorless), its odor (pure water is odorless), and its taste (pure water is tasteless). The following table summarizes some of the physical properties of water:

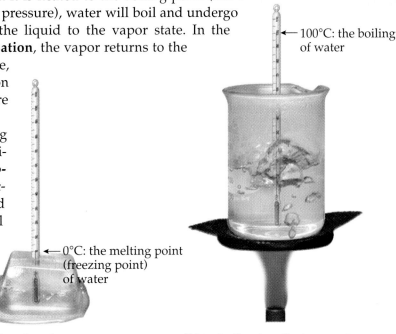

← 100°C: the boiling point of water

← 0°C: the melting point (freezing point) of water

Water melts when the temperature goes above 0°C; it freezes when the temperature goes below 0°C.

Water boils when the temperature goes above 100°C; it condenses when the temperature goes below 100°C.

Physical Properties of Pure Water (at Normal Atmospheric Pressure)

Melting point	0°C
Boiling point	100°C
Color	None
Odor	None
Taste	None

Each pure substance has its own unique set of physical properties. For example, pure ethanol, the component of beer and wine that gets you drunk, is a colorless liquid that looks just like water. However, it melts at −117°C and boils at 78°C. Thus, if you are presented with a colorless liquid and asked to determine whether it is water or ethanol, you can place a thermometer in it and heat it to boiling. If the thermometer reads 78°C, then the liquid is ethanol.

← 78°C

At normal atmospheric pressure, this liquid boils at 78°C, so it could be ethanol.

At normal atmospheric pressure, this liquid boils at 100°C—probably water.

← 100°C

Another important physical change is *sublimation*. **Sublimation** occurs when a pure substance goes directly from the solid to the gas state without passing through the liquid state. For example, when you heat an ice cube, it will first melt to liquid water before boiling to reach the vapor state. But carbon dioxide, another pure substance, behaves quite differently. Carbon dioxide exists as a solid below −78°C, but if you warm it above this temperature, it slowly vanishes into thin air without ever forming a puddle or any obvious wetness. For this reason, solid carbon dioxide is called "dry ice." It goes directly from the solid state to the gas state—it sublimes. Unlike water, carbon dioxide cannot exist in the liquid state unless a great deal of pressure is applied to it.

At room temperature (25°C) . . .

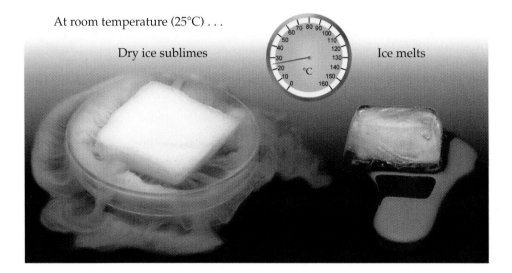

Dry ice sublimes   Ice melts

PRACTICE PROBLEMS——————————————————————————————

1.7 Multiple choice: Ice cubes slowly vanish from their tray while stored in a freezer. This is an example of:
(a) Evaporation
(b) Condensation
(c) Melting
(d) Sublimation

Answer: This is an example of sublimation, since the ice cubes slowly "vanish" by going directly from the solid to the gaseous state.

1.8 Multiple choice: Molten iron cooling into solid iron is an example of:
(a) Sublimation
(b) Condensation
(c) Freezing
(d) Melting

1.9 True or false? After being heated to beyond its boiling point, ethanol is no longer ethanol, but is something else. Explain.

1.10 You are presented with a block of metal and told that it is pure gold, but you have your doubts. How can you use a thermometer to determine whether the metal is pure gold?

MATTER AND ITS CHEMICAL TRANSFORMATIONS 1.4

So far we have been concentrating on the physical properties of pure substances. It's time now to switch to chemical properties. The **chemical properties** of a pure substance are the ways it behaves when it is combined with other pure substances. For example, sodium (a soft shiny metal with the elemental symbol Na, from the Latin *natrium*) catches fire when it is combined with water. This is a chemical property of sodium. A chemical property of chlorine (a greenish gas with the formula Cl_2) is that it "burns" most substances it comes into contact with. If you mix sodium and chlorine together, a rather incredible thing happens. A tremendous amount of heat and light are

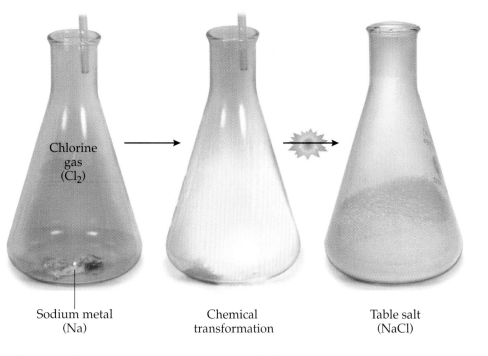

Chlorine gas (Cl_2)

Sodium metal (Na)

Chemical transformation

Table salt (NaCl)

given off. When things finally calm down enough to take a look, both the sodium and chlorine are gone, replaced by an entirely new pure substance, the compound sodium chloride (formula NaCl).

The new compound has chemical properties entirely different from those of either of the pure substances from which it was made. It no longer catches fire in water, and it no longer burns things it comes into contact with. It's ordinary table salt. What you have witnessed is a chemical transformation. A **chemical transformation** occurs when new substances, with new physical and chemical properties, are formed from some starting substance or substances. The starting substances are called **reactants**, and the new substances that are formed are called **products**. The process that goes on during the conversion of reactants into products is called a **chemical reaction**.

Chemists write chemical reactions using a symbolic representation, with the reactant substances on the left side of an arrow and the product substances on the right side. The arrow itself means "react to give." A chemist would write the chemical reaction between sodium and chlorine as

$$2\,\text{Na} + \text{Cl}_2 \longrightarrow 2\,\text{NaCl}$$
$$\text{Reactants} \qquad\qquad \text{Products}$$

In English this simply says "sodium and chlorine react to give sodium chloride." The matter has undergone a chemical transformation. Often, chemists will include the notation (g) for gaseous, (l) for liquid, or (s) for solid to represent the physical state of each substance in the reaction. Thus, we could have written

$$2\,\text{Na}(s) + \text{Cl}_2(g) \longrightarrow 2\,\text{NaCl}(s)$$

Of all the chemical transformations that occur, relatively few are carried out by chemists. Most are carried out by nature. The rusting of iron, be it in a nail or on your car, is an example of a chemical transformation in which iron reacts with oxygen in the presence of water to give the new compound iron oxide (formula $Fe_2O_3 \cdot H_2O$), more commonly known as rust.

Even more important to humans are the chemical transformations that occur during the digestion of food. Some of the food you eat gets converted to carbon dioxide and water, which you exhale. Your very thought processes involve many chemical reactions.

In spite of the stereotype of the crazed scientist madly mixing test tubes of chemicals for the joy of it, that's not the reason chemists mix substances together. Chemists are constantly searching for new compounds with new chemical and physical properties that can extend our knowledge about how matter behaves and can benefit humankind. New compounds to combat disease, to protect metal from corrosion, to whiten and brighten socks, . . . , the list goes on and on. Chemical transformation is what makes chemistry so interesting and worthy of study.

$$4\,\text{Fe} + 3\,\text{O}_2 + 2\,\text{H}_2\text{O} \longrightarrow 2\,\text{Fe}_2\text{O}_3 \cdot \text{H}_2\text{O}$$

Chemistry got its start hundreds of years ago in the medieval period. Back then, alchemists, the forerunners of modern-day chemists, were busy trying to turn base metals such as lead into gold. They didn't succeed, but what they discovered while trying laid the foundation for the science of chemistry. Today we know that the alchemists could never have succeeded, for while turning lead into gold is an example of transforming matter from one substance into another, it is not an example of a chemical transformation. In a chemical transformation, none of the elements that you start with may change. If your reactants have carbon, hydrogen, and oxygen in them, then your products must have only carbon, hydrogen, and oxygen in them and no other elements. A chemical transformation can't convert one element into another element, so turning lead into gold by chemical means is hopeless. Modern-day scientists, however, can turn lead into gold via a *nuclear transformation*, which converts one element into another; you'll learn more about this in Chapter 14.

Refer to the periodic table inside the front cover for names of elements that are not yet familiar to you.

1.11 Which of the following represent a chemical transformation?
   (a) $4\,P(s) + 5\,O_2(g) \longrightarrow 2\,P_2O_5(s)$
   (b) $H_2O(g) \longrightarrow H_2O(l)$
   (c) $3\,O_2 \longrightarrow 2\,O_3$

Answer: (a) and (c) are both examples of chemical transformations, since the products are different from the reactants; (b) is an example of a physical change, since we have water on both sides of the arrow.

1.12 Water (H_2O) and carbon dioxide (CO_2) are produced in a chemical reaction when methane (CH_4) and oxygen (O_2) are combined and heated. What are the reactants and what are the products?

1.13 Multiple choice: What, if anything, is wrong with the following chemical reaction?

$$Cl_2(g) + Br_2(l) \longrightarrow 2\,HCl(g) + 2\,HBr(g)$$

   (a) It is not a chemical reaction. It is only a physical transformation.
   (b) The product includes elements not in the reactants.
   (c) There is nothing wrong with the chemical reaction.

HOW SCIENCE IS DONE: THE SCIENTIFIC METHOD 1.5

Concepts such as what atoms are made of and how they react are essential to our understanding about chemistry. Many of the basic ideas that underlie chemistry were developed within the last 150 years, but the process that scientists used to develop these concepts has been around for much longer. In fact, any curious person seeking some answers to a question might, and probably has, used the *scientific method*. For example, consider Albert. It is a commonly held belief that scientific types have little or no romantic life. This is, in fact, not true. In his younger days, Albert found himself newly arrived at a university, seeking dates.

Being of a curious nature, he decided that he would study the topic of dating. He went on many dates, making observations, collecting data, and recording it all in his diary (his "laboratory notebook"). These dates were his *experiments*. **Experiments** are procedures that scientists carry out to study some phenomenon. For Albert, some dates were successful and some were not. On each date he would vary one aspect of his appearance (a racy new tie, a bold set of socks, a black eye patch), observe its effect on the outcome, and diligently record the results.

After several months of collecting data from many dates, he came to the conclusion that changes in his appearance above his neckline had great influence on the outcome of his dates, while modifications below his neckline had little or no effect. This led him to postulate the following Law of Dating: "For a successful date, look as good as you can from the neck up." A **law** is a conclusion based on the experimental data. At first, Albert thought this law a bit odd, but after much thought, he developed a *theory* that explained why it was true. A **theory** is a hypothesis that attempts to explain why a law is true. He theorized that in any stressful situation between two people, they watch each other's faces intensely for visual cues of how things are going. This Stress-Induced Intense Facial Watching Theory explained the Law of Dating.

The process that Albert used to satisfy his curiosity about dating—collecting data from experiments, using it to fashion a law, then postulating a theory to account for the law—is called the **scientific method**.

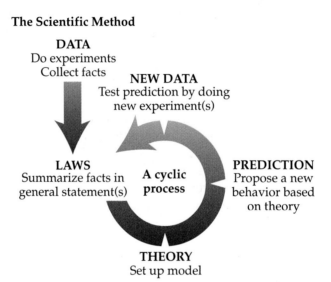

The Scientific Method

DATA
Do experiments
Collect facts

NEW DATA
Test prediction by doing
new experiment(s)

LAWS
Summarize facts in
general statement(s)

A cyclic process

PREDICTION
Propose a new
behavior based
on theory

THEORY
Set up model

The scientific method is meant to be a cyclical process in which scientists keep gathering data, postulating new laws, and modifying or discarding old ones. Based on these new or revised laws, theories or models are further tested and, if necessary, are refined or completely revamped. In repeatedly going through this cycle, sometimes theories and laws are proved to be wrong, and sometimes they hold up. This is how science is supposed to work. Even the great Sherlock Holmes knew this (from *The Yellow Face* by Sir Arthur Conan Doyle):

1.14 How does a theory relate to a law?

Answer: A theory attempts to explain why *a law is correct.*

1.15 How does a law relate to experimental data?

1.16 What is good evidence that a theory is correct?

If only it were so simple. Unfortunately (or, sometimes fortunately), this logical process is carried out by sometimes illogical, emotional (human) scientists. Sometimes, investigators believe so strongly in a pet theory that they enter the cycle from the wrong direction and attempt to interpret all they see in terms of their personal bias. **Bias** is a strong preference or inclination that inhibits impartial judgment. For example, today scientists believe that atoms are among the fundamental building blocks of matter. As we mentioned earlier, this is not a new idea. Democritus was battling Aristotle over the possibility of the existence of atoms over 2000 years ago. Now we know that Aristotle, who rejected atoms, was wrong, but we can't be too hard on him—he had no ability to examine matter on a submicroscopic level. The beliefs of the ancient Greeks were founded not on verifiable experimental evidence but on their opinions and convictions about the natural world. These personal biases can really get in the way of scientific development. Indeed, Aristotle went to his grave convinced that an adult man had more teeth in his mouth than an adult woman. As the renowned philosopher Bertrand Russell once commented, "All Aristotle had to do was ask Mrs. Aristotle to open her mouth," and he would have discovered the truth.

The fact that Aristotle never looked into a woman's mouth to check his theory—that he would not even conceive of it being necessary—is an ideal illustration of why scientists must always be on guard not to let personal bias, ambition, politics, ideology, or theology divert them from the scientific method. Scientists must be neutral to the point that they are willing to throw out all they were taught and replace it with something else if the scientific method so demands. That is not always easy. It has been observed quite often that "A new theory in science is only really accepted when the last of its opponents dies off."

Without the ability to gather data and do experiments, you can't employ the scientific method. Democritus was unable to support his original atomic theory. In contrast, some of the most beautiful examples of the application of the scientific method and the replacement of old theories with new ones come from the development of modern atomic theory, beginning with John

Dalton in the early 1800s. When you read about this development in Chapters 3 and 4 keep the scientific method in mind, and you will understand why theories came and went.

1.6 LEARNING CHEMISTRY WITH THIS BOOK

A common misconception is that chemistry is all math, calculations, and numerical problem-solving. In all honesty, chemistry does have that side to it. But when a chemist is presented with a question about matter, the first thing that comes to mind is *not* a complicated mathematical formula. Instead, chemists use a basic set of fundamental concepts, often best represented with pictures instead of mathematical equations. For example, a chemist and a nonchemist picture the concept of melting differently:

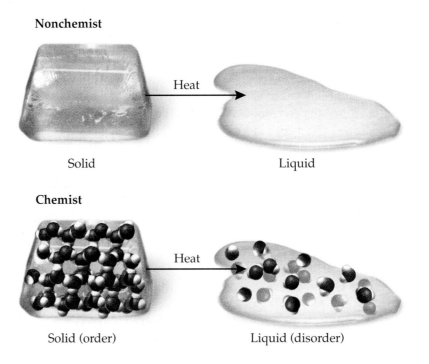

Nonchemist

Solid   Heat →   Liquid

Chemist

Solid (order)   Heat →   Liquid (disorder)

These fundamental concepts and images are the tools chemists use in answering questions about matter and the transformations it undergoes. Though this book will tackle the numerical side of chemistry, it will focus on the fundamental concepts, illustrating them with pictures and common everyday experiences. Even chemists forget mathematical expressions and memorized equations, but the fundamental concepts, stored as images in their minds, are with them all their lives.

Finally, we want to give you some advice about how to use this textbook. This advice comes from our own experience (after all, we were beginning chemistry students once). Read each chapter slowly, with a pad of paper and a pencil in hand so you can take notes, draw pictures, and write down questions on points you are not sure about. We have done three things to encourage you to do this. First, each chapter includes a number of Work-Patches marked with a stop sign. When you reach one of these, *stop* reading and try to do the problem or sketch the concept. *Don't go on until you can answer the question.* Check your answer against the answer given at the end of the chapter. Second, we have clustered practice problems throughout each chapter. In each cluster, the worked-out solution is given for the first practice problem. The other problems in the cluster are similar to the first. In some cases we will provide simple answers to these problems, but it is up to you to actually work out the solution. Blue numbers indicate practice problems for which answers are given in the back of the book; complete solutions for these problems are provided in the *Student Study Guide and Solutions Manual* that accompanies this text (see the Preface). Now, you could skip over the WorkPatches and practice problems, but honestly, you would not be doing yourself a favor. The best time to find out if you understand what you read is just after you read it. In addition, working the practice problems and Work-Patches as you encounter them will help you retain the material. Keep that paper and pencil handy, and do the WorkPatches and practice problems *as you encounter them.* Finally, we have ended each chapter with a section called "Have You Learned This?", which provides a list of the key concepts followed by questions that represent the essence of the chapter. Go over the list of key concepts, and if something doesn't quite ring a bell, go back to the indicated page and read that section again. When you can answer "yes" to the question "Have You Learned This?", then you will be ready to go on.

Have You Learned This?

Science (p. 2)

Technology (p. 2)

Chemistry (p. 3)

Matter (p. 3)

Pure substance (p. 3)

Mixture (p. 3)

Homogeneous mixture (p. 4)

Solution (p. 4)

Heterogeneous mixture (p. 4)

Element (p. 5)

Atom (p. 6)

Elemental substance (p. 6)

Compound (p. 6)

Chemical formula (p. 6)

Physical transformations of matter (p. 8)

States of matter (p. 8)

Freezing (p. 8)

Melting (p. 8)

Evaporation (p. 9)

Condensation (p. 9)

Physical properties of matter (p. 9)

Sublimation (p. 10)

Chemical properties of matter (p. 11)

Chemical transformation of matter (p. 11)

Reactant (p. 11)

Product (p. 11)

Chemical reaction (p. 11)

Experiment (p. 13)

Law (p. 14)

Theory (p. 14)

Scientific method (p. 14)

Bias (p. 15)

SCIENCE AND TECHNOLOGY

Answers to problems with blue numbers appear in the back of the book. Complete solutions for these problems are provided in the Student Study Guide and Solutions Manual that accompanies this text (see the Preface).

1.17 What is the difference between science and technology? Use an example to demonstrate what you mean.

1.18 Define chemistry.

1.19 Give an example (not from this book) of a piece of scientific information and the "good" and "bad" results of its technological application.

1.20 In the United States, most chemical research (science) is funded by the government, whereas most money spent on applying scientific knowledge (technology) comes from the chemical industry. Why do you think the chemical industry is reluctant to fund research?

1.21 Suppose from study of the human brain and brain waves comes an inexpensive pocket-sized technological device that allows for both the downloading (accurate mind reading) and uploading (instant learning) of information into the mind. One extreme opinion is to make the device freely accessible. The other extreme opinion is to outlaw the device and destroy the plans for creating it. Where would you stand on this issue? Give your reasons, and then elaborate on who should decide.

MATTER AND ITS TRANSFORMATIONS

1.22 Define matter.

1.23 Is it possible to have a mixture of compounds? If your answer is yes, then give an example.

1.24 How does a solution differ from a heterogeneous mixture?

1.25 Flour is mixed with powdered sugar and ground to a fine dust. Is this an example of a solution or a heterogeneous mixture? Explain.

1.26 Powdered aluminum is mixed with powdered iron.
   (a) The powders are ground together. Is this an example of a solution or a heterogeneous mixture? Explain.
   (b) The powders are ground together and then the mixture is heated above both their melting points to give molten metal. Is this an example of a solution or a heterogeneous mixture? Explain.
   (c) The molten solution from part (b) is allowed to cool and solidify. Is the resulting solid an example of a solution or a heterogeneous mixture? Explain.

1.27 The air you breathe is properly called a solution. Explain why.

1.28 What is an element? How many elements are known?

1.29 If H is the symbol for the element hydrogen, and O is the symbol for the element oxygen, then why is the symbol for sodium Na and the symbol for iron Fe?

1.30 What is the smallest piece of an element called?

For Problems 1.31–34, refer to the periodic table inside the front cover.

1.31 What are the elemental symbols for lead, molybdenum, tungsten, chromium, and mercury?

1.32 What are the elemental symbols for sulfur, chlorine, phosphorus, magnesium, and manganese?

1.33 Give the names for the following elements: Ti, Zn, Sn, He, Xe, Li

1.34 Give the names for the following elements: U, Pu, Cs, Ba, F, Si

1.35 How does an elemental substance differ from a compound?

1.36 What does a chemical formula tell you about a pure substance?

1.37 Compounds have properties that are different from those of their constituent elements. Discuss a real example to support this statement.

1.38 Which of the following are compounds, and which are elemental substances?

F_2, $BrCl_3$, P_4, C_2H_2, HCl, Ar, Al_2O_3, Al

1.39 Indicate whether the following pure substances are elemental substances or compounds.
(a) Chlorine (formula Cl_2)
(b) Octane (formula C_8H_{18})
(c) Sulfur (formula S_8)
(d) Neon (formula Ne)

1.40 The smallest portion of the compound hydrogen peroxide has 2 hydrogen atoms and 2 oxygen atoms. Write the chemical formula for hydrogen peroxide, writing hydrogen first, then oxygen.

1.41 The smallest portion of the compound nonane, a component of gasoline, has 9 carbon atoms and 20 hydrogen atoms. Write the chemical formula for nonane, writing carbon first, then hydrogen.

1.42 The smallest portion of the simple sugar glucose has twice as many hydrogen atoms as it has oxygen or carbon atoms. Given that glucose has 6 carbon atoms, write the chemical formula for glucose (write carbon first, then hydrogen, and then oxygen).

1.43 The air you breathe is made up mostly of nitrogen (formula N_2).
(a) What can you say about the smallest piece of nitrogen possible?
(b) Is nitrogen a compound? Explain.

1.44 What are the states of matter?

1.45 What is the term for the process in which matter goes directly from a solid to a gas?

1.46 What is the term for the process that is the opposite of evaporation?

1.47 Propane has a boiling point that is below room temperature, whereas hexane has a boiling point that is above room temperature. How could you distinguish propane from hexane by just looking at them at room temperature?

1.48 At normal atmospheric pressure, the freezing point of ethanol is $-117.3°C$, and the boiling point is $78.5°C$. What is the melting point of ethanol?

1.49 Solid mothballs work by filling a closet with chemical vapor, and they seem to shrink and disappear slowly with time. What is going on here?

1.50 What are the melting point and boiling point of water in °C and °F (at normal atmospheric pressure)?

1.51 O_2 is a breathable gas, but O_3 is a toxic gas. How do chemists explain this?

1.52 H_2O is a drinkable liquid, but H_2O_2 is a poison. How can this be possible, given that both are compounds of hydrogen and oxygen?

1.53 When sugar is added to water, it dissolves. Has the sugar undergone a physical change or a chemical change? Explain.

1.54 When a sample of ethanol burns, it is converted into carbon dioxide and water. Has the ethanol undergone a physical change or a chemical change? Explain.

1.55 What are chemists representing when they write chemical reactions?

1.56 When white phosphorus (formula P_4) is exposed to oxygen in the air, it catches fire and produces the compound P_4O_{10}. Is this a description of a physical property or a chemical property of phosphorus? Explain.

1.57 Is the tarnishing of silver a physical change or a chemical change? (The formula for the tarnish on silver is Ag_2O.)

1.58 Nitrogen (N_2) combines with hydrogen (H_2) to give ammonia (NH_3). Nitrogen and hydrogen are colorless, odorless gases that are not very soluble in water. Ammonia is also a gas but it has an extremely powerful odor and is very soluble in water. How can you explain this?

HOW SCIENCE IS DONE

1.59 How does a law differ from a theory?

1.60 If a theory is false, how will the scientific method reveal this?

1.61 Consider some event or activity for which you have collected data or observations over the years (washing clothes, cooking, dating, studying, etc.).
(a) State the activity and some question about it that your data might answer.
(b) Postulate a law regarding your observations.
(c) Postulate a theory.
(d) How could you test your theory?

1.62 The scientific method is often called foolproof and incapable of giving wrong answers, and yet history shows that scientists are sometimes wrong. What is it that often derails the scientific method?

WORKPATCH SOLUTIONS

1.1 (a) Heterogeneous mixture; different regions can easily be seen in the wood.
(b) Pure substance; made of a single element, iron (Fe).
(c) Heterogeneous mixture; the rust is a layer of the compound iron oxide over an iron nail.
(d) Solution; the dye is homogeneously distributed throughout the water.
(e) Heterogeneous mixture; no amount of mechanical grinding can result in a homogeneous mixture.
(f) Homogeneous solution; melting allows different compounds to mingle and establish a homogeneous mixture.

Chapter 2

The Numerical Side of Chemistry

NUMBERS IN CHEMISTRY—PRECISION AND ACCURACY 2.1

While we promised in Chapter 1 to dwell on fundamental concepts, chemistry does have its numerical side. Numbers can be very helpful in expressing, understanding, and applying chemical concepts. For example, a question currently under debate is whether modern society's love for burning fossil fuel is causing global warming. Burning fuel such as gasoline in cars or natural gas for heat produces carbon dioxide. This carbon dioxide helps trap the sun's warmth in our atmosphere. It is possible that we are producing so much carbon dioxide that we are affecting the planet's temperature, causing it to rise to the point where weather patterns are changing in undesirable ways. The question is, are we really doing this, or is our output of carbon dioxide negligible compared to natural producers of globally warming gases (decomposing vegetation, volcanic activity, etc.)? Only by a careful numerical analysis, at both the level of the chemical reactions involved and the global level, can we hope to find an answer to this question.

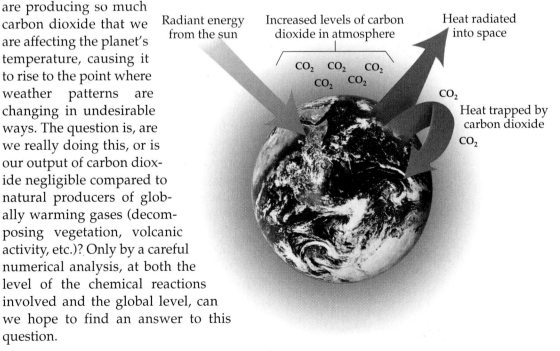

Radiant energy from the sun

Increased levels of carbon dioxide in atmosphere

Heat radiated into space

CO_2 CO_2 CO_2
CO_2 CO_2

CO_2

Heat trapped by carbon dioxide
CO_2

Let's tackle this numerical side of chemistry right now so we will be able to use it when we need to. You may already have a feel for some of the fundamental concepts that govern chemistry's numerical side. If you think about it a bit, numerical quantities are presented to you in one of two ways, either as an *exact number* or as the result of some *measurement*. For example, if someone tells you that there are 7 days in a week, the numerical quantity 7 is an exact number. It is an **exact number** because the number of days in a week is *universally defined* to be 7, and thus there is no uncertainty associated with this number. Also, the number of coins in your pocket is an exact number because you can't have a partial coin. That is, if you can *count* the number of items, the result will be an exact number.

On the other hand, some numerical quantities are the result of a **measurement**. For example, we could use a ruler to measure the diameter of a coin:

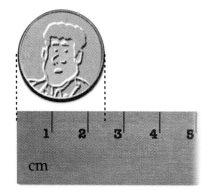

Let's agree that 2.5 cm is the true diameter of the coin. Next, we divide six students into two teams and ask each student to measure the coin once. Here are the results:

Team A	Team B
2.0 cm	2.2 cm
3.0 cm	2.2 cm
2.5 cm	2.2 cm

It looks like only the last student in team A measured the coin correctly. How can we explain the other students' results? Well, maybe one of the students in team A was tired, and another had difficulty locating the diameter. In team B, perhaps two of the team members were swayed by the third, who was absolutely convinced that the diameter was 2.2 cm. Bad measurements or not, these are the data presented by the teams, so let's work with them.

One thing you can do with repetitive measurements is calculate an average by summing the measurements and then dividing by the total number of measurements. For example, (2.0 cm + 3.0 cm + 2.5 cm)/3 = 2.5 cm is the average diameter for team A. The average diameters for both teams are shown below.

	Team A	Team B
	2.0 cm	2.2 cm
	3.0 cm	2.2 cm
	2.5 cm	2.2 cm
Average diameter:	2.5 cm	2.2 cm

We can use these data to discuss two important aspects of measured numerical quantities, *precision* and *accuracy*. **Precision** refers to the closeness of a series of measurements of the same object to one another. In our case, team B was precise because their measurements were all the same and thus as close to each other as possible. Team A was not precise because their measurements were all over the place, ranging from 2.0 cm to 3.0 cm. The term precision has meaning only when repetitive, identical measurements are made. We could not speak of precision if we measured the diameter just once.

If we look at the averages of the measurements, however, team A was more accurate. **Accuracy** refers to the closeness of a measured result to the true value, and team A's average diameter was closer to the true diameter (2.5 cm) than team B's. Unlike precision, the term accuracy can be applied if just a single measurement is made. For example, the third member of team A made the most accurate single measurement. We can summarize team B's results as being precise but not accurate, whereas team A's results were accurate but not precise. Thus, precision and accuracy do not have to go together. Of course, the goal is to achieve both accuracy and precision in measurements.

--PRACTICE PROBLEMS

2.1 A group of four people listening to a 2800 word speech tried to count the number of words in the speech as they were listening. They came up with the following results.

> Fred: 2736 words   Wilma: 2810 words
> Barney: 2792 words   Betty: 2734 words

(a) Which person was most accurate?
(b) Which person was most precise?
(c) Would you judge the group's average word count as being accurate or inaccurate?
(d) A different group of four people heard the same speech and came up with the following word counts: 2722, 2724, 2719, and 2723. Is this group more or less accurate than the first? Is this group more or less precise than the first?

Answer:
(a) *Barney was most accurate, since the number 2792 is closest to 2800, the actual number of words in the speech.*
(b) *Since each person made only one determination of word count, the term "precision" doesn't apply.*
(c) *The average word count for the group is (2736 + 2810 + 2792 + 2734)/4 = 2768. This is 32 less than the actual word count of 2800. Since 32 is quite small compared to 2800, the group's determination is pretty accurate.*
(d) *The average word count for the second group is (2722 + 2724 + 2719 + 2723)/4 = 2722. This is farther (78) from the true word count, so the second group is less accurate than the first. As for precision, the first group ranges from a high of 2810 to a low of 2734 (a spread of 76), whereas the second group ranges from a high of 2724 to a low of 2719 (a spread of 5). The smaller spread makes the second group more precise.*

2.2 Two students count the number of pieces of uncooked rice in a small cup. Both students repeat this measurement four times, with the following results:

Mike: 256, 263, 262, 266   Ike: 250, 242, 270, 278

The actual number of pieces of rice in the cup is 260. Which student is more accurate? Which student is more precise? Explain your answers.

2.3 Two students attempt to measure out a quart of water into a bucket. Jack has a half-quart container and Jill has a 10 gallon container. Which student will probably be more accurate at putting a quart of water into the bucket? Explain.

2.2 NUMBERS IN CHEMISTRY—UNCERTAINTY AND SIGNIFICANT FIGURES

Now, let's take an even closer look at numerical quantities that are the results of some measurement. For example, suppose you are told that a dinosaur bone is sixty-five million and twenty-three years old (65,000,023 years old). The number 65,000,023 has eight digits, but all of these figures do not have the same importance. Some, in fact, have no importance at all. The following discussion will give you some insight into why.

The problem with this reasoning is that the original age of 65 million years was approximate. Dating techniques used to determine the age of fossils are accurate to plus or minus a few million years at best. When compared with this uncertainty, the additional 23 years since the dating was carried out mean nothing. They are simply not significant relative to the error in the original measurement.

Like the fossil age of 65 million years, most numbers in chemistry come from some kind of measurement. You should always be aware that no measuring device can measure without some uncertainty. If you spend a lot of money on the best measuring device, you may lower the uncertainty, but there will always be some uncertainty associated with any measurement. For example, earlier we measured the diameter of a coin with a ruler marked only in centimeters:

We can see that the diameter is somewhere between 2 and 3 cm. The dashed line on the right seems to fall halfway between 2 and 3 cm, so we might record the diameter as 2.5 cm. The .5 is an estimate. Someone else might have estimated the diameter to be 2.4 cm. Yet another person might have estimated 2.6 cm. There is uncertainty in the last digit. Since the last digit is in the tenths position, we say that the uncertainty in the measurement is ±0.1 cm. You know that the uncertainty in the measurement is ±0.1 cm because you saw the ruler that was used to make the measurement. But what if you had never seen it and someone told you that the coin had a measured diameter of 2.5 cm? Without seeing the ruler, how would you know where the uncertainty in the measured number was? Perhaps a ruler with divisions of millimeters was used (as shown below, a millimeter is one-tenth of a centimeter). This would make the .5 a certainty instead of an estimate.

Without seeing the ruler you have no way of knowing how uncertain the measurement 2.5 cm is unless there is some agreed upon convention that tells you. Well, not surprisingly, there is just such a convention. The standard convention is that *the last digit written in a number is assumed to be where the uncertainty lies*. Thus, a measurement of 2.5 cm is uncertain in the tenths place, and the number is assumed to be uncertain to ±0.05 cm; that is, the actual distance could be anything from 2.45 to 2.55 cm. The ±0.05 is arrived at by placing a 1 in the position of the uncertainty (the tenths place) and dividing by 2 (0.1 ÷ 2 = 0.05). Using the ruler graduated in millimeters, we would report the coin diameter as 2.50 cm and not as 2.5 cm, since the uncertainty now lies one place farther to the right, in the hundredths position. By this same convention, the measured number 102 cm would be considered uncertain to ±0.5 cm because the last digit is in the ones position.

2.1 WORKPATCH   What is the uncertainty associated with the second ruler shown on page 27?
(a) ±0.5 cm   (b) ±0.05 cm   (c) ±0.005 cm

Explain your answer.

PRACTICE PROBLEMS

2.4 A desk is weighed and reported to be 185 lb. What is the assumed uncertainty in this weight?

Answer: The uncertainty is ±0.5 lb, since the last digit in the reported number is in the ones column.

2.5 A volume of liquid is measured as 16.0 gallons. What is the assumed uncertainty in this measured volume?

2.6 The voltmeter below is used to measure the voltage of a battery. Would the uncertainty in the measured voltage be:
(a) ±0.5 volt?   (b) ±0.05 volt?   (c) ±0.005 volt?

Now let's get back to the estimated coin diameter of 2.5 cm found by using the ruler marked only in centimeters. Suppose you are asked to calculate the radius of the coin. To do this you need to divide the diameter by 2.

Diameter measured with this ruler: Radius = Diameter divided by 2:

$$r = \frac{d}{2} = \frac{2.5 \text{ cm}}{2}$$

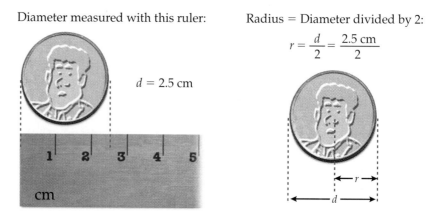

$d = 2.5$ cm

Your calculator would give 1.25 cm as the result. But wait! The last digit, 5, is in the hundredths column, implying that the radius is known to ±0.005 cm. How can this be? The ruler used to measure the diameter was accurate to only ±0.05 cm. Dividing the diameter by 2 didn't make the ruler more accurate. Even though your calculator gives a radius of 1.25 cm, the 5 is beyond the ruler's ability to estimate. Instead, the correct value for the radius, rounding up, would be 1.3 cm. This value is once again assumed to be uncertain to ±0.05 cm.

The moral is that you can't always keep all the digits a calculator produces. You can only keep the *significant* ones—the ones that are not beyond the accuracy of the measuring device. These digits are called **significant figures** (often shortened to **sig figs**) or **significant digits**.

ZEROS AND SIGNIFICANT FIGURES 2.3

The most important concept from the preceding section is that when you are presented with a measurement and don't know what measuring device was used, you should assume that the uncertainty lies in the last written digit. For example, if you are told that a board is 103.6 inches long, you should assume that the uncertainty is ±0.05 inch (since the 6 is in the tenths place) and all the digits are significant (there are four significant figures). But what about the number 0.06 inch? Are all three digits, the two 0's and the 6, significant? The answer is no. There is only one significant figure (the 6). The zeros are just placeholders, called *leading zeros*.

0.06 inch

Leading (placeholder) zeros are not significant

Leading zeros are all the zeros that precede the first nonzero digit. They are generally found in numbers whose absolute value is less than one. Our rule for where the uncertainty in the number is has not changed. The last written digit, 6, is in the hundredths column, so the uncertainty in the measurement 0.06 inch is ±0.005 inch. Notice that while the zeros in 0.06 inch are not significant, the zero in our first example, 103.6 inches, *is* significant. That's because the zero in 103.6 inches is not a leading zero. Leading zeros come before the first nonzero digit, never after it. Consider the following two numbers:

0.001 02   0.001 020

Both have three leading zeros (they are underlined). The first number therefore has only three significant figures, while the second number has only four (the digits that are not underlined).

Rule 1: **Leading zeros (all zeros before the first nonzero digit) are not significant.**

That last zero in the number 0.001 02<u>0</u> is given a special name, a *trailing zero*. **Trailing zeros** are those that appear to the right of the last nonzero digit in a number:

0.001 02   No trailing zeros
0.001 02<u>0</u>   One trailing zero

Trailing zeros are significant, and as such have meaning. The last zero in 0.001 020 is in the millionths column, and so the number 0.001 020 is known to an uncertainty of ±5 ten-millionths. The trailing zero gives the number 0.001 020 a smaller uncertainty than 0.001 02.

> **Rule 2: Trailing zeros (all zeros that follow the last nonzero digit *after* the decimal point) are significant.**

Try now to determine the number of trailing zeros and significant figures in each of the following numbers. Don't go on until you get the correct answers.

2.2 WORKPATCH

	Number of trailing zeros	Number of significant figures
20.201	?	?
20.210	?	?
20.0002	?	?
20.0	?	?
120.	?	?

The last number in the WorkPatch is an interesting case. It has one trailing zero, and its uncertainty is ±0.5. Now, you might think it a bit ridiculous to write the number one hundred twenty with a decimal point at the end. Why not just write it without the decimal point? If you wanted to buy one hundred twenty eggs, you would write 120 eggs on your shopping list, not 120. eggs. In fact, writing 120 eggs on your list is OK, since it is an exact (counted) number. However, if the number is a measurement, then leaving out the decimal point is the wrong thing to do. Without the decimal point, there are *two* ways to interpret the number: as 120 ±5 (that is, the zero is just a placeholder and is not significant), or as 120 ±0.5 (that is, the zero is a significant, trailing zero). With the decimal point included, there is no ambiguity. The decimal point indicates that any zeros to the left of the decimal point *are significant*. Since chemistry is a quantitative science, chemists must make sure that the numbers they write communicate a single clear meaning.

PRACTICE PROBLEMS

2.7 How many interpretations are there for the number 600?

Answer: With no decimal point at the end, 600 could be interpreted as having an uncertainty of ±50, ±5, or ±0.5. There is no way to know.

2.8 How would you express 600 inches ±0.5 inch?

2.9 Fill out the following table.

	Number of significant figures	Uncertainty
10.0	?	?
0.004 60	?	?
123	?	?

This brings us to an interesting question. Suppose you had measured the length of a building as 120 feet using a ruler that was marked only every 100 feet. This puts the uncertainty in the tens column, which means the measurement is uncertain to ±5 feet.

From the picture it looks as if the building is about 120 feet long. Of course, this is an estimate. Since the ruler is marked in only hundreds of feet, it is safe to say that you could be off by about 5 feet. Another person might read the ruler as 115 feet, and yet another might read it as 125 feet. That is, the length of 120 feet is uncertain to ±5 feet. Now, how are you supposed to write the number 120 and at the same time communicate to the world that the uncertainty in this measured number is ±5? You could not write 120. because that implies an uncertainty of ±0.5 foot. You could not write 120 because that is ambiguous. You could put a line under the middle digit (12̲0) and say "this is the last significant figure," but that is not how it is done. To accomplish this, we write the number in a different way, using *scientific notation*.

SCIENTIFIC NOTATION 2.4

The accepted way to unambiguously indicate uncertainty in a measured number is to use **scientific notation**—writing any numerical quantity as a number (A) multiplied by 10 raised to an exponent (x).

The general form of a numerical quantity written in scientific notation

$$A \times 10^x \qquad \textbf{Example:} \quad 1.2 \times 10^2$$

A represents some number   x is an exponent (or power) of 10

$A = 1.2$ $x = 2$

The thing to do now is *not panic*. Using scientific notation is easier than it might first appear. First, you should always include a decimal point in the number (A) part. The 10^x part is simply an instruction that tells you what to do with the decimal point in order to understand the real value of the number A. This principle works so long as the exponent (x) is a whole number, which it will always be in this book. That is, x will be 1, 2, 3, . . . , or 0, or -1, -2, -3, . . . , but not something like 1.28. When the exponent x is a positive number, you move the decimal point in A to the right x places. When the exponent is negative, you move the decimal point x places to the left. If the exponent is zero, then leave the decimal point alone. Below are shown some numerical quantities written in scientific notation and also in the normal way.

Scientific notation	Move decimal point	Normal notation
1.23×10^2	2 places to right	123.
1.23×10^{-1}	1 place to left	.123
1.23×10^0	Don't move it	1.23

How can you remember this? Look at what happened to the number 1.23. When the exponent was *positive* (10^2), the number 1.23 got *bigger* (it became 123.), because $10^2 = 100$. When the exponent was *negative* (10^{-1}), the number 1.23 got *smaller* (it became 0.123), because $10^{-1} = \frac{1}{10}$. When the exponent was zero (10^0), then 1.23 remained unchanged, because $10^0 = 1$.

Try the following practice problems now.

PRACTICE PROBLEMS

2.10 Convert 4.68×10^{-1} into normal notation.

Answer: The exponent is negative so the number must get smaller. Move the decimal point one place to the left: 0.468

2.11 Convert 47.3×10^{-2} into normal notation.

2.12 Convert 47.325×10^3 into normal notation.

Sometimes when you move the decimal point to convert a number from scientific notation to normal notation you will find that the decimal point moves right off the number. For example, consider the number 4.6×10^4. The exponent is positive, so we are supposed to move the decimal point *four places to the right* to make the number larger. If the exponent was negative (4.6×10^{-4}), the decimal point would move *four places to the left*. But the starting number has only two digits. What are we to do in these cases? Simple—we let the decimal point move off the number. Each time it moves farther off the number, it creates an empty column into which we place a zero. Look closely at how we handle this for our two examples.

What to do when the decimal point moves off the number on the right side

4.6×10^4 means move the decimal point four places to the right:

Three empty columns

4.6▼▼▼.

Moving the decimal four places to the right creates three empty columns without digits. Fill these empty columns with zeros:

46,000

What to do when the decimal point moves off the number on the left side

4.6×10^{-4} means move the decimal point four places to the left:

Three empty columns

.▼▼▼4.6

Moving the decimal four places to the left creates three empty columns without digits. Fill these empty columns with zeros:

.000 46

PRACTICE PROBLEMS

Write the following numbers in normal notation.

2.13 0.400×10^{-6}

Answer: 0.000 000 400

2.14 2.35×10^{-3}

2.15 6.0×10^{3}

Whichever way the decimal point moves, the procedure is the same. Simply fill the newly created empty columns with zeros.

We have been converting numbers from scientific into normal notation. You should also be able to go the other way and convert a number in normal notation into scientific notation. To do this, you first have to find the decimal point in the number. This is not a problem for a number like 62.8, where the decimal point is clearly written. When no decimal point is written, as in 125, assume there is a decimal point at the far right end.

Once you have located the decimal point, move it to the right of the first nonzero digit. In the number 125., the first nonzero digit is 1, so we move the decimal point two places to the left, making the number 1.25. Next, we must multiply 1.25 by the 10^{x} that would undo what we just did (in order to maintain the same value). Since we moved the decimal point two places to the left, we set $x = 2$ because 10^{2} would move the decimal point back two places to the right, giving the original number. So, in scientific notation, 125. is written as 1.25×10^{2}. Likewise, the number 0.000 50 is written as 5.0×10^{-4}. (We move the decimal four places to the right to place it after the first nonzero digit, and then multipy by 10^{-4} to undo this.)

Convert the following numbers from normal into scientific notation.   WORKPATCH  2·3
   (a) 123   (b) 0.000 06   (c) 0.000 060   (d) 1002.0

Did you notice that (b) and (c) in the WorkPatch look very similar? There's an important difference: the number in (c) has a significant trailing zero. Because this zero is significant, it must show up when the number is written in scientific notation. We never throw away significant figures, even if they are zero.

Now, if you recall, we started this discussion of scientific notation by wanting a way to write the number 120 feet ±5 feet. We can do this using scientific notation, but first we need to tell you one more critical thing. When converting from scientific to normal notation, *any zeros created by filling in empty columns due to the decimal moving off the number are not significant.* For example, consider the measured quantity 1.2×10^2 feet. Let's convert it to normal notation. The 10^2 tells us to move the decimal point two places to the right, creating one empty column which we must fill with a zero to arrive at 120 feet:

Empty column, fill with a zero

$$1.2 \times 10^2 \text{ feet} = 1.2\_ \text{ feet} = 120 \text{ feet}$$

But that added zero is not significant, as explained. The last significant digit, which is the one that always carries the uncertainty, is the 2. Thus, by writing the number 120 as 1.2×10^2, we are indicating that we mean 120 ±5.

Last significant digit

$$120 \pm 5 \qquad \text{is written in scientific notation as} \qquad 1.2 \times 10^2$$

What if we had used a more finely divided ruler graduated in feet? Then we would write 1.20×10^2 feet. The 10^2 still moves the decimal two places to the right, and we still get 120. feet, but now the last significant digit is the zero. Why? Because we included the zero (a significant, trailing zero) to begin with to indicate the reduced level of uncertainty.

Last significant digit

$$120 \pm 0.5 \qquad \text{is written in scientific notation as} \qquad 1.20 \times 10^2$$

Scientific notation gives us a way to indicate unambiguously where the uncertainty in a measured number is. This is one reason why we use it.

You try the next problem. Below, the diameter of a coin is measured with a ruler marked in centimeters and millimeters (each small division). Answer the following questions.

2.4 WORKPATCH

What is the uncertainty associated with this ruler?
(a) ±0.5 cm
(b) ±0.05 cm
(c) ±0.005 cm

Which choice for the diameter of the coin is correct?
(a) 5 cm
(b) 5.0 cm
(c) 5.00 cm
(d) 5.000 cm

Check your answers to the WorkPatch and make sure you understand before going on.

Another reason for using scientific notation is to write numbers that are huge or tiny more easily. For example, suppose you measure the length of some tiny particle with a powerful microscope and find it to be 0.000 000 120 meter long. All the zeros to the left of the digit 1 are placeholders and are not significant (the last zero is a trailing zero and is significant). To avoid having to write all those zeros, a scientist would write this number in scientific notation as 1.20×10^{-7}.

Stop, use pencil and paper, and prove to yourself that 1.20×10^{-7} is the same as

0.000 000 120

Writing 1.20×10^{-7} does away with all the placeholder zeros. This also works for very large numbers. Suppose a particular asteroid is measured to be 600,000,000,000 (six hundred billion) miles from us, and that the uncertainty in our measurement is ± 5 million miles. The arrow below points to the 10 millions column:

600,000,000,000

↑
10 millions column

This is where the uncertainty is. Thus, we want to indicate that all the zeros that come after it are not significant and are simply placeholders. To do this, we write this number in scientific notation as 6.0000×10^{11}.

With pencil and paper, prove to yourself that if you were given the number 6.0000×10^{11}, you (a) could come up with the normal notation 600,000,000,000, and (b) would be able to explain why the uncertainty is ± 5 million.

If you cannot do this WorkPatch problem, examine the solution at the end of the chapter.

─────PRACTICE PROBLEMS

Write the following numbers using scientific notation (to prove to yourself that it sometimes saves a lot of effort writing zeros).

2.16 Write the number 0.000 000 000 020 using scientific notation.

Answer: 2.0×10^{-11}

2.17 Write the number 47,100,000,000,000, using scientific notation. (Assume the uncertainty in the number to be ± 0.5 million.)

2.18 Again, write the number 47,100,000,000,000 using scientific notation. (This time, assume the uncertainty in the number to be ± 5 million.)

2.5 HOW TO HANDLE SIGNIFICANT FIGURES AND SCIENTIFIC NOTATION WHEN DOING MATH

When you do mathematical calculations with measurements, you must follow two rules to ensure that you don't end up with nonsignificant digits in your result. One rule applies for addition and subtraction. The other rule applies for multiplication and division. We'll cover the multiplication/division rule first.

Recall the earlier example of calculating the radius of a coin by dividing the measured diameter of 2.5 cm by 2. We couldn't keep the digit 5 in the resulting radius of 1.25 cm because it was not significant (it implied an uncertainty in the hundredths place, which was beyond the uncertainty of the ruler used to make the measurement). The rule determining the number of significant figures allowed in the result of multiplying or dividing two measured numbers is simple.

> **Rule 1: The result of multiplying or dividing two numbers cannot have more significant figures than the number with the fewest significant figures.**

For example, consider the following two multiplications:

$$\begin{array}{r} 2.0 \text{ cm} \quad \longleftarrow 2 \text{ sig figs} \\ \underline{\times\ 2 \qquad} \quad \longleftarrow 1 \text{ sig fig} \\ 4 \text{ cm} \quad \longleftarrow \text{Answer can have only 1 sig fig} \end{array}$$

BUT

$$\begin{array}{r} 2.0 \text{ cm} \quad \longleftarrow 2 \text{ sig figs} \\ \underline{\times\ 2.0 \qquad} \quad \longleftarrow 2 \text{ sig figs} \\ 4.0 \text{ cm} \quad \longleftarrow \text{Answer can have 2 sig figs} \end{array}$$

Sometimes the only way to keep your answer from having too many significant figures in it is to round it off. For example:

$$\begin{array}{r} 2.0 \text{ mm} \quad \longleftarrow 2 \text{ sig figs} \\ \underline{\times\ 8 \qquad} \quad \longleftarrow 1 \text{ sig fig} \\ 16 \text{ mm} \quad \longleftarrow \text{But the answer can have only 1 sig fig} \end{array}$$

Because you must round up to one significant figure, the correct answer is

20 mm   or, indicating one significant figure, 2×10^1 mm

You must always round off your answer if that is the only way to get the right number of significant figures.

PRACTICE PROBLEMS

2.19 27.5 sec/2.0 = ?

Answer: 27.5 sec/2.0 = 13.75 sec, but this is not correct. The answer can have only two significant figures in it, so the answer is rounded to 14 sec or 1.4 × 10¹ sec.

2.20 22.0 hr × 2.0 = ?

2.21 220. hr × 3 = ?

Now try this one.

What is the result of 222 ft/2.0 written with the correct number of significant figures?

Obviously, 222 ft/2.0 is 111 ft, but your result can have only two significant figures in it. So you need to write the answer using scientific notation. The answer is 1.1×10^2 ft. It is rounded off (1.1×10^2 ft = 110 ft), and it uses only two significant figures. The moral here is, use scientific notation to get you out of these significant figure jams.

Practice Problems

Answer the following problems to the correct number of significant figures (consider all numbers to be the result of some measurement):

2.22 1222 lb/2 = ?

Answer: 6×10^2 lb (Your calculator will give you 611, but your answer can have only one significant figure.)

2.23 (a) 1222 lb/2.0 = ?   (b) 1222 lb/2.00 = ?

2.24 9 × 9 = ?

The rule for how many significant figures are allowed in the result of an addition or subtraction is different from the multiplication/division rule.

> **Rule 2: When adding or subtracting a series of measurements, the result can be no more certain than the least certain measurement in the series.**

For example, consider adding the following measurements:

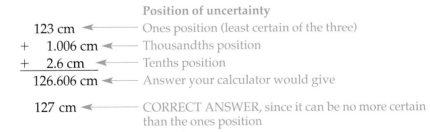

The least certain measurement in the series is 123 cm, which is uncertain in the ones position. The correct answer of 127 cm comes from rounding up the calculated result, 126.606, to the column where the uncertainty is, the ones position.

Try the following practice problems now.

PRACTICE PROBLEMS

Answer the following problems to the correct number of significant figures:

2.25 $1555 + 0.001 + 0.2 = ?$

Answer: 1555

2.26 $1555 + 0.001 + 0.8 = ?$

2.27 $142 - 0.48 = ?$

If you follow the rules given for multiplication/division and addition/subtraction, you will know how to round off the results your calculator gives you. Ignoring the rules of significant figures would put more certainty into your results than justified.

You will use a scientific calculator to do most of your calculations with numbers expressed in scientific notation. You can simply enter numbers directly in scientific notation and then add, subtract, multiply, or divide by pressing the appropriate keys. The calculator will handle the exponents for you. On most scientific calculators, to enter a number in scientific notation you first punch in the digits, then the $\boxed{\text{EE}}$ (or $\boxed{\text{EXP}}$) key, then the value of the exponent (x), and then the $\boxed{\text{+/−}}$ key if you want to make the exponent negative. Some examples are shown below:

Number	Key in as follows	How it would look on calculator
1.23×10^3	$\boxed{1}\boxed{.}\boxed{2}\boxed{3}\boxed{\text{EE}}\boxed{3}$	*1.23 E 03*
1.23×10^{-3}	$\boxed{1}\boxed{.}\boxed{2}\boxed{3}\boxed{\text{EE}}\boxed{3}\boxed{\text{+/−}}$	*1.23 E −03*

If you have trouble working with scientific notation on your calculator and the instruction manual doesn't help, see your instructor.

Finally, we need to mention how to keep track of significant figures when working with exact numbers. At the beginning of the chapter we defined an exact number as a number that does not come from a measuring device. Rather, it is an exactly known or defined quantity that has no uncertainty associated with it. For example, there are exactly 3 feet in a yard because that is how a yard is defined. The number 3 in this case is an exact number. The rule regarding exact numbers is that they have no effect on the number of significant figures in a calculated result. For example, suppose you measure the length of a house to be 60.50 feet (four significant figures). We could convert this to yards by dividing it by 3 (which is the same as multiplying it by 1/3), as shown below:

$$60.50 \text{ feet} \times \frac{1 \text{ yard}}{3 \text{ feet}} = 20.17 \text{ yard}$$

Measurement   Exact number

The answer has the same number of significant figures (four) as the original measurement. That is because 3 is an exact number. It has no uncertainty at all and therefore has no effect on the number of significant figures in the result.

NUMBERS WITH A NAME—UNITS OF MEASURE 2.6

A measurement makes no sense unless you know its units. For instance, it makes no sense to be told you can get to the airport in 3. Three minutes? Three hours? Three gallons of gas? You need the units.

Deciding which unit of measurement to use can be complicated because there are many possibilities for each type of measurement. Length can be measured in the system of inches, feet, yards, and miles, or it can be measured in the metric system of centimeters, meters, and kilometers. Volume can be measured in the system of quarts and gallons or in the metric system of liters. Weight can be measured in a variety of units. So there's potential for confusion in measurement, sometimes to a disastrous degree. Indeed, an airline crew once experienced serious problems because they knew their fuel load in one unit (pounds) but the airport crew in another country knew only metric units. A miscalculation was made in converting from one system to another and they ran out of fuel during the flight, making a forced (but safe) landing at an abandoned military field. To eliminate confusion, scientists always use metric units. In fact, scientists have gone even further and have adopted a standardized system of metric units called **SI units.** (SI stands for *Système Internationale*.)

Table 2.1 presents the basic SI units used in chemistry.

Table 2.1   Common SI Units Used in Chemistry

Physical quantity	Name of unit	Symbol
Length	meter	m
Mass	kilogram	kg
Time	second	s
Temperature	kelvin	K
Amount of substance	mole	mol

The basic SI unit of length is the meter (m). A meter is about $3\frac{1}{3}$ inches longer than a yard (a yard equals 3 feet).

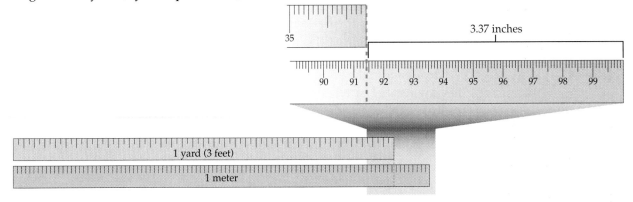

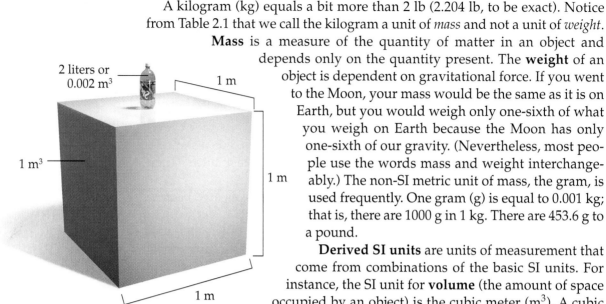

A kilogram (kg) equals a bit more than 2 lb (2.204 lb, to be exact). Notice from Table 2.1 that we call the kilogram a unit of *mass* and not a unit of *weight*. **Mass** is a measure of the quantity of matter in an object and depends only on the quantity present. The **weight** of an object is dependent on gravitational force. If you went to the Moon, your mass would be the same as it is on Earth, but you would weigh only one-sixth of what you weigh on Earth because the Moon has only one-sixth of our gravity. (Nevertheless, most people use the words mass and weight interchangeably.) The non-SI metric unit of mass, the gram, is used frequently. One gram (g) is equal to 0.001 kg; that is, there are 1000 g in 1 kg. There are 453.6 g to a pound.

Derived SI units are units of measurement that come from combinations of the basic SI units. For instance, the SI unit for **volume** (the amount of space occupied by an object) is the cubic meter (m^3). A cubic meter of volume is the space that would be enclosed by a box whose length, width, and depth were all 1 m. One cubic meter equals approximately 264 gallons.

Other derived SI units are listed in Table 2.2.

Table 2.2   Derived SI Units

Physical quantity	Name of unit	Symbol
Volume	cubic meter	m^3
Pressure	pascal	Pa
Energy	joule	J
Electrical charge	coulomb	C

Energy is a measure of the capacity to do work. The more energy a system has, the more work it can do. Such systems could be as diverse as a battery, a coiled spring, or you. Many chemical reactions can also release energy that can be used to do work—as when the burning of gasoline powers your car. The derived SI unit of energy, the joule (J), is a small amount of energy on an everyday scale. If you directly applied this much energy to the palm of your hand, you would barely feel it. It takes 4.184 J of energy to warm 1 g of water by 1°C. Burning a gallon of gasoline releases millions of joules of energy. A commonly used non-SI unit of energy is the calorie, which is a little more than four times larger than a joule (1 cal = 4.184 J). Nutritionists measure the energy of food in terms of Calories (Cal, with a capital C). A Calorie is equal to 1000 cal or 4184 J.

We often find the SI units in Tables 2.1 and 2.2 inconvenient to use because they are too big or too small for the job at hand. For example, suppose you wanted to measure out a teaspoon of liquid and express this volume in cubic meters. Such a small volume of liquid is only a tiny fraction of a cubic meter (1 teaspoon = 0.000 005 m^3). Would you want to have to ask for 0.000 005 m^3 of cough syrup every time you needed a teaspoon of it? And suppose you

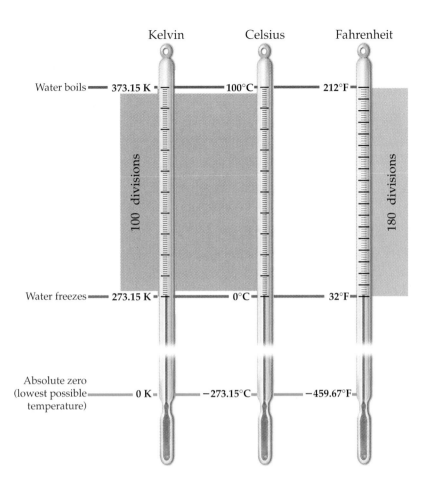

In the United States, people are a bit stubborn about using the Fahrenheit scale. Schools teach that water freezes at 32°F and boils at 212°F. Our normal body temperature is 98.6°F, and a pleasant day is about 75°F. Tell an average person in the United States that it is 35 degrees outside and the reaction would probably be, "35 degrees! I'd better wear a warm coat." Unfortunately, if that was 35°C, they would be in for a bit of a shock, since 35°C is the same as 95°F!

You should get used to the Celsius scale. Remember, a degree Celsius is about twice as large as a degree Fahrenheit, so if you go from a pleasant 24°C (75°F) to 30°C, you are increasing the temperature by 6°C, or almost 12°F. This takes us from 75°F to almost 86°F, a tad on the warm side. You can see the exact conversion below:

$$°F = 32 + \frac{9}{5} \times (30°C)$$
$$= 32 + \left(\frac{9 \times 30}{5}\right)$$
$$= 32 + 54$$
$$= 86°C$$

If you're used to units such as pounds, miles, quarts, gallons, and degrees Fahrenheit because you've worked with them all your life, it will take some practice to get used to the new units of measure. As you get practice using SI and metric units, you will get more of a feel for them also. Try the following WorkPatch to see if you are getting a feel for the metric volume units.

2.8 WORKPATCH   Suppose you want to fill a 2.5 L bucket with water using a 100 mL beaker. How many times will you have to fill your 100 mL beaker?

Could you do the WorkPatch in your head? The trick is to realize that 100 mL is one-tenth of a liter. Now try the following practice problems.

PRACTICE PROBLEMS

2.31 How many milliliters are there in 1.000 L?

Answer: Since a milliliter is 1/1000 of a liter, there are 1000 mL in a liter.

2.32 How many milliliters are there in 2.500 L?

2.33 How many milliliters are there in 246.7 cm^3 of water?

2.34 The temperature outside is 263.5 K. What is the temperature in degrees Celsius and in degrees Fahrenheit?

2.7 DENSITY: A USEFUL PHYSICAL PROPERTY OF MATTER

Density is a derived property that essentially tells you how much matter there is in a given volume. The "how much" is expressed as mass, usually in grams. Mathematically, density is defined as the mass per unit volume of the material in question:

$$\text{Density} = \frac{\text{Mass}}{\text{Volume}} = \frac{m}{V}$$

There is much less mass per unit volume of cloud than per unit volume of lead. A cloud has a much lower density than a piece of lead.

Is there more matter per given volume in clouds or in lead?

Unit volume

Unit volume

Chemists usually express densities in units of g/mL or g/cm^3. Table 2.4 lists the densities of some common substances. Notice how the last three entries in the table, all gases, have much lower densities than the other entries in the table, all liquids and solids.

Table 2.4   Densities of Some Common Substances*

Substance	Density (in g/mL)	Substance	Density (in g/mL)
Gold	19.3	Butter	0.9
Mercury (liquid)	13.6	Gasoline	0.7
Lead	11.4	Oxygen	0.001 43
Water (liquid)	0.997	Air	0.001 30
Water (solid ice at 0°C)	0.917	Helium	0.000 18

*Densities of liquids and solids are measured at 25°C; densities of gases are measured at 0°C and 1 atm pressure.

Try the following WorkPatch to see if you have a good feel for density.

WORKPATCH 2.9

Imagine that you have two solids A and B made of A atoms and B atoms, respectively. Assume that A and B atoms have the same mass. Suppose solid B has twice the density of solid A. Explain the different densities of A and B by drawing two small cubes, adding A atoms to one cube and B atoms to the other.

The density of liquid water is 0.997, or essentially 1.00 g/mL. This simply means that 1.00 mL of water weighs 1.00 g. Mercury, a liquid metal used in thermometers, has a density of 13.6 g/mL. This means that 1.00 mL of mercury weighs 13.6 g. Thus, 1.00 mL of mercury is 13.6 times heavier than the same volume of water.

Density can be very useful in identifying and characterizing an unknown substance. For example, suppose you were given a small cube of golden yellow metal that measured 2.00 cm × 2.00 cm × 2.00 cm and needed to determine if it was gold or painted lead. One way to do this would be to weigh the cube of metal and then divide this weight by the volume of the block to get its density. You could then compare this density to the entries in Table 2.4.

Volume of block = 2.00 cm × 2.00 cm × 2.00 cm = 8.00 cm^3 or 8.00 mL

Weight of block = 91.2 g

$$\text{Density} = \frac{91.2 \text{ g}}{8.00 \text{ mL}} = \frac{11.4 \text{ g}}{\text{mL}}$$

Since lead has a density of 11.4 g/mL and gold has a density of 19.3 g/mL, the yellow metal block must be lead painted to look like gold. Now you have a very accurate way to determine if the gold someone is selling you is really gold, and it is totally independent of the size of the sample. This is because density is an **intensive property**, a property that does not depend on the amount of material. Had you been given a larger or smaller block of metal than the one we just analyzed, the density would have been the same. The opposite of an intensive property is an **extensive property,** one that does depend on the amount present. Examples of extensive properties of matter are mass and volume. Since density equals mass/volume, we see that an intensive property—density—can be derived from two extensive properties.

Measuring the density of an object requires determining its mass and its volume. Determining mass is usually the easy part. That just requires weighing it. Determining volume, however, can be difficult if the object is irregularly shaped. For example, how could you easily measure your own volume? You could not simply multiply your length times your width times your depth as we do for a box, since your width and depth change as you go from feet to head. An easy way to measure the volume of an irregularly shaped object is to immerse it completely in a container filled to the very top with water. That will cause a volume of water equal to the volume of the object to be displaced and spill from the container. Collecting this overflow water and measuring its volume will give you the volume of the object. This is called determining the volume of an object by displacement.

Measuring the volume of an irregularly shaped object by displacement

PRACTICE PROBLEMS

2.35 Which is more dense, a 200 lb block of lead or a tiny, 0.1 g piece of gold?

Answer: The tiny piece of gold is more dense (see Table 2.4). The size of the piece does not matter since density is an intensive property.

2.36 A small cube that is 10.0 mm × 10.0 mm × 10.0 mm weighs 4.70 g. What is its density in grams per milliliter?

2.37 A small statue that weighs 500.0 g displaces 150.5 mL of water. What is its density in grams per milliliter?

2.8 DOING CALCULATIONS IN CHEMISTRY—UNIT ANALYSIS

Now we want to review a general strategy for solving many of the numerical problems that come up in chemistry. Often the biggest problem for students is knowing where to start. It can often appear that a blizzard of numbers has blown your way. For example, consider the following problem: If you weigh 2.00 g of the liquid metal mercury (Hg), what volume will it occupy in milliliters given that the density of mercury is 13.6 g/mL?

We have a measured mass of mercury (2.00 g) and a density of mercury (13.6 g/mL), and we need to find a volume in milliliters. Where should we begin? The answer comes from looking very closely at the units associated with the numerical quantities in the problem. The density of mercury has two different units, grams and milliliters. The mass of mercury, 2.00 g, has just one unit. The general strategy for solving a numerical problem makes use of this information in a two-step approach called **unit analysis**.

The first step of unit analysis is to write down the given measurement.

Step 1: Write down the given measurement.

2.00 grams mercury

The second step will be to use the density of mercury, but we first want you to look at density in a new way—as a conversion factor. A **conversion factor** can be recognized because *it will always have two different units in it, one over the other*. For example, you know that there are 60 seconds in a minute,

60 seconds = 1 minute

This is a conversion factor. It is often written in ratio form, as:

$$\frac{60 \text{ seconds}}{1 \text{ minute}} \quad \text{or} \quad \frac{1 \text{ minute}}{60 \text{ seconds}}$$

60 seconds per 1 minute   1 minute per 60 seconds

You can write it either way. The left ratio says there are 60 seconds per 1 minute (60 sec/min) while the right ratio says there is 1 minute per 60 seconds (min/60 sec). Both are true. Both ratios have two units associated with them, seconds and minutes, one over the other. *This is the hallmark of a conversion factor.*

Since the density of mercury is 13.6 g/mL, we can write

1 mL mercury = 13.6 g   *Note:* The "1" in "1 mL mercury" is treated as an exact number in all calculations.

and we can write this in ratio form as

$$\frac{13.6 \text{ g mercury}}{1 \text{ mL mercury}} \quad \text{or} \quad \frac{1 \text{ mL mercury}}{13.6 \text{ g mercury}}$$

Either form is correct, although traditionally density is written as mass over volume.

So, step 1 was to write down the measurement (2.00 g mercury). Step 2 is to multiply the measurement by the appropriate conversion factor in such a way that the unit of the measurement cancels out. For our example this means we do the following:

Step 2: Multiply the measurement by a conversion factor so that the unit of the measurement cancels.

$$2.00 \text{ g} \times \frac{1 \text{ mL}}{13.6 \text{ g}}$$

The conversion factor is written as 1 mL/13.6 g because this way, the units of grams will cancel out. Identical units cancel out when they appear in both the top (numerator) and the bottom (denominator) of a ratio because anything over itself equals 1.

Identical units on top and bottom cancel

$$2.00\ \cancel{g} \times \frac{1\ mL}{13.6\ \cancel{g}} = 0.147\ mL$$

Had we written the density "conversion factor" the other way (with grams on top and milliliters on bottom), the unit grams would appear twice on top and would not cancel. Step 2 says to multiply by the conversion factor *so as to make the unit of the number cancel!* The unit that survives is milliliter, a unit of volume. This is what the question asked us to solve for (the volume occupied by the mercury), so we are done. The answer to this question is just 2.00/13.6 or 0.147 mL.

PRACTICE PROBLEMS

2.38 Which of the following can be thought of as conversion factors?
  (a) 49.3 kg
  (b) 4.184 J/°C
  (c) 350 miles per hour
  (d) 12 eggs per dozen
  (e) 1 dozen grams

Answer: (b)–(d) are all conversion factors.

2.39 Write two conversion factors (two different ratios) to express the fact that there are 24 hours in a day.

2.40 Given a speed of 600.0 miles per hour and a measured distance of 50.0 miles, multiply the conversion factor by the measured distance to make identical units cancel. Calculate the result and include its unit.

2.41 Given a speed of 600.0 miles per hour and a measured time of 50.0 hours, multiply the conversion factor by the measured time to make identical units cancel. Calculate the result and include its unit.

One of the benefits of unit analysis is that you can use step 2 over and over again, multiplying by conversion factor after conversion factor, to put the final answer into any units you desire. For example, let's go back to our question about the volume occupied by 2.00 g of mercury. Suppose we wanted the answer to this question in liters instead of milliliters. All we would have to do is multiply by another conversion factor, as shown below.

If you are not at the units you want, keep multiplying by conversion factors

$$2.00\ \cancel{g} \times \frac{1\ \cancel{mL}}{13.6\ \cancel{g}} \times \frac{1\ L}{1000\ \cancel{mL}} = 0.000\ 147\ L$$

The new conversion factor comes from the fact that there are 1000 mL per liter. Note that it is written in such a way as to cancel out mL and make liters appear in the answer (mL on the bottom and L on the top). You can string out as many conversion factors as necessary to come up with the unit the problem is asking for. To do the math when many conversion factors are strung together, simply multiply everything on top, then divide by everything on the bottom. For this problem, you would key the following into your calculator: $2.00 \div 13.6 \div 1000 =$.

The next example will show you the power of stringing out conversion factors. Suppose it took you a total of 3.5 weeks to read this book. How many seconds did it take you to read it? You could use the following conversion factors:

> 60 seconds = 1 minute
> 60 minutes = 1 hour
> 24 hours = 1 day
> 7 days = 1 week

Once again we use the two-step approach of unit analysis: Step 1, write down the measured time it took you to read the book; step 2, multiply it by conversion factors to cancel out its unit and arrive at the unit you want. With respect to significant figures, you can treat all of the above conversion factors as exact, since they derive not from some measurement but from a definition (for example, a minute is defined as having 60 seconds in it). Remember, when you are doing math, exact quantities have no effect on the number of significant figures that will be in the answer.

How long did it take to read this book in seconds if it took you 3.5 weeks to read it?

$$3.5 \text{ weeks} \times \frac{7 \text{ days}}{1 \text{ week}} \times \frac{24 \text{ hours}}{1 \text{ day}} \times \frac{60 \text{ minutes}}{1 \text{ hour}} \times \frac{60 \text{ seconds}}{1 \text{ minute}} = 2,116,800 \text{ seconds}$$

Since the starting number has only two significant figures, the answer can have only two significant figures, so the answer is

2.1×10^6 sec

The method of unit analysis, starting with the number and then multiplying by conversion factors to make units cancel, is very simple and yet very powerful. Many conversion factors useful for doing numerical calculations in chemistry are given inside the back cover of this book. Try your hand now at the following practice problems.

PRACTICE PROBLEMS

2.42 The density of gold is 19.3 g/cm^3. What volume in milliliters will 20.0 g of gold occupy? [*Hint:* Don't be fooled. Remember that $1 \text{ cm}^3 = 1 \text{ mL}$.]

Answer: $20.0 \text{ g} \times \dfrac{1 \text{ mL}}{19.3 \text{ g}} = 1.04 \text{ mL}$

2.43 The density of air is $0.001\,30 \text{ g/mL}$. How much will 500.0 L of air weigh in grams? In kilograms?

2.44 The measured density of lead is 11.4 g/mL. What volume in milliliters will 1.50 lb of lead occupy? [1 lb = 453.6 g]

2.45 It takes 6 cups of flour to bake 1 cake; exactly 1 cup of flour weighs 120.0 g. If you have 6955 g of flour, how many cakes can you bake? [*Hint:* Two conversion factors are given to you in this problem. Find them and write them down first in ratio form. Then start with the number.]

2.9 ALGEBRAIC MANIPULATIONS

When you work with mathematical equations it is often desirable to rearrange them (rewrite them) to make it easier to arrive at the answer. This rearranging of an equation is called **algebraic manipulation** (or **algebra**, for short). You can often avoid having to do this by using unit analysis; however, sometimes it will be to your advantage to do some algebra. The fundamental rule when rearranging equations algebraically is simple: *When rearranging an equation, you are not allowed to do anything that changes the meaning of the equation; you can only change its appearance.*

A good example for demonstrating algebraic manipulation is density. We have seen that density is defined as mass over volume ($d = m/V$). Like any equation, $d = m/V$ can be algebraically rearranged at will. Let's return to an earlier problem that we already solved by unit analysis. If you have 2.00 g of liquid mercury, what volume will it occupy if its density is 13.6 g/mL? In this problem we are being asked to find a volume. This means we want to algebraically rearrange the density equation $d = m/V$ so that it ends up looking like $V =$ something. This is called solving the equation for volume. How can we do this? By following the fundamental rule: Do anything you want to this equation so long as it doesn't change its meaning. This means you can add anything to the equation, subtract anything from it, multiply it by something, or divide it by something, so long as you don't change its meaning. The way to ensure that the meaning isn't changed is *to do exactly the same thing to both sides of the equation.* If we add something on the left side of the equal sign, we must add exactly the same thing on the right side of the equal sign. If we multiply by something on the left side, we must multiply by exactly the same thing on the right side. Let's try this now for our density equation. Suppose we multiply both sides of the equation by volume (V), the quantity we are being asked to solve for:

$$d = \frac{m}{V}$$

Multiply both sides by V:

$$V \times d = \frac{m}{V} \times V$$

In a mathematical equation, anything over itself is equal to 1. On the right side of the equation we now have V on the top and V on the bottom. This is V/V, which equals 1. Follow the diagram at the top of the next page to see how this changes the appearance (but not the meaning) of our equation.

$$V \times d = \boxed{\frac{m}{V}} \times V \quad \begin{array}{l} V \text{ on top} \\ V \text{ on bottom} \end{array}$$

$$V \times d = m \times \frac{V}{V} \qquad V/V = 1, \text{ and } m \times 1 = m$$

$$V \times d = m$$

We now have $V \times d = m$. We are almost at our target of $V = $ something. We just have to remove d from the left side. To do this, we divide both sides of the equation by d. Notice how this further changes the appearance of our equation:

$$\frac{V \times d}{d} = \frac{m}{d} \qquad \begin{array}{l} \text{Dividing both sides by } d \text{ gives us} \\ d/d \text{ on the left side, which equals 1,} \\ \text{and } V \times 1 = V \end{array}$$

$$V = \frac{m}{d}$$

We have done it! By algebraic manipulation, we have taken the equation for density $d = m/V$ and rewritten it as $V = m/d$ without changing its meaning. We have solved it for V. Now, to calculate the volume that the question asked for, we simply take the mass and divide it by the density:

$$V = \frac{m}{d} = \frac{2.00 \text{ g}}{13.6 \text{ g/mL}} = 0.147 \text{ mL}$$

This is exactly the same answer we arrived at by the method of unit analysis.

Another way to look at algebraic manipulations is to consider yourself as an eliminator of variables. The variables are just the quantities in the equation [in our last example, mass (m), volume (V), and density (d) were the variables]. Your job is to eliminate the ones you don't want from one side of an equation, leaving behind only the one you want. For example, given the equation for density $d = m/V$, how could you quickly solve it for m? The answer is to eliminate the V from the right side and get m all by itself. The question then becomes, what do you have to do to eliminate V from the right side? The answer is to multiply both sides by V. This would give $Vd = Vm/V$, and $Vm/V = m$ (the V's cancel). You end up with $Vd = m$, and you have solved for m.

Sometimes you have to add or subtract instead of multiply or divide. For example, if you are given the equation $p = q + r$ and you are asked to solve it for q, then you need to get rid of the r from the right side. Since r stands alone (that is, it is not multiplied or divided by anything else), simply subtract r from both sides (*always do the same thing to both sides*). This gives $p - r = q + r - r$. On the right side, $r - r = 0$ (anything minus itself is zero), and so the r's are eliminated from the right side. We end up with $p - r = q$. We have solved the equation for q.

Try your hand at algebraic manipulations in the following problems now. Do whatever it takes to eliminate unwanted variables from one side of an equation. Remember that whatever you do to one side of the equation you must also do to the other. The first two problems involve multiplying or dividing. The last problem involves adding or subtracting.

Practice Problems

Carry out the following algebraic manipulations. Show how you solved each problem, as is done in the answer to Problem 2.46.

2.46 Given $p/q = r$, solve for p.

Answer:　To solve $\dfrac{p}{q} = r$

for p, we must eliminate q from the denominator of the left side. To do this, multiply both sides by q:

$$\frac{\cancel{q}p}{\cancel{q}} = rq$$

q cancels from the left side because it is in both the top and the bottom.

2.47 Given $p/q = r$, solve for q.

Answer:　$q = \dfrac{p}{r}$

2.48 Given $P + Q = z$, solve for P.

Answer: $P = z - Q$

As we said earlier, you can't avoid the numerical side of chemistry if you hope to study it in any detail. This chapter has prepared you for dealing with this numerical side, and it is important. However, the fundamental concepts of chemistry and the pictures, models, and stories that are discussed in the remainder of this book are just as important, if not more so. After all, you can't use the mathematical tools that you just learned to solve a problem until you understand what the problem is about.

Have You Learned This?

Exact number (p. 24)

Measurement (p. 24)

Precision (p. 25)

Accuracy (p. 25)

Uncertainty (p. 26)

Significant figures or significant digits (p. 29)

Leading zeros (p. 29)

Trailing zeros (p. 30)

Scientific notation (p. 31)

SI units and derived SI units (pp. 39, 40)

Mass (p. 40)

Weight (p. 40)

Volume (p. 40)

Energy (p. 40)

Density (p. 44)

Extensive and intensive properties (p. 46)

Unit analysis (p. 47)

Conversion factor (p. 47)

Algebraic manipulation (p. 50)

NUMBERS IN CHEMISTRY—PRECISION AND ACCURACY

2.49 In one case you are told that there are 3 feet in a yard. In another case you are told that a piece of wood is 3 feet long. What is the fundamental difference between the value of 3 feet in these cases?

2.50 What is the difference between precision and accuracy? Do they ever mean the same thing?

2.51 Suppose in making repetitive measurements you could have only a precise result or an accurate result, but not both. Which would you choose, and why?

2.52 Three people measure the distance from Main Street to Market Avenue using their car odometers. Their data are as follows: 2.3 miles, 2.6 miles, 3.1 miles. Survey charts show the actual distance to be 2.6 miles. Characterize the collected data in terms of accuracy and precision.

2.53 Two people attempt to measure the length, in feet, of a parking lot that they know is about 100 feet long. One person uses a 6 inch plastic ruler; the other uses a 120 foot tape measure (both graduated in 1/16 inch). Which person is likely to make the more accurate measurement? Explain why.

NUMBERS IN CHEMISTRY—UNCERTAINTY AND SIGNIFICANT FIGURES

2.54 A ruler is marked in 1/8 inch intervals. Is the uncertainty associated with the ruler $\pm 1/8$ inch or $\pm 1/16$ inch? Explain.

2.55 How are we to tell what the uncertainty is in a measurement?

2.56 Why do all measurements have some uncertainty?

2.57 What is the assumed uncertainty for each of the following measured numbers?
(a) 12.60 cm   (b) 12.6 cm   (c) 0.000 000 03 inch   (d) 125 feet

2.58 Indicate the number of significant figures for each of the numbers in Problem 2.57.

2.59 Using radioactive elements, scientists determined the age of a dinosaur skeleton to be 78.5 million years old. In years, what is the uncertainty in this measurement?

2.60 Underline any trailing zeros in the following measurements.
(a) 12.202 km   (b) 0.01 mL   (c) 205°C   (d) 0.010 g

2.61 For the measurements in Problem 2.60, circle all the zeros that are significant.

2.62 What is the uncertainty associated with each of the measurements in Problem 2.60?

2.63 The measurement 30 feet is ambiguous, but the measurement 30. feet is not. Explain what the ambiguity is and how adding the decimal point eliminates the ambiguity.

2.64 Give all the possible interpretations of the measurement 2200 feet.

SCIENTIFIC NOTATION

2.65 Convert the following measurements from scientific notation to normal notation. For each one, indicate the number of significant figures present.
(a) 5.60×10^1 kg   (b) 2.5×10^{-4} m
(c) 5.600×10^6 miles   (d) 0.02×10^2 feet

2.66 What is the uncertainty for each of the measurements in Problem 2.65?

2.67 Using scientific notation, write the measurement 30 ft as having an uncertainty of:
(a) ± 5 ft   (b) ± 0.5 ft   (c) ± 0.05 ft

2.68 Using scientific notation, write the measurement 2200 ft as having an uncertainty of ±50 feet.

2.69 Convert the following numbers from normal notation to scientific notation.
(a) 226   (b) 226.0   (c) 0.000 000 000 50   (d) 0.3   (e) 0.30
(f) 900,000,574 (with an uncertainty of ±0.5 million)
(g) 900,000,574 (with an uncertainty of ±50)

2.70 How many significant figures are there in the following measurements, and what is the uncertainty in each of the measurements?
(a) 0.001 kg   (b) 0.000 10 m   (c) 102 L
(d) 2.600×10^{-3} m   (e) 1.1×10^6 km

HOW TO HANDLE SIGNIFICANT FIGURES AND SCIENTIFIC NOTATION WHEN DOING MATH

2.71 To the correct number of significant figures, what is the result of adding the following measurements: 100. inches + 2 inches + 0.001 inch? What is the uncertainty in the result?

2.72 To the correct number of significant figures, what is the product of each of the following multiplications? Use scientific notation when necessary. (No units means the number is exact.)
(a) 2.3 cm × 2   (b) 2.3 m × 2.0 m   (c) 1000 J × 10   (d) 0.1 mm × 124 mm

2.73 A student walks 20,450.2 feet to school every day. A mile is defined as 5280 feet. How many miles does the student walk to school each day? Make sure your answer is written to the correct number of significant figures.

2.74 Do the following calculations using a scientific calculator. Watch your significant figures.
(a) 3.33×10^4 km + 2.22×10^5 km   (b) 2.444×10^9 J/2.444×10^{-9} J
(c) 2.34×10^2 m − 2.34×10^1 m   (d) 4.00×10^4 L + 6.00×10^{-1} L

NUMBERS WITH A NAME—UNITS OF MEASURE

2.75 What is the basic SI unit of length? Of volume?

2.76 What are two metric but non-SI units of volume, and why are they more often used than the basic SI unit of volume?

2.77 When is it correct to use cm^3 instead of mL? Explain your answer.

2.78 Why was the SI unit system developed by scientists?

2.79 Write the following measurements in units of meters using scientific notation (watch your significant figures).
(a) 2.31 gigameters (Gm)   (b) 5.00 micrometers (μm)
(c) 1004 millimeters (mm)   (d) 5.00 picometers (pm)
(e) 0.25 kilometer (km)

2.80 Which is larger, a degree Celsius or a degree Fahrenheit? Explain why.

2.81 Which temperature scales can have negative temperatures and which can't? For those that can't, explain why not.

2.82 Convert the following temperatures as requested.
(a) 22.5°C (convert to Fahrenheit and Kelvin)
(b) −3.00°F (convert to Celsius and Kelvin)
(c) 0.0 K (convert to Celsius and Fahrenheit)
(d) 65.1°C (convert to Fahrenheit and Kelvin)

2.83 How cold does it have to be for water to freeze in °F? In °C? In kelvins?

2.84 Where does the 9/5 ratio come from in the formula for converting °C into °F?

2.85 You are filling a bottle with water by pouring into it the contents of the two grad-uated cylinders shown below.

(a) What is the uncertainty in each graduated cylinder?
(b) What is the final volume of water in the bottle after you pour the contents of both graduated cylinders into it? What is the uncertainty of this volume? Explain your answers.

2.86 A student measures the length of one side of a perfect cube with the ruler below, which is marked in centimeters. She then calculates the volume, V, by cubing the length, l ($V = l^3 = 1 \times 1 \times 1$). What volume should she report? Give the answer to the proper number of significant figures and with the proper units of volume for this problem.

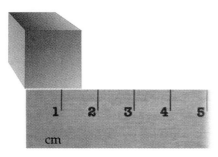

2.87 Using a ruler marked in centimeters and millimeters, a student measures the diameter of a ball to be 1.5 cm. His partner measures the same ball with the same ruler and comes up with 1.50 cm. The lab instructor scolds one of them for using the ruler incorrectly. Which one gets the scolding and why?

2.88 The students measure another ball with the ruler in Problem 2.87 and determine that its diameter is 2.55 cm. What is the radius of the ball to the correct number of significant figures?

DENSITY

2.89 Define density, and explain why units of density are called derived SI units.

2.90 How much would 1000.0 mL of liquid water at 25°C weigh in grams (consult Table 2.4)?

2.91 How much would 2.0 L of mercury at 25°C weigh in grams (consult Table 2.4)?

2.92 How much would a stick of butter whose dimensions are 10.0 cm × 10.0 cm × 10.0 cm weigh in grams (consult Table 2.4)?

2.93 Explain how you would measure the density of a pumpkin.

2.94 Two students attempt to measure the density of gold. One works with a 100 g bar of pure gold. The other works with a 200 g bar of pure gold. Which student will measure the larger density? Explain your answer fully.

2.95 It is a fact that an object will sink in a liquid if it is denser than the liquid, but it will float on the liquid if it is less dense than the liquid. Knowing this, consult Table 2.4 and describe a quick flotation test to distinguish lead from gold.

UNIT ANALYSIS

2.96 Suppose it takes you 1.25 days to drive to your aunt's house. How many seconds does it take you to get there? Use the unit analysis approach and show your work.

2.97 A train traveling at 45.0 miles per hour has to make a trip of 100.0 miles. How many minutes will it take for the train to make the trip? Use the unit analysis approach and show your work.

2.98 You have a great job in which you earn $25.50 per hour. How much do you earn per second? Use the unit analysis approach and show your work.

2.99 Gold has a density of 19.3 g/mL. Suppose you have 100.0 glonkins of gold. What volume in liters will the gold occupy? Here are some conversion factors to help you: There is 0.911 ounce per glonkin. There are 28.35 grams per ounce. Use the unit analysis approach and show your work. Treat conversion factors as exact.

2.100 One liter is equal to 0.264 gallon. Suppose you have 1.000×10^3 cm^3 of water. How many gallons of water do you have? Use the unit analysis approach and show your work. Treat conversion factors as exact.

2.101 You measure the dimensions of a rectangular block to be 10.2 cm × 43.7 cm × 95.6 cm. What is its volume in liters?

2.102 The block of Problem 2.101 weighs 2.43×10^2 kg. What is the density of the block in grams per milliliter?

2.103 You measure the side of a cube using a meter stick marked in centimeters. Unfortunately, the cube is longer than a meter. You mark where a meter ends with a pen and then, using a 15 cm ruler marked in millimeters, you measure the remaining distance to be 1.40 cm.
(a) What is the length of the side of the cube in centimeters?
(b) What is the volume of the cube in cubic centimeters? The lengths of the sides of a cube are all equal. Watch your significant figures. Use scientific notation if you have to.
(c) You weigh the cube on a scale. It weighs 111 kg. What is the density of the cube in grams per milliliter? Watch your significant figures.

ALGEBRAIC MANIPULATIONS

2.104 Why can't you multiply just one side of an equation by something when algebraically rearranging it?

2.105 Solve the following equation for x: $y = z/x$

2.106 Solve the following equation for x: $y = z - x$

2.107 Solve the following equation for x: $y = (z/x) + 2$

2.108 The density of an unknown liquid is 1.15 g/mL. How many grams of it are necessary to fill a volume of 50.00 mL? Do this problem by the method of algebraic manipulation, showing all steps.

2.109 Repeat Problem 2.108 using the method of unit analysis. Show all steps. Do you get the same answer as for Problem 2.108?

WorkPatch Solutions

2.1 (c) ±0.005 cm

2.2

	Number of trailing zeros	Number of significant figures
20.201	None	5
20.210	1	5
20.0002	None	6
20.0	2	3
120.	1	3

2.3 (a) 1.23×10^2 (b) 6×10^{-5} (c) 6.0×10^{-5} (d) 1.0020×10^3

2.4 Each small division represents 0.1 cm. You can estimate in between these lines, or to 0.01 cm. This gives an uncertainty of (c) ±0.005 cm. The diameter of the coin is (c) 5.00 cm.

2.5 In 1.20×10^{-7}, the exponent -7 instructs us to move the decimal point 7 places to the left, filling in the empty columns we generate with zeros:

Fill with zeros

.↓↓↓↓↓↓↓1.20 = .000 000 120

2.6 (a) In 6.0000×10^{11}, the exponent instructs us to move the decimal point 11 places to the right, filling any empty columns we generate with zeros:

Last significant digit — Fill with zeros ⌐Last significant digit

6.0000↓↓↓↓↓↓↓↓↓↓↓ = 600,000,000,000

(b) The last significant digit is the fourth zero from the 6. It is in the 10 millions column, so the number is ±5 million.

2.7 1.1×10^2 ft

2.8 Since it takes 1000 mL to get a liter, you would have to fill your 100 mL beaker 10 times to get 1 L. To get 2.5 L, you would have to fill your 100 mL beaker 25 times.

2.9 The diagrams below show that solid B has twice as many atoms per unit volume as does solid A. Since we said that atoms A and B have the same mass, this would give solid B twice the density (twice the mass per unit volume) of A.

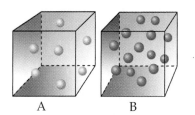

A   B

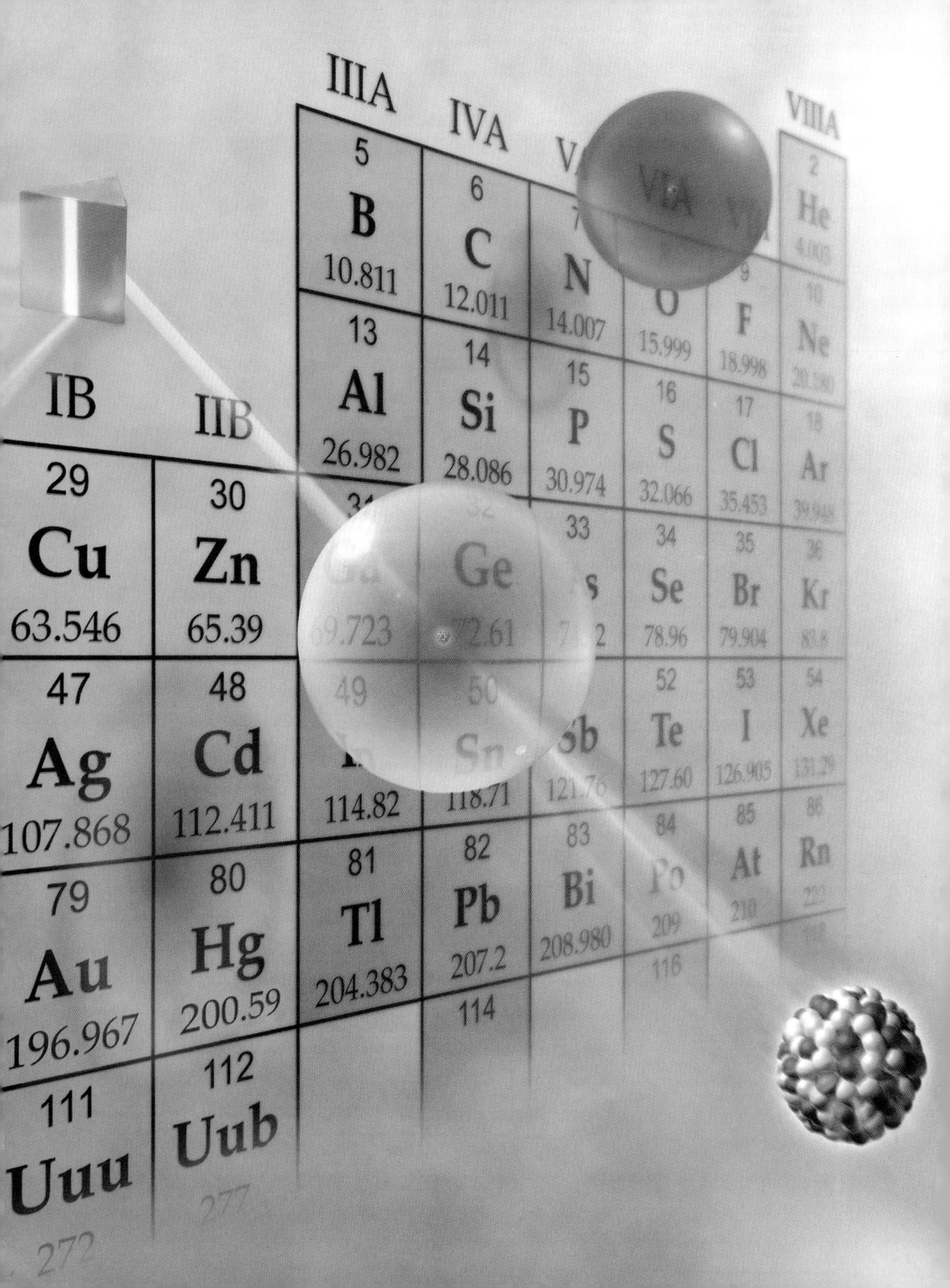
IIIA
IVA
VIIIA
5
B
10.811
6
C
12.011
N
14.007
15.999
F
18.998
2
He
Ne
IB
IIB
13
Al
26.982
14
Si
28.086
15
P
30.974
16
S
32.066
17
Cl
35.453
18
Ar
39.948
29
Cu
63.546
30
Zn
65.39
Ge
33
34
Se
78.96
35
Br
79.904
36
Kr
47
Ag
107.868
48
Cd
112.411
114.82
118.71
52
Te
127.60
53
I
126.905
54
Xe
79
Au
196.967
80
Hg
200.59
81
Tl
204.383
82
Pb
207.2
83
Bi
208.980
84
209
85
At
210
86
Rn
114
116
111
Uuu
112
Uub

The Evolution of Atomic Theory

DALTON'S ATOMIC THEORY 3.1

Recall from Chapter 1 the battle between the philosophies of Democritus and Aristotle that took place around 400 BC. Aristotle thought that matter was continuous and could be infinitely subdivided into smaller and smaller particles. Democritus postulated that matter was ultimately made up of tiny, indivisible particles which he called *atoms*. Aristotle's view won the day, and it took over 2000 years before the atomic theory of matter made a comeback. Historians generally credit the beginning of modern atomic theory to the work of a mild-mannered English schoolteacher named John Dalton in the early 1800s, and it is here that we will begin.

John Dalton's atomic model was just that, a model in the true spirit of the scientific method. It was formulated as an explanation or rationalization for a number of "laws" about how matter behaves in a chemical reaction. These laws were the result of careful experimental work done in the latter part of the 1700s and were based on observations of a great number of chemical substances and their reactions. These laws are the foundation on which modern atomic theory is based.

The first of them is the **law of conservation of matter**:

Law of conservation of matter: When a chemical change (reaction) takes place, matter is neither created nor destroyed.

Today we accept this as fact (with minor modifications), but it took a long time for this law to become established since there are a number of chemical reactions that *seem* to violate it.

The subject was set straight in 1775 by the French chemist Antoine Lavoisier, along with his wife Marie, who worked with him in his lab. They made sure that no matter, including gases, entered or left their reaction vessels. Under these conditions, they found no measurable changes in mass during a reaction. Today, we know that the weight gained by a rusting nail is due to oxygen gas taken up from the atmosphere to form rust, and that a glass of water and Alka-Seltzer loses weight due to the carbon dioxide gas that is given off.

The second law that led Dalton to his atomic hypothesis is called the **law of definite proportions (or constant composition):**

Law of definite proportions (or constant composition): When two elements react to form a compound, the total amount (mass) of the compound formed is determined by the characteristic composition of the compound and not by the masses of the elements used.

For example, suppose you wanted to make the compound aluminum iodide. This can be done by combining the pure elements aluminum and iodine. If you combine just the right amounts of each, both will be entirely consumed by the reaction, leaving you with pure aluminum iodide and no leftover aluminum or iodine:

A nail gains weight as it rusts.

A glass of water with an Alka-Seltzer tablet loses weight as the tablet dissolves.

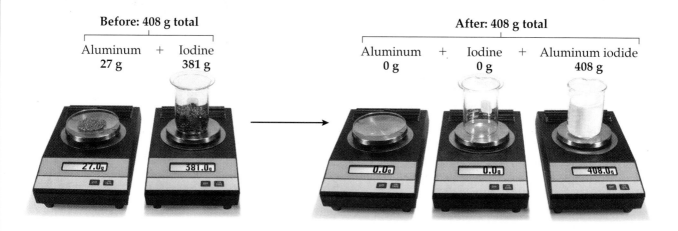

Before: 408 g total

Aluminum + Iodine
27 g   381 g

After: 408 g total

Aluminum + Iodine + Aluminum iodide
0 g   0 g   408 g

Since all the aluminum and iodine ended up in the new compound aluminum iodide, it is a simple matter to calculate the percent by mass of the compound that is aluminum and the percent by mass that is iodine:

The percent by mass of aluminum is:

$$\frac{\text{Mass of Al in compound}}{\text{Total mass of compound}} = \frac{27\text{ g}}{408\text{ g}} \times 100 = 6.6\% \text{ aluminum}$$

The percent by mass of iodine is:

$$\frac{\text{Mass of I in compound}}{\text{Total mass of compound}} = \frac{381 \text{ g}}{408 \text{ g}} \times 100 = 93.4\% \text{ iodine}$$

We now can say that aluminum iodide consists of approximately 7% (by mass) aluminum and approximately 93% iodine.

Now let's make some more aluminum iodide. This time, however, we will reduce the amount of aluminum used by half. Before Dalton, it was believed that any quantities of starting materials would combine and react to form product. That is, the reaction of 13.5 g of aluminum with 381 g of iodine would consume all the reactants and produce 394.5 g of aluminum iodide. If this were true, the percent by mass of aluminum and iodine in that sample of aluminum iodide would be 3.4% aluminum and 96.6% iodine. These percentages are not the same as we determined previously, and this result would violate the law of definite proportions. The actual results of this experiment are shown below:

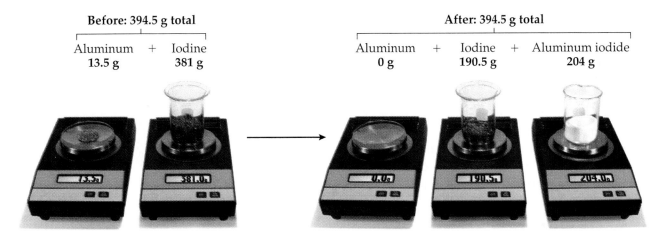

Before: 394.5 g total		After: 394.5 g total		
Aluminum + Iodine		Aluminum +	Iodine +	Aluminum iodide
13.5 g 381 g		0 g	190.5 g	204 g

This time 190.5 g of iodine are left over, and we formed only half as much aluminum iodide as in the first experiment. What can we make of this?

We started with a total of 394.5 g of material and we ended up with 394.5 g of material. No mass was gained or lost, so matter was conserved. But is the aluminum iodide we made in this reaction the same as in the first batch? After all, if you bake a cake with only half as much flour as the recipe calls for, it probably will not taste the same. To check, we can calculate the percentages of aluminum and iodine in the new batch of aluminum iodide:

The percent by mass of aluminum is:

$$\frac{13.5 \text{ g}}{204 \text{ g}} \times 100 = 6.6\% \text{ aluminum}$$

The percent by mass of iodine is:

$$\frac{381 \text{ g iodine used} - 190.5 \text{ g iodine left over}}{204 \text{ g}} \times 100 = 93.4\% \text{ iodine}$$

The percentages (proportions) are the same as before. This suggests that any sample of aluminum iodide—no matter what amount of it is created or how it is made—always contains 6.6% aluminum and 93.4% iodine. That is what the law of definite proportions means. Even though we combined 13.5 g of aluminum with 381 g of iodine in our second reaction, we did not get 394.5 g (13.5 g + 381 g) of aluminum iodide. We got only 204 g of aluminum iodide, and this aluminum iodide had the same percent composition as our first batch.

This law is true no matter what compound is created. For example, water is a compound of hydrogen and oxygen. No matter how water is made, no matter how much hydrogen and oxygen are combined with each other, the water produced always has the same percentage of oxygen and hydrogen. In the following WorkPatch, water is made by combining just the right amounts of hydrogen and oxygen (so that they are both used up), and then again by cutting the amount of hydrogen in half.

3.1 WORKPATCH Use the given data to calculate the percentages of oxygen and hydrogen present in water, and prove the law of definite proportions to yourself.

$$\text{Hydrogen} + \text{Oxygen} \longrightarrow \text{Water}$$
$$100.0 \text{ g} \qquad 793.6 \text{ g} \qquad 893.6 \text{ g}$$

(a) Is the law of conservation of matter obeyed?
(b) What is the percent hydrogen in water?
(c) What is the percent oxygen in water?

$$\text{Hydrogen} + \text{Oxygen} \longrightarrow \text{Water}$$
$$50.0 \text{ g} \qquad 793.6 \text{ g} \qquad 446.8 \text{ g} + 396.8 \text{ g leftover oxygen}$$

(d) Is the law of conservation of matter obeyed?
(e) What is the percent hydrogen in water?
(f) What is the percent oxygen in water?

Did you get the same percentages for both WorkPatch problems? You should have.

To Dalton, constant composition was very good evidence that the elements did, indeed, exist in the form of atoms. When you formed a chemical compound, he reasoned, you were actually combining atoms of the constituent elements. An atom of one element would always combine with a fixed number of atoms of another element, the number depending on the identity of the elements and the compound being made. As usual, not everyone agreed with Dalton, but a third law made it nearly impossible to ignore atomic theory, the **law of multiple proportions**:

Law of multiple proportions: When two elements (A and B) can combine to form more than one compound (say, AB and A₂B), then for a fixed weight of A, the weights of B in two different compounds always form a ratio that is expressible in small *whole* numbers.

This law nearly screams that atoms must exist. Again, the best way to understand the law is with an example.

Consider compounds 1 and 2, both gases, and both composed entirely of carbon and oxygen. One of these gases is biologically inert (unreactive) while the other is a deadly poison. If we make a sample of both of these compounds using a fixed amount of one of the elements, say 12 g of carbon for each, then we can look at both products to see how much oxygen was used:

We see that different amounts of oxygen are needed for the two different compounds. To make compound 1, 12 g of carbon combined with 32 g of oxygen. To make compound 2, the same amount of carbon combined with half as much oxygen—only 16 g. Is the law of multiple proportions at work here? We used a fixed weight of carbon (12 g) to make both compounds, so the amounts of oxygen incorporated into the two compounds should be expressible as a ratio of small whole numbers.

To make both compounds, 12 g of carbon was used.

32 g of oxygen was needed to make compound 1.
16 g of oxygen was needed to make compound 2.

The ratio of 32 g O/16 g O can be reduced to a ratio of small whole numbers:

$$\frac{32 \text{ g O}}{16 \text{ g O}} = \frac{2}{1}$$

Thus, compound 1 must have twice as many oxygen atoms per carbon atom as compound 2. Compound 1 is carbon dioxide, CO_2, a gas we exhale. Compound 2 is carbon monoxide, CO, a highly toxic component of automobile exhaust.

The law of multiple proportions is very powerful evidence in support of the existence of atoms. It could only be true if atoms are indeed indivisible particles, so that all chemical compounds would have to be made from *whole numbers* of atoms.

Now it's your turn to give the law of multiple proportions a try. The formula for water is H_2O. Another compound of hydrogen and oxygen is hydrogen peroxide, used as an antiseptic and for bleaching hair. According to Dalton, if we prepare both compounds using the same amount of hydrogen for each, then the ratio of the amounts of oxygen needed to make these compounds should be expressible as a ratio of whole numbers. Using the data in the following WorkPatch, determine the ratio of the amounts of oxygen needed.

3.2 WORKPATCH Both water and hydrogen peroxide are made from the same two elements, hydrogen and oxygen. A student analyzes both compounds by decomposing them to their elements and finds:

A sample of water produces 2 g of hydrogen and 16 g of oxygen.

A sample of hydrogen peroxide produces 2 g of hydrogen and 32 g of oxygen.

(a) What is the ratio $\dfrac{\text{g O in hydrogen peroxide}}{\text{g O in water}}$?

(b) If the formula for water is H_2O, what must be the formula for hydrogen peroxide?

Do you see how this kind of result argues for the existence of atoms? Every time Dalton did experiments like this he found a ratio of small whole numbers (2/1 for the above WorkPatch). Matter must therefore exist as indivisible, whole atoms.

PRACTICE PROBLEMS

3.1 A student makes two different compounds of sulfur and oxygen using the amounts of each elemental substance shown below. In both cases, all of the elemental substances are completely used up.

Compound 1:   5.00 g sulfur + 4.99 g oxygen
Compound 2:   5.00 g sulfur + 7.48 g oxygen

(a) What is the percent sulfur in compound 1?
(b) If the formula of compound 1 is SO_2, what is the formula of compound 2?

Answer: (a) %S = (5.00 g S/9.99 g compound 1) × 100 = 50.1% S
(b) Use the law of multiple proportions. The substance present in a fixed amount is sulfur, so determine the ratio of the amounts of oxygen:

(g O in compound 2)/(g O in compound 1) = 7.48 g/4.99 g = 1.50 = 3/2 (a ratio of
whole numbers)

Therefore if compound 1 has the formula SO_2, then compound 2 has the formula SO_3.

3.2 A student makes two different compounds of nitrogen and oxygen using the amounts of each elemental substance shown below. In both cases, all of the elemental substances are completely used up.

Compound 1:   10.0 g nitrogen + 11.42 g oxygen
Compound 2:   10.0 g nitrogen + 22.84 g oxygen

(a) What is the percent oxygen in each compound?
(b) If the formula of compound 1 is NO, what is the formula of compound 2?

Answer: (a) Compound 1, 53.3% oxygen; compound 2, 69.5% oxygen   (b) NO$_2$

3.3 101.96 g of a compound of aluminum and oxygen is 47.1% by mass oxygen.
(a) What is the percent by mass aluminum in this compound?
(b) Of the 101.96 g of this compound, how many grams of it are aluminum?

Answer: (a) 52.9% aluminum   (b) 53.9 g aluminum

Based on the law of conservation of matter, the law of constant composition, and the law of multiple proportions, Dalton formulated his atomic theory, which can be summarized in five short statements.

Dalton's atomic theory

1. All matter is made up of atoms (small, indivisible, indestructible, fundamental particles).
2. Atoms can neither be created nor destroyed (they persist, unchanged for all eternity).
3. Atoms of a particular element are alike (in size, mass, properties).
4. Atoms of different elements are different (different sizes, shapes, masses, properties).
5. A chemical change (reaction) involves the union or separation of individual atoms.

Molecules of the chemical compound we call water are always formed from one atom of oxygen and two atoms of hydrogen, H$_2$O. Why do atoms of one element combine with only a certain number of atoms of another element in forming chemical compounds? Dalton wanted a physical model, a picture of the atom that fit his theory and helped to explain it. Chemists of the time ultimately envisioned a "ball-and-hook" model for atoms that allowed atoms to combine with each other in particular ways. They described atoms of different elements as balls of different sizes with a characteristic number of hooks embedded in them. The number of hooks in each type of atom would account for its chemical reactivity and explain why elements of certain atoms characteristically combined with only a certain number of other atoms. They "reacted" with each other until their hooks were filled.

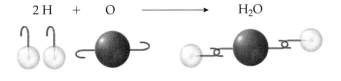

Today we know that atoms don't have hooks, but for the early 1800s, this was a good model. It made visual sense of the behaviors summarized by the three laws. Dalton claimed that if you accept the five statements of his theory as a viable model of reality, then the laws of conservation of matter, constant composition, and multiple proportions could be easily explained and accounted for.

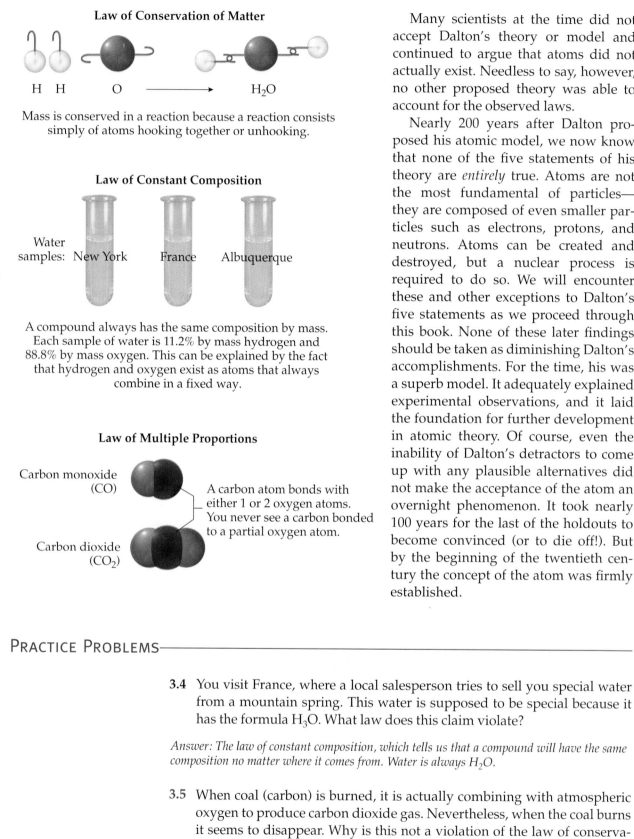

Law of Conservation of Matter

H   H   O $\longrightarrow$ H_2O

Mass is conserved in a reaction because a reaction consists simply of atoms hooking together or unhooking.

Law of Constant Composition

Water samples: New York   France   Albuquerque

A compound always has the same composition by mass. Each sample of water is 11.2% by mass hydrogen and 88.8% by mass oxygen. This can be explained by the fact that hydrogen and oxygen exist as atoms that always combine in a fixed way.

Law of Multiple Proportions

Carbon monoxide (CO)

Carbon dioxide (CO_2)

A carbon atom bonds with either 1 or 2 oxygen atoms. You never see a carbon bonded to a partial oxygen atom.

Many scientists at the time did not accept Dalton's theory or model and continued to argue that atoms did not actually exist. Needless to say, however, no other proposed theory was able to account for the observed laws.

Nearly 200 years after Dalton proposed his atomic model, we now know that none of the five statements of his theory are *entirely* true. Atoms are not the most fundamental of particles—they are composed of even smaller particles such as electrons, protons, and neutrons. Atoms can be created and destroyed, but a nuclear process is required to do so. We will encounter these and other exceptions to Dalton's five statements as we proceed through this book. None of these later findings should be taken as diminishing Dalton's accomplishments. For the time, his was a superb model. It adequately explained experimental observations, and it laid the foundation for further development in atomic theory. Of course, even the inability of Dalton's detractors to come up with any plausible alternatives did not make the acceptance of the atom an overnight phenomenon. It took nearly 100 years for the last of the holdouts to become convinced (or to die off!). But by the beginning of the twentieth century the concept of the atom was firmly established.

PRACTICE PROBLEMS

3.4 You visit France, where a local salesperson tries to sell you special water from a mountain spring. This water is supposed to be special because it has the formula H_3O. What law does this claim violate?

Answer: The law of constant composition, which tells us that a compound will have the same composition no matter where it comes from. Water is always H_2O.

3.5 When coal (carbon) is burned, it is actually combining with atmospheric oxygen to produce carbon dioxide gas. Nevertheless, when the coal burns it seems to disappear. Why is this not a violation of the law of conservation of matter?

3.6 Using Dalton's ball-and-hook atomic models, sketch an explanation of how it is possible for hydrogen to combine with oxygen to form two different compounds, water (H_2O) and hydrogen peroxide (H_2O_2).

DEVELOPMENT OF THE STRUCTURE OF THE ATOM 3.2

Once the existence of atoms was accepted, the obvious next question was, "What do atoms look like?" A series of extremely clever experiments done by physicists J. J. Thomson, James Chadwick, and others, decisively showed that atoms were not the fundamental particles of matter but were themselves composed of even smaller particles.

In 1897, J. J. Thomson discovered the first subatomic particle to be identified, the **electron**. He found that all atoms contained electrons and that the electrons from all atoms were identical. These electrons were very small and light particles indeed, having only 1/1836 the mass of a hydrogen atom, the smallest and lightest of all atoms. In addition, the electron had a negative electrical charge, which for convenience can be assigned a numerical value of −1.

Just 10 years later, in 1907, Thomson and E. Goldstein found another subatomic particle that was present in all atoms, the **proton**. The proton was much heavier than the electron, with a mass almost equal to that of a hydrogen atom. It was found to have a positive electrical charge equal in magnitude but opposite in sign to that of the electron. This charge is assigned a value of +1.

Twenty-five years later Sir James Chadwick demonstrated the existence of a third subatomic particle, the **neutron**. It had about the same mass as a proton but lacked any charge (it was electrically neutral, hence the name neutron). The lack of a charge made the neutron much more difficult to study and helps to explain why its discovery came so much later.

By the mid 1930s it was well established that atoms were made up of these fundamental particles, summarized in Table 3.1.

Table 3.1   The Subatomic Particles That Make Up Atoms

Subatomic particle	Discoverers	Properties
Electron (e)	Thomson, 1897	Present in all atoms
		Extremely light (1/1836 mass of H atom)
		Possesses negative charge, assigned −1
Proton (p)	Thomson and Goldstein, 1907	Present in all atoms
		About the same mass as H atom
		Has positive charge equal in magnitude but opposite in sign to electron, assigned +1
Neutron (n)	Chadwick, 1932	About the same mass as a proton
		Has no charge (is electrically neutral)

Once it was established that all atoms were composed of the same three fundamental particles, the path of investigation turned to the exploration of the internal structure of the atom. How could only three different subatomic pieces be put together to make all the different atoms in the periodic table?

The first model was proposed by Thomson shortly after his discovery of the electron. He knew two basic facts about the atom: (*1*) atoms contain small, negatively charged particles called electrons, and (*2*) the atoms of each element behave as if they have no electrical charge at all—they are electrically neutral. Thomson reasoned that there must be something in the atom that carries a positive charge to neutralize the electrons (protons had yet to be discovered). Thomson's model for the atom consisted of a "cloud of positive electricity" in which the negatively charged electrons were embedded. This became known as the "plum pudding" model since it reminded many of a classic English pudding.

It may seem a somewhat silly model today, but it fulfilled all the requirements of a model in the scientific method given the data available at that time. It was consistent with all the known facts about atoms and successfully accounted for their neutral behavior. What helped banish this model to the trash heap was a series of experiments performed just a few years later, in 1909, by the British physicist Ernest Rutherford.

Thomson "plum pudding" model of the atom

Negative electrons

Positive "pudding"

3.3 THE NUCLEUS

Rutherford's experiment, like many of the great experiments that have shaped modern chemistry and physics, was beautifully simple in its concept. Rutherford had been studying **alpha (α) particles**, small chunks of positively charged matter that are spontaneously and randomly given off by many naturally occurring radioactive elements such as radium, polonium, and radon. We now know that α particles are fast-moving helium atoms from which two electrons have been removed, giving them a +2 charge. Rutherford did not know this precisely, but he did know that α particles were about 7000 times more massive than electrons, that they had a positive charge, and that they were ejected from certain types of matter at very high speeds—about 9300 miles/second (1.5×10^4 km/sec), or approximately one-twentieth the speed of light. Rutherford thought of α particles as positively charged "bullets" that could be fired at various targets. The targets he chose were thin foils of pure metals. One was made of pure gold, a metal so soft and malleable that it could be hammered into a thin film only a couple of thousand atoms thick. He decided to "fire" the α particles at the thin gold foil and "see" what happened.

Of course, he could not actually see either the α particles or the individual atoms of gold in the foil target. However, a screen made of glass that was coated with zinc sulfide was known to glow (fluoresce) with a tiny point of green light when struck by an α particle. Rutherford simply positioned a zinc sulfide screen around his gold target so that any α particle that passed through the gold target would strike the screen, cause it to glow, and thus reveal its flight path. Since the "plum pudding" model of the atom was in vogue at the time, Rutherford expected the α particles to simply cruise right through the gold atoms pretty much undisturbed. After all, the electrons were much too small and light to cause any resistance, and the positive charge in the atom was spread out in a thin cloud-like blanket throughout the entire volume of the atom. If this model of the atom were correct, the α particles should go straight through the gold foil and hit the zinc sulfide screen in a straight-line trajectory from the radioactive source. To make sure that he and his students missed nothing, they sat in a pitch black room for an hour so that their eyes would be ultra-sensitive to any green flashes on the fluorescent screen. What they saw was *almost* what they expected.

Rutherford's α-Particle Scattering Experiment

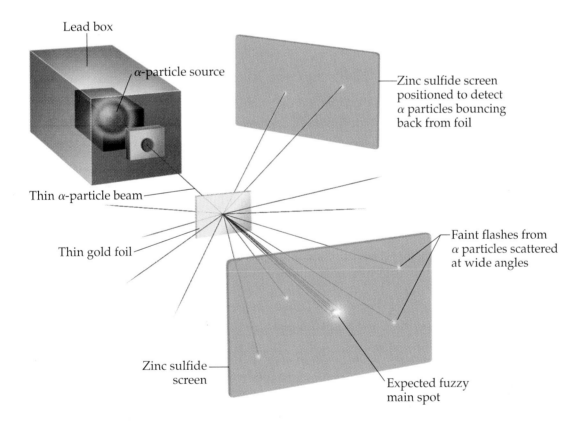

Most of the α particles did indeed go straight through the foil as predicted. But a few, barely noticeable as green flashes on the zinc sulfide screen, were deflected at large angles. Some α particles even bounced right back toward the source! This was incredible. In Rutherford's own words, "It was . . . as if you fired a 15-inch shell at a piece of tissue paper and it came back and hit you." If the results were to be believed, then the "plum pudding" model of how the atom was put together could not possibly be correct. A new model for the internal structure of the atom must be created to account for the results of Rutherford's experiment. In particular, two observations needed to be explained: (1) Most of the α particles went straight through the foil, and (2) a few (about 1 in 20,000) of the α particles were deflected from a straight-line path, some by very extreme angles.

The fact that *any* of the fast-moving α particles were deflected by the gold atoms in the target led Rutherford to postulate that there must be something small and massive inside each gold atom. Indeed, it had to be very compact and massive, since some of the α particles actually bounced back from it. In addition, since the α particles carry a positive charge, then it made sense that this massive center had a positive charge (since like charges repel each other). What about the fact that most of the α particles did go straight through? Rutherford reasoned that for this to be true, the positively charged, massive "something" inside the atom must also be extremely tiny so that there was little chance of an α particle striking or even coming near it. In other words, the atom must be mostly empty space, letting most of the α particles pass right through.

Thus, in attempting to verify the "plum pudding" model of the atom, Rutherford ended up having to discard it and replace it with an atom that was mostly empty space but had something that was tiny, massive, and positively charged inside it. This "something" he called the atom's **nucleus**, which we now know holds an atom's protons and neutrons. The "empty space" that makes up most of the volume of an atom is filled with the much lighter electrons that must also be present in the atom.

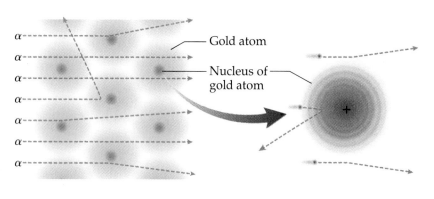
Gold atom
Nucleus of gold atom

3-3   WORKPATCH

Suppose one of Rutherford's contemporaries argued that the nucleus of an atom is negative. How would you use Rutherford's observations to argue that he must be wrong?

Think about the few positively charged α particles that are bounced straight back on hitting the foil, and you should be able to answer the WorkPatch.

Based on the number of α particles deflected and their angles of deflection, Rutherford was able to calculate the relative sizes of the "electron space" and the nucleus. He calculated that the nucleus was about 10^{-13} cm in diameter, while the diameter of the entire atom was about 10^{-8} cm. Thus, the nucleus and the atom differ in size by about 10^5—that is, the atom is 100,000 times larger than the nucleus. Rutherford's model of the atom was mostly empty space. At the center of the spherical atom was a tiny, dense nucleus; the rest of the volume was occupied by the electrons, which moved around the nucleus in some fashion. A contemporary comparison to Rutherford's model might be that of a Ping-Pong ball suspended in the middle of the Houston Astrodome. The Astrodome represents the total volume of the atom with the Ping-Pong ball as its nucleus. The electrons would be equivalent to a few mosquitoes flying around inside the stadium.

The Houston Astrodome

It certainly was not easy for Rutherford to completely discard an accepted model. Another, less thorough scientist might have been tempted to ignore the few faint flashes of green light that showed up at large angles, or might even have neglected to design the experiment to look for them, since "obviously" all the action would be in a straight-line path. In this case, however, Rutherford and the scientific method triumphed, replacing an old model (theory) with a new one, and inviting new sets of experiments to challenge and further refine it.

Rutherford himself realized that more work needed to be done. For instance, the electrons presented a problem. Rutherford assumed they had to be continuously moving, because, if they stopped, they would be sucked into the nucleus by its strong positive charge (opposite charges attract each other, even on the atomic level). But *how* were they moving? According to classical physics, every time a moving charged particle—like an electron—changes speed or direction, it must radiate energy. Thus, the electrons moving about inside an atom should quickly lose their energy, slow down, and spiral into the nucleus. In other words, the principles of classical physics demand that a Rutherford type of atom should collapse in on itself within a fraction of a second. Since it was pretty clear that all matter in the universe was in no immediate danger of collapsing, Rutherford knew that either something was wrong with his model or something was wrong with classical physics. He had difficulty accepting either proposition and so, as you might imagine, he put forth his model of the atom with a great deal of trepidation.

Rutherford knew that his model was in need of further refinement and modification. What he did not know was that the solution to the problem was going to be so revolutionary, so inconceivable, that even the scientists who developed it found it almost too difficult to accept. Albert Einstein went to his grave not fully believing in it and working feverishly to find the flaw in the analysis. Classical physics, used for centuries to calculate the force of gravity, the motions of the planets, and the trajectory of missiles, just did not work when applied to particles the size of an atom. To solve the problem of the atom, classical physics itself, the foundation of modern science, would have to be tossed away and replaced with something else. Over the next twenty years (1910–1930), a revolution in physics and quantum mechanics was born.

3.4 THE STRUCTURE OF THE ATOM

Let's take stock of where we are with regard to the sub-atomic pieces that go into making an atom. There are the relatively massive, positively charged (+1) protons (p) that reside at the center of the atom in a tiny nucleus. There are the extremely light, negatively charged (−1) electrons (e) that move about outside the positive nucleus. A fundamental law of physics (quantum as well as classical) is that oppositely charged particles (positive and negative) attract one another. The significance of this is that the electrons can't just leave the atom, at least not without the input of a good deal of energy. They are attracted to the nucleus and so are kept from wandering away.

The third subatomic particle is the neutron (n). This particle has no charge; it is neutral. Neutrons have approximately the same mass as protons.

ATOMIC NUMBER

The number of protons inside the nucleus is called the **atomic number** (abbreviated with the letter **Z**) of that atom. The atomic number is extremely important because it determines the identity of the atom. An atom with one proton (atomic number 1) is always hydrogen; an atom with six protons (atomic number 6) is always carbon; and so on.

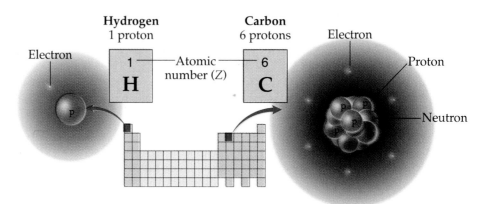

The atomic number is so important that the periodic table of the elements is arranged according to increasing atomic number. Examine the atoms in the figure at the bottom of page 72 and then answer the following WorkPatch.

WORKPATCH 3.4

From the figure, what do you think is the rule for how many electrons an atom has? What is the result of this rule on the overall charge of an atom?

Your answer to the WorkPatch should help you understand how the overall neutrality of an atom depends on the number of protons and electrons it possesses. (If you are not sure about this, check the answer at the end of the chapter before going on.)

Now, what about neutrons? Neutrons are fairly massive subatomic particles, so they must also be part of the nucleus since Rutherford determined that nearly all the atom's mass was concentrated there. Most hydrogen atoms have no neutrons; all other elements have at least one. Carbon atoms, with six protons, usually have six neutrons, but there is no simple relationship between the number of neutrons and protons in the nucleus of an atom. This is especially true for the heavier atoms in the periodic table. A close look at a typical uranium nucleus, with 92 protons (atomic number 92), shows 146 neutrons.

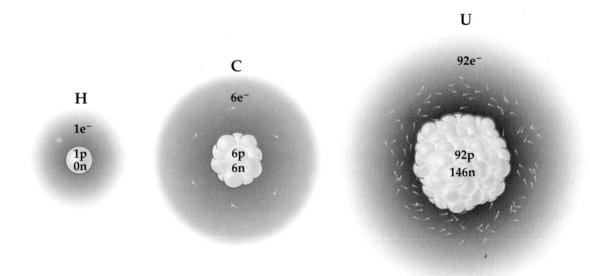

MASS NUMBER

Along with the atomic number (Z), chemists have defined another atomic quantity called *mass number*. An atom's **mass number** is simply the number of protons plus the number of neutrons in its nucleus. The mass numbers for H, C, and U are shown below.

$$\begin{array}{c} 1\,p \\ +\ 0\,n \\ \hline \end{array}$$

Mass number $\longrightarrow$ 1

$$\begin{array}{c} 6\,p \\ +\ 6\,n \\ \hline \end{array}$$

Mass number $\longrightarrow$ 12

$$\begin{array}{c} 92\,p \\ +\ 146\,n \\ \hline \end{array}$$

Mass number $\longrightarrow$ 238

Let's see if you're catching on to this. Complete the following WorkPatch and then we'll discuss it.

3·5 WORKPATCH

Sketch the carbon atom in the following three ways, showing protons, neutrons, and electrons, as we've done on page 73. All three of these carbon atoms should be electrically neutral. Be sure to indicate the number of neutrons in each nucleus.

(a) Make the mass number equal to 12.
(b) Make the mass number equal to 13.
(c) Make the mass number equal to 14.

If you did things correctly, then all three sketches should have 6 protons in the nucleus (since they are all carbon) and 6 electrons outside of the nucleus (to keep each atom electrically neutral). The only difference, therefore, must be in the number of neutrons in the nucleus: (a) should have 6 neutrons, (b) should have 7 neutrons, and (c) should have 8 neutrons. The question is, what should we call these three slightly different versions of the same element? The answer is carbon. They must all be carbon, because they all have an atomic number of 6. It is the atomic number and only the atomic number that determines the elemental identity of an atom. And yet, these three carbon atoms are not identical; there is a difference between them: (b) has an extra neutron, and (c) has two extra neutrons. Atoms (a), (b), and (c) are examples of *isotopes*. **Isotopes** are different versions of the same element (same atomic number) that contain a different number of neutrons in their nuclei (different mass numbers). Because they are the same element, they have identical chemical properties (there are some subtle differences but they are beyond the scope of this book). They differ only in their weights.

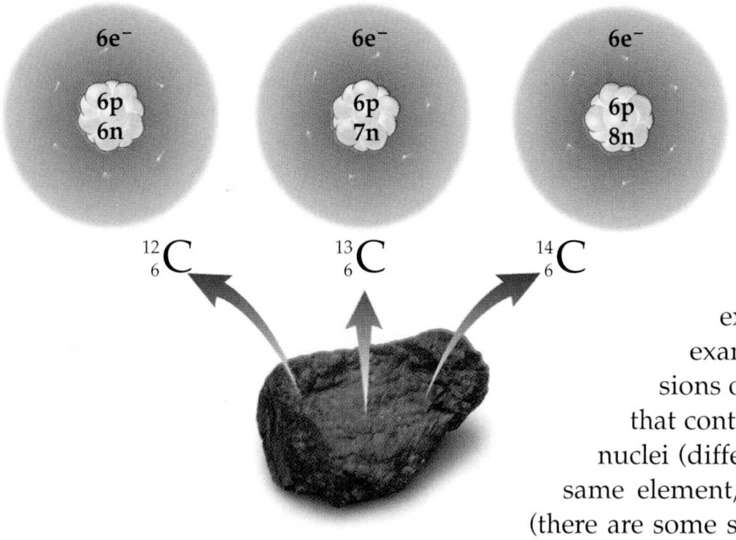

Chemists symbolize the different isotopes of an element by using full atomic symbols like the ones shown at left. In these symbols, the mass number is written at the upper left and the atomic number at the lower left:

$$\text{Mass number} \longrightarrow {}^{12}_{6}\text{C} \longleftarrow \text{Atomic number}$$

Very often, the atomic number is omitted from this symbol. Thus, the three isotopes of carbon may be written simply as ^{12}C, ^{13}C, and ^{14}C. These same isotopes might also be written as carbon-12, carbon-13 and carbon-14. When read aloud, these isotope symbols, ^{12}C for example, are pronounced "C-twelve" or "carbon-12."

All three of these carbon isotopes exist in nature, although not in equal abundance. On this planet 98.89% of all the carbon is ^{12}C, 1.11% is ^{13}C, and only a trace of the ^{14}C isotope exists in nature. The percent abundances of the isotopes would most likely be different on some other planet. Analysis of the percent of each isotope in a rock is one way chemists can determine if a rock is terrestrial in nature or if it comes from some extraterrestrial meteorite.

Nearly all the elements exist as two or more naturally occurring isotopes. Even the simplest of the elements, hydrogen, comes in three isotopic forms. Although they are really all hydrogen atoms (they all have one proton in their nucleus), they are given different names, as shown at right. (This is not done for isotopes of any of the other elements.)

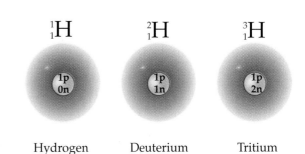

Hydrogen   Deuterium   Tritium

While isotopes of an element have essentially identical chemical properties, there is a single important difference between them. Since neutrons and protons carry most of an atom's mass, and isotopes differ in their number of neutrons, isotopes have different weights. Deuterium (^{2}H) is twice as heavy as ^{1}H, and tritium (^{3}H) is even heavier than deuterium. For this reason, ^{2}H is sometimes referred to as heavy hydrogen, and D_2O (the special symbol D is used for a deuterium atom) is called heavy water.

Together, the mass number and atomic number tell you quite a bit about the atom, but what if you are given only one of these numbers? Try the following WorkPatch.

Suppose you are told that an atom has a mass number of 16. Is this enough information to identify the element? Why?

WORKPATCH 3.6

If you are not sure of the answer, think about what you need to know to determine an atom's identity. Suppose you are given the additional information in the next WorkPatch.

Suppose you are told that an atom with a mass number of 16 also possesses 9 neutrons. Can the atom be identified now? What element is this?

WORKPATCH 3.7

This time you should be able to determine the atom's elemental identity (check the answer at the end of the chapter). The moral of this story is that to identify an atom you need to know its atomic number, and you can determine this number in a variety of different ways provided sufficient information is given.

PRACTICE PROBLEMS

3.7   Give the full atomic symbol for an atom that has 16 neutrons and whose atomic number is 15.

Answer: The atomic number of 15 identifies the element as phosphorus (P). The mass number is 15 + 16 = 31. The full atomic symbol is $^{31}_{15}$P.

3.8   Bromine (Br) has two abundant isotopes, one with 44 neutrons and the other with 46 neutrons. Give the full atomic symbols for both isotopes.

3.9   How many electrons do atoms of each of the isotopes in Practice Problem 3.8 have?

3.10 Fill in the following table:

	$^{14}_{7}\text{N}$	$^{24}_{12}\text{Mg}$	$^{24}_{11}\text{Na}$	$^{59}_{26}\text{Fe}$
Mass number	?	?	?	?
Atomic number	?	?	?	?
Number of protons	?	?	?	?
Number of neutrons	?	?	?	?
Number of electrons	?	?	?	?

ATOMIC WEIGHT

We have looked closely at atomic number and mass number. But neither of these numbers tells us how much an atom weighs. When it comes to reporting the weight (or mass) of an atom we turn to yet another quantity called **atomic weight**. Below we list the exact atomic weights for specific isotopes of the elements hydrogen, carbon, and magnesium.

Some atoms and their exact atomic weights

^{1}H	Atomic weight	1.007 83	^{12}C	Atomic weight	12
^{2}H	Atomic weight	2.014 10	^{13}C	Atomic weight	13.003 35
^{3}H	Atomic weight	3.016 05	^{24}Mg	Atomic weight	23.985 04

Do you notice something about the atomic weights? Look closely. There are no units associated with them. An atom of hydrogen-1 weighs 1.007 83, but 1.007 83 what? Grams? Pounds? Why aren't there any units? There are no units because the atomic weight scale is a *relative* scale. In this case, the word relative means *compared to something*. All atomic weights are given relative to the weight of an atom of ^{12}C, the principal isotope of carbon. By universal agreement among chemists, an atom of ^{12}C has been *assigned* an atomic weight of exactly 12. This is the only isotope whose atomic weight and mass number are equal. All the rest of the atoms are assigned weights relative to how heavy (or light) they are compared to ^{12}C. For example, ^{1}H has an atomic weight of 1.007 83. This means that a single ^{1}H atom is 1.007 83/12 or roughly one-twelfth the weight of a ^{12}C atom. Since ^{24}Mg has an atomic weight of 23.985 04, this means that ^{24}Mg is 23.985 04/12 or roughly twice as heavy as a ^{12}C atom. For this reason, the atomic weight scale is also called the *relative atomic weight scale*.

3.8 **WORKPATCH**

An atom is determined to be 4.015 times heavier than ^{12}C. What is the atomic weight of this atom?

We have one more topic to cover before we end our discussion of atomic weights. We said that an atom of the ^{12}C isotope weighs exactly 12 by universal agreement. But if you take a look at the periodic table, you will see that carbon's atomic weight is listed as 12.011, not 12. This atomic weight takes into account the fact that when you find carbon atoms in nature, they are not all ^{12}C atoms. As we have seen, carbon consists predominantly of two isotopes, ^{12}C (98.89% natural abundance) and ^{13}C (1.11% natural abundance).

6
C
12.011

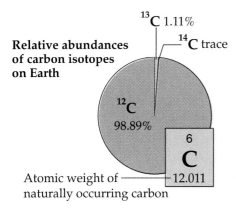

Relative abundances of carbon isotopes on Earth

^{13}C 1.11%

^{14}C trace

^{12}C 98.89%

6
C
12.011

Atomic weight of naturally occurring carbon

Any naturally occurring sample of carbon atoms will contain both isotopes in these abundances. The atomic weight of carbon reported in the periodic table is a so-called *weighted average* of the atomic weights of these two isotopes. Each isotope's atomic weight is multiplied by its percent abundance (in decimal form, the percent is divided by 100) and then the results are summed.

Weighted average atomic weight of naturally occurring carbon

$$= \overbrace{(12)(0.9889)}^{^{12}_{6}\text{C portion}} + \overbrace{(13.003\ 35)(0.0111)}^{^{13}_{6}\text{C portion}}$$
$$= 12.011$$

All the atomic weights listed in the periodic table of the elements are calculated in this way and are thus average weights for the naturally occurring mixture of isotopes. If you want the atomic weight of an individual isotope (other than ^{12}C), you need to consult a reference book such as the *Handbook of Chemistry and Physics* (CRC Press).

PRACTICE PROBLEMS

3.11 Suppose we fuse 16 identical bowling balls together. We call this our standard object and claim that it weighs exactly 16 (no units). Next, it is determined that you are 1.50 times heavier than the standard object. What is your relative weight?

Answer: Since you weigh 1.50 times as much as the standard object, your relative weight is (1.50)(16) = 24.0.

3.12 Referring to Practice Problem 3.11, the relative weight of a mystery object is 48.0. Is it lighter or heavier than the standard object? By how much?

3.13 Chlorine exists as two isotopes in nature, $^{35}_{17}$Cl (atomic weight = 34.969, % abundance = 75.77%) and $^{37}_{17}$Cl (atomic weight = 36.966).
(a) What is the % abundance of the $^{37}_{17}$Cl isotope?
(b) What is the atomic weight of naturally occurring chlorine?
(c) How many times heavier is $^{37}_{17}$Cl than $^{35}_{17}$Cl?

THE LAW OF MENDELEEV—CHEMICAL PERIODICITY 3.5

Chemists were keeping very busy during the nineteenth century. A major effort was underway to isolate and characterize all of the elements. Chemists throughout the world set out to decompose various chemical compounds into their component elements in order to determine the properties of these fundamental chemical building blocks. By 1860, as a result of these efforts, nearly 70 of the 115 elements we know about today had been isolated and studied. In the thousands of different chemical compounds and mixtures that chemists dissected, each with its own unique physical and chemical properties, only these 70 elements were found. This represented a great simplification for the science of chemistry. Any object in the universe could now, at least in principle, be understood in terms of the relatively few elements that made it up.

As elements were discovered and their properties examined, it became necessary to organize the data in some useful way that would help to make sense of it all. One of the greatest advances toward this goal was made in the mid-1800s by a Russian chemist named Dmitri Mendeleev. He listed the elements and their properties on individual cards and then experimented with different arrangements of these cards to look for any patterns. The breakthrough came when he arranged the cards by putting the elements in order of increasing atomic weight (as they were known at the time):

The first 20 elements arranged from lightest to heaviest

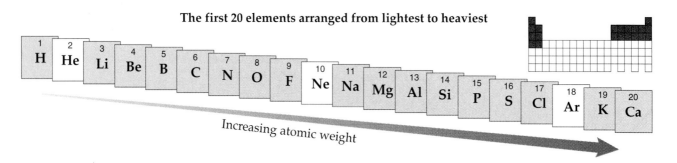

[We have inserted the elements helium (He), neon (Ne), and argon (Ar), which were completely unknown in Mendeleev's time.] With the elements arranged this way, Mendeleev noted that their chemical properties repeated in a regular way. For example, consider the properties of the element sodium (Na).

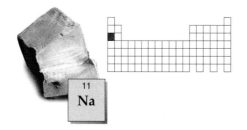

Elemental sodium is too reactive to be found in nature. However, chemists managed to isolate pure sodium from its compounds. It is a soft metal, silvery in appearance, with a low density and a low melting point (for a metal). Like most metals, it is a good conductor of electricity. It is also highly reactive, which can be readily demonstrated by dropping a piece of sodium metal in water. It reacts violently with the water, releasing flammable hydrogen gas which is often ignited. In addition, if a plant dye called litmus is added to the water, it turns a dark blue. An analysis of the reaction products shows that the sodium metal reacted with water to form NaOH, a compound known to make litmus turn blue.

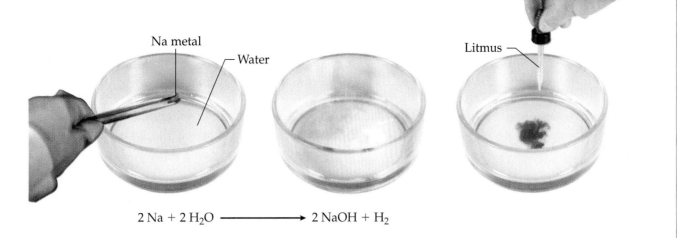

Na metal

Water

Litmus

$$2\,Na + 2\,H_2O \longrightarrow 2\,NaOH + H_2$$

Like most chemists of the time, Mendeleev knew all this, and by itself it was nothing to get excited about. But when he examined his arrangement of cards, looking for elements with chemical properties similar to those of sodium, he noticed something interesting. Eight elements to the left and right of sodium in his list were elements with almost identical chemical and physical properties. Eight elements to the right was potassium (K), and eight elements to the left was lithium (Li).

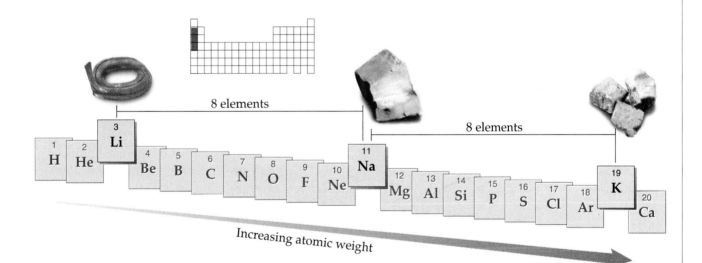

8 elements

8 elements

Increasing atomic weight

Both of these elements react with oxygen to form oxides (Li_2O and K_2O) with similar formulas to that of sodium oxide (Na_2O), and with water to form hydroxides (LiOH and KOH) in a manner similar to sodium (NaOH). All three are silvery metals that are too reactive to exist in their elemental form in nature. All are good electrical conductors, all three react vigorously with water releasing flammable hydrogen gas, and all three produce a blue solution with litmus.

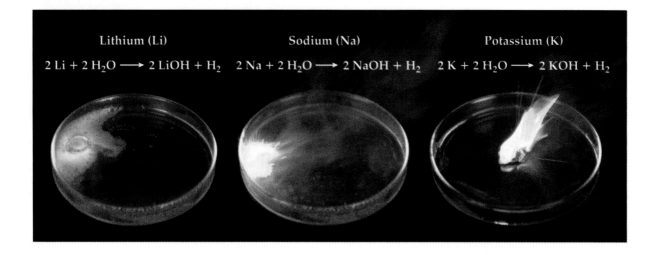

Lithium (Li)   Sodium (Na)   Potassium (K)

$2\,Li + 2\,H_2O \longrightarrow 2\,LiOH + H_2$ $2\,Na + 2\,H_2O \longrightarrow 2\,NaOH + H_2$ $2\,K + 2\,H_2O \longrightarrow 2\,KOH + H_2$

Was this just coincidence? Hardly! Mendeleev observed this same pattern of 8 for other elements in his uniquely arranged list. Magnesium (Mg) reacts with oxygen to give a (1:1) oxide of the formula MgO. Eight elements away in both directions (heavier and lighter) was another element that gave an oxide of similar formula (beryllium oxide, BeO, and calcium oxide, CaO). Also, as we now know, neon (Ne) refuses to react with oxygen at all. Sure enough, eight elements away in both directions are two other elements, helium (He) and argon (Ar) that also refuse to react with oxygen. There seems to be something almost "magical" about the number 8. This pattern was sometimes referred to as the "law of octaves."

When Mendeleev rearranged his cards so that elements with similar properties were put into the same vertical columns, he had eight distinct columns. No doubt about it. There had to be something special about this number 8. (While the elements of column 8 were unknown at the time, Mendeleev's table did have 8 columns.)

This repeating behavior of the chemical properties of the elements is called **chemical periodicity** or **periodic behavior** (thus the name periodic table). In recognition of his efforts, this discovery is often referred to as the **law of Mendeleev**. This same relationship was discovered independently, and at about the same time, by the German chemist Lothar Meyer, but Mendeleev received most of the credit for the periodic table.

Column I							Column VIII
1 **H** H_2O	Column II	Column III	Column IV	Column V	Column VI	Column VII	2 **He** no oxide
3 **Li** Li_2O	4 **Be** BeO	5 **B** B_2O_3	6 **C** CO_2	7 **N** N_2O_6	8 **O**	9 **F** F_2O	10 **Ne** no oxide
11 **Na** Na_2O	12 **Mg** MgO	13 **Al** Al_2O_3	14 **Si** SiO_2	15 **P** P_2C_6	16 **S** SO_2	17 **Cl** Cl_2O	18 **Ar** no oxide
19 **K** K_2O	20 **Ca** CaO						

Law of Mendeleev: Properties of the elements recur in regular cycles (periodically) when the elements are arranged in order of increasing atomic weight.

Mendeleev was fortunate that his periodic table worked as well as it did. The modern periodic table is arranged according to increasing atomic number, not atomic weight. Fortunately, there are only a few inversions that occur between Mendeleev's table and the modern one.

It was later discovered that eight was not the only magic number with respect to periodic behavior. In portions of the periodic table, properties repeat every 18 or 32 elements, as we shall soon see.

When Mendeleev began work on his table, only 70 elements were known. Indeed, when Mendeleev arranged the elements into groups of similar chemical properties, there were some "holes" in his table. For example, the column containing the element carbon (C) also contained the elements silicon (Si) and tin (Sn), but there was a "hole" where an element with a weight somewhere between that of Si and Sn should be.

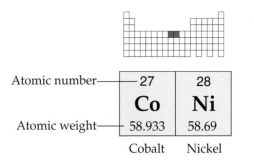

Cobalt comes before nickel in the modern periodic table. Mendeleev would have incorrectly put Co after Ni because it is heavier.

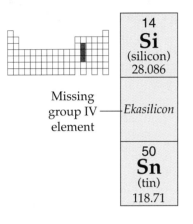

With his newly constructed table, Mendeleev was able to do what no chemist before him could do—predict the properties of elements that had yet to be discovered! This was quite a trick. Mendeleev reasoned that the missing element, which he called *ekasilicon* ("eka" meaning next in order) should have properties that were the average of the properties of the elements above and below it. This told chemists what properties to look for, and led to the rapid discovery of the missing element. Today we call ekasilicon (symbolized below by R) germanium, Ge. Look how close his predictions about germanium actually were.

	Predicted properties (average of Si and Sn)	Observed properties (measured after discovery)
Atomic weight	72	72.61
Density	5.5 g/cm^3	5.32 g/cm^3
Melting point	825°C	938°C
Oxide formula	RO_2	GeO_2
Density of oxide	4.7 g/cm^3	4.70 g/cm^3
Chloride formula	RCl_4	$GeCl_4$
Chloride boiling point	100°C	86°C

Life on our planet is based on the element carbon, but science fiction writers often like to conjure up silicon-based life forms. Why have these writers picked on the element Si to create alien life?

WORKPATCH 3·9

Mendeleev's insights greatly advanced our understanding of the chemistry of the elements but they also presented a new dilemma. Any chemically useful model of the atom would have to account for the existence of periodic properties in the chemical behavior of the elements.

Rutherford's model did not even attempt to address the chemical behavior of the elements, and that nasty classical physics problem of why electrons don't spiral down into the nucleus still wasn't explained. Modifying Rutherford's model was not going to be an easy task. The solution was to come with the new quantum physics and the resulting modern (or quantum) model of the atom, which we'll see in the next chapter. In the rest of this chapter we'll take a closer look at the modern periodic table, and then examine some of the clues that pointed the way to the quantum model of the atom.

3.6 THE MODERN PERIODIC TABLE

The modern periodic table is elegant in its simplicity, and yet it says so much. Its neat horizontal rows and vertical columns underlie a detailed understanding of the subatomic structure and chemical properties of the elements. The modern periodic table places the elements in vertical columns, called **groups**, and horizontal rows, called **periods**. Both the groups and the periods are numbered. There is no controversy on how to number the periods. They are numbered 1 through 7. There are, however, a few different ways to number or label the groups; we will use the most common method, as shown below.

Periodic Table of the Elements

Representative (main group) elements

IA	IIA	IIIB	IVB	VB	VIB	VIIB	VIIIB			IB	IIB	IIIA	IVA	VA	VIA	VIIA	VIIIA
1 H 1.0079																	2 He 4.003
3 Li 6.941	4 Be 9.012											5 B 10.811	6 C 12.011	7 N 14.007	8 O 15.999	9 F 18.998	10 Ne 20.180
11 Na 22.990	12 Mg 24.305											13 Al 26.982	14 Si 28.086	15 P 30.974	16 S 32.066	17 Cl 35.453	18 Ar 39.948
19 K 39.098	20 Ca 40.078	21 Sc 44.956	22 Ti 47.88	23 V 50.942	24 Cr 51.996	25 Mn 54.938	26 Fe 55.845	27 Co 58.933	28 Ni 58.69	29 Cu 63.546	30 Zn 65.39	31 Ga 69.723	32 Ge 72.61	33 As 74.922	34 Se 78.96	35 Br 79.904	36 Kr 83.8
37 Rb 85.468	38 Sr 87.62	39 Y 88.906	40 Zr 91.224	41 Nb 92.906	42 Mo 95.94	43 Tc 98	44 Ru 101.07	45 Rh 102.906	46 Pd 106.42	47 Ag 107.868	48 Cd 112.411	49 In 114.82	50 Sn 118.71	51 Sb 121.76	52 Te 127.60	53 I 126.905	54 Xe 131.29
55 Cs 132.905	56 Ba 137.327	57 La 138.906	72 Hf 178.49	73 Ta 180.948	74 W 183.84	75 Re 186.207	76 Os 190.23	77 Ir 192.22	78 Pt 195.08	79 Au 196.967	80 Hg 200.59	81 Tl 204.383	82 Pb 207.2	83 Bi 208.980	84 Po 209	85 At 210	86 Rn 222
87 Fr 223	88 Ra 226.025	89 Ac 227.028	104 Rf 261	105 Db 262	106 Sg 263	107 Bh 262	108 Hs 265	109 Mt 266	110 Uun 269	111 Uuu 272	112 Uub 277	114		116		118	

Transition metals

Rare earth elements

Lanthanides

58 Ce 140.115	59 Pr 140.908	60 Nd 144.24	61 Pm 145	62 Sm 150.36	63 Eu 151.964	64 Gd 157.25	65 Tb 158.925	66 Dy 162.5	67 Ho 164.93	68 Er 167.26	69 Tm 168.934	70 Yb 173.04	71 Lu 174.967

Actinides

90 Th 232.038	91 Pa 231.036	92 U 238.029	93 Np 237.048	94 Pu 244	95 Am 243	96 Cm 247	97 Bk 247	98 Cf 251	99 Es 252	100 Fm 257	101 Md 258	102 No 259	103 Lr 262

The groups are labeled with roman numerals followed by either the letter A or the letter B. The colors indicate that the periodic table separates the elements into three broad classes: Gray groups IA–VIIIA are called the **representative or main group** elements. Tan groups IB–VIIIB are known as **transition**

metals. Finally, the green elements at the bottom are called the **lanthanides** (or **rare earths**) and **actinides**. The symbol of each element is accompanied by its atomic number above and its atomic weight below. The representative, or A group, elements are the ones on which much of early chemistry was based. They also show the strongest periodic relationships in the table. Using only the representative elements, Mendeleev's law of octaves works magnificently well. Picking any element and then moving away eight elements in either direction brings you to another element whose properties are similar to those of the starting element.

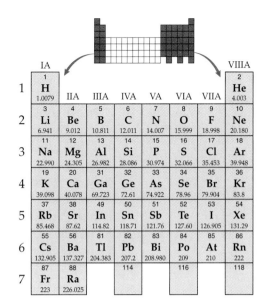

A periodic table showing only the representative or main group elements is eight columns wide. Chemical periodicity occurs every eight elements.

Look again at the entire periodic table and pay close attention to how the groups are numbered. At first glance, the numbering system looks a bit bizarre. There is a big block of B groups jammed in between the A groups. Even worse, the B groups start with number III, go to VIII (three columns are assigned this number), and then end with I and II. There are good reasons for this strange group numbering system. Nevertheless, you should be aware that a newer, more straightforward group numbering system has recently been adopted by IUPAC (the International Union of Pure and Applied Chemistry). It simply starts with the leftmost column and sequentially numbers the columns from 1 to 18.

Our periodic table includes useful information about the phase (solid, liquid, or gas) in which the pure element exists under standard conditions (25°C and 1 atmosphere of pressure). If the element's symbol is solid black, then the element exists as a solid. If it is red, then it exists as a gas. If it is blue, then it is a liquid. Finally, if the symbol is green, the element does not exist in nature and must be created by some type of nuclear reaction.

Now we will take a brief tour of the periodic table, starting with the representative elements (groups IA–VIIIA). These groups contain solid, liquid, and gaseous elements. In fact, all the elements that exist as gases at standard temperature (25°C) and pressure (1 atm) are representative (or main group) elements, including the two that make up most of the air we breathe, nitrogen and oxygen.

The first representative elements that we will examine are those of group VIIIA. This is a very special column of the periodic table. It is the only group whose members are all gases at standard temperature and pressure. In addition, all the elements in this group are extremely unreactive. For this reason the group VIIIA elements are sometimes called the *inert gases*, and for some time chemists actually believed that they were completely chemically unreactive. In 1962, however, the British chemist Neil Bartlett challenged this assumption and prepared the first compound of the gas xenon (Xe). Today we more correctly refer to this group as the **noble or rare gases**, but they are still among the most unreactive of the elements. With the exception of helium, they are all found in trace amounts in the atmosphere. The Earth's gravitational pull is not strong enough to keep helium in the atmosphere from escaping into space, and as a result all naturally occurring helium comes from deposits trapped underground.

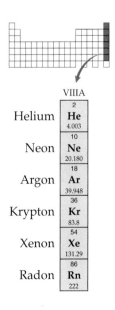

Radon (Rn), the only noble gas that is radioactive, has received some bad press lately. It has been discovered seeping into basements from naturally occurring underground sources. Sustained exposure is thought to be responsible for an increase in the incidence of lung cancer, and many people have purchased radon detectors for their basements.

As we continue to examine the representative elements, we find yet another way to divide them: They can be classified as either *metals*, *nonmetals*, or neither (the latter are called *metalloids* or *semi-metals* and have properties intermediate between those of a metal and those of a nonmetal). This division is indicated in the following illustration. The red stair-step line that starts at boron separates the metals from the nonmetals. The elements bordering this line (except for aluminum and polonium) are the metalloids (shown in green). Of the 44 representative elements, 7 are metalloids, 17 are nonmetals (including the noble gases), and the remaining 20 are metals.

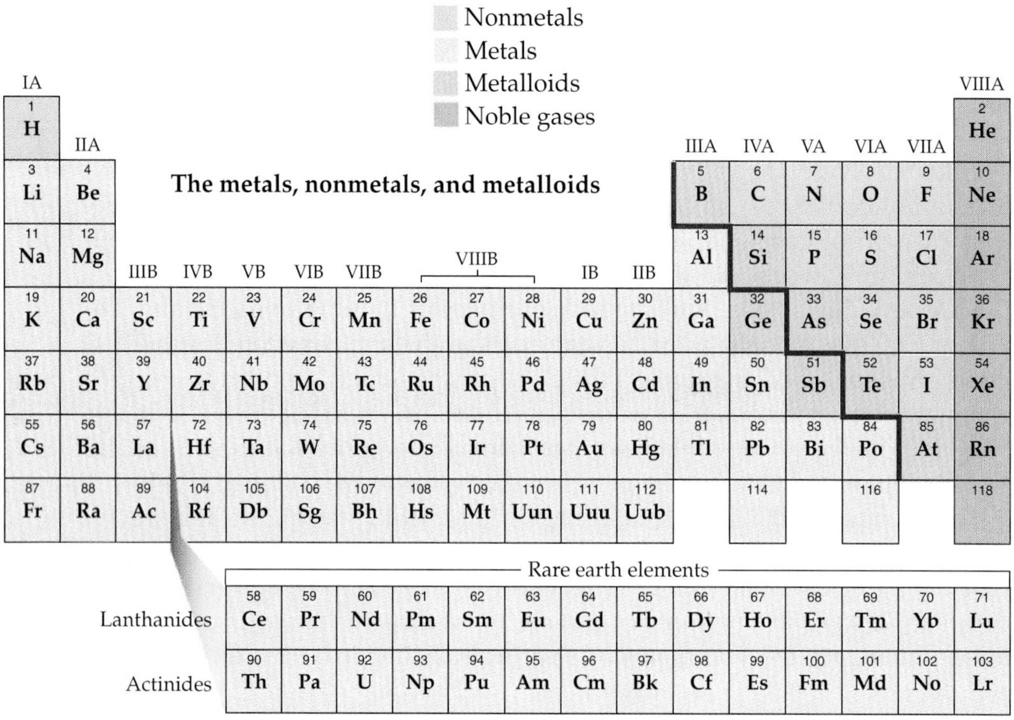

The metals, nonmetals, and metalloids

While we have not yet formally defined the terms metal and nonmetal, you probably have a pretty good feel for them based on everyday experience. **Metals** tend to be shiny solids that are bendable, malleable, and conduct heat and electricity well. **Nonmetals**, on the other hand, tend to be brittle and do not conduct heat or electricity well (they are insulators, the opposite of conductors). Lead (Pb), a representative element in group IVA once used for plumbing, is an obvious metal. Diamond, a pure form of carbon (C), is a nonmetal. For now, these everyday "definitions" of metal and nonmetal will suffice. Later we'll give more chemically accurate definitions of these terms that will relate to their tendency to gain or lose electrons when forming chemical compounds.

Some representative (or main group) metals such as aluminum (Al) and tin (Sn) are familiar to most people. You've certainly come across aluminum foil and tin cans ("tin" cans are actually steel coated with tin). The metals of groups IA and IIA are less familiar because they are much too reactive to be found in nature in their elemental forms. However, as parts of compounds, some of them are also quite familiar. The compound sodium chloride (table salt) has the group IA metal sodium (Na) in it. Many common antacids such as Milk of Magnesia or Tums contain compounds of the group IIA metals magnesium (Mg) and calcium (Ca).

The **metalloids or semi-metals** can act, depending on circumstances, like either a metal or a nonmetal. The elements silicon (Si), germanium (Ge), and arsenic (As) are good examples of elements that demonstrate classic metalloid character. With respect to conducting electricity, they neither conduct as well as metals nor do they insulate as well as nonmetals. They are somewhere in-between and are thus called *semiconductors*. These elements are used to make the sophisticated electronic chips that are the "brains" of computers. Indeed, the entire electronics industry owes its existence to the metalloid properties of these elements.

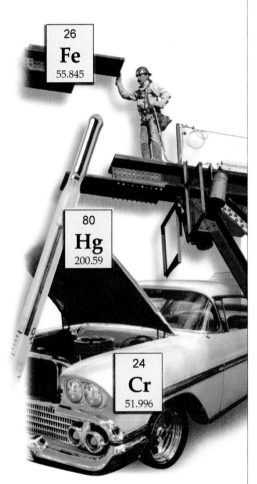

The rest of the elements in the periodic table belong to the lanthanides, actinides, or transition metals (the B groups). These form a bridge between the two parts of the representative elements. Perhaps the most familiar transition metal to you is iron (Fe), the major component of steel, used in everything from bicycles to skyscrapers. Others such as nickel (Ni), silver (Ag), gold (Au), copper (Cu), and zinc (Zn) are also familiar to most people. Mercury (Hg), commonly used in thermometers, is the only one of the transition metals that is a liquid at room temperature. Chromium (Cr) is used for plating other metals, as is done for car bumpers. Titanium (Ti) is used to strengthen bicycle frames. Platinum (Pt), used in pollution-control catalytic converters on cars, is even more expensive than gold. While the remaining transition metals may be less familiar to you, each has a wide number of uses in our technological society.

Of the roughly 115 known elements, approximately 75% are metals and only 16% are nonmetals! Nevertheless, all life that we know of is based on a nonmetallic element, carbon. Had nature built you out of one of the majority elements, a metal, you might have to worry more about rusting than about getting wrinkles in your old age.

PRACTICE PROBLEMS

3.14 How many groups constitute the main group elements?

Answer: 8

3.15 Does the stair-step metal–nonmetal boundary line ever enter into the transition metal portion of the periodic table?

3.16 What do the elements in a group of the periodic table have in common with one another?

Mendeleev's arrangement of the elements in the periodic table by weight resulted in the elements in a particular group (column) having similar chemical properties. Exactly why it worked out this way was not understood at the time. Nevertheless, these similarities invited giving each group a special name. Of the eight representative A groups, five have commonly used names, the first two and the last three. When metals from either group IA or group IIA are placed in water, they react with it to make the water alkaline (basic), hence the names **alkali metals** and **alkaline earth metals**, respectively. The group VIA elements, which include oxygen and sulfur, are called the **chalcogens**. The name comes from the Greek words *chalkos* for copper and *genes* for born. Most copper-containing minerals contain oxygen or sulfur. The group VIIA elements, which include chlorine, bromine, and iodine, are called the **halogens** (*halos* is Greek for salt). Many salts have a halogen in them, such as sodium chloride. And group VIIIA elements are called the noble or rare gases, as mentioned earlier. You should commit these main group names to memory.

Alkali metals | Alkaline earth metals | Chalcogens | Halogens | Noble gases

IA	IIA	IIIB	IVB	VB	VIB	VIIB	VIIIB			IB	IIB	IIIA	IVA	VA	VIA	VIIA	VIIIA
1 H 1.0079																	2 He 4.003
3 Li 6.941	4 Be 9.012											5 B 10.811	6 C 12.011	7 N 14.007	8 O 15.999	9 F 18.998	10 Ne 20.180
11 Na 22.990	12 Mg 24.305											13 Al 26.982	14 Si 28.086	15 P 30.974	16 S 32.066	17 Cl 35.453	18 Ar 39.948
19 K 39.098	20 Ca 40.078	21 Sc 44.956	22 Ti 47.88	23 V 50.942	24 Cr 51.996	25 Mn 54.938	26 Fe 55.847	27 Co 58.933	28 Ni 58.69	29 Cu 63.546	30 Zn 65.39	31 Ga 69.723	32 Ge 72.61	33 As 74.922	34 Se 78.96	35 Br 79.904	36 Kr 83.8
37 Rb 85.468	38 Sr 87.62	39 Y 88.906	40 Zr 91.224	41 Nb 92.906	42 Mo 95.94	43 Tc 98	44 Ru 101.07	45 Rh 102.906	46 Pd 106.42	47 Ag 107.868	48 Cd 112.411	49 In 114.82	50 Sn 118.71	51 Sb 121.76	52 Te 127.60	53 I 126.905	54 Xe 131.29
55 Cs 132.905	56 Ba 137.327	57 La 138.906	72 Hf 178.49	73 Ta 180.948	74 W 183.85	75 Re 186.207	76 Os 190.2	77 Ir 192.22	78 Pt 195.08	79 Au 196.967	80 Hg 200.59	81 Tl 204.383	82 Pb 207.2	83 Bi 208.980	84 Po 209	85 At 210	86 Rn 222
87 Fr 223	88 Ra 226.025	89 Ac 227.028	104 Rf 261	105 Db 262	106 Sg 263	107 Bh 262	108 Hs 265	109 Mt 266	110 Uun 269	111 Uuu 272	112 Uub 277		114		116		118

Given all we have said so far, you might wonder why hydrogen is included in group IA, the alkali metals. After all, hydrogen is not a metal. In fact, Mendeleev would not have put it there. We will wait until the next chapter to explain why it is there. Suffice it to say that while hydrogen is shown in group IA, nobody thinks of it as an alkali metal.

As we have seen, the structure of the periodic table reflects what we have learned about the chemistry of the various elements. Mendeleev's law of octaves applies very satisfactorily when you are dealing only with the representative elements. The chemical properties of these elements repeat every eight elements, and this part of the table contains eight columns to allow for this. But the law needs some modification if we are to include the rest of the elements in the table. If we continue to place elements into columns based primarily on their similar chemical behaviors, then we end up with the form of the periodic table as we know it today.

The introduction of the transition elements into the table begins with row (period) 4 and causes the number of columns to swell to 18 (8 representative elements + 10 transition metals per row). In this part of the table, therefore, chemical properties repeat every 18 elements and not every 8.

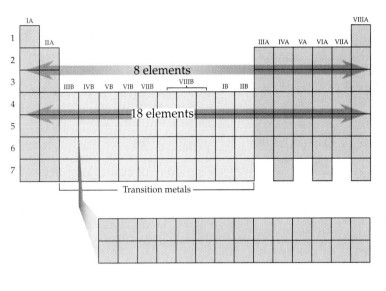

As we continue down the rows of the table we come to period 6, where we must now accommodate the rare earth elements. So far, we have shown these elements as a separate block below the main part of the table. That's just a convenience to allow the table to fit better on the printed page. Really, the lanthanide and actinide elements should fit right into periods 6 and 7, like this:

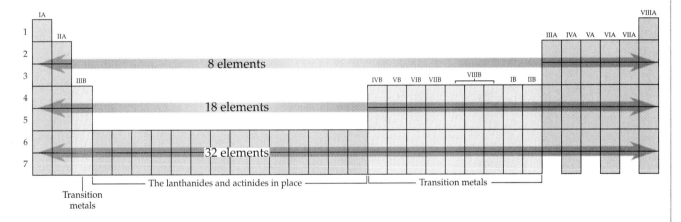

Thus, in periods 6 and 7, the periodic table becomes 32 columns wide, and the chemical properties of the elements in these periods repeat with a cycle of 32, rather than 8 or 18. It's much easier to see how the periodicity changes from 18 to 32 at period 6 looking at the periodic table with the lanthanide and actinide elements inserted in it.

PRACTICE PROBLEMS

3.17 What is the period and group number of the element with atomic number 15?

Answer: The element with atomic number 15 is phosphorus (P). It is in period 3 and group VA.

3.18 What elements with an atomic number greater than 40 can be expected to have chemistry similar to the element bromine (Br)? What is the periodicity with respect to these elements?

3.19 What would the periodicity be if the transition metals, lanthanides, and actinides were removed from the periodic table?

3.7 OTHER REGULAR VARIATIONS IN THE PROPERTIES OF THE ELEMENTS

Chemical behavior is not the only property of the elements that varies in a systematic way. Two others are *atomic size* and *ionization energy*. We'll look at the periodic trend in atomic size first.

ATOMIC SIZE

If we think of atoms as spheres, we can characterize each by its radius, called the **atomic radius**. Atomic radii can actually be measured by experimental techniques such as X-ray diffraction. They are often reported in angstroms (Å, 10^{-10} meter), nanometers (nm, 10^{-9} meter), or picometers (pm, 10^{-12} meter). The periodic table below illustrates the relative sizes of the representative elements:

Relative Atomic Sizes of the Representative Elements

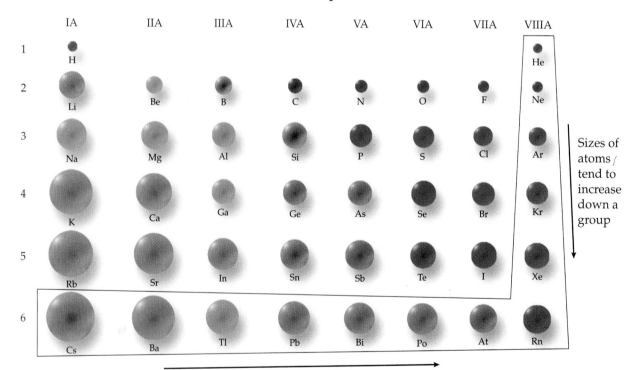

Sizes of atoms tend to increase down a group

Sizes of atoms tend to decrease across a period

Notice that the sizes of the atoms show regular variations in both the groups (columns) and the periods (rows). As you go down any group the atoms get larger (the atomic radius increases). Though there are some exceptions to this when you look at the full periodic table, generally, atoms get larger as you proceed down a group. By contrast, as you go across a period from left to right, the general trend is for the atoms to get smaller. We will see the reasons for this in the next chapter.

IONIZATION ENERGY

Another property of atoms that varies regularly in the periodic table is *ionization energy*. Atoms are neutral because the number of negatively charged electrons outside the nucleus equals the number of positively charged protons inside the nucleus. An atom in which this electron–proton balance is not maintained would have a net charge and is called an **ion**. The only practical way to create an ion is to add or remove electrons from a neutral atom. If you take a neutral lithium atom, for instance, and add one electron to it, the resulting ion has one more electron than it has protons and therefore has a net charge of -1. Such negatively charged ions are called **anions**. Likewise, if you remove one electron from a neutral lithium atom, the resulting ion will have one less electron than it has protons and a net positive charge of $+1$. Ions with a positive charge are called **cations**.

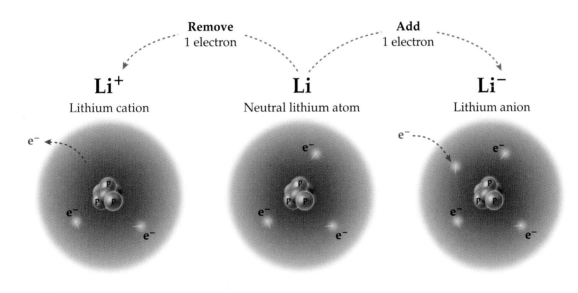

Now, what if we take another atom, say magnesium (Mg, atomic number 12), and remove two electrons from the neutral atom? There will now be an excess of two protons and we will end up with a cation with a $+2$ charge. This ion is written as Mg^{2+}. If we instead added two electrons to a magnesium atom (which for magnesium is actually very difficult to do), we would get an ion with a -2 charge which would be written as Mg^{2-}. In both cases, the usual convention in writing ions is to give the numerical value of the charge before the $+$ or $-$ sign. In the case of ions with a single charge, the 1 is usually omitted and the ion is written simply as, for example, Li^+ or Cl^-. An element symbol written without any charge (or with a zero, for example, Fe^0 or Cu^0) represents a neutral atom of that element.

3.10 WORKPATCH   Fill in the table for the following cations and anions:

	$^{14}_{7}N^{3-}$	$^{24}_{12}Mg^{2+}$	$^{23}_{11}Na^{+}$	$^{56}_{26}Fe^{3+}$
Mass number	?	?	?	?
Atomic number	?	?	?	?
Number of protons	?	?	?	?
Number of neutrons	?	?	?	?
Number of electrons	?	?	?	?

The above WorkPatch is meant to drive home an important point. The only way you can turn a (neutral) atom into an ion is by adding or removing electrons. You do not adjust the number of protons. If the number of protons were changed, the atom would not be the same element anymore.

Now let's look more closely at the removal of electrons from neutral atoms to form cations. Electrons are held in an atom due to their attraction to the positive nucleus. To remove an electron from the influence of the nucleus, you either have to pull it out or "kick" it with enough energy to free it. In more formal terms, you must expend energy to remove an electron from an atom. The minimum amount of energy required to remove the outermost electron from an atom is called the first **ionization energy (IE)**.

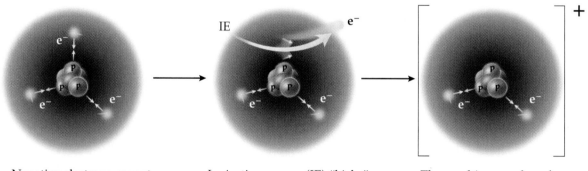

Negative electrons cannot leave the atom unless energy is supplied to overcome their attraction to the nucleus.

Ionization energy (IE) "kicks" an electron out of the atom.

The resulting surplus of one proton gives the atom an overall +1 charge. The atom has been ionized and is now a cation.

The periodic at the top of the facing page lists the ionization energies of the elements in units of electron volts (eV), a unit of energy commonly used by physicists. Look at the table closely. Can you see any trends in the ionization energies down a group or left to right across a period?

Just as there are periodic trends in atomic size (radius) there are also periodic trends in ionization energies. In general:

Ionization energy decreases as you go down a group.

Ionization energy increases as you go from left to right in a period.

This means that it becomes easier (takes less energy) to "kick" an electron out of an atom as you go down a group, and it takes more energy to remove an electron as you go from left to right across a period. If we plot the ionization energies as shown at the bottom of the facing page, these trends become even more apparent.

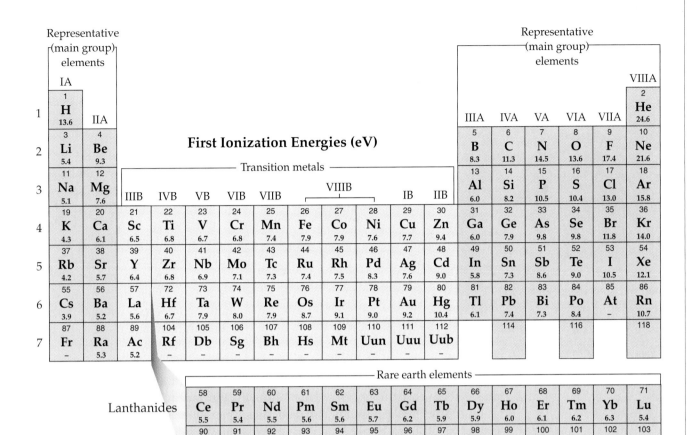

First Ionization Energies (eV)

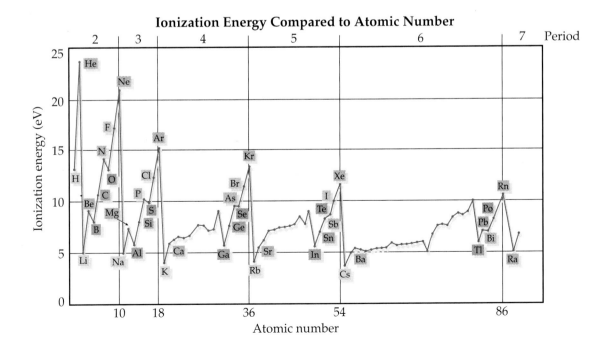

Ionization Energy Compared to Atomic Number

If we focus on the elements He, Ne, Ar, Kr, Xe, and Rn, you can see how the ionization energy decreases as you go down a group. If we then focus on the elements Li, Be, B, C, N, O, F, and Ne, you can see how the ionization energy generally increases as you go across a period from left to right.

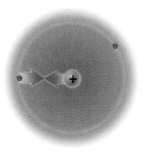

As atom gets larger, electrons are farther from the nucleus and are less strongly attracted

We can understand these period and group behaviors of the ionization energy by looking again at the trends in atomic size. As you go down a group, atoms get bigger and, at the same time, it becomes easier (requires less energy) to remove an electron. In a larger atom, the outermost electrons are, in general, farther away from the nucleus. This increased distance combined with a partial shielding from the nucleus of the outermost electrons by the inner ones reduces the attraction of the nucleus for the outermost electrons and thus makes them easier to remove. This means the ionization energy is lower.

Of course, the opposite is also true. As atoms get smaller (either going up a group or left to right across a period), it is more difficult to remove an electron, and the ionization energy increases.

These regular variations (periodic trends) in atomic size and ionization energy can be a great help to chemists in explaining the chemical behavior of the elements. For example, the alkali metals in group IA have relatively low ionization energies. This means that an element like sodium will easily lose an electron. Now, if you start at sodium and travel across period 3 from left to right, you come to a much smaller atom with a relatively large ionization energy, chlorine (Cl). It takes a large amount of energy for chlorine to lose an electron, and it generally does not do this when it reacts. In fact, just the opposite tends to occur; that is, Cl tends to gain an additional electron when it reacts. Bring sodium and chlorine together and this is exactly what happens. Sodium atoms lose electrons, becoming Na^+ cations, while the chlorine atoms pick up these electrons, and become Cl^- anions. The result is the spontaneous and rather spectacular chemical transformation of sodium and chlorine into the compound sodium chloride (NaCl), table salt.

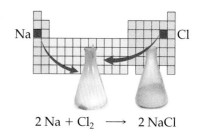

$$2\,Na + Cl_2 \longrightarrow 2\,NaCl$$

PRACTICE PROBLEMS

3.20 The alkali metals (group IA) all react with water by giving an electron to it. Based on their relative positions in the periodic table, which do you think would be more reactive with water, lithium or sodium?

Answer: Sodium. It is further down the group and therefore has a lower ionization energy, meaning it is easier for sodium to give up an electron. Since giving up an electron is how it reacts with water, this should make sodium the more reactive metal.

3.21 Arrange the following atoms from smallest to largest: Mg, Sr, S, O, Rb

3.22 Give the full symbol for the atom or ion that has 26 protons and 30 neutrons in its nucleus and 23 electrons outside its nucleus. Also give the number of the group this element is in.

3.23 Give the full symbol for the atom or ion that has 8 protons and 8 neutrons in its nucleus and 10 electrons outside its nucleus. Also give the name of the group this element is in.

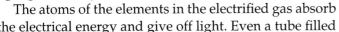

SEEING THE LIGHT—A NEW MODEL FOR THE ATOM 3.8

By 1870, chemists had conclusively demonstrated that the chemical properties of the elements repeated periodically. Rutherford's nuclear model of the atom could not explain this. Physicists had difficulty with the violations of classical physics that the nuclear model required. Just about everyone, including Rutherford, knew that his atomic model needed improvement.

The beginnings of an improved model came from a closer examination of the properties of light. Not just any light, but the light given off when gaseous atoms of the elements were provided with large amounts of energy. For example, glass tubes filled with neon gas or sodium vapor (produced by heating sodium metal) give off light of characteristic colors when an electric current is passed through them. We are all familiar with the orange-red glow produced by neon lights and the bright yellowish light given off by sodium street lights.

The atoms of the elements in the electrified gas absorb the electrical energy and give off light. Even a tube filled with hydrogen gas will glow with a bluish-pink color when electrified. What eventually gave scientists a valuable clue to the true nature of atomic structure was what they observed when they passed the light emitted by these electrified atoms through a prism or a diffraction grating. But before we can show you what they saw, we must first say a little about what light itself is.

We are all familiar with light. Flick a switch in a darkened room and an electrical current begins to flow through a tungsten filament until it glows white hot inside a glass bulb. The resulting white light illuminates the room.

Light moves through space at the incredible speed of 3.00×10^8 meters per second. Called *c*, the speed of light, this equals 186,000 miles per second or 671 million miles per hour! If you were traveling at the speed of light, you could travel around the world at its equator $7\frac{1}{2}$ times in a single second. But what exactly *is* light? Modern science tells us that light is two different things at once. It consists of particles or packets of energy called *photons*, and at the same time it acts like a *wave of energy*. Here we want to focus on the view of light as a wave.

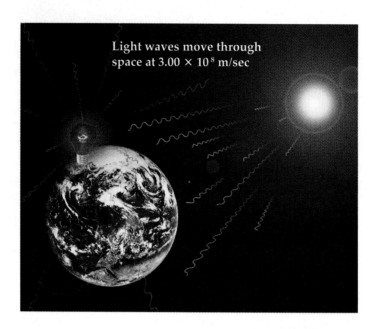

Light waves move through space at 3.00×10^8 m/sec

Being a wave, light can be characterized by its *wavelength*, symbolized by the Greek letter λ (lambda). **Wavelength** is simply the distance between identical adjacent points on the wave—that is, the distance between one crest (or trough) in the wave and the next.

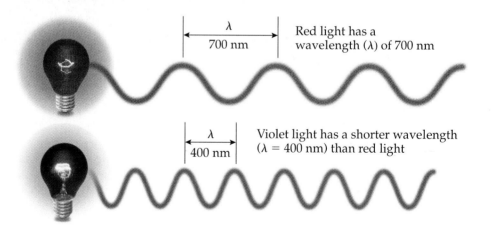

λ
700 nm
Red light has a wavelength (λ) of 700 nm

λ
400 nm
Violet light has a shorter wavelength (λ = 400 nm) than red light

Our brains interpret light of different wavelengths as different colors. Light with a wavelength between 650 and 750 nm (nanometers, 10^{-9} m) is seen as red in color. Violet light has shorter wavelengths, around 380–430 nm. The wavelengths in between are interpreted by our brains as the different colors of the rainbow. The wavelengths of light between 380 and 750 nm are collectively called visible light because these are the wavelengths that our eyes are designed to detect. When all these wavelengths impinge upon our eyes at once, we see white light. A prism or a diffraction grating will take this white light and separate or disperse it into its different wavelengths, giving rise to a *continuous spectrum*, the familiar rainbow of colors. It is called continuous because the colors smoothly blend into one another without any breaks.

Though our world can seem very bright and colorful at times, the fact of the matter is that we actually do not see most of the light that hits our eyes. Outside the 380–750 nm visible range there are many other wavelengths that are invisible to our eyes. The visible and invisible wavelengths are called **electromagnetic radiation**, since they consist of both an electric component and a magnetic component. The entire spectrum of electromagnetic radiation is known as the **electromagnetic spectrum**.

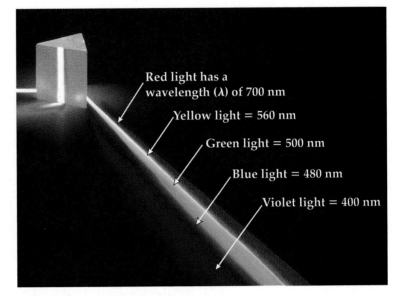

Red light has a wavelength (λ) of 700 nm

Yellow light = 560 nm

Green light = 500 nm

Blue light = 480 nm

Violet light = 400 nm

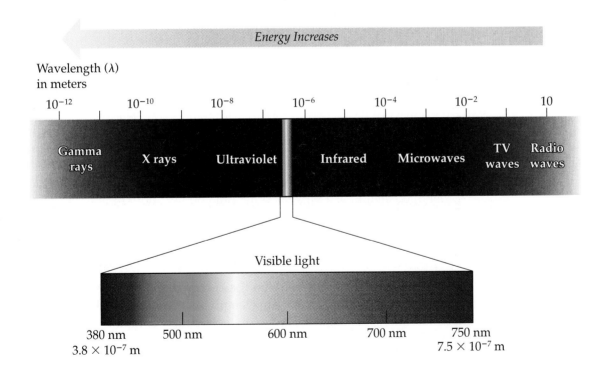

Energy Increases

Wavelength (λ) in meters

10^{-12} 10^{-10} 10^{-8} 10^{-6} 10^{-4} 10^{-2} 10

Gamma rays   X rays   Ultraviolet   Infrared   Microwaves   TV waves   Radio waves

Visible light

380 nm
3.8×10^{-7} m

500 nm

600 nm

700 nm

750 nm
7.5×10^{-7} m

While we cannot see the electromagnetic radiation outside the narrow visible region, it can affect us. This is because all electromagnetic radiation has energy associated with it, and the amount of energy depends on its wavelength. The equation relating the wavelength of a particular beam of electromagnetic radiation to its energy content is given at the top of the next page.

Equation to calculate the energy of electromagnetic radiation

$E = hc/\lambda$ where E = Energy of electromagnetic radiation
$c = 3.00 \times 10^8$ m/sec (the speed of light)
$h = 6.626 \times 10^{-34}$ J · sec (called Planck's constant)
λ = Wavelength of electromagnetic radiation

To use this equation you simply plug in the values for h and c (both constants) and the wavelength λ of the light. Watch your units, though! Since the speed of light c has units of meters/second, your wavelength must be expressed in units of meters. Since Planck's constant h has units of joules × seconds, your calculated energy will have units of joules.

PRACTICE PROBLEMS

3.24 What is the energy of blue light with a wavelength of 450.0 nm?

Answer: $E = \dfrac{hc}{\lambda} = \dfrac{6.626 \times 10^{-34} \text{ J} \cdot \text{sec})(3.00 \times 10^8 \text{ m/sec})}{(450.0 \text{ nm})(1 \times 10^{-9} \text{ m/nm})} = 4.42 \times 10^{-19}$ J

3.25 What is the energy of red light with a wavelength of 660.5 nm?

Answer: 3.01×10^{-19} J

3.26 What is the color of light that has an energy of 3.50×10^{-19} J? [*Hint:* Use algebraic manipulation to solve the light energy equation for λ. You will get an answer in meters. Convert it to nanometers (10^{-9} nm = 1 m) and consult the electromagnetic spectrum shown on page 95.]

Answer: 568 nm, yellow

3.27 What kind of electromagnetic radiation has a wavelength roughly of the order of the height of a person? What can you say about its energy?

The equation $E = hc/\lambda$ tells you that the energy of electromagnetic radiation is inversely proportional to its wavelength. As the wavelength of electromagnetic radiation gets smaller, the energy gets larger. Compare the answers to Practice Problems 3.24 and 3.25 to convince yourself of this. Red light of 660.5 nm has a larger (longer) wavelength than blue light with a wavelength of 450.0 nm, and thus the red light will have the smaller energy.

Invisible to us are all forms of electromagnetic radiation with wavelengths shorter than violet light (ultraviolet, X rays, and gamma rays). Because they contain so much energy, ultraviolet rays can cause sunburn, X rays can cause cancer, and gamma rays can kill. Also invisible to us are the forms of electromagnetic radiation with wavelengths longer than red light (infrared, microwaves, television waves, and radio waves), although we can feel infrared radiation as heat.

Now, what does light have to do with the revolution that led to the current model of the atom? In the late nineteenth century, physicists were examining the light given off by electrified gases of elements such as neon and hydrogen. When they passed this light through a prism they saw something remarkable. Rather than the continuous rainbow of colors seen when sunlight passes through a prism, physicists saw spectra like these:

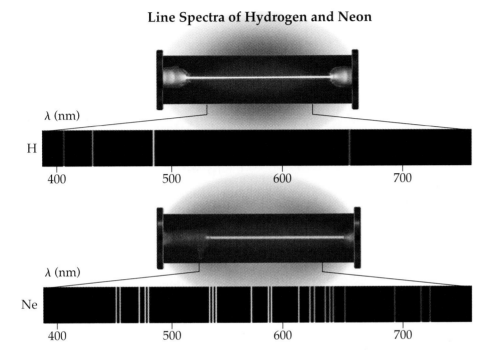

Line Spectra of Hydrogen and Neon

These are called **line spectra** because they are made up of separate lines of certain colors with regions of total blackness between. And as if this wasn't disconcerting enough, when the physicists did the same thing with other elements like sodium and mercury, they again obtained line spectra, but with lines of different colors and wavelengths. Each element produced its own characteristic line spectrum.

Since the energy of light depends on its wavelength (or color), each differently colored line that an atom emits must correspond to a different energy. This meant that something inside the atom was controlling the amounts of energy that were emitted when these atoms were energetically excited. Could this also mean that something inside the atom was only allowed to possess certain energies? Was this something the electrons? The answers to these questions came quickly in an avalanche of theoretical advancements during the 1920s and 1930s. During this time, quantum theory and the modern quantum model of the atom were born. The quantum model would not only explain the spectral properties of atoms but would also provide the "fix" for Rutherford's model and account for Mendeleev's observation of periodic chemical behavior. Chapter 4 will introduce you to this quantum model.

HAVE YOU LEARNED THIS?

Law of conservation of matter (p. 59)

Law of definite proportions or constant composition (p. 60)

Percent by mass (p. 60)

Law of multiple proportions (p. 62)

Dalton's atomic theory (p. 65)

Electron, proton, and neutron (p. 67)

Alpha (α) particles (p. 68)

Nucleus (p. 70)

Atomic number (Z) (p. 72)

Mass number (p. 73)

Isotopes (p. 74)

Atomic weight (p. 76)

Chemical periodicity (p. 80)

Law of Mendeleev (p. 80)

Periodic table (p. 82)

Groups and periods (p. 82)

Representative or main group elements (p. 82)

Transition metals (p. 82)

Lanthanides or rare earths (p. 83)

Actinides (p. 83)

Noble or rare gases (p. 83)

Metals and nonmetals (p. 84)

Metalloids or semi-metals (p. 85)

Alkali metals (p. 86)

Alkaline earth metals (p. 86)

Chalcogens (p. 86)

Halogens (p. 86)

Atomic radius (p. 88)

Ion (p. 89)

Anion and cation (p. 89)

Ionization energy (IE) (p. 90)

Wavelength (p. 94)

Electromagnetic radiation (p. 95)

Electromagnetic spectrum (p. 95)

Line spectrum (p. 97)

DALTON'S ATOMIC THEORY

3.28 When wood burns, the remaining ash weighs less than the original wood. Yet, the law of conservation of matter says that mass cannot be created or destroyed in a chemical reaction. How do you reconcile the result of burning with this law?

3.29 Volcanoes spew off hydrogen sulfide, a poisonous, bad-smelling gas. A 100.0 g sample of hydrogen sulfide gas was obtained from some strange volcanic planet. Analysis showed that 94.08 g of it are the element sulfur. The rest of the sample consists of the element hydrogen.
(a) What are the percentages by mass of sulfur and hydrogen in this sample?
(b) Hydrogen sulfide from Earth has exactly the same percent compositions for S and H as hydrogen sulfide from the strange volcanic planet. Which of Dalton's laws accounts for this?

3.30 If a scientist during Dalton's time found that hydrogen sulfide always had the formula H_2S, how would this have been explained?

3.31 Two compounds of iodine (I) and chlorine (Cl) are analyzed. Compound A consists of 126.9 g of I and 35.45 g of Cl. Compound B consists of 126.9 g of I and 106.4 g of Cl.
(a) What is the percent composition of each element in compound A?
(b) What is the percent composition of each element in compound B?
(c) Demonstrate that the law of multiple proportions is obeyed using compounds A and B.
(d) Assume that the formula for compound A is ICl. Based on your answer to part (c), postulate a formula for compound B.

3.32 Suppose 12.0 g of carbon (C) react with 70.0 g of sulfur (S) to give 76.0 g of the compound carbon disulfide (CS_2). In the process, all the carbon gets used up, but some elemental sulfur is left over.
  (a) For the law of conservation of matter to be obeyed, how much sulfur is unused?
  (b) What is the percent C in CS_2?
  (c) What is the percent S in CS_2?
  (d) What is the sum of the %C and %S in CS_2?

3.33 How does the law of multiple proportions suggest that matter can only exist as whole atoms and not as parts (halves, thirds) of atoms?

3.34 Nitrogen and hydrogen form the compound NH_3. How many hooks would Dalton put on the nitrogen atom to explain this compound?

3.35 Two different compounds, both consisting of sodium (Na) and oxygen (O), were analyzed. The data are given below.

Compound	Weight of sample analyzed	Weight of O present	Weight of Na present
A	19.50 g	8.00 g	?
B	61.98 g	16.00 g	?

  (a) Fill in the last column of the table.
  (b) Calculate the %Na and %O for both compounds.
  (c) Demonstrate the law of multiple proportions using the data in the table. [*Hint*: Divide all the weights for compound B by some number so as to adjust the weight of oxygen to be present in a fixed amount for both compounds.]

STRUCTURE OF THE ATOM

3.36 How did Rutherford interpret the observation that only a very few of the α particles were scattered by the gold foil?

3.37 How did Rutherford know that the nucleus wasn't negatively charged?

3.38 What was wrong with Rutherford's model of the atom as far as classical physics was concerned?

3.39 Suppose all the α particles in Rutherford's experiment went straight through the gold foil with absolutely no deflections. What would this say about the structure of the atom?

3.40 Why isn't an atom's mass number sufficient to determine an atom's elemental identity?

3.41 Does knowing how many electrons a neutral atom has tell you its elemental identity? Explain.

3.42 Physicists are fond of saying that an atom is mostly empty space. What justifies this statement?

3.43 Fill in the following table for four neutral atoms:

	$^{15}_{8}O$	?	?	?
Mass number	?	16	37	?
Atomic number	?	8	?	?
Number of protons	?	?	?	?
Number of neutrons	?	?	?	12
Number of electrons	?	?	17	11

3.44   Uranium exists mainly as two isotopes in nature, possessing mass number 235 and 238, respectively. Write the full atomic symbols for both isotopes.

3.45   What is wrong with this symbol?

$${}^{12}_{7}\text{C}$$

3.46   Give the full atomic symbol for the following atom:

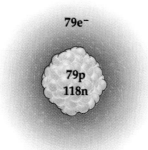

3.47   Write the full atomic symbol for the only atom that has its mass number equal to its atomic number.

3.48   Suppose you wanted to turn atoms of lead into atoms of gold. What would you have to do to the nucleus of the lead atoms?

3.49   What is the difference between an atom's atomic weight and its mass number?

3.50   Why is the atomic mass scale also called the relative atomic mass scale?

3.51   What is so special about the mass number and atomic mass of ${}^{12}_{6}\text{C}$?

3.52   How much heavier is an "average" oxygen atom than a ${}^{12}_{6}\text{C}$ atom? [Use the periodic table for the atomic weight of oxygen.]

3.53   How much heavier is an "average" titanium (Ti) atom than a ${}^{12}_{6}\text{C}$ atom? [Use the periodic table for the atomic weight of titanium.]

3.54   How much heavier is an "average" oxygen atom than a He atom? [Use the atomic weights from the periodic table.]

3.55   Why is the atomic weight of carbon not listed as 12 on the periodic table if the universal standard is that ${}^{12}_{6}\text{C}$ has an atomic weight of exactly 12?

3.56   Suppose we wanted to use one of the heavier atoms, like ${}^{235}_{92}\text{U}$, as the standard reference for the atomic weight scale.
(a) What would be the universally agreed upon atomic weight of the ${}^{235}_{92}\text{U}$ isotope?
(b) Why don't we use ${}^{235}_{92}\text{U}$ instead of ${}^{12}_{6}\text{C}$?

3.57   Bromine exists as only two isotopes in nature, ${}^{79}_{35}\text{Br}$ (atomic weight 78.918 336, % natural abundance = 50.69%), and ${}^{81}_{35}\text{Br}$ (atomic weight 80.916 289).
(a) What is the % natural abundance for ${}^{81}_{35}\text{Br}$?
(b) Calculate the atomic weight of the naturally occurring mixture of isotopes.

3.58   Naturally occurring hydrogen on Earth has an atomic weight of 1.0079. Suppose you were on another planet and found the atomic weight of hydrogen to be 1.2000. How would you explain this? (The atomic weight of ${}^{1}_{1}\text{H}$ is 1.007 825 2; the atomic weight of ${}^{2}_{1}\text{H}$ is 2.104 102 2.)

3.59 While an atom's mass number and its atomic weight are not the same, they are often quite close to one another. Consider the following: Uranium-235 has an atomic weight of 235.043 93 and a % abundance of 0.73%. Uranium-238 has an atomic weight of 238.0508 and a % abundance of 99.27%. Without doing any calculations, which of the following do you think represents the atomic weight of naturally occurring uranium?

(a) 234.04   (b) 236.03   (c) 237.03   (d) 238.03   (e) 238.07

Explain how you made your choice.

THE PERIODIC TABLE

3.60 (a) In what unique way did Mendeleev order the elements to make his discovery?
(b) What was his discovery?
(c) How does his ordering differ from the modern ordering?

3.61 How did Mendeleev's discovery aid in discovering additional elements?

3.62 Define what is meant by chemical periodicity with respect to chemical properties.

3.63 If one considers just the representative elements, how many groups would the periodic table have?

3.64 Considering just the representative elements, what percentage of them are metals, nonmetals, and metalloids? How do your results compare to the same percentages derived using the entire periodic table?

3.65 Give the names associated with groups IA, IIA, VIA, VIIA, and VIIIA.

3.66 What element seems poorly placed in group IA, the alkali metals? Why?

3.67 How does a group differ from a period in the periodic table?

3.68 As you take one step to the right in any period, how many electrons and protons are being added to the atom?

3.69 Magnesium (Mg) reacts with chlorine (Cl) to form the compound $MgCl_2$.
(a) Predict the formulas for the compounds formed when all the other alkaline earths react with chlorine.
(b) Predict the formulas for the compounds formed when all the alkaline earths react with bromine.
(c) Explain what principle you used to make your predictions.

3.70 State two things that are unique about the noble gases.

3.71 Name and give symbols for five of the transition metals.

3.72 How many elements wide are the transition metal and rare earth portions of the periodic table?

3.73 Why does chemical periodicity change from 8 to 18 and then again to 32?

3.74 Metals are excellent conductors of electricity, whereas nonmetals are poor conductors of electricity (insulators). Silicon (Si) is somewhere in-between and is referred to as a semiconductor. Why do you think this is so for Si?

OTHER PERIODIC TRENDS—ATOMIC SIZE AND IONIZATION ENERGY

3.75 Explain what is responsible for the size trend as you go across a period from left to right.

3.76 Order the following atoms from smallest to largest, judging from their relative positions in the periodic table: Cs, Fe, Ti, Hf

3.77 Order the following atoms from smallest to largest, judging from their relative positions in the periodic table: F, S, Br, Ar

3.78 How many electrons are in a $^{24}_{12}Mg$ cation that has a difference of 2 between the number of electrons and the atomic number? Give the full atomic symbol for this ion.

3.79 How many electrons are in the $^{15}_{7}N$ anion that has a difference of 3 between the number of electrons and the atomic number? Give the full atomic symbol for this ion.

3.80 Two students are studying using a plastic model of the atom to which they can easily add or remove electrons, protons, and neutrons. They build a model for the neutral $^{14}_{6}C$ atom. Their next job is to make the +2 ion of this carbon isotope. Student X removes 2 electrons from the atom. Student Y adds 2 protons to the atom's nucleus. Did both students successfully construct a +2 carbon-14 cation? Explain.

3.81 Fill in the following table:

	$^{15}_{8}O^+$	?	?	?
Mass number	?	27	?	58
Atomic number	?	?	15	?
Number of protons	?	13	?	?
Number of neutrons	?	?	16	30
Number of electrons	?	?	?	27
Charge on ion	?	+3	−3	+1

3.82 What is meant by the term "ionization energy"?

3.83 True or false? It always takes energy to remove an electron from a neutral atom. Explain your answer.

3.84 What is the trend in ionization energy as you go down a given group in the periodic table? As you go across a period from left to right?

3.85 Explain the trends in ionization energy in terms of the trends in atomic size.

3.86 Of the atoms Na, Mg, and Al, which should be the most difficult to ionize? Which should have the smallest ionization energy? Explain fully.

3.87 Of the atoms Na, Mg, and K, which should be the most difficult to ionize? Which should have the smallest ionization energy? Explain fully.

3.88 Refer to the chart shown at the bottom of page 91. Do the halogens follow the expected trend in ionization energies? Explain fully.

3.89 Sodium (Na) metal rapidly reacts with chlorine (Cl), a nonmetal.
(a) What reason is given for this in this chapter?
(b Would you predict the reaction of lithium (Li) with bromine (Br) to be more or less reactive than that between sodium and chlorine? Explain your choice.
(c) Would you predict the reaction of potassium (K) with fluorine (F) to be more or less reactive than that between sodium and chlorine? Explain your choice.

3.90 In the reaction of lithium (Li) with nitrogen, 3 lithium atoms react with 1 nitrogen atom to give the compound Li_3N. Atoms of one of these elements lose 1 electron; atoms of the other element gain 3 electrons.
(a) Which element gains the 3 electrons? Explain your choice.
(b) Is the element that gained 3 electrons a cation or an anion?
(c) Give the full atomic symbol for the ion of part (b), assuming it has 7 neutrons in its nucleus.

LIGHT

3.91 Electromagnetic radiation has a wavelength λ associated with it. Draw two waves of electromagnetic radiation, one representing green light and the other X rays, showing in a relative way how they differ in their wavelength.

3.92 X rays have a higher energy than green light. How can you prove this without doing any calculations?

3.93 White light from a heated tungsten filament (incandescent) light bulb gives a continuous visible spectrum upon being passed through a prism. What is meant by the word "continuous"?

3.94 Why are X rays and gamma rays so dangerous?

3.95 Light travels extremely rapidly ($c = 3.00 \times 10^8$ m/sec). Suppose you had to travel 30 miles to work every day. If you traveled at the speed of light for the entire trip, how long would it take you to get to work (in seconds)? [1 mile = 1.6093 km]

3.96 The Earth is 9.3×10^7 miles from the Sun. How long does it take the Sun's visible light to reach us (in minutes)? Also, given that the Sun also emits higher-energy gamma rays, how long does it take the gamma rays to reach us (in minutes)? [$c = 3.00 \times 10^8$ m/sec; 1 mile = 1.6093 km]

3.97 According to the equation for the energy of light, which statement is true? (1) The energy of light increases as its wavelength increases. (2) The energy of light decreases as its wavelength increases. Explain how the equation for the energy of light tells you which is true.

3.98 The unit of nanometers (nm) is commonly used for the wavelength of visible light. What does 1.00 nm equal in meters? What does it equal in inches? [There are 2.54 cm per inch.]

3.99 Exposure to gamma rays can kill you, while exposure to radio waves is not harmful. Why is this so?

3.100 Suppose a radio wave has a wavelength of 10 meters. What is the energy of this radiation (in joules)?

3.101 Suppose an X ray has a wavelength of 10 pm. What is the energy of this radiation (in joules)? How many times more energetic is this radiation than the radio wave in the previous problem? [1 pm = 10^{-12} meter]

3.102 A heated gas made up of individual atoms gives off a line spectrum when the emitted light is passed through a prism. How does a line spectrum differ from a continuous spectrum?

3.103 Upon electrification, hydrogen produces the following line spectrum:

Color	Wavelength (λ), nm
Indigo	410.1
Blue	434
Blue-green	486
Red	656.3

What are the energies (in joules) of each of the lines?

3.104 The fact that electrified atoms emit only certain colors of light in sharp lines and not all colors blended together tells us what about an atom?

WORKPATCH SOLUTIONS

3.1 (a) The law of conservation of matter is obeyed since 100.0 g + 793.6 g = 893.6 g.
(b) The percent hydrogen is (100.0 g/893.6 g) × 100 = 11.19% H.
(c) The percent oxygen is (793.6 g/893.6 g) × 100 = 88.81% O.
(d) The law of conservation of matter is obeyed since 50.0 g + 793.6 g = 446.8 g + 396.8 g.
(e) The percent hydrogen is (50.0 g/446.8 g) × 100 = 11.19% H.
(f) The percent oxygen is (396.8 g/446.8 g) × 100 = 88.81% O. [Notice that the amount of oxygen (396.8 g) in the calculation is the amount actually used, which is the starting amount minus the leftover amount.]

3.2 (a) The ratio of (g O in hydrogen peroxide)/(g O in water) is 32 g/16 g = 2/1. This ratio has meaning because of the fixed amount of hydrogen (2 g) used to make both compounds.
(b) The ratio in part (a) tells us that hydrogen peroxide has twice as much oxygen in it as water per given amount of hydrogen. Since the formula for water is H_2O, hydrogen peroxide must be H_2O_2.

3.3 The nucleus can't be negatively charged. If it were, then positively charged α particles would be attracted towards it and tend to stick. None would be deflected through large enough angles to appear to bounce back.

3.4 From the figure, it is obvious that an atom has as many electrons outside the nucleus as it has protons inside the nucleus. Since each electron has a −1 charge and each proton has a +1 charge, the overall atom must be neutral since an equal number of +1 and −1 charges cancel each other.

3.5 (a)   (b)   (c)

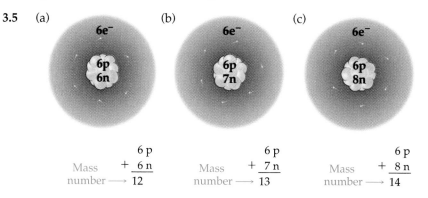

3.6 No. The mass number tells you the number of protons + neutrons. To identify the element you need to know the number of protons (atomic number) in the nucleus.

3.7 Now you can figure out the atomic number. If the atom has a mass number of 16 and it possesses 9 neutrons, then the number of protons is 16 − 9 = 7. The element with atomic number 7 is nitrogen (N).

3.8 If an atom weighs 4.015 times more than $^{12}_{6}C$, then its relative atomic weight is 4.015 × 12 = 48.18. (Remember that 12 is an exact number, so we don't need to round off.)

3.9 Because Si is in the same group as carbon, and thus should have similar chemical properties to C.

3.10

	$^{14}_{7}N^{3-}$	$^{24}_{12}Mg^{2+}$	$^{23}_{11}Na^{+}$	$^{56}_{26}Fe^{3+}$
Mass number	14	24	23	56
Atomic number	7	12	11	26
Number of protons	7	12	11	26
Number of neutrons	7	12	12	30
Number of electrons	10	10	10	23

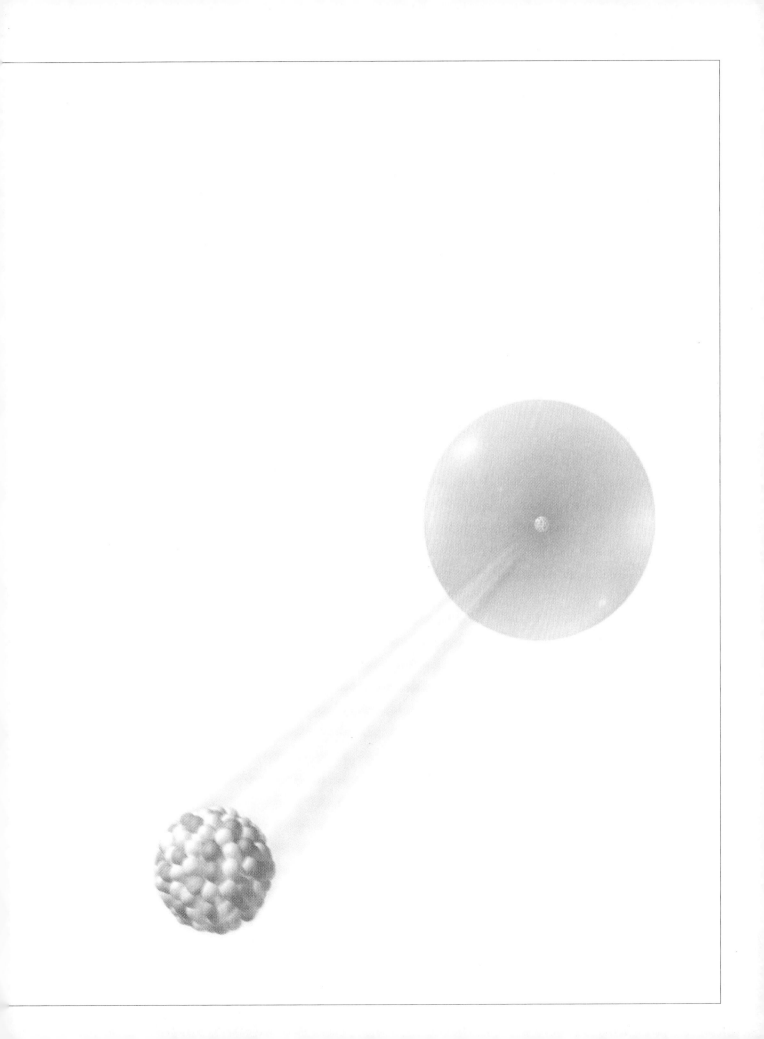

Chapter 4

The Modern Model of the Atom

A NEW KIND OF PHYSICS—ENERGY IS QUANTIZED 4.1

Newton gave us his three laws of motion in the 1680s. These were wonderful laws that allowed us to explain the motion of objects from cannon balls to planets. By 1900, Newton's laws had become the holy grail of physics—unchallengeable, the very foundation of modern science. You simply didn't question Newton's laws. Imagine then what it must have been like when scientists, trying to explain how electrons behaved in atoms, did just that. By 1930, atomic scientists had discarded the classical physics of Newton and replaced it with an entirely new kind of physics. They had to, because for the first time, Newton's laws had failed miserably. They simply could not explain the behavior of electrons in atoms. Rutherford had set the stage. Recall his model of the atom, with negative electrons buzzing around a tiny, massive, positive nucleus. Why don't the electrons fall into the nucleus? Why is the chemical behavior of the elements periodic? Why do electrifying gases of the elements give rise to line spectra, with their distinctive lines of color?

In 1913 the Danish physicist Niels Bohr proposed an answer that was revolutionary. He said that the energies of the electrons inside atoms are *quantized*. By **quantized** he meant that the electrons could have only particular, allowed energies. Such an assertion went against all the teachings of Newton's classical physics. Classical physics and everyday observation tells us that an object can have any amount of energy. Consider a tennis ball, for example. A tennis ball can have any amount of kinetic energy, depending on how hard you hit it. It can have low energy (a gentle lob over the net), high energy (fired out of a cannon), or anything in between.

But imagine a tennis ball that was allowed to have only certain kinetic energies so that it could travel at only certain allowed speeds. It could never travel at a speed in between the allowed ones no matter how it was hit. That would be quite a tennis ball! Well, tennis balls *do* have quantized energies; we just can't detect this fact (read on). However, we *can* detect that electrons in an atom have quantized energies.

Before we go on, we want to remind you of the scientific method discussed in Chapter 1. The scientific method often results in a theory being set aside and replaced with another one. Most often, a major change of this kind is not the result of one person's work. For example, when Bohr proposed that the energy of electrons in atoms is quantized, he was drawing on the work of the German physicist Max Planck. Planck had earlier proposed energy quantization in his efforts to explain the energy characteristics of light that was emitted by heated objects. Planck saw this only as a mathematical trick, something that allowed him to arrive at the correct answer in his calculations. He did not believe that energy was actually quantized in any real physical system. Bohr was aware of Planck's work, and he took the concept one step further by applying it to the electrons in an atom. That was a bold thing to do, particularly since Bohr had no explanation for why electrons should exhibit quantization. He just knew that it explained certain observations. The explanation of why quantization occurred would come about 10 years later from yet another scientist, Louis de Broglie. This is the scientific method in action.

Today, physicists recognize two kinds of physics, classical and quantum. **Classical physics** is the physics of Newton that says objects can have any energy. It works extremely well for large objects such as tennis balls, cannon balls, rockets, and planets. **Quantum physics** says that objects can have only certain, particular energies. It is the physics that correctly explains the behavior of tiny particles the size of an atom or smaller, such as electrons. In fact, quantum physics is really the more general of the two. Earlier we said that the motion of a tennis ball is quantized; we just can't detect the quantization. That is because the differences between the allowed energies for a large object like a tennis ball are so small that the energies seem to blend into one another to yield a smooth continuum. For example, consider heating a large amount of water on a stove. As you add heat (energy) to the water, the water's temperature (a measure of its energy content) appears to increase in a smooth, continuous fashion.

Even with the world's most sensitive and accurate thermometer, you would not observe any temperature jumps. That is not because there are no temperature jumps, but rather because the step size is too small to be measured. The step size is on the order of what is known as Planck's constant, about 10^{-34} of a degree. No thermometer could ever detect such small changes. However, the differences between the allowed energies for a tiny electron inside an atom are much larger, and are therefore detectable. We can therefore get away with using Newtonian physics for large objects, but we must use quantum physics if we want to understand how an electron behaves in an atom.

4.1 WORKPATCH   Imagine that cars were the size of atoms. What would traffic look like on the highway? How would it differ from stop-and-go traffic jams that you have experienced?

THE BOHR THEORY OF ATOMIC STRUCTURE 4.2

Bohr proposed that the electrons in an atom travel around the nucleus in circular orbits much like the planets orbit the Sun. The revolutionary part of his hypothesis was that only certain orbits, with certain fixed distances from the nucleus, were allowed. By imposing this restriction, he was actually saying that electrons in atoms could have only certain allowed energies. To understand the relation between the size of an electron's orbit and the energy of the electron, you need only remember that the nucleus is positively charged and an electron is negatively charged, so they are attracted to each other. Think of this attraction as a spring that connects the electron to the nucleus. When the electron is closest to the nucleus, the spring is relaxed and the atom is at its lowest possible energy. To move the electron away from the nucleus to a larger orbit requires putting energy into the atom to overcome the electron–nucleus attraction. In our spring model, that is equivalent to stretching the spring. The energy we put in is stored in the stretched spring. The stretched spring (and thus the atom itself) now contains more energy than before and therefore can be considered to be in a higher-energy, less stable situation. Unless there is something preventing it, the electron will spontaneously snap back into the lower orbit and release the energy that was stored in the stretched spring.

Electron ⎯⎯
Relaxed ⎯⎯
spring
Nucleus ⎯⎯

Energy added to stretch spring

Energy released when spring relaxes

Atom has lowest possible energy

Atom has higher energy

Atom returns to lowest energy

Therefore, the closer an electron is to the nucleus, the lower the electron's energy and the more stable the atom:

Electron farther from nucleus; attraction weaker; arrangement less stable

Electron close to nucleus; attraction strong; arrangement stable

So by saying that an electron could only be at certain distances from the nucleus, Bohr was really saying that the electrons in an atom could only have certain (quantized) energies—as if the electrons were connected to specific orbits.

In Bohr's model of the atom the electrons can jump from one orbit to another (these are called quantum jumps), but they can never be found between the orbits. In this view, electrons position themselves around the nucleus in an atom much like a person

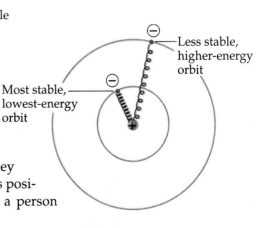

Less stable, higher-energy orbit

Most stable, lowest-energy orbit

standing on a ladder. You can stand on the first rung, the second rung, and so on, but you can't stand between two rungs. The same is true for electrons in atoms in Bohr's model.

Bohr took his quantized electron orbits, which were called **shells**, and assigned them a **principal quantum number (n)**, starting with $n = 1$ for the first orbit, $n = 2$ for the next, and so on as shown below:

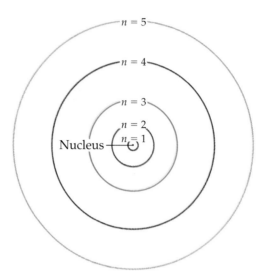

The shells continue with $n = 6, 7, 8,$ and so on up to infinity, getting larger and larger, taking the electron farther from the nucleus, and thus giving it a higher and higher energy. (The relative sizes of the orbits shown in the preceding illustration are correct for hydrogen, but they are not correct for other elements. From here on, therefore, we will show the orbits more schematically.)

Ideally, all the electrons in an atom would like to occupy the $n = 1$ shell, since it is the shell of lowest energy (maximum electron–nucleus attraction). But electrons also repel each other, and crowding them into a single shell would quickly raise the energy of the atom dramatically. In order to deal with this problem, and also to explain line spectra and periodic chemical behavior, Bohr maintained that each shell could hold a certain maximum number of electrons. The larger the shell, the more electrons it could hold, since a larger shell provides more room for the electrons to spread out. Bohr maintained that each shell could hold a maximum number of $2n^2$ electrons, where n is the principal quantum number of the shell. Here, then, are the main features of the Bohr model:

1. Orbits (shells) get larger (their radius or distance from the nucleus increases) as the orbit number n, called the principal quantum number, increases.
2. Electrons in the $n = 1$ orbit are the most stable and have the lowest energy. Electrons in the $n = 2$ orbit are the next most stable and are somewhat higher in energy. The $n = 3$ electrons are even less stable and higher in energy, and so on.
3. Since each orbit or shell can hold a maximum of $2n^2$ electrons:
   the $n = 1$ shell can hold a maximum of $2 \times 1^2 = 2$ electrons,
   the $n = 2$ shell can hold a maximum of $2 \times 2^2 = 8$ electrons,
   the $n = 3$ shell can hold a maximum of $2 \times 3^2 = 18$ electrons,
   the $n = 4$ shell can hold a maximum of $2 \times 4^2 = 32$ electrons, and so on.

Maximum electron capacity of the first four shells:

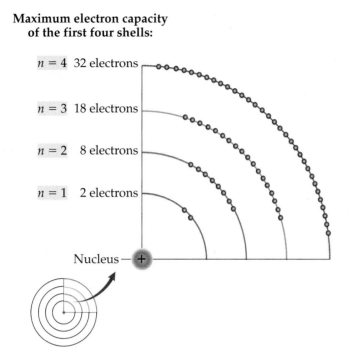

4. When building an atom, orbits are filled with electrons from the innermost shell (the one lowest in energy) on out. Electrons are added to each orbit until it is filled to capacity ($2n^2$ electrons) before moving up to the next level. For example, the lithium (Li) atom, with atomic number 3, has three protons in its nucleus and three electrons in its electron shells. Two of the electrons go into the most stable $n = 1$ shell, and the third electron must go into the less stable $n = 2$ shell.

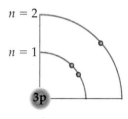

Lithium (Li)

This is the essence of the Bohr atomic model. It is not the most current model of the atom, which we will see at the end of this chapter, but it is valuable. Its major concepts, such as energy quantization and the increase in the electron's energy with increasing distance from the nucleus, are also important parts of the modern model. In addition, we can use the Bohr model to show you why atoms give line spectra and why the elements behave in a periodic fashion.

PRACTICE PROBLEMS

4.1 What is wrong with the following Bohr model of the beryllium (Be) atom (atomic number 4)?

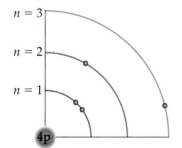

Answer: Don't begin filling the $n = 3$ orbit until the $n = 2$ orbit is full. Don't skip shells!

4.2 Draw a Bohr model for an atom of sulfur (S). How many additional electrons could be fit into the valence shell?

4.3 Why is an electron in an orbit with a low value of n in a more stable arrangement than one with a higher value of n?

4.4 How many electrons could the $n = 5$ shell in an atom hold?

4.3 PERIODICITY AND LINE SPECTRA EXPLAINED

Explaining the periodic behavior of the chemical properties of the elements was one of the major triumphs of Bohr's model. Consider the elements lithium (Li) and sodium (Na), both in group IA of the periodic table and both possessing similar chemical properties. According to Bohr, lithium, with 3 electrons, and sodium, with 11 electrons, would have the following structures:

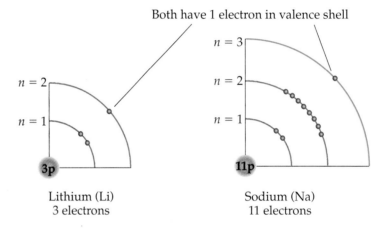

Lithium (Li)
3 electrons

Sodium (Na)
11 electrons

If you look closely at these diagrams you should be able to see a similarity between lithium and sodium. They both have completely filled inner shells and only a single electron in the outermost occupied shell. This outermost occupied shell of an atom is called the **valence shell**. Perhaps these elements have similar chemical properties because they have identical valence shell configurations. Does this hold true throughout the periodic table?

Let's look at another pair of elements from the other side of the periodic table, fluorine (F) and chlorine (Cl). They are both in group VIIA, so we know they have similar chemical properties. What about their valence shells? When the electrons are put into the orbits following Bohr's prescription, both elements end up with 7 electrons in their valence shells. This,

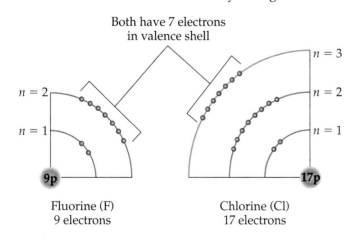

Fluorine (F)
9 electrons

Chlorine (Cl)
17 electrons

according to the Bohr model, is why F and Cl have similar chemical properties and belong in the same group of the periodic table.

The electron shell configurations for the first 18 elements in the periodic table, filled according to Bohr's rules, are shown in the following chart. Look at the outermost shell in each group to convince yourself that elements with similar chemical properties have identical valence shell configurations. Only helium (He) seems not to follow this rule. In a way, though, it does. Both helium and the atom below it, neon (Ne), have a completely filled valence shell. The difference is that the valence shell in helium is the $n = 1$ shell and can only accommodate 2 electrons. All atoms after Ne in this group contain 8 electrons in their valence shell.

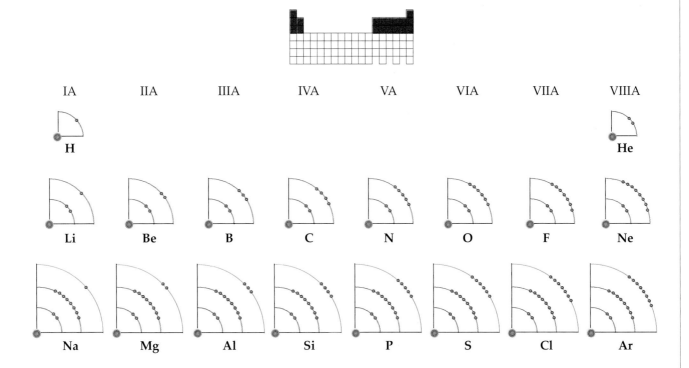

Careful examination of this chart can reveal a shortcut for determining the number of electrons in the valence shell of an atom. Hydrogen, lithium, and sodium each have 1 valence shell electron in the Bohr model, and they are also in group IA of the periodic table. Fluorine and chlorine both have 7 valence electrons, and are in group VIIA. For the representative elements, the group number on the periodic table gives the number of valence shell electrons.

The number of electrons in the valence shell of an atom is equal to the group number for the representative (A group) elements.

So, according to Bohr's model, the chemical properties of the elements are periodic because, as successive shells of electrons get filled, the valence shell configuration repeats.

The other success of the Bohr model was its ability to explain the line spectrum of hydrogen. Not only did the model explain why line spectra exist, but the simple mathematical calculations that Bohr carried out based on the model's assumptions reproduced exactly the experimental line spectrum of hydrogen. From the physicist's perspective this was the real success of the Bohr model. Bohr's equations could be used to calculate the energy of an electron in any allowed shell of the hydrogen atom. A scaled version of the energy values he calculated for the first six shells of the hydrogen atom is shown at the right in an **energy level diagram**. As we shall see, the absolute energies of the shells are not important. It is the differences between them that is all important. Thus, we have scaled the energies so that the energy of the lowest-energy ($n = 1$) shell is 1.0 electron volt (eV).

Notice how the energies of the shells get closer together as n increases. This fact will become important later on. Keep in mind that it is the energies of the shells that get closer together, not the shells themselves.

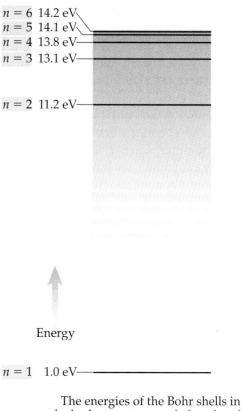

The energies of the Bohr shells in the hydrogen atom scaled so that the energy of the $n = 1$ shell is 1.0 eV

4.2   WORKPATCH

Examine the energy level diagram for hydrogen shown above, and answer the following questions:
   (a) If an electron is in the $n = 1$ shell, what is its scaled energy (in eV)?
   (b) If an electron is in the $n = 2$ shell, what is its scaled energy (in eV)?
   (c) How much energy would it take to transfer an electron from the $n = 1$ shell to the $n = 2$ shell?
   (d) Would it take the same, more, or less energy than your answer to part (c) to take an electron from the $n = 2$ shell to the $n = 3$ shell? Why?

Were you able to intuitively answer parts (c) and (d) of the WorkPatch? You know the electron's energy when it is in the $n = 1$ shell, and you know the electron's energy when it is in the $n = 2$ shell. The difference between these two energies, therefore, is the energy it will cost you to move the electron from the $n = 1$ to the $n = 2$ shell. This is a very important concept, so make sure you understand it. As for part (d), the energies of the shells get closer together with increasing n. Clearly then, the energy cost will be less to go from $n = 2$ to $n = 3$ than to go from $n = 1$ to $n = 2$.

To predict the line spectrum of hydrogen, Bohr assumed that the atom was in its *ground state*. The **ground state** of an atom is defined as the arrangement of electrons that is of lowest total energy. Thus, it is the configuration where all the electrons are placed in the lowest possible energy shells. For a hydrogen atom, which has just one electron, the ground state would appear as shown on the left below (we use a Bohr diagram to show you where the electron is and an energy level diagram to indicate the energy of the electron).

All hydrogen atoms spend most of their time in the ground-state electron configuration under normal conditions. However, when enough energy is added to the atom (for instance, by heating it or passing an electric current through it), the electron can move ("jump") into a higher-energy shell. When that occurs, we say the energy was absorbed by the atom and the atom is now in an *excited state*. In an **excited state**, one or more of the electrons in an atom are located in higher-energy shells although there is still room for more electrons in the lower-energy shells. A possible excited state for the hydrogen atom is shown on the right below.

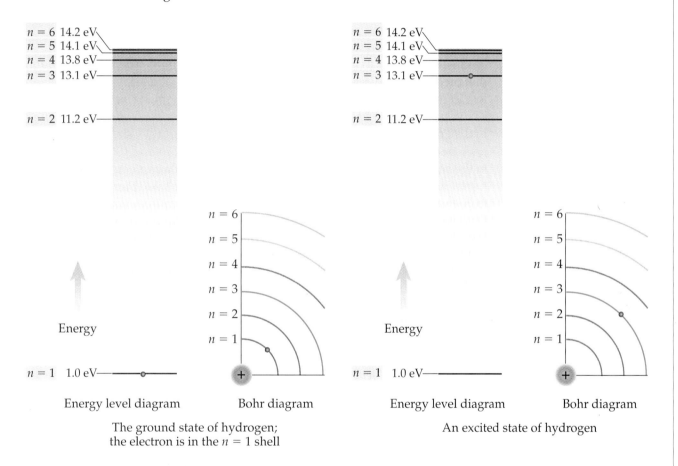

Energy level diagram   Bohr diagram   Energy level diagram   Bohr diagram

The ground state of hydrogen;   An excited state of hydrogen
the electron is in the $n = 1$ shell

This energy level diagram represents an excited state because there are vacant, lower-energy shells below the electron (levels have been skipped before they are full). Another possible excited state would have placed the electron in the $n = 2$ shell, and yet another in the $n = 4$ shell, and so on. We can create any of these excited states by starting with the ground-state hydrogen atom and adding enough energy to make the electron jump from the $n = 1$ shell to an upper shell.

What is the lowest-energy excited state of the hydrogen atom?   WORKPATCH **4·3**

The WorkPatch points out that all excited states are not created equal. For instance, the hydrogen excited state with the electron in the $n = 2$ shell has less energy than the excited state with the electron in the $n = 3$ level, because it takes more energy to raise the electron up to $n = 3$ than to $n = 2$.

4·4 WORKPATCH   Starting from the ground state, how much more energy does it take to create the $n = 3$ excited state of hydrogen than the $n = 2$ excited state?

Check your WorkPatch answer against ours. When you can do the Work-Patch, then you are becoming skilled at understanding and using a quantized energy level diagram.

When a tube of hydrogen gas is electrified, the electricity supplies the energy to cause various excited states. To account for the line spectrum produced by such a gas, we must consider what happens *after* the excited states are created. As it turns out, atoms do not usually remain in an excited state for long. Since there are vacancies in lower-energy levels, the electrons can drop back down into them. That is desirable from an energy point of view. To create the excited state, the atom absorbs energy, and when the atom returns to the ground state, it loses the same energy.

One way the atom can lose energy is to emit it as light. The light created in this process (called *relaxation*) has the same amount of energy as the electron lost in making its energy level jump. This light is characterized by a particular wavelength, and this is where line spectra come from. When an excited hydrogen atom relaxes from the $n = 2$ state to its ground state, the electron goes from an energy level of 11.2 eV down to 1.0 eV, losing 10.2 eV of energy in the process. This energy is released as light with a wavelength corresponding to 10.2 eV in energy. Using Table 4.1, we can see that this particular electron transition results in ultraviolet light being emitted. Since ultraviolet light is invisible to our eyes, we can't see this particular line in the hydrogen line spectrum.

Table 4.1   Energy and Wavelength Ranges for Ultraviolet and Visible Light

		Energy in electron volts (eV)	Wavelength (λ) in nanometers (nm)
Ultraviolet		> 3.74	< 380
Indigo/violet		3.34–2.95	380–430
Blue		2.95–2.64	430–480
Green		2.64–2.24	480–565
Yellow		2.24–2.10	565–620
Orange		2.10–1.95	620–650
Red		1.95–1.69	650–750

However, the hydrogen spectrum does contain four lines in the visible region of the electromagnetic spectrum—one red line, one green line, one blue line, and one indigo line. According to Table 4.1, red corresponds to light with an energy in the range of 1.69–1.95 eV. A close inspection of the energy level diagram shows that there is one and only one jump that corresponds to energy in this range. An electron in an excited hydrogen atom relaxing from the $n = 3$ level to the $n = 2$ level would produce light with exactly 1.9 eV of energy (red light).

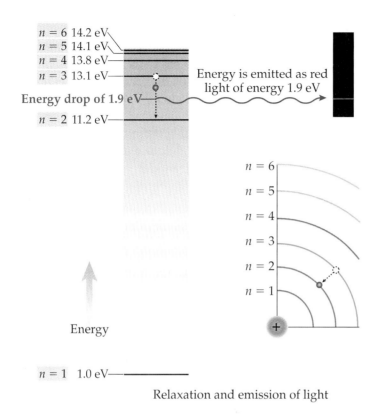

Relaxation and emission of light

Thus we can conclude that the electron in an excited atom does not necessarily return to its ground state in a single jump. It can also reach the ground state by making a succession of smaller downward jumps.

The hydrogen atom has three other visible lines in its line spectrum: green, blue, and indigo. Use the energy level diagram for hydrogen and Table 4.1 to determine the electron transitions or jumps that are responsible for these lines.

WORKPATCH 4.5

Indigo Blue   Green   Red
$n = 3 \longrightarrow n = 2$

Were you able to do the WorkPatch? Since, of these four lines, indigo has the highest energy, the jump corresponding to the indigo line should be the largest. Check your answer to see that it is.

Bohr's model accounts for the existence of line spectra in atoms because the electrons in an atom are only allowed certain, fixed energies and never anything in between. Since the levels are fixed, the differences between them are also limited, and only certain wavelengths (energies) of light can be produced. The energy levels for atoms of other elements have different numerical values and hence different line spectra. The rapid acceptance of Bohr's model was due mainly to its success in exactly reproducing the line spectrum of the hydrogen atom. Not only could Bohr correctly calculate the exact position and energy of the four lines in the visible spectrum, but he could also match exactly the lines detected in other regions of the spectrum (ultraviolet and infrared). Nevertheless, Bohr's model of the atom did have some problems.

The lines Bohr predicted for the helium spectrum did not match up at all with the real spectrum. In fact, his calculations failed to predict the correct spectrum for any atom other than hydrogen. In less than 20 years, Bohr's model would be replaced with the modern model of the atom. However, his postulate that the energy levels for electrons in atoms are quantized and his rules for the maximum number of electrons that each shell can accommodate are still correct. Therefore we can continue to use these aspects of simple Bohr theory to explain many things in chemistry. Before we do, however, we must look at an important modification to the Bohr model.

PRACTICE PROBLEMS

4.5 The ground state for the lithium (Li) atom and the scaled energies of its shells are shown below. Draw a Bohr model for the lowest-energy excited state of Li.

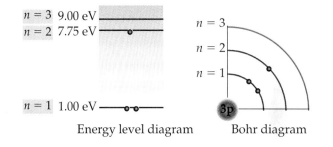

Energy level diagram   Bohr diagram

Answer: Two excited states are drawn below, but only (a) correctly answers the question. Remember, both are created from the ground state. To produce (a), an electron must be lifted from the n = 2 shell to the n = 3 shell. This takes 1.25 eV (1.25 eV of energy must be put into the atom). To produce (b), an electron is lifted from the n = 1 shell to the n = 2 shell. This takes 6.75 eV. Since it took more energy to create (b), (b) is a higher-energy excited state than (a). Thus, (a) is the lowest-energy excited state that can be prepared from the ground-state Li atom.

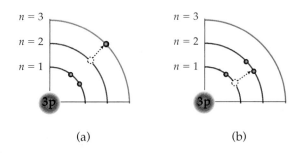

(a)   (b)

4.6 Draw a Bohr model for the highest-energy excited state of Li, using just the three shells shown in Practice Problem 4.5 and exciting only a single electron.

4.7 Draw a Bohr model for a Li⁺ cation in its ground state.

4.8 An atom has atomic number 6 and has 8 electrons.
(a) What element is this?

(b) Is this atom neutral, a cation, or an anion? If it is an ion, what is its charge?

(c) Draw a Bohr model for this atom in its ground state.

4.9 An F⁻ anion has a total of 10 electrons. A student draws the following Bohr model for this anion. There are two things wrong with this diagram regarding the electrons. What are they?

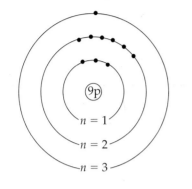

SUBSHELLS AND ELECTRON CONFIGURATION 4.4

Now we need to look at an important refinement to the Bohr model, the essence of which is also important in the modern model. During Bohr's time, instruments improved enough to show that the simple line spectra generated by excited atoms were not so simple after all. In many cases, what was originally thought to be a single line actually turned out to be a number of very closely spaced lines of nearly identical color (energy). Bohr struggled with these findings for several years, and tried various modifications to his basic model in an attempt to account for them. What he and others eventually proposed was, in essence, that some of the electron shells in an atom actually consist of a set of **subshells**, which are very closely spaced in energy and size.

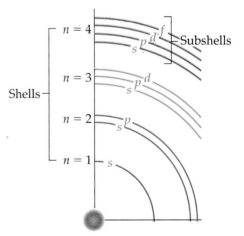

Each shell n has n subshells that are close to each other in size and energy.

As shown at the right, these subshells are given single-letter designations (s for the first subshell within a level, p for the second, d for the third, and f for the fourth). Not every level contains all these subshells. The first level ($n = 1$) contains just an s subshell. Every level we move out from $n = 1$ gets an additional subshell, so the $n = 2$ level consists of two subshells (s and p), the $n = 3$ level has three subshells (s, p, and d), and so on. In addition, as shown at the top of the next page, each subshell has a different maximum electron capacity. An s subshell can have a maximum of 2 electrons, a p subshell can hold a maximum of 6 electrons, a d subshell can hold up to 10 electrons, and an f subshell can hold 14 electrons. These subshells also increase in energy and size from $s \rightarrow p \rightarrow d \rightarrow f$.

Electron capacity
of shells $(2n^2)$

Electron capacity
of subshells

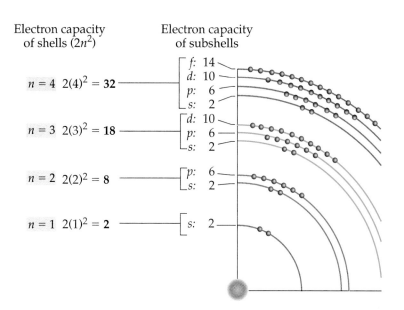

$n = 4$ $2(4)^2 = 32$

$f:$ 14
$d:$ 10
$p:$ 6
$s:$ 2

$n = 3$ $2(3)^2 = 18$

$d:$ 10
$p:$ 6
$s:$ 2

$n = 2$ $2(2)^2 = 8$

$p:$ 6
$s:$ 2

$n = 1$ $2(1)^2 = 2$

$s:$ 2

Notice that the maximum electron capacity of each principal shell is unchanged from the original Bohr model. For instance, the $n = 3$ shell still holds a maximum of 18 electrons.

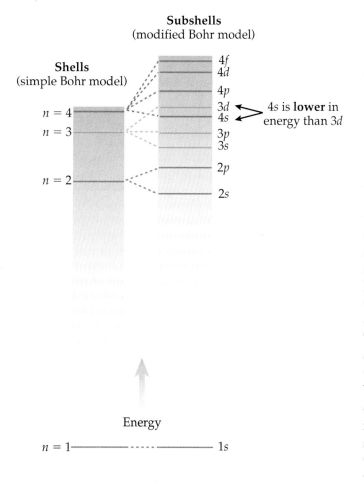

Shells
(simple Bohr model)

Subshells
(modified Bohr model)

$n = 4$

$n = 3$

$n = 2$

4f
4d

4p
3d
4s
3p
3s

4s is **lower** in
energy than 3d

2p

2s

Energy

$n = 1$ —————————— 1s

When we first introduced the energy level diagram for the hydrogen atom, we pointed out that the energies of the shells get closer and closer together as n increases. We also said that this would become important later on. Now is the time to show you why. In part because the energy levels get closer together as n increases, the energies of some subshells from adjacent shells end up crossing. The diagram at left illustrates this.

As we've said, the subshells within a shell increase in energy in the order $s \rightarrow p \rightarrow d \rightarrow f$. For example, within the $n = 2$ shell, the 2s subshell has the lowest energy, followed by the 2p subshell. Within the $n = 3$ shell, the order is 3s, then 3p, then 3d. And all of the $n = 3$ subshells are higher in energy than the $n = 2$ subshells. But look carefully at the 4s subshell. It is *lower* in energy than the 3d subshell, even though it has a larger value of n. Additional crossings occur as we proceed beyond $n = 4$. The consequences of these crossings are extremely important! Remember, Bohr stated that the ground state of an atom is achieved by filling the shells in order of increasing energy, starting with the $n = 1$ shell and working up. The same rule holds when we take subshells into account. This solves a problem that initially befuddled Bohr—the so-called potassium problem.

Potassium (K) is a group IA atom possessing 19 electrons. Group IA atoms are supposed to have one valence electron, but that's not what the simple Bohr model gives us. Look at what happens if we add 19 electrons to the simple Bohr energy level diagram (the energies are different for K than for H, but the diagrams are similar):

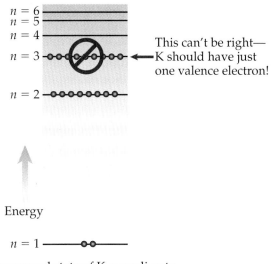

This can't be right—K should have just one valence electron!

Energy

The ground state of K according to the simple Bohr model (shells only)

The inner $n = 1$ and $n = 2$ shells have been filled to capacity, as expected. But the valence $n = 3$ shell has 9 electrons in it instead of the expected 1 electron for this group IA atom. The modified subshell model fixed this problem. Demonstrate this for yourself now.

Using 19 dots to represent potassium's 19 electrons, fill the subshells to arrive at the ground-state K atom.

WORKPATCH **4.6**

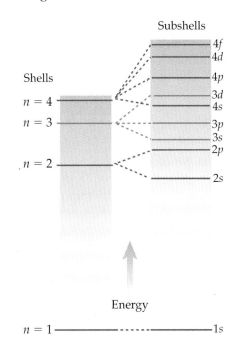

In the WorkPatch, you should have arrived at 1 electron in the valence $4s$ subshell. You have just solved the potassium problem! We are about to discuss this energy level diagram in detail, so check your answer against the one at the end of this chapter.

ELECTRON CONFIGURATION

The exact way that the electrons are arranged in the various subshells of an atom is called the **electron configuration**. An energy level diagram can accurately depict the electron configuration of an atom; however, it is a tedious way to represent it. Chemists have therefore developed a notation to indicate an atom's electron configuration. Each occupied subshell, starting with the lowest-energy subshell, is indicated with a lowercase letter (s, p, d, f) preceded by a number indicating its principal shell (n). The number of electrons occupying a subshell is indicated as a superscript. Thus, we can replace the modified Bohr diagram for potassium with the following to indicate its electron configuration:

$$1s^2 2s^2 2p^6 3s^2 3p^6 4s^1$$

This electron configuration notation is a lot easier to write, and it communicates the same information. It tells us how the electrons are arranged among the various subshells. In addition, it is an easy matter to identify the valence electrons. In general, the valence electrons are considered to be those in the subshells of highest quantum number n. For potassium, the highest quantum number is 4, and if we sum up the electrons present in the $n = 4$ subshells, we get 1, the number of valence electrons in the potassium atom.

Valence electron
$$1s^2 2s^2 2p^6 3s^2 3p^6 \overbrace{4s^1}$$

A slightly more difficult example is bromine. Bromine's 35 electrons fill the subshells to yield the ground-state electron configuration shown below:

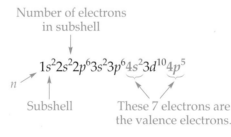

If we sum up the electrons in the highest n subshells, we get 7 valence electrons. To emphasize the valence electrons, chemists often write them last:

$$1s^2 2s^2 2p^6 3s^2 3p^6 3d^{10} 4s^2 4p^5$$

PRACTICE PROBLEMS

4.10 Write the electron configuration of silicon (Si) using the electron configuration notation. How many valence electrons does it have?

Answer: Silicon has atomic number 14, so it has 14 electrons. Using the subshell energy level diagram, we start by putting the first 2 electrons in the 1s subshell, the next 2 in the 2s subshell, the next 6 in the 2p subshell, the next 2 in the 3s subshell, and the remaining 2 in the 3p subshell:

$$1s^2 2s^2 2p^6 3s^2 3p^2$$

The highest quantum number n is 3, so summing up the number of electrons in the n = 3 subshells gives us 4, the number of valence electrons in Si.

4.11 Write the electron configuration of arsenic (As) using the electron configuration notation. Does the number of valence electrons agree with the group number?

4.12 Write the electron configuration of scandium (Sc) using the electron configuration notation.

The subshell energy level diagram that we have presented only went up to $n = 4$. As we proceed beyond this, additional subshell energy crossings occur. You will be pleased to know that you do not need to memorize any of these crossings. That's because the periodic table can be used as a quick and easy guide for deducing the electron configuration of any element. To do this, we think of the periodic table as being constructed from four blocks, the s block, the p block, the d block, and the f block:

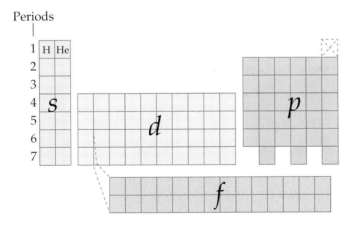

Notice that the s block is two elements wide. That is because an s subshell can hold a maximum of 2 electrons. The p block is six elements wide, since a p subshell can hold up to 6 electrons. Likewise, the d and f blocks are 10 and 14 elements wide, respectively, due to the maximum number of electrons those subshells can hold. Also, notice that we moved helium (He) from its usual position to a position in the s block, adjacent to hydrogen.

The designation (s, p, d) of each block represents the subshell that is being filled. The period tells you what level (n) is currently filling, *unless you are in the d or f block*. In the d block, the value of n for the d subshell being filled is 1 less than the period number. If you are in the *f* block, then the value of *n* for the *f* subshell being filled is 2 less than the period number. This takes into account the subshell crossings. A few examples will help to make this clear. Suppose we want to determine the electron configuration of sulfur (S). Keep glancing at the periodic table below and follow along the arrows as we discuss how to do this. We start at hydrogen (as we always will) and move across the first period from left to right until we reach the end of the period at helium. Each element we encounter on our way adds one more electron to the atom. Helium is the second element, so we have 2 electrons so far. They are both in the 1*s* subshell because we have moved along a path in the first period

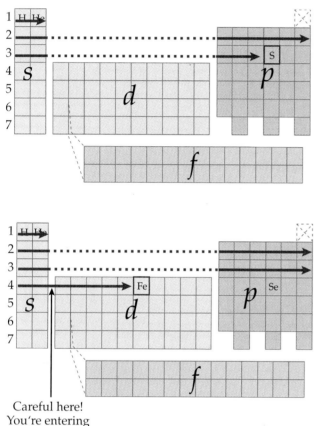

and we are in the *s* block. So far we have $1s^2$. Next, we move to the beginning of the next period. Here we move through both the *s* and *p* blocks of the second period, gaining $2s^2$ and $2p^6$. Having reached the end, we move on to the third period, going through the *s* block of period 3 ($3s^2$), and then going four elements deep into the *p* block to reach our target, sulfur. Four deep into the *p* block of period 3 gives us 4 electrons, or $3p^4$. The complete result for S is $1s^22s^22p^63s^23p^4$. As expected, there are 6 valence electrons in the $n = 3$ shell for this group VIA element.

Now let's try iron (Fe), a *d* block element. Again we start at H. The only catch is to remember that when you plunge into the *d* block, you must subtract 1 from the period to get the correct *n* (to account for the subshell crossings). This means that the final 6 electrons should be listed as $3d^6$, and not as $4d^6$. Thus, the electron configuration for iron is $1s^22s^22p^63s^23p^64s^2\,3d^6$.

If we were looking for the electron configuration of selenium (Se), then our arrow would continue past iron, out of the *d* block, and into the *p* block. Once we leave the *d* block and enter the *p* block we must bump *n* back up by 1. The electron configuration of selenium (Se) therefore is $1s^22s^22p^63s^23p^64s^2\,3d^{10}4p^4$.

Careful here!
You're entering
the *d* block!

The ground-state electron configuration obtained by using either the subshell energy diagram or the periodic table is not always correct, because there are exceptions to the rules we have been following. Practice Problem 4.13 will show you one of them.

PRACTICE PROBLEMS

4.13 Using the periodic table as a guide, determine the ground-state electron configuration of copper (Cu).

Answer: You should have arrived at $1s^22s^22p^63s^23p^64s^23p^9$, since Cu is nine deep into the d block of the fourth period. However, Cu is one of those exceptions, and its true electron configuration is $1s^22s^22p^63s^23p^64s^13p^{10}$.

4.14 Using the periodic table as a guide, determine the ground-state electron configuration of krypton (Kr). Also, explain why it is proper for it to be in group VIIIA.

4.15 Using the periodic table as a guide, determine the ground-state electron configuration of palladium (Pd).

So far we have been neglecting the f-block elements—that is, the elements beyond lanthanum (La) and actinium (Ac). Lanthanum and actinium can be thought of as detour signs to the f block. Remember that the value of n in the f block is 2 less than the period you are in. Once you get through the f block, it's back into the d block (where the value of n is only 1 less than the period). For example, let's determine the electron configuration of gadolinium (Gd):

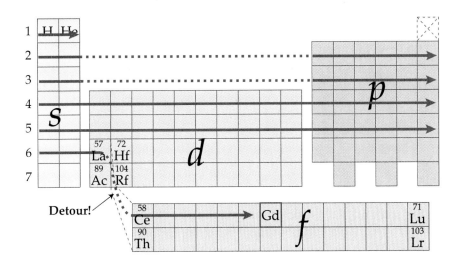

Follow the arrows. We get $1s^2$ from the first row, $2s^22p^6$ from the second row, $3s^23p^6$ from the third row, $4s^23d^{10}4p^6$ from the fourth row, $5s^24d^{10}5p^6$ from the fifth row, and finally $6s^25d^14f^7$ from the last arrow (we are only 1 deep into the d block and 7 deep into the f block). The result is $1s^22s^22p^63s^23p^64s^23d^{10}4p^65s^24d^{10}5p^66s^25d^14f^7$. We must point out that for the f-block elements there are more exceptions to this method of filling the subshells than there are elements that obey it. For most, you must transfer the single d electron into the f orbital that is being filled. This is shown in the abbreviated electron configurations listed below:

Electron configurations of the highest-energy subshells (the ones being filled)

58	59	60	61	62	63	64	65	66	67	68	69	70	71
Ce	**Pr**	**Nd**	**Pm**	**Sm**	**Eu**	**Gd**	**Tb**	**Dy**	**Ho**	**Er**	**Tm**	**Yb**	**Lu**
$6s^24f^15d^1$	$6s^24f^3$	$6s^24f^4$	$6s^24f^5$	$6s^24f^6$	$6s^24f^7$	$6s^24f^75d^1$	$6s^24f^9$	$6s^24f^{10}$	$6s^24f^{11}$	$6s^24f^{12}$	$6s^24f^{13}$	$6s^24f^{14}$	$6s^24f^{14}5d^1$
90	91	92	93	94	95	96	97	98	99	100	101	102	103
Th	**Pa**	**U**	**Np**	**Pu**	**Am**	**Cm**	**Bk**	**Cf**	**Es**	**Fm**	**Md**	**No**	**Lr**
$7s^26d^2$	$7s^25f^26d^1$	$7s^25f^36d^1$	$7s^25f^46d^1$	$7s^25f^6$	$7s^25f^7$	$7s^25f^76d^1$	$7s^25f^9$	$7s^25f^{10}$	$7s^25f^{11}$	$7s^25f^{12}$	$7s^25f^{13}$	$7s^25f^{14}$	$7s^25f^{14}6d^1$

Writing electron configurations for atoms that possess many electrons, such as the *f*-block elements, can get pretty tedious. Chemists have developed an additional shortcut in which the noble gas atom that directly precedes an element is used to represent its own electron configuration. For example, the noble gas that directly precedes gadolinium (Gd, atomic number 64) is xenon (Xe, atomic number 54). The first 54 electrons in Gd can thus be replaced with the Xe symbol, and the electron configuration of Gd can be written as $[Xe]6s^25d^14f^7$. (In this notation, the symbol for the noble gas is always enclosed in square brackets.)

PRACTICE PROBLEMS

4.16 Write the electron configurations of chlorine (Cl) and lutetium (Lu), using both the full notation and the noble gas abbreviated notation.

Answer: Cl: $1s^22s^22p^63s^23p^5$; $[Ne]3s^23p^5$
Lu: $1s^22s^22p^63s^23p^64s^23d^{10}4p^65s^24d^{10}5p^66s^25d^14f^{14}$; $[Xe]6s^25d^14f^{14}$

4.17 Write the electron configuration of radium (Ra), using both the full notation and the noble gas abbreviated notation.

4.18 Write the electron configuration of uranium (U), using both the full notation and the noble gas abbreviated notation.

Finally, we can turn this around and ask questions about an atom once we've determined its electron configuration. For example, arsenic (As) has the ground-state electron configuration $1s^22s^22p^63s^23p^64s^23d^{10}4p^3$ (or $[Ar]4s^23d^{10}4p^3$). We could now ask, "Where does this atom belong in the periodic table? What period is it in? What group is it in?" You should be able to answer these questions. Since the outermost occupied (valence) shell is $n = 4$, we expect to find As in the fourth period. In addition, As has 5 valence electrons, which we can emphasize by rewriting the electron configuration as $1s^22s^22p^63s^23p^63d^{10}4s^24p^3$. Five valence electrons puts it in group VA.

Here are a few additional hints. If an element has a partially filled *d* subshell, then it is a transition metal (a *d*-block element). If the element has a partially filled *f* subshell, then it is an *f*-block element. A partially filled *s* or *p* subshell identifies the element as an *s*-block element or a *p*-block element, respectively, and thus as a representative element (although there are exceptions).

PRACTICE PROBLEMS

4.19 In what period and group is the element with the electron configuration $[Ar]4s^23d^{10}4p^5$? What element is this?

Answer: Rewriting the electron configuration as $[Ar]3d^{10}4s^24p^5$ to emphasize the valence electrons, we see that the highest value of n is 4, so this atom is in the fourth period. There are a total of 7 valence electrons and the p subshell is partially filled, so this is a group VIIA element. It is bromine (Br).

4.20 In what period is the element with the electron configuration $1s^2 2s^2 2p^6 3s^2 3p^6 4s^2 3d^3$? What element is this?

4.21 In what period and group is the element with the electron configuration $[Xe]6s^1$? What element is this?

COMPOUND FORMATION AND THE OCTET RULE 4.5

The simple Bohr model can be very useful in predicting and understanding the chemical compositions of many compounds. For example, it can be used to explain why the compound sodium chloride has the formula NaCl and never Na_2Cl, $NaCl_2$, or any other variation. The one-to-one ratio of sodium to chlorine in this compound can be understood in terms of an early chemical principle known as the *octet rule*. The octet rule arose from the observation of the unique nature of the elements in group VIIIA of the periodic table, the noble gases. As we saw in Chapter 3, the elements in this group are exceptionally unreactive. This lack of "desire" to combine with other elements was interpreted as meaning that the elements in group VIIIA are exceptionally stable. The Bohr model provided another unifying feature for these elements—they all have 8 electrons in their valence shells (with the exception of helium, which has 2). The obvious implication is that an element will become unreactive (unusually stable) if it has 8 valence electrons. Thus was born the octet rule. According to the **octet rule** (from the Latin *octo*, meaning eight), the reason the other elements in the periodic table (groups IA–VIIA) participate in chemical reactions is to ultimately end up with 8 electrons in their valence shells, thus gaining a stability similar to that of the noble gases.

> **The Octet Rule: Elements react to form compounds in such a way as to put 8 electrons in their outermost (valence) shell, giving them a valence electron configuration identical to a noble gas and making them exceptionally stable.**

There are many exceptions to this rule. For example, hydrogen achieves stability with only 2 electrons (a duet), becoming like helium in electronic structure. The elements beryllium (Be) and boron (B) often prefer 4 and 6 electrons, respectively, instead of 8. Many atoms in the third period and beyond often end up with more than an octet, and the transition metals usually don't obey the octet rule. Even with these exceptions, the octet rule can be quite useful, particularly for carbon, which obeys it well. We can use it for sodium and chlorine to understand why the formula of sodium chloride is NaCl and not something else.

Let's consider the formation of sodium chloride from sodium and chlorine atoms in terms of the octet rule. We begin with a chlorine atom, which has a total of 17 electrons and the simple Bohr structure shown at right.

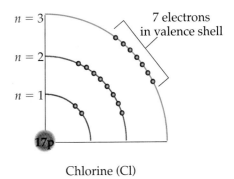

Chlorine (Cl)

Since chlorine is a member of group VIIA, chlorine atoms have 7 electrons in their outermost shell before the reaction. To understand how chlorine atoms react, focus on the valence shell and apply the octet rule by asking, "How can a chlorine atom get 8 electrons in this shell?" Since the valence shell is only 1 electron short of an octet, the obvious answer is for each chlorine atom to pick up 1 more electron, which would produce a chloride ion with a −1 charge. But where will this electron come from? To answer this, let's look at a sodium atom. Sodium atoms have a total of 11 electrons and the following Bohr structure:

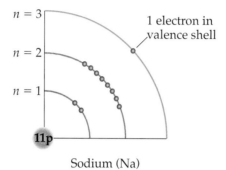

Sodium (Na)

Sodium atoms have only a single electron in their valence shell. Again we apply the octet rule and ask, "What is the best way for a sodium atom to end up with 8 electrons in its valence shell?" Like chlorine, sodium could do this by picking up more electrons, but each atom would have to gain 7 electrons. This would result in a sodium ion with a −7 charge, and that's quite a bit of excess negative charge for the sodium nucleus to hold onto. An easier way for sodium to get its octet is to lose its single valence electron from the $n = 3$ shell, leaving this shell empty. The $n = 2$ shell becomes the new valence shell, and this shell already has 8 electrons in it! In addition, if sodium loses 1 electron, then this electron is now available for a chlorine atom.

Recall that Cl needed to gain 1 electron to satisfy the octet rule. Now we know where it might come from. In reacting with chlorine, each sodium atom loses 1 electron to satisfy the octet rule and becomes a cation with a +1 charge. Each chlorine atom gains 1 electron and becomes an anion with a −1 charge. In doing so, both of the newly formed ions satisfy the octet rule.

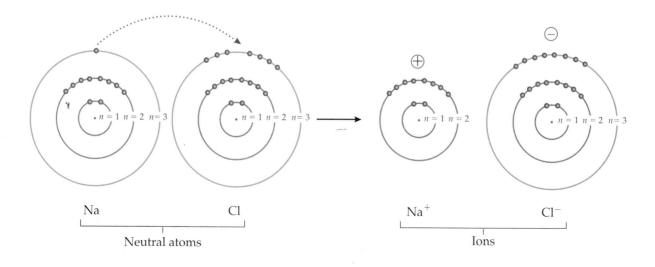

In essence, 1 electron is exchanged between the atoms. Note that in this exchange the atom that loses the electron, sodium, is a metal. The atom that gains the electron, chlorine, is a nonmetal. This turns out to be the typical behavior for metals and nonmetals in general. We can also now explain why the composition of sodium chloride is NaCl and not $NaCl_2$, etc. Each sodium atom loses 1 electron, and each chlorine atom picks up 1 electron. In order to balance this electron exchange and not end up with a deficit or excess of electrons, these elements combine in a 1:1 ratio. That is not the case for all elements. Let's try to predict the formula of the compound magnesium fluoride. We start by drawing the Bohr structures for atoms of Mg (group IIA, metal) and F (group VIIA, nonmetal):

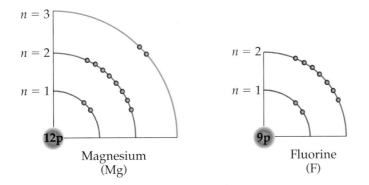

Now consider the best way for each atom to achieve an octet in its valence shell. Like chlorine, each nonmetal fluorine atom needs to gain only 1 electron, and in doing so will form the anion F^-. Magnesium, a metal atom, can most easily arrive at an octet by losing electrons, but it must lose 2 electrons and will therefore form a cation with a +2 charge (Mg^{2+}). This has a profound effect on the formula of the compound they form. Since Mg loses 2 electrons and F gains only 1, it takes two F atoms to react with one Mg atom and balance the electron exchange. The +2 charge and the two −1 charges of the resulting ions add up to zero, making the compound neutral. The formula of magnesium fluoride is thus MgF_2. Pictorially, here is how this might happen:

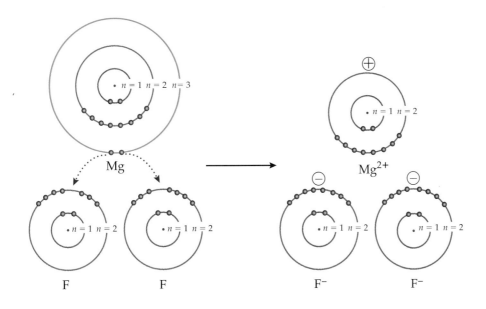

You might realize by now that it is not really necessary to draw the Bohr structures of the atoms in order to determine how each achieves its octet. There is a shortcut, at least for the representative (group A) elements. Reconsider the magnesium fluoride example. Magnesium is in group IIA of the periodic table, so we know that its valence shell contains 2 electrons without resorting to a Bohr diagram. We also know that Mg is a metal and metals tend to lose electrons when they react, so it's a good bet that Mg will lose both its valence electrons to satisfy the octet rule. Fluorine is a group VIIA element and thus has 7 electrons in its valence shell. Fluorine is also a nonmetal, and nonmetals tend to gain electrons when they react, so it is likely that F will gain 1 electron to satisfy the octet rule. Thus, instead of having to draw Bohr structures to determine formulas, we just need to consider the group the atoms are in and whether they are metals or nonmetals. Table 4.2 sums up the behavior of the representative elements for the metals in groups IA–IIIA and for the nonmetals in groups VA–VIIA. You can use this table to predict the most likely formula for the compound that will result from the reaction between any representative metal and nonmetal.

Table 4.2   Summary of the Chemistry of the Representative Elements

Group	Valence shell	To achieve octet in valence shell		To become
IA		Will lose 1 electron	$\longrightarrow$	+1 cation
IIA		Will lose 2 electrons	$\longrightarrow$	+2 cation
IIIA		Will lose 3 electrons	$\longrightarrow$	+3 cation
VA		Will gain 3 electrons	$\longrightarrow$	−3 anion
VIA		Will gain 2 electrons	$\longrightarrow$	−2 anion
VIIA		Will gain 1 electron	$\longrightarrow$	−1 anion

Notice that the elements of group IVA are not included in this table. We purposely left them out because they can be a bit "schizophrenic." Each of these elements has 4 valence shell electrons and therefore would need to either gain or lose 4 electrons to achieve an octet. Neither alternative is very satisfactory and thus, these elements tend to react in an entirely different way. We will discuss this in more detail in the next chapter.

PRACTICE PROBLEMS

4.22 Predict the formula Na_xO_y of the compound that results from reaction between the elements Na and O.

Answer: Na is a group IA metal and will thus lose 1 electron to become Na^+.
O is a group VIA nonmetal and will gain 2 electrons to become O^{2-}.
To arrive at charge neutrality, the formula must be Na_2O.

4.23 Draw the reaction above in terms of Bohr models as was done for MgF_2 in the text.

4.24 Predict the formula of the compound that results from the reaction between the elements Ba and F.

4.25 Predict the formula of the compound that results from the reaction between the elements Al and O.

You can see from these practice problems that the octet rule is a powerful predictive tool.

In Chapter 3 we introduced the terms metal and nonmetal and defined them according to their observable, physical properties and appearance. Now we are in a position to revisit these terms and redefine them according to their chemical reactions.

Chemical definition of metals and nonmetals

A **metal** is an element that tends to lose its valence electrons in chemical reactions, becoming a cation in the process.

A **nonmetal** is an element that tends to gain valence electrons in chemical reactions, becoming an anion in the process.

To a chemist, this tendency to lose or gain valence electrons in a reaction is the defining property that makes an element a metal or a nonmetal. It is a definition to remember.

ATOMIC SIZE REVISITED 4.6

In the last chapter we discussed the trends in atomic size, moving across and down the periodic table. This trend is illustrated again in the figure on page 132. As you can see, atoms get larger going down a group and smaller going from left to right across a period.

Relative Atomic Sizes of the Representative Elements

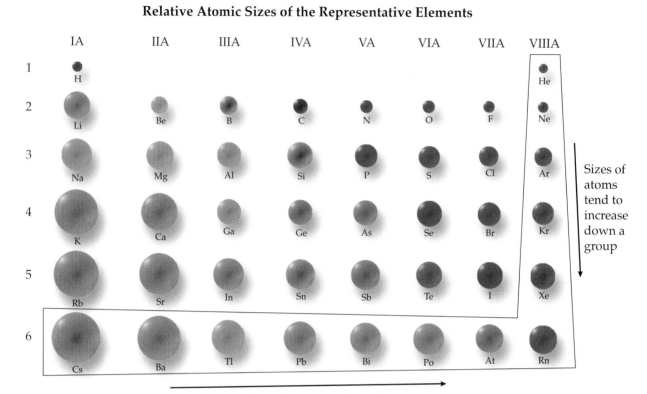

Sizes of atoms tend to increase down a group

Sizes of atoms tend to
decrease across a period

We are now in a position to explore why size varies in this way. Let's start with the trend within a group. Atoms get larger (their atomic radii increase) going down any group. The following figure shows the Bohr structure for the first three elements in group IA:

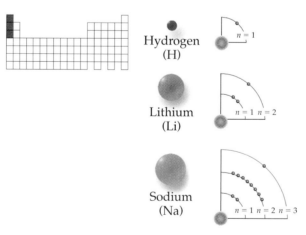

Hydrogen
(H)

Lithium
(Li)

Sodium
(Na)

One of the consequences of Bohr's model is that shells get larger as n increases. Therefore, sodium, with its outermost electron in the $n = 3$ shell, is a larger atom than lithium, which has its valence electron in the $n = 2$ shell. Moving down any group, each step places the valence electron(s) in a larger shell, thus increasing the size of the atom. This explains the size trend down a group.

What about the size trend across a period? Atoms get smaller going from left to right across a period. The following figure shows the Bohr structures for the first four elements in period 2:

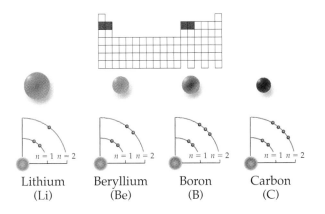

Lithium
(Li)

Beryllium
(Be)

Boron
(B)

Carbon
(C)

Since we are in period 2, the valence shell is the $n = 2$ shell, and it does not change as we move across the period. Therefore, you might expect that the sizes of these atoms would be roughly the same. But remember, as we go across a period from left to right, each step adds an additional electron to the valence shell and one more proton to the nucleus. The added proton makes the nucleus more positive, which causes an increased attraction for all its electrons, including the valence electrons. This results in all the electrons being pulled in more tightly and shrinks the shells, thus shrinking the atom. This combination of the same valence shell and an increased nuclear charge is responsible for the observed size trend within a period.

Redraw the Bohr structures of the previous figure to make them fit the size trend. In each nucleus, write the total number of protons present. Which atom would be hardest to ionize?

WORKPATCH

What happens to the sizes of all the shells in your WorkPatch drawings as you go from left to right? Check your drawings carefully against the answers at the end of the chapter. Good drawings make the answer to the ionization question obvious.

THE MODERN QUANTUM MECHANICAL MODEL OF THE ATOM 4.7

The Bohr model began with the assumption that the electrons in an atom could possess only certain allowed energy values (their energies were quantized). It is an understatement to say that many scientists of the day, having been trained in classical physics, had difficulty adjusting to this concept. But there was little time to complain. In less than 10 years, the new quantum physics came along with a description of the electrons in an atom so unusual that even the scientists who gave birth to the field weren't very happy about it. Bohr himself said, "Anyone who is not shocked by quantum theory has not understood it." Erwin Schroedinger, who developed the fundamental equation on which much of quantum physics is based, has been quoted as saying, "I don't like it, and I'm sorry I ever had anything to do with it." Much of this discomfort was generated not so much by the concept of quantized energy levels but by the other bizarre behaviors that quantum physics demanded of electrons. For example, the equations of quantum physics predict that electrons can exhibit a form of behavior known as *tunneling*. To best understand this behavior, let's assume that a baseball follows the same laws of physics as an electron. The tunneling theory says that if you place a baseball in a solid steel box and weld a cover onto the box, you can come back to the box sometime later and find the baseball sitting outside the still sealed metal box, even though the baseball lacks sufficient energy to penetrate the box. No magic or trickery is involved here. It's just possible (according to the laws of quantum physics) for the baseball to pass through the solid walls of the box under the right circumstances. Baseballs are large objects and therefore follow the laws of classical physics, but electrons are small enough that quantum effects are important, and electron tunneling is experimentally observed. In fact, a number of electronic devices on the market today (for example, the scanning tunneling electron microscope and Josephson junctions, used to make superfast switches for supercomputers) actually make use of this strictly quantum mechanical behavior. But this was not the worst of it.

One of the basic postulates of quantum physics is the *uncertainty principle*, put forth by Werner Heisenberg in the 1930s. The essence of this principle is that in dealing with particles the size of an electron, it is impossible to know exactly where the electron is or even where it is going with any real accuracy. In addition, the more carefully and precisely you measure the location of an electron, the less well you will be able to predict where it will be in the next instant. This is exactly the opposite of classical physics, where, for example, careful determination of the present position and velocity of the Moon allows us to know exactly where it will be days or weeks (even centuries) later so we can aim a rocket and land on it.

The uncertainty principle has some direct consequences when dealing with the structure of an atom. Since we cannot know an electron's location within an atom, it cannot be correct to think of electrons as traveling in specific, well-defined, fixed orbits around a nucleus. The best that quantum mechanics allows us to do is to put a "fence" around the nucleus of an atom and calculate the *probability* of finding one of its electrons somewhere inside the fence.

For some physicists, this was too much to accept. Einstein himself refused to believe fully in quantum mechanics and its probability-based calculations, insisting that "God does not play dice with the universe." Today most physicists and chemists have been forced by the weight of an incredible number of experimental results to subscribe to the quantum mechanical description of the atom. Others have been "dragged, kicking and screaming," all the way into a

begrudged but inevitable acceptance of this new brand of physics. Even now, first exposure of students to the field of quantum mechanics (or quantum physics) does not usually meet with enthusiastic acceptance.

The fence that quantum mechanics places around an atom according to a specified level of probability is not rectangular. Depending on the subshell (or energy) assigned to a particular electron in an atom, quantum mechanics can tell us something about the size and shape of the area inside the fence where we most probably will find the electron. These probable areas are referred to as *orbitals*. **Orbitals** are regions of space around the nucleus of an atom in which it is reasonably probable that we will find the electron assigned to that orbital (or subshell). The shapes of these orbitals are defined by drawing a boundary around the nucleus within which there is a 90–95% chance of finding the electron. A few of these orbitals are shown below:

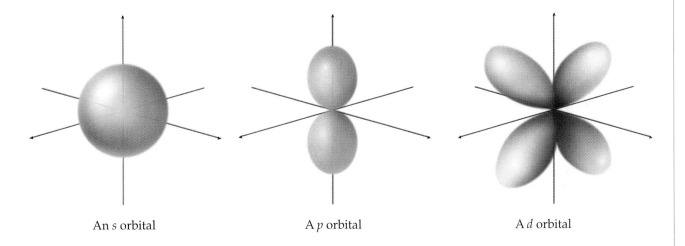

An *s* orbital   A *p* orbital   A *d* orbital

These pictures show the shapes of the areas inside which the electrons assigned to that particular subshell spend most of their time. All *s* orbitals are spherical, and all *p* orbitals are shaped like a figure eight. Like the Bohr orbits, these orbitals get larger as *n* increases. So an electron assigned to a 1*s* subshell or orbital in an electron configuration will be found (at least 90% of the time) within a sphere of a particular radius drawn around the nucleus of the atom. An electron in a 2*s* orbital can also be found (with a 90% probability) in a spherical volume surrounding the nucleus, but the volume of the 2*s* sphere would be significantly larger than that drawn for the 1*s* electron.

s orbitals

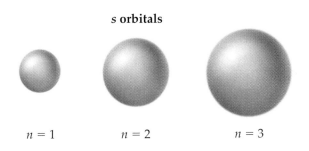

$n = 1$ $n = 2$ $n = 3$

Another way to visualize these orbitals is to imagine taking a picture of a particular electron as it moves about the nucleus of an atom. The picture is taken with the shutter left open for a long period of time. This produces a "picture" of the electron's movement around the nucleus—it is somewhat fuzzy but shows the overall region of space occupied in the electron's travels. Dark areas of the picture indicate greater electron densities than those of lighter areas; that is, the electron spends more time in the darker areas than in the lighter areas. We can draw a boundary around this fuzzy cloud to outline the area where the electron spends most of its time, and this boundary will have the same shape as the orbitals drawn from quantum theory. This is what a picture of the $1s$ electron in a hydrogen atom would look like:

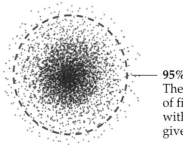

95% boundary:
There is a 95% probability
of finding the electron
within this area at any
given time

While we can't actually photograph an electron in an orbital, we can calculate the shapes, energy, and sizes of orbitals using the key mathematical equation of quantum physics, known as the *Schroedinger equation.*

So, how do we think of the atom today? The most current model retains Rutherford's tiny, massive, positive nucleus and Bohr's proposal about the quantized energy of electrons, but it replaces the straightforward picture of an orbit with one of a fuzzy orbital (sometimes called an *electron cloud*), which can only describe where it is most probable for an electron to be. If you continue on with more advanced courses in chemistry you will study the modern quantum mechanical description of the atom in greater depth. However, what you have learned in this chapter will serve you well for understanding chemical bonding, which comes next.

HAVE YOU LEARNED THIS?

Quantized energy (p. 107)

Classical physics (p. 108)

Quantum physics (p. 108)

Bohr model of the atom (p. 109)

Shells (p. 110)

Principal quantum number (n) (p. 110)

Valence shell (p. 112)

Energy level diagram (p. 114)

Ground state (p. 114)

Excited state (p. 115)

Subshells (p. 119)

Electron configuration (p. 122)

Octet rule (p. 127)

Metals and nonmetals (p. 131)

Orbitals (p. 135)

ENERGY IS QUANTIZED

4.26 If you exhibited quantum behavior with respect to your location on a racetrack, describe how you would appear to others as you ran in a race.

4.27 Why don't we see everyday objects exhibiting quantum behavior?

4.28 What is meant by the term "quantized energy"?

4.29 Which is more general, classical physics or quantum physics? Explain your answer.

THE SIMPLE BOHR MODEL

4.30 How does the energy of an object relate to its stability?

4.31 The following drawings represent charged particles. Which situation is most stable? Which situation is least stable? Explain your answers.

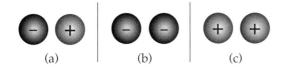

4.32 The following drawings represent charged particles. Which situation is most stable? Which situation is least stable? Explain your answers.

4.33 What happens to both the energy of an electron in an atom and its distance from the nucleus as n increases?

4.34 What type of physics could be used to describe an electron in an atom if it could have *any* energy?

4.35 What is another name for a Bohr orbit?

4.36 Is energy required or is energy released when an electron is moved farther away from its nucleus? Explain.

4.37 What are two things that happen to an electron in an atom as n increases?

4.38 What is the maximum number of electrons a Bohr orbit can accommodate?

4.39 Explain why we construct a Bohr model of the atom by first filling a lower shell to capacity before going to an upper shell.

4.40 Does the following Bohr model represent a ground or excited state? Explain your answer.

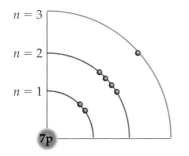

4.41 Using Bohr's rules, draw the lowest-energy configuration for the atom represented in Problem 4.40.

4.42 Why can't an atom's electrons ever be between orbits?

4.43 Explain why saying that an electron can only be at certain distances from the nucleus is the same thing as saying that the electron can have only certain energies.

PERIODICITY AND LINE SPECTRA EXPLAINED

4.44 What is meant by an atom's valence shell of electrons?

4.45 According to Bohr, why do atoms in the same group in the periodic table have similar chemical properties?

4.46 Explain how the Bohr model of the atom accounts for the existence of atomic line spectra.

4.47 What is a ground-state electron configuration? What is an excited-state electron configuration? Explain in words and also by drawing a Bohr model of the Mg atom in both the ground state and an excited state.

4.48 Draw Bohr models and use them to explain why phosphorus (P) and arsenic (As) have similar chemical properties.

4.49 (a) How much energy would be necessary to take a hydrogen from its ground state to an excited state in which the $n = 3$ orbit is occupied?
(b) What would be the energy of light emitted when the excited atom relaxed back to the ground state?
(c) What kind of electromagnetic radiation would be emitted? (Consult Table 4.1.)

4.50 What would happen to the electron in a ground-state hydrogen atom if it was given 5.1 eV of energy?

4.51 In order for hydrogen atoms to give off continuous spectra, what would have to be true?

4.52 Draw the Bohr model for a Cl atom and for a Cl^- ion. In both, identify the valence shell.

4.53 Draw a Bohr model for an Al atom and for an Al^{3+} ion. In both, identify the valence shell.

4.54 For each of the diagrams below do the following:
(a) Determine whether anything is wrong with the diagram. Elaborate.
(b) Give a full atomic symbol for the diagram if there is nothing wrong with it.
(c) For those diagrams that have nothing wrong with them, indicate whether they represent a ground state or an excited state.

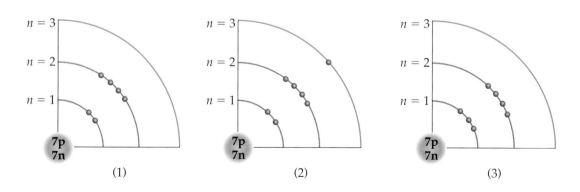

(1)   (2)   (3)

4.55 What is a very useful rule regarding valence electrons for representative elements?

4.56 Referring to Table 4.1, draw a wave of red light and a wave of blue light to show (relatively) how the wavelengths compare. Which light has the higher energy? Which light would be associated with relaxation to the ground state from a higher-energy excited state rather than from a lower-energy excited state?

4.57 According to the Bohr model, why do atoms increase in size going down any group in the periodic table?

4.58 (a) According to the Bohr model, why might someone expect that atoms would not change in size on going from left to right across a period?
(b) In fact, they do change in size. What is the trend and why?

4.59 The simple Bohr model (the model without subshells) works well up to the nineteenth element, potassium (K). Explain how it fails at K.

SUBSHELLS AND ELECTRON CONFIGURATION

4.60 What was the experimental evidence that supported the existence of subshells? Explain how this evidence suggested subshells.

4.61 (a) What is the numbering system used to label shells?
(b) What is the lettering system used to label subshells?
(c) How many subshells are there in a given shell?

4.62 How many electrons can a given subshell hold before it is considered full?

4.63 Bohr solved the potassium problem by putting its last electron where? How did he justify this?

4.64 You have seen that the 4s subshell fills before the 3d subshell. Since this is so, what would be wrong with drawing the Bohr diagram as follows (what would it suggest that is in fact not true)?

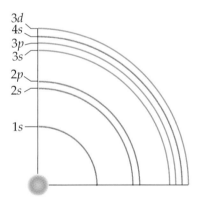

4.65 Write the electron configuration for the following elements without using the noble gas abbreviated form (use the periodic table to assist you).
(a) B   (b) Sc   (c) Co   (d) Se   (e) Ru

4.66 Repeat Problem 4.65, but use the noble gas abbreviated form this time.

4.67 Write the electron configuration for the following elements without using the noble gas abbreviated form.
(a) Ba   (b) W   (c) Pb   (d) Pr   (e) Pa

4.68 Repeat Problem 4.67, but use the noble gas abbreviated form this time.

4.69 In which group and period in the periodic table are these atoms found?
(a) $1s^22s^22p^3$ (b) $1s^22s^22p^63s^1$ (c) $1s^22s^22p^63s^23p^64s^23d^{10}4p^5$

4.70 A student has written what he thinks are some ground-state electron configurations. Which ones have something wrong with them? What is wrong?
(a) $1s^22p^63s^1$ (b) $1s^22s^63s^23p^64s^24p^6$ (c) $2s^22p^63s^23p^64s^23d^7$
(d) $1s^22s^22p^73s^33p^6$ (e) $1s^22s^22p^53s^1$
(f) $1s^22s^22p^63s^23p^64s^23d^{10}4p^65s^14d^4$

4.71 How many valence electrons does each of these atoms have?
(a) $1s^22s^22p^3$ (b) $1s^22s^1$ (c) $1s^22s^22p^63s^23p^64s^23d^7$
(d) $1s^22s^22p^63s^23p^6$ (e) $1s^2$

4.72 What does knowing the period of an atom tell you about the atom's valence electrons?

4.73 Write electron configurations for O, O^{2+}, and O^{2-}. Which form would you expect to find in most compounds of oxygen? Why?

4.74 When using the periodic table to assign electron configurations, what is the rule for the n quantum number when you are in the d block? When you are in the f block?

4.75 Why are the s, p, d, and f blocks in the periodic table 2, 6, 10, and 14 blocks wide, respectively?

COMPOUND FORMATION AND THE OCTET RULE

4.76 What is the octet rule, and what is the justification behind it?

4.77 How do metal atoms usually attain an octet in chemical reactions?

4.78 How do nonmetal atoms usually attain an octet in chemical reactions?

4.79 Predict the formula of the compound that forms when sodium atoms react with sulfur atoms. Completely explain your reasoning.

4.80 Predict the formula of the compound that forms when lithium atoms react with nitrogen atoms. Completely explain your reasoning.

4.81 Why are group numbers for the representative elements useful in predicting how many electrons an atom will gain or lose in a chemical reaction?

4.82 Consider the following Bohr diagrams for two reactants:

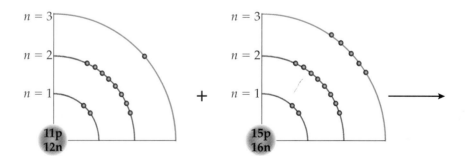

(a) Which elements are reacting with one another? Give full atomic symbols.
(b) Which is the metal, and which is the nonmetal?
(c) What is the formula of the resulting compound?
(d) Draw Bohr diagrams for the ions formed.

4.83 When a group IIA element X reacts with a group VIIA element Y, what will be the formula? Why?

4.84 Explain what is meant by this statement: The element hydrogen is an exception to the octet rule and yet it obeys it in principle.

4.85 Aluminum forms compounds with both sulfur and oxygen. Why are the formulas of the resulting compounds Al_2S_3 and Al_2O_3, respectively? Why are they similar?

4.86 True or false? It is impossible for a cation to exist in an excited state. Justify your answer with a Bohr diagram using a real element as an example.

4.87 True or false? It is impossible for the H^+ cation to exist in an excited state. Justify your answer.

4.88 True or false? The O^{2-} and F^- anions have identical electron configurations. Justify your answer with Bohr diagrams.

4.89 Which part of the following statement is true and which part is false? Since Mg^{2+} and Na^+ have identical electron configurations, they also have similar properties. Explain your answer fully.

4.90 How can you tell how many electrons a representative metal is likely to lose? What, in general, will be the charge of the cation it forms?

4.91 How can you tell how many electrons a representative nonmetal is likely to gain? What, in general, will be the charge of the anion it forms?

4.92 When a cation and an anion join to form a compound, how does knowing the charge of the ions help to determine the formula?

ATOMIC SIZE REVISITED

4.93 Which atom has a smaller $1s$ subshell, lithium (Li) or beryllium (Be)? Justify your answer.

4.94 Which atom has a smaller valence subshell, lithium (Li) or sodium (Na)? Explain your answer.

4.95 Which atom is larger, lithium (Li) or beryllium (Be)? Explain your answer.

4.96 Which atom is larger, lithium (Li) or sodium (Na)? Explain your answer.

4.97 Why do atoms shrink as you proceed left to right across a period?

4.98 Why do atoms get larger as you proceed down a group?

4.99 Rank the following atoms from smallest to largest: Si, Mg, Rb, Na

THE MODERN QUANTUM MECHANICAL MODEL OF THE ATOM

4.100 What is wrong with Bohr's planetary model of atomic electrons according to modern quantum mechanical theory? [*Hint:* Use Heisenberg's uncertainty principle in your answer.]

4.101 Quantum mechanical tunneling is one consequence of quantum theory. What is quantum mechanical tunneling?

4.102 What is an orbital?

4.103 Draw an s orbital, a p orbital, and a d orbital.

4.104 Suppose you behaved like a quantum mechanical particle. Describe what a picture of you taken by a camera might look like.

4.105 Why were physicists originally so upset with the quantum mechanical model of the atom?

WORKPATCH SOLUTIONS

4.1   Atom-sized cars in stop-and-go traffic would only be allowed to travel at certain allowed speeds and never any speed in-between. Thus, cars would instantaneously change from stopped to one of a set of allowable speeds. If all the cars didn't switch to the same allowable speed, a major pile-up would occur!

4.2   (a) 1.0 eV
(b) 11.2 eV
(c) The difference between them, (11.2 eV − 1.0 eV) = 10.2 eV
(d) Less energy, (13.1 eV − 11.2 eV) = 1.9 eV

4.3   The hydrogen atom with the electron in the $n = 2$ shell

4.4   12.1 eV − 10.2 eV = 1.9 eV more energy

4.5   Green corresponds to an electron jump from $n = 4$ to $n = 2$.
Blue corresponds to an electron jump from $n = 5$ to $n = 2$.
Indigo corresponds to an electron jump from $n = 6$ to $n = 2$.

For example, the green line corresponds to light of energy between 2.24 and 2.64 eV (Table 4.1). The only relaxation that will produce this energy is from the $n = 4$ level (13.8 eV) to the $n = 2$ level (11.2 eV): 13.8 eV − 11.2 eV = 2.6 eV.

4.6

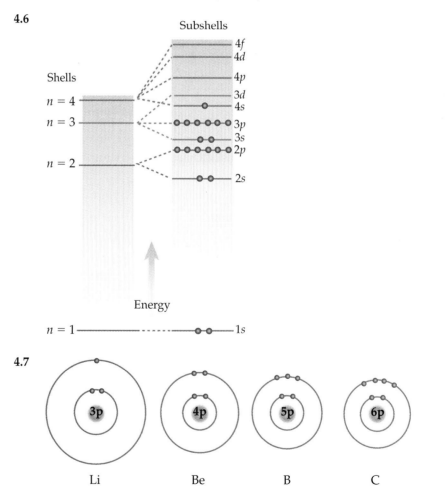

4.7

Carbon would be the hardest to ionize since its shells are closest to the nucleus, increasing the force of attraction between the electrons and the nucleus. The shells are smallest for C since it has the most positive nucleus of the atoms in this period.

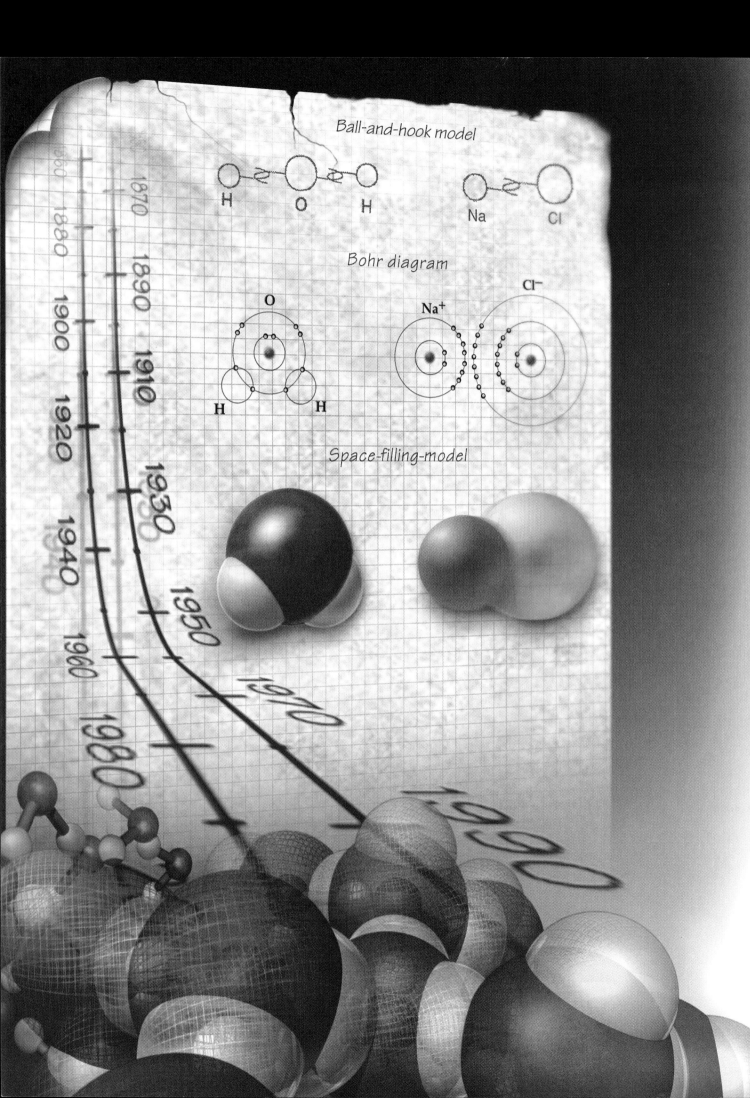

Chapter 5

Bonding Between Atoms

MOLECULES: WHAT ARE THEY? WHY ARE THEY? 5.1

Take a deep breath. Everyone knows that the air filling your lungs contains gaseous oxygen from the atmosphere, but exactly what is it that you are breathing? The gas you are inhaling is not made up of single, individual oxygen (O) atoms. Oxygen atoms are extremely reactive, and your lungs would be instantly seared by them. Instead, you are actually breathing O_2 *molecules*.

A **molecule** is defined as a collection of atoms that are bound together. In the case of oxygen, two oxygen atoms are all that is needed to form a molecule. An example of a molecule that is made from more than two atoms is methane. Methane is the main component of natural gas, the fuel burned in home furnaces, kitchen stoves, and the bunsen burners in your chemistry lab. The methane molecule, CH_4, is a collection of five atoms consisting of a single carbon atom and four hydrogen atoms:

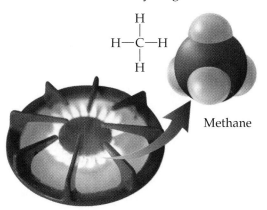

Methane

Note that while methane and oxygen both exist as molecules, only CH_4 is considered to be a *compound* since it consists of more than one element. Molecular O_2 is an elemental substance.

One of the things that makes molecules so interesting is that they have chemical and physical properties that are very different from those of the atoms from which they are made. For example, a simple change from individual O atoms to O_2 molecules changes the properties from that of a toxic poison to a life-sustaining gas. Another example is the molecule glucose, one of the simplest kinds of sugar. Much of the food you eat either contains glucose or gets converted into glucose molecules during the process of digestion. The glucose molecule is assembled from 24 atoms (6 carbon atoms, 12 hydrogen atoms, and 6 oxygen atoms) connected together. Whereas individual carbon, hydrogen, and oxygen atoms have no particular nutritional value, the glucose molecule, $C_6H_{12}O_6$, is food for the cells in your body.

It is precisely this change in properties that results when atoms combine to form molecules that makes molecules so interesting and worthy of study. Huge corporations invest large amounts of money to design and produce new molecules with specific, tailor-made properties. Two examples you are probably familiar with are the compounds aspartame (NutraSweet), which has the property of being sweet without providing significant calories, and taxol, which has been in the news lately due to its ability to inhibit some forms of cancer. There are literally millions upon millions of different molecules, ranging from simple two- or three-atom molecules, such as oxygen (O_2) or water (H_2O), to huge supermolecules containing thousands of atoms, such as DNA.

Natural sugars are broken down into glucose in the body

5.1 WORKPATCH

Consider the two molecules H_2O and H_2O_2 shown below. Would you expect them to have nearly identical properties?

$$H{\diagup}\!\!\!\!O\!\!\!\!{\diagdown}H \qquad H{\diagup}\!\!\!\!O\!\!\!\!{\diagdown}O{\diagup}\!\!\!\!H$$

$$H_2O \qquad\qquad H_2O_2$$

Did you answer the WorkPatch correctly? The answer (see the end of the chapter) emphasizes what we have been discussing about the chemical properties of molecules.

5.2 HOLDING MOLECULES TOGETHER—THE COVALENT BOND

One of the most fundamental and important questions of chemistry is, "What holds a molecule together?" Atoms aren't sticky, and they don't come with hooks attached. What is it that prevents molecules from coming apart into separate atoms?

To answer this question let's consider the element hydrogen. In the periodic table, an atom of hydrogen is represented as H. However, a tank of hydrogen gas is not filled with H atoms. It is filled with H_2 molecules. Like oxygen, hydrogen is more stable as a diatomic (two-atom) molecule. Indeed, if you could get your hands on a tank full of H atoms, you would not have it for very long. The gas in the tank would rapidly transform into a collection of H_2 molecules. To discover why hydrogen prefers to exist this way let's begin by doing a thought experiment. Imagine that you have been shrunk down to the size of a hydrogen atom (about 10^{-10} m, also called 1 angstrom, 1 Å). In front of you are two H atoms. Each H atom consists of a positively charged nucleus and a single, negatively charged, valence electron. Each electron is held firmly to its own nucleus by the attraction between their opposite electrical charges. Let's begin with the two H atoms far from each other (~10 Å). At this distance, they just sit there, each unaware that the other one exists.

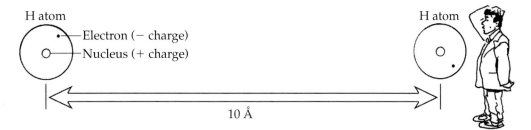

Now let's imagine moving the H atoms closer to each other, a little bit at a time. At 9 Å, they still just sit there. At 8 Å, still no change.

But when the atoms are moved to within 5 Å of each other, something interesting happens. Suddenly, the atoms spontaneously start to move toward each other! Their movement is slow at first, but as they get closer they begin to move faster. Soon they're racing toward each other!

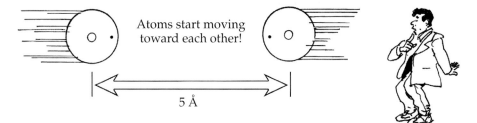

As these two atoms get closer and closer together, you turn away in anticipation of a messy crash.

When you look, you discover the two atoms are still intact, with their nuclei separated by a distance of 0.74 Å. At the same time, the temperature in the vicinity of the new molecule has increased. An H_2 molecule has been created.

H$_2$ molecule

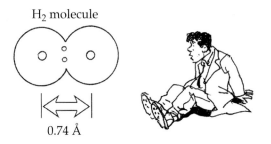

0.74 Å

What happened? Why did the H atoms start moving toward each other when they got close, and why did they stop at 0.74 Å? Where did the temperature increase come from?

There is much to explain. Let's start with the first observation. When the H atoms were brought to within 5 Å of each other they began to move toward each other. This was due to the same force that holds each electron close to its own nucleus. Opposite charges attract. When the two H atoms got close enough to one another, the negative electron of one atom began to be attracted to the positive nucleus of the other atom, and vice versa. This force of attraction began to pull the atoms toward each other.

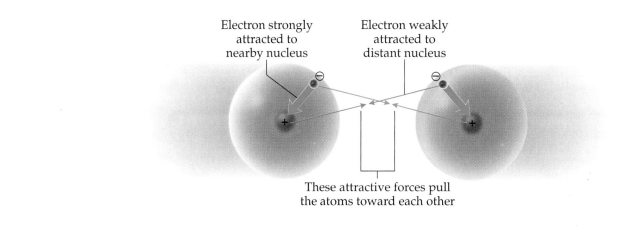

Electron strongly attracted to nearby nucleus

Electron weakly attracted to distant nucleus

These attractive forces pull the atoms toward each other

This electron is attracted to both nuclei

So is this one

As the atoms got closer, this attractive force became stronger, and the atoms picked up speed. When they stopped and formed the H_2 molecule, there was greater total positive–negative attraction in the H_2 molecule than there was in the two isolated H atoms because each electron was then attracted to two nuclei.

Remember, attractive interactions stabilize particles, so this extra positive–negative attractive force means that the molecule consisting of two H atoms is more stable than the two isolated H atoms. It is this extra attractive force that holds the molecule together. We call this additional attractive force a *covalent bond*.

A **covalent bond** is the extra force of attraction that results from valence electrons being *attracted* to two nuclei. It holds molecules together.

In order to break the H_2 molecule apart into isolated hydrogen atoms, you would have to provide enough energy to overcome this additional attractive force.

Why did the atoms stop at 0.74 Å? Once again the reason is a force between charges, but this time, it's between like charges. Like charges repel each other, and the nuclei of both atoms are positive. When the nuclei are 0.74 Å apart, the repulsion between the two nuclei becomes great enough to keep the H atoms from getting any closer. Likewise, the electrons are now close enough for their repulsions to become significant, and this also keeps the atoms apart. A covalent bond is a delicate balance between these attractive and repulsive forces.

Nuclei repel each other

Electrons repel each other

0.74 Å

Finally, as the two H atoms formed a molecule, the temperature increased. Why? The H_2 molecule is more stable than two isolated H atoms because of the extra attraction of the covalent bond. In this case, the energy released by the increased attractions, along with the energy of the collision itself, had to be carried off by collisions between the newly formed H_2 molecule with other atoms or molecules around it. This caused an increase in the speed of the surrounding molecules and the resulting increase in temperature. You would have to put this same amount of heat (energy) back into the molecule to separate it into isolated atoms—that is, to break the covalent bond.

In the H_2 molecule, both electrons will spend their time "buzzing around" both nuclei, but when we draw a representation of the molecule, we always show the electrons occupying the region between the nuclei. This is a reasonable place to put them since they are attracted equally to both nuclei, and this is in fact where they are located most of the time.

The electrons spend most of their time between nuclei . . .

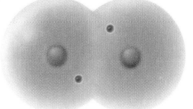

Drawn this way it looks as if the nuclei are sharing the two electrons. In fact, that is exactly what a chemist would say is happening. This leads to another, slightly different definition of a covalent bond:

although they spend some time in other places

A **covalent bond** is the extra force of attraction that results from valence electrons being *shared* between two nuclei. It holds molecules together.

Co- is a prefix that means partner, as in somebody you share with; *valent* refers to the fact that it is valence (outer-shell) electrons that are being shared and are responsible for bonding the H atoms to each other. So the name *covalent* stresses a very important point: When atoms come together to form molecules, it is always the valence electrons that are involved. An atom's inner-shell electrons (also called *core electrons*) are almost never involved in bonding, because they are too close to their own nucleus and therefore too strongly attracted to it to be shared with another nucleus.

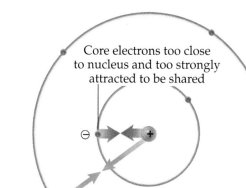

Core electrons too close to nucleus and too strongly attracted to be shared

Valence electrons can be shared with another nucleus to form covalent bonds

We didn't have to worry about this for hydrogen, since a hydrogen atom has only one electron. However, as we consider bonding between other elements, we will concern ourselves only with the valence electrons.

There are several ways to represent atoms connected together by covalent bonds. Some of these are shown below:

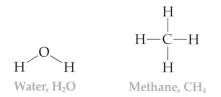

H:H   or   H··H   or   H—H

The line drawn between the H atoms stands for the two shared electrons that give rise to the covalent bond. A line is easier to draw than two dots, and it is the representation most often used. Every time you see a line drawn between atoms in a molecule it stands for two shared electrons between two nuclei and the covalent bond that results. Take another look at the water and methane molecules. In water there are two covalent bonds (two lines); in methane there are four:

Water, H₂O   Methane, CH₄

Though we will be using lines to represent covalent bonds throughout the book, you should remember that they are just a chemist's device. If you could

see an individual water molecule, you would not see any lines. You would see only one oxygen atom and two hydrogen atoms that always stay close to each other.

Covalent bonds are quite strong; in fact, energies involved in this type of bonding are among the largest that chemists encounter. The O–H covalent bonds in the water molecule are the reason it doesn't fly apart into atoms even when you boil it. The steam that rises from a pot of boiling water is made up of intact H₂O molecules. To break the O–H bonds in water would take far more energy than that required to boil it.

To summarize, energy is always released when a covalent bond forms between two atoms. There are no exceptions to this statement. The converse is also true. It always takes energy to break a covalent bond.

Covalent bonds are very strong.

Covalent bonds

Covalent bonds

To further explore covalent bonding, let's return to the hydrogen atom/ molecule case. Recall that when isolated and apart from each other, each hydrogen atom has only one valence electron. What about after they come together to form a molecule of H_2? Does each hydrogen atom in the molecule still have only one valence electron? The answer depends on how we choose to do our electron counting. In the case of a covalent bond between atoms, chemists have decided to "double-count" the shared electrons. That is, in H_2, each H atom in the molecule is considered to have two electrons.

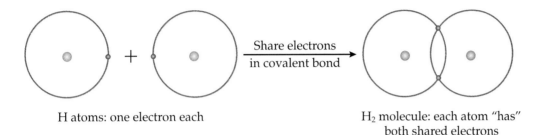

H atoms: one electron each H_2 molecule: each atom "has"
                                                both shared electrons

If you had a joint bank account with a friend and you did your accounting this way, you could quickly get into some financial trouble. Obviously, a shared $100 bank account does not give each of you $100 to spend. Nevertheless, this is how we count shared valence electrons. This means that sharing electrons is one way for atoms to get more electrons. The covalent bond in the H–H molecule effectively gives each hydrogen atom one more electron than it started with. Every covalent bond an atom forms will increase its number of valence electrons by 1. One justification for this counting method is that it helps us to predict how many atoms will combine when they come together, as we will see shortly.

We are leading up to something important here regarding bonding between atoms, and the best way to introduce you to it is to consider bringing two helium (He) atoms together to attempt to form a molecule:

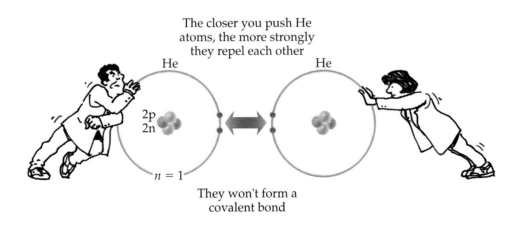

The closer you push He
atoms, the more strongly
they repel each other

They won't form a
covalent bond

A very different scenario now takes place compared to the two hydrogen atoms. The closer two He atoms get to one another, the more they repel one another. Indeed, they refuse to form a covalent bond between them. Why the difference from hydrogen? A hint comes from examining the behavior of the other elements in helium's group in the periodic table, group VIIIA (the noble gases). All the remaining elements in group VIIIA have 8 electrons in their valence shell, and all of them strongly resist forming covalent bonds. Does the octet rule play a role in determining when covalent bonds will form and how many of them will form? The answer is an emphatic yes.

 5.2 WORKPATCH   How many valence electrons does each atom in the methane molecule have? "Double-count" shared electrons.

$$
\begin{array}{c}
\text{H} \\
| \\
\text{H—C—H} \\
| \\
\text{H}
\end{array}
$$

You should have found an octet on one of the atoms in the WorkPatch if you did your counting correctly.

5.3 WORKPATCH   How many valence electrons does a carbon atom normally have? How many valence electrons does it have in the methane molecule? How many electrons did it gain, and how many covalent bonds to hydrogen did it form?

Are you starting to get an insight into why methane is CH_4 and not CH_5 or something else?

5.3 MOLECULES, DOT STRUCTURES, AND THE OCTET RULE

What determines whether an element forms covalent bonds or not? How many bonds is an element likely to form? Why is a water molecule always H_2O? Why not H_3O or H_4O or HO_2? The answers to these questions come from applying the octet rule, which says that all the elements want to be like the group VIIIA (noble gas) atoms and possess 8 electrons in their valence electron shells (or 2 electrons if their valence shell is the $n = 1$ shell). This rule often holds true when atoms are part of a molecule, where they achieve an octet by sharing electrons with other atoms.

By understanding how many electrons an atom "wants" to share, we can better understand why water has the formula H_2O and not something else, as illustrated in the adjacent figure. The best way to begin is to determine the number of valence electrons a neutral, isolated atom has before it becomes

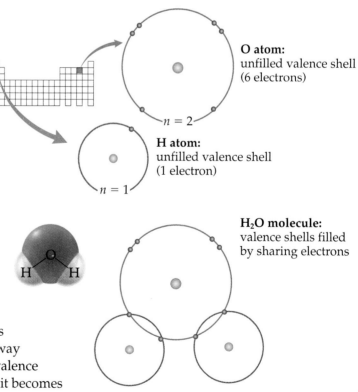

O atom:
unfilled valence shell
(6 electrons)

H atom:
unfilled valence shell
(1 electron)

H_2O molecule:
valence shells filled
by sharing electrons

part of a molecule. This is easy. Recall that for a representative element, the group number tells you the number of valence electrons. While the noble gases in group VIIIA have 8 valence electrons (He, with 2, is the only exception), all the other representative elements in the periodic table have fewer than 8 valence electrons—that is, less than an octet. For example, every group IVA element is short of an octet by 4 electrons.

One way group IVA elements can get 8 electrons is to share all 4 of their valence electrons with other atoms. For example, remember the methane (CH_4) molecule we looked at earlier? Carbon starts with only 4 valence electrons and needs 4 more to complete an octet. Therefore, it forms four covalent bonds with other atoms.

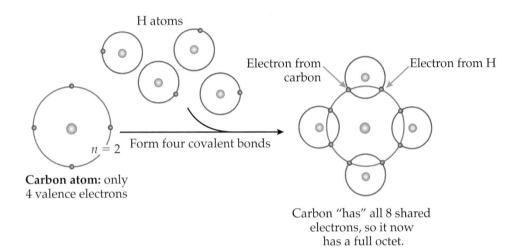

Since the group number tells you the number of valence electrons an atom has, it is a simple matter to figure out how many additional electrons it needs to get 8. This number is also the number of covalent bonds it must form to get them. Table 5.1 summarizes the requirements of each group of nonmetal atoms (we will save the metals for later). Examine it closely; then we'll use it to determine the formula of a molecule.

Table 5.1 Number of Covalent Bonds an Atom Will Form

Nonmetal atoms	Group number	Valence electrons	Electrons needed to achieve an octet* and also the number of covalent bonds it will form
H	IA	1	1*
C, Si, Ge	IVA	4	4
N, P, As, Sb	VA	5	3
O, S, Se, Te	VIA	6	2
F, Cl, Br, I, At	VIIA	7	1

*Duet in the case of a hydrogen atom.

The number in the last column of the table is the number of covalent bonds the atom must form to achieve an octet. For example, let's look at ammonia, a molecule made from nitrogen and hydrogen. Find nitrogen in Table 5.1. It is in group VA and therefore has 5 valence electrons. It needs 3 more electrons to achieve its octet. In ammonia, these 3 electrons come from forming three covalent bonds to H atoms. That is why the formula for ammonia is NH_3—not NH_2 or NH_4.

PRACTICE PROBLEMS

5.1 Predict the formula of the compound that forms between carbon and chlorine.

Answer: From Table 5.1, carbon needs to form four bonds and chlorine will form one bond. The formula is therefore CCl_4.

5.2 Predict the formula of the compound that forms between phosphorus (P) and hydrogen.

5.3 Predict the formula of the compound that forms between silicon (Si) and bromine (Br).

Table 5.1 is fine, but to build more complex molecules like glucose and DNA we need a scheme for drawing atoms that can tell us at a glance how many covalent bonds each atom will form. A chemist named G. N. Lewis invented just such a drawing system for molecules back around 1916. His drawings are called **dot diagrams**, or sometimes **Lewis dot diagrams**. Dot diagrams are used to show an atom's valence electrons. In a dot diagram, the symbol of an element is surrounded with dots, one dot for each valence electron.

Dot diagrams for the period 2 elements—proceed around the atom symbols, adding the valence electrons as shown

	Li	Be·	Ḃ·	·Ċ·	·N̈·	·Ö:	·F̈:	:N̈e:
Group ⟶	IA	IIA	IIIA	IVA	VA	VIA	VIIA	VIIIA

In every case, the number of dots is equal to the group number for these representative elements. Notice how the electrons are added. The first 4 electrons (Li through C) are positioned on the four sides of the symbol. It is not until we have a fifth electron, in N, that we begin to pair the electrons. Thus, carbon is surrounded by 4 single electrons, whereas nitrogen has 3 single electrons and one pair. We refer to the pairs as **electron pairs** (or **paired electrons**), and we call the single electrons **unpaired electrons**. Thus, nitrogen has 3 unpaired electrons and one electron pair. As we will see, the difference between paired and unpaired electrons is important.

Electron pair ⟶ ·N̈·
Unpaired electron ⟶

There is some flexibility with respect to how you arrange the unpaired and paired electrons. For example, all the dot diagrams for oxygen shown below are correct:

All these forms are equivalent

·Ö: :Ö·   ·Ö·

Because the number of valence electrons determines what the dot diagram for an atom will look like, all atoms with the same number of valence electrons will have identical dot configurations. Thus, if we generate dot diagrams for periods 3–7 of the representative elements, the dots will be in the same positions as those shown on the preceding page for period 2. We won't concern ourselves with dot diagrams for the transition metals, lanthanides, or actinides. This leaves just two elements, hydrogen and helium, both of which are somewhat unique. Their dot diagrams are as follows:

H·   He:

Note that helium's 2 valence electrons are always paired.

We now have simple diagrams for all the representative elements. These dot diagrams can help us determine the bonding that can occur between them. If you examine all the dot diagrams that we have considered, only the group VIIIA atoms (He, Ne, Ar, Kr, Xe, Rn) have all their valence electrons paired up and possess a complete octet (or duet for He). All the other representative elements have unpaired valence electrons and thus less than an octet. The octet rule states that they need 8 valence electrons (duet for H) to achieve stability. To get 8 valence electrons, they share their unpaired electrons with other atoms.

Let's see how this works by using fluorine as an example. Fluorine is in group VIIA so each atom has 7 valence electrons, three pairs and one unpaired electron:

The fluorine atom needs one more valence electron to achieve an octet. If two F atoms get together and share their unpaired electrons, then both can have 8. In doing so, a single covalent bond forms between them. No more bonds will form, as both atoms now have octets. This is why fluorine exists as a diatomic molecule F_2—not as F or F_3, etc.

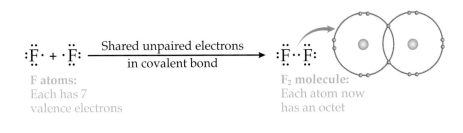

Why do two He atoms refuse to form the molecule He_2 when brought close to one another?   **WORKPATCH**

How did you answer the WorkPatch? You could have said that there were no unpaired electrons. Or you could have answered in terms of the octet rule. Either way would have been correct.

PRACTICE PROBLEMS

5.4 In forming molecules, atoms share unpaired electrons in order to achieve an octet in their valence shell, except hydrogen. What number of electrons does hydrogen "want," and why is it not eight?

Answer: Hydrogen "wants" 2 electrons (a duet instead of an octet). This is because hydrogen is a first period atom, so its valence shell is the n = 1 shell. This shell is filled when there are 2 electrons in it.

5.5 What is the formula of the compound that forms between H atoms and F atoms? Justify your answer with dot diagrams.

5.6 Why is water H_2O and not H_3O or something else? Justify your answer with dot diagrams.

The strategy of sharing unpaired electrons and aiming for octets works for all sorts of molecules. For example, the molecules NF_3 and CH_4 are formed from atoms as follows:

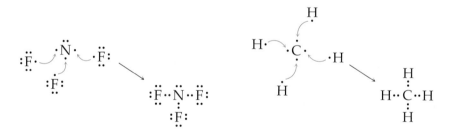

Drawing all these dots gets tedious, so we usually just draw a line for each shared pair of electrons:

As we said previously, a line between two atoms is understood to represent two shared electrons, which constitute a covalent bond. Therefore, shared pairs of electrons are often called **bonding pairs**. The unshared electron pairs on the F atoms and on the N atom in NF_3 are called **lone pairs** since they belong to only a single atom.

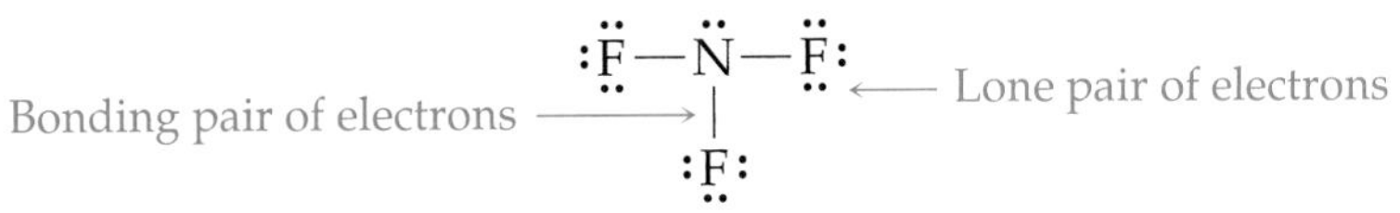

$$:\ddot{F}—\ddot{N}—\ddot{F}:$$

Bonding pair of electrons ⟶ $|$ ⟵ Lone pair of electrons

$$:\ddot{F}:$$

Note that not all atoms in a molecular dot diagram have lone pairs (such as the carbon atom in methane or hydrogen atoms in any molecule). Because they are not shared, lone pairs are not involved in bonding. Nevertheless, they are important, and should never be left out of a molecular dot diagram when present.

How many valence electrons are present on each atom in the NF_3 molecule shown above?

WORKPATCH 5·5

Did you count the lone pairs once, but double-count the shared pairs when doing your electron counting for the WorkPatch? This demonstrates the importance of the lone pairs with respect to the octet rule.

With dot diagrams in hand you are now ready to build more complicated molecules.

Consider ethane (C_2H_6), a minor component of natural gas. What does this molecule look like? Start with the atoms as dot diagrams and try putting them together so that each C ends up with 8 valence electrons and each H ends up with 2 electrons. Then replace every pair of shared electrons with a line. When you are done, compare your attempt with the answer at the end of the chapter. [*Hint:* Since hydrogen comes into every bonding situation with 1 electron and needs only 1 additional electron to complete its valence shell, hydrogens are usually found as the terminal ends (outside) of molecules.]

WORKPATCH 5·6

There should be no lone pairs in your dot diagrams for this WorkPatch. Remember, hydrogen never has a lone pair in a molecule, since 2 electrons are all it needs to fill its valence shell, and these will come from the covalent bond. As for carbon, since it is initially surrounded by 4 unpaired electrons, it also will have no lone pairs if it can form bonds to four other atoms to gain 4 more electrons, which is the case in ethane. Will carbon ever have lone pairs in a molecule? As we will see in the next section, the answer is, it's possible but rare.

───────────────────────────── PRACTICE PROBLEMS

5.7 Draw a dot diagram for the molecule hydrazine, N_2H_4, sometimes used in rocket fuel.

Answer: Start with the pieces and put them together in a way that satisfies all octets (duets for H):

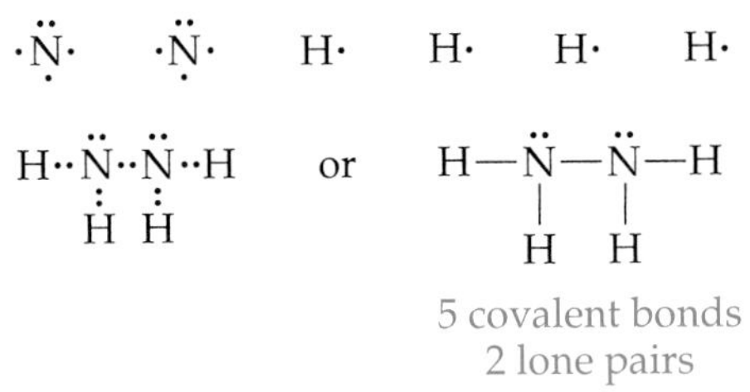

·$\ddot{N}$·   ·$\ddot{N}$·   H·   H·   H·   H·

H··$\ddot{N}$··$\ddot{N}$··H   or   H—$\ddot{N}$—$\ddot{N}$—H
  $\overset{..}{H}$ $\overset{..}{H}$ $|$ $|$
                          H   H

5 covalent bonds
2 lone pairs

5.8   Draw a dot diagram for hypochlorous acid, HClO.

5.9   Draw a dot diagram for the molecule propane, C_3H_8, used as a fuel for torches and heating.

5.4 MULTIPLE BONDS

Sometimes two atoms will share more than one pair of electrons with each other. Such a situation is called a **multiple covalent bond**, or **multiple bond** for short. For example, let's reconsider those diatomic O_2 molecules you are breathing. To understand the bonding in this molecule, we start as usual by drawing a dot diagram for one oxygen atom:

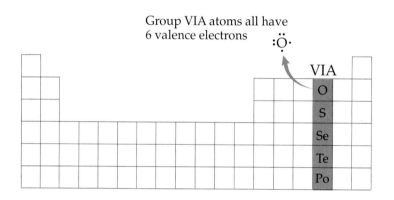

Group VIA atoms all have 6 valence electrons

A group VIA oxygen atom has 2 unpaired valence electrons. If two oxygen atoms get close, they will form a covalent bond, with each sharing one of its unpaired electrons:

Each oxygen atom has   Each oxygen atom has
6 valence electrons   7 valence electrons

Where do we go from here? Both oxygen atoms now have 7 electrons, not yet a complete octet. Each oxygen atom still has one unpaired electron left, and unpaired electrons are for sharing. So don't stop. Share the remaining unpaired electrons:

Each oxygen atom has   Each oxygen atom has
7 valence electrons   8 valence electrons

Look what this did for each oxygen atom. Each is now surrounded by 8 valence electrons.

The octet rule has been satisfied for both oxygen atoms by the formation of a multiple bond. In this case, there are two shared pairs of electrons between the atoms, or two covalent bonds. We refer to this as a **double bond**. When there is just one covalent bond between two atoms it is called a **single bond**. As you might expect, a double bond is stronger than a single bond. This is generally true.

The air you breathe contains another diatomic gas, nitrogen (N_2). To draw this molecule, we start with dot diagrams for the isolated nitrogen atoms, then bring them together to share electrons until both atoms end up with 8 valence electrons:

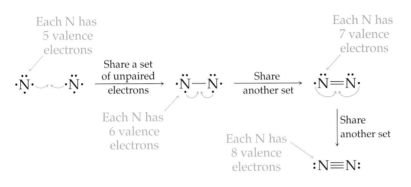

We arrive at a **triple bond**, three shared pairs between the N atoms, a very strong bond indeed.

Related to the C_2H_6 (ethane) molecule of WorkPatch 5.6 is the acetylene molecule, C_2H_2, used in oxyacetylene torches that cut through steel. This molecule also contains a multiple bond. Determine what kind in the following WorkPatch.

Draw a dot diagram for the acetylene molecule, C_2H_2.

WORKPATCH 5·7

You should have found that there is really only one way to put the acetylene molecule together. If you put both hydrogens on the same carbon atom, then you would end up violating the octet rule and leaving unpaired electrons on the other carbon atom.

PRACTICE PROBLEMS

5.10 Draw a dot diagram for the carbon dioxide molecule (CO_2).

Answer: :Ö·⤑⟵·Ċ·⤑⟵·Ö: ⟶ :Ö═C═O:

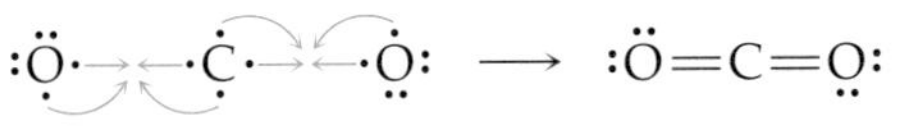

5.11 Draw a dot diagram for the propyne molecule, C_3H_4. One carbon has three hydrogens bound to it, one has one hydrogen bound to it, and one has no hydrogens bound to it.

5.12 Draw a dot diagram for the hydrogen cyanide molecule, HCN. The hydrogen is attached to the carbon.

5.13 Draw a dot diagram for the acetone molecule, C_3H_6O, the active ingredient in nail polish remover. The oxygen atom is bound to only one carbon atom and no hydrogens.

Drawing dot diagrams is a great way to learn about covalent bond formation in molecules, but it is tedious and doesn't always work. There is a simpler, four-step method that we want to show you now. We will demonstrate this method using two examples, carbon monoxide (CO) and carbon dioxide (CO_2).

Step 1: Determine the *total* number of valence electrons (dots) that will be in the final diagram. To do this, simply add up the group numbers of all the atoms in the molecular formula.

Step 1: Determine the total number of dots that will be in the diagram.

Step 2: Connect the atoms with single bonds. If you are not told how the atoms are connected to one another, then assume that the first atom in the formula is the central atom and all other atoms are attached to it. (This works only for simple molecules.) Remember that every time you connect two atoms with a single bond you are using 2 electrons.

Step 2: Connect the atoms using single covalent bonds.

C—O   O—C—O

2 dots used   4 dots used

Step 3: Put in the remaining dots two at a time as lone pairs, first on the terminal atoms of your structure and then, if there are any left, on the central atoms. Keep adding electrons to an atom until its octet is filled before moving on to the next atom. You must stop putting in lone pairs when you run out of electrons. This is very important. Do not use more electrons than you calculated in step 1.

Step 3: Put in remaining dots as lone pairs, satisfying octets as you go.

If at the end of this step, every atom has 8 valence electrons around it (and every hydrogen has 2 electrons), then you are done, and the molecule will have only single bonds. In our two examples, we are not done, since carbon has only 4 valence electrons around it in both cases. When this happens, we must go on to step 4.

Step 4: If there are atoms that do not have an octet, then send lone pairs to the rescue by moving them into a bonding position between the atoms. The lone pairs must come from an *adjacent* atom (one that is attached to the octet-deficient atom).

Carbon monoxide (CO)

$$:C \overset{\frown}{} \ddot{O}: \longrightarrow :C{\equiv}O:$$

Carbon dioxide (CO$_2$)

$$:\ddot{O} \overset{\frown}{-} C \overset{\frown}{-} \ddot{O}: \longrightarrow :\ddot{O}{=}C{=}\ddot{O}:$$
(a)

OR

$$:\ddot{O} \overset{\frown}{-} C {-} \ddot{O}: \longrightarrow :O{\equiv}C{-}\ddot{O}:$$
(b)

OR

$$:\ddot{O} {-} C \overset{\frown}{-} \ddot{O}: \longrightarrow :\ddot{O}{-}C{\equiv}O:$$
(c)

The solution above reveals that the carbon monoxide (CO) dot diagram contains a triple bond between the carbon and oxygen atoms. Each atom also contains one lone pair, and each atom has a complete octet. This was accomplished using exactly the 10 valence electrons we started with. By following steps 1–4 we came up with a valid molecular dot diagram. For carbon dioxide (CO$_2$), we drew three dot diagrams, all equally valid with respect to the octet rule. Convince yourself of this by examining the diagrams above and answering the following WorkPatch.

How did we get three different dot diagrams for carbon dioxide?

WORKPATCH 5.8

Do you understand why we arrived at more than one valid dot diagram for CO$_2$? If you do, then you will know when to expect this situation. But all three dot diagrams for CO$_2$ can't be correct, can they? Well, actually they are all correct. They are identical in their arrangement of atoms (carbon in the middle with oxygens attached to it). The only difference is the arrangement of their electrons. Valid dot diagrams that differ from one another only in their arrangement of electrons are called **resonance forms**. The actual molecule is considered by chemists to be some combination of these three forms, with all the forms not necessarily equally weighted. Experiments and theoretical considerations show that carbon dioxide is mostly like resonance form (a), with perhaps just a tad of (b) and (c) mixed in.

Up to now we have been generating dot diagrams for neutral (uncharged) molecules. Before we present some practice problems we want to modify the first step in the above procedure so that it can be extended to cover *polyatomic ions*. **Polyatomic ions** are just what they sound like, ions made from many atoms. Some examples are nitrate (NO$_3^-$), used to preserve meats; phosphate (PO$_4^{3-}$), found in old-fashioned soda pop; and ammonium (NH$_4^+$), often found in soaps and detergents. These are all ions because they have an overall charge. Other than that, we can treat them as normal molecules and follow steps 1–4 to generate dot diagrams. However, the charge affects the total number of dots in the diagram, so step 1 has to be modified to account for this. We must consider how to convert a neutral molecule into a cation or an anion. An anion is negative because it has a surplus of electrons. The surplus is equal to the charge of the anion. For example, the neutral molecule NO_3 has 23 valence electrons (one group VA atom and three group VIA atoms). The anion NO_3^-, with its -1 charge, must therefore have 24 valence electrons. The neutral molecule PO_4 has 29 valence electrons (one group VA atom and four group VIA atoms). The anion PO_4^{3-}, with its -3 charge, would therefore have 32 valence electrons. A cation is positive because it has a deficiency of electrons. For example, the molecule NH_4 has 9 valence electrons (one group VA atom and four group IA atoms). The NH_4^+ cation, with its $+1$ charge, would therefore have 8 valence electrons.

> **Modified Step 1:** Determine the total number of dots that will be in the diagram. To do this, simply add up the group numbers of all the atoms in the molecular formula. *If you are dealing with a polyatomic ion, adjust the dot count according to the charge (subtract electrons for cations, add them for anions).*

You are now ready for some practice problems. Some will have resonance forms, some will not—it's up to you to decide. When resonance forms exist, draw them all.

PRACTICE PROBLEMS

5.14 Draw a dot diagram for the carbonate anion, CO_3^{2-}.

Answer:

Step 1: CO_3^{2-} has a total of 24 dots:

$$C \text{ is group IVA} = 4 \text{ dots}$$
$$O \text{ is group VIA (there are three of them)} = 6 \times 3 = 18 \text{ dots}$$
$$-2 \text{ charge gives an additional 2 electrons} = \underline{2 \text{ dots}}$$
$$24 \text{ dots}$$

Step 2: Assume the first atom written is the central atom and all other atoms are attached to it. Connect them with single bonds.

The three bonds consume 6 electrons, leaving 18 to go.

Step 3: Put in the remaining electrons as lone pairs, satisfying octets as you go.

All 24 electrons are in. Are we done? No. Carbon does not have an octet. We must proceed to step 4.

Step 4: If all octets are not satisfied, send adjacent lone pairs to the rescue.

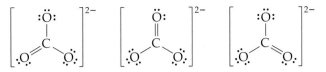

We get three resonance forms, depending on which lone pair we use to create the double bond. Note that if we are dealing with a polyatomic ion, it is traditional to put the ion in square brackets and indicate the overall charge as a superscript.

5.15 Draw a dot diagram for SO_3.

5.16 Draw a dot diagram for $SO_3{}^{2-}$.

5.17 Draw a dot diagram for NO^+.

5.18 Do the dot diagrams shown below represent resonance forms? Explain completely.

$$\ddot{O}=C=\ddot{O} \qquad :C\equiv O-\ddot{O}:$$

Ionic Bonding: Bring on the Metals 5.5

In our discussion of molecules we haven't mentioned metal atoms. In general, metal atoms don't share their valence electrons with other atoms because they have little ability to attract additional electrons to themselves. In fact, when metal atoms form compounds with nonmetals, they tend to completely give up, or lose, their valence electrons instead of sharing them. We introduced this concept in the previous chapter, but we can now look at it again in terms of the bonding that occurs.

Consider the compound sodium chloride (NaCl), table salt. It is a stable compound in which group IA Na atoms are bound to group VIIA Cl atoms. But this bonding is not the result of shared electrons as was the case in H_2. Look at what happens to a sodium metal atom when it wanders close to a nonmetal chlorine atom:

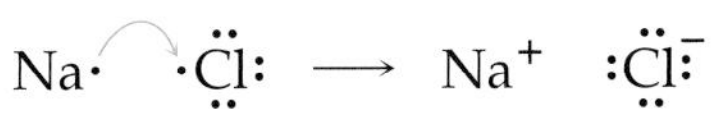

$$Na\cdot \quad \cdot\ddot{\underset{\cdot\cdot}{Cl}}: \longrightarrow Na^+ \quad :\ddot{\underset{\cdot\cdot}{Cl}}:^-$$

There is no electron sharing. Chlorine takes an electron from sodium. We say that the metal's electron has been transferred to the nonmetal. Stripped of its single valence electron, the Na atom is now a positive Na^+ cation. Having gained the electron, the Cl atom is now a negative Cl^- anion. This agrees with what we said in Chapter 4, that metals tend to lose electrons and nonmetals tend to gain them. When a metal and a nonmetal get together, this is generally what happens.

If sodium and chlorine don't share electrons, then they can't bond together covalently. But clearly, the Na⁺ and Cl⁻ ions in table salt stick together (bond) somehow. What holds these ions together is the fact that they have become oppositely charged. Since opposite charges attract, the Na⁺ and Cl⁻ ions literally "stick to each other" via a strong electrostatic force of attraction.

This attractive force holding NaCl together is similar to a covalent bond in that it arises from the attraction between something positive and something negative. It is different in that there is no sharing of electrons. This time, the attractive force arises from oppositely charged ions that were formed by a transfer of electrons. Since this bond is formed by an attraction between ions, it is called an **ionic bond**.

In general, ionic bonds form between a metal and a nonmetal, while covalent bonds form between nonmetals. Even though these bonds arise in different ways, they are similar in strength, since both are due to the attractions between relatively large positive and negative charges—between electrons and nuclei in covalent bonds or between oppositely charged ions in ionic bonds. Finally, we should point out that simple ionic compounds like NaCl, made from one nonmetal and one metal, do not exist as molecules. Rather, they exist as a large, ordered, three-dimensional network of positively and negatively charged ions called an **ionic lattice**. In this lattice, the ions are arranged such that positive ions are adjacent to negative ions (resulting in ionic bonds), but similarly charged ions, which would repel, are kept farther apart. A portion of the NaCl lattice is shown here. This entire lattice is held together by strong, ionic bonds between oppositely charged ions.

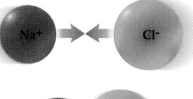

Strong attraction between oppositely charged ions

Ions stick together

The sodium chloride crystal lattice

5.9 WORKPATCH What two words best represent the essential difference between covalent and ionic bonding?

Understanding the difference the WorkPatch refers to means that you are prepared for the next section. Nature is not so simple as to have only ionic or covalent bonds. There are also bonds that lie somewhere between.

5.6 EQUAL VERSUS UNEQUAL SHARING OF ELECTRONS— ELECTRONEGATIVITY AND THE POLAR COVALENT BOND

When two atoms come near enough to one another, a bond may form between them. The bond occurs because their valence electrons redistribute themselves in such a way as to give rise to an attractive force between the atoms. The redistribution involves either a sharing (covalent bond) or a complete transfer (ionic bond) of electrons. We can represent this pictorially. Look at the two atoms, A_1 and A_2 shown below. We've given each a single, unpaired valence electron:

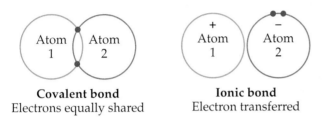

Two things can happen when these two atoms get close to one another. The valence electrons can be shared or transferred. Examine the pictures closely:

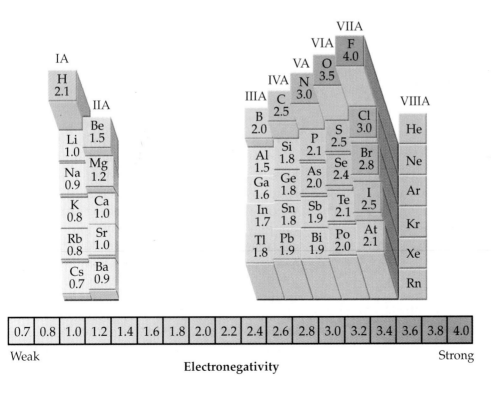

Covalent bond
Electrons equally shared

Ionic bond
Electron transferred

Whether the electrons are shared or transferred, the result is the same in that a strong attractive force—a chemical bond—arises. But are equal sharing or complete transfer the only possibilities? To gain further insight into this question, we must turn to a defined property of atoms called *electronegativity*. **Electronegativity (EN)** is a numerical rating of an atom's ability to attract the shared electrons in a covalent bond to itself. The chart below lists the electronegativity values of the representative elements.

| 0.7 | 0.8 | 1.0 | 1.2 | 1.4 | 1.6 | 1.8 | 2.0 | 2.2 | 2.4 | 2.6 | 2.8 | 3.0 | 3.2 | 3.4 | 3.6 | 3.8 | 4.0 |

Weak   Strong
Electronegativity

The electronegativity scale ranges from 0.7 to 4.0. The higher the value, the greater is an atom's ability to attract shared electrons to itself. Fluorine, with an assigned electronegativity of 4, is the best at attracting shared electrons to itself. All other electronegativity values are scaled relative to fluorine = 4.0. Francium, at 0.7, is the worst. In general, metal atoms tend to have low electronegativity values and nonmetal atoms tend to have high electronegativity values.

There is also a periodic trend in the electronegativity values, which is emphasized by color shading in the chart on page 165. The higher values are in the upper right-hand corner (shown in orange), and the lower values are in the lower left-hand corner (shown in yellow). The four elements nitrogen (N), oxygen (O), fluorine (F), and chlorine (Cl) are the most electronegative elements in the periodic table. These are the electron "hogs."

A useful rule of thumb is that the closer you are on the periodic table to these four elements, the more electronegative the element will be. This means that as you proceed up a group or go from left to right across a period, electronegativity generally increases. Notice that the noble gas atoms of group VIIIA are not assigned any electronegativities. This is because they don't typically form bonds to other atoms, so their ability to attract shared electrons is a moot point.

We can use these electronegativity values to predict whether a chemical bond between two specific atoms is going to be ionic or covalent, or maybe something else. Consider bringing together two atoms, both of which have high electronegativity values. This means they both have a strong attraction for additional electrons, so neither will be willing to transfer its valence electrons to the other. As a result, the atoms will share their electrons and the bond will be covalent. What if we bring together two atoms that both have low electronegativity values? In this case, the bond that forms will still be covalent. As long as the electronegativity values are roughly equal, then the abilities of the atoms to attract shared electrons will be roughly equal, and the electrons will be shared.

When atoms having similar electronegativities join . . .

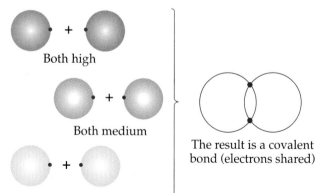

Both high

Both medium

Both low

The result is a covalent bond (electrons shared)

When would two atoms come together and form an ionic bond? Remember, we said previously that ionic bonds tend to form between metals and nonmetals. We also saw that nonmetals have high electronegativity values and metals have lower electronegativity values. Therefore, we can conclude that ionic bonding results when the two elements involved have very different electronegativity values. This makes sense. The atom with the high electronegativity is a powerful attracter of valence electrons, while the atom with the low electronegativity is a poor attracter of valence electrons and is therefore more willing to give them up. This will result in a transfer of valence electrons from the less electronegative atom to the more electronegative atom.

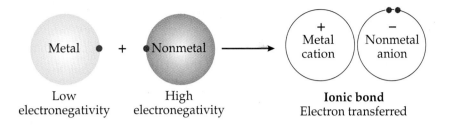

When we use electronegativities to predict the type of bond that will form between two atoms, the important thing is the difference between the electronegativity values and not the individual values themselves. This difference can be expressed as ΔEN, where EN stands for electronegativity:

$$\Delta\text{EN} = (\text{EN of atom 1}) - (\text{EN of atom 2})$$

We always subtract the smaller EN from the larger EN so that this difference is always positive.

If two atoms have similar electronegativities, then ΔEN will be small, and the atoms will share electrons and form covalent bonds. If one atom has a substantially different electronegativity (higher or lower) than the other, the ΔEN will be large and the atoms will bond ionically.

What do we mean by small and large values for ΔEN? The ΔEN values can range from 0 (when identical atoms are bonded to one another) to 3.3 (4.0 − 0.7, the difference in electronegativities between the highest and lowest values in the table). We can define a somewhat arbitrary dividing line near the midpoint of this range (2.0) and say that a ΔEN value below 2.0 is small enough to result in a covalent bond, while a ΔEN value above 2.0 is large enough to produce an ionic bond. This is shown below in a graphical fashion using a ΔEN line along with the calculated values of ΔEN for three examples (Cl_2, HCl, and NaCl). Since Cl_2 and HCl have ΔEN values to the left of the dividing line, they are covalent compounds; NaCl, to the right of the dividing line, is considered ionic.

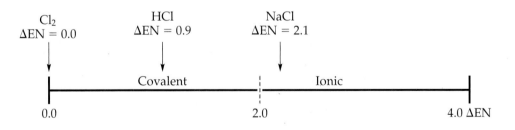

Now what about that dividing line between ionic and covalent at ΔEN = 2.0? Is it really like an on/off switch? Is everything to the left of the line purely covalent, with equal sharing of electrons until you cross the line, and then, bang! electrons are transferred and all bonding is ionic? Not at all. Nature is rarely that black and white. While all bonds in the region from 0 to 2.0 are considered covalent, there is a gradual change in the nature of the covalent bond as we move toward the dividing line. Electrons are still being shared between the atoms, but as we get closer to the dividing line, the ability of one of the atoms to attract the shared electrons to itself is constantly increasing. We eventually end up in a situation where, while the electrons are still being shared,

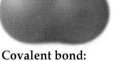

Covalent bond:
electrons shared equally

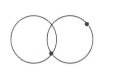

$\delta+$ $\delta-$

Polar covalent bond:
electrons shared unequally

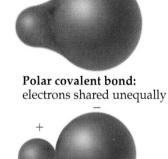

Ionic bond:
electron transferred

they are not being equally shared. The bonding electrons will tend to spend more time near the atom with the higher electronegativity, giving that atom a slight excess of the shared electrons and giving that atom a partial negative charge. The other atom, the one with the lower electronegativity, has lost a bit of the shared electrons and as a result ends up with a partial positive charge. These partial charges are represented by using the lowercase Greek letter delta, $\delta+$ for partial positive and $\delta-$ for partial negative charges, and the bond that results is called a *polar covalent bond*. A **polar covalent bond** is a covalent bond in which the electrons are shared unequally. The three types of bonds (covalent, polar covalent, and ionic) can be shown either with dot diagrams or more realistically by what are known as electron cloud (orbital) diagrams, where the blue color indicates where the valence electrons spend most of their time.

Polar covalent bonding occurs when the electronegativity of one atom differs from the electronegativity of the other atom, but not by enough to result in electron transfer. Now let's go back and modify our original ΔEN line to account for this new classification of bonds. We will add another dividing line at 0.4 to separate pure (nonpolar) covalent bonds from polar covalent bonds:

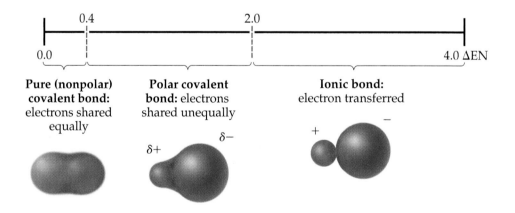

With this modification, we can now reclassify the three examples (Cl_2, HCl, and NaCl) as having either covalent, polar covalent, or ionic bonds:

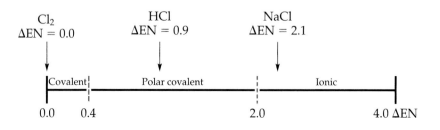

The molecule Cl_2 is still covalent, but HCl is now called polar covalent. The electrons in this bond are not equally shared, they spend more time around the Cl atom than the hydrogen:

Polar covalent bond

$$\overset{\delta+}{H}\text{————}\overset{\delta-}{Cl}$$

EN = 2.1   EN = 3.0
        ΔEN = 0.9

Notice the resulting partial charges ($\delta+$ and $\delta-$) over the atoms. These partial charges are not nearly as strong as the full +1 and −1 charges for the Na^+ and Cl^- ions in NaCl. The Na^+ and Cl^- charges result from the complete transfer of an electron. In HCl the electrons are still shared, they are just shared somewhat unequally. The more electronegative chlorine atom gets the "lion's share," giving it a partial negative charge and the hydrogen a partial positive.
    You try it now.

PRACTICE PROBLEMS

For Practice Problems 5.19–5.21, use the electronegativity values in the chart on page 165 to calculate ΔEN and predict whether the bonds (Ba–Cl, C–H, and Si–O) are covalent, polar covalent, or ionic.

5.19 $BaCl_2$

Answer:

EN Ba = 0.9;   EN Cl = 3.0; ΔEN = 2.1

The bond between Ba and Cl is ionic.

5.20 CH_4

5.21 SiO_2

5.22 Without ever knowing the actual electronegativity values, a student claims $BaCl_2$ is more ionic than $BeCl_2$. All she has access to is a periodic table. How does she know she is right?

    In summary, the ΔEN line shows you that ionic and covalent bonds are really not so different. As you move along the line, covalent blends into ionic, and many compounds exist with bonding that lies somewhere between these two extremes. Stable substances like Cl_2 or HCl or NaCl do not fly apart because valence electrons are shared equally, shared unequally, or transferred between the atoms. All three situations give rise to an attractive electrostatic force between the atoms that we call a chemical bond (covalent, polar covalent, or ionic). Valence electrons truly are the "glue" that bonds atoms to one another.

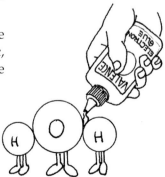

5.7 NOMENCLATURE: NAMING CHEMICAL COMPOUNDS

By international agreements over the years, chemists have arrived at a standardized system for naming chemical compounds. There are two main reasons for this. First, there are so many different chemical compounds that it would be impossible to give each its own common name (such as water for H_2O, ammonia for NH_3, calomel for Hg_2Cl_2, and so on). We would quickly run out of names. Second, common names such as water or calomel give no hint as to the composition of the compound. You simply have to memorize that a name like calomel indicates a compound made from two mercury atoms and two chlorine atoms. Instead, chemists have devised a naming system that never runs out of names, gives a unique name for each different compound, and also gives information as to the composition of the compound. Some of the rules of this nomenclature system are presented below.

NAMING BINARY IONIC COMPOUNDS

Binary compounds are those made up of only two elements. **Binary ionic compounds** are made up of two ions, a positive cation (usually a metal) and a negative anion (usually a nonmetal). The formula is always written with the elemental symbol for the cation preceding the symbol for the anion. For example, sodium chloride is written as NaCl and never as ClNa. Like the formula, the name of a binary ionic compound identifies the cation first, followed by the anion. The cation is given the name of the element itself, while the anion has the suffix "ide" added to its elemental name. An ionic compound made from fluorine and lithium would be written as LiF and would be called lithium fluor*ide*. Here are a few more examples of binary ionic compounds:

Naming binary ionic compounds

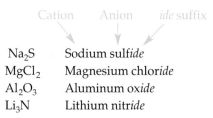

Na_2S	Sodium sulf*ide*
$MgCl_2$	Magnesium chlor*ide*
Al_2O_3	Aluminum ox*ide*
Li_3N	Lithium nitr*ide*

Notice that each name tells you what elements are present in the compound and even identifies the anion (the one with the *ide* ending), but does not seem to tell you how many of each ion are in the formula. In fact, it does. Just by looking at the name magnesium chloride, you should be able to determine that the formula is $MgCl_2$. You already have the knowledge to do this. You learned that the metal in an ionic compound will lose electrons. Since magnesium is in group IIA, it will lose 2 electrons to become Mg^{2+}. You have also learned that nonmetals gain electrons. Since chlorine is in group VIIA, it will gain 1 electron and become Cl^-. Finally, in order for magnesium chloride to be electrically neutral, there must be two Cl^- ions for every Mg^{2+} ion, making the formula $MgCl_2$. So, just from seeing the name magnesium chloride, you can deduce the formula. This is the beauty of this naming system.

5.10 WORKPATCH

Thallium (Tl) forms a compound with oxygen called thallium oxide. Based on their locations in the periodic table, what is the expected formula?

Check your answer to the WorkPatch at the end of the chapter. You should be able to go from formula to name or from name to formula for any binary ionic compound.

5.23 What is wrong with the formula MgCl?

Answer: To have this formula, the ions would need to have charges that were equal in magnitude but opposite in charge, such as +1 and −1, or +2 and −2, etc., but Mg forms Mg^{2+} cations and Cl forms Cl^- anions.

5.24 Fill in the blanks in the table.

Formula	Cation	Anion	Name
CaF_2	?	?	?
?	?	?	Cesium bromide
?	Al^{3+}	S^{2-}	?
K_2O	?	?	?

5.25 What do all the formulas of the group IIA bromides have in common?

The transition metals and some of the heavier representative metals throw a monkey wrench into this simple nomenclature system. These elements can form more than one kind of cation. For example, unlike sodium, which always exists in compounds as Na^+, iron commonly exists both as Fe^{2+} and Fe^{3+}. This means that the name iron chloride could represent the formula $FeCl_2$ and/or $FeCl_3$. This is not good. We can't allow ambiguity in our naming system. Fortunately, a simple addition to the system will fix this problem. When dealing with a transition metal or one of the heavier representative metals, a roman numeral (in parentheses) is included in the name after the cation to indicate the magnitude of the positive charge. For example, iron(II) chloride (pronounced as "iron two chloride") implies Fe^{2+} and would have the formula $FeCl_2$, while iron(III) chloride would be $FeCl_3$ (Fe^{3+}). This takes care of the problem and allows us to go from name to formula or from formula to name.

5.26 Name the compound $TiCl_2$.

Answer: Cl is in group VIIA and so will gain 1 electron to become a Cl^- anion. Since there are two of these, then titanium (Ti) must have a +2 charge to maintain electrical neutrality. The name is therefore titanium(II) chloride.

5.27 Name the compound $TiCl_3$.

5.28 Give the formula of tin(II) fluoride.

You should be aware of an older but still commonly used system for naming these types of compounds. Instead of using a roman numeral, the name of the cation is changed by adding a suffix to indicate its charge. For the element iron, the suffix "ous" is used for Fe^{2+} (ferrous), and the suffix "ic" is used for Fe^{3+} (ferric). However, the suffixes "ous" and "ic" do not necessarily mean +2 and +3 respectively. Rather, "ous" is used for the lower of the possible positive charges and "ic" is used for the higher charge. For example, tin (Sn) commonly exists in two cationic forms, +2 and +4. The compounds formed between tin and fluorine could be either SnF_2 or SnF_4. In this case, SnF_2, tin(II) fluoride, would be referred to as stannous fluoride, and SnF_4, tin(IV) fluoride, would be called stannic fluoride. Think about this the next time you are brushing your teeth with fluoride toothpaste. The active ingredient is stannous fluoride.

SnF_2

NAMING COMPOUNDS THAT CONTAIN POLYATOMIC IONS

Earlier we introduced polyatomic ions as essentially small molecules with a charge. Some polyatomic ions are very stable and show up over and over again in a variety of compounds. Hypochlorite, for example, is one that you have probably used many times without ever realizing it. The ClO^- ion is the active ingredient in liquid laundry bleach. Table 5.2 lists the most common polyatomic ions, along with their charges and names. You should commit the names, formulas, and charges of these ions to memory.

Table 5.2   Some Common Polyatomic Ions

Cation		Anions		Anions		Anions	
NH_4^+	Ammonium	OH^-	Hydroxide	HCO_3^-	Bicarbonate	SO_4^{2-}	Sulfate
		NO_3^-	Nitrate	ClO^-	Hypochlorite	CO_3^{2-}	Carbonate
		NO_2^-	Nitrite	CH_3COO^-	Acetate	PO_4^{3-}	Phosphate

Once you know these polyatomic ions, naming compounds that contain them is easy. As an example, consider the compound sodium phosphate. What would the formula for this compound be? From Table 5.2 we know that the phosphate ion has the composition PO_4 and a −3 charge. Since sodium, a group IA element, forms ions with a charge of +1, the overall formula must be Na_3PO_4 to achieve overall charge neutrality.

$$\underbrace{PO_4^{3-}}_{-3} + \underbrace{\begin{matrix} Na^+ \\ Na^+ \\ Na^+ \end{matrix}}_{+3} \longrightarrow \text{Neutral compound, } Na_3PO_4$$

What is the formula for magnesium phosphate? The phosphate ion still has a −3 charge, but now the magnesium cation, a group IIA element, has a +2 charge. To arrive at an overall neutral compound, we need to combine two phosphates with three Mg^{2+} ions:

$$\underbrace{\begin{array}{c} PO_4{}^{3-} \\ PO_4{}^{3-} \end{array}}_{+6} \; + \; \underbrace{\begin{array}{c} Mg^{2+} \\ Mg^{2+} \\ Mg^{2+} \end{array}}_{-6} \longrightarrow \text{Neutral compound, } Mg_3(PO_4)_2$$

The formula of magnesium phosphate is therefore $Mg_3(PO_4)_2$. Notice that we put parentheses around the phosphate; otherwise, the formula would have looked like this: Mg_3PO_{42}. That would imply that there are 42 oxygen atoms in the compound. Whenever there is more than one polyatomic ion in a formula it must be surrounded by parentheses.

PRACTICE PROBLEMS

5.29 Write the formula for ammonium phosphate.

Answer: Since ammonium is $NH_4{}^+$, and phosphate is $PO_4{}^{3-}$, the formula must be $(NH_4)_3PO_4$ to obtain charge neutrality.

5.30 Write the formula for sodium carbonate.

5.31 Write the name for $Ca(ClO)_2$.

5.32 Write the formula for magnesium bicarbonate.

NAMING BINARY COVALENT COMPOUNDS

Binary covalent compounds generally consist of two different nonmetals bound to one another by covalent or polar covalent bonds. They are named in much the same way as the binary ionic compounds with one slight difference. In the case of ionic compounds, the name magnesium chloride was enough to tell us that the formula was $MgCl_2$ because we could count on the magnesium always being +2 and the chlorine always being −1. With binary covalent compounds this is no longer the case. When two nonmetallic elements combine, a range of possibilities exist. This means that we need to insert prefixes into the name to explicitly tell us how many of each element are present in the formula. The prefixes that we use are shown in Table 5.3 and are derived from early Greek names for the numbers.

Table 5.3   Greek Prefixes

1	mono	6	hexa
2	di	7	hepta
3	tri	8	octa
4	tetra	9	nona
5	penta	10	deca

Also, since both elements in these compounds are nonmetals, we treat the less electronegative element as a metal, list it first, and use the name of the element. The more electronegative element is listed second and has the suffix "ide" appended to its name. For example, the compound HCl is called hydrogen chloride.

(5.11) WORKPATCH

Why does it make sense to give the more electronegative element in a binary covalent compound the suffix "ide"?

Let's consider a few examples. Carbon and oxygen can combine to form two different binary compounds, CO and CO_2. These compounds are called carbon *mon*oxide and carbon *di*oxide. We use prefixes *mon(o)* and *di* to indicate that one or two atoms of these elements are present in the compound. These prefixes can also be used with the first element. For example, the compounds NO_2 and N_2O_3 are named nitrogen dioxide and dinitrogen trioxide, respectively. Notice that NO_2 is *not* called mononitrogen dioxide. If there is only one atom of the least electronegative element in the formula, the mono is generally left out.

PRACTICE PROBLEMS

5.33 Name the compound whose formula is N_2O.

Answer: Dinitrogen monoxide

5.34 Name the compound NO.

5.35 Name the compound PCl_5.

5.36 Give the formula of the compound whose name is tetraphosphorus decaoxide.

Finally, some binary covalent compounds have been around for so long that the common names are the only ones used. For example, NH_3 is always referred to as ammonia and not nitrogen trihydride. And absolutely nobody calls H_2O dihydrogen monoxide! To prove this to yourself, the next time you go to a restaurant see how clever your waiter thinks you are when you ask for some dihydrogen monoxide, but this time in a clean glass.

Molecule (p. 145)

Covalent bond (p. 149)

Dot diagram (p. 154)

Electron pair (or paired electrons) (p. 154)

Unpaired electrons (p.154)

Bonding pairs (p. 156)

Lone pairs (p. 156)

Multiple covalent bonds (p. 158)

Double bond (p. 158)

Single bond (p. 159)

Triple bond (p. 159)

Resonance forms (p. 161)

Polyatomic ions (pp. 161, 172)

Ionic bond (p. 164)

Ionic lattice (p. 164)

Electronegativity (p. 165)

Polar covalent bond (p. 168)

Binary compounds (p. 170)

Binary ionic compounds (p. 170)

Binary covalent compounds (p. 173)

MOLECULES

5.37 Give a general definition of a molecule.

5.38 Are all molecules also compounds? Explain.

5.39 What is it about molecules that makes them worth preparing? Give some examples.

5.40 What common diatomic molecules exist in our atmosphere? What triatomic molecules exist in our atmosphere? Give the formulas for all your answers.

5.41 Ethanol (C_2H_6O) is consumed in alcoholic beverages, but another alcohol, methanol (CH_4O), is toxic. What explanation can you give for this?

5.42 Can a molecule also be an elemental substance? If so, give some examples.

THE COVALENT BOND

5.43 Define a covalent bond.

5.44 What is wrong with saying that a covalent bond is just "two shared electrons"?

5.45 Why does the sharing of two electrons between two atoms bond the atoms to each other?

5.46 In terms of energy, how is an H_2 molecule more stable than two isolated H atoms?

5.47 Why is an H_2 molecule more stable than two isolated H atoms?

5.48 The bond distance in an H_2 molecule is 0.74 Å. Why isn't it shorter than this? Why isn't it longer than this?

5.49 Using what you know about how charged particles interact, explain why a covalent bond between two nuclei does not involve three shared electrons.

5.50 Why are an atom's valence electrons the only electrons involved in bonding?

5.51 Suppose one of the electrons from the covalent bond in H_2 suddenly vanished. Why would the bond between the atoms weaken?

5.52 A student decides to boil water to produce hydrogen gas and oxygen gas. Will this work? Explain your answer.

5.53 What do we mean by "double-counting" when it comes to counting the electrons around an atom in a molecule?

5.54 What is the accepted shortcut for drawing a shared pair of electrons (a covalent bond) in a molecular drawing?

5.55 How many electrons does an atom gain for each covalent bond that it forms in a molecule?

MOLECULES, DOT STRUCTURES, AND THE OCTET RULE

5.56 A student draws the following dot diagrams for N, O, and F. What, if anything, is wrong with each?

$$\cdot \overset{\cdot}{\underset{\cdot \cdot}{N}} \cdot \qquad \overset{\cdot \cdot}{\underset{\cdot \cdot}{O}} \qquad \cdot \overset{\cdot \cdot}{\underset{\cdot \cdot}{F}} \cdot$$

5.57 How many dots will a dot diagram for a representative element have around it?

5.58 Draw dot diagrams for the following atoms:
(a) S   (b) I   (c) He   (d) B

5.59 Chlorine (Cl), neon (Ne), and helium (He) all exist as gases, but only one of them is diatomic. Which is it, and why is it diatomic while the others are monatomic?

5.60 Electrons in an atomic dot diagram can exist in either of two ways. What are they?

5.61 Why isn't the formula for water HO?

5.62 Hydrogen (H) and sulfur (S) form the toxic compound hydrogen sulfide, a gas that smells like rotten eggs and is spewed from volcanoes. Predict the formula of hydrogen sulfide starting with atomic dot diagrams.

5.63 Phosphorus (P) and bromine (Br) form a compound. Predict the formula of this compound starting with atomic dot diagrams.

5.64 Nitrogen (N) forms a compound with bromine (Br) similar to that in Problem 5.63. How is it similar? Explain why this is so.

5.65 Ethers are compounds of C, H, and O that are often used as solvents. One particular ether molecule has the formula C_2H_6O. The structure is such that both carbons are attached to the oxygen atom. Starting with atomic dot diagrams, draw a dot diagram for this ether molecule. Report on the total number of bonding pairs and lone pairs of electrons.

5.66 Ethanol, the alcohol found in alcoholic beverages, has the same formula as the ether of Problem 5.65. However, it has a different structure, in which only one carbon atom is bound to the oxygen atom. Starting with atomic dot diagrams, draw a dot diagram for the ethanol molecule. Indicate the total number of bonding pairs and lone pairs of electrons. Also comment on what this says about the role of structure in determining chemical properties of a molecule.

5.67 The molecule HCl is known, but the molecule HeCl is not. Explain why this is so.

5.68 In the dot diagram for a helium atom, why is it important to draw the two valence electrons as a lone pair?

5.69 Oxygen, in almost all of its compounds, has two bonds to it. Explain why this is so.

5.70 How many bonds will an atom from group VA generally form? Explain why this is so.

MULTIPLE BONDS

For all the problems in this section, you must draw all the resonance forms when they exist.

5.71 Draw a dot diagram for ethylene, C_2H_4.

5.72 Is the bond between the carbon atoms in the ethylene molecule of Problem 5.71 stronger or weaker than the bond in acetylene, C_2H_2 (WorkPatch 5.7)? Explain fully.

5.73 Explain what is meant by resonance forms.

5.74 Draw a dot diagram for the molecule SO_2, sulfur dioxide, a gas that comes from burning coal and is responsible for acid rain. [*Hint:* Sulfur is in the middle.]

5.75 Draw a dot diagram for the molecule O_3, ozone, the molecule in our upper atmosphere that protects us from the Sun's harmful ultraviolet electromagnetic radiation.

5.76 Should there be any similarities between the dot diagrams of Problems 5.74 and 5.75? Explain your answer.

5.77 A student claims that the bonds in ozone are really not double bonds or single bonds, but somewhere between (roughly 1.5 bonds between each oxygen). Justify this statement.

5.78 Experiments show that it takes more energy to break the bond between oxygen atoms in the O_2 molecule than in the O_3 molecule. How can you explain this?

5.79 Is anything wrong with the following dot diagram for carbon monoxide (CO)? If so, explain what it is and fix it.

$$:\ddot{C}-\ddot{O}:$$

5.80 Is anything wrong with the following dot diagram for acetone (C_3H_6O)? If so, then fix it.

5.81 How many covalent bonds do you think would form between the phosphorus atoms in the molecule P_2? Explain your answer and draw a dot diagram.

5.82 Draw a dot diagram for the nitrate (NO_3^-) ion.

5.83 Draw a dot diagram for the hypothetical O_2^{2+} ion.

5.84 Acetaldehyde has the formula C_2H_4O. Draw a dot diagram for it given the following atomic connections.

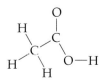

5.85 Acetic acid, a major component of vinegar, has the formula $C_2H_4O_2$. Draw a dot diagram for it given the following atomic connections.

5.86 The acetate ion has the formula $C_2H_3O_2^-$. Draw a dot diagram for it given the following atomic connections. [*Hint:* Watch for resonance forms.]

5.87 Draw a dot diagram for perchloric acid, $HClO_4$. The chlorine is the central atom to which all the oxygens are attached, and the hydrogen is attached to one of the oxygens.

5.88 Draw a dot diagram for the NO_2^+ cation.

5.89 Consider benzene, a pleasant-smelling but carcinogenic (cancer causing) liquid. Benzene molecules have the formula C_6H_6, with the atoms connected as shown. Complete the dot diagram for this molecule and include any resonance forms.

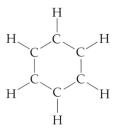

5.90 How many valence electrons are present on each carbon atom and each hydrogen atom in the benzene molecule of Problem 5.89?

IONIC BONDING

5.91 Draw atomic dot diagrams for the following simple ions:
(a) I^- (b) O^{2-} (c) Cl^- (d) H^+

5.92 What is an ionic bond? How does it differ from a covalent bond, and how is it similar to a covalent bond?

5.93 What must occur between atoms for an ionic bond to form?

5.94 Between what types of elements is ionic bonding most likely to occur?

5.95 Draw dot diagrams for Cl_2 and for $MgCl_2$ (put Mg in the middle) that agree with the type of bonding occurring in each. (That is, show electrons as being shared or transferred, and if they are transferred, show the charges of the resulting ions.)

5.96 Predict the formulas of the ionic compounds that result from combining:
(a) Mg (solid metal) with Br_2 (liquid)   (b) Be (solid metal) with O_2 (gas)
(c) Na (solid metal) with I_2 (solid)

5.97 Predict the formula of the ionic compound formed by each of the following pairs of elements:
(a) Ca, I   (b) Ca, O   (c) Al, S   (d) Ca, Br

5.98 What is wrong with saying that NaCl exists as a molecule, just like H_2O does?

ELECTRONEGATIVITY AND THE POLAR COVALENT BOND

5.99 There is one very important word missing from the following definition of electronegativity. What is it? "Electronegativity is an indication of an atom's ability to attract electrons to itself."

5.100 Why did we call the elements N, O, Cl, and F electron "hogs" in Section 5.6?

5.101 What is the most electronegative element in the periodic table? What is the least electronegative element? What kind of compound would form if these two elements were brought together?

5.102 How do the categories metal and nonmetal relate to electronegativity?

5.103 What are the trends for electronegativity in the periodic table:
(a) Down a group?   (b) Across a period from left to right?
(c) Going from the bottom left corner to the upper right corner?

5.104 What was electronegativity used to predict in this chapter, and how was this done?

5.105 How does a polar covalent bond differ from a covalent bond? Give two examples of diatomic molecules that exhibit these types of bonds.

5.106 Would you classify the bonds in ethane (C_2H_6) as covalent, polar covalent, or ionic? Explain.

5.107 Lithium is a metallic element. Consider the hypothetical species dilithium, Li_2. Predict whether the bonding in such a species would be covalent, polar covalent, or ionic. Explain fully.

5.108 Which molecule has bonds that are the most polar covalent?

$$H_2, \quad CO, \quad H_2S, \quad H_2O$$

5.109 Chemists sometimes think of molecules with polar covalent bonds as being part covalent and part ionic. How can a bond be both covalent and ionic?

5.110 In the following molecule XY, one of the atoms is more electronegative than the other. Which is more electronegative, and how do you know?

X   Y

5.111 In the following molecule XY, one of the atoms is more electronegative than the other. Which is more electronegative, and how do you know?

$$\overset{\delta-}{X}\!-\!\overset{\delta+}{Y}$$

5.112 How would you change the electronegativities of the atoms at either end of a polar covalent bond to make the bond ionic? To make the bond covalent?

5.113 Electronegativity is not about measuring an electron's negativeness. Given your knowledge of what electronegativity measures, think of a new word or expression that could replace this misleading term and also communicate what it measures.

5.114 Without knowing the electronegativity values, in what situation can you be absolutely sure that the bonding between two atoms will be purely covalent?

5.115 Classify the bonds in each of the following as ionic, covalent, or polar covalent. Explain each choice.
(a) Br_2 (b) PCl_3 (c) $LiCl$ (d) ClF (e) $MgCl_2$

NOMENCLATURE

5.116 What is a binary compound? Is atmospheric oxygen an example of a binary compound?

5.117 What does the suffix "ide" mean, and which element in the name of a binary ionic compound gets it?

5.118 Which element in a binary covalent compound gets the suffix "ide"? Why?

5.119 Any ionic compound has an overall neutral charge. How does this fact help to determine its formula?

5.120 Consider the two binary compounds Al_2O_3 and N_2O_3.
(a) Which is ionic and which is covalent? Explain.
(b) Al_2O_3 is properly named aluminum oxide, but N_2O_3 is named dinitrogen trioxide. Explain fully why one name uses Greek prefixes and the other does not.

5.121 Name the following binary ionic compounds:
(a) Ca_3N_2 (b) AlF_3 (c) Na_2O (d) CaS

5.122 Give the formulas of the following binary ionic compounds:
(a) Calcium bromide   (b) Sodium sulfide
(c) Potassium nitride   (d) Lithium oxide

5.123 What "monkey wrench" do transition metals throw into the picture when naming binary compounds? What do we include in the nomenclature system to accommodate them?

5.124 Name the following binary compounds of transition metals using both the modern and older systems.
(a) CuCl and $CuCl_2$ (b) $Fe(OH)_2$ and $Fe(OH)_3$

5.125 Name the following compounds:
(a) Na_2SO_4 (b) $(NH_4)_3PO_4$ (c) KClO   (d) $CaCO_3$ (e) $Al(NO_3)_3$

5.126 Give the formulas for the following compounds:
(a) Ammonium acetate   (b) Ammonium carbonate
(c) Iron(II) nitrate   (d) Ferric hydroxide
(e) Calcium hypochlorite

5.127 Name the following binary covalent compounds:
(a) PCl_3 (b) SO_2 (c) N_2O_4 (d) P_5O_{10}

WorkPatch Solutions

5.1   The molecules H_2O (water) and H_2O_2 (hydrogen peroxide) have very different chemical properties. For example, you drink water, but H_2O_2 is a harsh antiseptic and bleach. Different formulas mean different compounds with different properties, even when the elements in the formula are the same.

5.2   The carbon atom in the methane (CH_4) molecule has 8 valence electrons (2 from each bond). Each hydrogen atom has 2 valence electrons (from the bond that attaches it to carbon).

5.3   Carbon normally has 4 valence electrons (it is a group IVA atom). In methane it has 8 valence electrons, so it has gained 4 electrons. It also forms four bonds to hydrogen, so carbon gains one more valence electron with each bond to hydrogen.

5.4   Helium already has a filled valence shell (a duet), so it has no "desire" to share electrons with other elements to gain more electrons.

5.5   Every atom in the NF_3 molecule has an octet of valence electrons (if the bonding pairs are double counted).

5.6

$$
\begin{array}{ccc}
 & H & H \\
 & | & | \\
H- & C- & C-H \\
 & | & | \\
 & H & H
\end{array}
$$

5.7   H—C≡C—H

5.8   We were able to generate three different dot diagrams for CO_2 by moving lone pairs into bonding positions from different atoms.

5.9   Sharing versus transfer

5.10   Thallium (Tl) is a metal in group IIIA, so it will lose 3 electrons to become Tl^{3+}. Oxygen is a nonmetal in group VIA, so it will gain 2 electrons to become the oxide ion, O^{2-}. To obtain a neutral compound, we must combine two Tl^{3+} ions (for a +6 charge) with three O^{2-} ions (for a −6 charge). The formula is therefore Tl_2O_3.

5.11   The more electronegative element in a binary covalent compound will have the partial $\delta-$ charge, and the suffix "ide" is associated with the anionic or negative part of a compound.

EQUAL
SWEETENER
WITH
NUTRASWEET
As sweet as 2 tsp sugar
QUADE
Suggest Highlighting A BLUE POCKET
NUTRASWEET ("EQUAL") & THE
Asportame
N & O will give
(STRUCTURAL,
peptide bond
H2NCH CNHCHCOCH3
HOOCCH2
CH2
Asp
Phe

The Structure of Molecules

WHY IS THE SHAPE OF A MOLECULE IMPORTANT? 6.1

In the last chapter we discussed the forces that hold a molecule like methane together and why it has the formula CH_4. Now we'll take a closer look at how we've been drawing molecules. The dot structures we've been using were usually drawn as if they were flat, or two-dimensional. But dot structures are meant to show only the bonding in a molecule and not its true three-dimensional shape. Methane is really not a flat molecule.

What does the methane molecule actually look like? In this chapter we will answer this question and learn how to predict the three-dimensional shape of a molecule.

At this point you might be wondering why we should care about the shape of a molecule. After all, we can't see individual molecules. What possible relevance could the shape have? Quite simply, the shape of a molecule could save your life, and most likely it already has. Chemists figured this out shortly before World War II began. In all wars up to that point, the greatest number of deaths were caused not by the wounds inflicted in battle, but by the infections that set in shortly thereafter. A young soldier with a stomach full of shrapnel would lie in a makeshift field hospital recovering from abdominal surgery. The operation would go well, but there was just no way to keep a few microscopic bacteria from entering his bloodstream and reproducing wildly. It was an all too common scenario. The soldier's body would try to fight the infection, but eventually he would die. During World War I, medics could do almost nothing to stop this, and hundreds of thousands died. But World War II was different. For the first time, a medic could intervene with an effective defense—a drug called sulfanilamide. Sulfanilamide was a trap, meant to fool the bacteria. Its molecular structure is similar to another molecule found in the bloodstream, *para*-aminobenzoic acid (PABA). You can see the similarities below:

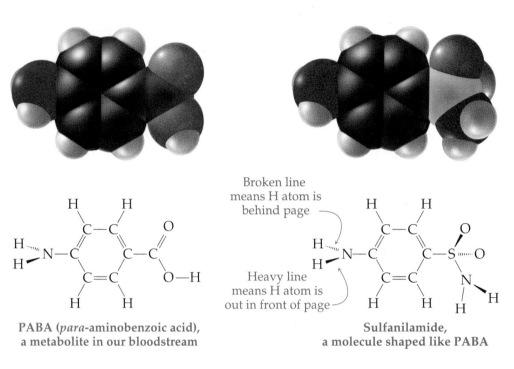

PABA (*para*-aminobenzoic acid),
a metabolite in our bloodstream

Broken line means H atom is behind page

Heavy line means H atom is out in front of page

Sulfanilamide,
a molecule shaped like PABA

We all have lots of PABA in our bloodstream. PABA is a metabolite—that is, a natural product of our body's metabolism. Bacteria can use PABA to their own advantage by combining it with two other molecular fragments also present in our bloodstream. Once connected together, these three fragments form one of the B vitamins—folic acid, the structure of which is shown below:

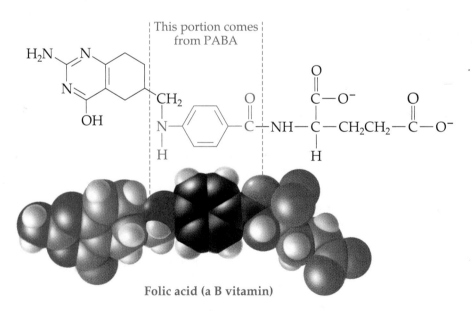

This portion comes from PABA

Folic acid (a B vitamin)

This vitamin is essential for the health and continued functioning of the bacteria. Humans also need this vitamin in order to live, but we have lost the ability to make it from these three parts and must ingest it whole in the food we eat. It is this difference in the source of folic acid used by bacteria and their human hosts that makes sulfanilamide and many other antibiotics the wonder drugs they are.

Folic acid is created inside the bacterial cell by a specific molecule called an *enzyme*. Enzymes are large protein molecules that serve as catalysts in living systems. They increase the speed of specific chemical reactions within the cell. Enzymes have precisely shaped pockets (called *active sites*) on their surface which they use to bind with other, smaller molecules that have complementary shapes. This is often referred to as the "lock and key" model of enzyme action. The pocket or active site is the lock, and the complementary molecule that fits perfectly into the pocket is the key. The enzyme in bacteria that constructs folic acid from its three molecular pieces has a pocket which is complementary in shape to the PABA molecule.

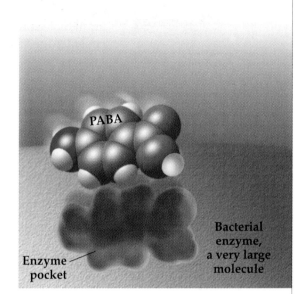

Efficiently manufacturing their own vitamin B, the bacteria can thrive and reproduce in the human bloodstream. But not when sulfanilamide molecules are waiting. Since sulfanilamide has almost the same shape as the PABA molecule, it fits nicely into the pocket of the bacterial enzyme, which is tricked into making a defective form of folic acid, incorporating the sulfanilamide instead of PABA. Starved for the real B vitamin, the bacteria soon die, and the infection subsides. Sulfanilamide thus saved thousands of lives in World War II, and was the precursor to a number of modern antibiotics called sulfa drugs.

So, is the shape of a molecule important? Unquestionably! You can be sure that the new antibacterial and antiviral agents that have yet to be discovered—including the drugs that will one day free humanity of herpes, AIDS, hepatitis, meningitis, and cancer—will have molecular shape at the core of their ability to cure.

VALENCE SHELL ELECTRON PAIR REPULSION (VSEPR) THEORY 6.2

Of course, no one can build a molecule with a specific shape or even predict what the shape of a molecule will be without first understanding what it is that determines a molecule's shape. **Valence shell electron pair repulsion (VSEPR) theory** is the model most often used to predict molecular shape. It is based on the simple concept that the valence electrons in a molecule will repel each other. Let's go back to methane to investigate this. We'll start by drawing a dot diagram for methane where lines are used to represent bonding pairs of electrons:

All electrons are negatively charged, so each bonding pair of electrons represents a region of concentrated negative charge. Since like charges repel, the electrons in the four bonds of methane will repel each other. The bonding pairs of electrons will get as far apart from each other as possible in order to minimize this repulsion. The flat, square version of methane shown on page 185, with its 90° angles, would seem to accomplish this pretty well. Certainly we can envision other ways of attaching four hydrogens to a central carbon atom that would make the repulsions between the bonds worse. For example, consider the following arrangement:

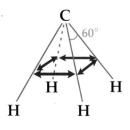

A methane molecule would never assume this shape. With such small H–C–H angles, the bonding pairs of electrons would be practically on top of one another, causing strong repulsion between the bonding pairs.

While we can easily generate structures that are worse than our original square planar shape, the real question is, "Can we do any better?" If we no longer insist that the methane molecule be flat, the answer is "Yes." There is a better arrangement, one that opens up the H–C–H angles beyond 90° and fully maximizes the distance between the four bonding pairs of electrons. That shape is shown below:

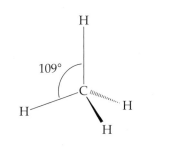   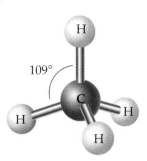

— Bond is out in front of page
⋯⋯ Bond goes back behind page
— Bond is in the plane of the page

Ball-and-stick representation

Pay particular attention to the drawing on the left. Two of the C–H bonds are drawn as lines, one as a solid wedge, and one as a dashed wedge. These are meant to communicate three-dimensionality, as explained in the diagram. The methane molecule is said to have a *tetrahedral* shape or geometry. A **tetrahedron** is a regular four-sided polygon in which the four sides are identical equilateral triangles. The following figure shows a methane molecule inscribed within a tetrahedron:

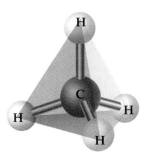

The H–C–H angles are all roughly 109°, the standard tetrahedral angle. This is as large as they can be. When methane adopts this shape, the four bonding pairs are as far apart from one another as possible. This minimizes the repulsion between the electron pairs and thus maximizes the stability of the molecule.

What we have said so far about methane is not specific to that molecule. Any time there are four pairs of electrons around an atom in a molecule, a tetrahedron is the preferred arrangement. This is simply the best way to keep the four groups of electrons as far apart from each other as possible. Once you understand this, you understand one of the most critical factors that determines molecular shape.

Of course, we will not always be dealing with four pairs of electrons. Let's look at two other organic molecules, formaldehyde (used to preserve biological specimens) and acetylene (a flammable gas used by metal workers in high-temperature oxyacetylene torches).

Formaldehyde, CH_2O Acetylene, C_2H_2

The dot structures for these molecules have been drawn flat, with nice, neat right angles. These structures are meant to show only the bonds and lone pairs in a molecule, not its actual shape. Once the dot structure has been determined, the question of how the bonds and atoms are actually oriented in space can be answered. We must arrange the electrons in such a way as to minimize the repulsion between them. We start just as we did for methane, by counting the number of bonds around a particular carbon atom, but there is a new twist here. Both of these molecules have multiple bonds (a double bond in formaldehyde, a triple bond in acetylene). The rule for counting multiple bonds for the purpose of VSEPR electron counting is to *treat multiple bonds as if they were single bonds*. This is done because all the electrons in a multiple bond occupy roughly the same region of space. This means that surrounding the carbon atom in formaldehyde there are just three "pairs" of electrons, two single bond pairs and a double bond "pair." For each carbon atom in acetylene, there are just two groups of electrons around it, a single bond pair and one triple bond "pair." The word "pair" is enclosed in quotes because the double bond actually consists of 4 electrons and the triple bond of 6 electrons. From now on we will use the term "electron group" instead of pair to avoid this confusion. Once the counting is done, the task of determining the shape of the formaldehyde molecule comes down to answering the question, "How can three groups of electrons best arrange themselves around a central (carbon) atom?" In the case of acetylene, it becomes, "How will two groups of electrons best arrange themselves around a central (carbon) atom?" The answer in both situations involves a flat arrangement of the bonds. Can you determine the best angles between these bonds? Try it in WorkPatch 6.1. If you can, then you've caught on to VSEPR theory.

(6.1) WORKPATCH   Redraw the structures shown to get the electron groups as far apart as possible. What angles did you use?

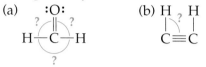

Be sure to check your answers to the WorkPatch at the end of the chapter before going on.

So far we have looked at three specific molecules and used VSEPR theory to predict their shapes. In methane there were four electron groups around the carbon atom, in formaldehyde there were three, and in acetylene there were two (counting multiple bonds as if they were single bonds). But there is nothing special about these molecules. The same results can be expected for any molecule that contains an atom with the same number of electron groups around it, as shown in Table 6.1.

Table 6.1   VSEPR Arrangements of Electron Groups Around an Atom

Number of electron groups	Name of molecular shape (geometry)	Structure and bond angle	Examples
Two	Linear	 180° — A —	O=C=O 
Three	Trigonal planar  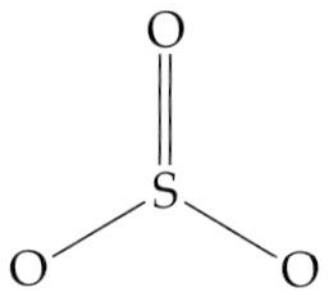 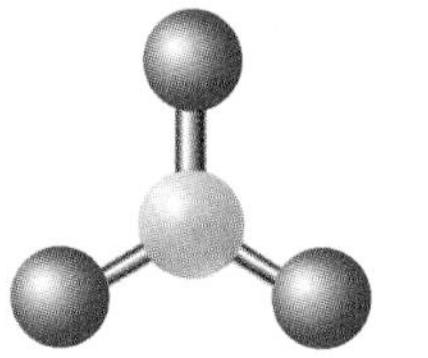	120°	O=S, O, O
Four	Tetrahedral 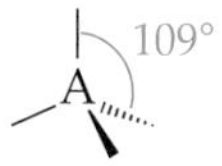	109°	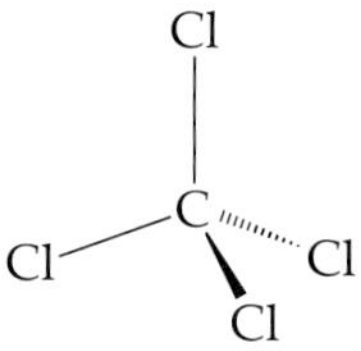 Cl—C(Cl)(Cl)Cl 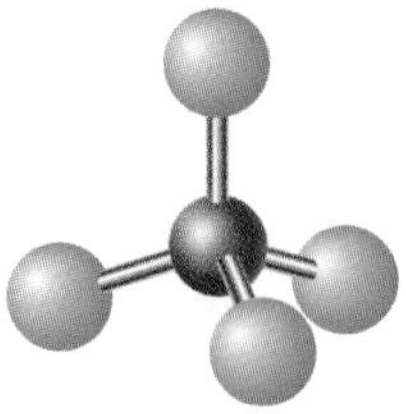

Since determining the shape of a molecule requires knowing the number of electron groups that surround each atom, you really can't get started until you have a good dot diagram in front of you. This means that when you are asked to determine the shape of a molecule, you must first draw a dot diagram if one has not been supplied.

For all these problems, name the shape, draw the shape, and indicate the bond angles.

6.1 What is the shape of the CO_2 molecule?

Answer: First draw a valid dot diagram, then examine it via VSEPR.

:O:
‖ ?
C═Ö $\xrightarrow{\text{Best arrangement for two groups is linear (180°)}}$ Ö═C═Ö   180°

Valid dot diagram
Two electron groups
about the carbon

Molecule is linear
180° bond angle

Note: Don't worry about the arrangement of the electron pairs about the peripheral (outer) atoms of a molecule. Only the electrons on the inner atoms affect the shape.

6.2 What is the shape of the ammonium (NH_4^+) cation?

6.3 What is the shape of the C_2Cl_2 molecule (connected in the following order: Cl–C–C–Cl)?

6.4 What is the shape of the $SOCl_2$ molecule? [*Hint:* All atoms are connected to S, and there is a sulfur–oxygen double bond in the molecule.]

So far, none of the molecules we have considered had lone pairs of electrons on the central atom. It's now time to consider these molecules as well. A good example is the ammonia molecule, NH_3. Ammonia is an interesting substance. It exists as a gas at room temperature, but you are probably more familiar with it dissolved in water as a household cleaner. It is the fifth most commonly produced chemical in the United States, used in greater quantities as an agricultural fertilizer than as a household grease cutter. What is its shape? To answer this question we have to decide how the electron groups are arranged in space around the central nitrogen atom, so we will need a valid dot diagram to begin.

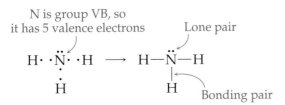

This dot diagram is valid but, as usual, it is drawn with an arbitrary planar shape using right angles. Now focus on the central nitrogen atom and count the total number of electron pairs that surround it. The nitrogen atom in ammonia is surrounded by four pairs of electrons: three bonding pairs and one lone pair. The lone pair must be included in the total count. It is negatively charged, just like a bonding pair, and will thus repel the other electron groups. Therefore, much like the carbon in methane, or any other atom with four electron groups around it, the electron groups on the ammonia N will adopt a tetrahedral arrangement in order to minimize their repulsions.

> Count all electrons when determining the shape

> Ignore the lone pairs when naming the shape

So, the shape of the ammonia molecule is tetrahedral, right? Well, not quite. Ammonia is not considered to be a tetrahedral molecule. The above drawing is perfectly correct, but we describe the shape of a molecule differently when lone pairs are present. The convention is to include lone pairs in our initial count of electron groups in determining the overall arrangement of these electron groups, *but when it is time to describe the shape we ignore the lone pairs*. This is done because ultimately we want the name of the shape to indicate the arrangement of the *atoms* in a molecule, not the electrons.

How do we describe the shape of the ammonia molecule? Ignoring the lone pair of electrons, the ammonia molecule is basically a tetrahedron with the top part removed. What's left looks likes a pyramid, so the ammonia molecule is said to have a pyramidal shape.

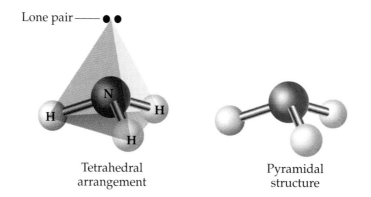

Lone pair

Tetrahedral arrangement

Pyramidal structure

Let's take this one step further. What is the H–N–H angle in ammonia? To answer this question, think about where we started, before we ignored the lone pair. The four electron pairs surrounded the N atom in a tetrahedral arrangement. Therefore, the H–N–H angles in ammonia should be close to the standard 109° tetrahedral angle:

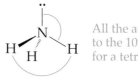

All the angles should be close to the 109° value expected for a tetrahedral geometry

The experimentally determined angles in ammonia are actually a little bit smaller than the ideal 109° tetrahedral angle because the electrons in a lone pair take up more space than electrons in a bonding pair. This tends to squeeze down on the bonding pairs, compressing the angles. A useful rule of thumb is that each lone pair on a second-period atom compresses the remaining bond angles about that atom by approximately 2°. Thus, we predict the H–N–H bond angle in ammonia to be approximately 107°, which is in good agreement with the experimental results.

The actual H–N–H angles are 107°, close to the tetrahedal value

"Squeeze play"

In summary, we follow steps 1–4 to apply VSEPR theory to determine the shape of a molecule:

Step 1: Draw a valid electron dot structure for the molecule.

Step 2: Count the number of electron groups around the central atom. Remember to include lone pairs and to count a multiple bond as one group.

Step 3: Using the count from step 2, decide on the overall electron arrangement around the central atom. This arrangement includes the lone pairs.

Step 4: Finally, ignore the lone pairs and describe the resulting shape.

Let's see if you've got it. We will work partway through the determination of the shapes of the water (H_2O) and ozone (O_3) molecules, and then we'll ask some questions. We begin as usual with electron dot diagrams:

H—Ö—H   Ö=Ö—Ö:

Water   Ozone

Focus on the central atom in each of these molecules, and count the total number of electron groups around it (remember that multiple bonds count only once). The oxygen atom in water is surrounded by four electron groups (two bonding pairs and two lone pairs). The central oxygen atom in ozone has three electron groups around it (a double bond, a single bond, and one lone pair). How will these groups arrange themselves in space? VSEPR theory tells us:

Four electron groups arranged around oxygen tetrahedrally

Three electron groups arranged around oxygen as a planar triangle

WORKPATCH 6.2

Now, how would you describe the shapes of the water and ozone molecules? Follow the convention! Ignore the lone pairs on the central atoms. Cover them up and think of words or phrases that describe their shapes.

What did you call the shapes of these molecules? Certainly they are not linear. They should not be called trigonal planar, because we reserve that term for cases where there are three atoms surrounding a central atom. Here they are again without the lone pairs:

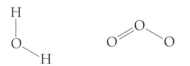

Check the WorkPatch answer at the end of the chapter to see how the shapes of these molecules are most commonly described.

Finally, you should have noticed that the water and ozone molecules have essentially the same shape, but there is an important difference between them. The H–O–H angle in water is quite different from the O–O–O angle in ozone. If you can predict these two angles, then you are ready to go on.

(6.3) WORKPATCH

In the structures preceding WorkPatch 6.2, we indicated the idealized (approximate) bond angles in the water (H_2O) and ozone (O_3) molecules. Recalling that there is roughly a 2° compression from the ideal VSEPR angle for every lone pair on the central atom, predict a more accurate bond angle for each of these molecules.

Applying steps 1–4 for determining the shape of a molecule to all possible cases of four, three, and two electron groups, we obtain Table 6.2.

Table 6.2   VSEPR Shape Table

Total number of electron groups	Structure	Electron group arrangement	Bond angles (idealized)*	Molecular shape
Four				
4 bonding groups no lone pairs		Tetrahedral	109°	Tetrahedral
3 bonding groups 1 lone pair		Tetrahedral	~107° (109°)*	Pyramidal
2 bonding groups 2 lone pairs		Tetrahedral	~105° (109°)*	V-shaped or bent
Three				
3 bonding groups no lone pairs		Trigonal planar	120°	Trigonal planar
2 bonding groups 1 lone pair		Trigonal planar	~118° (120°)*	V-shaped or bent
Two				
2 bonding groups no lone pairs		Linear	180°	Linear

*If central atom is in second period, −2° for each lone pair.

To make sure you understand how Table 6.2 was generated, try the next WorkPatch and the following practice problems.

(6.4) WORKPATCH

We left three possible situations out of Table 6.2:

1. 4 electron groups: 1 bonding group and 3 lone pairs
2. 3 electron groups: 1 bonding group and 2 lone pairs
3. 2 electron groups: 1 bonding group and 1 lone pair

These are considered to be trivial cases. Why is this so? What shape(s) would these bonding situations give rise to?

PRACTICE PROBLEMS

Determine the shapes of the following molecules and polyatomic ions, some of which have lone pairs on the central atom.

6.5 Consider PBr_3. Draw and name its shape, and estimate the bond angles.

Answer: First draw a dot diagram (there will be a total of 26 electrons):

$$:\overset{..}{\underset{..}{Br}} - \overset{..}{P} - \overset{..}{\underset{..}{Br}}:$$
$$|$$
$$:\overset{}{\underset{..}{Br}}:$$

Next, count electron groups about the central atom. There are four (three single bonds and one lone pair), indicating a tetrahedral arrangement of these groups about the P atom. However, ignoring the lone pair on P, the molecule is pyramidal in shape, with a predicted Br–P–Br angle of less than 109° (actual bond angle is 102°).

P
Br Br Br   Pyramidal

6.6 Consider the sulfate ion, SO_4^{2-}. Draw and name its shape, and estimate the bond angles.

6.7 Consider the molecule hydrogen cyanide, HCN (hydrogen and nitrogen are attached to carbon). Draw and name its shape, and estimate the bond angles.

6.8 Consider sulfur dioxide, SO_2, a pollutant that comes from burning coal contaminated with sulfur. Draw and name its shape, and estimate the bond angles.

6.9 Consider nitric acid, HNO_3. The hydrogen is bound to one of the oxygens and all of the oxygens are bound to nitrogen. Draw and name its shape (to simplify you can ignore the hydrogen when you name the shape). Estimate *all* the bond angles.

POLARITY OF MOLECULES, OR WHEN DOES 2 + 2 NOT EQUAL 4? 6.3

We have already seen how the shape of a molecule could be used to trick bacteria. Molecular shape is important in many ways, since the chemical and physical properties of molecules are often dependent on shape. Molecular "stickiness" is one property that is very dependent on molecular shape. Molecules can "stick" to each other so tightly that they cannot easily be pulled apart. Try it! The next time you have a drink with some ice in it, take out an ice cube, hold it in both hands, and try ripping it in half. You can't. The trillions of individual water molecules that make up your ice cube are stuck together as if

someone had used Super Glue. This simple experiment is proof of the existence of a large net attractive force among all the water molecules in the ice cube. Such forces among molecules are called **intermolecular attractive forces**, or *intermolecular forces* for short. If the intermolecular forces among the water molecules could somehow be turned off, the ice cube would come apart and instantly turn into a gas (water vapor). It is because of these intermolecular forces that molecular substances can exist in condensed phases (liquid and solid). Where do these intermolecular forces come from? What factors determine how strong or weak they are? Again the answer can be found by examining both the composition and the shapes of the molecules involved.

To examine how and why molecules "stick" to one another, let's look at the specific examples shown in Table 6.3.

Table 6.3   Some Examples to Consider

Molecule	Formula	Shape	Structure
Hydrogen	H_2	Linear	H—H
Hydrogen fluoride	HF	Linear	H—F
Hydrogen chloride	HCl	Linear	H—Cl
Water	H_2O	Bent	H⌄O⌄H 105°
Dichloroacetylene	C_2Cl_2	Linear	Cl—C≡C—Cl
Tetrachloromethane (carbon tetrachloride)	CCl_4	Tetrahedral	Cl₃C structure 109°
Phosgene	Cl_2CO	Trigonal planar	O=C(Cl)Cl 120°

We'll start with the two simplest molecules, H_2 and HF. The biggest difference between these two diatomic molecules, besides their elemental composition, is in their bonding. The H_2 molecule is held together by a covalent bond (equal sharing of electrons). In HF, the very different electronegativities of hydrogen and fluorine give rise to a polar covalent bond (unequal sharing of electrons). This unequal sharing means that the bonding electrons spend more time around the F atom and less time near the H atom. That produces a molecule with a partial negative charge ($\delta-$) on the more electronegative F atom and a partial positive charge ($\delta+$) on the less electronegative H atom. If you are not sure about this, stop and review electronegativity and polar covalent bonding in Chapter 5.

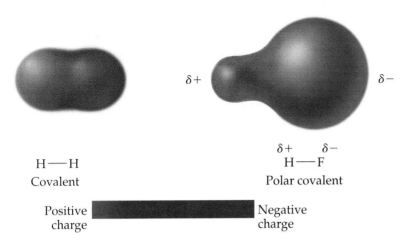

$\delta+$ $\delta-$

H——H

Covalent

$\delta+$ $\delta-$
H——F

Polar covalent

Positive   Negative
charge   charge

The presence of partial charges in HF causes these molecules to behave in some significantly different ways from H_2 molecules. One of these differences can be seen when samples of these molecules are placed between two oppositely charged metal plates. Here is what happens:

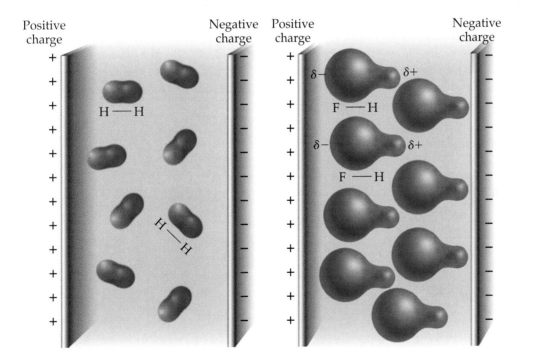

The H_2 molecules are not noticeably affected by the charged plates, and they orient themselves in a random fashion between them. The situation is different for the HF molecules. Because the partially negative fluorines are attracted to the positive plate and the partially positive hydrogens are attracted to the negative plate, the molecules all assume the same orientation. Based on this behavior, HF is called a **polar** (or **dipolar**) molecule and H_2 is called a **nonpolar** molecule. (This is an

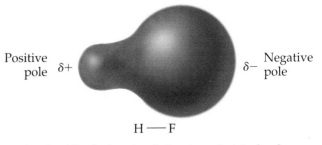

A polar (dipolar) molecule has two electrical poles

operational definition; that is, a definition based on how something behaves in a particular situation.)

Diatomic molecules like N_2 and O_2, the two most abundant molecular substances in our atmosphere, are both nonpolar molecules because of the equal sharing of electrons between the atoms within each molecule. Other diatomics like HF and HCl are polar molecules due to the polar covalent nature of their bonds. Molecules that are polar do not necessarily have the same degree of polarity. Look at the electronegativities of the atoms in HF and HCl:

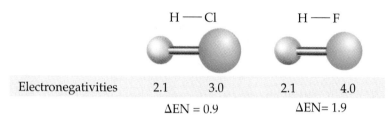

| Electronegativities | 2.1 | 3.0 | 2.1 | 4.0 |

$$\Delta EN = 0.9 \qquad \Delta EN = 1.9$$

Both these molecules are polar (or dipolar), but HF is more polar than HCl. They both exhibit unequal sharing of the bonding electrons, but fluorine is more electronegative than chlorine (recall that fluorine is the most electronegative element that there is). The greater value of ΔEN for HF tells us that fluorine really "hogs" those shared electrons. Thus, the degree of unequal electron sharing in HF is greater than in HCl. This gives rise to the development of larger partial charges in HF than in HCl, and this increased charge separation makes the molecule more polar.

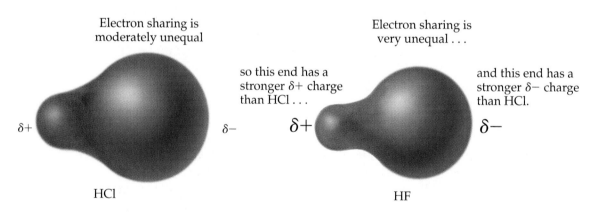

The accepted convention for indicating the polarity of a bond (and even, as we shall see, of the molecule as a whole) is to use an arrow that points from the $\delta+$ to the $\delta-$ charge. The arrowhead points to the negative end, and the arrow is crossed at the positive end. When the polarity of a molecule is measured experimentally, we get its **dipole moment**, which is symbolically represented by the same type of arrow. Sometimes, the length of the arrow is used to indicate the degree of polarity in a bond or molecule. The longer the arrow, the greater the dipole moment, and the more polar the molecule is. Physicists refer to these arrows as *vectors* (a **vector** is a number that has a direction associated with it, usually represented by an arrow).

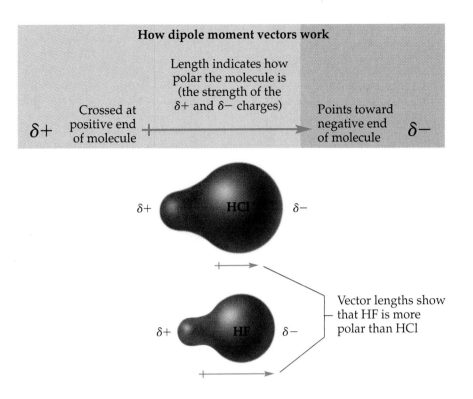

When will a bond have a zero dipole moment associated with it?

WORKPATCH 6.5

Notice that the WorkPatch asked you about the dipole moment (polarity) associated with a bond. However, when we started discussing polarity, we talked about the H_2 and HF *molecules*. For these simple molecules, considering the polarity of the bond is the same as considering the polarity of the molecule because there is only one bond in the molecule. But what about molecules with more than one bond? In these cases, determining whether the entire molecule is polar becomes a bit more involved.

Let's consider two molecules that have more than one polar bond in them, water and dichloroacetylene. Notice a few things about these examples. As always, the bond dipole vectors point from the $\delta+$ to the $\delta-$ end of the bond. Notice that the end of the arrow over the $\delta+$ atom is crossed, like a plus sign. There is no bond dipole vector over the carbon–carbon triple bond in dichloroacetylene because this bond is nonpolar (electrons in this bond are shared equally). Both molecules contain two polar covalent bonds: the O–H bonds in water and the C–Cl bonds in dichloroacetylene. So, are both molecules

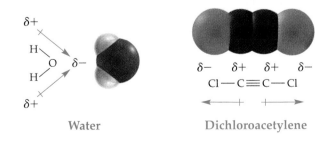

Water   Dichloroacetylene

polar? The only way to truly know is to put both molecules between oppositely charged electrical plates and see what happens. The figure below shows the result of just such an experiment.

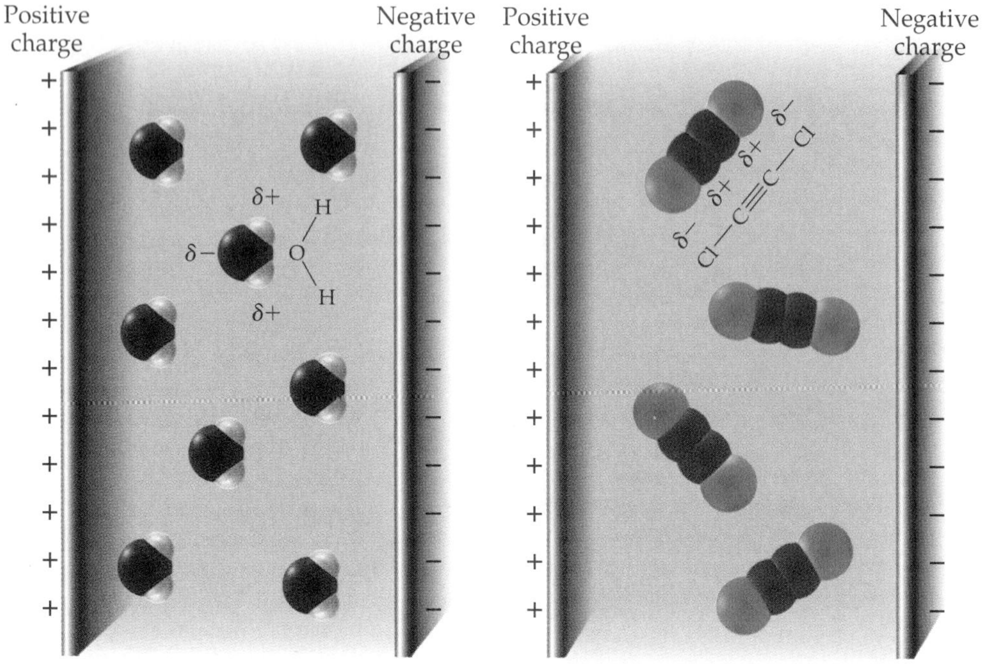

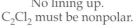

They line up.
H_2O must be polar.

No lining up.
C_2Cl_2 must be nonpolar.

Notice how the water molecules all assume the same orientation. The dichloroacetylene molecules, on the other hand, remain randomly oriented. By our operational definition of polar molecules, water is considered a polar molecule whereas dichloroacetylene is nonpolar. Nonpolar? How can dichloroacetylene be nonpolar when it contains polar covalent bonds with $\delta+$ and $\delta-$ ends? The trick here is to recall that each individual bond dipole moment is a vector, a number with a direction attached to it (the direction of the arrow). When you add the numbers 2 + 2 you always get 4. But when you add two vectors, each of length 2, the answer you get depends on which way the two vectors are pointing relative to each other. In vector arithmetic, 2 + 2 can equal anything from 0 to 4, as shown below.

Vectors pointing in the same direction add

Vectors pointing in opposite directions cancel (subtract)

Vectors at an angle to each other add to give a vector having a length between zero and the sum of the vector lengths (depending on the angle)

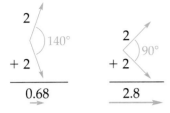

If the vectors have the same value but point in opposite directions, then they will cancel each other out and give zero as a result, as shown on page 198. This is exactly what happened in dichloroacetylene. The individual bond dipole vectors have equal magnitude but point in opposite directions. They completely cancel and give a value of zero for the overall molecular dipole moment. Even though two of the bonds in the dichloroacetylene molecule are polar, the molecule as a whole is nonpolar.

A case where individual bond dipole vectors cancel

In water, the two individual O–H vectors do not cancel each other. Instead, they add to give a vector that points in the direction shown below—between the two original vectors. This is the **net molecular dipole moment (vector)** for the entire molecule. Since this vector is not zero, the molecule has a dipole moment and is considered a polar molecule.

A case where they do not cancel

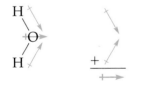

So molecular shape is important once again. For molecules with more than two atoms, just having polar bonds is not enough for the entire molecule to be polar. For these molecules, molecular polarity also depends on molecular shape. Try the following WorkPatch to see if you are catching on to these concepts.

Which molecule is more polar? [Be careful!]

(a) $\ddot{O} = C = \ddot{O}$ (b) $\ddot{S} = C = \ddot{S}$

WORKPATCH 6.6

Now, which of the following molecules is more polar?

(c)   O   (d)   S
     H   H   H   H

The molecules in (a) and (b) of the WorkPatch are a bit of a trick question. Did you catch it? The molecules in (c) and (d) should have you thinking about the size of the vectors (bond dipole moments) over each bond.

In more complicated molecules with more than three atoms, a quick way to test whether all the individual bond dipoles cancel is to do a "tug-of-war" test. We can try this for two molecules, phosgene, Cl_2CO (a highly poisonous gas used in World War I), and tetrachloromethane, CCl_4 (also called carbon tetrachloride, a common solvent):

Slightly stronger pull

Individual bond dipole vectors add up to give . . .

The overall dipole moment for the entire molecule

Phosgene—trigonal planar   Tetrachloromethane—tetrahedral

The geometries and individual bond dipoles are shown above. Both of these molecules have polar bonds, but is either molecule polar? To answer this, imagine that you are at the central atom of each molecule and that the bond dipoles are ropes that are tugging on you. Then ask the question, with ropes tugging in that particular arrangement, "Would you move?" If the answer is yes, then the bond dipoles do not cancel and the molecule is polar. This is the case for phosgene. Standing where carbon is, you would be pulled upward, the result of combining one stronger pull up with the two weaker pulls down, one to the left and one to the right (the angles between your legs or your leg and head will be approximately 120° by VSEPR). You would move, so phosgene is a polar molecule.

6.7 WORKPATCH

Why is the pull upward in phosgene slightly stronger than either of the downward pulls? What would happen if the three pulls were equal in strength?

This WorkPatch is extremely important. Check both of your answers and don't go on until you understand them.

What about tetrachloromethane? It has four polar bonds. Is the whole molecule polar? Now we have four equal tugs pulling toward the corners of a tetrahedron. Would you move? Looking at the tetrahedron the way it is usually drawn, it is not so easy to determine if the pulls cancel. It helps to rotate the molecule into a different view, as shown. In this case, the four pulls would cancel, and you would not move. Tetrachloromethane therefore isn't polar, even though its individual bonds are. Convince yourself that this is so. Build a model if you can't see this.

The "tug-of-war" test can be very useful in determining the polarity of larger molecules. However, the test is useless unless you can first determine the shape of the molecule and estimate the relative strengths of the different tugs. Try the following practice problems, keeping in mind that for nonpolar covalent bonds—those for which the difference in electronegativities (ΔEN) is less than 0.4—the dipole moment is zero. This is true of any C–H bond in a molecule (EN C = 2.5; EN H = 2.1; so ΔEN = 0.4).

The four individual bond moments add up to zero (they cancel)   Different view

6.10 Is trichlorophosphine (PCl_3) a polar molecule? If it is, then draw the dipole moment vector for the entire molecule, and show where the $\delta+$ and $\delta-$ regions for the molecule are.

Answer: Yes, it is a polar molecule. PCl_3 should have 26 electrons.

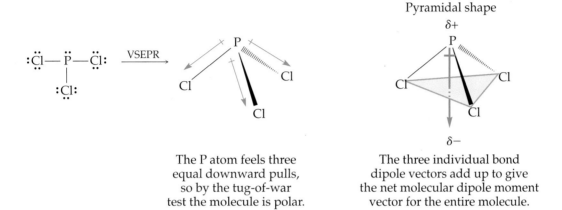

The P atom feels three equal downward pulls, so by the tug-of-war test the molecule is polar.

The three individual bond dipole vectors add up to give the net molecular dipole moment vector for the entire molecule.

6.11 Is chloroform ($CHCl_3$) a polar molecule? If it is, then draw the dipole moment for the entire molecule and show where the $\delta+$ and $\delta-$ regions for the molecule are.

6.12 Show how the molecules of Problems 6.10 and 6.11 would arrange themselves in an electric field.

When we started this section we asked, "What makes molecules 'stick' to one another (why can't you rip an ice cube in half)?" And we promised you that the shapes of molecules play a role in answering this question. This is because the way molecules are attracted to one another depends on whether the molecules are polar, and as we have just seen, molecular polarity depends on molecular shape. We are now prepared to take a closer look at intermolecular forces.

INTERMOLECULAR FORCES—DIPOLAR INTERACTIONS 6.4

How do water molecules "stick" to one another to form liquid water or solid ice? We now know enough about the water molecule to begin answering this question.

First of all we know that the water molecule is V-shaped (bent), and because of this and its polar covalent bonds, the molecule is polar. The dipole moment vector for the molecule shows a partially negative area located at the oxygen atom and a partially positive portion located on the $\delta+$ hydrogens. What happens if we bring a group of water molecules together? Since opposite charges attract, the $\delta+$ portion of one molecule is attracted to the $\delta-$ portion of another, causing them to come close to one another, as shown on the next page.

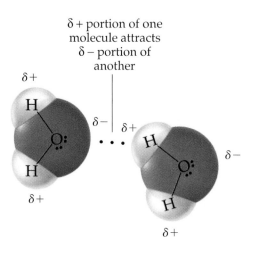

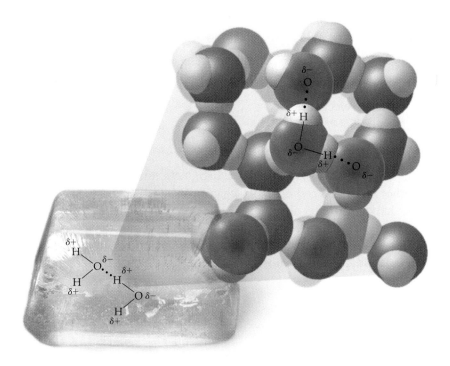

This is the origin of the attractive force between water molecules—the reason these molecules "stick" together so strongly in ice. In a piece of ice, billions of water molecules are attracted to one another by dipolar *intermolecular forces*.

Intermolecular forces between polar molecules are called **polar forces**, or **dipole–dipole forces**, indicating an attraction between the dipoles of different molecules. On a per atom basis, dipole–dipole forces are among the strongest that exist between molecules. Even so, these forces are nowhere near as strong as the ionic and covalent bonds, sometimes referred to as *intramolecular forces* (*intra* meaning within), that we discussed earlier.

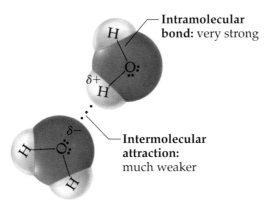

Intramolecular bond: very strong

$\delta+$

Intermolecular attraction: much weaker

$\delta-$

This was just a first look at the much larger topic of intermolecular forces. What we have told you about the water molecule is incomplete, and needs to be further refined. In addition, there are other types of intermolecular forces besides dipole–dipole attractions. For example, the hydrogen molecule, H_2, is nonpolar. It has no partial charges, yet NASA fills fuel tanks on the space shuttle with *liquid* hydrogen. The H_2 molecules must "stick" to each other to form a liquid. But how? We will take a more in-depth look at intermolecular forces in Chapter 9.

For now, you shouldn't lose sight of the importance of molecular shape in all of this. If water were a linear molecule instead of bent, then it would have no dipole moment (the individual bond dipoles would cancel). Water molecules would no longer have a strong polar force of attraction between them, and all H_2O on this planet would exist as vapor. Life, at least as we now know it, could not exist if this were so.

You should get comfortable with being able to go from a molecular formula like H_2O or PH_3 to the shape of the molecule. You can't answer questions about molecular polarity and intermolecular forces unless you know the shape of the molecule. And you can't determine a molecule's shape without a valid dot diagram.

The process for going from formula to polarity

Formula

Determine the total number of valence electrons (use group numbers to help you).
Connect atoms using single bonds.
Put in remaining dots as lone pairs to satisfy octets.
If all octets are not satisfied, send lone pairs to the rescue (multiple bonds).

Dot diagram

Apply the concepts of VSEPR to the interior atoms.

Shape

Draw all bond dipole moments (consider the electronegativities).
Determine the net molecular dipole moment.

Molecular polarity, intermolecular forces

The dot diagram is the key starting point in this process. Start with this, stick to the outline above, and keep practicing.

HAVE YOU LEARNED THIS?

VSEPR theory (p. 185)

Tetrahedron (p. 186)

Intermolecular attractive force
(p. 194)

Dipolar (polar) molecule (p. 195)

Nonpolar molecule (p. 195)

Dipole moment (p. 196)

Vector (p. 196)

Net molecular dipole moment
(vector) (p. 199)

Dipole–dipole forces (p. 202)

SHAPES OF MOLECULES AND VSEPR

6.13 Why is it better for methane to have 109° bond angles rather than 90° bond angles?

6.14 What do you think a bacterium might have to do to evolve and become resistant to a drug like sulfanilamide that is designed to kill it?

6.15 Why is the theory that governs the shape of molecules called VSEPR and not just EPR?

6.16 Draw the methane molecule true to its tetrahedral shape using wedges and dotted lines to show three-dimensionality.

6.17 Draw a methane molecule inscribed inside a tetrahedron such that the H atoms touch the vertices of the tetrahedron. Why is this shape called a tetrahedron?

6.18 Shown below are correct electron dot diagrams for some simple molecules.

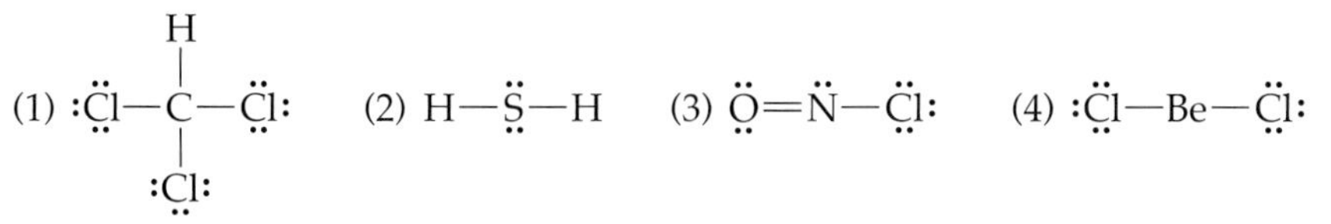

(1) $:\ddot{C}l-C-\ddot{C}l:$ with H on top and $:\ddot{C}l:$ on bottom   (2) $H-\ddot{S}-H$ (3) $\ddot{O}=\ddot{N}-\ddot{C}l:$ (4) $:\ddot{C}l-Be-\ddot{C}l:$

(a) Draw each molecule according to its true shape. Use wedges and dotted lines as necessary. Indicate all bond angles.
(b) Give the name that describes the shape of each molecule.

6.19 Shown below are correct electron dot diagrams for some simple molecules and polyatomic ions.

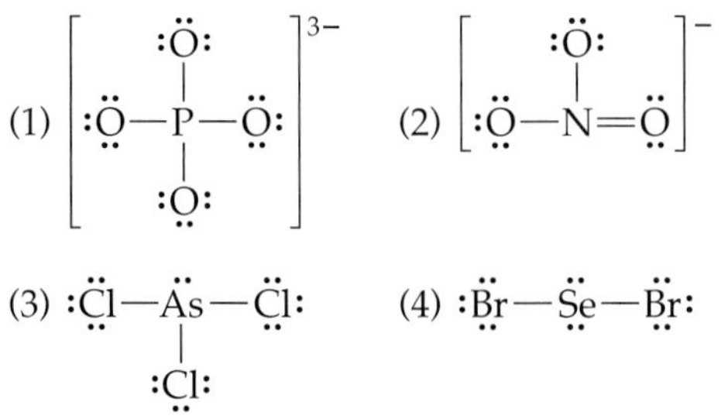

(1) $\left[:\ddot{O}-P-\ddot{O}:\right]^{3-}$ with $:\ddot{O}:$ on top and $:\ddot{O}:$ on bottom   (2) $\left[:\ddot{O}-N=\ddot{O}\right]^{-}$ with $:\ddot{O}:$ on top

(3) $:\ddot{C}l-\ddot{A}s-\ddot{C}l:$ with $:\ddot{C}l:$ on bottom   (4) $:\ddot{B}r-\ddot{S}e-\ddot{B}r:$

(a) Draw each molecule according to its true shape. Use wedges and dotted lines as necessary. Indicate all bond angles.
(b) Give the name that describes the shape of each molecule.

6.20 What gives us the right to treat multiple bonds as if they were one electron pair in VSEPR shape determination?

6.21 Two molecules may be correctly described as bent or V-shaped and yet one can have a bond angle of 120° while the other has a bond angle of 109°. How is this possible?

6.22 Why are the bond angles in ammonia smaller than the expected tetrahedral value?

6.23 Ammonia has four pairs of electrons around the central nitrogen atom, and yet we don't call it a tetrahedral molecule. Why not?

6.24 Consider the following molecule. The diagram shows how the atoms are connected, but it is not a complete dot diagram.

$$
\begin{array}{ccc}
H & H & O \\
| & | & | \\
C = C - C \\
| & & | \\
H & & H
\end{array}
$$

(a) Complete the dot diagram.
(b) Redraw the molecule showing its true shape. Use wedges and dotted bonds if necessary.
(c) Indicate all bond angles.
(d) Why can't we describe the shape of this molecule in one word, as we can for molecules like H_2O?

6.25 Consider the following molecule. The diagram shows how the atoms are connected, but it is not a complete dot diagram.

$$
\begin{array}{ccc}
& & H \\
& & | \\
H & & O \\
| & & | \\
C = C - C - H \\
| & | & | \\
H & H & H
\end{array}
$$

(a) Complete the dot diagram.
(b) Redraw the molecule showing its true shape. Use wedges and dotted bonds if necessary.
(c) Indicate all bond angles.

6.26 Why is it necessary to first obtain a correct electron dot diagram before trying to predict the shape of a molecule?

6.27 A hexagon has angles of 120° inside the ring and is flat. The following two molecules are often drawn as flat hexagons, but only one is truly flat. Which one is it and why?

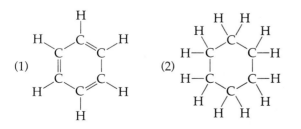

6.28 *Proteins* are long-chain molecules that make up our skin, hair, muscles, and enzymes, among other things. Shown below is a dot diagram of the part of a protein known as a *peptide bond*. Protein chains often adopt a coiled structure that is partly responsible for their biological function. In the diagram below indicate what the actual geometry is around each of the three carbon atoms and the nitrogen atom in the peptide bond. Using wedges and dotted bonds, redraw this peptide bond as it really appears.

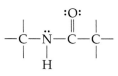

POLARITY AND INTERMOLECULAR FORCES

6.29 In terms of an operational definition, when is a molecule considered to be polar?

6.30 Is it possible to predict whether a molecule will be polar? Explain completely what you would need to know about the molecule.

6.31 True or false? All molecules that contain polar bonds must be polar. Explain your answer.

6.32 Consider the molecules HCl and HBr.
(a) Which molecule has the larger bond dipole moment? Explain why.
(b) Which molecule is more polar? Explain why.

6.33 What does the magnitude of a bond dipole moment (the length of the arrow) tell you about the bond?

6.34 Why do we use an arrow to represent a bond's dipole moment? Why not just use a number?

6.35 What does electronegativity have to do with a bond's dipole moment?

6.36 How is a bond dipole moment drawn?

6.37 The molecule CO_2 has very polar bonds, and yet the molecule itself is nonpolar. Explain how this is possible.

6.38 The polyatomic ion NO_2^- has polar bonds and is polar. Why the difference from CO_2 (Problem 6.37)?

6.39 Consider all the hydrogen halide molecules HX, where X is a group VIIA atom.
(a) Which is the most polar? Why?
(b) Which is the least polar? Why?
(c) Draw all these molecules, showing their relative bond dipole moments.

6.40 Consider the two molecules CO and CO_2. They are both made of the same elements and yet only one is polar. Why is this so?

6.41 Consider the following molecules. For those that are polar, draw the net molecular dipole moment.
(a) $CHCl_3$ (b) CH_3Cl (c) H_2S (d) ONCl   (e) $BeCl_2$

6.42 What is a dipole–dipole interaction? Give an example.

6.43 Are dipole–dipole interactions between molecules as strong as the interactions between oppositely charged ions? Explain.

6.44 What do we mean by intermolecular forces? What evidence is there that they exist?

6.45 Draw two HCl molecules and show how they would be attracted to one another. Show the partial charges and dipole moment vectors for both molecules, and orient the molecules properly with respect to each other.

6.46 Draw two ammonia molecules true to shape and show how they would be attracted to one another. Show the partial charges and dipole moment vectors for both molecules, and orient the molecules properly with respect to each other.

6.47 Why is it important to always show the lone pairs in an electron dot diagram?

6.48 What would be the consequences of coming up with an incorrect electron dot diagram for a molecule?

6.49 Consider the molecule $SiCl_4$.
(a) Draw a correct electron dot diagram.
(b) Draw the molecule true to shape, and label all bond angles.
(c) Using one or two words, describe the shape of this molecule.
(d) Draw in the individual bond dipole moments.
(e) Is the molecule polar? If yes, draw the net molecular dipole moment vector.

6.50 Consider the molecule AsF_3.
(a) Draw a correct electron dot diagram.
(b) Draw the molecule true to shape, and label all bond angles.
(c) Using one or two words, describe the shape of this molecule.
(d) Draw in the individual bond dipole moments.
(e) Is the molecule polar? If yes, draw the net molecular dipole moment vector.

6.51 Consider the molecule SO_3.
(a) Draw a correct electron dot diagram.
(b) Draw the molecule true to shape, and label all bond angles.
(c) Using one or two words, describe the shape of this molecule.
(d) Draw in the individual bond dipole moments.
(e) Is the molecule polar? If yes, draw the net molecular dipole moment vector.

6.52 The atoms in the molecule HSCN are connected to each other in the order given in the formula.
(a) Draw a correct electron dot diagram.
(b) Draw the molecule true to shape, and label all bond angles.
(c) Draw in the individual bond dipole moments.
(d) Is the molecule polar? If yes, draw the net molecular dipole moment vector.

6.53 Consider the molecule hydrazine, N_2H_4, used in rocket fuel. The atoms are connected as follows: each nitrogen has two hydrogens on it, and the nitrogens are bound to each other.
(a) Draw a correct electron dot diagram.
(b) Draw the molecule true to shape, and label all bond angles.
(c) Using one or two words, describe the shape of this molecule.
(d) Draw in the individual bond dipole moments.
(e) Is the molecule polar? If yes, draw the net molecular dipole moment vector.

6.54 Consider the molecule HNF_2 (N is the central atom in the molecule).
(a) Draw a correct electron dot diagram.
(b) Draw the molecule true to shape, and label all bond angles.
(c) Using one or two words, describe the shape of this molecule.
(d) Draw in the individual bond dipole moments.
(e) Is the molecule polar? If yes, draw the net molecular dipole moment vector.

6.55 Consider the molecule N_2O (connected in the following order: N–N–O).
(a) Draw a correct electron dot diagram.
(b) Draw the molecule true to shape, and label all bond angles.
(c) Using one or two words, describe the shape of this molecule.
(d) Draw in the individual bond dipole moments.
(e) Is the molecule polar? If yes, draw the net molecular dipole moment vector.

6.56　Some molecules pose special challenges to the rules for obtaining a correct electron dot diagram. Consider the molecule NO_2 (N is the central atom).
(a) What challenge does it present?
(b) Suppose you were allowed to violate the octet rule for the N atom. What would the electron dot diagram look like?
(c) Based on your answer to part (b), what is the shape of such a molecule? Would it be polar?

6.57　Consider the polyatomic ion ammonium, NH_4^+.
(a) Draw a correct electron dot diagram.
(b) Draw the polyatomic ion true to shape, and label all bond angles.
(c) Using one or two words, describe the shape of this polyatomic ion.
(d) Draw in the individual bond dipole moments.

6.58　Imagine that you had a molecule in which you could vary the magnitude of the dipole moment. What should happen to its boiling point as the dipole moment is increased? Explain.

6.59　PH_3 is a nonpolar molecule. Why is this so? [*Hint:* Consider electronegativities.]

WorkPatch Solutions

6.1　(a)

Formaldehyde　　　　Acetylene

6.2　Both molecules would be called bent or V-shaped.

6.3　Expect a roughly 4° compression from 109° due to two lone pairs on oxygen:

Expect a roughly 2° compression from 120° due to one lone pair on oxygen:

6.4　These are all trivial because they all represent one central atom with one other atom attached (only one bonding group in each case). When a molecule consists simply of two atoms attached to one another it can only be linear. No other shape is possible.

6.5　A bond is guaranteed to have a zero dipole moment when the electron sharing in the covalent bond between two atoms is absolutely equal. This occurs when both atoms have identical electronegativities (that is, when both atoms are the same element).

6.6　CO_2 has individual bonds that are more polar due to the greater electronegativity difference between C and O than between C and S. However, since both molecules are linear, these bond dipole moments cancel. Therefore, neither molecule is polar.

(a) $\overset{..}{\underset{..}{O}}{=}C{=}\overset{..}{\underset{..}{O}}$　　　　(b) $\overset{..}{\underset{..}{S}}{=}C{=}\overset{..}{\underset{..}{S}}$

Since H_2O has the more polar bonds, H_2O will be the more polar molecule.

(c)

6.7   The pull upward in the $COCl_2$ molecule is slightly stronger because the C–O bond is more polar than the C–Cl bond. This is because the electronegativity difference between C and O is greater than the electronegativity difference between C and Cl. If the three pulls were equal, and assuming the bond angles were 120° (trigonal planar), then the pulls would cancel and the molecule would be nonpolar.

Chapter 7

Chemical Reactions

What Is a Chemical Reaction?   7.1

What exactly is a chemical reaction? We touched briefly on this subject in Section 1.4, but now we want to answer this question in more detail. In the mid-1980s, chemists combined just the right amounts of two different metal oxides [yttrium oxide, Y_2O_3, and copper(II) oxide, CuO] with a metal carbonate (barium carbonate, $BaCO_3$), and heated them in a stream of oxygen (O_2). The result was a new ceramic compound, $YBa_2Cu_3O_7$, that possessed extraordinary properties. For example, when this compound is cooled to 90 K ($-183°C$, or $-298°F$), a magnet placed above it will float as if defying gravity!

This new ceramic compound is called a *superconductor*. A **superconductor** is a material that conducts electricity with zero resistance. In addition, magnetic field lines cannot penetrate a superconductor as they do ordinary materials. The result is that a magnet will float above a superconductor. Magnetically levitated (maglev) trains are just one of the uses envisioned for this phenomenon. Just put some strong magnets on the bottom of a train and replace the steel railroad tracks with tracks made from this new compound. Then, cool the tracks to 90 K with liquefied nitrogen, a relatively cheap industrial commodity, and

$$BaCO_3 + CuO + Y_2O_3$$
$$\text{Heat} \downarrow O_2$$
$$YBa_2Cu_3O_7$$

An experimental Japanese maglev train floats above superconducting tracks

the train will float virtually frictionless above the tracks. This is the near future, brought to you by a chemical reaction that turned a few piles of otherwise not very useful metal oxides into a brand new compound.

It would appear from our example that a chemical transformation has taken place—that is, a chemical reaction produces something new, just as we said in Section 1.4. In a **chemical reaction**, one or more substances are converted (or changed) into new substances that have different compositions and properties from the starting substances. We call the original starting substances **reactants** and the final new substances **products**. We write reactions with the reactants on the left side of the arrow and the products on the right side:

$$\text{Reactants} \longrightarrow \text{Products}$$

Remember, the arrow itself means "reacts to form." That is, reactants *react* with each other *to form* products. It is important that you understand what constitutes a chemical reaction and what doesn't. Try the WorkPatch to see whether you do.

7.1 WORKPATCH   Of the three possibilities below, which are chemical reactions?

(a) $H_2O(l) \longrightarrow H_2O(g)$
[Remember, (l) means "liquid" and (g) means "gas."]
(b) $CO_2 + H_2O \longrightarrow H_2CO_3$
(c) $CO_2 + O_2 \longrightarrow O_2 + CO_2$

Did you get the correct answer? There has been a change in part (a) of the WorkPatch, but a simple phase change from liquid to gas is not a chemical reaction. To be a chemical reaction, at least one new substance must be formed as a product, different from any of the reactants. But how does this happen? How do reactants react to form products? It is this *how* that we are now prepared to consider.

7.2 How Are Reactants Transformed into Products?

Let's start with a much simpler reaction than the one that makes a superconductor. Imagine that you have a flask filled with two different diatomic molecules, hydrogen (H_2) and iodine (I_2). Hydrogen is a colorless gas. Iodine is a dark solid at room temperature, but if we heat the flask, we can drive the iodine into the gas phase and produce a purple vapor consisting of iodine molecules. The hydrogen and iodine molecules in the gas phase react with each other to give a colorless gas made up of hydrogen iodide (HI) molecules.

This product has completely different chemical and physical properties from those of the reactants, as shown in the figure below:

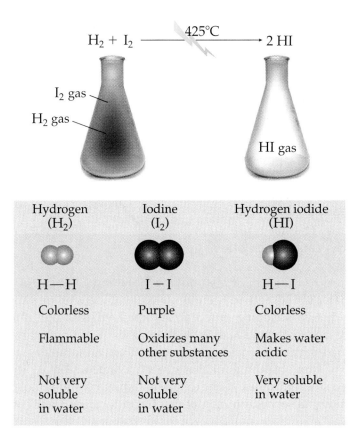

Since both of our reactants are gases, we say that the reaction between them is a **gas-phase reaction**. How does a mixture of H_2 and I_2 molecules turn itself into HI molecules? To get some insight, let's change our point of view. Imagine you are so small that you can see individual atoms and molecules. At the very beginning of the reaction, you can clearly see the H_2 and I_2 molecules, each with a strong single covalent bond between the atoms. An instant later the reaction is over. In the blink of an eye, all the H–H and I–I covalent bonds have broken and new H–I covalent bonds have formed.

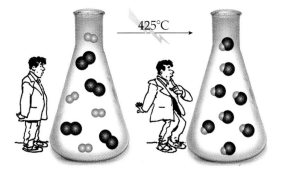

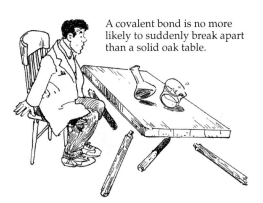

A covalent bond is no more likely to suddenly break apart than a solid oak table.

At first you might be puzzled. Why did the H–H and I–I bonds break? It couldn't be that they were weak and ready to break anyway, because, as we've seen, covalent bonds are among the strongest chemical forces. And strong bonds certainly would not break for no reason at all. That would be like the strong legs on a solid oak table suddenly breaking off for no reason. (When was the last time you saw this happen?) You should be just as surprised to see all those H–H and I–I covalent bonds break, so why did they? Once again, let's change our point of view. This time, put yourself *on* an H_2 or I_2 molecule in the gas-filled flask. You'd better hold on tight!

At 425°C, these molecules are moving at high speeds

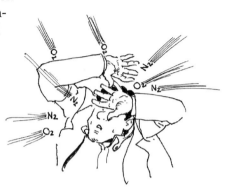

Molecules like H_2 and I_2 are not just sitting still. At room temperature you would be zipping about at hundreds of miles per hour!

This is true for all small molecules in the gas phase. For example, at room temperature, the oxygen (O_2) and nitrogen (N_2) molecules in the air you are breathing are traveling along at speeds close to 500 miles per hour. They travel even faster if the temperature is increased. So, why don't you feel like you are in a hurricane at this very moment? The answer is that while the molecules in the air are moving rapidly, they are moving in different, random directions. They are crashing into you at high speeds, each pushing you a little bit, but in different directions. The net result is that all the pushes cancel each other, and the air feels still.

Now, let's return to the flask of fast-moving H–H and I–I reactant molecules. Sooner or later, an H_2 and an I_2 molecule will collide with one another. As with two speeding automobiles involved in a collision, the energy associated with their motion (their kinetic energy) gets absorbed by the colliding molecules. We are all familiar with the result of cars absorbing all that energy. Windows break, tires fly off, and steel crumples. In the case of colliding H_2 and I_2 molecules, the energy absorbed on impact can begin to break their covalent bonds. The fundamental concept in this scenario is that:

H—H   I—I

Collision course!

It always takes energy to break a chemical bond, covalent or ionic.

There are no exceptions to this rule—ever. The opposite is also true. Energy is always released when a new bond forms.

So, for our gas-phase $H_2 + I_2$ reaction, the energy needed for bond breaking in the reactant molecules comes primarily from collisions between the fast-moving molecules themselves. Immediately after these collisions, however, the scenario differs radically from a crash between automobiles. When automobiles crash and break apart, they stay that way. (At least, no one has ever observed a car spontaneously reassemble itself after impact.) In the case of the collision between H_2 and I_2 molecules, new H–I covalent bonds can form. A new substance is formed. A chemical reaction has occurred.

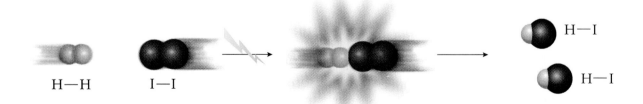

H—H   I—I

This description of a chemical reaction is a simplified and idealized one. For example, it is possible to reform the original bonds and restore the H_2 and

I_2 reactant molecules. In Chapter 12 we will deal with this possibility. For now, we will focus on cases in which collisions between reactant molecules result in the formation of product molecules.

Imagine a collision between an oxygen (O_2) molecule and a hydrogen (H_2) molecule to give the product hydrogen peroxide (H_2O_2). Draw dot diagrams for all the molecules and then indicate which bonds must be broken and which bonds must form. [*Hint:* The atoms in hydrogen peroxide are connected H–O–O–H.]

WORKPATCH 7·2

Did you notice in the above WorkPatch that two reactant molecules collide to give just one molecule of product? This was not the case for the $H_2 + I_2$ reaction. This observation sets us up for what comes next. So far we have been talking about what a chemical reaction is in terms of bond breaking and bond making, and what causes the bonds to break. We will spend the rest of this chapter dealing with another critical aspect of chemical reactions: How much reactant forms how much product?

BALANCING CHEMICAL EQUATIONS 7.3

If we combine a certain amount of H_2 with a certain amount of I_2, how much HI will we make? It is easy to answer this question on a molecular level because we can "see" it in the illustration on page 214. If we take one H_2 molecule and bang it into one I_2 molecule with sufficient force to break bonds, we can make two HI molecules. We can write this using a *balanced equation*:

$$H_2 + I_2 \longrightarrow 2\,HI$$

This is shorthand for "one molecule of H_2 and one molecule of I_2 react to give two molecules of HI." A **balanced equation** is one in which the law of conservation of matter is obeyed. If we start with an unbalanced equation, we have to add numerical coefficients in front of the reactants and/or products to balance it. (If no numerical coefficient is written in front of a reactant or product, the coefficient is assumed to be 1.) An equation is balanced when the numbers of atoms of each element are equal on both sides of the arrow. Check this:

$$
\begin{array}{ccccc}
H_2 & + & I_2 & \longrightarrow & 2\,HI \\
H—H & & I—I & & H—I \\
& & & & H—I \\
\underbrace{\hspace{2em}}_{\text{2 H atoms}} & & \underbrace{\hspace{2em}}_{\text{2 I atoms}} & & \underbrace{\hspace{3em}}_{\substack{\text{Total of 2 H atoms} \\ \text{and 2 I atoms}}}
\end{array}
$$

On each side of the arrow there are 2 H atoms and 2 I atoms. This means we start with 2 H atoms and 2 I atoms, and we end up with 2 H atoms and 2 I atoms. Nothing is lost or gained. Following Dalton's law of conservation of matter, a reaction must be balanced.

Here is another version of the HI reaction. Is it balanced?

$$2\,H_2 + 2\,I_2 \longrightarrow 4\,HI$$

It is certainly not the same as the equation above, but it is still balanced. There are now four of each type of atom on both sides of the arrow. There is nothing really wrong with writing the balanced equation this way, but it is traditional to use the smallest possible whole numbers.

What if you come across an equation that is unbalanced, like this one?

$$H_2 + I_2 \longrightarrow HI$$

You can add or adjust the numerical coefficients (also called balancing coefficients) in order to balance the equation. In this case, you simply add the balancing coefficient 2 in front of the HI:

$$H_2 + I_2 \longrightarrow 2\,HI$$

To balance an equation you can always add coefficients. What you must never do is adjust the subscripts within a compound's formula. For example, you would never balance the HI equation this way:

$$H_2 + I_2 \longrightarrow \cancel{H_2I_2}$$

We changed the subscripts from HI to H_2I_2, so the reaction is now balanced (it obeys Dalton's law of conservation of matter). But what we did is absolutely wrong, because the reaction of H_2 and I_2 makes HI, not H_2I_2, which would be a totally different compound with a whole different set of chemical and physical properties. When balancing a chemical equation, *never* change the subscripts within a formula. If you do, then you will have changed the identity of the substance and the nature of the chemical reaction.

Many of the equations that we deal with in this book can be balanced by the method of inspection, which is just a formal way of saying a trial-and-error approach. The best way to proceed is to consider one element at a time, adding or adjusting the coefficients in front of every formula that contains that element so that there are equal numbers of atoms of that element on both sides of the arrow. Then move on to another element in the equation and repeat the procedure.

To illustrate this method, let's balance the following equation for the production of ammonia from nitrogen and hydrogen:

Unbalanced:

$$N_2 + H_2 \longrightarrow NH_3 \quad \text{Ammonia}$$

In scanning this equation from left to right, the first element we come to is nitrogen (N). There are 2 nitrogen atoms on the left, so we must have 2 nitrogen atoms on the right. We can achieve this by putting a coefficient of 2 in front of the NH_3:

$$N_2 + H_2 \longrightarrow 2\,NH_3$$

Now the nitrogen atoms are balanced. The next element to balance is hydrogen. There are 2 H atoms on the reactant (left) side of the equation. How many hydrogen atoms are there on the right side? If you said 3, you need to look again. There are 3 H atoms in each NH_3 molecule, but there are now 2 ammonia molecules on the right side, so there are a total of 6 H atoms on the product (right) side. We need to fix this by increasing the total number of hydrogen atoms on the left side to 6. To do this, we simply place a 3 in front of the H_2:

Balanced:

$$N_2 + 3\,H_2 \longrightarrow 2\,NH_3$$

Now there are 6 hydrogen atoms on both sides. Both the nitrogen and hydrogen are balanced, so the equation is now balanced.

Try balancing the following unbalanced chemical equations. Here's a helpful hint: If the equation has a pure elemental substance in it, such as H_2, N_2, O_2, Fe, etc., save this element for last when balancing by inspection. That will sometimes make the job of balancing a bit easier.

──**PRACTICE PROBLEMS**

Balance the following chemical equations.

7.1 $CH_4 + O_2 \longrightarrow CO_2 + H_2O$

Answer: Oxygen appears as a pure elemental substance, so we will save it for last. Starting with carbon (an arbitrary choice), we note that there is 1 carbon on both sides of the equation, so C is already balanced. Moving on to hydrogen, there are 4 hydrogen atoms on the left and only 2 on the right. To fix this, we put a balancing coefficient of 2 in front of the H_2O:

$$CH_4 + O_2 \longrightarrow CO_2 + 2\,H_2O$$

This balances the hydrogen. Now it's time to balance the oxygen. There are 2 oxygen atoms on the left (O_2) and a total of 4 on the right (2 from CO_2 and 2 from 2 H_2O's). To balance the equation, just put a 2 in front of the O_2:

$$CH_4 + 2\,O_2 \longrightarrow CO_2 + 2\,H_2O$$

7.2 $Fe + O_2 \longrightarrow Fe_2O_3$

7.3 $C_6H_{12}O_6 + O_2 \longrightarrow CO_2 + H_2O$

7.4 $HCl + Na_2CO_3 \longrightarrow NaCl + CO_2 + H_2O$

Once a chemical equation is balanced, it is a valuable tool for figuring out all sorts of "how much?" questions. How much of each reactant must I mix together to make a certain amount of product? How much product will I make if I mix certain amounts of reactants together? Will any of the reactants be left over (unused), and if so, which ones and how much? We should point out that a number of chemical equations are extremely difficult, if not impossible, to balance by the trial-and-error method of inspection. A variety of methods have been developed to balance such reactions, but these methods are beyond the scope of this book. As we come across these reactions in later chapters we will present their equations in balanced form.

7.4 STOICHIOMETRY: WHAT IS IT?

Let's go back to our reaction of hydrogen with iodine to make hydrogen iodide:

$$H_2 + I_2 \longrightarrow 2\,HI$$

Suppose you wanted to go into the laboratory and actually make 10 grams of HI. How much H_2 should you use? How much I_2? How could you do the reaction in a "balanced" way, using just the right amounts of H_2 and I_2 so that when the reaction was over, both of the reactants were used up and you were left with only 10 grams of HI? These "how much" questions are the subject of **stoichiometry** (stoy-key-om'-e-tree), the study of the quantitative aspects of a chemical reaction. The word is derived from two Greek roots, *stoicheon* (for element) and *metron* (to weigh or measure). We use stoichiometry to calculate things like how much reactant we need to make a certain amount of product, or how much product we can expect to make given a certain amount of reactant. We are going to show you how to do these stoichiometry calculations. The key to doing them will be the balanced equation for the reaction. However, before we look at chemical stoichiometry calculations, we want to convince you that this is something you are already familiar with.

Every time you follow a recipe, you are using stoichiometry. For example, the recipe shown requires 5 eggs to make one cheesecake. If you wanted to bake two cheesecakes, you would simply use the recipe for baking one cake and double it (that is, you would use 10 eggs). You'd do this without much thinking at all, but in using 10 eggs, you'd be doing a stoichiometry calculation in your head. Now try doing these other stoichiometry calculations in your head.

Cheesecake
—————————
3 blocks cream cheese
1 cup sugar
5 eggs

PRACTICE PROBLEMS

Do the following stoichiometry problems based on the cheesecake recipe.

7.5 You have only 3 cups of sugar. How many cakes can you make?

Answer: Since 1 cup of sugar makes one cake, then 3 cups can make three cakes.

7.6 If you use 3 cups of sugar, how many eggs will you need?

7.7 Given 100 cups of sugar, 100 blocks of cream cheese, and 10 eggs, how many cheesecakes can we make?

The last question was a bit more complicated than the others because you first had to realize that one of the ingredients—eggs—was in short supply. However, once you realize this, it is easy to figure in your head that the answer is two cakes. Chemists would call the eggs the *limiting reactant* in Practice Problem 7.7, while the sugar and cream cheese are the *excess reactants*. A **limiting reactant** is one that is present in short supply relative to the other reactants, and is therefore the one that ultimately determines how much product you can make. There will not always be a limiting reactant present. For example, combining 6 blocks of cream cheese, 2 cups of sugar, and 10 eggs is just enough to make two cheesecakes. Nothing is present in excess and there is no limiting reactant. We will take a closer look at limiting reactant problems later in the chapter.

Now we want to get back to doing stoichiometry calculations in your head. Suppose you had to teach someone how to do these stoichiometry calculations on paper. How might you do it? As a chemist you would first rewrite the cheesecake recipe so that it looked more like a chemical equation:

$$3 \text{ blocks cream cheese} + 5 \text{ eggs} + 1 \text{ cup sugar} \longrightarrow \text{Cheesecake}$$

It may look less appetizing this way, but it still says the same thing. Now, on paper, how do we solve the following problem: "Given that two cakes were made, how many eggs were required to bake them?" We have seen this type of problem before in Chapter 2. This is just a unit conversion problem, much like converting feet to meters. In this case, we want to convert from the given units of cakes to the units of the answer, eggs. We begin just as we did in Chapter 2, by writing down the number of cakes that are given, and then multiplying by the appropriate conversion factor to convert to the desired units of eggs:

$$2 \text{ cakes} \times \underbrace{\frac{5 \text{ eggs}}{1 \text{ cake}}}_{\substack{\text{Conversion} \\ \text{factor}}} = 10 \text{ eggs}$$

$\underbrace{\phantom{2 \text{ cakes}}}_{\text{Number}}$

The conversion factor 5 eggs/1 cake came directly from the balanced cheesecake equation (the recipe). Of course, we could have written the conversion factor upside down, as 1 cake/5 eggs, but we want the units of cakes to cancel. Remember that identical units cancel only when they are on top and bottom. What is interesting is that the conversion factor came from the balancing coefficients of the cheesecake equation. Let's go back and revisit the stoichiometry practice problems. This time, don't do them in your head.

PRACTICE PROBLEMS

Now rework Practice Problems 7.5–7.7 by the method of unit conversion. Write out your solution in full as is done above, making sure to show your conversion factor and which units cancel.

7.5 You have plenty of cream cheese and eggs but you have only 3 cups of sugar. How many cakes can you make?

Answer: Sugar is the limiting reactant so we use it to answer the question:

$$3 \text{ cups sugar} \times \underbrace{\frac{1 \text{ cake}}{1 \text{ cup sugar}}}_{\substack{\text{Conversion} \\ \text{factor}}} = 3 \text{ cakes}$$

$\underbrace{\phantom{3 \text{ cups sugar}}}_{\text{Number}}$

7.6 If you use 3 cups of sugar, how many eggs will you need?

7.7 Given 100 cups of sugar, 100 blocks of cream cheese, and 10 eggs, how many cheesecakes can we make?

The balanced cheesecake equation is actually a source of conversion factors for doing stoichiometry problems. All you have to do is write down the number that the problem gives you, multiply it by the appropriate conversion factor from the balanced equation, and you have the answer. That's basically all there is to solving stoichiometry problems.

Before we move on and apply these techniques to an actual chemical reaction, let's try one more cheesecake stoichiometry problem: Suppose you are given 24 ounces of cream cheese. How many cakes can you make?

$$3 \text{ blocks cream cheese} + 5 \text{ eggs} + 1 \text{ cup sugar} \longrightarrow \text{Cheesecake}$$

This last question was a curveball. While the balanced cheesecake equation is a source of many conversion factors, it does not tell us how many ounces there are to a block. We need a conversion factor that tells us how many ounces are in a block. If you know that there are 8 ounces of cream cheese per block, you can do the problem. With the conversion factor

$$\frac{1 \text{ block cream cheese}}{8 \text{ ounces cream cheese}}$$

we can use unit conversion to translate into the language of the recipe:

$$24 \text{ ounces cream cheese} \times \frac{1 \text{ block cream cheese}}{8 \text{ ounces cream cheese}} = 3 \text{ blocks cream cheese}$$

Now we can finish the problem. We take the translated amount of cream cheese and multiply it by the appropriate conversion factor from the balanced equation:

$$3 \text{ blocks cream cheese} \times \frac{1 \text{ cake}}{3 \text{ blocks cream cheese}} = 1 \text{ cake}$$

We have our answer. You can make one cheesecake with 24 ounces of cream cheese.

This last stoichiometry problem was a little more complicated because we were given the amount of cream cheese in units that were different from the balanced equation. This will always happen when we stop baking cheesecakes and start mixing chemicals. That's because balanced chemical equations *always* speak in terms of numbers of atoms or molecules, but we will usually be dealing with units of grams. Therefore, before we can use a balanced chemical equation as a recipe and source of conversion factors for solving stoichiometry problems, we have to learn how to translate back and forth between *grams* of atoms and molecules to *numbers* of atoms and molecules. This is where we get some help from the *mole*.

The mole is used to convert from the language of numbers of molecules to grams of molecules. No, not this mole.

Suppose you need to prepare 10 grams of HI. The "recipe" below doesn't directly tell you how.

$$H_2 + I_2 \longrightarrow 2\,HI$$

It does tell you how to prepare 10 *molecules* of HI. According to the balanced reaction, you would need 5 H_2 molecules and 5 I_2 molecules to prepare 10 HI molecules. But the question was, how do you prepare 10 *grams* of HI? By itself, this recipe isn't much help, and a recipe that requires you to count out individual molecules would be pretty worthless in any case. Not only can't you see individual H_2 and I_2 molecules, the number that you would need to count to prepare 10 grams of HI would be astronomical.

one billion and one, one billion and two . . .

How can we use a balanced chemical equation that "speaks" in numbers of molecules but works in terms of grams? This is where we get assistance from the mole. In concept, a mole is similar to a dozen, which always stands for 12 of anything. A mole is just a whole lot larger. A **mole** is 6.02×10^{23} of anything. This is also called **Avogadro's number**, in honor of Amadeo Avogadro (1776–1856), who first explained how equal volumes of different gases have the same number of molecules. Just as you can have a dozen of something (donuts, dollars, molecules), you can have a mole of something (or Avogadro's number of something). A mole of anything (donuts, dollars, molecules) is always equal to 6.02×10^{23} of them. Whereas a dozen is a conveniently sized number for things like doughnuts, a mole is a conveniently sized number when dealing with atoms and molecules. Because atoms and molecules are so tiny, you need a tremendous number of them to have a useful amount. You can get a feel for just how large a mole is by writing it in normal notation:

602,000,000,000,000,000,000,000

This is almost a trillion trillion. Let's put this in perspective. If you counted every grain of sand on every beach and ocean floor on this planet, you would only come close to 1 mole. If you had a mole of dollars and you spent a billion dollars a second for your entire life, you could spend only about 0.001% of your money before you died.

While Avogadro's number is huge, atoms and molecules are so tiny that it takes a lot of them to make a dent. A mole of water molecules is barely a swallowful, and a mole of sugar molecules will fit in the palm of your hand. It might interest you to know that Avogadro was in his grave for many years before Avogadro's number was experimentally determined with reasonable accuracy. He never knew the value of the quantity named after him, but you should never forget it.

How is the mole going to help us with our translation problem? Let's look at the balanced equation for making HI again. Instead of reading the reaction as "1 molecule of H_2 and 1 molecule of I_2 react to give two molecules of HI," we can also read it as "1 **mole** of H_2 molecules and 1 **mole** of I_2 molecules react to give 2 **moles** of HI molecules."

$$H_2 + I_2 \longrightarrow 2\,HI \text{ says:}$$

1 molecule of H_2 and 1 molecule of I_2 react to give 2 molecules of HI

OR

1 mole of H_2 molecules and 1 mole of I_2 molecules react to give 2 moles of HI molecules

This is a fundamental concept in chemistry. Any balanced chemical equation can always be read in terms of moles of molecules instead of individual molecules. This should make sense to you. After all, if the balanced equation says

1 molecule of H_2 reacts with 1 molecule of I_2 to give 2 molecules of HI

then it also says

1 dozen H_2 molecules react with 1 dozen I_2 molecules to give 2 dozen molecules of HI

or

1 mole of H_2 molecules react with 1 mole of I_2 molecules to give 2 moles of HI molecules

I NEED ONE TRILLION HI MOLECULES. PLEASE TELL HER HOW MANY GRAMS OF HI THAT IS.

BUT OF COURSE! THAT'S MY JOB!

This by itself doesn't help us, since we're still speaking in terms of numbers of molecules, and we don't count molecules in a lab, we weigh them. We go to the storage room, find a bottle of iodine, and weigh out 10 grams of it. If we use 10 grams of I_2, then how many grams of H_2 will we need? How many grams of HI will we make? To answer these questions we need to learn one more thing about the mole. Just as we needed to translate from ounces of cream cheese to blocks of cream cheese before we could use the cheesecake recipe, the mole will allow us to translate from grams to numbers so that we can use a balanced chemical equation.

The value of the mole, 6.02×10^{23}, is defined as the number of carbon atoms in exactly 12 grams of $^{12}_{6}C$. Experiments over the years determined that there are 6.02×10^{23} atoms of $^{12}_{6}C$ in 12 grams. Simply knowing this fact allows us to count $^{12}_{6}C$ atoms by weighing them! To "count out" 6.02×10^{23} (one mole) of $^{12}_{6}C$ atoms, all you have to do is weigh out exactly 12 grams.

The mole lets us count atoms by weighing.

$$12 \text{ g of } {}^{12}_{6}C = 1 \text{ mole of } {}^{12}_{6}C$$

$$= 6.02 \times 10^{23} \text{ atoms of } {}^{12}_{6}C$$

12.000g

In fact, the mole allows you to count atoms of any element by weighing. All we need to do is use the relative atomic weights that we discussed in Chapter 3. For example, what if we wanted 1 mole of magnesium (Mg) atoms? The relative atomic weight of naturally occurring magnesium is 24.305, as shown in the periodic table. The relative atomic weight tells us that magnesium is 24.305/12, or 2.025, times as heavy as $^{12}_{6}C$. So, if a mole of $^{12}_{6}C$ weighs 12 grams, then a mole of magnesium will weigh 2.025 times that, or 24.305 grams. In other words, the concept of moles takes the relative atomic weights, which were unitless, and gives them all units of grams! Consider the following elements and their relative atomic weights:

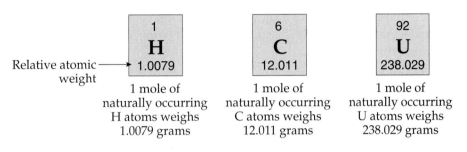

Relative atomic weight

| 1 mole of naturally occurring H atoms weighs 1.0079 grams | 1 mole of naturally occurring C atoms weighs 12.011 grams | 1 mole of naturally occurring U atoms weighs 238.029 grams |

A mole of any atom weighs its relative atomic weight in grams. We call the weight of a mole of an atom or molecule its **molar mass**. Therefore, the molar mass of hydrogen is 1.0079 g/mole, the molar mass of naturally occurring carbon is 12.011 g/mole, and the molar mass of uranium is 238.029 g/mole. Once you understand this concept, you can count atoms of any element by weighing them. Said another way, you can convert back and forth between numbers of atoms and grams of atoms. Try the following WorkPatch.

Consider the element sulfur, a yellow solid. Suppose you weighed 64.132 g of it.

WORKPATCH 7.3

   (a) What is the molar mass of sulfur?
   (b) How many moles of sulfur do you have?
   (c) How many atoms of sulfur do you have?

Check your answers against ours at the end of the chapter, and do not go on until you agree with it. Understanding what follows depends on understanding this fundamental mole concept.

PRACTICE PROBLEMS

Try to do the following problems in your head.

7.8 Suppose you have half a mole of S atoms.
   (a) How many S atoms do you have?
   (b) How much will it weigh in grams?

Answer:
(a) Half a mole is half of 6.02×10^{23}, so you have 3.01×10^{23} S atoms.
(b) A mole of S weighs 32.066 g, so half a mole weighs 16.033 g.

7.9 Suppose you have a tenth of a mole of uranium atoms.
(a) How many U atoms do you have?
(b) How much will it weigh in grams?

7.10 Suppose you have 120.11 g of naturally occurring carbon atoms.
(a) How many moles of carbon atoms do you have?
(b) How many carbon atoms do you have?

The understanding that a mole of atoms of an element weighs its relative atomic weight in grams will be critical to your being able to do chemical stoichiometry problems.

We can calculate the weight of a mole of molecules in the same way. For instance, how much would a mole of CO_2 molecules weigh in grams? The molecular formula tells us that 1 molecule of carbon dioxide is made from 1 carbon atom and 2 oxygen atoms. Following the line of reasoning we developed earlier, it must also be true that 1 mole of CO_2 molecules is made from 1 mole of carbon atoms and 2 moles of oxygen atoms. This is very important! All the subscripts in a formula not only say how many atoms there are in one molecule, they also say how many moles of atoms there are in 1 mole of the molecule.

What the subscripts say:

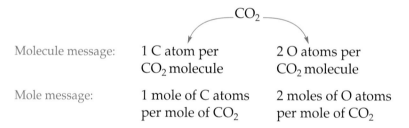

	CO_2	
Molecule message:	1 C atom per CO_2 molecule	2 O atoms per CO_2 molecule
Mole message:	1 mole of C atoms per mole of CO_2	2 moles of O atoms per mole of CO_2

This is similar to the balancing coefficients in a chemical equation, which we've already seen can always be interpreted as moles. To see if you are catching on, try the following WorkPatch. You should be able to do parts (a) through (c) in your head, but a calculator will help for (d).

7.4 WORKPATCH

Suppose you are given a mole of methane, CH_4.
(a) How many moles of carbon atoms do you have?
(b) How many moles of hydrogen atoms do you have?
(c) How many atoms of carbon are there in a mole of methane?
(d) How many atoms of hydrogen are there in a mole of methane?

[*Hint for (c) and (d)*: Remember that a mole of anything is 6.02×10^{23} of it.]

Once you understand the principle behind the answers to the WorkPatch, it is a simple matter to figure out the molar mass of a molecule—that is, how much a mole of it weighs in grams. Let's do this for CO_2. Since 1 mole of CO_2 contains 1 mole of carbon atoms and 2 moles of oxygen atoms, and we know how much a mole of each of these atoms weighs, then all we have to do is add up the molar masses as shown below:

	Molar mass
1 mole C atoms	12.011 g/mole
1 mole O atoms	15.999 g/mole
1 mole O atoms	15.999 g/mole
1 mole of CO_2 molecules	44.009 g/mole

Or, since there are 2 moles of O atoms, we could have written:

	Molar mass
1 mole C atoms	12.011 g/mole
2×1 mole O atoms	2×15.999 g/mole
1 mole of CO_2 molecules	44.009 g/mole

The molar mass of CO_2 is 44.009 g/mole. If you weigh out 44.009 g of CO_2, then you have 1 mole of it, or 6.02×10^{23} molecules of it. Once again you have counted by weighing.

Using the molar masses of the atoms from the periodic table, we can calculate the molar mass for any chemical compound so long as we know its formula. Here are some more examples:

Formula	Molar mass
HI	$H(1.0079) + I(126.905) = 127.912$ g/mole
H_2	$2 \times H(1.0079) = 2.0158$ g/mole
I_2	$2 \times I(126.905) = 253.81$ g/mole
H_2O	$2 \times H(1.0079) + O(15.999) = 18.015$ g/mole

The mole allows us to go back and forth between a number of atoms or molecules and their weight in grams.

PRACTICE PROBLEMS

Consider propane, C_3H_8, used for gas grills and torches.

7.11 (a) If you have 1 mole of propane, how many propane molecules do you have?

    (b) If you have 1 mole of propane, how many hydrogen atoms do you have?

    (c) How much does a mole of propane weigh in grams?

Answer:
(a) 1 mole of propane molecules means that you have 6.02×10^{23} propane molecules. That is the meaning of 1 mole.

(b) 1 mole of propane contains 8 moles of hydrogen atoms (this is what the subscript 8 tells you in the formula C_3H_8). We know that there are 6.02×10^{23} atoms of H per mole of them, so

$$8 \text{ moles H atoms} \times \frac{6.02 \times 10^{23} \text{ H atoms}}{1 \text{ mole H atoms}} = 4.82 \times 10^{24} \text{ H atoms}$$

(c) $\underbrace{(3 \times 12.011 \text{ g})}_{3 \text{ moles C}} + \underbrace{(8 \times 1.0079 \text{ g})}_{8 \text{ moles H}} = \underbrace{44.096 \text{ g}}_{\substack{\text{Molar mass of propane} \\ \text{(the weight of 1 mole of } C_3H_8)}}$

7.12   How much does 2 moles of propane weigh in grams?

7.13   How many moles of carbon atoms are in 2 moles of propane?

7.14   Propane reacts with oxygen (burns) to produce CO_2 and H_2O, as shown in the balanced equation

$$C_3H_8 + 5 O_2 \longrightarrow 3 CO_2 + 4 H_2O$$

Write this reaction in English using the word mole(s) four times.

7.6 REACTION STOICHIOMETRY

We are now ready to use the mole for doing reaction stoichiometry problems. Let's go back to our HI reaction:

$$H_2 + I_2 \longrightarrow 2 HI$$

Viewed as a recipe this says, "Combine 1 mole of H_2 molecules with 1 mole of I_2 molecules and they will react to give 2 moles of HI molecules." We can finally go into the lab and carry out this reaction, because we can now count molecules by weighing them. One mole of H_2 molecules weighs 2.0158 g (2 × 1.0079 g/mole, the molar mass of H), and 1 mole of I_2 molecules weighs 253.81 g (2 × 126.905 g/mole, the molar mass of I). If we combine these amounts of H_2 and I_2, the reaction even tells us how many grams of product to expect. Since the molar mass of HI is 127.912 g/mole, then we should produce 2 × 127.912 = 255.824 g of HI. This is called the **theoretical yield** of the reaction, the maximum amount of product you can hope to make for a given amount of reactants.

Interestingly, reactions do not always produce the theoretical yield, for a variety of reasons. Many reactions have competing side reactions that consume some of the reactant to produce unwanted impurities. Or, when it is time to recover the product, difficulties arise in collecting it or purifying it, causing some of it to be lost. Whatever the reason, it is very common to get something less than the theoretical yield of product. The amount of product that you actually end up with is called the **actual yield** of the reaction. Suppose that, for example, we were able to isolate only 225.10 g of HI instead of the predicted theoretical yield of 255.824 g. The actual yield of the reaction would be 225.10 g of HI.

The actual yield allows us to define yet another quantity, the *percent yield* of the reaction, which is just what it sounds like. The **percent yield** of a reaction is the percentage of the theoretical yield that we were able to isolate. It is calculated as shown below:

$$\% \text{ yield} = \frac{\text{Actual yield}}{\text{Theoretical yield}} \times 100$$

For example:

$$\% \text{ yield} = \frac{225.10 \text{ g HI}}{255.824 \text{ g HI}} \times 100 = 87.990\%$$

Our percent yield is almost 88%, which is not bad. Some reactions done by the chemical and pharmaceutical industry have smaller percent yields and are thus quite wasteful of their starting reactants. You end up absorbing this in the price you pay for pharmaceuticals and other useful chemicals.

—PRACTICE PROBLEMS

7.15 Consider the balanced reaction

$$CH_4 + 2 O_2 \longrightarrow 2 H_2O + CO_2$$

(a) Write this reaction in English, using the word mole(s) wherever appropriate.
(b) To produce 1 mole of CO_2 from this reaction, how many grams of CH_4 and O_2 must you combine?
(c) What is the theoretical yield of H_2O for this reaction?
(d) Suppose you recovered 30.0 g of H_2O. What would be the percent yield of this reaction?

Answer:
(a) 1 mole of methane molecules and 2 moles of oxygen molecules react to give 2 moles of water molecules and 1 mole of carbon dioxide molecules.
(b) The molar mass of methane is (12.011 g C) + (4 × 1.0079 g H) = 16.043 g CH_4. This is the weight of 1 mole of methane, which is what the reaction calls for. The molar mass of oxygen (O_2) is (2 × 15.999 g O) = 31.998 g. This is the weight of 1 mole of O_2. The reaction calls for 2 moles, so we multiply by 2 to get 63.996 g O_2.
(c) The most we can hope to form is 2 moles of water. The molar mass of water is (2 × 1.0079 g H) + (15.999 g O) = 18.015 g H_2O. This is the weight of 1 mole of water. So the theoretical yield is just twice this, or 36.030 g H_2O.
(d) % yield = (30.0 g/ 36.030 g) × 100 = 83.3%

7.16 Consider the unbalanced reaction

$$\underset{\text{Benzene}}{C_6H_6} + \underset{\text{Hydrogen}}{H_2} \longrightarrow \underset{\text{Cyclohexane}}{C_6H_{12}}$$

(a) Balance the reaction by inspection.
(b) Write this reaction in English, using the word mole(s) wherever appropriate.
(c) To produce 1 mole of C_6H_{12} from this reaction, how many grams of C_6H_6 and H_2 must you combine?
(d) What is the theoretical yield of C_6H_{12} for this reaction?
(e) Suppose only 24.0 g of C_6H_{12} was recovered. What would be the percent yield of this reaction?

7.17 Consider the unbalanced reaction

$$C_6H_{12}O_6 + O_2 \longrightarrow CO_2 + H_2O$$
Glucose

(a) Balance the reaction by inspection.
(b) Write this reaction in English, using the word mole(s) wherever appropriate.
(c) To produce 6 moles of H_2O from this reaction, how many grams of glucose and O_2 must you combine?
(d) What is the theoretical yield of CO_2 for this reaction?
(e) Suppose only 196.0 g of CO_2 was recovered. What would be the percent yield of this reaction?

Up to now we have been using a balanced equation like a recipe, following its instructions exactly. But recall that when we were baking cheesecakes, we sometimes wanted to depart from the recipe, doubling it to make two cakes, etc. The same holds true for chemical reactions. Our balanced reaction for hydrogen iodide is $H_2 + I_2 \longrightarrow 2\,HI$, and it tells us how to make 2 moles of HI. But we rarely want to produce exactly 2 moles of HI (255.824 g, twice its molar mass). Suppose we want to make only 10.0 g of HI. How do we cut back on the recipe? The answer comes from a concept we covered in Chapter 2—conversion factors. Both a balanced chemical equation and a molecular formula are sources of conversion factors that are useful for doing stoichiometry problems. We simply read them directly from the balanced equation or the formula, like this:

Conversion factors that come from the balanced equation
$H_2 + I_2 \longrightarrow 2\,HI$

$\dfrac{2 \text{ moles HI}}{1 \text{ mole } H_2}$ 2 moles of HI are produced per 1 mole of H_2 used

$\dfrac{1 \text{ mole } H_2}{2 \text{ moles HI}}$ 1 mole of H_2 is used per 2 moles of HI produced

⎫ Same, just inverted

$\dfrac{2 \text{ moles HI}}{1 \text{ mole } I_2}$ 2 moles of HI are produced per 1 mole of I_2 used

$\dfrac{1 \text{ mole } I_2}{2 \text{ moles HI}}$ 1 mole of I_2 is used per 2 moles of HI produced

⎫ Same, just inverted

$\dfrac{1 \text{ mole } I_2}{1 \text{ mole } H_2}$ For every 1 mole of I_2 consumed, 1 mole of H_2 is consumed

$\dfrac{1 \text{ mole } H_2}{1 \text{ mole } I_2}$ For every mole of H_2 consumed, 1 mole of I_2 is consumed

⎫ Same, just inverted

Conversion factors that come from the formulas

$$\frac{2 \text{ moles H atoms}}{1 \text{ mole } H_2 \text{ molecules}} \quad \text{From the formula } H_2$$

$$\frac{2 \text{ moles I atoms}}{1 \text{ mole } I_2 \text{ molecules}} \quad \text{From the formula } I_2$$

$$\frac{1 \text{ mole H atoms}}{1 \text{ mole HI molecules}} \quad \text{From the formula HI}$$

$$\frac{1 \text{ mole I atoms}}{1 \text{ mole HI molecules}} \quad \text{From the formula HI}$$

This concept is so important that you need to practice it now.

Consider the balanced reaction

WORKPATCH 7·5

$$\underset{\text{Aluminum}}{4 \text{ Al}} \quad + \quad \underset{\text{Oxygen}}{3 \text{ O}_2} \quad \longrightarrow \quad \underset{\substack{\text{Aluminum} \\ \text{oxide}}}{2 \text{ Al}_2\text{O}_3}$$

(a) Write the reaction in English, using the word moles for every reactant and product.
(b) Write every single conversion factor that you can think of from the balanced chemical equation.
(c) Write all the conversion factors from the formula Al_2O_3.

Did you get six conversion factors from the reaction? Check your answers before going on, because being able to get these conversion factors is essential to doing general stoichiometry problems.

Now, how do we use the balanced HI reaction to determine how many grams of H_2 and I_2 we need to make 10.0 g of HI? Since the reaction speaks in moles, the first thing we must do is translate the 10.0 g of HI into moles of HI. This is the job of the mole and the molar mass. The molar mass of HI is 127.912 g/mole; in other words, 1 mole of HI weighs 127.912 g. Thus, we can write the molar mass of HI as if it were a conversion factor:

$$\text{Molar mass of HI} = \frac{127.912 \text{ g HI}}{1 \text{ mole HI}} \quad \text{and its inverse:} \quad \frac{1 \text{ mole HI}}{127.912 \text{ g HI}}$$

You can write it either way. Which fraction you use depends on whether you want to translate from moles to grams or from grams to moles. The first step in doing this stoichiometry problem is to convert from grams of HI to moles of HI, using the second fraction as a conversion factor:

$$10.0 \text{ g } \cancel{HI} \times \underbrace{\frac{1 \text{ mole HI}}{127.912 \text{ g } \cancel{HI}}}_{\substack{\text{Inverse of molar} \\ \text{mass of HI}}}$$

We used the form of the conversion factor that will make the units of grams cancel top and bottom. At this point we have changed over to units of moles of HI (it is the only unit not crossed out). The second step is to use the balanced chemical equation to find out how many moles of H_2 and I_2 we need. Let's tackle H_2 first. We find the conversion factor from the balanced chemical equation that takes us from units of moles of HI to moles of H_2, and then just string it on to our calculation:

$$10.0 \text{ g HI} \times \frac{1 \text{ mole HI}}{127.912 \text{ g HI}} \times \underbrace{\frac{1 \text{ mole } H_2}{2 \text{ moles HI}}}_{\text{From balanced equation}}$$

This cancels out the units of moles of HI and leaves us with units of moles of H_2. Since the question is "how many grams of H_2 do we need?" we want our final answer in grams. So, the third and final step is to use the molar mass of H_2 to convert back to grams. Notice that the molar mass conversion factor for H_2 is written so that it makes the proper units cancel:

$$10.0 \text{ g HI} \times \frac{1 \text{ mole HI}}{127.912 \text{ g HI}} \times \frac{1 \text{ mole } H_2}{2 \text{ moles HI}} \times \underbrace{\frac{2.0158 \text{ g } H_2}{1 \text{ mole } H_2}}_{\text{Molar mass of } H_2}$$

Surviving unit

We are done. The only units not crossed out are the units of the question we were asked, grams of H_2. All we have to do is perform the math and attach the surviving unit to it. Here is what you would enter into your calculator.

$$10.0 \times 2.0158 \div 127.912 \div 2 = 0.0788 \text{ g } H_2$$

This is the answer we have been looking for. To make 10.0 g of HI, we need 0.0788 g of H_2. We'll leave the calculation of the amount of I_2 to you (in Work-Patch 7.6, on the next page).

To summarize, here is the three-step method we just used to solve our typical stoichiometry problem. In each step we are multiplying by an appropriate conversion factor that comes from a molar mass or from the balanced chemical equation.

Start with grams of some reactant or product.

Step 1: Convert grams to moles using a molar mass conversion factor.

Convert to moles.

Step 2: Convert to moles of the reactant or product the question is asking about by using an appropriate conversion factor from the balanced equation.

Convert to grams of what the question is asking about.

Step 3: Convert moles to grams using a molar mass conversion factor.

ANSWER

Now, it's your turn to try this.

Using the three-step procedure, determine how many grams of I_2 are required to make 10.0 g of HI. Consult the balanced equation and completely write out your method, crossing out units that cancel.

Your first step should have been identical to the first step we used when solving for the amount of H_2. Check your work against ours to see how you did.

WORKPATCH **7.6**

PRACTICE PROBLEMS

Glucose reacts with O_2 to give CO_2 and H_2O. The balanced reaction and molar masses are given below.

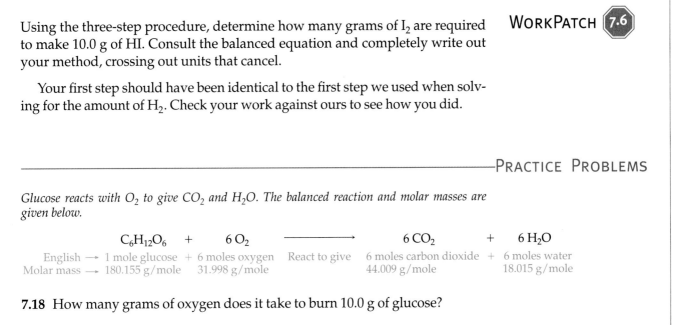

$$C_6H_{12}O_6 \quad + \quad 6\,O_2 \quad \longrightarrow \quad 6\,CO_2 \quad + \quad 6\,H_2O$$

English → 1 mole glucose + 6 moles oxygen   React to give   6 moles carbon dioxide + 6 moles water
Molar mass → 180.155 g/mole   31.998 g/mole   44.009 g/mole   18.015 g/mole

7.18 How many grams of oxygen does it take to burn 10.0 g of glucose?

Answer:

$$10.0 \text{ g glucose} \times \frac{1 \text{ mole glucose}}{180.155 \text{ g glucose}} \times \frac{6 \text{ moles } O_2}{1 \text{ mole glucose}} \times \frac{31.998 \text{ g } O_2}{1 \text{ mole } O_2} = 10.7 \text{ g } O_2$$

7.19 How many grams of water will burning 10.0 g of glucose produce? [*Hint*: Start by writing 10.0 g glucose and convert it to moles of glucose. Then get a conversion factor from the balanced equation and convert to moles of water. Finish by converting back to grams.]

7.20 What is the theoretical yield of carbon dioxide if 10.0 g of glucose are burned? [*Hint*: Start by writing 10.0 g glucose and convert it to moles of glucose. Then get a conversion factor from the balanced equation and convert to moles of carbon dioxide. Finish by converting back to grams.]

7.21 If 10.0 grams of carbon dioxide were produced, how many grams of glucose must have been burned? [*Hint:* Start by writing 10.0 g carbon dioxide and convert it to moles of carbon dioxide. Then get a conversion factor from the balanced equation and convert to moles of glucose. Finish by converting back to grams.]

Though we have stressed following the three-step method, it can be modified to suit the occasional oddball question. For example, the answer to Practice Problem 7.18 is that it takes 10.7 g of O_2 to burn 10.0 g of glucose. You could have been asked to go further, and calculate how many O atoms are in 10.7 g of O_2. Our knowledge of the mole (6.02×10^{23} of anything) and our knowledge of what the formula O_2 means (each oxygen molecule consists of 2 oxygen atoms) gives us more conversion factors, which we use to find the answer in Practice Problem 7.22.

PRACTICE PROBLEMS

7.22 How many O atoms are in 10.7 g of oxygen (O_2) molecules?

Answer: Don't be afraid to string out conversion factors. They may come from a balanced equation, a formula, or your knowledge of the mole, depending on the question you are trying to answer.

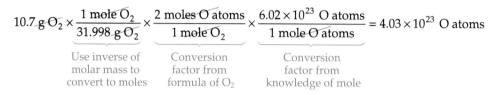

$$10.7 \text{ g } O_2 \times \frac{1 \text{ mole } O_2}{31.998 \text{ g } O_2} \times \frac{2 \text{ moles } O \text{ atoms}}{1 \text{ mole } O_2} \times \frac{6.02 \times 10^{23} \text{ O atoms}}{1 \text{ mole } O \text{ atoms}} = 4.03 \times 10^{23} \text{ O atoms}$$

Use inverse of molar mass to convert to moles	Conversion factor from formula of O_2	Conversion factor from knowledge of mole

7.23 How many aluminum atoms are there in 10.0 g of aluminum oxide (Al_2O_3)?

7.24 How many molecules of water are there in 10.0 grams of water?

With the proper conversion factors, you can use balanced equations to calculate "how much" when running a reaction at any scale. Our final advice is this. Since reactions speak in the language of moles, you can bet that for almost every stoichiometry problem, you will be converting to moles. If in doubt, convert to moles.

7.7 FORMULAS AND PERCENT COMPOSITIONS FROM COMBUSTION ANALYSIS

When we do reactions and stoichiometry problems on paper, we always know what the products are. In reality, reactions sometimes give totally unexpected products, or products that are contaminated with impurities. For example, suppose you have a headache and a friend offers you a tablet, saying, "This is aspirin, I just synthesized it in a chemical reaction. Have some."

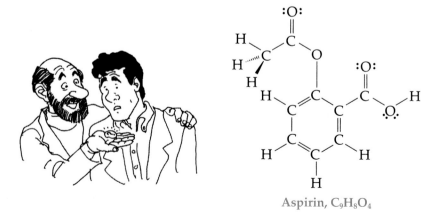

Aspirin, $C_9H_8O_4$

If enough impurities or unexpected products were in the aspirin, it could cure your headache forever! Before it is considered fit for consumption, a chemist would have to confirm both the aspirin's identity and its purity. One good test would be to determine the formula of the compound that the reaction produced. Hopefully, it's $C_9H_8O_4$, the formula for aspirin. One of the best ways to determine the formula of an unknown compound is to take an accurately weighed amount of the compound, and—believe it or not—burn it. This is called **combustion analysis**. Burning something is reacting it with oxygen, so when you burn a compound, all the elements in it combine with oxygen. For example, the combustion reaction for aspirin is

$$C_9H_8O_4 + 9\,O_2 \longrightarrow 9\,CO_2 + 4\,H_2O$$

All the carbon ends up in CO_2 and all the hydrogen ends up in H_2O. You can think of the carbon dioxide and the water as "garbage cans" where all the carbon and hydrogen end up.

In the lab, a combustion analyzer like the one diagrammed below is commonly used to burn unknown substances and determine their formulas. Different materials in the analyzer absorb the H_2O and the CO_2 that are produced. The absorbants are weighed before and after the combustion. The mass gain of each is equal to the mass of water and carbon dioxide produced.

What combustion analysis does

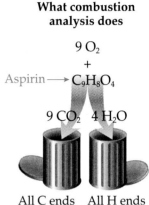

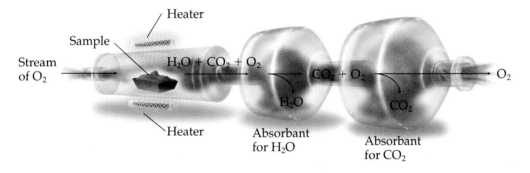

Here is the experimental data from the combustion analysis of a 0.100 g sample of the uncertain "aspirin":

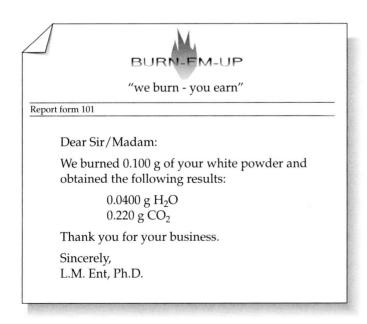

BURN-EM-UP

"we burn - you earn"

Report form 101

Dear Sir/Madam:

We burned 0.100 g of your white powder and obtained the following results:

0.0400 g H_2O
0.220 g CO_2

Thank you for your business.

Sincerely,
L.M. Ent, Ph.D.

Our job now is to turn the data into a formula. Right now, all we know is that the formula of this white powder is $C_xH_yO_z$; we don't know the values of any of the subscripts. We do know, however, that they can be thought of as moles. Since all of the aspirin sample's hydrogen went into making water, if we figure out how many moles of hydrogen there are in the 0.0400 g of water, we will know how many moles of hydrogen there were in the sample. This is just a simple application of stoichiometry, writing down the number 0.0400 g H_2O and applying conversion factors until we get moles of H:

$$0.0400 \text{ g } H_2O \times \underbrace{\frac{1 \text{ mole } H_2O}{18.015 \text{ g } H_2O}}_{\substack{\text{Inverse of} \\ \text{molar mass}}} \times \underbrace{\frac{2 \text{ moles H atoms}}{1 \text{ mole } H_2O}}_{\substack{\text{From formula} \\ \text{of water}}} = \underbrace{0.004\ 44 \text{ mole H atoms}}_{\substack{\text{This is the value} \\ \text{of } y \text{ in } C_xH_yO_z.}}$$

We can also do this for the carbon. Since all of the aspirin sample's carbon went into making carbon dioxide, if we figure out how many moles of carbon there are in the 0.220 g of carbon dioxide, we will know how many moles of carbon there were in the sample.

$$0.220 \text{ g } CO_2 \times \underbrace{\frac{1 \text{ mole } CO_2}{44.009 \text{ g } CO_2}}_{\substack{\text{Inverse of} \\ \text{molar mass}}} \times \underbrace{\frac{1 \text{ mole C atoms}}{1 \text{ mole } CO_2}}_{\substack{\text{From formula} \\ \text{of } CO_2}} = \underbrace{0.005\ 00 \text{ mole C atoms}}_{\substack{\text{This is the value} \\ \text{of } x \text{ in } C_xH_yO_z.}}$$

So far we have the formula $C_{0.005\ 00}H_{0.004\ 44}O_z$ for our aspirin sample. Don't worry that the subscripts are not whole numbers. We will take care of that shortly. Now we have to find the value of the z subscript for oxygen. We have a small problem, though. We don't know where the oxygen in the combustion products CO_2 and H_2O came from. Some may have come from the sample. Some may have come from the "stream of O_2" used in the combustion analyzer. The only way to find out is to calculate the weight of the 0.005 00 mole of C and 0.004 44 mole of H in our sample, and subtract this from the weight of the aspirin sample (0.100 g). Anything left is the weight of oxygen in the sample.

$$0.004\ 44 \text{ mole H} \times \underbrace{\frac{1.0079 \text{ g H}}{1 \text{ mole H}}}_{\text{Molar mass of H}} = 0.004\ 48 \text{ g H}$$

$$+$$

$$0.005\ 00 \text{ mole C} \times \underbrace{\frac{12.011 \text{ g C}}{1 \text{ mole C}}}_{\substack{\text{Molar mass} \\ \text{of C}}} = 0.0600 \text{ g C}$$

$$\underline{0.0645 \text{ g}} \leftarrow \text{Due to C and H}$$

Of the 0.100 g sample of aspirin, 0.0645 g is due to C and H. If the only elements in the compound are carbon, hydrogen, and oxygen, then the rest must be oxygen:

$$
\begin{array}{ll}
0.1000 \text{ g} & \text{Sample} \\
-\ 0.0645 \text{ g} & \text{Due to C and H} \\
\hline
0.0355 \text{ g O} &
\end{array}
$$

If the result here were zero, then we could conclude that the compound had no oxygen in it. Since our compound does have oxygen, let's convert the grams of oxygen to moles to find the subscript z in the formula $C_xH_yO_z$:

$$0.0355 \text{ g O} \times \frac{1 \text{ mole O}}{15.999 \text{ g O}} = 0.002\ 22 \text{ mole O}$$

Inverse of
molar mass of O

Notice that we used the molar mass of O (not O_2) in our calculation. Oxygen is not diatomic in compounds.

We now have the complete formula of the sample that we burned: $C_{0.005\ 00}H_{0.004\ 44}O_{0.002\ 22}$. OK, so you don't like it. Neither do we. That's because we want whole numbers of moles in our formula. To fix this, we divide all the subscripts by the smallest one (z, which is 0.002 22):

$$\frac{C_{0.005\ 00}H_{0.004\ 44}O_{0.002\ 22}}{0.002\ 22 \quad 0.002\ 22 \quad 0.002\ 22} \longrightarrow C_{2.25}H_{2.00}O_{1.00} = C_{2.25}H_2O$$

Well, it almost worked. Many times you will be done at this point, but sometimes not, as is the case here. We are allowed to change any subscripts that are *very close* to whole numbers into whole numbers, but the subscript 2.25 on C is not very close to 2 or to 3. In this case we need to find a number that we can multiply 2.25 by to make it a whole number. This takes a little thought. With 4 we get $2.25 \times 4 = 9$, a whole number. We must be careful, though. If we multiply 2.25 by 4, we must multiply *all* the subscripts by 4. The rule is, whatever we do to one subscript, we must always do to the others, otherwise we will change the relationships among them.

$$C_{2.25 \times 4}H_{2.00 \times 4}O_{1.00 \times 4} \longrightarrow C_9H_8O_4$$

We are done. We have gone from the data of a combustion analysis to a formula. The analysis indicates that the sample does indeed have the formula of aspirin (assuming the sample was pure). Of course, this does not prove that the sample is aspirin, since different compounds can have the same formula. For example, both ethyl ether, CH_3OCH_3, and ethanol, CH_3CH_2OH, have the same formula, C_2H_6O. Let's summarize the steps we used:

Determining a formula from combustion analysis data:

Step 1: Convert grams CO_2 $\longrightarrow$ moles C   This is subscript x in $C_xH_yO_z$
Convert grams H_2O $\longrightarrow$ moles H   This is subscript y in $C_xH_yO_z$

Step 2: Convert moles C from step 1 $\longrightarrow$ grams C
                                                              +
Convert moles H from step 1 $\longrightarrow$ grams H
                                              grams of C and H

Step 3: Find grams of O:

grams of original sample that was burned
− grams of C and H from step 2

grams of O $\longleftarrow$ If it's zero, subscript z is zero.
                             If it's not zero, convert to moles O.
                             This is subscript z in $C_xH_yO_z$.

Step 4: Divide all the subscripts by the smallest one, and if they are very close to whole numbers, then round them to whole numbers.

If some subscripts are still very far from whole numbers, find some number to multiply by that makes them whole. Then multiply all the subscripts by this number.

The results from a combustion analysis can be used to calculate other things besides the formula. We can also calculate the *mass percent* of each element present in the aspirin. For example, the human body is roughly 70% by mass water, meaning roughly 70% of your weight is due to water. We could do the same sort of thing for aspirin. Given a sample of aspirin, what percent of the sample's weight is due to carbon? What percent is due to hydrogen? What percent is due to oxygen? The answers are that aspirin is 60.0% by mass carbon, 4.48% by mass hydrogen, and 35.5% by mass oxygen. We calculate these mass percents from the combustion analysis results. To do this, we simply take the weight of each element present in the sample, divide it by the weight of the sample (0.100 g), and multiply by 100:

$$\%C = \frac{0.0600 \text{ g C}}{0.100 \text{ g aspirin}} \times 100 = 60.0\% \text{ C}$$

$$\%H = \frac{0.004\,48 \text{ g H}}{0.100 \text{ g aspirin}} \times 100 = 4.48\% \text{ C}$$

$$\%O = \frac{0.0355 \text{ g O}}{0.100 \text{ g aspirin}} \times 100 = 35.5\% \text{ O}$$

If we add these mass percents together (and obey the rules of significant figures), we should always get 100%. Now try the following practice problems.

PRACTICE PROBLEMS

7.25 A compound is known to contain carbon and hydrogen. It might also contain oxygen. A 0.250 g sample of the compound is burned to produce 0.561 g H_2O and 0.686 g CO_2.
(a) What is the formula for this compound?
(b) What is the percent by mass for each element in this compound?
(c) Write the balanced combustion reaction (reaction with O_2) for this compound.

Answer (not worked out on purpose—you do it):
(a) CH_4
(b) 74.9% C; 25.1% H
(c) $CH_4 + 2\,O_2 \longrightarrow 2\,H_2O + CO_2$

7.26 A compound is known to contain carbon and hydrogen. It might also contain oxygen. A 0.250 g sample of the compound is burned to produce 0.293 g H_2O and 0.478 g CO_2.
(a) What is the formula for this compound?

(b) What is the percent by mass for each element in this compound?

(c) Write the balanced combustion reaction (reaction with O_2) for this compound.

The combustion analysis problems we have been doing so far have been chosen to avoid a complication that we want to deal with now. Let's burn a weighed sample of one more compound, but this time we will tell you the formula right at the start. We will burn a 0.500 g sample of the simple sugar glucose, formula $C_6H_{12}O_6$. The combustion analysis data are shown here. We already know the formula for this combustion analysis sample, so if we take the data from the combustion report, we ought to be able to come up with the formula $C_6H_{12}O_6$. Let's see if we can. The problem is worked out for you below. Follow it through carefully.

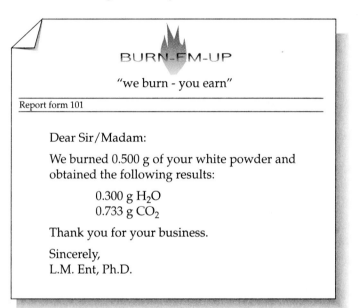

BURN-EM-UP

"we burn - you earn"

Report form 101

Dear Sir/Madam:

We burned 0.500 g of your white powder and obtained the following results:

0.300 g H_2O
0.733 g CO_2

Thank you for your business.

Sincerely,
L.M. Ent, Ph.D.

Step 1: Convert g CO_2 to moles C and g H_2O to moles H:

$$0.733 \text{ g CO}_2 \times \frac{1 \text{ mole CO}_2}{44.009 \text{ g CO}_2} \times \frac{1 \text{ mole C}}{1 \text{ mole CO}_2} = 0.0166 \text{ mole C}$$

$$0.300 \text{ g H}_2\text{O} \times \frac{1 \text{ mole H}_2\text{O}}{18.015 \text{ g H}_2\text{O}} \times \frac{2 \text{ moles H}}{1 \text{ mole H}_2\text{O}} = 0.0333 \text{ mole H}$$

Formula so far is $C_{0.0166}H_{0.0333}O_z$.

Step 2: Figure out how much the moles of C and H in the burned sample weigh:

$$0.0166 \text{ mole C} \times \frac{12.011 \text{ g C}}{1 \text{ mole C}} = 0.199 \text{ g C}$$

$$+$$

$$0.0333 \text{ mole H} \times \frac{1.0079 \text{ g H}}{1 \text{ mole H}} = \frac{0.0333 \text{ g H}}{0.232 \text{ g}} \leftarrow \text{Due to C + H}$$

Step 3: Determine if there is O in the compound. If there is, convert it to moles.

$$
\begin{array}{ll}
0.500 \text{ g} & \text{Sample burned} \\
-\ 0.232 \text{ g} & \text{Due to C + H} \\
\hline
0.282 \text{ g} & \text{Due to O}
\end{array}
$$

$$0.282 \ \cancel{\text{g O}} \times \frac{1 \text{ mole O}}{15.999 \ \cancel{\text{g O}}} = 0.0168 \text{ mole O}$$

Formula so far is $C_{0.0166}H_{0.0333}O_{0.0168}$.

Step 4: Divide subscripts through by the smallest subscript:

$$\underset{0.0166 \quad 0.0166 \quad 0.0166}{C_{0.0166}H_{0.0333}O_{0.0168}} \longrightarrow C_{1.00}H_{2.006}O_{1.01} \longrightarrow \underset{\text{Formula}}{CH_2O}$$

What went wrong? The calculations based on the combustion analysis give us a formula of CH_2O, but we know that glucose has the formula $C_6H_{12}O_6$. The answer is nothing went wrong. In general, a combustion analysis does not give the actual formula of a compound, it gives the *empirical formula*. An **empirical formula** (also called the *simplest formula*) gives the ratio of the elements present in a compound, whereas a **molecular formula** gives you the actual composition of the compound. Thus, CH_2O is the empirical formula of glucose. It tells us that there are twice as many H atoms in the compound as there are C or O atoms, and that the C and O atoms are present in equal numbers. There are cases where the empirical formula is also the molecular formula (this was the case in all the examples and problems we did up to this point), but you can't count on this. Most often, the empirical formula and the molecular formula are not the same.

Since a combustion analysis always gives the empirical formula, this presents us with a problem. How do we know if it is also the molecular formula? And if it is not, then how can we figure out what the molecular formula is? To answer both of these questions we need to know the molar mass of the substance. But wait a minute—if you don't know the molecular formula in the first place, then you can't calculate the molar mass. Therefore, when you need a molar mass to go from an empirical formula to a molecular formula, we will give it to you (based on values from experimental results). For glucose, the molar mass is 180.155 g/mole. Now it's easy. We calculate the molar mass of the empirical formula and compare it to the molar mass of the compound:

Empirical formula: CH_2O

$$
\begin{array}{l}
\text{Molar mass} \\
\text{of } CH_2O
\end{array}
= \underbrace{12.011 \text{ g/mole}}_{C} + \underbrace{(2 \times 1.0079 \text{ g/mole})}_{H} + \underbrace{15.999 \text{ g/mole}}_{O}
$$

$$= 30.026 \text{ g/mole}$$

Molecular formula: $C_6H_{12}O_6$

Molar mass of $C_6H_{12}O_6 = 180.155$ g/mole

These numbers are not the same, so the empirical formula cannot be the molecular formula.

If the molar mass of the actual compound and that of the empirical formula are the same, then the empirical formula is also the molecular formula. When they are different, then simply divide the molar mass of the empirical formula into that of the actual compound. The result tells you how many empirical formula units go into making the actual compound. Then multiply the subscripts in the empirical formula by this result to get the molecular formula. Here is how it works for glucose:

Converting an empirical formula into a molecular formula

Step 1: Divide the molar mass of the actual compound by the molar mass of the empirical formula:

$$\frac{\text{Molar mass of glucose}}{\text{Molar mass of empirical formula}} = \frac{180.155 \text{ g / mole}}{30.026 \text{ g / mole}}$$

$$= 6 \qquad \text{Approximately}$$

Step 2: To get the subscripts for the molecular formula, multiply all the subscripts in the empirical formula by the number from step 1:

$$C_{1\times6}H_{2\times6}O_{1\times6} = C_6H_{12}O_6 \qquad \text{Molecular formula}$$

Now try the following practice problems.

PRACTICE PROBLEMS

7.27 A compound is known to contain carbon and hydrogen. It is uncertain if it also contains oxygen. A 0.500 g sample of the compound is burned to produce 0.409 g H_2O and 0.999 g CO_2.
   (a) What is the empirical formula for this compound?
   (b) What is the mass percent of each element in this compound?
   (c) The molar mass of this compound is 132.159 g/mole. What is the molecular formula for this compound?
   (d) Write the balanced combustion reaction (reaction with O_2) for this compound.

Answer (not worked out on purpose—you do it):
(a) C_2H_4O
(b) 54.6% C; 9.16% H; 36.2% O
(c) $C_6H_{12}O_3$
(d) $C_6H_{12}O_3 + O_2 \longrightarrow CO_2 + H_2O$
To balance the equation, look at C and H first. Then balance the elemental substance (O_2) last:

$$C_6H_{12}O_3 \quad + \quad O_2 \quad \longrightarrow \quad 6\,CO_2 + 6\,H_2O$$

This give us   We need another   18 O atoms on this side
3 O atoms   15 O atoms here to
match the 18 on the right

How do we make O_2 provide us with 15 O atoms? The answer is by multiplying it by 7.5 (that is, $\frac{15}{2}$):

$$C_6H_{12}O_3 + 7.5\,O_2 \quad \longrightarrow \quad 6\,CO_2 + 6\,H_2O$$

$7.5 \times 2 = 15$ O atoms

This is a perfectly correct balanced equation, but if you prefer whole numbers, then you can multiply all the coefficients by some number that makes them whole (in this case, 2):

$$2\,C_6H_{12}O_3 + 15\,O_2 \longrightarrow 12\,CO_2 + 12\,H_2O$$

7.28 A compound is known to contain carbon and hydrogen. It is not certain if it also contains oxygen. A 1.000 g sample of the compound is burned to produce 1.284 g H_2O and 3.137 g CO_2.
(a) What is the empirical formula for this compound?
(b) The molar mass of this compound is 28.054 g/mole. What is the molecular formula for this compound?
(c) Write the balanced combustion reaction (reaction with O_2) for this compound.

7.8 GOING BACK AND FORTH BETWEEN FORMULA AND PERCENT COMPOSITION

In this section we are going to exercise your developing stoichiometry skills. You just saw how combustion analysis can give you both the formula and the elemental mass percents for a compound. It is also possible to be given a formula and calculate the elemental mass percents, or to be given the elemental mass percents and calculate a formula. Let's continue to use glucose, $C_6H_{12}O_6$, as our example. To go quickly from this (or any) formula to the mass percent of each element in the compound, all we have to do is assume that we have 1 mole of the compound. The weight of 1 mole of any compound will be its molar mass in grams.

If we have 1 mole of glucose, then we have 6 moles of C atoms, 12 moles of H atoms, and 6 moles of O atoms (this comes straight out of the formula $C_6H_{12}O_6$). To get the mass percents, all we have to do is figure out how much these different moles of atoms weigh, divide each by the total weight of 1 mole of glucose (180.155 g/mole, calculated earlier), and multiply by 100. The key to getting started is to assume that you have 1 mole of the compound.

Going from the formula to the elemental mass percents:

Step 1: Assume you have 1 mole of the compound (glucose in this case). The formula tells you how many moles of each element you have. Figure out what everything weighs in grams.

$$6 \text{ moles C} \times \frac{12.011 \text{ g C}}{1 \text{ mole C}} = 72.066 \text{ g C}$$

$$12 \text{ moles H} \times \frac{1.0079 \text{ g H}}{1 \text{ mole H}} = 12.0948 \text{ g H}$$

$$6 \text{ moles O} \times \frac{15.999 \text{ g O}}{1 \text{ mole O}} = 95.994 \text{ g O}$$

Step 2: Divide the weight of each element by the total weight of the mole of compound (its molar mass) and multiply by 100. The molar mass of glucose ($C_6H_{12}O_6$) is 180.155 g.

$$\%C = \frac{72.066 \text{ g C}}{180.155 \text{ g}} \times 100 = 40.00\% \text{ C}$$

$$\%H = \frac{12.0948 \text{ g H}}{180.155 \text{ g}} \times 100 = 6.71\% \text{ H} \qquad \text{Arbitrarily rounded to the hundredths place}$$

$$\%O = \frac{95.994 \text{ g O}}{180.155 \text{ g}} \times 100 = \underline{53.28\% \text{ O}}$$

$$99.99\%$$

The percents should add up to 100. After rounding, it will.

We just went from formula to elemental mass percents. We can also go from elemental mass percents to formula. All the elemental mass percents for a compound add up to give 100%. This time the key to getting started is to assume that we have 100 g of the compound. This allows us to reinterpret the elemental percents as grams. We can then convert the grams of each element to moles for use as formula subscripts. This is done for you below, using the percents we just calculated for glucose.

Going from the elemental mass percents to the formula:

Step 1: Assume you have 100 g of the compound (glucose in this case). Convert the percents to grams, and then convert all the grams to moles.

$$40.00\% \text{ C} \longrightarrow 40.00 \text{ g C} \times \frac{1 \text{ mole C}}{12.011 \text{ g C}} = 3.33 \text{ moles C}$$

$$6.71\% \text{ H} \longrightarrow 6.71 \text{ g H} \times \frac{1 \text{ mole H}}{1.0079 \text{ g H}} = 6.66 \text{ moles H}$$

$$53.28\% \text{ O} \longrightarrow 53.28 \text{ g O} \times \frac{1 \text{ mole O}}{15.999 \text{ g O}} = 3.33 \text{ moles O}$$

Step 2: Use the moles of each element as formula subscripts and divide them by the smallest one to get whole numbers:

$$C_{\frac{3.33}{3.33}}H_{\frac{6.66}{3.33}}O_{\frac{3.33}{3.33}} \longrightarrow CH_2O$$

Notice that this method gives you the empirical formula. If you want the molecular formula, you need to know the molar mass of the compound.

You should now be able to quickly go from formula to elemental mass percents and vice versa. Just remember what initial assumption to make for each procedure.

PRACTICE PROBLEMS

7.29 What are the percents by mass of each element in hydrogen peroxide, H_2O_2?

Answer: 5.93% H, 94.07% O

7.30 What are the percents by mass of each element in trinitrotoluene, also known as TNT? The formula is $C_7H_5N_3O_6$.

7.31 A compound is found to have the following elemental mass percents:

$$\%Cl = 89.09 \qquad \%C = 10.06 \qquad \%H = 0.84$$

The molar mass of the compound is 119.378 g/mole. What are the empirical and molecular formulas for this compound?

7.32 A compound is known to contain C and H. It might also contain O. The compound is analyzed for C and H only, yielding the following elemental mass percents:

$$\%C = 54.53 \qquad \%H = 9.15$$

The molar mass of the compound is 88.106 g/mole. What are the empirical and molecular formulas for this compound?

7.9 DEALING WITH A LIMITING REACTANT

In Section 7.4 we introduced the concept of the limiting reactant. There are two ways to run a reaction. One way is in a stoichiometric (balanced) fashion, making sure that the reactants are present in exactly the ratio dictated by the balanced equation. When a reaction is run in a balanced fashion, then it doesn't matter which reactant you use to calculate the theoretical yield of product. The other way to run a reaction is to have a short supply of one reactant and an excess of all the others. The one in short supply is called the limiting reactant. The limiting reactant is important because it determines the amount of product that can be produced. Recall our earlier example using the cheesecake recipe:

Cheesecake recipe

3 blocks cream cheese

5 eggs

1 cup sugar

If you have 100 blocks of cream cheese, 100 cups of sugar, and 10 eggs, then the eggs are in short supply. The eggs are the limiting reactant, and the theoretical yield of cheesecakes is two.

When a reaction is run with a limiting reactant, then you must use the limiting reactant to calculate the theoretical yield of product. Unfortunately, when it comes to a chemical reaction, it won't always be so obvious if a reaction is being run in a stoichiometric or limiting fashion or which reactant is limiting. You need to learn how to determine these things. Let's work with the simplest kind of reaction first, one where all the reactants in the balanced equation are used in the same molar amount. Our HI reaction is a good example.

$$H_2 + I_2 \longrightarrow 2\ HI$$

Both reactants used at
the same mole amount
(1 mole of each)

If we ran this reaction by combining 10 moles of H_2 with 10 moles of I_2, the reaction would be balanced, since both reactants are present in the same molar amount, just what the balanced reaction calls for. If we combined 10 moles of H_2 with 11 moles of I_2, then we would be running the reaction in a limiting fashion. The H_2 would be the limiting reactant and the I_2 the excess reactant. There would be 1 mole of I_2 left at the end of the reaction. More importantly, if we wanted to know the theoretical yield of HI, we would have to use the H_2 to find out.

Using the limiting reactant to determine the theoretical yield

$$10\ \text{moles}\ H_2 \times \frac{2\ \text{moles HI}}{1\ \text{mole}\ H_2} = 20\ \text{moles HI}$$

Limiting   From balanced   Theoretical
reactant   equation   yield

To determine whether this reaction was being run in a stoichiometric or limiting fashion and which reactant was limiting, all we had to do was compare moles of H_2 to moles of I_2 and see if one was smaller than the other. The one present in a smaller molar amount was the limiting reactant. In actuality, you are much more likely to be told how much of each reactant you are starting with in grams rather than moles. If you are told you have 10.0 g H_2 and 10.0 g I_2, it is no longer obvious if the reaction is being run in a balanced or limiting fashion because the amounts are in grams whereas the reaction speaks in moles. It's that old language problem again, but all we need to do is to translate the gram amounts into moles using the appropriate molar mass conversion factors:

$$10.0\ g\ H_2 \times \frac{1\ \text{mole}\ H_2}{2.0158\ g\ H_2} = 4.96\ \text{moles}\ H_2$$

Inverse of molar mass of H_2

$$10.0\ g\ I_2 \times \frac{1\ \text{mole}\ I_2}{253.81\ g\ I_2} = 0.0394\ \text{mole}\ I_2$$

Inverse of molar mass of I_2

Now it's easy to see that the I_2 is present in a far smaller molar amount and is the limiting reactant. The theoretical yield of HI would be twice the number of moles of I_2.

For simple reactions where all the reactants are used in the same molar amount in the balanced equation, finding the limiting reactant is just a matter of converting everything to moles and then looking for the reactant present in the smallest molar amount. For reactions where the reactants are used in different molar amounts, such as the combustion of propane shown below, this won't work.

$$C_3H_8 + 5\,O_2 \longrightarrow 3\,CO_2 + 4\,H_2O$$

Reactants used in different mole amounts

For example, suppose we try to run this reaction by combining 1 mole of propane (C_3H_8) with 4 moles of oxygen (O_2). If we blindly did things as we did for the HI reaction, we would notice that propane was present in the smaller molar amount and call it the limiting reactant. But that would be wrong, because the balanced equation says that each mole of propane requires 5 moles of oxygen. We have 1 mole of propane but we are only given 4 moles of oxygen. The oxygen is in short supply and will run out first, so oxygen is the limiting reactant.

For reactions that have unequal reactant balancing coefficients, you can't just look for which reactant is present in the smaller molar amount to determine which, if any, is limiting. You must also take the balanced equation into account. There is a simple way to do this. After you convert all the amounts of reactants to moles, determine the mole-to-coefficient ratios by dividing each molar amount by its corresponding coefficient from the balanced equation. The reactant that generates the smallest mole-to-coefficient ratio is the limiting reactant. If all the reactants generate the same number, then the reaction is being run in a balanced fashion. For example, suppose we combine 6 moles of propane with 29 moles of oxygen. Are we running the reaction in a balanced or limiting fashion? If it is limiting, which reactant is the limiting reactant? To find out, we divide each molar amount by its balancing coefficient and check the results:

$$C_3H_8 \quad + \quad 5\,O_2 \longrightarrow 3\,CO_2 + 4\,H_2O$$

$$\frac{6 \text{ moles } C_3H_8}{1} = 6 \qquad \frac{29 \text{ moles } O_2}{5} = 5.8$$

Smallest mole-to-coefficient ratio, so O_2 is the limiting reactant

The numbers generated are not equal, so the reaction is being run in a limiting fashion. The smallest mole-to-coefficient ratio is associated with O_2, so oxygen is the limiting reactant. This method always works, but be careful! *You must always do this with moles and never with grams.* If you are given the amounts of reactants in grams (which almost always will be the case), you must first convert everything to moles before proceeding with this method.

Now it's time for you to try some practice problems.

PRACTICE PROBLEMS

Determine whether the reaction is being run in a balanced or limiting fashion using the method of dividing the moles of each reactant by its balancing coefficient. If it is limiting and you are asked for the theoretical yield of product, remember to use only the limiting reactant.

7.33 Consider the balanced propane combustion reaction

$$C_3H_8 + 5\,O_2 \longrightarrow 3\,CO_2 + 4\,H_2O$$

Suppose 10.0 g of propane are combined with 10.0 g of oxygen. What is the theoretical yield in grams of CO_2?

Answer: First we calculate the molar masses, since we'll probably need them:

$$C_3H_8 = 44.096 \text{ g/mole} \qquad O_2 = 31.998 \text{ g/mole}$$
$$CO_2 = 44.009 \text{ g/mole} \qquad H_2O = 18.015 \text{ g/mole}$$

Next, we convert the amounts of reactants to moles using the molar masses:

$$10.0 \text{ g } C_3H_8 \times \frac{1 \text{ mole } C_3H_8}{44.096 \text{ g } C_3H_8} = 0.227 \text{ mole } C_3H_8$$

$$10.0 \text{ g } O_2 \times \frac{1 \text{ mole } O_2}{31.998 \text{ g } O_2} = 0.312 \text{ mole } O_2$$

Next, we find the mole-to-coefficient ratios by dividing each of these molar amounts by their balancing coefficients from the balanced equation:

$$\frac{0.227 \text{ mole } C_3H_8}{1} = 0.227 \qquad \frac{0.312 \text{ mole } O_2}{5} = 0.0624$$

Smallest mole-to-coefficient ratio indicates limiting reactant

Finally, we use the limiting reactant, O_2, to calculate the theoretical yield of CO_2:

$$0.312 \text{ mole } O_2 \times \frac{3 \text{ moles } CO_2}{5 \text{ moles } O_2} \times \frac{44.009 \text{ g } CO_2}{1 \text{ mole } CO_2} = 8.24 \text{ g } CO_2$$

7.34 Consider the balanced propane combustion reaction

$$C_3H_8 + 5 O_2 \longrightarrow 3 CO_2 + 4 H_2O$$

(a) How many grams of oxygen are necessary to burn 100.0 g of propane in a balanced fashion?
(b) What is the theoretical yield of water in grams?

7.35 Chemical treatment of zinc sulfide (ZnS) with oxygen (O_2) gives zinc oxide (ZnO) and sulfur dioxide gas (SO_2).
(a) Write a balanced equation for the reaction of ZnS with O_2.
(b) If 10.0 g ZnS is combined with 10.0 g O_2, what is the theoretical yield of both products in grams?
(c) How much of the excess reactant would be left over (in grams)?
(d) Suppose after running this reaction only 7.50 g ZnO was recovered. What would be the percent yield of ZnO?

Stoichiometry problems of the type we have been doing in this chapter are the heart of chemical calculations. "How much?" and "What did I make?" are always important questions, whether you are baking cheesecakes or producing aspirin. Just keep in mind that our recipes—the balanced equations of chemical reactions and the chemical formulas within them—speak in moles. To use them, you must speak in moles also. If in doubt, convert to moles. It's almost always a good first step.

HAVE YOU LEARNED THIS?

Superconductor (p. 211)   Avogadro's number (p. 221)

Chemical reaction (p. 212)   Molar mass (p. 223)

Reactants, products (p. 212)   Theoretical yield (p. 226)

Gas-phase reaction (p. 213)   Actual yield (p. 226)

Balanced chemical equation (p. 215)   Percent yield (p. 226)

Stoichiometry (p. 218)   Combustion analysis (p. 233)

Limiting reactant (p. 218)   Empirical formula (p. 238)

Mole (p. 221)   Molecular formula (p. 238)

CHEMICAL REACTIONS

7.36 Tap water contains dissolved oxygen gas, symbolized as O_2 (aq) [(aq) stands for *aqueous*, which means dissolved in water]. Adding heat to tap water causes the dissolved oxygen to leave. Which of the two reactions below represents a chemical reaction, and which does not? Defend your answer.
(a) $2 H_2O_2(l) \longrightarrow 2 H_2O(l) + O_2(g)$

(b) $H_2O(l) + O_2(aq) \xrightarrow{\text{Heat}} H_2O(l) + O_2(g)$

7.37 Someone claims that a substance has undergone a chemical reaction. What would you have to demonstrate to prove this?

7.38 Recall the reaction to make the superconductor at the beginning of this chapter. What could be done to prove that the product of the chemical reaction is truly a new compound and not just a heterogeneous mixture of the reactants?

7.39 Under what circumstances (if any) does it *not* take energy to break a chemical bond?

7.40 Consider the following reaction. Describe it in terms of which bonds must be broken and which bonds must be formed.

$$2\,Cl\!-\!Cl + \underset{\underset{H}{|}}{\overset{\overset{H}{|}}{H\!-\!C}}\!-\!\underset{\underset{H}{|}}{\overset{\overset{H}{|}}{C}}\!-\!H \longrightarrow \underset{\underset{Cl}{|}}{\overset{\overset{H}{|}}{H\!-\!C}}\!-\!\underset{\underset{Cl}{|}}{\overset{\overset{H}{|}}{C}}\!-\!H + 2\,H\!-\!Cl$$

7.41 For most gas-phase chemical reactions, where does the energy come from to begin to break covalent bonds in reactant molecules?

7.42 Cooling a gas-phase reaction slows it down. Cooling it enough can actually stop it. Why do you think this is so?

7.43 Write down how you would explain to someone how a chemical reaction changes one set of compounds into another.

7.44 For most gas-phase reactions, increasing the number of reactant molecules in a flask makes the reaction go faster. Why do you think this is so?

BALANCING CHEMICAL EQUATIONS

7.45 Why must a chemical equation be balanced?

7.46 What is wrong with changing the subscripts within the formula of a compound to balance a chemical equation?

7.47 Balance the following chemical equation by inspection using the proper balancing coefficients:

$$C_2H_4 + O_2 \longrightarrow CO_2 + H_2O$$

7.48 The following chemical equation is properly balanced. Translate it into an English sentence.

$$CH_4 + 2\,O_2 \longrightarrow CO_2 + 2\,H_2O$$

7.49 Balance the following chemical equation by inspection, adding the proper balancing coefficients:

$$Fe_2O_3 + C \longrightarrow Fe + CO_2$$

7.50 Balance the following chemical equation by inspection, adding the proper balancing coefficients:

$$Al_2Cl_6 + H_2O \longrightarrow Al(OH)_3 + HCl$$

7.51 Balance the following chemical equation by inspection, adding the proper balancing coefficients:

$$KClO_3 \longrightarrow KCl + O_2$$

7.52 Balance the following chemical equation by inspection, adding the proper balancing coefficients:

$$C_2H_2 + O_2 \longrightarrow H_2O + CO_2$$

7.53 Translate the balanced chemical equation from Problem 7.52 into English. The compound C_2H_2 is called acetylene and is used as a fuel for torches.

SIMPLE STOICHIOMETRY

7.54 How is a balanced chemical equation like a recipe?

7.55 What is meant by "limiting reactant"?

7.56 Suppose you run the reaction $A + 2\,B \longrightarrow C$, and you discover that the product C is contaminated with A. How would you explain this?

7.57 Consider the following recipe for a dozen cookies:

>**Recipe for one dozen sugar cookies**
>1 egg
>2 cups flour
>3 cups sugar
>$\frac{1}{4}$ pound butter
>1 cup milk

(a) Write at least five conversion factors from the above recipe that might be useful for solving stoichiometry problems. Include units.
(b) Using the method of unit conversion, calculate how many cups of flour are necessary for baking 30 cookies. Write down your method and cross out units that cancel.
(c) Why can't you answer the following question using just the information in the recipe: How many eggs are required to completely use up 1 container of milk?

(d) Given the additional conversion factor that 1 container of milk = 4 cups of milk, answer the question in part (c). Write down your method and cross out units that cancel.

(e) Write the recipe in chemical equation form.

(f) You want to make as many cookies as possible. You open the cupboard and discover that you have only 3 cups of flour and 3 cups of sugar (plus unlimited amounts of all the other ingredients).

  (i)   Which is the limiting reactant, the sugar or the flour?

  (ii) How many cookies can you make?

  (iii) How much of the excess reactant (flour or sugar) will be left over?

7.58 What is the difference between running a chemical reaction in a stoichiometric (balanced) versus nonstoichiometric (unbalanced) fashion?

7.59 The recipe for a cheesecake is:

> 3 blocks of cream cheese
>
> 5 eggs
>
> 1 cup sugar

(a) To bake 3 cakes, how many eggs do you need? Write down your method and cross out units that cancel.

(b) If you have 25 eggs, 9 blocks of cream cheese, and 4 cups of sugar, how many cakes can you make? Write down your method and cross out units that cancel.

(c) If you have 63 blocks of cream cheese, how many eggs do you need? Write down your method and cross out units that cancel.

THE MOLE

7.60 How is a mole similar to a dozen?

7.61 When is it allowed to insert the word mole into a chemical equation when translating it into English?

7.62 The human population of our planet is about 5 billion people. What percentage of a mole is 5 billion?

7.63 How many bicycles are there in 1 mole of bicycles? How many tires are there in 1 mole of bicycles?

7.64 How many O_2 molecules are there in 1 mole of O_2 molecules? How many O atoms are there in 1 mole of O_2 molecules?

7.65 How many pennies are there in 2.5 moles of pennies? How many dollars does this equal? Answer both questions by using conversion factors, and show which units cancel.

7.66 How many years are there in a mole of seconds? Answer this question by using conversion factors, and show which units cancel.

7.67 Translate the following balanced equation into English, first without using the word moles, then again with the word moles.

$$2 SO_2 + O_2 \longrightarrow 2 SO_3$$

7.68 Sometimes using the word moles when translating an equation into English can be very helpful. For example, consider the following *correctly balanced* equation. What difficulty do you run into when you try to translate it into English without using the word moles? How does using the word moles solve the difficulty?

$$C_2H_2 + \tfrac{5}{2}O_2 \longrightarrow 2 CO_2 + H_2O$$

7.69 In general, how much does 1 mole of atoms of any element weigh in grams?

7.70 In general, how much does 1 mole of any molecule weigh in grams?

7.71 How do you calculate the molar mass of a compound?

7.72 How many atoms of the $_6^{12}C$ isotope are in 12 g?

7.73 How many carbon atoms are in a sample of naturally occurring carbon that weighs 12.011 g?

7.74 If you have 1 mole of glucose ($C_6H_{12}O_6$):
(a) How many moles of carbon atoms do you have?
(b) How many moles of hydrogen atoms do you have?
(c) How many oxygen atoms do you have? (Note that we are not asking for moles here.)

7.75 Consider the ammonia (NH_3) molecule.
(a) If you have 1 mole of ammonia, how many moles of H atoms do you have?
(b) If you have 2 moles of ammonia, how many moles of H atoms do you have?
(c) If you have 2 moles of ammonia, how many N atoms do you have? (We are not asking for moles here.)

REACTION STOICHIOMETRY

7.76 What do we mean by the theoretical yield of a reaction?

7.77 What do we mean by the actual yield of a reaction?

7.78 Why is the actual yield of a reaction often not equal to the theoretical yield?

7.79 What do we mean by the percent yield of a reaction?

7.80 A student runs a reaction to prepare 40.0 g of aspirin and yet only recovers 15.5 g. What is the percent yield?

7.81 How does molar mass solve the "language problem" when using a chemical equation as a recipe?

7.82 Consider the unbalanced reaction

$$NO + O_2 \longrightarrow NO_2$$

(a) Balance the reaction.
(b) Translate this reaction into English using the word mole wherever you can.
(c) To produce NO_2 by the reaction you just wrote for part (b), how many grams of NO and O_2 must you combine?
(d) What is the theoretical yield of NO_2 from the reaction in part (b)?
(e) You carry out the reaction in part (b) and recover 22.5 g of NO_2. What is the percent yield?

7.83 Consider the unbalanced reaction

$$HCl + Zn \longrightarrow H_2 + ZnCl_2$$

(a) Balance the reaction.
(b) Translate this reaction into English using the word mole wherever you can.
(c) To run the reaction you just wrote for part (b), how many grams of HCl and Zn must you combine?
(d) What is the theoretical yield of H_2 from the reaction in part (b)?
(e) You recover 2.00 g of H_2 after carrying out the reaction in part (b). What is the percent yield?

7.84 Consider the unbalanced reaction

$$Na + Cl_2 \longrightarrow NaCl$$

(a) Balance the reaction.
(b) Translate this reaction into English using the word mole wherever you can.
(c) To run the reaction you just wrote for part (b), how many grams of Na and Cl_2 must you combine?
(d) What is the theoretical yield of NaCl from the reaction in part (b)?
(e) You carry out the reaction in part (b) and recover 45.50 g of NaCl. What is the percent yield?

7.85 The formula $C_6H_{12}O_6$ is a source of many conversion factors. Write at least three of them.

7.86 The balanced reaction $2\,AgBr \longrightarrow 2\,Ag + Br_2$ and the formulas of the substances in it are a source of many conversion factors. Write all that are possible, using the word mole in all of them.

7.87 The following reaction is unbalanced:

$$I_2 + Cl_2 \longrightarrow ICl_3$$

(a) Balance the equation.
(b) Write all the possible conversion factors that can come from the balanced equation and the formulas of the substances in it, using the word mole in all of them.

7.88 How many H atoms are there in 2.0158 g of H atoms?

7.89 How many moles of O_2 molecules are in 24.0 g of O_2 molecules?

7.90 How many moles of O atoms are in 24.0 g of O_2 molecules?

7.91 Consider sulfuric acid, H_2SO_4, used in car batteries.
(a) What is the molar mass of sulfuric acid?
(b) How much does 1 mole of H_2SO_4 weigh in grams?
(c) How much does 2.50 moles of H_2SO_4 weigh in grams?
(d) How much does 1000 molecules of H_2SO_4 weigh in grams? [*Hint*: Start by writing "1000 molecules H_2SO_4" and then apply conversion factors. Remember our admonition, if in doubt, convert to moles.]

7.92 How many O_2 molecules are there in 1.00 g of O_2 molecules?

7.93 Suppose you wanted one billion (1.00×10^9) water molecules and you didn't have time to sit and count them out. How many grams of water would you need to get one billion water molecules?

7.94 How many grams of glucose ($C_6H_{12}O_6$) would you need to get 5.00×10^{30} carbon atoms?

7.95 Consider the following balanced chemical reaction:

$$2\,H_2O_2 \longrightarrow 2\,H_2O + O_2$$

(a) Given 20.0 g of H_2O_2 (hydrogen peroxide), how many grams of water will you make?
(b) How many grams of hydrogen peroxide would you need to make 20.0 g of water?
(c) How many grams of hydrogen peroxide would you need to make 20.0 g of oxygen?

7.96 Consider the following balanced chemical reaction:

$$SCl_4 + 2\,H_2O \longrightarrow SO_2 + 4\,HCl$$

(a) How many grams of H_2O will react with 5.00 g of SCl_4?
(b) How many grams of SO_2 can you make from 10.00 g H_2O?

(c) Suppose you react 5.00 g of SCl_4 with the amount of H_2O you calculated in part (c). How many grams of HCl will you form?

7.97 Consider the following balanced chemical equation:

$$2 H_2 + O_2 \longrightarrow 2 H_2O$$

(a) How many grams of water can you make from 5.00 g of H_2 and an excess amount of O_2?
(b) How many grams of O_2 do you need to make 5.00 g of H_2O?
(c) Given 100.0 g H_2, how many grams of O_2 are required to run the reaction in a stoichiometric fashion?
(d) What is the theoretical yield of water upon combining 50.0 g O_2 with an excess amount of H_2?
(e) Express the answer to part (d) in terms of the number of water molecules.

7.98 Consider the following unbalanced equation:

$$P + O_2 \longrightarrow P_2O_5$$

(a) How many grams of phosphorus (P) are required to react completely with 20.0 g O_2?
(b) What is the theoretical yield if you combine the amounts of reactants in part (a)?

7.99 Consider the following unbalanced equation:

$$H_2 + N_2 \longrightarrow NH_3$$

(a) To run this reaction in a balanced fashion, how much nitrogen is required if you start with 10.0 g of H_2?
(b) How many grams of ammonia (NH_3) will you produce?
(c) How many molecules of ammonia will you produce?

FORMULAS AND PERCENT COMPOSITIONS FROM COMBUSTION ANALYSIS

7.100 A compound has the empirical formula C_2H_4O. Its molar mass is about 90 g/mole. What is its molecular formula?

7.101 An organic compound of carbon and hydrogen has the empirical formula CH. What is its molecular formula if its molar mass is:
(a) 26 g/mole?   (b) 52 g/mole?   (c) 78 g/mole?

7.102 A 1.540 g sample of a liquid burns in oxygen to produce 2.257 g CO_2 and 0.9241 g H_2O.
(a) What are the mass percents of the elements present in this sample?
(b) What is the empirical formula for this compound?
(c) The molar mass of this compound is determined to be about 30 g/mole. What is the molecular formula for this compound?

7.103 A 2.136 g sample of a solid burns in oxygen to produce 5.933 g of CO_2 and 1.227 g of H_2O.
(a) What are the mass percents of the elements present in this sample?
(b) What is the empirical formula for this compound?
(c) The molar mass of this compound is determined to be about 94 g/mole. What is the molecular formula for this compound?

7.104 A 1.000 g sample of a liquid burns in oxygen to produce 3.383 g CO_2 and 0.692 g H_2O.
(a) What are the mass percents of the elements present in this sample?
(b) What is the empirical formula for this compound?
(c) The molar mass of this compound is determined to be about 80 g/mole. What is the molecular formula for this compound?

7.105 A 1.000 g sample of a liquid that contains only carbon and hydrogen burns in oxygen to produce 1.284 g H_2O.
(a) What are the mass percents of the elements present in this sample?
(b) What is the empirical formula for this compound?
(c) The molar mass of this compound is determined to be about 72 g/mole. What is the molecular formula for this compound?

7.106 Can a compound have an empirical formula and a molecular formula that are identical? If so, give an example.

7.107 A compound used as an insecticide contains only C, H, and Cl. When a 3.200 g sample is burned in oxygen, 6.162 g of CO_2 and 0.9008 g of H_2O are produced. What are the mass percents of the elements present in this sample?

7.108 A compound used as an insecticide contains only C, H, and Cl. When a 3.000 g sample is burned in oxygen, 2.724 g of CO_2 and 0.5575 g of H_2O are produced.
(a) What are the mass percents of the elements present in this sample?
(b) What is the empirical formula for this compound?

GOING BACK AND FORTH BETWEEN FORMULA AND PERCENT COMPOSITION

7.109 Determine the simplest formula of the compound with the following mass percents of the elements present: 40.00% C; 6.71% H; 53.28% O.

7.110 Determine the simplest formula of the compound with the following mass percents of the elements present: 58.5% C; 4.91% H; 19.5% O; 17.1% N.

7.111 Determine the simplest formula of the compound with the following mass percents of elements present: 26.4% Na; 36.8% S. This compound also contains oxygen.

7.112 Determine the simplest formula of the compound with the following mass percents of elements present: 43.2% K; 39.1% Cl. This compound also contains oxygen.

7.113 Ethanol, the alcohol in beer and wine, has the formula C_2H_6O. Calculate the mass percent of each element in ethanol.

7.114 Ethylene glycol, used for antifreeze, has the formula $C_2H_6O_2$. Calculate the mass percent of each element in ethylene glycol.

7.115 Penicillin G, used as an antibiotic, has the formula $C_{16}H_{18}N_2O_4S$. Calculate the mass percent of each element in penicillin G.

7.116 The hormone thyroxine has the formula $C_{15}H_{11}NO_4I_4$. Calculate the mass percent of each element in thyroxine.

DEALING WITH A LIMITING REACTANT

7.117 A gaseous mixture containing 5.00 moles of H_2 and 7.00 moles of Br_2 reacts to form HBr.
(a) Write a balanced equation for this reaction.
(b) Which reactant is limiting?
(c) What is the theoretical yield for this reaction in moles?
(d) What is the theoretical yield for this reaction in grams?
(e) How many moles of excess reactant are left over at the end of the reaction?
(f) How many grams of excess reactant are left over at the end of the reaction?

7.118 Chlorine (Cl_2) and fluorine (F_2) react to form ClF_3. A reaction vessel is charged with 2.50 moles of Cl_2 and 6.15 moles of F_2.
(a) Write a balanced equation for this reaction.
(b) Which reactant is limiting?
(c) What is the theoretical yield for this reaction in moles?
(d) What is the theoretical yield for this reaction in grams?

(e) How many moles of excess reactant are left over at the end of the reaction?

(f) How many grams of excess reactant are left over at the end of the reaction?

7.119 5.00 g of solid sodium (Na) and 30.0 g of liquid bromine (Br_2) react to form NaBr.

(a) Write a balanced equation for this reaction.

(b) Which reactant is limiting?

(c) What is the theoretical yield for this reaction in grams?

(d) How many grams of excess reactant are left over at the end of the reaction?

(e) When this reaction is actually performed, 14.7 g of NaBr are recovered. What is the percent yield of the reaction?

7.120 Chlorine (Cl_2) and fluorine (F_2) react to form ClF_3. A reaction vessel is charged with 10.00 g Cl_2 and 10.00 g F_2. [*Hint:* Refer to Problem 7.118.]

(a) Write a balanced equation for this reaction.

(b) Which reactant is limiting?

(c) What is the theoretical yield for this reaction in grams?

(d) How many grams of excess reactant are left over at the end of the reaction?

(e) When this reaction is actually performed, 12.50 g of ClF_3 are recovered. What is the percent yield of the reaction?

7.121 Sodium (Na) reacts with hydrogen (H_2) to form sodium hydride (NaH). A reaction mixture contains 10.00 g Na and 0.0235 g H_2.

(a) Write a balanced equation for this reaction.

(b) Which reactant is limiting?

(c) What is the theoretical yield for this reaction in grams?

(d) How many grams of excess reactant are left over at the end of the reaction?

(e) When this reaction is actually performed, 0.428 g of NaH are recovered. What is the percent yield of the reaction?

7.122 Butane (C_4H_{10}), used as the fuel in disposable lighters, reacts with oxygen (O_2) to produce CO_2 and H_2O. Suppose 10.00 g butane is combined with 10.00 g O_2.

(a) Write a balanced equation for the combustion of butane.

(b) Which reactant is limiting?

(c) What is the theoretical yield of each product for this reaction in grams?

(d) How many grams of excess reactant are left over at the end of the reaction?

(e) How many additional grams of the limiting reactant are required to run this reaction in a balanced fashion?

WorkPatch Solutions

7.1 (a) Not a chemical reaction because the same compound exists on both sides of the arrow. This is simply a phase change.

(b) A chemical reaction because new compounds, different from the reactants, are produced.

(c) Not a chemical reaction since the same compounds appear on both sides of the arrow.

7.2 There are two correct answers to this question:

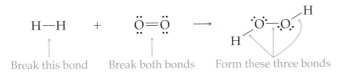

or

7.3 (a) 32.066 g/mole (This is just the relative atomic weight with the units grams/mole added.)

(b) Since 1 mole of S weighs 32.066 g and we have twice this (64.132 g), then we have 2 moles of S.

(c) 2 moles of S is $2 \times (6.02 \times 10^{23})$ sulfur atoms $= 1.20 \times 10^{24}$ sulfur atoms.

7.4 (a) $1 \text{ mole CH}_4 \text{ molecules} \times \dfrac{1 \text{ mole C atoms}}{1 \text{ mole CH}_4 \text{ molecules}} = 1 \text{ mole C atoms}$

Conversion factor
from formula CH_4

(b) $1 \text{ mole CH}_4 \text{ molecules} \times \dfrac{4 \text{ moles H atoms}}{1 \text{ mole CH}_4 \text{ molecules}} = 4 \text{ mole H atoms}$

Conversion factor
from formula CH_4

(c) $1 \text{ mole CH}_4 \text{ molecules} \times \dfrac{1 \text{ mole C atoms}}{1 \text{ mole CH}_4 \text{ molecules}} = \dfrac{6.02 \times 10^{23} \text{ C atoms}}{1 \text{ mole C atoms}}$

Same as part (a)

$= 6.02 \times 10^{23} \text{ C atoms}$

(d) $1 \text{ mole CH}_4 \text{ molecules} \times \dfrac{4 \text{ moles H atoms}}{1 \text{ mole CH}_4 \text{ molecules}} = \dfrac{6.02 \times 10^{23} \text{ H atoms}}{1 \text{ mole H atoms}}$

Same as part (b)

$= 2.41 \times 10^{24} \text{ H atoms}$

7.5 (a) Four moles of aluminum and three moles of oxygen react to give two moles of aluminum oxide.

(b) Conversion factors from reaction:

$$\dfrac{4 \text{ moles Al}}{3 \text{ moles O}_2} \qquad \dfrac{3 \text{ moles O}_2}{4 \text{ moles Al}} \qquad \dfrac{4 \text{ moles Al}}{2 \text{ moles Al}_2\text{O}_3} \qquad \dfrac{2 \text{ moles Al}_2\text{O}_3}{4 \text{ moles Al}}$$

$$\dfrac{3 \text{ moles O}_2}{2 \text{ moles Al}_2\text{O}_3} \qquad \dfrac{2 \text{ moles Al}_2\text{O}_3}{3 \text{ moles O}_2}$$

(c) Conversion factors from the formula Al_2O_3:

$$\dfrac{2 \text{ moles Al atoms}}{1 \text{ mole Al}_2\text{O}_3} \qquad \dfrac{1 \text{ mole Al}_2\text{O}_3}{2 \text{ moles Al atoms}}$$

$$\dfrac{3 \text{ moles O atoms}}{1 \text{ mole Al}_2\text{O}_3} \qquad \dfrac{1 \text{ mole Al}_2\text{O}_3}{3 \text{ moles O atoms}}$$

7.6 $10.0 \text{ g HI} \times \dfrac{1 \text{ mole HI}}{127.912 \text{ g HI}} \times \dfrac{1 \text{ mole I}_2}{2 \text{ moles HI}} \times \dfrac{253.81 \text{ g I}_2}{1 \text{ mole I}_2} = 9.92 \text{ g I}_2$

Balanced   Molar mass
reaction   of I_2

Chapter 8

The Transfer of Electrons Between Atoms in a Chemical Reaction

WHAT IS ELECTRICITY? 8.1

Did you ever accidentally bite down on a piece of aluminum foil? Did it hurt?

The pain is due to a particular kind of chemical reaction, one that produces an electric jolt that goes straight to the nerve endings in your teeth. Electricity from a chemical reaction? Yes, some reactions can produce electricity. However, before we go on to study such reactions, let's quickly review what electricity is.

Electricity is simply the flow of electrons. In many ways, electricity is similar to water flowing through a hose. **Voltage** is analogous to water pressure. The higher the voltage, the greater the "pressure" or "push" behind the electrons that causes them to flow. (Remember, we symbolize an electron by e^-.)

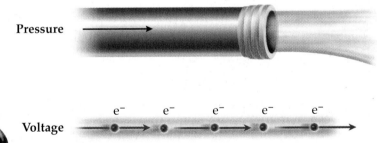

The "push" begins at a power plant where electricity is produced mechanically by a generator consisting of copper wire wound on a rotor that is spun between the poles of a magnet. The rotor is spun by a turbine that is usually powered by steam from burning oil or coal or by flowing water (hydroelectric).

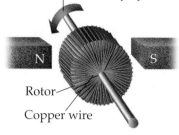

Axle turned mechanically by turbine

N   S

Rotor
Copper wire

As the copper wire moves through the magnetic field, electrons in the wire are pushed and made to flow, producing electricity. What you buy from the power company is the right to use their flowing electrons. They flow through your electronic devices and do work and then return to the power company to be pushed again.

Electrons can also be made to flow through a wire by chemical means. The chemical reactions that do this often come packaged in containers called *batteries*. In this chapter you will learn how some chemical reactions can generate electricity, and how to recognize this type of reaction.

8.2 ELECTRON BOOKKEEPING—OXIDATION STATES

While all chemical reactions convert reactants into products, some chemical reactions also transfer one or more electrons from one atom to another as part of the conversion process. For example, the reaction of calcium metal with acid (aqueous H^+ ions) occurs as follows:

$$Ca\!: + 2\,H^+ \longrightarrow Ca^{2+} + H\!\cdot\!\cdot H$$

This pair of valence electrons starts out belonging to Ca . . .

but ends up belonging to (and being shared between) the H atoms.

During the course of this reaction there must be a flow or transfer of electrons from the calcium atom to the hydrogen ions. Another way to say this is that the calcium atom lost some electrons while the H^+ ions gained them. It is this type of reaction that can be used to push electrons through a wire and produce electricity. These reactions are known as **electron transfer reactions**. Note that only valence electrons are transferred in these reactions. Core electrons are too tightly held to their respective nuclei to be easily transferred.

Not all reactions occur with electron transfer. Some proceed without any atom gaining or losing electrons. Since electron transfer reactions are the only ones that can be used to produce electricity, it is important that we be able to distinguish which reactions proceed with electron transfer and which do not. This is fairly easy if the reaction is written as shown above, with the valence electrons explicitly shown. Written this way, you can see that the calcium lost 2 electrons because its lone pair disappeared, and you can see that the hydrogens gained 2 electrons because a covalent bond appeared between the two H's. The problem is that you will rarely see reactions written this way. Rather, you will be presented with just the balanced reaction, which for our example is $Ca + 2\,H^+ \longrightarrow Ca^{2+} + H_2$. Now how can you tell that electrons are being transferred as reactants convert to products? Perhaps you feel that you still can tell because charges are changing (for example, calcium is going from neutral to +2). For calcium to go from neutral to +2 it must lose 2 electrons, so electrons must be transferred. But consider the reaction $H_2 + Cl_2 \longrightarrow 2\,HCl$. No charges are changing here. The reactants and product are all neutral molecules. Nevertheless, this reaction is also an electron transfer reaction.

So, given just a balanced reaction, how are we to tell whether electron transfer is occurring? What we need is electron ownership information. For example, suppose we wrote the $H_2 + Cl_2$ reaction in the following way:

$$H_2 + Cl_2 \longrightarrow 2\,HCl$$

Valence electrons:	Each H	Each Cl	H atom	Cl atom
ownership	atom owns	atom owns	owns 0	owns 8
information	1 electron	7 electrons	electrons	electrons

With the valence electron ownership information given below the balanced equation, we can now tell that electron transfer occurs. According to the ownership information, the H atoms in the H_2 molecule each own 1 valence electron, but they end up owning 0 electrons in the HCl molecules. Each H atom has lost an electron. The ownership information also indicates that each Cl atom has gained 1 electron. Evidently, electrons moved from H atoms to Cl atoms during the course of this reaction, so it can be classified as an electron transfer reaction.

Unfortunately, balanced chemical reactions don't come with electron ownership information written below every atom. We need a way, therefore, to determine the number of electrons owned by every atom on both the reactant and product sides of a chemical equation. This can be done in much the same way that we keep track of our personal finances, by doing some bookkeeping. Chemists have created an electron bookkeeping method called the *oxidation state method*. Its goal is to assign a number to every atom, called an **oxidation number** or **oxidation state**, that can tell us how many valence electrons each atom owns. By assigning these oxidations states, we will be able to look at a balanced reaction and tell whether electron transfer occurs.

The best way to understand how oxidation states are assigned to atoms in a molecule is to use electron dot diagrams. We'll demonstrate with a water molecule:

$$\overset{\displaystyle \cdot\overset{\cdots}{\underset{\cdots}{O}}\cdot}{H \qquad H}$$

There are 8 valence electrons (dots) in this diagram. Electron bookkeeping means asking the question, "How many electrons does each atom in the molecule own?" Since the answer depends on how the bookkeeping is done (how we define the word "own"), we need to agree on a few rules. The first rule states:

Lone pairs of electrons are assigned completely to the atom on which they are drawn in the dot diagram.

(This makes sense, since lone pairs are not shared with another atom.)

Oxygen owns its lone pairs completely

What about the bonding electrons? In our discussion of covalent bonding and the octet rule in Chapter 5, we double counted the bonding electrons in a

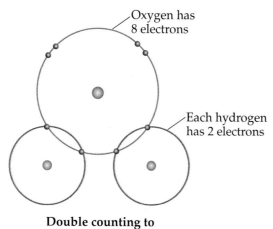

Oxygen has
8 electrons

Each hydrogen
has 2 electrons

**Double counting to
satisfy the octet rule**

dot structure. For example, consider one of the O–H bonds in water. The oxygen atom was said to own both shared electrons, and the hydrogen atom was also said to own both shared electrons. This is how we satisfied the octet rule for each atom (or the duet rule in the case of hydrogen).

The oxidation state method of bookkeeping is different. The rule for shared electrons in the oxidation state bookkeeping method is:

In a chemical bond, the more electronegative atom gets complete ownership of both the shared electrons.

Recall from Section 5.6 that electronegativity is a measure of an atom's ability to attract shared electrons to itself. It is often useful to remember that the four most electronegative elements in the periodic table are fluorine (EN = 4.0), oxygen (EN = 3.5), chlorine (EN = 3.0), and nitrogen (EN = 3.0), the electron "hogs." Hydrogen has only a moderate electronegativity value (EN = 2.1).

Let's now apply the oxidation state system of bookkeeping to the water molecule:

Electron bookkeeping the oxidation state way

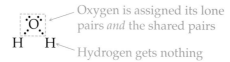

Oxygen is assigned its lone pairs *and* the shared pairs

Hydrogen gets nothing

Since oxygen is more electronegative than hydrogen, oxygen is assigned all the electrons in the O–H bonds. The hydrogen atoms get none of these electrons. Oxygen also gets its own lone pairs. So, by the oxidation state method of electron bookkeeping, oxygen owns all 8 of the electrons in the molecule, and the hydrogen atoms get none.

At least for now, the only way we can do this type of bookkeeping is to have a dot diagram to look at. In Chapter 6 we saw that it was necessary to have a dot diagram to be able to use VSEPR theory to predict molecular shape. Many times in chemistry you will find that the first step in understanding a problem is to draw a dot diagram.

8.1 WORKPATCH

Using the oxidation state method of electron bookkeeping, determine how many electrons each atom owns in the molecule NH_2F. [*Hint:* Start by drawing a dot diagram. Nitrogen is the central atom.]

You should have found that only one atom in the NH_2F molecule owns an octet by this method of bookkeeping. Was it the most electronegative atom? Check your answer at the end of the chapter.

At this point you might be puzzled by the existence of two different systems for electron bookkeeping. First we saw the double counting method, which we used to satisfy the octet rule. Now we've introduced the oxidation state method, which does not double count shared electrons, but instead assigns them to the most electronegative atom. The two counting methods come up with different answers for how many electrons each atom owns.

Which counting method is right? The answer is, both of them are "right." Which one you use depends on what you want to know. If you want to know whether a particular dot diagram is valid with respect to the octet rule, then double count. If you want to know whether electrons are transferred from atom to atom in a chemical reaction, then use the oxidation state method.

Now that we know how to use the oxidation state method to formally assign electrons to the atoms in a molecule, we are ready to calculate actual oxidation states for the atoms. We do this by comparing the number of electrons that each atom owns (using oxidation state bookkeeping) with the number of valence electrons the free atom of that element owns. For the representative elements, as we learned in Chapter 3, the number of valence electrons on a free atom is equal to the element's group number in the periodic table. For example, the number of valence electrons for an isolated oxygen atom (group VIA) is 6. The number of valence electrons for an isolated hydrogen atom (group IA) is 1.

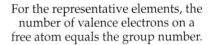

For the representative elements, the number of valence electrons on a free atom equals the group number.

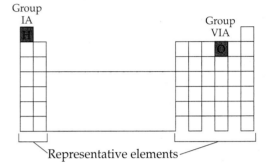

Representative elements

The **oxidation state** of an atom in a molecule is simply the difference between the number of valence electrons on a free atom of that element and the number of electrons that are assigned to the atom by oxidation state bookkeeping:

Number of valence electrons on a free atom of the element

− Number of electrons assigned to the atom by oxidation state bookkeeping

Oxidation state (or oxidation number)

For water,

	Oxygen	Hydrogen	
Number of valence electrons on a free atom of the element	6	1	
− Number of electrons assigned to the atom by oxidation state bookkeeping	− 8	− 0	
Oxidation state:	−2	+1	

Notice that the oxidation state can be either negative or positive. It is negative if the atom owns more electrons by oxidation state bookkeeping than it would as a free atom; it is positive if the atom owns fewer electrons than it would as a free atom. An oxidation state of zero means the atom owns the same number of electrons by oxidation state bookkeeping as it owns as a free atom. To summarize:

If the oxidation state is +n, then the atom owns n fewer electrons in the molecule or compound than it would as a free atom.

If the oxidation state is −n, then the atom owns n more electrons in the molecule or compound than it would as a free atom.

The oxygen atom in water has a -2 oxidation state because, by oxidation state bookkeeping, it owns 2 more electrons than it would as a free atom. The hydrogen atoms in water are in the $+1$ oxidation state because they each own 1 less electron than the free atom. At this point, it is very important to realize that these oxidation states are artificial numbers that result from arbitrary bookkeeping rules. They do not indicate charges. For example, the oxygen atom in a water molecule is partially negative ($\delta-$) due to the polar covalent O–H bond, but it does not carry a full -2 charge.

PRACTICE PROBLEMS

8.1 Consider the NH_2F molecule from WorkPatch 8.1. Assign oxidation states to each atom in the molecule.

Answer: In WorkPatch 8.1 you should have assigned 8 electrons to F, 6 to N, and none to either H atom, using the oxidation state method of electron bookkeeping. Now we just subtract from each of these numbers the number of valence electrons on the free atoms:

$$H_A-\ddot{N}-H_B$$
$$|$$
$$:\ddot{F}:$$

	Fluorine	Nitrogen	Hydrogen$_A$	Hydrogen$_B$
Number of valence electrons on a free atom of the element	7	5	1	1
$-$ Number of electrons assigned to the atom by oxidation state bookkeeping	$-$ 8	$-$ 6	$-$ 0	$-$ 0
Oxidation state:	-1	-1	$+1$	$+1$

8.2 Consider the carbon dioxide (CO_2) molecule.
  (a) Assign oxidation states to each atom in the molecule. [*Hint:* Draw a dot diagram first.]
  (b) How many electrons does the C atom own by the oxidation state method of electron bookkeeping?
  (c) How many more or fewer valence electrons are assigned to the carbon atom in CO_2 by the oxidation state method than are present on a free carbon atom?
  (d) Is it correct or incorrect to say that the C atom in CO_2 has a complete octet of valence electrons? Explain.

8.3 Consider the methane (CH_4) molecule.
  (a) Assign oxidation states to each atom in the molecule. [*Hint:* Draw a dot diagram first.]
  (b) How many electrons does the C atom own by the oxidation state method of electron bookkeeping?
  (c) How many more or fewer valence electrons are assigned to the carbon atom in CH_4 by the oxidation state method than are present on a free carbon atom?

8.4 Consider the chloroform ($CHCl_3$) molecule.
   (a) Assign oxidation states to each atom in the molecule. [*Hint:* Draw a dot diagram first.]
   (b) How many electrons does the C atom own by the oxidation state method of electron bookkeeping?
   (c) How many more or fewer valence electrons are assigned to the carbon atom in $CHCl_3$ by the oxidation state method than are present on a free carbon atom?
   (d) Is it correct or incorrect to say that the C atom in $CHCl_3$ has a complete octet of valence electrons? Explain.

SHORTCUT METHOD FOR ASSIGNING OXIDATION STATES

The good news about the method we have been using to assign oxidation numbers is that it always works. The bad news is that it requires a dot diagram, and drawing one takes time, if it can be done at all. For this reason, chemists have devised a shortcut for assigning oxidation states without the need for a dot diagram or even a table of electronegativities. However, what we gain in time we lose in accuracy, since this shortcut method may not always produce a correct answer. Nevertheless, it works reliably for many of the compounds that chemists routinely encounter.

Shortcut rules

Rule 1. The oxidation state of an atom in the most abundant naturally occurring form of an element typically is zero.

The atoms in the substances shown below, including those in diatomic molecules, are assigned an oxidation state of zero.

Element	Naturally occurring elemental form	
Hydrogen	H_2	
Oxygen	O_2	
Nitrogen	N_2	
Fluorine	F_2	
Chlorine	Cl_2	
Bromine	Br_2	
Iodine	I_2	Oxidation state of atoms: 0
Helium	He	
Neon	Ne	
Phosphorus	P_4	
Sulfur	S_8	
Iron	Fe	
Aluminum	Al	

The elemental form for nonmetals is usually (but not always) a diatomic molecule, while the noble gases are monatomic. For all metals, the elemental form is represented with just the symbol of the element. There are exceptions to these rules. For example, ozone (O_3) is another elemental form of oxygen; whereas the two O atoms in O_2 have a zero oxidation state, all the O atoms in O_3 do not.

8.2 WORKPATCH

Assign oxidation states for all the oxygen atoms in the ozone (O_3) molecule using the following special rule:

> **When two atoms of the same element (and therefore of identical electronegativity) share electrons, the oxidation state method of electron bookkeeping says to divide them equally between the atoms (each atom gets half of them).**

A dot diagram for ozone is shown below.

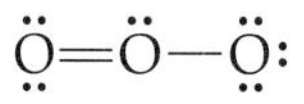

$$\ddot{\text{O}}=\ddot{\text{O}}-\ddot{\text{O}}:$$

WorkPatch 8.2 included our first mention of assigning oxidation states to molecules that contain electrons shared between atoms of identical electronegativity. Using the rule in the WorkPatch, you should have found that two of the oxygen atoms in ozone have nonzero oxidation states.

> **Rule 2. The oxidation state of an oxygen atom in a compound is almost always −2.**

For example, the oxidation state of the oxygen atoms in the compounds H_2O, CO, CO_2, $C_6H_{12}O_6$ (glucose), and $Fe_2O_3 \cdot H_2O$ (rust) is −2. Molecules called peroxides, which contain an O–O single bond, are exceptions to this rule. Ozone, which also has O–O bonds, is also an exception (see WorkPatch 8.2). The only other time this rule fails is when oxygen is bound to an element more electronegative than itself. Only fluorine (F) meets this criterion. Except for these situations, it is a safe bet that an oxygen atom in a compound has an oxidation state of −2.

> **Rule 3. The oxidation state of a hydrogen atom in a compound is assigned as +1.**

For example, the oxidation state of the hydrogen atoms in the compounds H_2O, $C_6H_{12}O_6$ (glucose), CH_4 (methane), and C_2H_2 (acetylene) is +1. Compounds in which H is bonded to a less electronegative atom (called hydrides) are an exception. We will not consider these cases.

> **Rule 4. The oxidation state of any group IA atom [alkali metals such as lithium (Li), sodium (Na), and potassium (K)] in a compound is always +1.**

That's *in a compound!* This means that the oxidation state of the lithium atoms in the compound lithium chloride, LiCl, or lithium oxide, Li_2O, is +1. The oxidation state of lithium metal Li, its elemental form, is zero by rule 1.

> **Rule 5. The oxidation state of any group IIA atom [alkaline earth metals such as magnesium (Mg), barium (Ba), and calcium (Ca)] in a compound is always +2.**

Again, that's *in a compound!* This means that the oxidation state of the magnesium atom in magnesium chloride, $MgCl_2$, is +2. The oxidation state of magnesium metal Mg, its elemental form, is zero.

Rule 6. The oxidation state of a monatomic (one-atom) ion is equal to the charge of the ion.

Some examples are shown below.

Monatomic ion	Oxidation state
Al^{3+}	+3
O^{2-}	−2
H^+	+1
Cl^-	−1
N^{3-}	−3

Finally, we need a way to assign oxidation states to all the other atoms in the periodic table that are not covered by rules 1–6. We use an extremely important rule for oxidation states that applies in all situations and to which there are no exceptions.

Rule 7. The sum of the oxidation states for every atom in a formula must add up to the overall charge of the formula.

If the formula represents a neutral molecule, like H_2O, it has no charge and the oxidation states must add up to zero. If you are dealing with a polyatomic ion such as sulfate (SO_4^{2-}), then all the oxidation states must add up to the charge on the ion (for sulfate, −2).

Which of the seven rules to apply and the order of their application can change from example to example, but we recommend that you always try applying rules 2 and 3 first, and save rule 7 for last. A few examples of using the shortcut rules are worked out below. Follow them through, but realize that we would have arrived at the same result if we had started with a dot diagram.

	Atom	Oxidation state	
H_2O	O	−2	Rule 2: O is −2
	H	+1	Rule 3: H is +1
	H	+1	
	Sum:	0	Oxidation numbers add up to zero, the charge on the water molecule (rule 7).
OH^-	O	−2	Rule 2: O is −2
	H	+1	Rule 3: H is +1
	Sum:	−1	Oxidation numbers add up to −1, the charge on the hydroxide ion (rule 7).

These were straightforward examples. Now we'll make it a bit more difficult by including an atom for which there is no rule. The trick is to apply all the other rules that you can, starting with rules 2 and 3 for O and H, respectively, and saving rule 7 for last.

	Atom	**Oxidation state**
$(SO_4)^{2-}$	O	-2
	O	-2
	O	-2 Rule 2: O is -2
	O	-2
	S	? There is no rule for S
	Sum:	-2 The sum must equal -2, the charge on the sulfate ion (rule 7).

Rule 7 says that all the oxidation numbers in a formula must add up to give the overall charge of the formula, which in this case is -2. We have four oxygen atoms at -2 each giving a total of -8. What do we have to add to -8 to equal the -2 charge of the sulfate ion? The answer is $+6$. The unknown oxidation number of the sulfur atom must therefore be $+6$.

We'll work through two more examples, each containing one element for which there is no shortcut rule.

	Atom	**Oxidation state**
$(NH_4)^+$	H	$+1$
	H	$+1$
	H	$+1$ Rule 3: H is $+1$
	H	$+1$
	N	? There is no rule for N
	Sum:	$+1$ The sum must equal $+1$, the charge on the ammonium ion (rule 7).

Rule 7 says that all the oxidation numbers in a formula must add up to give the overall charge of the formula, which in this case is $+1$. We have four hydrogen atoms with oxidation states of $+1$, giving a total of $+4$. What do we have to add to $+4$ to get $+1$, the charge on the ammonium ion? The answer is -3. The unknown oxidation number of the nitrogen atom must therefore be -3.

	Atom	**Oxidation state**
$(Cr_2O_7)^{2-}$	O	-2
	O	-2
	O	-2
	O	-2 Rule 2: O is -2
	O	-2
	O	-2
	O	-2
	Cr	? There is no rule for Cr
	Cr	?
	Sum:	-2 The sum must equal -2, the charge on the dichromate ion (rule 7).

Rule 7 says that all the oxidation numbers in a formula must add up to give the overall charge of the formula, which in this case is -2. We have seven oxygen atoms with oxidation states of -2 each, giving a total of -14. To get -2, the charge on the dichromate ion, we have to add $+12$. Now we have to be careful. The unknown oxidation number of Cr is not +12, because there are two Cr atoms in this compound. In cases like this, we can usually assume that the two Cr atoms have the same oxidation state. This means that each chromium atom must have an oxidation state of $+12/2 = +6$.

Before giving you some practice problems, we want to give you one more shortcut rule, which we call the halide rule:

If a halogen atom is in a compound, it will most likely have a −1 oxidation number.

Recall that the halogens are the group VIIA atoms (F, Cl, Br, I). So, for example, we can now assign the oxidation states to the atoms in bromoform, $CHBr_3$.

	Atom	Oxidation state	
$CHBr_3$	H	+1	Rule 3: H is +1
	Br	−1	
	Br	−1	Halide rule: Br is −1
	Br	−1	
	C	?	There is no rule for C
	Sum:	0	The sum must equal 0 (rule 7).

Rule 7 says that all the oxidation numbers in a formula must add up to give the overall charge of the formula, which in this case is zero. We have one hydrogen atom at +1 and three bromine atoms at −1 each, giving a total of −2. What do we have to add to −2 to get 0? The answer is +2. The unknown oxidation number of the carbon atom must therefore be +2.

You have to be a little careful with the halide rule. As in rule 2 for oxygen and rule 3 for hydrogen, the halide rule has some exceptions. Part (c) in Work-Patch 8.3 will test your knowledge of a basic concept that will enable you to determine when you can use the halide rule and when you cannot.

Suppose that a chlorine atom in a compound has an oxidation state of −1.
(a) Is it assigned more or fewer valence electrons by oxidation state bookkeeping than are present on a free chlorine atom?
(b) How many more or fewer?
(c) For a chlorine atom in a compound to have a −1 oxidation state, what must be true about the electronegativity of the atom it is bonded to?

WORKPATCH 8.3

Check your answers to the WorkPatch to make sure you understand these basic concepts. Now let's examine a case where the halide rule does not apply.

Consider the chlorate ion, ClO_3^-, in which all three oxygen atoms are bound to a central chlorine atom. The oxidation state of the Cl atom in this ion is not −1.
(a) Why doesn't the halide rule apply for the chlorate ion?
(b) What is the oxidation state of the Cl atom in chlorate?

WORKPATCH 8.4

These last two WorkPatches are very important. They are reminders that in oxidation state electron bookkeeping, it is the most electronegative atom that gets the shared electrons. Make sure you understand this principle, and then try the following practice problems using the shortcut rules.

PRACTICE PROBLEMS

Assign oxidation states to all the atoms using the shortcut rules.

8.5 $COCl_2$ (the oxygen and chlorine atoms are all bound to the central carbon atom).

Answer: O is −2 (rule 2). Cl is −1 (halide rule applies since Cl is more electronegative than C). C is +4 (the sum of the oxidation numbers for the O and two Cl's is −4).

8.6 $MgBr_2$

8.7 Fe_2O_3

8.8   HClO (the Cl and O are bound to each other).

8.9 CO_3^{2-}

8.10 Cu^{2+}

Chemists frequently use the shortcut rules to assign oxidation states, but when they are inadequate (for example, the rules don't cover a molecule like CS_2), or when they present some ambiguity, a dot diagram will always work. A good example of the ambiguity we mentioned is demonstrated by acetic acid, $C_2H_4O_2$. By the shortcut rules we would assign the following oxidation states:

	Atom	**Oxidation state**	
$C_2H_4O_2$	O	−2	Rule 2: O is −2
	O	−2	
	H	+1	Rule 3: H is +1
	H	+1	
	H	+1	
	H	+1	
	C	?	There is no rule for C.
	C	?	
	Sum:	0	The sum must equal 0 (rule 7).

Rule 7 says that all the oxidation numbers in a compound must add up to give the overall charge of the formula, which in this case is zero. Since the hydrogen and oxygen oxidation numbers sum up to give zero, then if we make the oxidation number of both carbons equal to zero, rule 7 will be obeyed. But we can also satisfy rule 7 by making one carbon +1 and the other −1, or +2 and −2, or +3 and −3, etc. There is no clear answer here. The only way out is a dot diagram.

Given the dot diagram for acetic acid below, assign each atom an oxidation number.

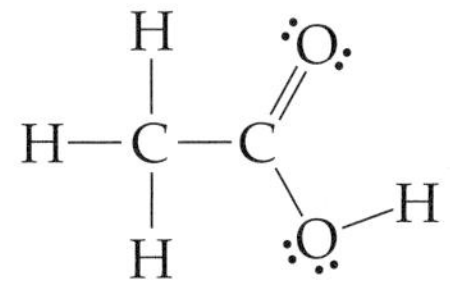

Using the dot diagram and applying the electronegativity rules for oxidation state electron bookkeeping, you should have arrived at an unambiguous set of oxidation states for the carbon atoms in acetic acid.

So far we have simply been assigning oxidation states to the atoms in a chemical compound. But remember that our original goal was to be able to determine whether any transfer of electrons has taken place among the atoms in a chemical reaction. Oxidation numbers were designed to help us do just this, and we are now ready to use them for this purpose.

RECOGNIZING ELECTRON TRANSFER REACTIONS 8.3

Since oxidation numbers tell you how many electrons each atom in a compound owns relative to the free atom, then *if the oxidation number of any atom changes during a reaction, a formal transfer of electrons must have occurred.* Look at the chemical reaction below, where the oxidation number of each atom is printed above the atom:

$$\overset{+1\,-1}{NaCl} + \overset{+1\,-1}{LiBr} \longrightarrow \overset{+1\,-1}{NaBr} + \overset{+1\,-1}{LiCl}$$

Consider each atom individually. Does the oxidation number of sodium (Na) change on going from reactant to product? Does the oxidation number of chlorine (Cl) change on going from left to right? What about lithium and bromine? In this example none of the oxidation numbers change during the course of the reaction. This means that even though a chemical reaction takes place, there is no transfer of electrons between any of the atoms. Each atom owns the same number of electrons before the reaction as after it.

The situation is quite different in the following reaction:

$$\overset{-4\,+1}{CH_4} + \overset{0}{2\,O_2} \longrightarrow \overset{+4\,-2}{CO_2} + \overset{+1\,-2}{2\,H_2O}$$

Carbon starts with an oxidation state of -4 and ends up as $+4$. Oxygen begins at zero and ends up as -2. This means that, according to the oxidation state method of electron bookkeeping, a transfer of electrons between these atoms must have taken place. In going from zero to -2, each of the oxygen atoms gains 2 electrons. When an atom gains electrons, we say that it has been **reduced** (or has undergone *reduction*); **reduction** is defined as a gain of electrons. This makes sense if you think about reduction not in terms of the number of electrons but in terms of the oxidation state of the atom, which is reduced to a lower value.

Reduction

- Gain of electrons (by oxidation state bookkeeping)
- Reduction in oxidation state (from a higher to a lower value)

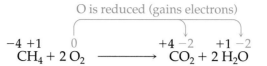

In the reaction above, the oxygen atoms gain electrons, and we say they are reduced. In general, an atom is reduced when its oxidation number *becomes more negative* on going from reactant to product.

If the oxygen atoms gain electrons in this electron transfer reaction, where do they come from? The key word here is *transfer*. It implies that for oxygen to have gained electrons, some other atom must have lost them. Electron transfer implies *gain* and *loss*—you can never have one without the other. To find the atom that lost the electrons, simply look for the one whose oxidation number increased (became more positive) during the course of the reaction. In this reaction, it is carbon. To go from an oxidation state of -4 to $+4$, each carbon loses 8 electrons. When an atom loses electrons we say that it has been **oxidized** (or has undergone *oxidation*); **oxidation** is defined as the loss of electrons. The carbon atoms are oxidized in this reaction.

Oxidation

- Loss of electrons (by oxidation state bookkeeping)
- Increase in oxidation state (from a lower to a higher value)

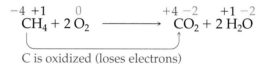

Now remember, by the octet rule method of electron bookkeeping, no atom has lost or gained electrons. For example, the carbon atom in both methane and carbon dioxide has an octet:

$$
\begin{array}{c}
\text{H} \\
| \\
\text{H}-\overset{\displaystyle |}{\underset{\displaystyle |}{\text{C}}}-\text{H} \qquad \ddot{\text{O}}=\text{C}=\ddot{\text{O}} \\
\text{H}
\end{array}
$$

It is only when we switch to the oxidation state method of electron bookkeeping that we can talk about the loss and gain of electrons.

Reduction (the gain of electrons) and oxidation (the loss of electrons) go together. One cannot occur in a chemical reaction without the other. We acknowledge this fact by referring to chemical reactions in which electrons are transferred as **redox** (pronounced ree-dox) reactions.

The burning of methane (CH_4) in oxygen in the reaction above is a redox reaction. The burning of any substance (generally referred to as combustion) is a redox reaction, including the multistep "burning" of glucose in the cells of your body. Many chemical reactions are redox reactions, but not all of them. To be classified as a redox reaction an electron transfer must take place—that is, oxidation numbers must change.

Indicate whether each reaction is a redox reaction. If it is, which atom gets oxidized, and which atom gets reduced? Consult the shortcut rules.

8.11 $P_4 + 6\,Br_2 \longrightarrow 4\,PBr_3$ [*Hint:* Br is more electronegative than P.]

Answer:

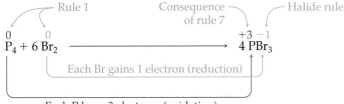

8.12 $4\,Fe + 3\,O_2 \longrightarrow 2\,Fe_2O_3$

8.13 $Ca + 2\,H^+ \longrightarrow Ca^{2+} + H_2$

8.14 $2\,NaBr + MgO \longrightarrow MgBr_2 + Na_2O$

We are getting close to an explanation of how we can use electron transfer (redox) reactions to produce electricity. But first we need to introduce a bit more terminology. To do this, let's go back to the combustion of methane:

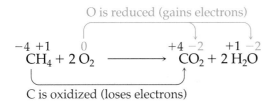

Up to now we have been indicating the individual atoms that get oxidized or reduced. The C atom in CH_4 is oxidized. The O atoms in O_2 are reduced. While this is completely correct, chemists usually don't think in terms of the oxidation or reduction of individual atoms in a compound. Instead, a chemist would say that the entire CH_4 molecule is oxidized, and the entire O_2 molecule is reduced. When the situation is described using whole molecules instead of individual atoms, we use the terms *oxidizing agent* and *reducing agent*. An **oxidizing agent** is a substance that oxidizes something else. A **reducing agent** is a substance that reduces something else. In our example, O_2 is an oxidizing agent, because it oxidizes CH_4. It does this by gaining the electrons that methane loses. This means that the oxidizing agent, O_2, is actually reduced during the course of the chemical reaction.

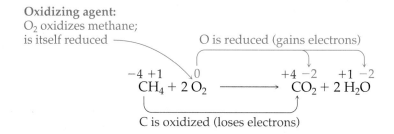

The same kind of logic works in identifying the reducing agent. In our example, CH_4 is a reducing agent, because it reduces O_2. It does this by providing the electrons that O_2 gains. This means that the reducing agent, CH_4, is actually oxidized during the course of the reaction.

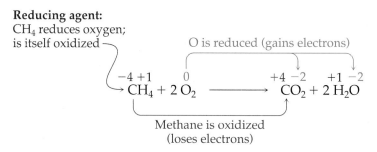

This terminology of oxidizing agent and reducing agent can be confusing, but it is something you will have to become comfortable with, and practice helps. The following are a repeat of Practice Problems 8.11–8.13, but this time you are asked to identify the oxidizing and reducing agents.

PRACTICE PROBLEMS

For each redox reaction, indicate which substance is the oxidizing agent and which is the reducing agent.

8.15 $P_4 + 6\,Br_2 \longrightarrow 4\,PBr_3$

Answer:

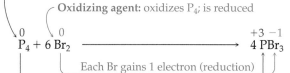

8.16 $4\,Fe + 3\,O_2 \longrightarrow 2\,Fe_2O_3$

8.17 $Ca + 2\,H^+ \longrightarrow Ca^{2+} + H_2$

Now let's carry out a real redox reaction. We will put a strip of zinc metal into a beautiful blue aqueous solution of Cu^{2+} ions. Immediately, a chemical reaction begins. Within minutes, the blue color begins to fade and the silvery zinc strip becomes plated with copper metal.

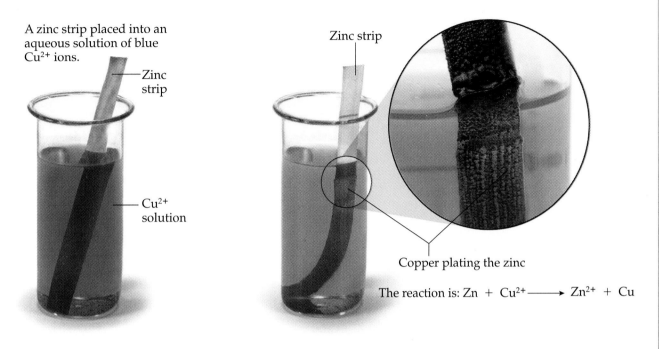

A zinc strip placed into an aqueous solution of blue Cu^{2+} ions.

Zinc strip

Zinc strip

Cu^{2+} solution

Copper plating the zinc

The reaction is: $Zn + Cu^{2+} \longrightarrow Zn^{2+} + Cu$

By now you should be able to look at the balanced equation and quickly determine that this is indeed a redox reaction. Just to be sure, complete the following WorkPatch.

WORKPATCH 8.6

Step 1: Assign oxidation states to the atoms and ions in the reaction (use shortcut rules 1–7).

$$Zn + Cu^{2+} \longrightarrow Zn^{2+} + Cu$$

Step 2: Check to see if any of the oxidation states have changed.

(a) Is this a redox reaction (yes or no)?
(b) What was oxidized?
(c) What was reduced?
(d) What is the oxidizing agent?
(e) What is the reducing agent?

Without a doubt, electrons are being transferred from Zn to Cu^{2+} in this reaction. Now let's go and harness some electricity from this redox reaction.

ELECTRICITY FROM REDOX REACTIONS 8.4

Here, again, is the redox reaction we have chosen to harness electricity from:

$$Zn + Cu^{2+} \longrightarrow Zn^{2+} + Cu$$

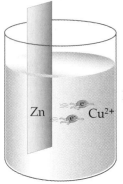

How do the electrons get from Zn to Cu²⁺? **Not this way!**

The zinc metal is oxidized, so the Zn atoms in the zinc strip each lose 2 electrons, while the copper ions (Cu^{2+}) are reduced, each picking up 2 electrons. But exactly how do the electrons get from the strip of Zn metal to the aqueous Cu^{2+} ions in solution? Because water is such a poor conductor of electricity, they can't just leap off the zinc strip and swim out to the Cu^{2+} ions. Instead, the Cu^{2+} ions must migrate to the zinc strip and actually touch it for the electron transfer from Zn to the Cu^{2+} to take place.

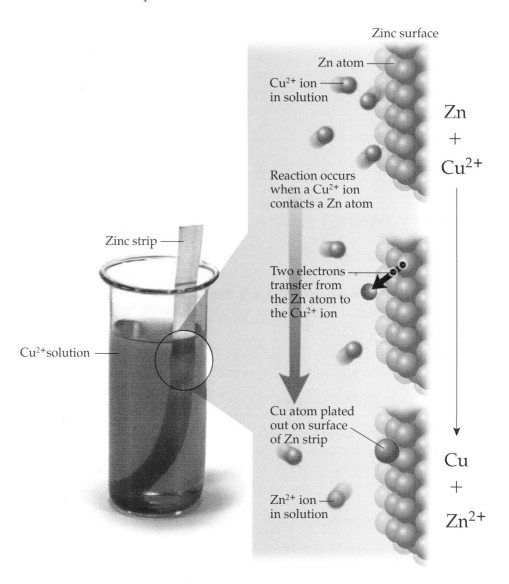

When a Cu^{2+} ion touches the zinc strip, it accepts 2 electrons and is reduced to elemental copper (Cu). That is where the copper metal that plates out on the Zn strip comes from. This also explains why the solution becomes less blue: Blue Cu^{2+} ions are being removed from solution, and at the same time, Zn atoms in the zinc strip are being oxidized to colorless Zn^{2+} ions. These newly formed Zn^{2+} ions enter the solution. If we could somehow get the electrons that transfer from the zinc to the copper ions to travel through a wire, then we would be creating electricity (recall our description of electricity as a flow of electrons in a wire). This would be electricity produced by a chemical (redox) reaction, and it would last as long as there are reactants converting into products.

One way to accomplish this is to separate the Zn strip from the Cu^{2+} ions so that they are no longer in direct contact with each other, and then provide a wire path for the electrons to flow across. A common laboratory setup for this is shown in the figure below. The setup, called a **galvanic cell** by chemists, is what you would call a **battery**. In this example, a strip of zinc is placed into an aqueous solution of Zn^{2+} ions (created by dissolving zinc sulfate, $ZnSO_4$, in water), and a strip of copper is placed into a solution of Cu^{2+} ions (created by dissolving copper sulfate, $CuSO_4$, in water). In other words, each strip of metal is placed into a solution of its own ions. Both solutions also have sulfate (SO_4^{2-}) ions in them. The metal strips are then connected with a wire that can conduct electrons between them. An additional component, called a *salt bridge*, is necessary to complete the circuit. This can be as simple as a piece of absorbent paper saturated with a solution of an ionic salt, or the tube shown in the figure. The ionic salt used in the salt bridge should be composed of *spectator ions*—that is, ions that are not directly involved in the redox reaction. A common salt is sodium sulfate, Na_2SO_4, which provides Na^+ and SO_4^{2-} spectator ions to the system. We will discuss the necessity for the salt bridge shortly.

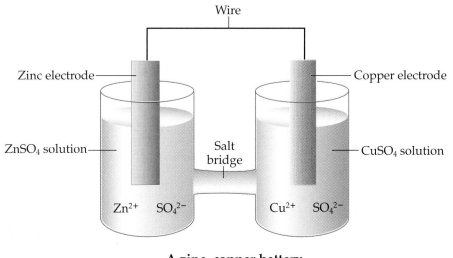

A zinc–copper battery

The metal strips of zinc and copper are called **electrodes**. No longer does a Cu^{2+} ion directly touch the zinc electrode, but they are in "electrical contact" with one another. There is an uninterrupted pathway for the electrons to travel between them. That pathway is through the wire. When a copper ion comes in contact with the copper electrode, the Cu^{2+} ion is, in effect, connected with the Zn electrode. The zinc metal can now give up electrons to the copper ions through the wire. The Cu^{2+} ions are reduced to copper metal, depleting the solution of Cu^{2+} ions. The newly formed copper metal (Cu) plates out on the copper electrode, causing it to increase in mass. The zinc atoms (Zn) in the zinc electrode are oxidized to Zn^{2+}, and these newly formed ions enter the solution around the electrode, increasing the amount of Zn^{2+}

ions in the solution. This causes the zinc electrode to be eaten away with time. All of this is shown diagrammatically in the figure. Follow the sequence carefully to see what is happening.

How a battery generates an electric current

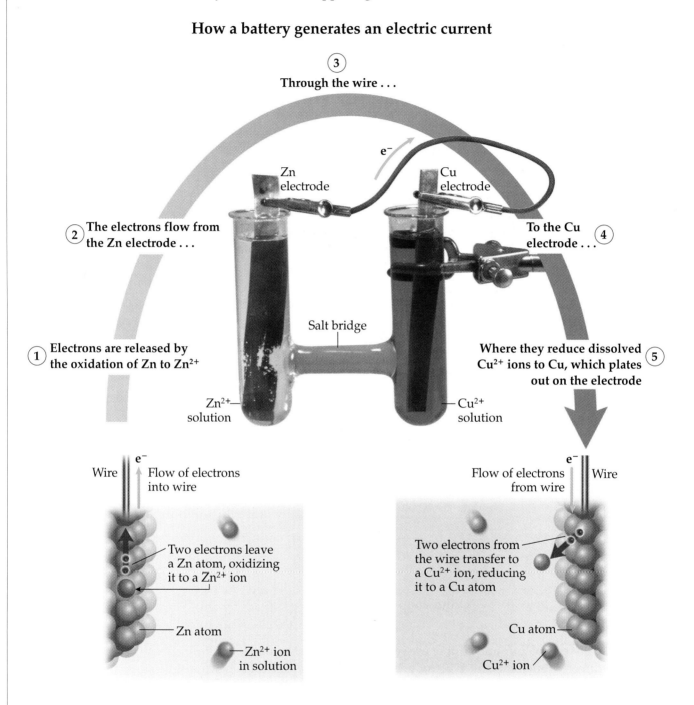

③ Through the wire . . .

② The electrons flow from the Zn electrode . . .

④ To the Cu electrode . . .

Zn electrode

Cu electrode

e⁻

① Electrons are released by the oxidation of Zn to Zn^{2+}

⑤ Where they reduce dissolved Cu^{2+} ions to Cu, which plates out on the electrode

Salt bridge

Zn^{2+} solution

Cu^{2+} solution

Wire

e⁻

Flow of electrons into wire

Flow of electrons from wire

e⁻

Wire

Two electrons leave a Zn atom, oxidizing it to a Zn^{2+} ion

Two electrons from the wire transfer to a Cu^{2+} ion, reducing it to a Cu atom

Zn atom

Zn^{2+} ion in solution

Cu atom

Cu^{2+} ion

Now we can better appreciate the presence of the salt bridge. Without it, the zinc side of the system (or cell) would become more and more positively charged as new Zn^{2+} ions are created. This would make it impossible for the negatively charged electrons to leave the zinc side. Likewise, the copper side would become increasingly negative as the Cu^{2+} ions are reduced and removed from solution, resulting in a relative excess of SO_4^{2-} ions. This would repel the electrons away from the copper side. This buildup of charge would cause the redox reaction to come to a halt, and electrons would no longer flow through the wire. The battery would not function. With the salt bridge in place, as the zinc side becomes more positively charged, the SO_4^{2-} ions in the tube can move into the solution and neutralize the charge. Likewise, the Na^+ ions from the salt bridge help neutralize the buildup of negative charge on the copper side. In other words, a salt bridge is necessary for this type of reaction in order to maintain a charge balance in both halves of the system.

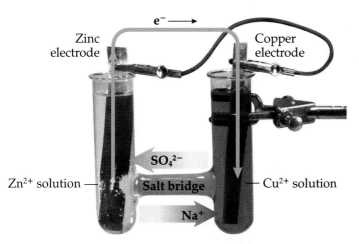

The function of the salt bridge

Oxidation reaction:	Reduction reaction:
$Zn \longrightarrow Zn^{2+} + 2e^-$	$Cu^{2+} + 2e^- \longrightarrow Cu$
As the oxidation reaction produces Zn^{2+} ions, negatively charged SO_4^{2-} ions flow in from the salt bridge, preventing a buildup of positive charge.	As the reduction reaction removes Cu^{2+} ions from the solution, Na^+ ions flow in from the salt bridge, preventing a buildup of negative charge.

The electrons must flow through a wire, so we are producing usable electricity. In addition, there is a measurable "pressure," or voltage, pushing the electrons along. This zinc–copper battery produces about 0.9 volt, enough to light a small bulb or run a motor to do some work. Until one of the reactants (Zn or Cu^{2+}) is used up, this battery will keep on going, and going, and going. In principle, any redox reaction can be used to make a battery. All we need to do is separate the oxidizing agent and the reducing agent from one another, and then connect them with a wire.

Consider the redox reaction between magnesium metal and silver ions:

WORKPATCH 8.7

$$Mg + 2\,Ag^+ \longrightarrow Mg^{2+} + 2\,Ag$$

Design and label a battery based on this reaction.

Astounding as it may seem, you can make a real battery using just strips of zinc and copper and a lemon or other citrus fruit. You can even use a potato. Here is the lemon battery we built for this book:

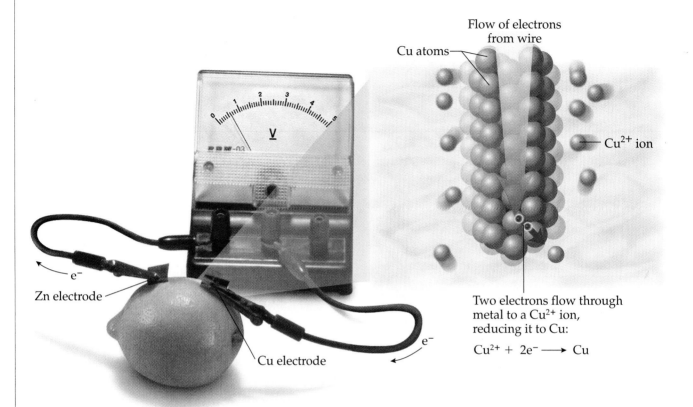

Flow of electrons from wire

Cu atoms

Cu^{2+} ion

e$^-$

Zn electrode

Cu electrode

e$^-$

Two electrons flow through metal to a Cu^{2+} ion, reducing it to Cu:

$$Cu^{2+} + 2e^- \longrightarrow Cu$$

The only real difference between this lemon battery and the zinc–copper battery we made earlier is that we haven't provided a solution of Cu^{2+} around the Cu electrode. Exactly what happens at the copper electrode is still an area of speculation. One reasonable explanation is that Cu^{2+} ions are already present on the electrode surface due to prior corrosion. Another is that the lemon itself causes some of the Cu atoms from the copper electrode to oxidize to Cu^{2+} ions. These ions remain in the vicinity of the copper electrode. As soon as the two electrodes are connected by a wire, zinc atoms begin to oxidize, releasing electrons. These electrons travel through the wire and reduce the Cu^{2+} ions back to Cu metal. The lemon itself serves as a salt bridge.

As we said, any redox reaction can potentially be used to make a battery if you can separate the substance that gets oxidized from the one that gets reduced and then connect them with a wire. Interestingly, you can simply place a zinc strip and a copper strip in the same beaker of salt water and run a low-current digital clock (a setup like this sits on the desk of one of the authors of this book). This and our lemon battery are a bit unconventional. Perhaps the ones in the photo look more familiar to you.

The standard "dry cell" is nothing more than a redox reaction in a can. A cross-section of one type is shown at the top of the next page. The reactants are manganese dioxide (MnO$_2$) and metallic zinc (Zn). As you can see from the diagram, these reactants are not in direct contact with each other. A semimoist paste that serves as a salt bridge separates them. The zinc metal is oxidized (loses 2 electrons) to become Zn^{2+} ions (Zn is the reducing agent):

Zinc is oxidized (loses 2 electrons).
Each Zn becomes Zn^{2+}.

The manganese dioxide (MnO_2) is changed into di-manganese trioxide (Mn_2O_3) during the reaction. The manganese atom (Mn) in manganese dioxide is in the $+4$ oxidation state but becomes $+3$ in Mn_2O_3. So the manganese is reduced, and MnO_2 is the oxidizing agent:

Manganese is reduced (gains 1 electron).
Each Mn^{4+} in MnO_2 becomes Mn^{3+} in Mn_2O_3.

The overall redox reaction for the battery is:

MnO₂ paste around graphite core

NH₄Cl and ZnCl₂ paste

Zinc metal can

Cross-section of a dry cell

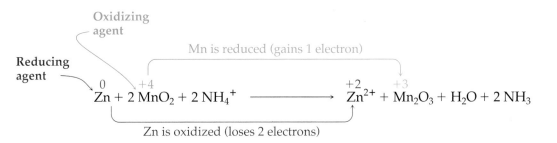

Oxidizing agent

Reducing agent

Mn is reduced (gains 1 electron)

$$\overset{0}{Zn} + 2\,\overset{+4}{MnO_2} + 2\,NH_4^+ \longrightarrow \overset{+2}{Zn^{2+}} + \overset{+3}{Mn_2O_3} + H_2O + 2\,NH_3$$

Zn is oxidized (loses 2 electrons)

Connect a wire between the two terminals and the reaction begins—electrons flow from Zn to MnO_2. Put a light bulb in the circuit and you've got a flashlight.

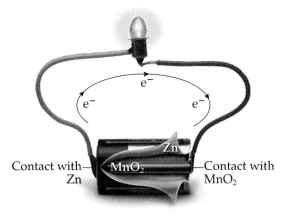

e^-

e^- e^-

Zn

MnO₂

Contact with Zn

Contact with MnO₂

Redox reactions are at the heart of all batteries, from the battery in your car to the tiny battery in a hearing aid. All involve at least one reactant that loses electrons (oxidation), while another reactant simultaneously gains them (reduction). You have probably noticed the $+$ and $-$ signs used to label the electrodes on a battery. In our zinc–manganese dry cell, which electrode would be marked with which sign? Recall that electrons are negatively charged and will therefore always be attracted to anything with a positive charge. In the zinc–manganese dry cell, the electrons flow from the Zn electrode toward the MnO_2 electrode. This means that the MnO_2 electrode is assigned the + sign.

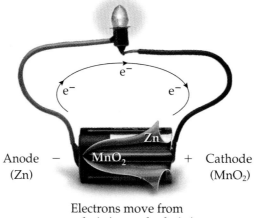

Electrons move from
anode (−) to cathode (+)

Not surprisingly, chemists have devised names for these two electrodes. The positive electrode is called the **cathode**, and the negative electrode is called the **anode**. To help you remember this, just recall the names of positive and negative ions. A positive ion is a *cat*ion and a negative ion is an *an*ion. For both batteries and ions, the prefixes *cat* and *an* mean positive and negative, respectively.

 WORKPATCH   Return to the drawing of the battery you made in WorkPatch 8.7 and add the labels +, −, cathode, and anode.

It's important that you be able to do WorkPatch 8.8 because it leads us to a very fundamental question. What ultimately decides which way the electrons flow? For the zinc–copper battery you were told that zinc is oxidized and feeds electrons to the Cu^{2+} ions. That is, we started with the balanced redox reaction written as $Zn + Cu^{2+} \longrightarrow Zn^{2+} + Cu$. But what if this were not the case? What if the diagram of the cell were put in front of you and you were asked, "Which way do the electrons flow? What gets oxidized? What is reduced? Why was the reaction written with electrons flowing from zinc to copper, and not the other way?"

 WORKPATCH   Multiple choice: In general, which way do electrons flow in a battery?
  (a) From what gets oxidized to what gets reduced
  (b) From what gets reduced to what gets oxidized
Explain how you made your choice.

This WorkPatch should remind you that the ultimate answer to the question "Which way do the electrons flow?" requires knowing what is oxidized and what is reduced. In the next section we will see how you can determine this. For now, try the following practice problems, where the necessary information is given.

8.18 Construct a battery from the redox reaction

$$Ni + Pb^{2+} \longrightarrow Ni^{2+} + Pb$$

To answer, draw a battery similar to the one you drew for WorkPatch 8.7. Make sure you label which way the electrons flow and label cathode, anode, +, and −.

Answer: The nickel electrode is the anode (−). The lead electrode is the cathode (+). Electrons flow from the nickel anode to the to lead cathode.

8.19 For the nickel–lead battery of Practice Problem 8.18:
   (a) What is the oxidizing agent?
   (b) What is the reducing agent?
   (c) Which electrode gets larger with time?
   (d) Which electrode is eaten away with time?

8.20 A battery is constructed from iron (Fe) and silver (Ag) by dipping a strip of each metal into a solution of its ions (Fe^{3+} and Ag^+, respectively). As the battery operates, the Fe^{3+} concentration increases, while the Ag^+ concentration decreases.
   (a) What is getting oxidized?
   (b) What is getting reduced?
   (c) Draw a battery similar to the one you drew for WorkPatch 8.7. Make sure you label which way the electrons flow and label the cathode, anode, +, and −.

Which Way Do the Electrons Flow? The EMF Series 8.5

Look back for a moment at the illustration on page 273. When we put a strip of zinc into the solution of Cu^{2+} ions, a redox reaction occurred spontaneously. Neutral Zn atoms spontaneously lost electrons to the Cu^{2+} ions in solution, causing copper metal to plate out onto the zinc strip. What would happen if we tried the opposite arrangement? Suppose that instead of putting a piece of Zn into a solution of Cu^{2+} ions, we put a strip of Cu metal into a solution of Zn^{2+} ions.

Would the Zn^{2+} ions oxidize (take electrons from) the Cu atoms? In other words, would we see the zinc metal plate out on the copper strip? The equation for this hypothetical reaction would be as follows:

$$Zn^{2+} + Cu \xrightarrow{?} Zn + Cu^{2+}$$

This is the opposite of the reaction that occurs in an actual zinc–copper battery:

$$Zn + Cu^{2+} \longrightarrow Zn^{2+} + Cu$$

Copper strip

Zn^{2+} solution

What will happen?

In fact, our hypothetical reaction doesn't occur; when you immerse a copper strip in a solution of Zn^{2+} ions, nothing happens. Clearly,

Zn metal gives up electrons to Cu^{2+} ions,
BUT
Cu metal does not give up electrons to Zn^{2+} ions.

Why is this so? Recall from Chapters 4 and 5 that the chemical definition of a metal is: "An element that easily loses its valence electrons." But, given the results summarized above, we can conclude that all metals are not created equally with respect to this definition. *Some metals lose electrons more readily than others.* In the competition we set up between the two systems Zn/Cu^{2+} versus Zn^{2+}/Cu, we saw that a spontaneous reaction occurs in only the first case, where Zn metal gives up electrons and copper accepts them. Because of this we say that zinc is a more *active* metal than copper. An **active metal** is one that more easily loses valence electrons. Zinc atoms have a greater tendency than copper atoms to lose their valence electrons, so zinc is the more active metal. Because this is so, zinc can "force" ions of Cu^{2+} to take its electrons, but Cu cannot "force" Zn^{2+} ions to take its electrons. The following diagram sums this up. In a zinc–copper system, the electron flow is always from zinc to copper due to zinc's greater tendency to lose electrons.

**Electrons always flow from the more active metal
to the less active metal**

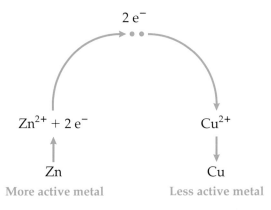

$2\,e^-$

$Zn^{2+} + 2\,e^-$ Cu^{2+}

Zn   Cu

More active metal   Less active metal

We should point out that there was no way to know whether Zn or Cu was more active the very first time these metals were studied. The result had to be determined by setting up both experiments described above and noting which one reacted spontaneously.

In principle we could compare any two metals to determine which one is more active. We could carry out this same experiment using zinc (Zn) and magnesium (Mg), for example, by dipping each metal into a solution of the other's ions and seeing which one reacts spontaneously. This experiment is shown at the top of the next page. We observe that nothing at all happens in the beaker with the zinc strip immersed in a solution of Mg^{2+} ions. However, the reverse situation, in the beaker on the right, is quite different. The magnesium quickly becomes coated with zinc metal.

Test to find out which metal is more active — Zn or Mg

From these experimental observations, we can conclude the following about the relative activities of magnesium and zinc:

Magnesium (Mg) metal will give its electrons to Zn^{2+} ions,
BUT
Zinc (Zn) metal cannot force its electrons onto Mg^{2+} ions.

Electrons flow from Mg (more active) to Zn (less active)

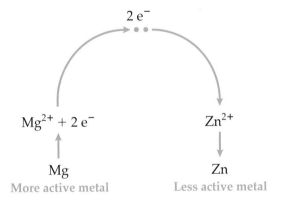

In other words, magnesium is a more active metal than zinc (its "desire" to lose electrons is even greater than zinc's). And since zinc is more active than copper, then magnesium must also be more active than copper. We can use a series of experiments of this type to determine the relative order of activity for a number of different metals. This relative ordering of metal activities is called the **electromotive force series** (or **EMF series**, for short). We have begun the list below:

An electromotive force series

Mg   Most active metal (loses electrons most easily)

Zn   Activity

Cu   Least active metal (loses electrons least easily)

EMF series

Most active
(loses electrons
most easily)

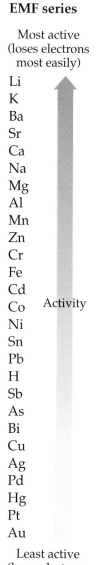

Li
K
Ba
Sr
Ca
Na
Mg
Al
Mn
Zn
Cr
Fe
Cd
Co   Activity
Ni
Sn
Pb
H
Sb
As
Bi
Cu
Ag
Pd
Hg
Pt
Au

Least active
(loses electrons
least easily)

We could extend this list by testing additional pairs of metals and metal ions, but chemists have done this for us over the years. A portion of the resulting EMF series is shown in the margin.

Remember that a more active metal will always give electrons to ions of a less active metal. Using the EMF series, we can predict which redox reactions will be spontaneous and which way the electrons will flow in a battery. For example, if we set up the experiment shown below, in which beaker will a spontaneous redox reaction occur?

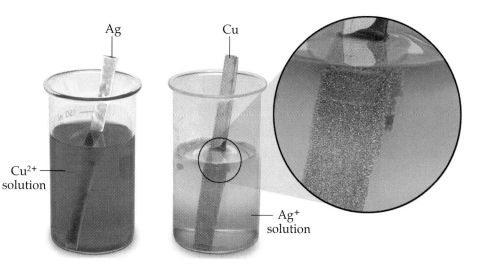

Copper is higher on the EMF list than silver (not by much, but it's higher). This means that copper is more active than silver, and so neutral copper, Cu, will give electrons to the silver ions, Ag^+. The redox reaction

$$Cu + 2\,Ag^+ \longrightarrow Cu^{2+} + 2\,Ag$$

will spontaneously occur in the beaker on the right. The Ag^+ ions will be reduced to elemental silver and plate out on the copper strip. As for the beaker on the left, absolutely nothing will happen. The reverse of the spontaneous redox reaction won't happen, because silver is less active than copper. In a copper–silver battery, electrons flow from the copper electrode (more active

metal) to the silver electrode (less active metal). This means that copper is the anode ($-$) and silver is the cathode ($+$). Copper is oxidized and Ag^+ ions are reduced. All these results can be predicted from a quick glance at the EMF series. We can even take advantage of this knowledge to make a beautiful Christmas decoration by bending a piece of copper wire into the shape of a Christmas tree and placing it into a solution of Ag^+ ions. The spontaneous redox reaction $Cu + 2\,Ag^+ \longrightarrow Cu^{2+} + 2\,Ag$ will occur, and pure silver metal will plate out on the copper tree. In addition, the Cu^{2+} ions formed will color the solution a bright blue.

8.21 What will happen if Zn is placed into a solution of Al^{3+} ions?

Answer: Absolutely nothing; Al is higher on the EMF series and is thus the more active metal. This means that Al could give its valence electrons to Zn^{2+} ions, but Zn will not spontaneously give its electrons to Al^{3+}.

8.22 Which of the following reactions would be spontaneous?
    (a) $3\,Zn^{2+} + 2\,Al \longrightarrow 3\,Zn + 2\,Al^{3+}$
    (b) $3\,Zn + 2\,Al^{3+} \longrightarrow 3\,Zn^{2+} + 2\,Al$

8.23 Using a diagram, show how you would construct a zinc–aluminum battery. Label the cathode, the anode, the + electrode, and the − electrode. Also, be sure to indicate the direction of electron flow and identify the beaker in which oxidation occurs and the beaker in which reduction takes place.

8.24 Consider a battery made from lead (Pb), copper (Cu), Pb^{2+} ions, and Cu^{2+} ions. In such a battery, what is the oxidizing agent, and what is the reducing agent? [*Hint:* Start by consulting the EMF series on page 284, and determine which metal oxidizes and which metal ion gets reduced.]

Understanding the EMF series helps us to predict which redox reactions are spontaneous and how to make batteries. It can also help us combat an enemy that costs us billions of dollars each year, an enemy that weakens our bridges, destroys our infrastructure, and even tried to tear down the Statue of Liberty! Who is this unpatriotic fiend? Corrosion.

ANOTHER LOOK AT OXIDATION: THE CORROSION OF METALS 8.6

We have defined oxidation as the loss of electrons. We are not going to change this definition, but we do want to embellish it a bit. Earlier in this chapter we talked about the burning of natural gas (methane, CH_4):

$$\overset{-4\,+1}{CH_4} + 2\,\overset{0}{O_2} \longrightarrow \overset{+4\,-2}{CO_2} + 2\,\overset{+1\,-2}{H_2O}$$

Methane is oxidized, since the oxidation state of the carbon atom changes from −4 to +4, signifying the loss of 8 electrons. However, there is an older definition of oxidation. One hundred years ago, before chemists even knew of the existence of electrons, they would have said that the methane was "oxidized" upon burning. At that time, oxidation literally meant "combining with oxygen," which is what happens when something burns. Many elements, including metals, react or combine with oxygen ("burn"). To see an example of this, just go to any junkyard and find a '65 Chevy. Rust! Rust is iron that has combined with oxygen, iron that has "burned":

$$\underset{\text{Iron metal}}{4\,Fe} + 3\,O_2 + 2\,H_2O \longrightarrow \underset{\text{Iron oxide (rust)}}{2\,Fe_2O_3 \cdot H_2O}$$

Of course, most cars don't burst into flames when they rust. This "burning" of metal, this combining with oxygen, occurs slowly. We call this slow burn **corrosion**.

Before we go on, let's get our definitions sorted out. We have the modern definition of oxidation, the loss of electrons. We also have the older definition of oxidation, combining with oxygen. Are these definitions compatible? Or, put another way, does an element lose electrons when it combines with oxygen ("burns")? The answer is yes if the element combining (forming a new bond) with oxygen is less electronegative than oxygen. Since oxygen is more electronegative than every element in the periodic table except fluorine, the answer in general is yes. When a metal (or any other element) forms new bonds with oxygen, it loses the bonding electrons according to the bookkeeping rules of the oxidation state method. Examine the oxidation numbers of the atoms in the reaction below, and you will see that iron is indeed oxidized by the modern definition when reacting with oxygen. Each iron atom loses 3 electrons. So, we see that the new and the old definitions of oxidation are truly one and the same.

$$\overset{0}{4\,Fe} + \overset{0}{3\,O_2} + 2\,H_2O \longrightarrow \overset{+3\,-2}{2\,Fe_2O_3 \cdot H_2O}$$

Iron metal   Iron oxide (rust)

Corrosion (rusting, oxidation) of iron or steel (which is iron combined with small amounts of other elements) is a major economic problem in any modern society, since so many things are built with steel. We try to prevent the corrosion of things like cars and bridges by covering the iron with paint or applying sealants, but this ultimately ends up a losing battle. We live in an atmosphere full of oxygen, and sooner or later, nature will claim your car. Sometimes, however, we can unwittingly make the situation worse! Such was the case with the Statue of Liberty. She was built with a "time bomb" ticking away inside, an unintentional design flaw involving a redox reaction that almost brought her down.

The Statue of Liberty is made from sculpted sheets of copper metal. These sheets are hung on a steel skeleton that provides the structural support. The statue sits on a small island, constantly exposed to the wind and spray of salt water. Needless to say, the exposed copper is subject to corrosion. Copper was chosen for the external skin of the statue because when it corrodes it becomes coated with a green material called a patina. The actual composition of the patina is a mixture of the hydroxy carbonate $Cu_2(OH)_2CO_3$ and the hydroxy sulfate $Cu_2(OH)_2SO_4$. The copper in these compounds is in the +2 oxidation state. This means that the exposed surface of the metallic copper has been oxidized to Cu^{2+}.

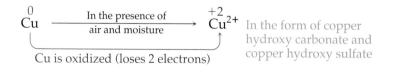

By itself, this is not a problem for the statue. Unlike iron oxide (rust), which flakes off and continually exposes fresh iron to further corrosion, oxidized copper (the patina) forms a tough, impenetrable coating over the underlying copper, protecting it from contact with oxygen and further corrosion. Given these properties of copper, the Statue of Liberty should not corrode after the initial patina has formed. For about 100 years, this was true. Then, almost overnight, it started to fall apart. The arm holding the torch was in serious danger of falling off!

The problem was actually inside the statue—the steel (iron) skeleton was reacting with the copper sheets. Consulting the EMF series on page 284, you can see that iron is a more active metal than copper. If iron metal is in contact with copper, and the copper is in its oxidized state (Cu^{2+}), then the iron will spontaneously oxidize and give electrons to the Cu^{2+} ions, reducing them back to elemental copper.

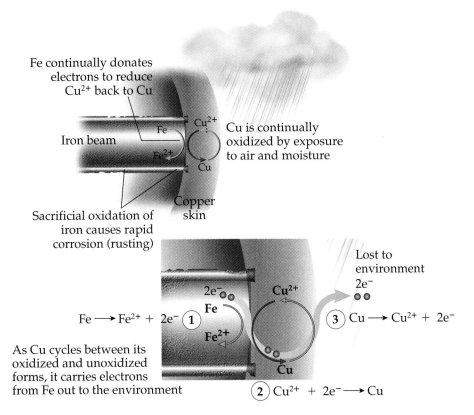

Fe continually donates electrons to reduce Cu^{2+} back to Cu

Iron beam

Cu is continually oxidized by exposure to air and moisture

Sacrificial oxidation of iron causes rapid corrosion (rusting)

Copper skin

As Cu cycles between its oxidized and unoxidized forms, it carries electrons from Fe out to the environment

Lost to environment $2e^-$

$Fe \longrightarrow Fe^{2+} + 2e^-$ ①

③ $Cu \longrightarrow Cu^{2+} + 2e^-$

② $Cu^{2+} + 2e^- \longrightarrow Cu$

More active metals lose electrons to ions of less active metals. This is great for the copper. Iron "sacrifices" itself to save it. As air and moisture re-oxidize the copper, the iron in the skeleton keeps losing its electrons to Cu^{2+}, reducing it back to elemental copper (Cu). The net result is that the inner steel skeleton oxidizes and crumbles very quickly.

Actually, the designers of the statue knew that this would happen. They put thick asbestos pads between the steel framework and the copper skin to insulate them from each other. Unfortunately, over the years, the asbestos padding rotted away, allowing the copper and iron to come into direct contact. At that point, the statue became one huge redox reaction, with her steel skeleton as one of the reactants. After ignoring the problem for a number of years, a massive effort was launched to save the statue. Many millions of dollars later, she now has a reinforced, corrosion-resistant, stainless steel skeleton incorporating high-tech plastics to insulate it from the copper.

Scaffolding surrounded the Statue of Liberty as corrosion damage was being repaired.

An active metal sacrificing itself for a less active one is also responsible for that rather painful sensation you experience when you bite down on a piece of aluminum foil. (Remember, we mentioned this at the beginning of this chapter.) Some of the tin (Sn) or silver (Ag) used in the mercury amalgam of your fillings has been oxidized to Sn^{2+} or Ag^+ by continual exposure to air and acids in your mouth. Aluminum, being a far more active metal than tin or silver (it is far above Sn and Ag in the EMF series), sacrifices itself on contact with your filling. The Al oxidizes to Al^{3+} and feeds electrons to the Sn^{2+} and Ag^+ ions, reducing them back to their elemental forms. This electron transfer gives the nerves in your teeth the equivalent of an electric shock.

The phenomenon of a sacrificial metal can actually be used in the never-ending battle to prevent corrosion. One way to protect iron (or steel) from corrosion is to coat the iron (plate it) with a more active metal such as chromium (Cr). Many of the active metals like chromium and aluminum form tough oxide coatings that strongly adhere to the metal underneath, protecting it from further corrosion. This works well for small pieces of iron, but chromium is an expensive metal. You wouldn't want to chrome plate the steel hulls of ocean-going vessels or large steel electrical utility poles. It would cost too much. What is often done instead is to attach a piece of a more active metal like magnesium to the steel. The more active magnesium (Mg) spontaneously oxidizes (to Mg^{2+}) and feeds its electrons to the iron in the steel, protecting it from oxidation. All we need to do is to periodically replace the piece of magnesium, the sacrificial (more active) metal.

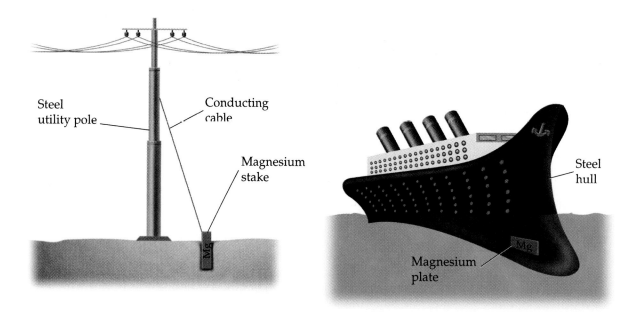

Steel utility pole

Conducting cable

Magnesium stake

Steel hull

Magnesium plate

The magnesium sacrifices itself to prevent corrosion of the iron (steel)

Of course, there are some instances, often symbolic, where we want metals to last a lifetime without any signs of corrosion.

If marriage is supposed to be "till death do you part," then what metal does the EMF series indicate should be used for making wedding rings? Explain your answer.

WORKPATCH 8.10

Certainly, a ring made of iron, which corrodes rapidly, would not be appropriate in this case.

From the "burning" of glucose in our cells to the batteries in our electronic devices to the combustion of methane, chemical reactions that transfer electrons from one atom to another can be found almost everywhere. Not all chemical reactions involve electron transfer, but you now know how to recognize the ones that do.

HAVE YOU LEARNED THIS?

Electricity (p. 257)

Voltage (p. 257)

Electron transfer reaction (p. 258)

Oxidation state (oxidation number) (pp. 259, 261)

Reduction (p. 269)

Oxidation (p. 270)

Redox (p. 270)

Oxidizing agent (p. 271)

Reducing agent (p. 271)

Battery (galvanic cell) (p. 275)

Electrodes (p. 275)

Anode (p. 280)

Cathode (p. 280)

Active metal (p. 282)

Electromotive force (EMF) series (p. 283)

Corrosion (p. 286)

ELECTRICITY AND ELECTRON BOOKKEEPING (OXIDATION STATE)

8.25 How is electricity similar to water flowing in a hose?

8.26 How is electricity mechanically produced in a power plant?

8.27 What is an electron transfer reaction?

8.28 Why did chemists find it necessary to invent oxidation states?

8.29 What do we mean by electron bookkeeping?

8.30 Which method of electron bookkeeping for shared electrons is correct, the double-counting method of the octet rule or the oxidation state method?

8.31 When assigning an oxidation state to an atom, why do we need to know how many valence electrons are present on a free atom of that element?

8.32 According to the oxidation state method of electron bookkeeping, which atom is assigned the shared electrons in a bond between two atoms?

8.33 Draw an electron dot diagram for NF_2H, and use the oxidation state method of electron bookkeeping to determine how many electrons each atom should be assigned.

8.34   Draw an electron dot diagram for H_2O_2 (hydrogen peroxide), and use the oxidation state method of electron bookkeeping to determine how many electrons each atom should be assigned.

8.35   Draw an electron dot diagram for NO_3^-, and use the oxidation state method of electron bookkeeping to determine how many electrons each atom should be assigned.

8.36   Assign oxidation states to every atom in the molecules in Problems 8.33–8.35.

8.37   Consider the following molecules. Without using the shortcut rules, determine the oxidation state for each atom in the molecules and show how you calculated it.
(a) H_2 (b) O_2 (c) Cl_2

8.38   What does the answer to Problem 8.37 teach you?

8.39   What must always be true when you add up all the oxidation states for the atoms in a molecule?

8.40   Suppose an atom has an oxidation state of $+3$.
(a) Would more or fewer electrons be assigned to this atom by oxidation state bookkeeping than are present on a free atom of that element?
(b) How many more or fewer?

8.41   Suppose an atom has an oxidation state of -2.
(a) Would more or fewer electrons be assigned to this atom by oxidation state bookkeeping than are present on a free atom of that element?
(b) How many more or fewer?

8.42   Suppose the shortcut rules can determine the oxidation state of every atom in a compound except one. How can you find the oxidation state of the remaining atom?

8.43   Use the shortcut rules to assign oxidation states to the atoms in the following molecules:
(a) PCl_3 (b) H_2S (c) MnO_4^- (d) HNO_3
(e) $HCOOH$ (f) $S_2O_3^{2-}$

8.44   Use the shortcut rules to assign oxidation states to the atoms in the following molecules:
(a) CO (b) CH_2Cl_2 (c) $HCOO^-$ (d) $PtCl_6^{2-}$

8.45   Use the shortcut rules to assign oxidation states to the atoms in the following ions and molecules:
(a) O^{2-} (b) Li_3N (c) $MgSO_4$ (d) MnO_2

8.46   Consider ClO^- and $AlCl_3$. For one of these substances the halide shortcut rule works. For the other it does not.
(a) Which one does it work for, and why?
(b) Why doesn't it work for the other?
(c) Assign oxidation states to all the atoms in both substances.

8.47   Assign oxidation states to the atoms in the molecule IF.

8.48   Consider the organic compound $C_3H_6O_2$, methyl acetate.
(a) What is the problem with using the shortcut method to assign oxidation numbers to the atoms in methyl acetate?
(b) The dot diagram of methyl acetate is shown below. Assign oxidation states to every atom.

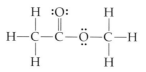

RECOGNIZING ELECTRON TRANSFER REACTIONS

8.49 What does the term "reduction" mean?

8.50 What does the term "oxidation" mean?

8.51 How are oxidation states useful in determining whether a reaction is a redox reaction?

8.52 What happens to an atom's oxidation state when it is reduced?

8.53 What happens to an atom's oxidation state when it is oxidized?

8.54 What does the word "transfer" imply about an electron transfer reaction?

8.55 Why can we always call an electron transfer reaction a redox reaction?

8.56 Which of the following is a redox reaction?
(a) $2 Na + 2 H_2O \longrightarrow 2 NaOH + H_2$
(b) $MgBr_2 + 2 NaF \longrightarrow MgF_2 + 2 NaBr$
(c) $2 CO + O_2 \longrightarrow 2 CO_2$
(d) $SO_2 + H_2O \longrightarrow H_2SO_3$

8.57 For those reactions in Problem 8.56 that are redox reactions:
(a) Indicate which atoms get oxidized and which atoms get reduced.
(b) Indicate which reactant is the oxidizing agent.
(c) Indicate which reactant is the reducing agent.

8.58 Which of the following is an electron transfer reaction?
(a) $2 CrO_4^{2-} + 2 H^+ \longrightarrow Cr_2O_7^{2-} + H_2O$
(b) $Fe + NO_3^- + 4 H^+ \longrightarrow Fe^{3+} + NO + 2 H_2O$
(c) $2 C_2H_6 + 7 O_2 \longrightarrow 4 CO_2 + 6 H_2O$
(d) $2 AgBr \longrightarrow 2 Ag + Br_2$

8.59 For those reactions in Problem 8.58 that are electron transfer reactions:
(a) Indicate which atoms get oxidized and which atoms get reduced.
(b) Indicate which reactant is the oxidizing agent.
(c) Indicate which reactant is the reducing agent.

8.60 The following reaction is responsible for producing electricity in your car battery (often called a lead storage battery):

$$Pb + PbO_2 + 2 H_2SO_4 \longrightarrow 2 PbSO_4 + 2 H_2O$$

(a) Assign oxidation states to all the atoms. [*Hint:* For $PbSO_4$, you can get the charge of the Pb if you remember that the charge of the sulfate ion is -2 (SO_4^{2-}). Then use shortcut rule 6 to get the oxidation state of the Pb in $PbSO_4$.]
(b) Identify the atom that gets oxidized and the atom that gets reduced.
(c) Identify the reactant that is the oxidizing agent and the reducing agent.

8.61 Hydrogen (H_2) gas burns very well in the presence of oxygen (O_2) to give water:

$$2 H_2 + O_2 \longrightarrow 2 H_2O$$

In principle, should it be possible to use this chemical reaction to produce electricity? Explain.

8.62 In the upper atmosphere, sunlight can convert oxygen into ozone:

$$2 O_2 \longrightarrow O_3 + O$$
Ozone

Is this a redox reaction? Completely justify your answer.

8.63 The hydroquinone molecule can be converted to the quinone molecule as shown below:

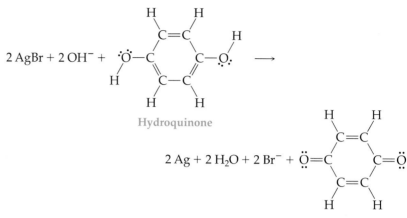

You may want to use a combination of shortcut rules and dot diagrams to assign oxidation numbers before answering the following questions:
(a) Which atom(s) are oxidized?
(b) Which atoms(s) are reduced?
(c) What is the oxidizing agent?
(d) What is the reducing agent?

ELECTRICITY FROM REDOX REACTIONS

8.64 If you put a piece of iron into an aqueous solution of blue Cu^{2+} ions, the spontaneous redox reaction

$$Fe + Cu^{2+} \longrightarrow Fe^{2+} + Cu$$

will occur. An aqueous solution of Fe^{2+} ions is red-brown in color.
(a) What is oxidized?
(b) What is reduced?
(c) What is the oxidizing agent?
(d) What is the reducing agent?
(e) What will happen to the piece of iron over time?
(f) Do the electrons move from the oxidizing agent to the reducing agent, or vice versa?
(g) How could you turn this into a battery? (Explain and show it with a diagram, using a nail and a penny as your source of iron and copper, respectively.)

In Problems 8.65–8.69 you will be given the spontaneous redox reaction (or enough information regarding it) so that you can construct and label a battery.

8.65 Construct a battery from the spontaneous redox reaction:

$$Cr + 3 Ag^+ \longrightarrow Cr^{3+} + 3 Ag$$

Draw a battery similar to the one you drew for WorkPatch 8.7. Make sure you label which way the electrons flow, and also label the cathode, anode, +, and −.

8.66 For the chromium–silver battery of Problem 8.65:
(a) What is the oxidizing agent?
(b) What is the reducing agent?
(c) Which electrode gains mass with time?
(d) Which electrode is eaten away with time?

8.67 Construct a battery from the spontaneous redox reaction:

$$Mg + Ni^{2+} \longrightarrow Mg^{2+} + Ni$$

Draw a battery similar to the one you drew for WorkPatch 8.7. Make sure you label which way the electrons flow, and also label the cathode, anode, +, and −.

8.68 For the magnesium–nickel battery of Problem 8.67:
(a) What is the oxidizing agent?
(b) What is the reducing agent?
(c) Which electrode gains mass with time?
(d) Which electrode is eaten away with time?

8.69 A battery is constructed from tin (Sn) and copper (Cu) by dipping strips of each metal into a solution of its ions (Sn^{2+} and Cu^{2+}, respectively). As the battery operates, the Sn^{2+} concentration increases, while the Cu^{2+} concentration decreases.
(a) What is getting oxidized?
(b) What is getting reduced?
(c) Draw a battery similar to the one you drew for WorkPatch 8.7. Make sure you label which way the electrons flow, and also label the cathode, anode, +, and −.

8.70 How does knowing which way the electrons flow allow you to assign + and − to the proper electrodes?

8.71 What trick can you use to help you remember the sign (+ or −) of the cathode and anode?

WHICH WAY DO THE ELECTRONS FLOW? THE EMF SERIES

8.72 Suppose you have two metals, A and B, and solutions of their ions, A^+ and B^+. Metal B is more active than metal A.
(a) Write the spontaneous redox reaction for this situation.
(b) What is the oxidizing agent?
(c) What is the reducing agent?
(d) Describe which way the electrons flow.

8.73 A zinc electrode and a copper electrode are used to make a battery, as shown:

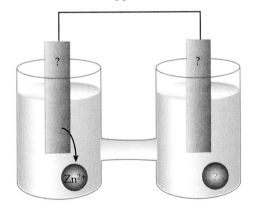

(a) Replace the question marks with appropriate labels.
(b) Indicate which way the electrons flow
(c) Write the spontaneous redox reaction.
(d) Add the labels +, −, anode, and cathode to the drawing.
(e) Write "oxidation occurs at this electrode" under the appropriate beaker.
(f) Write "reduction occurs at this electrode" under the appropriate beaker.
(g) Indicate in which beaker the concentration of ions increases with time.
(h) Indicate in which beaker the concentration of ions decreases with time.
(i) Indicate in which beaker the electrode appears eaten away with time.
(j) Indicate in which beaker the electrode gains mass with time.

8.74 Suppose you have three different metals, X, Y, and Z, each of which also exists as a +2 ion. Design an experiment to place them properly in order of their activity (a mini-EMF series).

8.75 What use is the EMF series?

8.76 What happens when you place an active metal into a solution of ions of a less active metal?

8.77 What happens when you place a less active metal into a solution of ions of a more active metal?

8.78 Using the EMF series on page 284, decide which of the following redox reactions is spontaneous. Explain your answer.
(a) $3K + Al^{3+} \longrightarrow Al + 3K^+$
(b) $Al + 3K^+ \longrightarrow 3K + Al^{3+}$

8.79 Using the EMF series on page 284, decide which of the following redox reactions is spontaneous. Explain your answer.
(a) $3Ag + Au^{3+} \longrightarrow Au + 3Ag^+$
(b) $Au + 3Ag^+ \longrightarrow 3Ag + Au^{3+}$

8.80 You are trapped on a desert island with plenty of water (both fresh and salt) a drinking glass, some wire, a radio, and no batteries. You do have a tin (Sn) cup, a tube of toothpaste containing stannous fluoride (SnF_2, a source of Sn^{2+} ions), a silver (Ag) pendant, and undeveloped black and white film (such film has silver bromide, AgBr, in it, a source of Ag^+ ions).
(a) How would you use the above materials to construct a battery? Show how with a diagram, including an arrow over the wire to show which way the electrons flow. (You can make a salt bridge by soaking a sock in salt water and then dipping the end in both cells.)
(b) Which metal would be eaten away? Explain.
(c) Which is the oxidizing agent?
(d) Which is the reducing agent?

8.81 Show with a diagram how you would construct a copper–nickel battery. Label the direction of flow of electrons, cathode, anode, +, and −. Also, indicate in which beaker oxidation occurs and in which beaker reduction occurs.

8.82 Recharging a battery means forcing a spontaneous redox reaction to run backwards—in the nonspontaneous direction—once all the reactants have been spent. In Practice Problem 8.24 you considered a battery made from lead and copper.
(a) Write the spontaneous redox reaction for this battery.
(b) Write the recharging reaction for this battery.

8.83 If the EMF series is based on activity, then which two elements from the EMF series on page 284 do you think would combine to give the battery with the highest voltage? Explain your answer.

THE CORROSION OF METALS

8.84 There are two definitions of oxidation, one involving oxygen, the other not.
(a) State the two definitions of oxidation.
(b) Are they compatible? Explain.
(c) When we speak of corrosion of metals, what chemical reaction are we usually talking about (use Fe as an example)?

8.85 How can attaching a piece of Mg to the iron hull of a ship protect it from rusting?

8.86 Suppose gold were not available for a wedding ring, but you still wanted a ring that would last "forever" and not corrode. According to the EMF series on page 284, what would be a good alternative metal to use?

8.87 Explain what almost brought down the Statue of Liberty, and what was done to keep the problem from happening again.

8.88 What problem might develop with the casing of the dry cell battery shown on page 279 if it is not made thick enough? Explain.

WorkPatch Solutions

8.1

N is assigned 6 electrons
N is more electronegative than H, but less electronegative than F

H ∴N∴ H ⟵ H is assigned 0 electrons
H is less electronegative than N

:F:

F is assigned 8 electrons
F is more electronegative than N

8.2

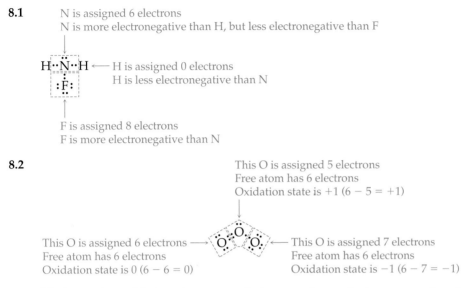

This O is assigned 5 electrons
Free atom has 6 electrons
Oxidation state is +1 (6 − 5 = +1)

This O is assigned 6 electrons ⟶
Free atom has 6 electrons
Oxidation state is 0 (6 − 6 = 0)

⟵ This O is assigned 7 electrons
Free atom has 6 electrons
Oxidation state is −1 (6 − 7 = −1)

8.3
(a) A negative oxidation state means that more valence electrons are assigned to the atom by oxidation state bookkeeping than are present on a free atom of the element.
(b) An oxidation state of −1 means that oxidation state bookkeeping assigns that chlorine atom 1 more valence electron than the number present on a free chlorine atom.
(c) Chlorine has 7 valence electrons as a free atom (group VIIA). To have an oxidation state of −1, and therefore 8 valence electrons, it must share its unpaired electron with an unpaired electron from another atom that is less electronegative than itself.

8.4
(a) The chlorine atom is bonded to atoms that are more electronegative than it is. Oxidation state bookkeeping always assigns the shared bonding electrons entirely to the more electronegative atom. Therefore, all the bonding electrons in this ion are assigned to the oxygen atoms. The chlorine atom gets none of them, so the halide rule does not apply.
(b) If each O atom has an oxidation state of −2 (rule 2), then the oxidation states of the three O atoms add up to −6. Therefore, the Cl atom must have an oxidation state of +5 for the sum of the oxidation states of all the atoms to add up to the −1 charge on the ClO_3^- ion.

8.5

This C is assigned 7 electrons
Free atom has 4 valence electrons
Oxidation state is −3 (4 − 7 = −3)

This O is assigned 8 electrons
Free atom has 6 valence electrons
Oxidation state is −2 (6 − 8 = −2)

H O
H ∴C∴C
H O H

This C is assigned 1 electron
Free atom has 4 valence electrons
Oxidation state is +3 (4 − 1 = +3)

All H's in this molecule are assigned 0 electrons
Free atom has 1 valence electron
Oxidation state is +1 (1 − 0 = +1)

This O is assigned 8 electrons
Free atom has 6 valence electrons
Oxidation state is −2 (6 − 8 = −2)

8.6 (a) Yes; $\overset{0}{Zn} + \overset{+2}{Cu^{2+}} \longrightarrow \overset{+2}{Zn^{2+}} + \overset{0}{Cu}$

(b) Zn changed to Zn^{2+}

(c) Cu^{2+} changed to Cu

(d) Cu^{2+} gets reduced (gains 2 electrons), so it is the oxidizing agent.

(e) Zn gets oxidized (loses 2 electrons), so it is the reducing agent.

8.7

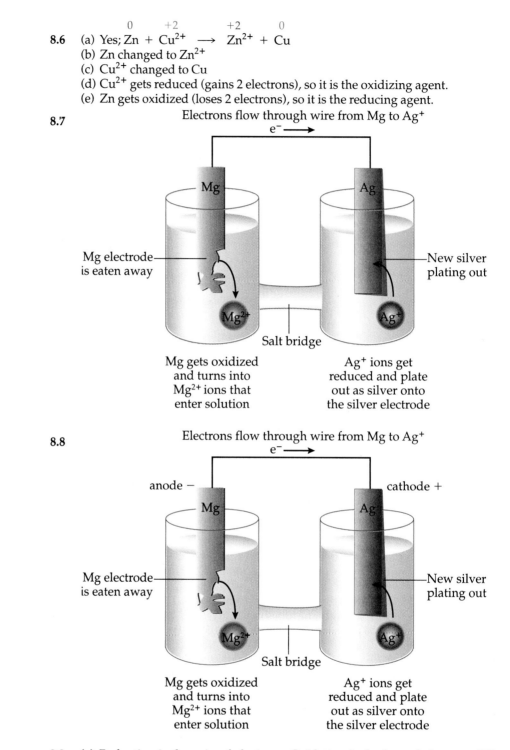

8.9 (a) Reduction is the gain of electrons. Oxidation is the loss of electrons. When something is oxidized, it loses electrons, which are then available to be gained by what is being reduced.

8.10 At the bottom of our abbreviated EMF series is gold (Au). It is the least active metal listed, and therefore is very resistant to oxidation and thus corrosion.

Chapter 9

Intermolecular Forces and the Phases of Matter

WHY DOES MATTER EXIST IN DIFFERENT PHASES? 9.1

It's the middle of July, 34°C, and you feel as if you're covered with a wet blanket. Sound familiar?

These are the sweltering, high-humidity days of summer when there is just too much water in the air. You can't see this water because it exists in the gas, or vapor, phase, but you can certainly feel it. It makes you feel miserable. The twentieth century solution to this problem is the air-conditioner, which cools the air by drawing it in with a fan and passing it over cold refrigeration coils. If you were to look at the coils inside the air-conditioner, you would notice that they are dripping wet. Cooling the humid air causes the water vapor in the air to *condense* into a liquid on the cold coils, thus drying the air. This is why air-conditioned air feels so good on a humid summer day—it's not just cooler, it's also drier. By setting the thermostat to its lowest setting, you can sometimes cause the refrigeration coils to get so cold that the liquid water on them actually freezes into solid ice. By simply cooling the air, an air-conditioner can drive atmospheric water vapor through three phases—from the *gas phase*, through the *liquid phase*, into the *solid phase*.

We touched briefly on this subject in Section 1.3. In this chapter, we will run through this scenario again, but from a different point of view. This time, we will look at what happens on a molecular level. Why should cooling the air cause the water in it to change phases? Exactly what are the water molecules doing in each phase? And let's not forget about the other gases present in the atmosphere, nitrogen (N_2, 78.1% by volume), oxygen (O_2, 20.1% by volume), argon (Ar, 0.934% by volume), and carbon dioxide (CO_2, 0.033% by volume). Unlike water, these gases don't change phase in the air-conditioner, even though they can exist in liquid and solid forms.

Liquid N_2 Liquid O_2 Solid CO_2 (dry ice)

Why do these substances remain in the gas phase while water vapor condenses and eventually freezes? What is so different about water molecules that could account for this difference in behavior?

MODEL OF A GAS

If you could see the individual molecules that make up the atmosphere (N_2, O_2, Ar, and CO_2), you would see that all these molecules are in constant motion. They travel in all directions, at speeds up to 500 miles per hour (as mentioned earlier in Chapter 7). The molecules are either colliding with each other or zipping past one another.

Now, what happens when we cool the air by passing it over the cold refrigeration coils in our air-conditioner? During this cooling process, we would observe something very interesting. The molecules in the air would slow down. *This is very important:* The average speed of the molecules in a gas depends on the temperature. When you add heat to a gas, the molecules in the gas move faster, on average. When you remove heat from a gas, the molecules in the gas move more slowly, on average.

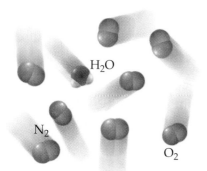

Molecules in hot air move very fast.

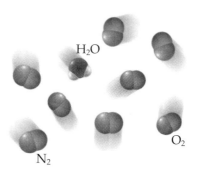

Molecules in cold air move more slowly.

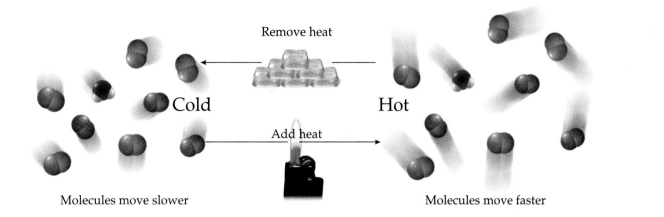

Remove heat

Cold Hot

Add heat

Molecules move slower Molecules move faster

MODEL OF A SOLID

So maybe the reason the water molecules condense out of the gas phase to become a liquid upon cooling is that they have slowed down. But this cannot be the whole answer. If it were, then the CO_2, O_2, and N_2 molecules and the Ar in the atmosphere would also condense, since they also slow down as heat is removed. There must be something else involved. Think back to Chapter 6, where we suggested you try ripping an ice cube in half.

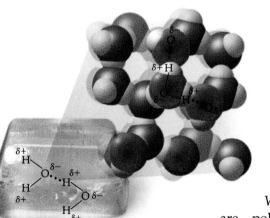

It can't be done because the water molecules in the ice cube are "stuck" too tightly to one another by the intermolecular forces of attraction between them. The $\delta+$ end of one polar water molecule is attracted to the $\delta-$ end of another. These forces are responsible for the existence of water in the solid phase (ice).

Water molecules are polar no matter what phase they are in. Even in the gas phase, they are attracted to each other when they are close. In a gas, however, these attractive forces are too weak to make the fast-moving molecules stick together. Their **kinetic energy** (the energy associated with their motion) is great enough to overcome their mutual attraction, and they shoot on past each other.

Attraction

$\delta+$

$\delta-$

Speed too great . . . attraction too weak . . . fast-flying molecules zoom on past each other

MODEL OF A LIQUID

However, as the gas is cooled (energy is removed), the molecules slow down. At some temperature the molecules will finally be moving slowly enough for the attractive intermolecular forces to overcome the

Attraction

Attraction draws
slow-moving
molecules together

$\delta-$ $\delta+$

These two
adhering water
molecules could
gather others,
becoming the
nucleus of a
liquid droplet.

Water molecules in the
liquid phase cling in
contact with each other,
but they move around in a
constant, random jostle.

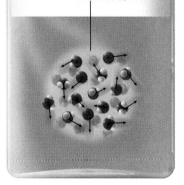

molecules' kinetic energy. At that temperature, the gas-phase molecules start to stick to each other and form the liquid phase.

When water changes from the gas phase to the liquid phase, we say that **condensation** has occurred. If you reverse this process—that is, start with liquid water and add heat—the molecules in the liquid will move faster and faster until they can finally overcome the attractions between them and leap into the gas phase. This reverse process, going from the liquid to the gas, is called **boiling** or **vaporization**.

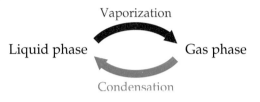

Vaporization

Liquid phase   Gas phase

Condensation

The temperature at which this occurs is called the **boiling point**. Since condensation is just the reverse of vaporization, the boiling point is also the temperature at which the condensation of gaseous water occurs.

Continued cooling of the liquid water (continued removal of energy) will eventually cause it to freeze to solid ice. To see why, let's take a close look at liquid water. A molecular view would show that the molecules in the liquid phase are still moving, even though they are all stuck to each other.

Even though the molecules in a liquid are in a much more crowded environment than they would be in a gas, they still have enough kinetic energy to jostle past each other, continually breaking the attractive force to one molecule and reestablishing it with another. You can come close to experiencing what it's like to be a molecule in the liquid phase by visiting a large shopping mall on the day after Thanksgiving. You can continue to move throughout the mall and get your shopping done all the while getting pushed, shoved, and bumped as you travel.

As the temperature of the liquid water drops, the molecules continue to slow down. Soon, they are no longer moving fast enough to even jostle past each other. At that temperature (called the **freezing point**), the molecules are effectively frozen in place. They have arrived at the solid phase.

As close as you'll probably come to
feeling like a molecule in a liquid

The structure of solid water (ice).
The molecules are locked in position
by the intermolecular attractions.

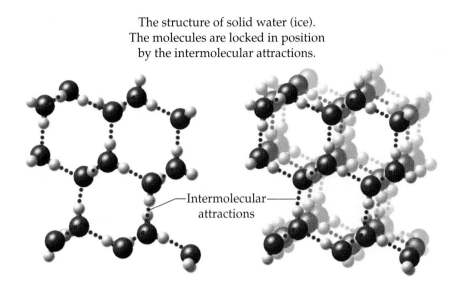

Intermolecular—
attractions

OVERVIEW OF THE THREE PHASES

A molecular view helps to understand why water exists in three phases and how it changes phases. The key points:

1. Heat causes molecules to move. The more heat, the faster they move, and the higher the temperature.
2. Intermolecular forces of attraction between molecules tend to stick the molecules together. If the molecules are moving fast enough, they have sufficient kinetic energy to overcome these forces and remain in the gas phase. Cooling can slow them down enough so that they stick together in the liquid phase, and further cooling allows them to form the solid phase.

In the gas phase, molecules are essentially independent of each other and are, on average, rather far apart from each other. In the liquid and solid phases, molecules are quite close to each other (they essentially "touch" each other). It is for this reason that the liquid and solid phases are often referred to as **condensed phases**.

Now that we understand what happens with water, we can go back and consider the other gases in the atmosphere, CO_2, N_2, Ar, and O_2, the gases that didn't change phase upon cooling. Why don't blocks of dry ice clog the coils of our air-conditioners? Why don't pools of liquid nitrogen and liquid oxygen form at our feet? One possibility is that these molecules have no attractive intermolecular forces between them. This is wrong, however. All molecules attract each other, all the time. They just don't do so with equal strength. The forces of attraction between two CO_2 molecules (or two N_2 molecules or two O_2 molecules) are much weaker than the forces of attraction between two H_2O molecules. Thus, in order to get these molecules to stick to one another, you have to cool them (slow them down) much more than water. It takes a temperature of $-78°C$ (at normal atmospheric pressure) to slow down CO_2 molecules enough so that the forces of attraction between them can begin to overcome their kinetic energy, allowing them to form a solid (normal atmospheric pressure is discussed in Section 9.5). It takes a temperature of −183°C (at normal atmospheric pressure) to slow down oxygen molecules enough to form a liquid. Only when the temperature gets down to −196°C (at normal atmospheric pressure) will nitrogen begin to form a liquid. This is extremely cold, so cold that materials like rubber, plastic, and bananas lose their usual flexibility and become as rigid and as fragile as glass.

9.1 WORKPATCH   The normal boiling point of ethanol, the alcohol in beer and wine, is 78°C. The normal boiling point of water is 100°C. (Normal means that the pressure at which the phase change occurs is normal atmospheric pressure.) What can you say about the strength of the intermolecular forces for ethanol relative to water?

The WorkPatch is a reminder that physical properties like boiling point and freezing point are related to the strength of the attractive forces between individual molecules.

PRACTICE PROBLEMS

9.1 Consider H–Cl molecules. Draw a picture showing how they are attracted to one another using dotted lines to represent intermolecular forces.

Answer:

$$\overset{\delta+}{H}-\overset{\delta-}{Cl}\cdots\overset{\delta+}{H}-\overset{\delta-}{Cl}$$
$$\overset{\delta+}{H}-\overset{\delta-}{Cl}$$

The dotted lines show the intermolecular attractive forces. Other arrangements are possible. Just make sure the δ+ end of one molecule approaches the δ− end of another.

9.2 Based just on the strength of the molecular dipole moments, which compound should have a higher boiling point, HF or HBr? Explain your choice.

9.3 Consider NH_3 molecules. Draw a picture showing how they attract one another.

9.4 Based just on the strength of the molecular dipole moments, which compound should have a higher boiling point, H_2S or H_2O? Explain your choice.

We stated earlier that "all molecules attract each other, all the time." Even though it requires extremely low temperatures, the fact that oxygen and nitrogen can exist as liquids at all is proof of this statement. For any substance to condense into a liquid phase, there must be some attractive force to stick the molecules together. The nature of this force in water is somewhat obvious. Water molecules are polar, each molecule possesses $\delta+$ and $\delta-$ ends. The charged portion of one molecule is attracted to the oppositely charged portion of another molecule. It is the nature, or the origin, of the attractive forces between other molecules that we must now explain. While water molecules are polar, molecules like CO_2, N_2, and O_2 are completely nonpolar. They do not have partially charged ends, so how are they attracted to each other? And why do some molecules attract each other strongly while others do so only weakly?

INTERMOLECULAR FORCES 9.2

In Chapter 6, we introduced the dipolar attractive forces that exist between polar molecules. But these are obviously not the only types of intermolecular forces. Let's consider carbon dioxide (CO_2). In Chapter 6 we determined that this molecule is linear and therefore nonpolar.

At a temperature of $-78°C$ (at normal atmospheric pressure), CO_2 molecules go directly from the gas phase into the solid phase. The fact that CO_2 solidifies at a much lower temperature than water (which freezes at $0°C$) indicates that the attractive forces between CO_2 molecules are considerably weaker than those between H_2O molecules. But the real question remains, why are CO_2 molecules attracted to each other at all if they are truly nonpolar and have no $\delta+$ and $\delta-$ ends? The answer comes from a model proposed over 100 years ago by the chemists Fritz London and J. D. van der Waals. Because of their work, the relatively weak attractive force between nonpolar molecules was referred to as a **London force** or a **van der Waals attraction**. (More recently, van der Waals force has taken on a broader meaning.) While this interaction is weak, there can be many such interactions in a large molecule, and the total attraction can be very strong. Nonetheless, on a per atom basis, it is the weakest kind of intermolecular force there is.

Basically, the original model used to explain London forces treats any atom or molecule as a spherical container of electrons. The electrons inside the

Nonpolar
(no $\delta+$ or $\delta-$ ends)

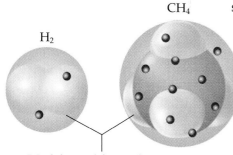

Models used for explaining
London forces

sphere are treated as being in constant, random motion. According to this simplified model, the hydrogen molecule (H_2) becomes a sphere containing 2 electrons; the methane molecule (CH_4) is a larger sphere that contains a total of 10 electrons (6 from the carbon and 1 each from the four hydrogen atoms).

Now imagine that due to the constant and random motion of the electrons and for the briefest of instants, the distribution of the electrons inside the sphere (molecule) becomes unbalanced:

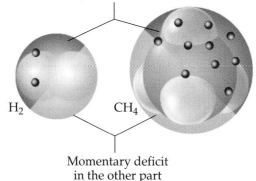

Momentary excess of electrons
in one part of molecule

Momentary deficit
in the other part

Atoms and molecules do, in fact, exhibit such brief and constantly shifting electron imbalances. Typically, these imbalances are slight (our diagrams exaggerate them). Nevertheless, when one part of a molecule has an excess of electrons, for that instant it will have a *slight* negative charge. The opposite side—the one briefly drained of electrons—will be *slightly* positive. In the next instant, some other part of the molecule may be slightly negative, and the opposite side slightly positive. These temporary charges are very small, smaller than the partial $\delta+$ and $\delta-$ charges in a dipolar molecule. To signify this, we write these charges as $\delta\delta+$ (partial partial positive) and $\delta\delta-$ (partial partial negative).

One instant . . .   The next instant . . .   The next instant . . .

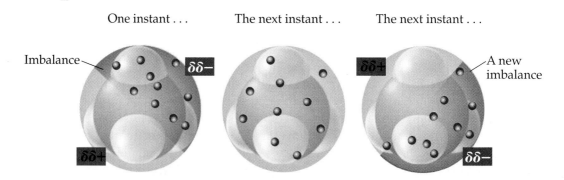

The constantly shifting distribution of electrons is what allows nonpolar molecules to attract one another. To understand this, we must consider more than one molecule at a time. Imagine a row of three methane molecules, as shown at the top of the next page.

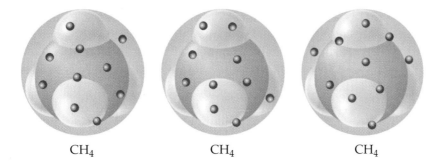

CH₄   CH₄   CH₄

Suppose the electrons in the central molecule undergo a shift to the left. How will the other molecules respond? A buildup of electrons (and negative charge) on the left side of the central molecule will repel the electrons in the atom on the left. Likewise, the partial positive charge on the right side of the central atom will attract the electrons in the atom on the right.

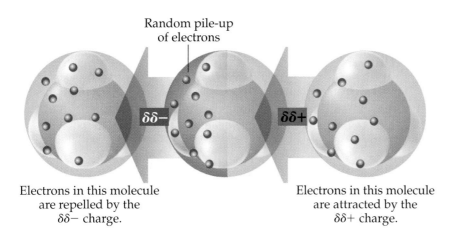

Random pile-up
of electrons

$\delta\delta-$ $\delta\delta+$

Electrons in this molecule
are repelled by the
$\delta\delta-$ charge.

Electrons in this molecule
are attracted by the
$\delta\delta+$ charge.

The shifting of electrons in the molecules on the left and right creates temporary partial charges in these molecules as well. The central molecule is said to have *induced* (caused) the adjacent molecules to become imbalanced in precisely the direction that brings about a net attraction between them (the $\delta\delta+$ end of one molecule ends up adjacent to the $\delta\delta-$ end of another):

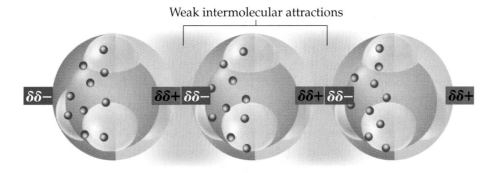

Weak intermolecular attractions

$\delta\delta-$ $\delta\delta+$ $\delta\delta-$ $\delta\delta+$ $\delta\delta-$ $\delta\delta+$

The original transient dipole in the central molecule created, or induced, dipoles in the adjacent molecules, resulting in a net attraction. The result is that the molecules "stick" to one another even though they are not polar—

One instant . . .

The next instant . . .

that is, even though they don't have permanent $\delta+$ and $\delta-$ ends. Of course, with billions of molecules instead of just the three shown, we get billions of instantaneous induced dipoles, resulting in overall attractive forces between the molecules. This pattern lasts for only an instant. In the next instant, the pattern of aligned dipoles is replaced by another aligned pattern that is equally effective at attracting molecules to each other.

London forces exist between all molecules. Polar molecules experience both dipolar and London intermolecular forces at the same time, whereas nonpolar molecules experience only London intermolecular forces.

How weak (or strong) are London forces? Remember, we have been exaggerating the electron imbalances in our diagrams. In actuality, these imbalances are very slight, and the partial charges that develop are very small. Thus, for molecules possessing few electrons (small molecules made from relatively light atoms, such as N_2, O_2, H_2O, or CH_4), London forces are extremely weak. The dipolar forces that exist between small dipolar molecules are quite a bit stronger than the London forces that exist between small molecules.

9.2 WORKPATCH   Why are the attractive intermolecular forces between water molecules much stronger than the attractive intermolecular forces between oxygen (O_2) molecules?

The answer to the WorkPatch gets right to the heart of intermolecular forces: They depend on the existence of partial charges. Even though water molecules have both dipolar and London intermolecular forces going for them, the London forces can be ignored because they are dwarfed by the dipolar forces. This is not the case for methane (CH_4), a nonpolar molecule. All it has pulling its molecules together are weak London forces. For this reason methane gas must be cooled to $-164°C$ (its normal boiling point) before it will liquefy. You really have to slow down methane molecules before the weak attractive forces between them have any effect. Liquefying hydrogen (H_2) requires even lower temperatures. The attractive forces between these molecules are so weak that almost any motion will overcome them. You must cool hydrogen gas to $-253°C$ at normal atmospheric pressure before it becomes a liquid.

On the other hand, London forces can be quite substantial for large molecules with many atoms and therefore many electrons. Thus, while methane is a gas at room temperature, nonpolar carbon tetrachloride (CCl_4), with its large chlorine atoms that possess 17 electrons each, is a liquid with a boiling point of $77°C$. This is more than $200°C$ higher than the boiling point of methane, attesting to the much stronger London forces present between CCl_4 molecules than between CH_4 molecules.

9.5 Carbon tetrabromide (CBr_4) has a normal melting point of 88–90°C. Explain why this makes sense when compared to the boiling points of methane and carbon tetrachloride.

Answer: Since CBr_4 has a melting point well above normal room temperature (~25°C), CBr_4 is a solid at room temperature. At room temperature, CCl_4 is a liquid and CH_4 is a gas. This makes perfect sense. All three molecules are tetrahedral in shape and nonpolar, so their only intermolecular forces are London forces. Since CBr_4 has the most electrons (146), it has the strongest London forces of the three molecules.

9.6 The molecule ethylene ($H_2C=CH_2$) is a gas at room temperature, but the molecule polyethylene, made from a large number of ethylene units bonded to each other, is a solid (used to make plastic food wrap, among other things). Explain why this is so.

9.7 Consider the molecules NH_3 and PH_3.
   (a) Draw a dot diagram for each.
   (b) Is either molecule polar? (The electronegativities of N, P, and H are 3.0, 2.1, and 2.1, respectively.)
   (c) In which substance would the London forces be stronger? Explain your answer.
   (d) Which substance would you expect to have the higher boiling point based just on London forces? Explain your answer.

A CLOSER LOOK AT DIPOLAR FORCES—HYDROGEN BONDING 9.3

We cannot complete a discussion of intermolecular forces without discussing a special type of dipolar attraction that exists between certain polar molecules. This attractive force is special because it is considerably stronger than an average dipolar attraction. To illustrate this, we will consider two polar molecules, acetone and water.

Acetone, the main ingredient in many brands of nail polish remover, has the molecular formula C_3H_6O. Because of the structure of the molecule and the large electronegativity difference between the carbon and oxygen atoms, the acetone molecule has a large permanent dipole:

Acetone

Dipole moment vector

The resulting attractive dipolar forces between acetone molecules make acetone a liquid at room temperature. It boils at 56°C (at normal atmospheric pressure).

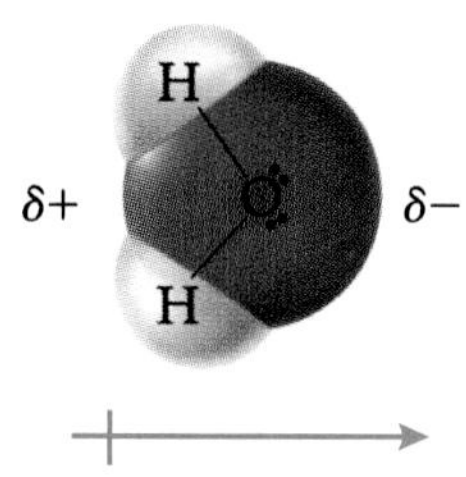

A water molecule is also polar, having a permanent dipole that is somewhat smaller than that of acetone. Yet, water boils at 100°C (at normal atmospheric pressure), an unusually high boiling point for such a small molecule. If water boiled at the temperature one would expect for a molecule of its size and composition (see Problem 9.47), the Earth would be a very different place indeed. Virtually all the water on this planet would be in the gas phase, making life as we know it impossible.

The fact that water boils at 100°C while acetone boils at 56°C tells us that the intermolecular attractive forces between water molecules are stronger than those between acetone molecules. Since the dipole moment of water is smaller than the dipole moment of acetone, there must be another reason for this difference. In the water molecule, the dipole is the result of the two hydrogen atoms bonded directly to an oxygen atom. While acetone also contains hydrogen atoms, they have no significant partial charges since they are bonded to carbon, an atom with a similar electronegativity. The dipole moment in acetone comes from the bond between oxygen and carbon.

Dipole moment vector comes
from polar O–H bonds

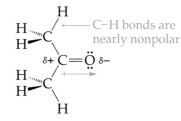

Dipole moment vector comes
from C=O bond

This structural variation makes all the difference in the world to the strength of the dipolar forces. To understand why, we need to examine the hydrogen and oxygen atoms close up. When a hydrogen atom is covalently bound to oxygen, the oxygen atom attracts the lion's share of the two shared electrons toward itself, and the hydrogen atom develops a sizable δ+ charge. This result is not unexpected. Just about any element bound to oxygen will develop a δ+ charge, since all atoms except fluorine have a lower electronegativity than oxygen. But there is something unique about the hydrogen atom. A hydrogen atom has no other electrons. When the two electrons of the O–H bond in water are pulled toward oxygen, the hydrogen atom looks almost like a naked proton sticking out in space. The picture at left shows how a water molecule might seem to another molecule. The oxygen atom is wrapped in a thick blue blanket of electron charge. Each hydrogen nucleus (a proton) is covered by only a thin veil of negative charge, since the shared electrons spend little time on the hydrogen atom.

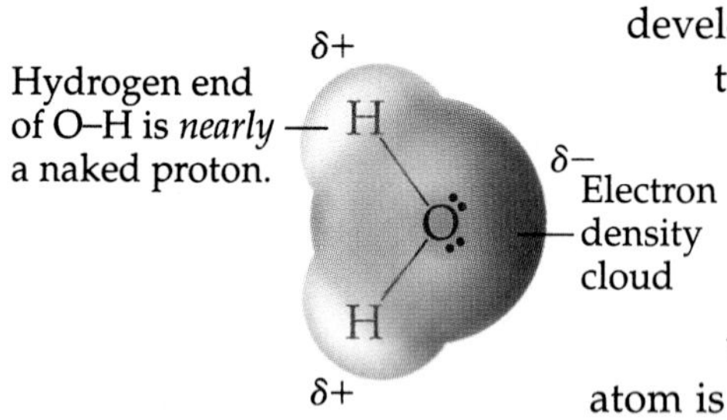

The nearly naked protons have a strong tendency to seek out electrons wherever they can find them. One obvious source would be a lone pair of electrons on the oxygen atom in another water molecule. The nearly naked proton can get in very close to the lone pair electrons since it has no other

Hydrogen end
of O–H is *nearly*
a naked proton.

Electron
density
cloud

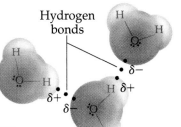

Hydrogen bonds

electrons to be repelled by them. This combination of the strong $\delta+$ charge on the hydrogen and the short distance between it and the oxygen lone pair in another molecule lead to an unusually strong attractive force between the two. These dipolar interactions are given a special name to indicate their uniqueness. They are called **hydrogen bonds** and are usually represented with a dotted line drawn between the atoms involved.

Calling this dipolar interaction a "bond" can be somewhat misleading. While this attractive force between water molecules is stronger than a normal dipole–dipole interaction, it is still nowhere near as strong as an actual covalent bond. The strength of the O–H covalent bond in a water molecule is over 400 kJ/mole (that is, it takes over 400 kJ of energy to break 1 mole of O–H covalent bonds). The hydrogen "bond" between two water molecules is only a fraction as strong—around 16 kJ/mole. This means that it would take roughly 16 kJ to separate a mole of water molecules that are stuck to each other by hydrogen bonds. Despite the dubious choice of the word "bond" in describing this interaction, it is still different enough from normal dipolar attractions (whose strength typically ranges from 4 to 8 kJ/mole) that it deserves to be treated as a special case.

It takes 400 kJ of energy to break a mole of O–H bonds . . .

But only 16 kJ to break a mole of H bonds . . .

And only 4–8 kJ to break a mole of ordinary dipolar interactions.

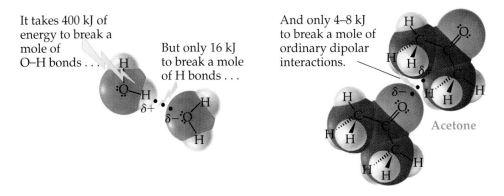

Acetone

Water is an excellent example of hydrogen bonding, but it doesn't tell the complete story. The fundamental requirement for the existence of a hydrogen bond is a nearly naked proton sticking out in space. This requirement is met whenever a hydrogen atom is directly bonded to an atom with a much higher electronegativity. Only oxygen, nitrogen (N), and fluorine (F) fill the bill. If there is an O–H, N–H, or F–H covalent bond in a molecule, then there can be hydrogen bonding between the molecules.

To form hydrogen bonds, molecules must have at least one of these covalent bonds:

$$\cdots\text{H}-\text{N}\Big/\Big\backslash \quad (\text{or}\quad \text{H}-\text{N}=)$$

$$\cdots\text{H}-\text{O}-$$

$$\cdots\text{H}-\text{F}$$

Recall that O, N, and F are three of the four "electron hogs" we discussed in Chapter 5 (chlorine is the fourth). These are the four most electronegative elements in the periodic table. So it's not so strange that when a hydrogen

atom is bonded to an atom of one of these elements, the hydrogen is practically stripped of its electrons, and conditions are set for hydrogen bonding.
Chlorine is usually not included in the set of elements that lead to hydrogen
bonding.

9.3 WORKPATCH   Two molecules are shown below. Both are liquids at room temperature. One
boils at 35°C and the other at 78°C.

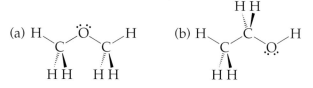

Which boils at which temperature? Why? If hydrogen bonds are involved in
either case, make a drawing using dotted lines to show them.

Both molecules in the WorkPatch have the same formula, C_2H_6O, so the
difference between their boiling points must be due to their different structures. What is it about their structures that leads to the difference? Make sure
you know before going on.

In a very real sense, hydrogen bonding is responsible for life itself. DNA—
the substance that carries the blueprint for living things—depends on hydrogen bonding to do its work. The DNA molecules in your cells encode the
information for building and operating most of the molecular structures and

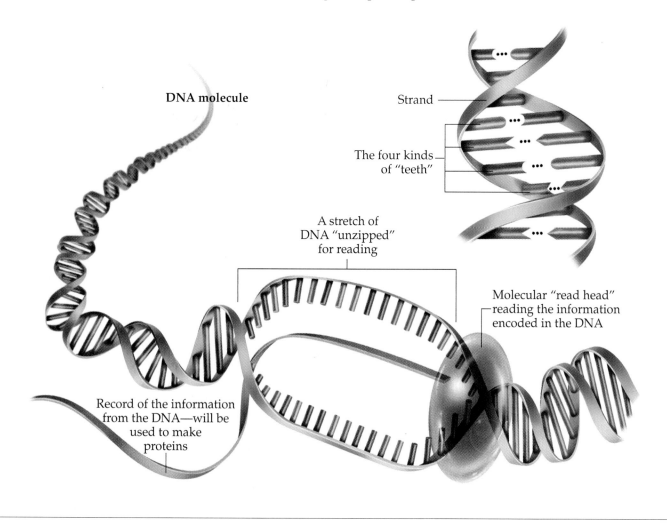

DNA molecule

Strand

The four kinds
of "teeth"

A stretch of
DNA "unzipped"
for reading

Molecular "read head"
reading the information
encoded in the DNA

Record of the information
from the DNA—will be
used to make
proteins

machinery in your body, from your toenails to your brain. (More precisely, DNA specifies the structure of protein molecules in the body, and these proteins take care of the rest.) Structurally, a DNA molecule is like a very long, twisted zipper, with two outer strands held together by a ladder of teeth. These teeth carry the information. There are four kinds of teeth, which can be read like the letters of a word. When a cell needs information from a portion of a DNA molecule, the two strands in that region are unzipped, and the sequence of teeth on one strand is read. Then the strands zip back up.

Like a real zipper, a DNA molecule has to be easy to zip and unzip, but when it is zipped, it should stay zipped. Unlike a real zipper, however, the teeth of DNA are held together by chemical interactions. Hydrogen bonds are perfect for this job, since covalent bonds would be too strong and require too much energy to break, whereas London forces would be too weak. Shown below is a small piece of a DNA molecule. Look for the hydrogen bonds (dotted lines) between the teeth. These hydrogen bonds are broken when the DNA is unzipped, and they form again when the DNA zips up.

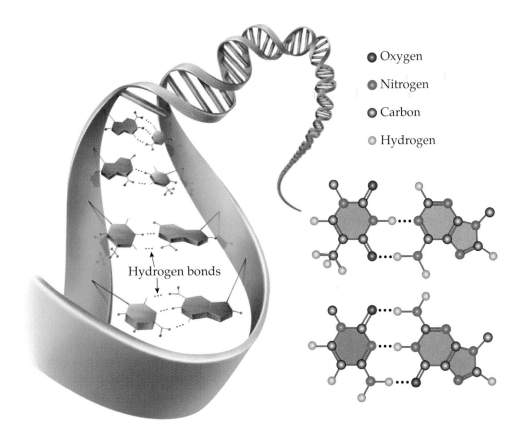

Oxygen
Nitrogen
Carbon
Hydrogen

Hydrogen bonds

We would like to end the discussion of intermolecular forces with a cautionary note. It's easy to get in the habit of thinking that hydrogen bonds are always the strongest intermolecular forces, dipolar forces always the next strongest, and London forces always the weakest. While this is generally true, it is not always true. For example, antimony trihydride (SbH_3) is an essentially nonpolar molecule because Sb and H have similar electronegativities. Nevertheless, SbH_3 has a higher boiling point than ammonia (NH_3), which forms hydrogen bonds. That is because antimony is a big atom, possessing 51 electrons. This results in London forces between SbH_3 molecules that are stronger than the hydrogen bonds between NH_3 molecules. So, as always, respect the general trend, but keep an open mind.

PRACTICE PROBLEMS

9.8   In Practice Problem 9.7 you should have predicted that, based only on London forces, PH_3 has a higher boiling point than ammonia, NH_3. In fact, it does not. Why is this so?

Answer: The answer to Practice Problem 9.7 was based only on London forces, which are greater for PH_3. However, ammonia meets the criterion for hydrogen bonding (H bound to N). The much stronger hydrogen bonds between ammonia molecules give it the higher boiling point.

9.9   Draw some ammonia (NH_3) molecules and show the hydrogen bonding between them using dotted lines.

9.10 The hydrogen halide molecules HF, HCl, HBr, and HI have the following order of boiling points: HF > HI > HBr > HCl.
(a) Why does HF have the highest boiling point?
(b) Why is the boiling point of HI greater than that of HBr or HCl?

9.4 NONMOLECULAR SUBSTANCES

So far we have seen that intermolecular forces (London forces, dipolar forces, or hydrogen bonding) are responsible for the existence of some substances in a condensed phase. This was the case for water. Water in its solid form (ice) is an example of a **molecular solid**, a solid made up of discrete individual molecules. If you could shrink yourself down to the size of a molecule and enter an ice cube, you would see an ordered arrangement of water molecules "locked" into place by hydrogen bonds. But not all solids are made up of individual molecules that owe their existence to relatively weak intermolecular forces. Some solids contain no molecules at all, but exist in a condensed phase due to the far stronger covalent or ionic bonding forces. Sodium chloride (NaCl, table salt) is an example of a *nonmolecular ionic solid*. As we have seen in Chapter 5, NaCl is made up of a vast array (lattice) of positive sodium ions (Na^+) and negative chloride ions (Cl^-). There is nothing you could point to and call a molecule. There is only a repeating pattern of positive and negative ions in the lattice. A solid that doesn't consist of discrete, individual molecules is called a **nonmolecular solid**.

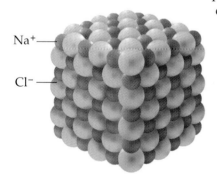

Na$^+$

Cl$^-$

A portion of the NaCl lattice

All simple ionic compounds are like this. Calcium chloride ($CaCl_2$), sodium bromide (NaBr), and magnesium oxide (MgO) are all lattices of ions. Since there is nothing in these solids that we can call a molecule, it would be incorrect to call the forces holding them together "intermolecular."

Intermolecular means "between molecules," and there are no molecules. Instead, these compounds are held together by the much stronger ionic bonds that we talked about in Chapter 5. Melting such compounds means breaking these ionic bonds. Overcoming these strong attractions requires tremendous amounts of energy (heat), so many ionic compounds have melting points in excess of 1000°C. Sodium chloride has a melting point (mp) of 801°C, very high when compared to the 0°C melting point of water.

What is wrong with the reasoning behind the following incorrect statement? "The fact that water melts at 0°C and NaCl melts at 801°C indicates that ionic bonds are much stronger than covalent bonds."

WORKPATCH 9.4

The incorrect statement in the WorkPatch is a common misconception among chemistry students. After all, water molecules do have covalent bonds in them. If you see why the statement is incorrect, you truly understand the fundamental difference between melting a molecular versus a nonmolecular solid. Check your answer at the end of the chapter.

There are other solids that do not contain molecules but are also not ionic. Some examples are silicon dioxide (quartz sand), SiO_2, from which glass is made, and diamond, a form of pure carbon. These are often called **network solids** or **network covalent substances** because they consist of a large network of atoms held together by covalent bonds. The structures of these nonmolecular substances are shown below.

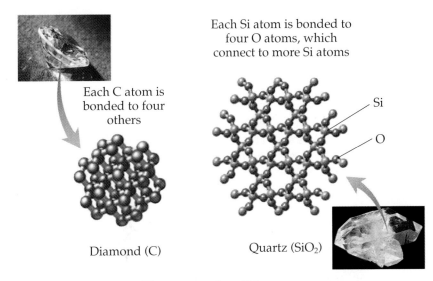

Each C atom is bonded to four others

Each Si atom is bonded to four O atoms, which connect to more Si atoms

Si

O

Diamond (C)   Quartz (SiO_2)

Two network solids

Diamond is a vast network of carbon atoms, with each atom covalently bonded to four other carbon atoms. Quartz is a vast network of silicon atoms, each covalently bonded to four oxygen atoms, which are in turn covalently bonded to more silicon atoms. Once again, there is nothing that stands out as an individual molecule. What we have now is a lattice of atoms instead of ions, all interconnected by very strong covalent bonds. Melting these compounds requires breaking covalent bonds, and consequently, requires adding a lot of heat. Diamond melts above 3550°C, and quartz sand requires temperatures in excess of 1700°C before it begins to liquefy.

It is not always easy to determine whether a particular solid substance is made up of molecules (like solid H_2O) or not (like SiO_2 or NaCl) just by looking at the formula. If the formula includes a group IA or IIA metal plus a nonmetal (like NaCl or $MgBr_2$), then it is a good bet that the solid is ionic. For other compounds, like SiO_2, the formula does not contain enough information to tell if its solid phase is molecular or nonmolecular. If we hadn't told you that quartz is a network (covalent) solid, you would have to determine its internal structure by experimental means. The melting point, however, can often provide a good hint, since it is directly related to the strength of the forces holding the solid together. If it is tremendously high, approaching 1000°C or beyond, the solid is most likely nonmolecular. If its melting point is much lower, weaker forces are present, and it is more likely to be molecular.

To melt quartz, you have to break strong covalent bonds.

To melt ice, you just have to break hydrogen bonds.

As usual, there are exceptions to these generalizations. One major exception is in the case of a solid, pure metal. Pure metals, like silver (Ag), iron (Fe), copper (Cu), and gold (Au), are nonmolecular solids. They are made up of a lattice of neutral metal atoms. The structure of pure gold is shown below:

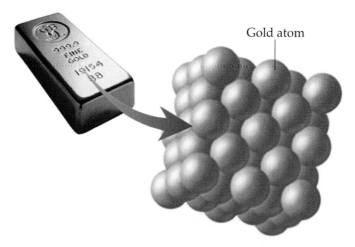

Gold atom

The atoms are held together in this nonmolecular solid by something called **metallic bonding**. Like covalent bonds, metallic bonds are the result of shared electrons. In a piece of metal, the valence electrons on the metal atoms are free to move about and be shared by *all* the nuclei, giving rise to extra attraction.

(This mobility of valence electrons explains why metals conduct electricity and heat so well.) Like covalent and ionic bonds, metallic bonds are true bonds, which means that, in general, they are very strong. Gold (Au) has a melting point of 1064°C. Iron (Fe) has a melting point of 1535°C. These are very high temperatures, reflecting the high strength of these metallic bonds. However, sodium metal (Na) has a melting point of only 98°C, and mercury (Hg) is a liquid at room temperature with a melting point of −39°C. These are two cases where metallic bonding is quite weak, even though these substances are nonmolecular solids.

PRACTICE PROBLEMS

9.11 The compound sodium oxide (Na_2O) is a white-gray powder that sub-limes (goes directly from solid to gas) at 1275°C. Is sodium oxide a molec-ular or nonmolecular solid?

Answer: First, sodium oxide is made from a group IA metal and a nonmetal. This practically ensures that the compound will be a nonmolecular, ionic solid. Second, this compound sublimes (goes directly from the solid to the gas) at a tremendously high temperature. This also indicates that the compound is nonmolecular (the high temperature is required to break ionic bonds in the ionic lattice of Na^+ and O^{2-} ions).

9.12 How is a piece of iron similar to a piece of sodium chloride in terms of the forces that must be overcome to melt it?

9.13 Elemental sulfur is a yellow solid material that melts at 113°C. Studies show sulfur to be a molecular solid consisting of individual S_8 molecules in a lattice. What forces must be overcome to melt the solid?

9.14 The following are solids at room temperature. Classify them as molecular, ionic, network covalent, or metallic (mp = melting point).
(a) Zirconium (Zr), mp = 1852°C
(b) Lead (Pb), mp = 328°C
(c) Calcium nitride (Ca_3N_2), mp = 1195°C
(d) Graphite form of carbon (C), sublimes at 3652°C
(e) Yellow phosphorus (P_4), mp = 44°C

DESCRIBING THE GAS PHASE 9.5

We began this chapter by examining the gas phase. Our model of a gas con-sists of a collection of molecules that are spaced far apart and that move around randomly, sometimes colliding with one another. Intermolecular forces have almost no effect on their behavior—the molecules are too far apart and are moving too fast to feel more than brief, faint touches of attraction and repulsion for each other. Our goal now is to quantify this model by developing a mathematical description (an equation) for the gas phase that will allow us

A gas—a perfect picture
of total chaos

to predict how a gas behaves under a variety of conditions. For example, if we have a tank of gas and we warm it up by 10°C, what will happen to the pressure in the tank? Will it increase or decrease, and by how much?

You might think that among all the phases of matter, deriving such a predictive mathematical equation would be most difficult for the gas phase. After all, in a gas, molecules are moving randomly at high speeds in different directions. The gas phase would seem to be an extremely chaotic situation.

The liquid phase seems so much neater, with molecules actually touching one another and moving much more slowly. It appears that it should be easier to derive a mathematical description of the liquid phase from its more ordered model. But, in fact, it is the very chaos associated with the gas phase that makes it easier to deal with.

To exactly describe the behavior of any phase, you must be able to accurately describe the interactions between the molecules in that phase. These forces are difficult to describe mathematically because they depend on many variables. For one thing, they depend on the identities of the molecules involved. Are they polar or nonpolar? Are they large or small, spherical or elongated? These forces can also depend on how far apart the molecules are and how they are oriented with respect to one another. In a condensed phase, they even depend on a molecule's location—that is, is the molecule on the surface or within the interior of the condensed phase? All of this is summarized below:

Factors that can influence the strengths of intermolecular forces

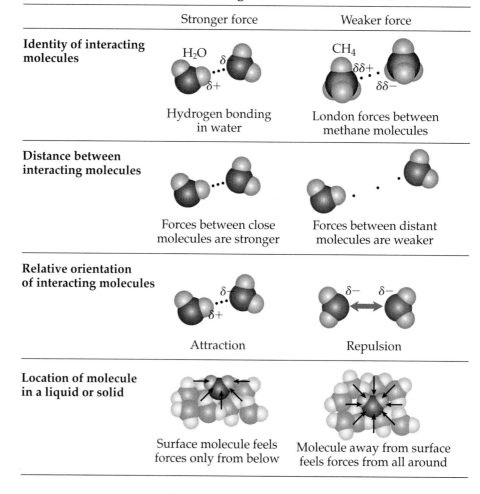

	Stronger force	Weaker force
Identity of interacting molecules	Hydrogen bonding in water	London forces between methane molecules
Distance between interacting molecules	Forces between close molecules are stronger	Forces between distant molecules are weaker
Relative orientation of interacting molecules	Attraction	Repulsion
Location of molecule in a liquid or solid	Surface molecule feels forces only from below	Molecule away from surface feels forces from all around

In a liquid—and, to some extent, in a solid—these interactions are constantly changing as the molecules move around, making it very difficult to describe them mathematically, even for a "simple" liquid like water. In the gas phase, however, the molecules are far enough apart that the intermolecular forces are extremely weak, so weak that they can be ignored. Ignoring them is an approximation, but it is a reasonable one.

Why is it never a reasonable approximation to ignore the intermolecular forces in the liquid and solid phases?

WORKPATCH 9.5

This approximation makes it possible to derive a mathematical description for a gas, but not for liquids or solids, where intermolecular interactions cannot be ignored. By treating a gas as a collection of independent, noninteracting particles, we no longer need to concern ourselves with the actual molecular identity of the gas itself. The model derived under this assumption will apply to any gas—a sort of "one size fits all" approach. In making this simplifying approximation of a gas, we are describing what chemists call an *ideal gas*. An **ideal gas** is one in which there are zero intermolecular forces between molecules. In an ideal gas, the collection of molecules still moves rapidly and in random directions, but only along straight line paths. They move in straight line paths because they don't attract or repel each other. The only time a gas molecule deviates from its course is when it collides with another gas molecule or with an object, such as the walls of a container the gas is in. When this happens, the molecules bounce off each other or off the container walls along some new straight line path.

No real gas behaves exactly like an ideal gas. However, in many cases, a real gas can be treated as though it were an ideal gas—that is, as though its molecules did not interact at all. Calculations based on this assumption usually give results that are very close to reality and good enough to be useful. With this simple model we can understand and explain quite a bit about the behavior of a gas. For example, we know from experience that a gas always expands to completely fill any container, whereas solids and liquids do not.

Some gas molecules

The same group of gas molecules

A gas expands to fill its container

Some liquid molecules

The same group of liquid molecules

A liquid does not

The ideal gas model explains why this is true. If the molecules don't attract each other, then they are free to travel until they hit something. That something is the wall of the container. Thus, a gas expands to fill its container. In a condensed phase, molecules are not free to wander all over the container.

They are attracted ("stuck") to each other, and cannot fill the container as it expands.

We can also understand gas pressure using the ideal gas model. Perhaps you've seen the demonstration in the photo. If you remove all the air inside a metal can (this can be done with a vacuum pump), the can will collapse.

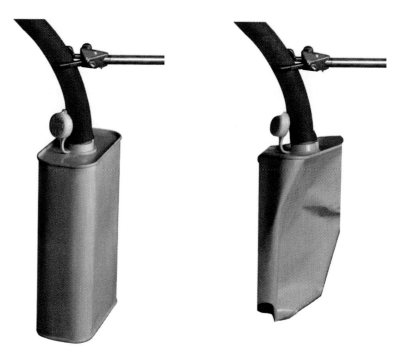

The fact that the can collapses tells us that there must be some external force present that is crushing it inward. But what is the source of this force? It is the billions of fast-moving air molecules in the atmosphere that are colliding with the outside walls of the metal can. Each collision imparts an inward force on the can. The force from these collisions is felt over the entire surface of the can and is referred to as air pressure.

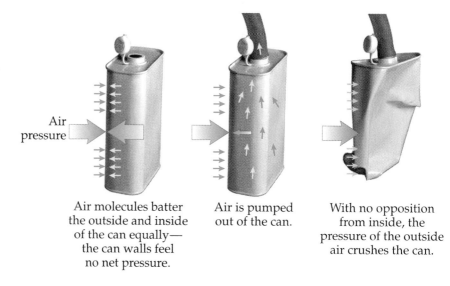

Air pressure

Air molecules batter the outside and inside of the can equally— the can walls feel no net pressure.

Air is pumped out of the can.

With no opposition from inside, the pressure of the outside air crushes the can.

Technically, **pressure** (abbreviated P) is defined as force per unit area ($P = F/A$). The same atmospheric pressure that crushed the can is, at this very moment, attempting to crush you. So, why aren't you collapsing? For the same reason the can didn't collapse until the air inside was pumped out. Before they were removed, the air molecules inside the can were also colliding with the walls, pushing them out with the same force the outside air molecules were pushing in. The net result is that nothing happened until the inside air was removed.

This same idea can be used to construct a device called a *mercury barometer*, used to measure the pressure of the atmosphere. To make a barometer, we can immerse one end of a glass tube into a beaker of liquid mercury (Hg) and then attach a vacuum pump to the other end and remove all the air from the tube. The mercury will rise above the level in the beaker to a specific height and then stop.

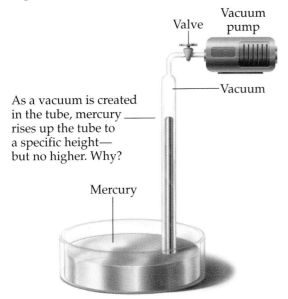

What makes the mercury rise up into the tube? The same thing that crushed the can, air pressure. Before we removed the air from inside the tube, molecules of air both outside and inside the tube collided with the mercury in the dish, pushing down on it with the same force:

When we removed the air from inside the tube, we eliminated the air pressure inside the tube. Thus, the mercury was pushed into the tube by the air pressure outside the tube.

The mercury is pushed up the tube by the outside force (atmospheric pressure) until the weight of the mercury column supplies enough opposing force to resist. On an average day at sea level, this will occur when the mercury column rises to a height of 760 mm. We therefore define **normal atmospheric pressure** as being equal to 760 mm of mercury (abbreviated 760 mm Hg) or 1 atmosphere (abbreviated 1 atm).

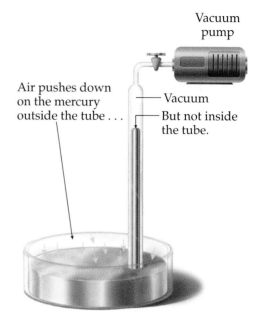

Vacuum pump

Air pushes down on the mercury outside the tube . . .

Vacuum

But not inside the tube.

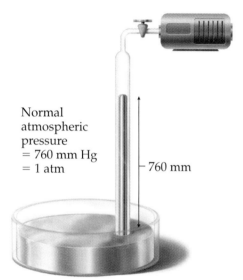

Normal atmospheric pressure
= 760 mm Hg
= 1 atm

760 mm

Our barometer on an average day at sea level

Weather forecasters use barometers to measure the pressure of the atmosphere, although they usually report it in inches of mercury instead of millimeters (760 mm = 29.9 inches; therefore, normal atmospheric pressure is 29.9 inches of mercury). When the atmospheric pressure drops, the column of mercury falls below its "normal" height. Then it's time to grab your umbrella, since low pressure fronts are often associated with rain.

Pressure is one property we can use to describe the state or condition of a sample of gas, but there are others. We can easily measure the volume (V) that a sample of gas occupies since it will always be equal to the volume of the container it is in. Likewise, we can measure the temperature (T) of the gas sample. It is conventional to report the temperature of a gas using the Kelvin or absolute temperature scale discussed in Chapter 2. This is a bit of an inconvenience, since most thermometers are calibrated in °F or °C, but it is a simple matter to measure a temperature using one scale and then convert it to another. Recall that if you measure the temperature of a gas in °C, you can convert it to kelvins by adding 273.15:

Temperature in K = (Temperature in °C) + 273.15

Along with pressure, volume, and temperature, there is one additional measurement that will prove useful in describing a gas sample: how much of it there is. This can be reported in a variety of ways. By convention, we specify the amount of gas present in moles, and we use the symbol n to represent this quantity. When we say that $n = 1$ for a gas, we mean that we have 1 mole, or 6.02×10^{23} molecules, of gas.

Measurable properties used to describe a gas

Pressure it produces (P)

Volume it occupies (V)

Temperature (T)   Specified in kelvins

Amount (n)   Specified as a number of moles

—PRACTICE PROBLEMS

9.15 An "empty" 0.500 gal milk container sits in a room at a temperature of 25.20°C. The barometric pressure is exactly equal to 1 atm.
  (a) What is the temperature of the gas in kelvins?
  (b) What is the pressure of the gas in millimeters Hg?
  (c) What is the volume of the gas in liters? [1 gal = 3.78 L]

Answer:
(a) 25.20°C + 273.15 = 298.35 K
(b) Since the pressure is exactly equal to 1 atm, P = 760 mm Hg.
(c) $0.500 \text{ gal} \times \dfrac{3.78 \text{ L}}{1 \text{ gal}} = 1.89 \text{ L}$

9.16 A container is evacuated with a vacuum pump and weighed. Next, it is filled with hydrogen (H_2) gas and reweighed. It is found to be 10.50 g heavier after filling. How many moles of H_2 gas are in the container? [*Hint:* You will need the molar mass of H_2.]

9.17 The gas in a pressurized container is at 5 times normal atmospheric pressure.
  (a) What is the gas pressure in millimeters Hg?
  (b) What is the gas pressure in inches of Hg? [2.54 cm = 1 inch]

9.18 A mercury barometer develops a leak, allowing some air to enter the glass tube. Will such a barometer read too high or too low a pressure? Explain your answer.

—THE MATHEMATICAL DESCRIPTION OF A GAS 9.6

Using the measurable quantities P, V, n, and T, we can characterize the state or condition of any sample of gas. If we phone some friends in another country, halfway around the world, and give them these values for a particular sample of gas, they can reconstruct a sample identical to ours. In fact, it turns out that it is not necessary to specify all four of these values to uniquely characterize a sample of gas, because the variables P, V, n, and T are not completely independent of each other. Changing one of these variables will cause changes in some or all of the others. It is the nature of the relationships between these variables that our mathematical equation will focus on. Since we will use the ideal gas model to develop this equation, the equation is called the *ideal gas equation* (or *ideal gas law*). We can begin to develop this equation by asking three questions about pressure:

  1. What happens to the pressure when we increase (or decrease) the *volume* of the container the gas is in?
  2. What happens to the pressure when we increase (or decrease) the *temperature* of the gas in a container?

3. What happens to the pressure when we increase (or decrease) the *amount* of gas in a container?

To predict the answers to these questions we will use nothing more then the simple model of an ideal gas. Let's begin with the first question. We've already seen that pressure comes about from gas molecules colliding with the container walls, as shown below. What happens to the pressure of a gas when we decrease the volume of the container? You can answer this question visually using the model of an ideal gas. Try it now in the following WorkPatch.

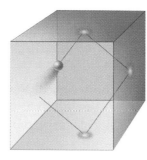

9.6 WORKPATCH   Draw two boxes like the one shown above, making one larger than the other. Put a single fast-moving ideal gas molecule inside each box (traveling at the same speed in both boxes). Show its path over the same interval of time for both boxes, long enough to allow for some collisions with the walls. Do you think the pressure in the small container will be the same as, less than, or greater than the pressure in the larger container?

What about the number of collisions with the walls in each box? There should be a difference. Do you see why? Check your answer against the one at the end of the chapter. This simple picture reveals why the pressure of an ideal gas always increases when the volume of its container decreases (and nothing else is changed). Your drawings should have revealed that the number of collisions per unit time increases as the volume of the box decreases. Thus, as the volume of the container gets *smaller*, the pressure exerted by the gas gets *larger*.

> **Rule 1: The pressure of a gas *increases* when the volume it occupies *decreases*.**

Examine rule 1 very closely. Pressure and volume behave oppositely. As one increases, the other decreases. Stated mathematically:

> **Rule 1: *P* is inversely proportional to *V***

These two statements of rule 1 for an ideal gas say exactly the same thing. We can, however, simplify the mathematical version a bit by dropping the word "inversely." This is easy to accomplish. If P is inversely proportional to V, that means it is directly proportional to $1/V$ (the inverse of V). This is a general mathematical fact for any two inversely related quantities.

> **Rule 1: *P* is proportional to $\dfrac{1}{V}$**

This rewording puts the volume (V) in the denominator of the expression. Now, as V gets smaller, the term $1/V$ gets bigger. Since P is proportional to

$1/V$, this means that P also gets bigger, which is what our model predicts. All three statements of rule 1 say exactly the same thing, that pressure will increase as volume decreases.

Now let's consider the second question: "What happens to the pressure when we increase the temperature of the gas in a container?" Once again we'll use our ideal model of a gas. As we heat the gas, we are putting energy into it. This causes the gas molecules to move faster, on average. Indeed, heating a gas is like stepping on the accelerator pedal in your car.

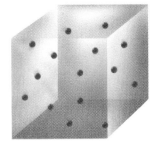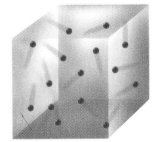

Higher temperature— molecules move faster

Since the molecules are moving faster, they will collide with the walls of the container more often per unit time, and they will also hit the walls with more force. Both effects result in more force being exerted on the walls per unit area, and so the pressure will increase.

Rule 2: The pressure of a gas *increases* when its temperature *increases*.

Look closely at rule 2. Unlike pressure and volume, which we found were inversely related, pressure and temperature are directly related. When one increases, so does the other. Now we can once again say exactly the same thing in a mathematical way:

Rule 2: *P* is proportional to *T*

The third question about pressure was: "What happens to the pressure when we increase the amount (the number of moles) of gas in a container?" In what is getting to be a repetitive theme, we'll once again use the model of an ideal gas to answer this question. The top box on the right has several gas molecules in it. The lower box has lots more. Which box has the higher pressure?

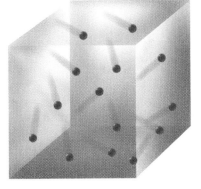

More molecules in the lower box mean more collisions with the walls per unit time. More collisions mean more force pushing outward—that is, more pressure. This gives us a third rule about the behavior of gases:

Rule 3: The pressure of a gas *increases* when the number of moles of gas *increases*.

Rule 3 states that pressure and the amount of gas present (in moles) are directly related. As one gets larger, so does the other. Mathematically:

Rule 3: *P* is proportional to *n*

We are done. Here is what we have concluded:

Rule 1: *P* is proportional to $\frac{1}{V}$

Rule 2: *P* is proportional to *T*

Rule 3: *P* is proportional to *n*

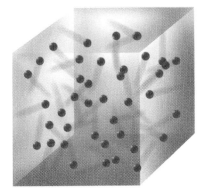

We can combine all three rules into one summary rule:

***P* is proportional to $\frac{nT}{V}$**

Notice that n and T are in the numerator (they are directly proportional to P), whereas V is still in the denominator (it is inversely proportional to P). We can turn this into a full-fledged mathematical equation by replacing the words "proportional to" with some mathematical operator, like an equals sign. To convert a proportion into an equality we need to introduce a constant of proportionality:

$$P = (\text{Constant}) \times \frac{nT}{V}$$

In principle, this constant is not difficult to determine. All we need to do is take a container of gas (one whose behavior is very close to ideal) and experimentally determine the values of all four variables (P, n, T, and V). Then we can use the equation above to determine a value for the constant. Of course, we don't have to do this at all since others before us have done it. This constant of proportionality is called the **ideal gas constant**, represented by R. Its value is given below:

Ideal gas constant $R = 0.0821$ L $\cdot$ atm/K $\cdot$ mole

When we include R in the summary rule, we get

$$P = R \times \frac{nT}{V} = \frac{nRT}{V}$$

This is the **ideal gas equation** (**ideal gas law**) that we have been seeking, the one that can predict how a gas will behave. This one mathematical equation summarizes everything that we know about an ideal gas.

Notice that R has units. *They are important!* When you use this value of R, you must make sure that you express volume in liters (L), pressure in atmospheres (atm), temperature in kelvins, and the amount of gas in moles. If you use any other units for these variables and attempt to use the ideal gas law, you will get a wrong answer.

Now that we have derived this ideal gas law, let's apply it. Suppose we take a closed 2.0 L glass bottle filled with 0.10 mole of air at 25°C and start heating it, all the while measuring the temperature of the air in the bottle.

At 600°C, the bottle explodes!

Why did this happen? Because heating the air molecules made them move faster, causing them to collide with the walls with more force and more often, creating more pressure. The pressure eventually increased to a point where the glass bottle could not contain it any longer, and the bottle exploded. During the experiment, we didn't measure the pressure inside the bottle, only the temperature. What was the pressure inside the bottle when it exploded? The bottle blew up when the air inside reached 600°C. We know the values of the variables n, T, and V, so we can use the ideal gas law to solve for the pressure. Here is what we know at the time the bottle blew up:

Amount of gas n = 0.10 mole
Volume of gas V = 2.0 L
Temperature of gas T = 600°C

But remember! Everything must have the units of the gas constant R. The temperature must be converted into kelvins:

$$K = °C + 273.15$$
$$= 600°C + 273.15$$
$$= 873.15 \text{ K}$$

Now we can plug in the values for n, V, T, and R, and solve for P using the ideal gas law:

$$P = \frac{RnT}{V}$$

$$P = \frac{0.0821 \frac{\cancel{L} \cdot atm}{\cancel{K} \cdot \cancel{mole}} \times 0.10 \, \cancel{mole} \times 873.15 \, \cancel{K}}{2.0 \, \cancel{L}} = 3.6 \text{ atm}$$

All the units cancel except the units of pressure (atm), so our answer is expressed in these units. The pressure when the bottle blew up was 3.6 atm, or 3.6 times greater than normal atmospheric pressure. The ideal gas law has allowed us to determine the pressure of a gas without ever directly measuring it.

Here is another problem that makes use of the ideal gas law. Suppose you work in a helium balloon factory, filling balloons with helium. One day your boss walks in and accuses you of putting too much helium into each balloon, needlessly cutting into corporate profits. She demands that you tell her how many grams of helium you are putting into each balloon. This is not an easy question to answer. You can't simply weigh a helium-filled balloon. It will float right off the balance! But then you remember the ideal gas law, $P = nRT/V$.

Your boss asked for the number of grams of helium in the balloon. The ideal gas law doesn't have grams in it, but it does have moles (n). You can use it to calculate the number of moles of helium in the balloon and then convert this to grams later. Rearranging the ideal gas equation to solve for n instead of P gives

$$n = \frac{PV}{RT}$$

You know the value of R, so now you need the P, V, and T of the helium gas in the balloon. A thermometer hanging on the wall reads 22°C as the room temperature. Since the balloon has been in the room for awhile, you reason that the gas in the balloon must be at the same temperature:

22°C + 273.15 = 295.15 K

A barometer in the room contains a column of mercury that is 748 mm high. The pressure of the gas in the balloon must be the same, or the balloon would expand or shrink until it was. (Actually, the pressure in the balloon is a little larger due to the elastic force of the rubber balloon, but we will ignore this slight difference.) To use the ideal gas equation you need the pressure in units of atmospheres (atm). Since 760 mm Hg equals 1 atm, you can use this as a conversion factor to convert the pressure into the desired units:

$$748 \ \text{mm Hg} \times \frac{1 \ \text{atm}}{760 \ \text{mm Hg}} = 0.984 \ \text{atm}$$

You now have P and T. That just leaves the volume, V, of the helium to be determined. Remembering that the volume of an object can be determined by water displacement (Chapter 2), you fill a bucket with water to the brim and submerge the balloon in the water. Collecting the overflowing water and placing it into a graduated cylinder produces a volume of 240 mL. That's the volume of the balloon. To use the ideal gas law, the volume must be in liters, not milliliters. Since there are 1000 mL in a liter, 240 mL converts to 0.240 L.

You now have the P, V, and T of the helium in the balloon, all in the appropriate units. Putting them into the ideal gas law and solving for n, the moles of helium in the balloon, gives:

$$n = \frac{PV}{RT}$$

$$n = \frac{0.984 \ \text{atm} \times 0.240 \ \text{L}}{0.0812 \dfrac{\text{L} \cdot \text{atm}}{\text{K} \cdot \text{mole}} \times 295.15 \ \text{K}} = 0.010 \ \text{mole He}$$

The balloon is filled with 0.010 mole of helium. But, the boss asked for the amount in grams. Your periodic table indicates that the atomic weight of the helium atom is 4.003. One mole of helium (He) atoms therefore weighs 4.003 g, and you can calculate the number of grams:

$$0.010 \text{ mole He} \times \frac{4.003 \text{ g He}}{1 \text{ mole He}} = 0.040 \text{ g He}$$

With answer in hand, you proudly march into your boss' office and report that each balloon is being filled with 0.040 g of helium. Unfortunately, you are fired. The company limit is 0.030 g per balloon.

Before leaving the ideal gas law we should mention that the answers obtained from its solution are not exactly correct. In both the exploding-bottle problem where we solved for pressure, and in the helium-balloon problem where we solved for amount, the answers calculated wouldn't exactly match the answers we would get if we actually measured these quantities. The calculated answers are close to the truth, but they will be off by a bit. This is because the ideal gas equation was developed using the model of an ideal gas, which assumes that the molecules have absolutely no attractive intermolecular forces between them. Ideal gases cannot be liquefied at any temperature. But real gas molecules do interact somewhat and can be liquefied. The small interactions are enough to make the behavior of a real gas deviate from that predicted by the ideal gas law, but usually the deviations are small, especially if low temperatures and high pressures are avoided.

PRACTICE PROBLEMS

9.19 According to the ideal gas law, what will happen to the pressure of a gas in a container if:
(a) You double the temperature in kelvins?
(b) You double the volume in liters?
(c) You double the amount of gas present in moles?
(d) You double the temperature in kelvins and you also double the volume in liters?

Answer:
(a) Since T is in the numerator of the ideal gas equation, if it doubles, so will P.
(b) Since V is in the denominator of the ideal gas equation, if it doubles, the pressure will be cut in half.
(c) Since n is in the numerator of the ideal gas equation, if it doubles, so will P.
(d) This question asks what happens when you do (a) and (b) simultaneously. Since (a) doubled the pressure and (b) halved it, the net result is that the pressure does not change.

9.20 Suppose that 3.00 g of gaseous nitrogen (N_2) are placed into a 2.00 L container. The pressure is measured to be 450.5 mm Hg. What is the temperature of the gas in °C? ($R = 0.0821$ L · atm/K · mole) [*Hint:* Solve the ideal gas equation for T.]

9.21 An automobile tire is filled with O_2 to a total pressure of 40.0 pounds per square inch (lb/in.^{2}). The temperature is 22.5°C. The inside volume of the inflated tire is 10.5 gal. How many grams of O_2 are in the tire? Note that 760 mm Hg = 14.696 lb/in.^{2}, and 1 gal = 3.785 L. (R = 0.0821 L $\cdot$ atm/K $\cdot$ mole) [*Hint:* Solve the ideal gas equation for n.]

We have covered quite a bit in this chapter. We examined a simple model of each of the three phases of matter (solid, liquid, and gas) at the molecular level. We looked at the intermolecular forces (London, dipolar, and hydrogen bonding) that are responsible for the existence of the condensed phases. We then focused on the gas phase and derived a mathematical equation (the ideal gas law) to predict how a gas behaves. We did this by examining the model of an ideal gas and considering what would happen to pressure as the other variables (amount, temperature, and volume) are changed. An important thing to remember at this point is that our understanding of these phases and their behaviors, as well as the mathematical equations that can be used to predict their behavior, are all based on simple, visual models. Chemists may forget equations and facts, but the underlying pictures and models are with them forever—this is where their true understanding comes from.

HAVE YOU LEARNED THIS?

Model of a gas (p. 300)

Model of a solid (p. 301)

Kinetic energy (p. 301)

Model of a liquid (p. 301)

Condensation (p. 302)

Boiling, vaporization (p. 302)

Boiling point (p. 302)

Freezing point (p. 302)

Condensed phases (p. 303)

London forces (p. 305)

Hydrogen bond (p. 311)

Molecular solid (p. 314)

Nonmolecular solid (p. 314)

Network solid (or network covalent substance) (p. 315)

Metallic bonding (p. 316)

Ideal gas (p. 319)

Pressure (p. 321)

Normal atmospheric pressure (p. 322)

Ideal gas law (pp. 323, 326)

Ideal gas constant (p. 326)

THE PHASES OF MATTER

9.22 Describe as completely as you can how small molecules like N_2 and O_2 behave at the molecular level in a warm room.

9.23 Explain why cooling a gas should eventually cause it to condense into a liquid.

9.24 What would be true of a gas if there were no such thing as intermolecular forces?

9.25 Is it incorrect to say that molecules are motionless in the liquid phase? Explain.

9.26 Explain in molecular terms how heating causes a liquid to change to the gas phase.

9.27 Water vapor liquefies when cooled below 100°C. Gaseous nitrogen liquefies when cooled below −196°C. What does this information tell you about the relative strengths of the intermolecular forces for these molecules?

9.28 Why are liquids and solids referred to as condensed phases?

9.29 Why does a gas expand to fill the container it is in, but a liquid and a solid do not?

9.30 Describe each phase of matter on the molecular level with respect to the amount of order present.

9.31 For this question you should consider only polarity. Molecule A–B has a smaller dipole moment than molecule C–D.
(a) Which molecule has the higher freezing point? Explain why.
(b) Which molecule has the higher boiling point? Explain why.

9.32 Draw a picture that shows how three relatively close, polar HBr molecules in the gas phase would attract one another. What kind of intermolecular forces are involved?

INTERMOLECULAR FORCES

9.33 Explain what gives rise to London forces and when they occur.

9.34 Would you expect CCl_4 or CBr_4 to have a higher boiling point? Explain your answer.

9.35 Chloromethane (CH_3Cl) has a much higher boiling point than methane (CH_4). Give two reasons for this.

9.36 Propane (C_3H_8) is a gas at room temperature, whereas octane (C_8H_{18}) is a liquid. Explain why this is so.

9.37 What is wrong with the statement "London forces are always weaker than dipolar forces"?

9.38 Consider the molecules HF and HCl. The electronegativities of the atoms involved are: H, 2.1; Cl, 3.0; F, 4.0.
(a) Which of the two molecules is more polar? Explain your answer.
(b) For which substance are the dipolar attractions between the molecules stronger? Explain your answer.
(c) Which gas would, upon cooling, liquefy first? Explain your answer. Also, indicate which compound would have the higher boiling point based on your answer.
(d) Do both molecules also experience London forces of attraction? If yes, which would have the greater London forces?

9.39 When discussing the intermolecular forces between methanol molecules, chemists usually ignore any London forces between them. Why are they justified in doing this?

Methanol

9.40 At room temperature, Cl_2 exists as a gas, Br_2 exists as a liquid, and I_2 exists as a solid. Completely explain why each element exists in that phase.

9.41 Long-chain hydrocarbon molecules of the type $CH_3(CH_2)_{20}CH_3$ are solids and are used for things like wax. The C–H bonds are essentially nonpolar. Why is wax a solid at room temperature?

9.42 How does the attraction between two molecules depend on the distance between them?

9.43 Why should the attraction between two water molecules depend on how they are oriented with respect to one another? Use drawings to illustrate your answer.

9.44 When can hydrogen bonding occur between molecules?

9.45 What is so special about hydrogen atoms that makes hydrogen bonding possible?

9.46 Two different compounds have the same elemental composition, C_3H_8O. One has a low boiling point and the other a much higher boiling point. What must be true about the structure of one of these compounds compared to the other?

9.47 Consider molecules of the type H_2A, where A is a group VIA atom (O, S, Se, Te). Their relative boiling points are shown below.

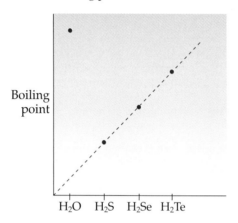

(a) Why do the boiling points of H_2S, H_2Se, and H_2Te steadily increase?
(b) The boiling point of H_2O is well above the line that the others are on. Explain why this is so.

9.48 What do we mean by "induced dipole" when discussing London forces?

9.49 Why do we use dotted lines rather than solid lines to represent hydrogen bonds?

9.50 Examine the diagram of DNA on page 313 to answer this question: Why can hydrogen bonds form in DNA?

9.51 Why would either covalent bonds or London forces be inappropriate for attaching the two strands of DNA to one another?

9.52 How is it possible for a nonpolar molecule to have a higher boiling point than a polar one?

NONMOLECULAR SUBSTANCES

9.53 What is the fundamental difference between a molecular and a nonmolecular substance?

9.54 Is it sometimes, always, or never possible to tell from a compound's formula whether it is a molecular or nonmolecular substance?

9.55 What do we mean by a "lattice of ions"?

9.56 In general, nonmolecular solids have much higher melting points than molecular solids. Why is this so?

9.57 What is a network covalent solid? Give an example.

9.58 What evidence is there that metallic bonds can be as strong as ionic or covalent bonds?

9.59 Carbon tetrachloride (CCl_4) is a liquid at room temperature, whereas carbon tetrafluoride (CF_4) is a gas that liquefies at $-128°C$ at normal atmospheric pressure. How can you explain this?

9.60 All of the following are solids at room temperature. Classify them as molecular, ionic, network covalent, or metallic.
(a) Potassium (K), mp = 64°C
(b) Potassium chloride (KCl), mp = 770°C
(c) Red phosphorus (P), mp = 590°C
(d) Boron triiodide (BI_3), mp = 50°C

DESCRIBING THE GAS PHASE

9.61 Why is it proper to think of the gas phase of matter as being more chaotic than either of the condensed phases?

9.62 What assumption is made for an ideal gas, and what gives us the right to make that assumption?

9.63 Consider a container of gas at the same pressure inside as the room pressure outside. If a tiny hole is punched in the side of the container, will the gas leak out? Explain your answer.

9.64 How does a gas produce pressure?

9.65 Why does liquid rise up a straw when you suck on it?

9.66 Describe how a mercury barometer works.

9.67 What is the definition of normal atmospheric pressure?

9.68 True or false? 1 atm = 76 cm Hg.

9.69 A weather forecaster reports the barometric pressure as 29.7 inches of mercury. [1 in. = 2.54 cm.]
(a) How many millimeters of Hg is this?
(b) How many atmospheres is this?

9.70 The pressure in a tank of oxygen is 2000.5 lb/in.^{2} (760 mm Hg = 14.696 lb/in.^{2}).
(a) How many millimeters of Hg is this?
(b) How many atmospheres is this?

9.71 How do you convert from °C to K? Convert room temperature (22.0°C) to kelvins.

9.72 Convert $-100.5°C$ to kelvins.

9.73 A balloon of methane (CH_4) gas has a temperature of $-2.0°C$ and contains 2.35 g of the gas. Express the temperature of the gas in kelvins and the amount in moles.

9.74 A tank of acetylene gas (C_2H_2) contains 48.5 lb of the gas and is at a pressure of 600.2 lb/in.^{2} (760 mm Hg = 14.696 lb/in.^{2}, 453.6 g = 1 lb). Express the pressure of the gas in atmospheres and the amount of gas in moles.

DESCRIBING A GAS MATHEMATICALLY

9.75 What happens to the pressure of a gas when the volume of its container is increased? Explain with pictures and words.

9.76 What happens to the pressure of a gas when its temperature is decreased? Explain with pictures and words.

9.77 What happens to the pressure of a gas when the amount of gas inside the container is increased? Explain with pictures and words.

9.78 Consider the following diagrams representing different gas samples, all at the same temperature, and answer the following questions:

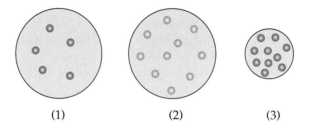

(1)   (2)   (3)

(a) Which gas is at the lowest pressure? Explain.
(b) Which gas has the highest pressure? Explain

9.79 Which of the two "gas" samples below is at the lower temperature? Explain, and also tell why we put the word "gas" in quotes.

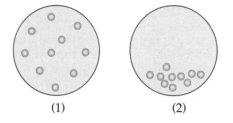

(1)   (2)

9.80 State how the pressure of a gas depends on its volume.

9.81 State how the pressure of a gas depends on its temperature.

9.82 State how the pressure of a gas depends on the amount of gas present.

9.83 What do we mean by inverse proportionality? By direct proportionality? Give an example of each using the way the pressure of a gas depends on something else.

9.84 Suppose the variable X is proportional to $1/Y$. What does this tell you about how X and Y are related?

9.85 "The older I get, the fewer hairs I have on my head." What kind of relationship (proportion or inverse proportion) exists between this gentleman's age and his hair? Explain your answer.

9.86 What kind of relationship (proportion or inverse proportion) exists between the strength of intermolecular forces and the distance between molecules? Explain your answer.

9.87 What are the units of the ideal gas constant R? Why are they important?

9.88 Rewrite the ideal gas law solving for V.

9.89 Rewrite the ideal gas law solving for T.

9.90 Rewrite the ideal gas law solving for n.

9.91 According to the ideal gas law, what should you get if you measure P, V, n, and T for a gas sample and then calculate the quantity PV/nT?

9.92 According to the ideal gas law, what would happen to the pressure of a gas if you doubled the amount of gas in a container while also cutting the volume of the container in half? Explain.

9.93 According to the ideal gas law, what would happen to the pressure of a gas if you doubled the amount of gas in a container while also tripling the kelvin temperature of the gas? Explain.

9.94 A student thinks he remembers reading that "If you double the temperature of an ideal gas, its pressure will also double." He is given a problem where he has an ideal gas at 25.0°C and its pressure is 2.5 atm. He is asked what the temperature must be raised to in order to double the pressure to 5.0 atm. He answers, "50.0°C, of course." Why is he wrong? What lesson should he learn about using the ideal gas law? What is the correct answer, in °C?

9.95 The gas inside a balloon is characterized by the following measurements: Pressure = 745.5 mm Hg; Volume = 250.0 mL, Temperature = 25.5°C. How many moles of gas are inside the balloon?

9.96 A gas is inside a container whose volume is variable. The container is in an ice bath at 0°C and there are 2.0 moles of gas in it. What must the volume, in liters, be adjusted to in order to produce a gas pressure of 2.5 atm?

9.97 What must the temperature (in °C) be if 2.0 moles of a gas in a 4.0 L steel container has a measured pressure of 100 atm?

9.98 According to the ideal gas law, what would be the volume of a gas at absolute zero (0 K)? Does the answer make sense? Describe what really happens to any gas as it is cooled toward this temperature.

9.99 An automobile tire at 22°C with an internal volume of 20.0 L is filled with air to a total pressure of 30 psi (pounds per square inch). [1 atm = 14.696 psi]
(a) How many moles of air (gas) are present in the tire?
(b) If the gas was entirely nitrogen (N_2), how many grams of it would be in the tire? How many pounds of it would be in the tire? [453.6 g = 1 pound]

9.100 Why are the results calculated using the ideal gas law not exactly equal to the "true" results obtained by an experimental measurement?

9.101 Suppose you want to carry out the chemical reaction

$$H_2(g) + Cl_2(g) \longrightarrow 2\,HCl(g)$$

in the lab. Someone gives you 1 mole of Cl_2 gas to start with.
(a) What volume of H_2 gas would you need (in liters) to have 1 mole of H_2, given that the H_2 pressure and temperature are 1.00 atm and 22.5°C, respectively?
(b) What would the volume of the product be if it were collected at 1.00 atm and 22.5°C?

WORKPATCH SOLUTIONS

9.1 Ethanol has a boiling point of 78°C, which is lower than the 100°C boiling point of water. This means that it takes less heat (energy) to boil ethanol. Therefore, the intermolecular forces between ethanol molecules must be weaker than those between water molecules.

9.2 Oxygen (O_2) is a nonpolar molecule. It has no permanent $\delta+$ and $\delta-$ partial charges on it that can create strong attractive forces between molecules. All it has are the very tiny, instantaneous $\delta\delta+$ and $\delta\delta-$ charges that result in weak London forces between molecules. Water, on the other hand, is a polar molecule with relatively large, permanent $\delta+$ and $\delta-$ charges that can produce relatively strong attractive forces between molecules.

9.3 Ethanol (b), boils at the higher temperature of 78°C, whereas ethyl ether (a) boils at 35°C. The reason can't be attributable to London forces, since both have the same formula (C_2H_6O) and therefore the same number of electrons to be unbalanced. Both molecules are polar, but only ethanol (b) has a hydrogen atom directly attached to an oxygen atom. This allows for the formation of hydrogen bonds between molecules, generally the strongest type of intermolecular attraction, giving it the higher boiling point.

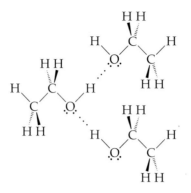

9.4   Although water molecules do have covalent bonds, melting ice doesn't involve breaking these bonds. Melting ice involves overcoming the much weaker hydrogen bonds between water molecules. The covalent bonds within the molecules remain completely intact. Thus, comparing the melting points of NaCl and water compares the strength of ionic bonds and hydrogen bonds, not covalent bonds.

9.5   In the condensed phases, the molecules are "touching," so the intermolecular forces (no matter what type) are relatively strong and thus can't be ignored. Remember, intermolecular attractive forces between molecules are stronger when the molecules are closer together.

9.6   Pressure in the smaller box should be greater, because there are more collisions per unit time with the walls in the smaller box than there are in the larger box.

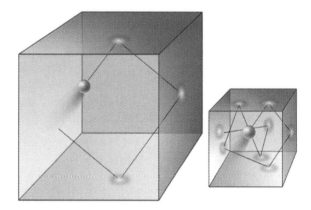

Chapter 10

Solutions

What Is a Solution? 10.1

The next time you're at a supermarket, take a look at the detergent aisle. You'll find a huge array of products, all claiming that when they're added to water, they have the muscle to remove dirt and stains. Actually, water by itself does a pretty good job on most household dirt and stains. Consider a puddle of grape juice that has dried on a kitchen counter. Strenuous wiping with a dry cloth will generally not remove it, but a gentle wipe with a water-dampened cloth does the trick. The purple color on the cloth reveals that the juice has dissolved into the water. It appears that water's stain-removing muscle is due to its ability to dissolve things. Of course, there are some stains that water alone will not remove. Recall that salad dressing stain on your new T-shirt, for example. In this chapter we are going to answer two questions: (1) What exactly happens when a substance dissolves? (2) Why is water so good at dissolving some things and so poor at dissolving others?

A **solution**, as we saw in Chapter 1, is a homogeneous mixture of two or more substances. By homogeneous we mean that the composition is the same throughout the solution. This means that when sugar, for example, is dissolved in water, every tiny volume of the solution has the same number of water and sugar molecules in it. Another way to say this is that the solution has the same *concentration* (the same amount of sugar per amount of water) everywhere. Technically speaking, the substance that is present in the greater amount (in our case, water) is called the **solvent**. The substance present in the lesser amount is called the **solute**. We can refine our definition now to say that a solution is a homogeneous mixture of a solute (or solutes) dissolved in a

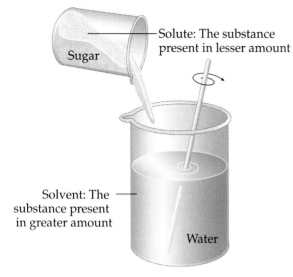

Solute: The substance present in lesser amount

Sugar

Solvent: The substance present in greater amount

Water

solvent. It is also important to note that once a solution is prepared, it will not separate back into its components.

The word solution often brings to mind a solid dissolved in a liquid, like sugar in water, but not all solutions are like that. Solutions can also be combinations of liquids. Vinegar is a solution of acetic acid in water, both clear, colorless liquids. Since vinegar is mostly water, water is the solvent and acetic acid is the solute.

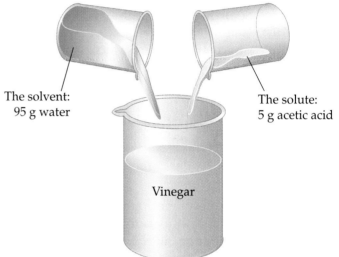

The solvent:
95 g water

The solute:
5 g acetic acid

Vinegar

A solution of a liquid in a liquid

Solutions can be more exotic yet. Many metal alloys, for instance, are solid solutions consisting of one metal evenly dissolved in another. The solutions are made when the metals are molten (that is, in their liquid phases). For example, sterling silver is a solid solution of a small amount of copper (the solute, Cu) dissolved in silver (the solvent, Ag). If the copper atoms are not evenly distributed among the silver atoms, the result is not sterling silver and is not a solution.

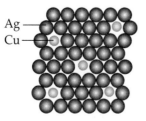

Ag
Cu

Sterling silver: a solid solution. The Cu atoms are evenly distributed among the Ag atoms.

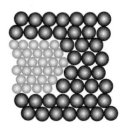

Not a solid solution. The two metals are not evenly distributed in each other.

The air you are breathing is a solution of gases, primarily O_2 dissolved in N_2. Nitrogen, at 78%, is considered the solvent, and oxygen (21%) is the principal solute; other solute gases are also present in smaller amounts. Like any solution, these mixtures are homogeneous.

For most of the solutions we will consider in this chapter, water is the solvent. Solutions in which water is the solvent are called **aqueous solutions** (from the Latin *aqua* for water).

Before we leave the definition of a solution, we want to be sure that you truly understand the meaning of the word "homogeneous" in this context. This topic was covered in Chapter 1, but it is worth another look. You could put a mixture of powdered quartz sand (silicon dioxide, SiO_2), sodium carbonate (Na_2CO_3), and calcium oxide (CaO) into a mortar and grind it with a pestle into a fine, pulverized dust. Even if you did this for a thousand years, the result would not be a solution. It would be a heterogeneous mixture, because there would still be tiny grains of pure sand, pure sodium carbonate, and pure calcium oxide. Only by melting the components and stirring well would you achieve a molten liquid solution that would become a solid solution (glass) if you allowed it to cool and solidify. The same is true for a mixture of salt and sugar. No amount of grinding these two solids together will yield a solution, but dissolve them both in the same beaker of water and you get a homogeneous mixture—a solution—of both solutes. Note that for both of these examples, we needed a liquid phase to achieve homogeneity. Simply mixing two solids together and grinding just won't do the job. You can also mix different gases together and get a homogeneous mixture—a solution—as we illustrated with air.

From what you learned about the different phases of matter in Chapter 9, why do you think it is necessary to employ a liquid or gaseous phase to obtain the necessary level of homogeneity to achieve a solution?

WORKPATCH 10.1

The models for a liquid and a gas in Chapter 9 should help you understand what it takes to create a homogeneous solution. The answer to WorkPatch 10.1 illustrates that chemistry works somewhat like a jigsaw puzzle, with the concepts and models you learn at different points fitting together to give you an understanding of the whole.

PRACTICE PROBLEMS

10.1 Classify the following as solutions or heterogeneous mixtures:
(a) A hot cup of instant coffee
(b) Chicken vegetable soup
(c) Blood
(d) Filtered blood
(e) A chromium-plated steel automobile bumper
(f) A stainless steel automobile bumper [Stainless steel is prepared by combining iron (Fe), up to 30% chromium (Cr), smaller amounts of nickel (Ni), and carbon (C), and heating them to molten.]

Answers: (a), (d), and (f) are solutions. (b), (c), and (e) are heterogeneous mixtures. (Unfiltered blood has cells in it, and solids will separate out of blood upon standing.)

10.2 Identify the solvent and solute(s) for each of the following solutions:
   (a) Nail polish remover (30% acetone in water)
   (b) Humid air
   (c) Stainless steel (see Practice Problem 10.1)
   (d) An aqueous solution of aspirin

10.3 For alcoholic drinks, the "proof" is equal to twice the percentage of alcohol; for example, a 50 proof drink is 25% alcohol in water. Vodka is normally sold between 80 and 100 proof, but suppose you came across a bottle of 135 proof vodka. Would you be justified in calling the alcohol the solvent and the water the solute? Explain why.

10.2 ENERGY AND THE FORMATION OF SOLUTIONS

Liquid water is an excellent solvent, capable of dissolving many substances (solutes). The human body is approximately 80% water. Blood and cellular fluids are made up of mostly water with small amounts of dissolved solutes (salts, gases, proteins, enzymes, etc.). So many different substances are soluble in water that it is often referred to as the *universal solvent*.

What is it about water that makes it such a good solvent? To answer this question, let's reexamine the water molecule. We've seen that water is a polar molecule, a consequence of the large difference in electronegativity between its oxygen and hydrogen atoms and its bent shape:

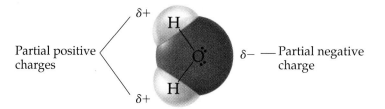

This polarity combined with the almost "naked proton" character of the H atoms leads to exceptionally strong dipolar attractions between water molecules—the hydrogen bond we introduced in Chapter 9.

We also saw in Chapter 9 that water molecules in the liquid state are in constant motion, moving in random directions, but always strongly attracted to one another, always touching one another, jostling past one another. Remember the analogy of a sale crowd in a shopping mall?

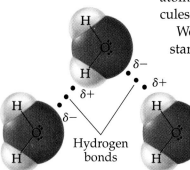

Now let's consider what happens when a solute is dissolved into the water. Since the ionic compound sodium chloride (NaCl) dissolves easily in water, we'll use it as our solute. We've seen that solid sodium chloride crystals are made up of relatively stationary Na^+ and Cl^- ions in a repeating three-dimensional pattern called a lattice. This lattice is held together by strong ionic bonds. This is why solid NaCl has to be heated to the extraordinarily high temperature of 801°C before it will melt, becoming a liquid made of mobile Na^+ and Cl^- ions. Yet, when we put this same compound into water at room temperature, the lattice comes apart easily to produce a solution of mobile Na^+ and Cl^- ions in water.

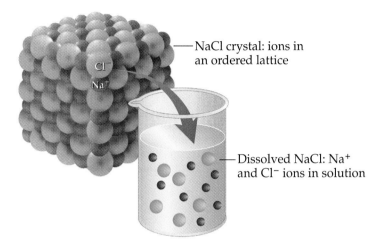

NaCl crystal: ions in an ordered lattice

Dissolved NaCl: Na^+ and Cl^- ions in solution

How can water succeed at breaking the ionic bonds of the crystal at room temperature (22°C), when it takes 801°C to do the same thing without the water? To answer this question, we need to look more closely at the dissolving process and break it down into its elementary steps.

Three steps must occur simultaneously for NaCl to dissolve in water. First, the NaCl lattice has to be broken apart into individual Na^+ and Cl^- ions. This means overcoming the attractive forces (ionic bonds) between the ions within the lattice. Since ionic bonds are quite strong, this step requires an input of energy.

Step 1: Freeing ions from the crystal lattice of the solute

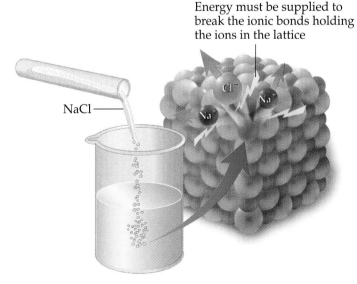

Energy must be supplied to break the ionic bonds holding the ions in the lattice

NaCl

Second, spaces must be opened up in the solvent to make room for the solute particles. Recall that the solvent molecules are attracted to one another and are touching one another. For a solute particle to dissolve, the solvent molecules have to make room and move away from one another. This requires overcoming the attractions between them, so this second step also requires an input of energy.

Step 2: Making room in the solvent

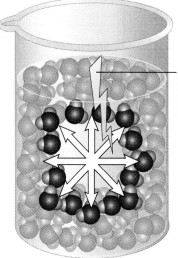

Energy must be supplied to overcome the attractive interactions between the solvent molecules

The third step places the individual Na⁺ and Cl⁻ ions into the spaces created for them in the solvent. Now the charged ions of the solute can interact with the polar water molecules that surround them, resulting in an attractive force called an **ion–dipole force**.

Step 3: Interaction between solvent and solute

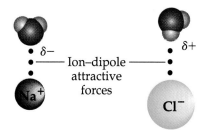

In the above figure, the water molecules are drawn so that an oxygen atom faces the Na⁺ ion, whereas a hydrogen atom faces the Cl⁻ ion. Explain why this is so.

Your WorkPatch answer explains why the Na⁺ and Cl⁻ ions are attracted to the water molecules. In fact, there is room for more than one water molecule about each ion. Both Cl^- and Na^+ surround themselves with up to six water molecules, making for many ion–dipole attractions per ion:

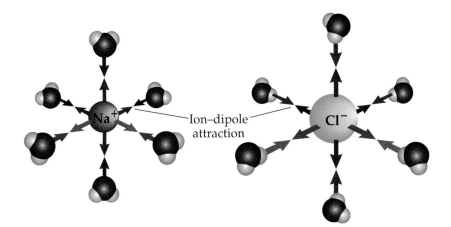

This means that each of the Cl^- and Na^+ ions is surrounded by a shell of water molecules, oriented as shown above. The surrounding of solute ions by many solvent molecules is called **solvation**. When the solvent is water, this phenomenon is called **hydration**. When we write the equation

$$NaCl(s) \xrightarrow{\text{H}_2\text{O}} Na^+(aq) + Cl^-(aq)$$

the symbol (*aq*), standing for aqueous, reminds us that each dissolved ion is surrounded by and attracted to many water molecules. The ions have been solvated (hydrated).

We have broken the process of forming a solution into three successive steps, but it is important for you to realize that all three steps actually occur at once. The final step—hydration—is different from the first two. The first two steps require overcoming attractive forces and thus require energy input. However, energy is released upon the formation of attractive interactions, so energy is released as the ions are hydrated. The energy released by this hydration is called the **hydration energy**. The fact that the third step in the dissolving process releases energy is very important, because it supplies the bulk of the energy that the first two steps require. However, hydration can supply sufficient energy only if the solvent–solute attractive forces are strong enough—that is, only if they liberate enough energy to do the job.

We can now finally address the question, "When will a solute dissolve in a solvent?" The answer comes from looking closely at the relative magnitudes of the energies involved in the three steps. The following problems will give you some practice in determining the magnitude of the hydration energy. Keep in mind that:

When two objects of opposite charge attract one another, the attractive force between them increases as the magnitude of the charges increases, or the distance between the objects decreases.

This fundamental law is called **Coulomb's law** and is very useful for explaining many phenomena, including hydration energy.

PRACTICE PROBLEMS

10.4 Consider the ionic compound magnesium chloride ($MgCl_2$). Do you think that the hydration energy for this compound will be greater than, less than, or about equal to that of NaCl?

Answer: The hydration energy released for $MgCl_2$ will be greater than that for NaCl. This is because $MgCl_2$ has Mg^{2+} ions in it. The attractive forces between oppositely charged particles increases as the magnitude of the charges increases. The +2 charge, which is greater than the +1 charge of the Na^+ ions in NaCl, will lead to stronger ion–dipole attractions with water molecules, releasing more hydration energy than for NaCl.

10.5 Imagine that you are trying to dissolve NaCl in liquid carbon tetrachloride (CCl_4) instead of water. Would the energy released in step 3 be greater than, less than, or about equal to that of NaCl in water? Explain your answer completely.

10.6 Based on the answer for Practice Problem 10.4, would the energy required for step 1, pulling apart the ionic lattice, be greater for NaCl or $MgCl_2$? Explain your answer.

Very often the combination of the solute and the solvent is referred to as a *system*. In considering the solution process for any particular solvent/solute pair, what is important is how the energy of the system changes for each of the three steps in the solution formation process. By examining the change in energy at each step, we will be able to determine whether a solute will dissolve in a solvent. The symbol that we will use to represent the system's energy changes is ΔE (the Greek letter delta, Δ, means change). As we have seen, the first two steps require an input of energy whereas the third step releases energy. We will assume that when a step requires energy, the total energy of the system must increase. We will represent this with a positive value for ΔE. Likewise, when a step releases energy, the total energy of the system must decrease. We will represent this with a negative value for ΔE. Now, read over the following step-by-step energy analysis for forming a solution, and compare it with the diagram at the top of the next page.

To form a solution:

Step 1: We must overcome the solute–solute attractive forces. This takes energy, so ΔE is positive. Note that if the solute is a liquid, the energy requirement is much less than for an ionic solid (where a lattice must be separated into ions); for a gas, the energy needed is essentially zero.

Step 2: We must overcome the solvent–solvent attractive forces (pry solvent molecules apart from one another to make room for the solute). This takes energy, so again ΔE is positive.

Step 3: Solute–solvent attractive forces must develop. This releases energy, so ΔE is negative.

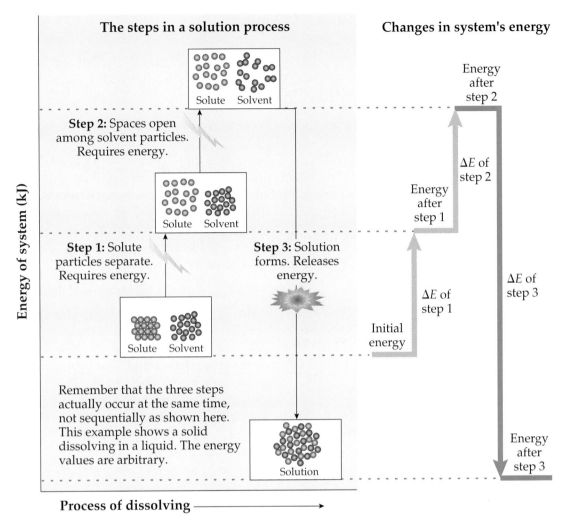

The steps in a solution process

Solute Solvent

Step 2: Spaces open among solvent particles. Requires energy.

Solute Solvent

Step 1: Solute particles separate. Requires energy.

Step 3: Solution forms. Releases energy.

Solute Solvent

Remember that the three steps actually occur at the same time, not sequentially as shown here. This example shows a solid dissolving in a liquid. The energy values are arbitrary.

Solution

Energy of system (kJ)

Process of dissolving ⟶

Changes in system's energy

Energy after step 2

ΔE of step 2

Energy after step 1

ΔE of step 1

Initial energy

ΔE of step 3

Energy after step 3

Whether or not a solute will dissolve in a solvent depends on the total energy change for the entire process. The total energy change involved, which we write as ΔE_{total}, is just the sum of the energy changes (ΔE) for the three steps:

$$\Delta E_{total} = \Delta E_{step\ 1} + \Delta E_{step\ 2} + \Delta E_{step\ 3}$$

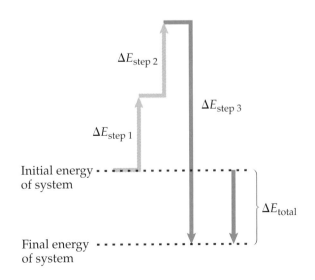

$\Delta E_{step\ 2}$

$\Delta E_{step\ 3}$

$\Delta E_{step\ 1}$

Initial energy of system

ΔE_{total}

Final energy of system

For example, for a solution of LiCl in water,

$$\Delta E_{step\ 1} = +853\ kJ \qquad \Delta E_{step\ 2} = +96\ kJ \qquad \Delta E_{step\ 3} = -986\ kJ$$

If we add these energy changes, we get ΔE_{total}:

$$\Delta E_{total} = (+853\ kJ) + (+96\ kJ) + (-986\ kJ) = -37\ kJ$$

We get a negative ΔE_{total}, which means that overall, energy is released. In other words, the system goes down in energy in step 3 more than it goes up in steps 1 and 2 combined, as shown in the energy diagram.

In this case, step 3 provides plenty of energy to drive steps 1 and 2. In fact, step 3 provides 37 kJ more energy than is needed. As a general rule, we can say:

If $\Delta E_{total} < 0$, the substance will dissolve.

In our example, dissolving LiCl, an ionic salt, in water resulted in an energy surplus, but this is not always the case. In the next figure, steps 1 and 2 go up in energy more than step 3 goes down. The difference is an energy deficit, so step 3 can't feed enough energy to steps 1 and 2 to make them occur. The result is that ΔE_{total} is positive (overall, energy is absorbed), and the substance will probably not dissolve.

$\Delta E_{step\ 2} = +96\ kJ$

$\Delta E_{step\ 3} = -986\ kJ$

$\Delta E_{step\ 1} = +853\ kJ$

Initial energy of system

Final energy of system

$\Delta E_{total} = (+853\ kJ)$
$+ (+96\ kJ)$
$+ (-986\ kJ)$
$= -37\ kJ$

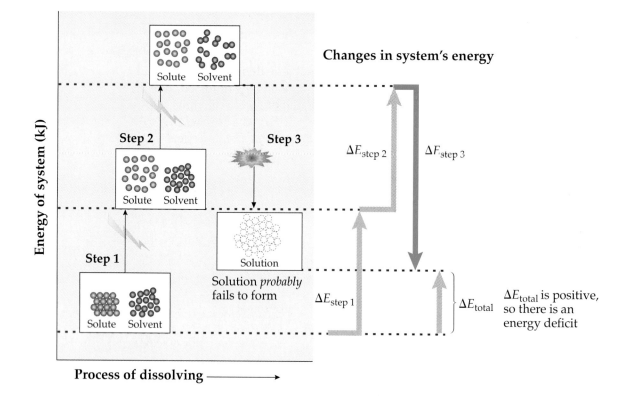

Changes in system's energy

Solute   Solvent

Step 2

Solute   Solvent

Step 3

$\Delta E_{step\ 2}$ $\Delta E_{step\ 3}$

Step 1

Solution

Solution *probably* fails to form

$\Delta E_{step\ 1}$

Solute   Solvent

ΔE_{total} ΔE_{total} is positive, so there is an energy deficit

Energy of system (kJ)

Process of dissolving ⟶

As a general rule, we can say:

When $\Delta E_{total} > 0$, the substance will probably not dissolve.

Why "probably"? We will answer that question in Section 10.4.

10.7 For a given solute in water, the energy changes are:

$$\Delta E_{step\ 1} = 835\ kJ \qquad \Delta E_{step\ 2} = 98\ kJ \qquad \Delta E_{step\ 3} = -805\ kJ$$

Will this solute dissolve in water? Explain your answer, both numerically and in terms of the three-step model for dissolving.

Answer: $\Delta E_{total} = (+835\ kJ) + (+98\ kJ) + (-805\ kJ) = +128\ kJ$

ΔE_{total} *is positive (there is an energy deficit), so the solute probably will not dissolve in water. In terms of the three-step model, the first two steps take energy (both energies are positive). The third step (solvation) releases energy (negative), but not enough (it's less than needed in the first two steps).*

10.8 Suppose we wanted to dissolve a gaseous solute in water instead of a solid solute like NaCl. Would you expect $\Delta E_{step\ 1}$ to be larger for dissolving the gaseous or the solid ionic solute? Explain your answer completely.

10.9 The more negative ΔE_{total} is, the more likely that a solute will dissolve. Explain why this is so.

10.10 Magnesium chloride, $MgCl_2$, is an ionic solid containing Mg^{2+} ions. Suppose you want to dissolve some $MgCl_2$ in water.
(a) Why do the magnesium cations have a +2 charge?
(b) $\Delta E_{step\ 1}$ for $MgCl_2$ is much more positive than $\Delta E_{step\ 1}$ for NaCl. Explain what this means and why it might be so.
(c) $MgCl_2$ is more soluble in water than NaCl. Explain how this is possible in light of the information given in part (b). [*Hint:* Consider the answer to Practice Problem 10.4.]

A general rule for determining whether a solute will dissolve is often expressed as "like dissolves like." The term "like" here refers to a similarity in polarity. What this rule means is that polar solutes dissolve best in polar solvents and nonpolar solutes dissolve best in nonpolar solvents. This is reasonable, since it is most likely for step 3 to release enough energy to feed steps 1 and 2 when the solvent and the solute molecules are similar in terms of their polarities. For example, NaCl, a very polar substance, dissolves in polar solvents such as water because the ion–dipole interactions of step 3 help compensate for the energy requirements of step 1 and 2. What if we tried to dissolve NaCl in a nonpolar solvent like carbon tetrachloride (CCl_4)? Try this now.

10.3 **WORKPATCH**   Imagine trying to dissolve NaCl in CCl$_4$. Considering the relative energies required in each step, choose the correct answers below. [*Hint:* Consider your answer to Practice Problem 10.5.]

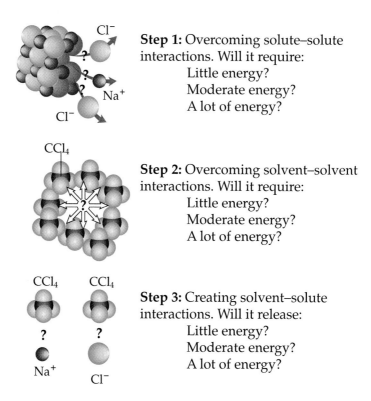

Step 1: Overcoming solute–solute interactions. Will it require:
    Little energy?
    Moderate energy?
    A lot of energy?

Step 2: Overcoming solvent–solvent interactions. Will it require:
    Little energy?
    Moderate energy?
    A lot of energy?

Step 3: Creating solvent–solute interactions. Will it release:
    Little energy?
    Moderate energy?
    A lot of energy?

You should have found in the WorkPatch that step 3 releases very little energy, whereas step 1 requires a great deal of energy. Step 3 can't compensate for steps 1 and 2; this hypothetical solution process would have a large energy deficit. Consequently, NaCl does not dissolve in CCl$_4$.

According to our general rule, "like dissolves like," nonpolar solutes should dissolve well in nonpolar solvents. Consider the forces involved in the three steps in this case:

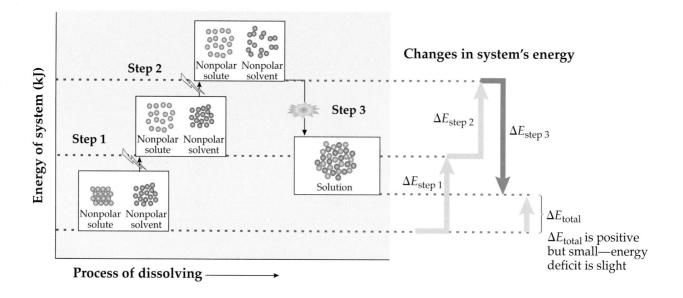

Changes in system's energy

In step 1, the weak London forces that hold together the nonpolar solute must be broken. In step 2, the weak London forces that hold together the solvent must be overcome. The solvent–solute interactions that form in step 3 also consist of London forces. Thus, steps 1 and 2 require little energy, and step 3 doesn't release much energy. Since the total energy consumed by steps 1 and 2 is small, and the energy released by step 3 is also small, ΔE_{total} will be very small, regardless of whether it's positive or negative.

In fact, even if ΔE_{total} is positive for such a process—that is, even if there's a slight energy deficit—a nonpolar solute is likely to dissolve well in a nonpolar solvent. That is because there is another factor besides the energy difference, ΔE_{total}, that can help a solute to dissolve. This factor, called *entropy*, can help tip the balance in favor of dissolving when the energy deficit is small.

ENTROPY AND THE FORMATION OF SOLUTIONS 10.3

Entropy can be thought of as a measure of the amount of disorder or randomness in a situation. For example, consider your room. When all your clothing is folded and put away, you have a neat and ordered room. This is referred to as an ordered, or low-entropy, situation. If you take all your clothes and randomly throw them about the room, you have a messy, disordered, or high-entropy, situation.

Order — Low Entropy   **Disorder — High Entropy**

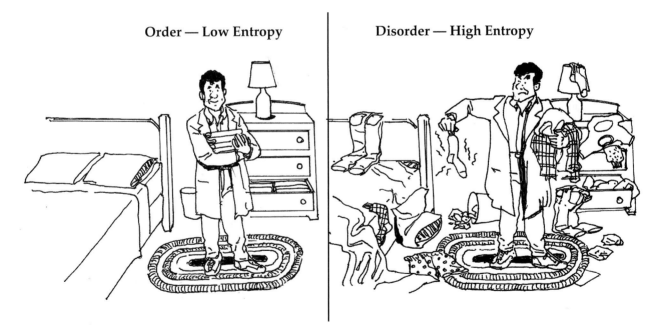

A fundamental law of nature called the **second law of thermodynamics** states that:

For anything to happen spontaneously anywhere in the universe, the entropy of the universe must increase.

In other words, the universe "prefers" a more random situation. There is a simple reason for this. A random situation is a lot more probable than an ordered one. For example, let's return to your room. Imagine you have just

A highly unlikely event

done the laundry and are ready to put it away into the proper drawers. You stand over the open drawers and toss the clothes into the air. Now, there is a chance that the clothes will fall, neatly folded, into the proper drawers. While this could certainly happen, the chances of obtaining such an ordered arrangement of your clothes is minuscule. That's because there are only a limited number of ways to order your clothes neatly and into the proper drawers, but there are an infinite number of ways to get a disordered, chaotic arrangement. The chances of getting a random arrangement are therefore much greater.

Similarly, when solid NaCl dissolves, the ordered rows of Na^+ and Cl^- ions are replaced by a much more random situation in which the ions are scattered throughout the solution. The result is that the entropy of the universe has increased, and the second law of thermodynamics has been satisfied.

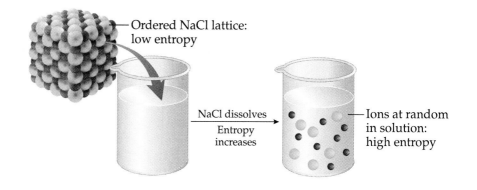

— Ordered NaCl lattice: low entropy

NaCl dissolves
Entropy increases

— Ions at random in solution: high entropy

This argument has been simplified for greater impact. Actually, other factors go into determining the entropy change for the universe when a solute dissolves. However, considered by itself, the decomposition of the ordered solute lattice into scattered solute particles always contributes to an increase in the entropy of the universe. According to the second law of thermodynamics, this contributes to the spontaneity of the dissolving process. This entropy increase is generally a weak effect, but it can sometimes tip the balance in favor of dissolving. Whether it can depends on ΔE_{total} for the dissolving process. For example, suppose we have a substance that dissolves in a solvent with a large energy deficit (ΔE_{total} much greater than zero). Chances are that the entropy increase upon dissolving wouldn't be large enough to make a dent, and the solute will not dissolve. This is the case we ran into when we tried to dissolve polar NaCl in nonpolar carbon tetrachloride (CCl_4). But in cases where the energy deficit is not too large, the entropy increase can tip the balance in favor of solubility. This is the case (discussed earlier) for dissolving a nonpolar solute in a nonpolar solvent. In general, the entropy increase is a small effect compared to ΔE_{total} and is thus important only when the energy deficit is relatively small. However, it is generally true that dissolving nonpolar solutes in nonpolar solvents almost always results in a small energy deficit. Thus, the entropy increase explains why nonpolar substances are almost always soluble in nonpolar solvents.

Consider two glass bulbs containing two different gases, A and B, connected but separated by a closed valve. Assume that ΔE_{total} for their mixing is zero. Upon opening the valve, the gases will spontaneously mix given enough time. The reverse process will never happen, no matter how long you wait. Explain why this is so.

WORKPATCH 10.4

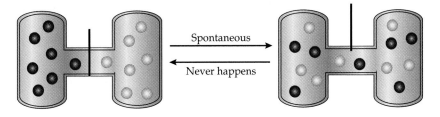

SOLUBILITY, TEMPERATURE, AND PRESSURE 10.4

You now know what must occur for a solute to dissolve in a solvent. But just knowing that a particular substance will dissolve in a solvent is only part of the story. When chemists are working with solutions they need to know how much solute can be dissolved in a given amount of solvent. The maximum amount of a solute that can be dissolved in a given amount of solvent is called its **solubility**. Solubility is often reported as the maximum number of grams of solute that will dissolve in 100 g of the solvent. An interesting aspect of solubility is that it can be changed. For example, try to dissolve 200 g of sugar (sucrose) in 100 g of ice-cold (0°C) water. Even with lots of stirring, some of the sugar just sits at the bottom of the glass and refuses to dissolve. But heat the water up to 100°C and it all dissolves. Evidently, temperature is one variable that can be used to control the solubility of sugar in water. As the temperature of the water increases, so does the solubility of the sugar, as shown in the bar chart.

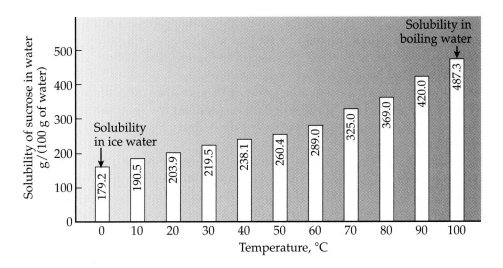

For most molecular solids (like sucrose), an increase in temperature generally causes an increase in solubility. The same is true for most ionic solids, as you can see from the following figure:

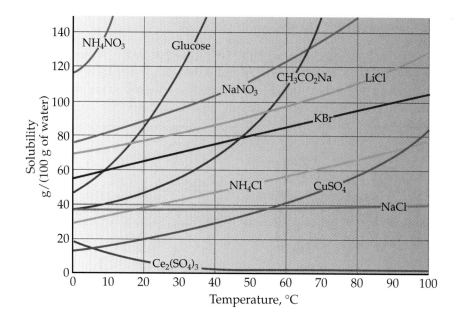

You can see from the graphs that the increase in solubility with increasing temperature is small for some compounds (NaCl), while the increase is quite dramatic for others (CH$_3$CO$_2$Na, sodium acetate). Cerium sulfate, Ce$_2$(SO$_4$)$_3$, is one of the exceptions. It actually becomes less soluble with increasing temperature.

While it might be tempting to relate the way solubilities change with temperature to the energy changes that take place in the dissolving process, this should not be done. For example, LiCl dissolves with an energy surplus whereas NH$_4$NO$_3$ dissolves with an energy deficit, and yet both compounds increase in solubility with an increase in temperature. Therefore, the best way to determine the effect of temperature on solubility is by experiment.

On the other hand, in systems in which a gas dissolves in a liquid, solubility consistently decreases with increasing temperature. Many familiar aqueous solutions fit this profile. The familiar fizz in soft drinks is due to dissolved CO$_2$ gas. Raising the temperature of the solvent decreases the solubility of the gaseous solute. Think of what it's like to drink warm soda pop. It tastes flat because most of the dissolved CO$_2$ has escaped from the solution.

Another variable that can affect the solubility of a gaseous solute in a liquid solvent is pressure. Increasing the pressure of a gas above a liquid will always force more gas into the liquid—that is, it increases the solubility of the gas. One way to do this is by pushing on a piston above the gas, as shown at the top of the next page.

Pushing down on the piston decreases the gas volume and thus increases its pressure (recall the ideal gas law from Chapter 9). The gas is essentially being squeezed or forced into the solution. This is how carbonated beverages are made. Beverage makers increase the solubility of CO$_2$ by forcing it into solution using high pressure (2–5 atm), well above the normal pressure of 1 atm. When you open a can of soda, the pressure above the solution

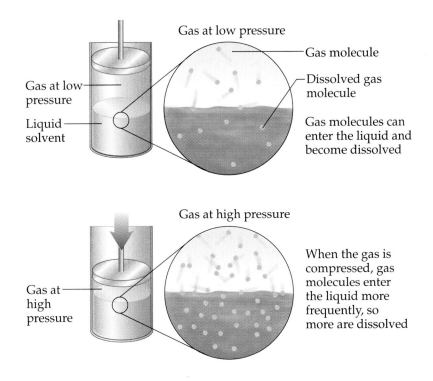

immediately drops back to 1 atm, drastically decreasing the solubility of the CO_2, which comes bubbling out of solution. If you leave the can open long enough, most of the CO_2 will escape and the drink will taste flat.

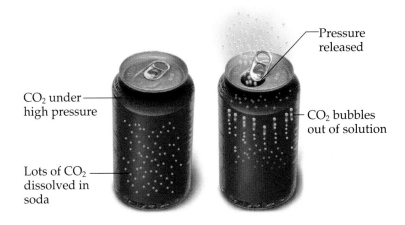

The fact that gases become more soluble as pressure increases is known as **Henry's law**.

A similar thing happens when divers rise too quickly from a deep-sea dive. Deep under water, the increased pressure squeezes more of the gas the divers are breathing into their blood, increasing the solubility of the gas in their blood. This gas is mostly N_2 if they are breathing compressed air. Coming up too rapidly is like popping the top on a can of soda pop. The rapid decrease in pressure lowers the solubility of the nitrogen in their blood. The nitrogen actually bubbles out of the blood as it quickly escapes from solution. This condition, known as the bends, can be serious enough to be fatal.

The most effective treatment is to place the afflicted diver as quickly as possible in a high-pressure chamber (called a *hyperbaric chamber*). The pressure in the chamber is raised to a level that will cause the gas to redissolve in the blood, temporarily curing the bends. Then the pressure is reduced slowly enough that the extra dissolved gas can come out of solution through the diver's lungs instead of forming bubbles in the blood.

Solubility values like those in the graphs on pages 353 and 354 represent the maximum amount of solute that will dissolve in a given amount of the solvent. When the maximum amount of solute is dissolved in solution, we say that the solution is **saturated**. Knowing the solubility enables us to calculate how much solute it takes to saturate a given amount of solvent. For example, suppose we have 250.0 g of water and we want to use it to prepare a saturated sugar solution at 20.0°C. How much sugar will we need? From the bar chart on page 353, we find that the solubility of sucrose at 20.0°C is 203.9 g sucrose/100 g water. To answer the question, we just use the solubility as a conversion factor and multiply it by the amount of solvent we have:

$$250 \text{ g water} \times \frac{203.9 \text{ g sucrose}}{100.0 \text{ g water}} = 509.8 \text{ g sucrose}$$

Try the following practice problems to gain some experience using solubility data.

PRACTICE PROBLEMS

10.11 How many grams of sucrose will it take to saturate 1 ton of water at 20.0°C? [1 ton = 2000 lb; 453.6 g/lb]

Answer:

$$1.00 \text{ ton water} \times \frac{2000 \text{ lb water}}{1 \text{ ton water}} \times \frac{453.6 \text{ g water}}{1 \text{ lb water}} \times \frac{203.9 \text{ g sucrose}}{100.0 \text{ g water}} = 1.85 \times 10^6 \text{ g sucrose}$$

10.12 How many grams of potassium chloride (KCl) will it take to prepare a saturated solution in 500.0 mL of boiling water? [The solubility of KCl at 100°C is 56.7 g/100 g water; assume the density of water is 1.00 g/mL.]

10.13 With regard to steps 1–3 for the formation of a solution, what is the biggest difference between dissolving a gas and dissolving a solid?

GETTING UNLIKES TO DISSOLVE: SOAPS AND DETERGENTS 10.5

We've been talking so far about like dissolving like. But much of the dirt that accumulates on our bodies and our clothing—such as grease and oil—is of a nonpolar nature, so washing ourselves and our clothes with polar water won't accomplish much in the way of cleaning. In order to dissolve away these nonpolar substances we need a nonpolar solvent. We could use a nonpolar solvent like carbon tetrachloride, but the expense and toxicity of CCl_4 make this somewhat impractical to do at home. This is, however, effectively what happens when you send your clothes off to be dry cleaned. Your clothes are immersed in a nonpolar liquid called TCE (trichloroethane) in which the nonpolar grease and oil will dissolve. Dry cleaning is "dry" only in the sense that no water is used.

"Dry" cleaning

Water—abundant, cheap, nontoxic, and recyclable—is still the solvent of choice for household cleaning. But how do we make nonpolar dirt dissolve in polar water? Let's go back to the detergent aisle of the supermarket. Getting "unlikes" to dissolve is the job of soaps and detergents. Each soap or detergent molecule consists of a long nonpolar "tail" of carbon and hydrogen (called a *hydrocarbon tail*) and a highly polar "head":

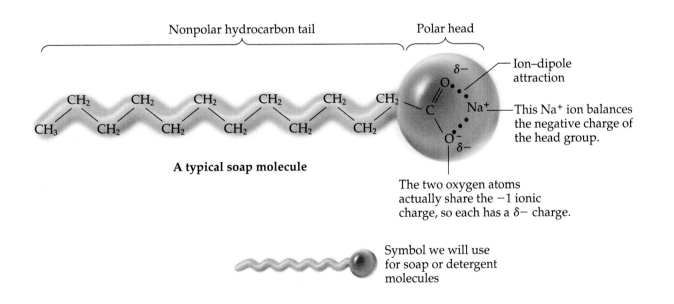

Nonpolar hydrocarbon tail   Polar head

A typical soap molecule

Ion–dipole attraction

This Na^+ ion balances the negative charge of the head group.

The two oxygen atoms actually share the −1 ionic charge, so each has a $\delta-$ charge.

Symbol we will use for soap or detergent molecules

Being polar, like water, the head of a soap or detergent molecule is soluble in water, and is called **hydrophilic** (meaning "water-loving"). The nonpolar hydrocarbon tail is insoluble in water, and is referred to as **hydrophobic** (meaning "water-fearing").

When soap molecules are added to water, an interesting spherical structure called a *micelle* spontaneously forms:

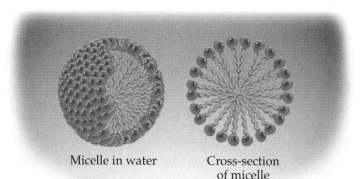

Micelle in water   Cross-section
                              of micelle

The outside of the micelle is formed from the polar heads and is hydrophilic. The inside is made up of the nonpolar tails, all attracted to one another via London forces, and is hydrophobic. A micelle therefore presents a polar surface to the water molecules which, being polar themselves, are attracted to the micelle via intermolecular forces. The polar heads possess partially negative ($\delta-$) O atoms, which form hydrogen bonds with water molecules. These attractions to water result in the micelle being soluble in water.

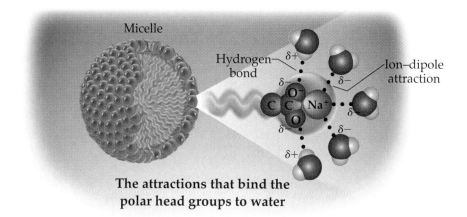

**The attractions that bind the
polar head groups to water**

It's quite a different story inside the micelle. Inside, among the conglomeration of nonpolar tails, the environment is strictly hydrophobic. This is just the right kind of environment to dissolve nonpolar grease molecules (like dissolves like). Grease and oil molecules that encounter the micelle quickly dissolve into its interior. Since the grease dissolves into the micelle and the micelle is dissolved in water, the result is that the detergent dissolves the grease in the water—a clever way of getting unlikes to dissolve in one another.

The first soaps were prepared from animal fats cooked with potash (KOH). The potash came from the ashes of wood fires, stirred with water. Animal fat and bacon grease provided the hydrocarbon tail, while the potash helped to create the polar head.

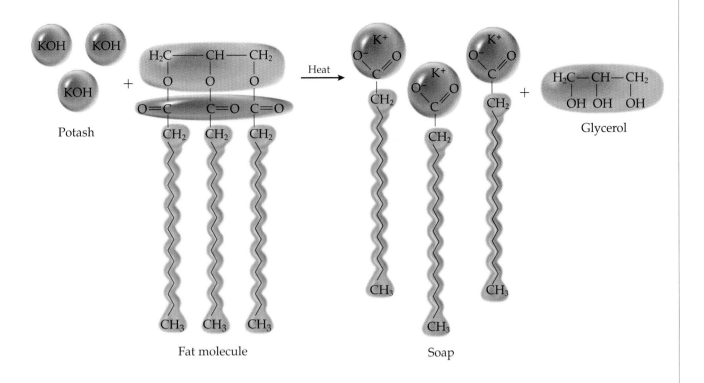

Today, the chemical industry spends a great deal of time trying to improve soaps and detergents, making grease dissolve even better in water, attempting to whiten and brighten your socks until your feet practically glow when you put them on.

Suppose you put detergent into a dry-cleaning machine that used nonpolar carbon tetrachloride (CCl_4). Assuming micelle-type structures still form:

WORKPATCH **10.5**

    (a) Make a diagram of the micelle-type structures that would form in CCl_4, and explain your diagram.

    (b) Explain what kind of dirt they would be good at removing.

Indeed, there are many dry-cleaning detergents. The micelle you drew to answer the WorkPatch explains how they might work.

CONCENTRATION: HOW MUCH SOLUTE IS DISSOLVED IN THE SOLVENT—MOLARITY 10.6

The solubility charts we looked at earlier tell us the maximum amount of a particular solute we can dissolve in a fixed amount of solvent. For example, at 20°C we can dissolve a maximum of 203.9 g of table sugar in 100 g of water. This sugar–water solution is *saturated* and would be quite sweet. We could also make a less sweet sugar–water solution by dissolving less than the maximum, say 20 g of sugar, in 100 g of water. The **concentrations** of these two solutions, the amount of solute per amount of solvent or solution, would be different. In this and the next section we are going to look at the quantitative ways to describe the concentration of a solution. But first we'll take a brief look at ways to describe concentration qualitatively. For example, you take a drink of some of the watery, light-brown liquid from the snack bar coffee machine and mutter, "This coffee's pretty weak." Or, you sip some dark, expensive cappuccino at a classy restaurant and proclaim, "This is *strong*!" Weak and strong are informal qualitative terms describing how much coffee (solute) is dissolved in the water (solvent). Instead of "weak" and "strong" you could have said "dilute" and "concentrated," other terms often used to describe concentration. These terms are also qualitative, because they don't describe the actual composition of the solution.

There are times when actually knowing the concentration of a solution takes on a whole new level of importance

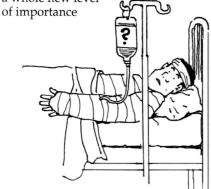

There will be times when you need more accurate, quantitative information as to exactly how much solute is dissolved in a given sample of solution. For example, suppose you have been in an auto accident and find yourself in immediate need of an intravenous saline (salt) solution. Too concentrated a solution will kill you, and too dilute a solution won't help. It's at a time like this that you seriously hope the attending health professional knows something more about concentration than just "weak" or "strong." Considering that your life depends on it, an accurate, quantitative numerical value of the concentration would be nice.

Saturated is a quantitative term. As we mentioned earlier, a solution is said to be **saturated** when the maximum amount of solute has been dissolved in it. If you add more solute, it just settles to the bottom of the container and doesn't dissolve. In fact, a good way to prepare a saturated solution is to add an excess of solute so that some doesn't dissolve, and then filter the solution.

The easy way to make a saturated solution

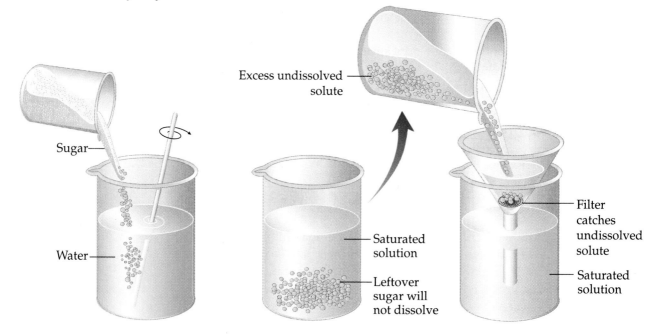

At 0°C, you can dissolve a maximum of 35.7 g of NaCl in 100 g of water. No more will dissolve. The word saturated therefore has a quantitative meaning, equal to the solubility of the solute at the temperature the solution was pre-pared. Unfortunately, just saying the word saturated doesn't tell you directly how much salt is dissolved in a saturated NaCl solution. You would have to look up the solubility value in a table or chart (like the one on page 354). It would be much more useful to have some quantitative description of the concentration that could instantly tell you the amount of solute present in the solvent. One such description commonly used by chemists is called *molarity*. From the name you can probably guess what's coming.

When describing the amount of solute dissolved, chemists most often think in terms of moles and express the concentration in terms of molarity. (If you need to review the concept of the mole, look back at Section 7.5.) The **molarity** of a particular solution is defined as the number of moles of solute that are present (dissolved) in 1 liter of solution:

Did you think you were rid of me?

$$\text{Molarity (M)} = \frac{\text{Moles of solute}}{\text{1 L of solution}}$$

For example, if you dissolve 1 mole (58.443 g) of NaCl in enough water to give exactly 1 L of solution, you will have a 1 molar (1 M) solution of NaCl:

How to prepare a 1 molar NaCl solution

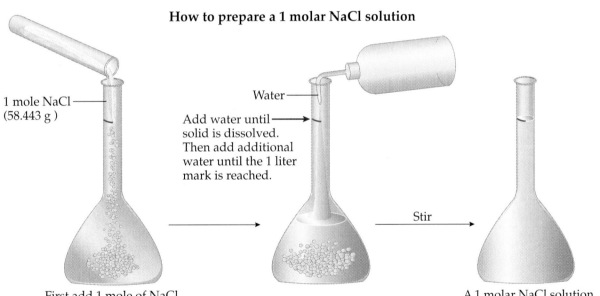

1 mole NaCl (58.443 g)

Water

Add water until solid is dissolved. Then add additional water until the 1 liter mark is reached.

Stir

First add 1 mole of NaCl . . .

A 1 molar NaCl solution

Now we want to warn you about a mistake that many students make (but you never will). A 1 M NaCl solution is not made by taking 1 mole of NaCl and adding 1 L of water to it. It is made by taking 1 mole of NaCl and adding enough water so that you end up with a total of 1 L of solution. The denomi-nator in the molarity expression says "1 L of solution," *not* "1 L of solvent." Make sure you understand this difference.

How NOT to prepare a 1 molar NaCl solution

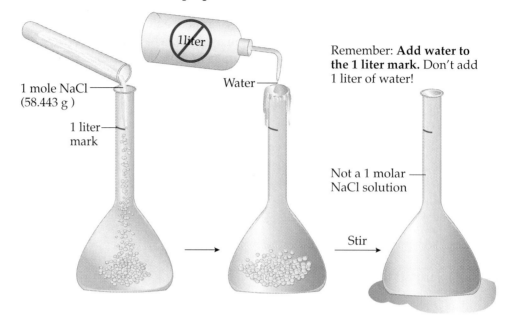

1 mole NaCl
(58.443 g)

1 liter
mark

Water

Remember: **Add water to
the 1 liter mark.** Don't add
1 liter of water!

Not a 1 molar
NaCl solution

Stir

Molarity gives us what we wanted. If someone hands you a bottle of saline solution labeled 1 M, then you quantitatively and absolutely know its concentration. No longer is it weak or strong, or saturated or unsaturated. It's 1 M. Every liter of that solution contains exactly 1 mole (58.443 g) of NaCl. A 2 M NaCl solution is twice as concentrated as a 1 M solution—every liter of 2 M solution contains 2 moles (116.886 g) of NaCl. Likewise, a 0.5 M solution of NaCl is half as concentrated as the original—every liter of a 0.5 M solution contains 0.5 mole (29.222 g) of NaCl.

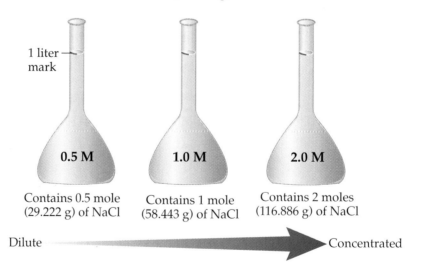

1 liter
mark

0.5 M

1.0 M

2.0 M

Contains 0.5 mole
(29.222 g) of NaCl

Contains 1 mole
(58.443 g) of NaCl

Contains 2 moles
(116.886 g) of NaCl

Dilute   Concentrated

200.0 mL

1.00 M NaCl
solution

**How
much
NaCl?**

How many moles of
NaCl are in the flask?
How many grams?

In fact, knowing the molarity, we can quickly determine how much solute we have for any volume of solution. We just use molarity as a conversion factor. For example, suppose you have a flask that contains 200.0 mL of a 1.00 M NaCl solution. If all the water evaporated, solid sodium chloride would remain. How much solid sodium chloride would there be in the flask? In other words, how much NaCl is contained in 200.0 mL of this 1.00 M solution?

We can find the answers to these questions by using unit analysis (Section 2.8). We will treat the molarity (1.00 M) as a conversion factor since it has two units (moles of solute over liters of solution). We will treat the volume (200.0 mL), which has just one unit (mL), as the measurement. This is not to say that the molarity is not a measured quantity. It is, but it can also serve as a conversion factor. We are now ready to solve the problem.

Step 1: Write down the measurement: 200.0 mL solution

Step 2: Multiply it by the appropriate conversion factor to make the starting unit (mL solution) cancel. We will use molarity, but first we have to convert the 200.0 mL to L:

$$200.0 \text{ mL solution} \times \underbrace{\frac{1 \text{ L solution}}{1000.0 \text{ mL solution}}}_{\substack{\text{Conversion factor} \\ \text{to change mL to L}}} \times \underbrace{\frac{1.00 \text{ mole NaCl}}{1 \text{ L solution}}}_{\substack{\text{Molarity} \\ \text{conversion factor}}} = 0.200 \text{ mole NaCl}$$

The answer is that 200.0 mL of a 1.00 M NaCl solution contains 0.200 mole of NaCl. To convert the answer to grams, we multiply by the appropriate conversion factor, the molar mass of NaCl:

$$0.200 \text{ mole NaCl} \times \underbrace{\frac{58.443 \text{ g NaCl}}{1 \text{ mole NaCl}}}_{\substack{\text{Molar mass} \\ \text{conversion factor}}} = 11.7 \text{ g NaCl}$$

Thus, there are 11.7 g of NaCl in our 200.0 mL of 1.00 M salt solution. Of course, we did not have to do this last calculation separately. We could have just strung the molar mass conversion factor onto the end of the previous calculation.

Armed with molarity as a conversion factor, you are now ready to practice some numerical problems. Remember, you can always use the inverse of a conversion factor to make units cancel.

———PRACTICE PROBLEMS

10.14 How many milliliters of a 1.500 M solution of NaCl do you need to obtain 100.0 g of NaCl?

Answer: As usual, we write the measurement and start multiplying it by appropriate conversion factors:

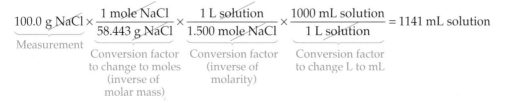

$$100.0 \text{ g NaCl} \times \underbrace{\frac{1 \text{ mole NaCl}}{58.443 \text{ g NaCl}}}_{\substack{\text{Conversion factor} \\ \text{to change to moles} \\ \text{(inverse of} \\ \text{molar mass)}}} \times \underbrace{\frac{1 \text{ L solution}}{1.500 \text{ mole NaCl}}}_{\substack{\text{Conversion factor} \\ \text{(inverse of} \\ \text{molarity)}}} \times \underbrace{\frac{1000 \text{ mL solution}}{1 \text{ L solution}}}_{\substack{\text{Conversion factor} \\ \text{to change L to mL}}} = 1141 \text{ mL solution}$$

Notice that the first thing we did was convert to moles. Remember our general advice for "amount" problems in chemistry. Convert to moles as quickly as you can. It's almost always the right thing to do.

10.15 How many milliliters of a 2.55 M solution of glucose ($C_6H_{12}O_6$, molar mass = 180.155 g/mole) do you need to obtain 25.0 g of glucose?

10.16 How many grams of ethanol (C_2H_6O) are there in 200.0 mL of a 2.00 M aqueous solution of ethanol?

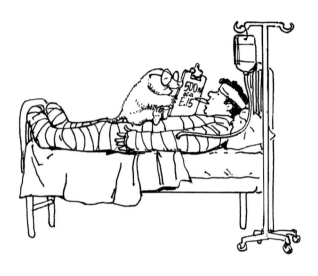

There is one other type of problem involving molarity that you should know how to do, a practical problem that asks: "How do you prepare a specific amount of a solution with a specific molarity?" For example, we last observed our assistant in the hospital on a bottle of intravenous saline solution. He notices the bottle is just about empty and so he tells the nurse, "You need to prepare 500.0 mL of a 0.15 M NaCl solution to refill this bottle." Realizing that the nurse studied introductory chemistry quite a while ago, our panicked assistant reminds him how to do this. His explanation follows.

There are two ways to prepare this life-saving solution. One is to weigh out the proper amount of NaCl and dissolve it in the proper amount of water. We'll call this preparing a solution from scratch. A second way would be to take a more concentrated stock solution of NaCl off the shelf and dilute it to the proper concentration. Both ways are commonly used, so we'll go over both.

Method 1: Preparing a solution from scratch

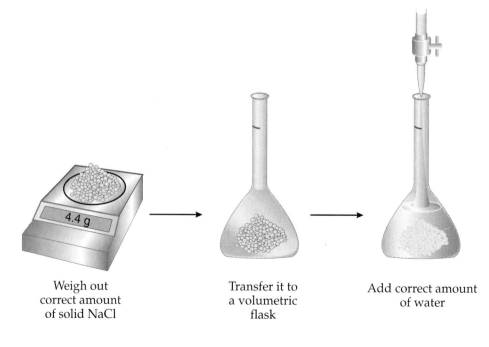

Weigh out
correct amount
of solid NaCl

Transfer it to
a volumetric
flask

Add correct amount
of water

Method 2: Preparing a solution by diluting a stock solution

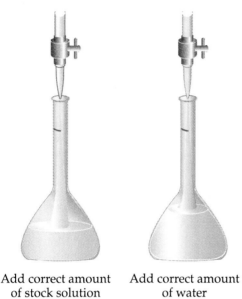

Add correct amount
of stock solution

Add correct amount
of water

Method 1: Preparing a solution from scratch. Our task is to prepare 500.0 mL
of a 0.15 M NaCl solution. We must first calculate how much NaCl
we need. Just use the usual method of unit analysis:

$$500.0 \text{ mL solution} \times \frac{1 \text{ L solution}}{1000 \text{ mL solution}} \times \frac{0.15 \text{ mole NaCl}}{1 \text{ L solution}} = 0.075 \text{ mole NaCl}$$

Measurement   Change units   Molarity
conversion factor

The answer is in moles. Since we weigh things in grams, we still
need to convert to grams:

$$0.075 \text{ mole NaCl} \times \frac{58.443 \text{ g NaCl}}{1 \text{ mole NaCl}} = 4.4 \text{ g NaCl}$$

Molar mass conversion factor

So to prepare 500.0 mL of a 0.15 M NaCl solution from scratch,
take 4.4 g of NaCl, and add enough water to get exactly
500.0 mL of solution. (Remember, do *not* add
500 mL of water!) Finally, stir well while
you are adding the water.

Water

500.0 mL

4.4 g

Weigh out 4.4 g
of solid NaCl

Transfer it to
a volumetric flask

Add water to
make 500.0 mL

Method 2: **Preparing a solution by diluting a more concentrated stock solution.** On the shelf is a large bottle of 0.50 M NaCl solution. This is our "reservoir" of solution, what is called a stock solution. We can use it to prepare the 500.0 mL of 0.15 M solution that we need. Since the stock solution is more concentrated than the solution we need, we have to dilute some amount of it with water. How much stock solution should we use, and how much water do we add to dilute it? At first glance this might seem more difficult than preparing the solution from scratch since we have to consider two solutions, the stock solution and the one we want to prepare. How much NaCl do we need to prepare 500.0 mL of a 0.15 M saline solution? This is exactly the same question we asked in method 1, so we know that the answer is 0.075 mole of NaCl. In method 1 we converted this amount to grams because we were preparing the solution from scratch (from solid NaCl). But this time we will get our 0.075 mole of NaCl from the stock solution, so the question becomes, "What volume of 0.50 M stock solution do we need to get 0.075 mole of NaCl?" Look closely at the question. There is a measurement (0.075 mole NaCl) and a conversion factor (0.50 M). It's time to use the unit analysis approach again:

$$0.075 \text{ mole NaCl} \times \underbrace{\frac{1 \text{ L stock solution}}{0.50 \text{ mole NaCl}}}_{\substack{\text{Conversion factor} \\ \text{(inverse of molarity} \\ \text{of stock solution)}}} = 0.15 \text{ L stock solution}$$

This calculation tells us that exactly 0.15 L of stock solution will contain 0.075 mole of NaCl. Notice that in this calculation we used the inverse of the molarity of the stock solution as the conversion factor.

To prepare our solution, we put 0.15 L (150 mL) of stock solution into a flask. This puts 0.075 mole of NaCl into the flask. Then,

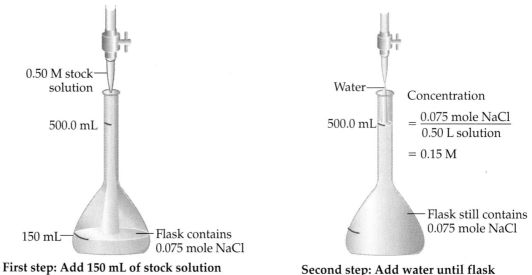

First step: Add 150 mL of stock solution

Second step: Add water until flask contains 500.0 mL of solution

we simply add enough water to get a total of 500.0 mL of solution. Once again we have 0.075 mole of NaCl dissolved in a total volume of 500.0 mL (0.50 L) of solution, or a 0.15 M NaCl solution.

Here is a summary of what we just did:

How to prepare a solution of desired molarity

Step 1: Determine the moles of solute that you need to prepare the solution. This will always be equal to the desired volume in liters times the desired molarity:

Moles of solute needed = (Volume of solution) × (Molarity of solution)

Step 2: If preparing the solution from scratch, convert moles of solute to grams by using a molar mass conversion factor.

OR

Step 2: If preparing the solution from stock solution, convert moles of solute to volume of stock solution by multiplying by the inverse of the molarity of the stock solution.

Step 3: For both methods, add enough water to reach the desired volume while stirring well.

Preparing specific volumes of solutions of particular molarities, both from scratch and by diluting more concentrated stock solutions, are routine practices carried out every day in chemical laboratories, medical offices, and hospitals throughout the world. Now it's time to convince yourself that you can do it.

PRACTICE PROBLEMS

You need 400.0 mL of a 2.00 M salt (NaCl) solution. On the shelf is pure solid NaCl and a 3.00 M stock solution of NaCl. Figure out how to prepare your solution by both methods.

10.17 From scratch:

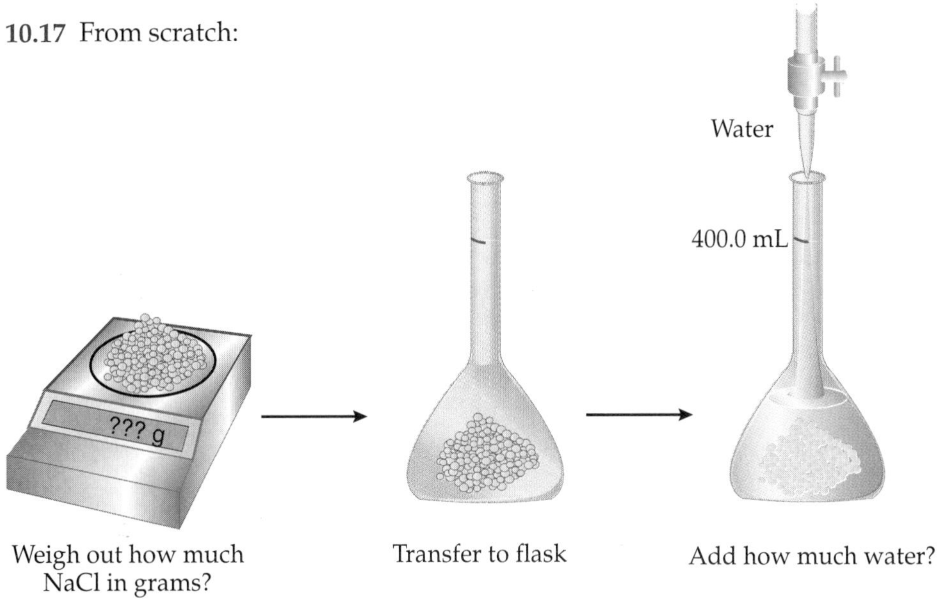

Water

400.0 mL

Weigh out how much
   NaCl in grams?

Transfer to flask

Add how much water?

10.18 From stock solution:

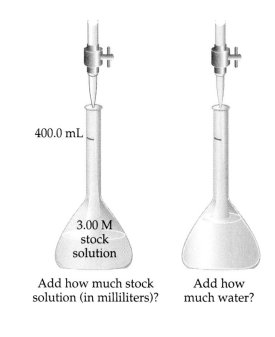

400.0 mL

3.00 M
stock
solution

Add how much stock
solution (in milliliters)?

Add how
much water?

10.7 PERCENT COMPOSITION FOR SOLUTIONS

Molarity is an excellent way to specify a solution's concentration quantitatively, and chemists use it routinely. However, solutions encountered outside of a chemical laboratory, in hospitals, doctors' offices, and grocery stores, are often likely to have concentrations expressed by *percent composition*. In a fundamental sense, percent composition is similar to molarity. In fact, most of the different ways of numerically reporting the concentration of a solution are based on the following definition:

$$\text{Concentration} = \frac{\text{Amount of solute}}{\text{Amount of solution}}$$

For molarity, the numerator (amount of solute) is expressed in moles, and the denominator (amount of solution) is expressed in liters of solution. Percent composition uses the same definition of concentration, but with different quantities in the numerator and the denominator. In fact, there are three types of **percent composition**, depending on whether the solute and solution are weighed in grams or measured by volume.

Percent by mass (wt %)	Percent by volume (vol %)	Percent mass/volume (wt/vol %)
$\dfrac{\text{Grams of solute}}{\text{Grams of solution}} \times 100$	$\dfrac{\text{Volume of solute}}{\text{Volume of solution}} \times 100$	$\dfrac{\text{Grams of solute}}{\text{Volume of solution}} \times 100$

Percent by mass (abbreviated as wt %, where wt stands for weight) is often used to describe the concentration of a solution that was prepared by dissolving a solid in some solvent. Let's try one example. If you dissolve 10.0 grams of solid sucrose in enough water so that the entire solution weighs 100.0 g, then the percent by mass concentration of the solution is 10.0 wt %, as demonstrated in the illustration:

Percent by mass

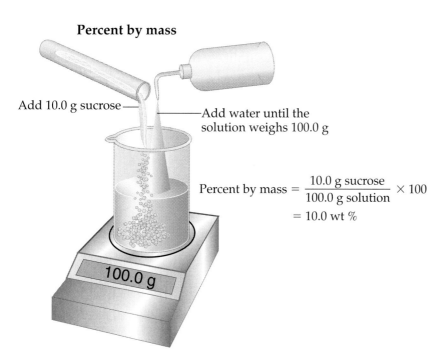

Add 10.0 g sucrose
Add water until the solution weighs 100.0 g

$$\text{Percent by mass} = \frac{10.0 \text{ g sucrose}}{100.0 \text{ g solution}} \times 100$$
$$= 10.0 \text{ wt } \%$$

100.0 g

When the solute is a liquid, percent by volume (vol %) is often used. For example, dissolving 25.0 mL of the liquid ethanol in enough water to give a total solution volume of 100.0 mL produces a solution with a concentration of 25.0 vol %, as shown below:

Percent by volume

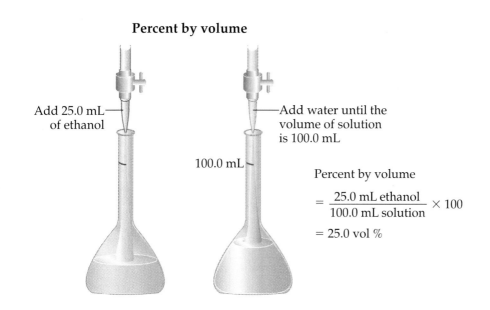

Add 25.0 mL of ethanol

Add water until the volume of solution is 100.0 mL

100.0 mL

Percent by volume

$$= \frac{25.0 \text{ mL ethanol}}{100.0 \text{ mL solution}} \times 100$$
$$= 25.0 \text{ vol } \%$$

Interestingly, you cannot prepare the above 25.0 vol % solution by adding 75 mL of water to the 25 mL of alcohol. When you mix different liquids together, the volumes are usually not exactly additive. In the case of ethanol and water a solution made by mixing 25 mL and 75 mL, respectively, will have a total volume that is measurably less than 100 mL. That is why the diagram says "Add water until the volume of solution is 100 mL" instead of "Add 75 mL of water."

The examples that we used to calculate percent compositions were a bit unrealistic because we set the denominator equal to 100.0 to make the calculations easy. In real situations, it's unlikely that you will have exactly 100.0 g or 100.0 mL of a solution. Suppose you have 273.0 g of a salt solution that you know contains 35.0 g of NaCl. What is its percent composition by mass? You use exactly the same equation that we used for our simpler example and calculate the percent composition exactly the same way:

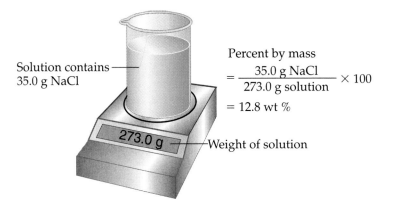

Solution contains 35.0 g NaCl

$$\text{Percent by mass} = \frac{35.0 \text{ g NaCl}}{273.0 \text{ g solution}} \times 100$$

$$= 12.8 \text{ wt } \%$$

273.0 g — Weight of solution

The percent composition for a solution can serve as a conversion factor, just as molarity can. For example, suppose you have 265.5 mL of an aqueous ethanol solution that is 30.0 vol % ethanol. How much ethanol (in milliliters) is in the bottle?

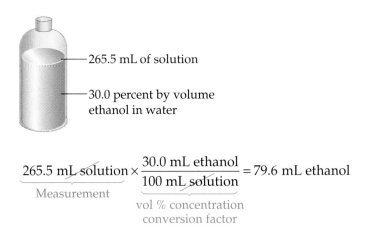

265.5 mL of solution

30.0 percent by volume ethanol in water

$$\underbrace{265.5 \text{ mL solution}}_{\text{Measurement}} \times \underbrace{\frac{30.0 \text{ mL ethanol}}{100 \text{ mL solution}}}_{\substack{\text{vol \% concentration} \\ \text{conversion factor}}} = 79.6 \text{ mL ethanol}$$

Any concentration (wt %, vol %, wt/vol %, molarity) is a potential conversion factor.

The density of pure ethanol is 0.7893 g/mL. How could you use this information to answer the question above in grams of ethanol instead of milliliters of ethanol?

WORKPATCH 10.6

The WorkPatch is one more reminder that many quantities in chemistry can serve as conversion factors.

Finally, although it's less common, we can also use percent mass/volume (wt/vol %) to characterize the concentration of a solution. In this case, we use units of grams for amount of solute and volume (mL) for amount of solution. Because there are three types of percent concentration, it is always important to state which definition you are using. For example, a bottle of an ethanol–water solution labeled 5% ethanol in water does not give you enough information about the solution's composition. If the 5% is percent by mass (5 wt %), then 100 g of the solution contains 5 g of ethanol. If the 5% is percent by volume (5 vol %), then 100 mL of the solution contains 5 mL of ethanol, which is only 3.9 g of ethanol. This ambiguity is one reason chemists prefer to use molarity.

PRACTICE PROBLEMS

10.19 A solution is prepared by dissolving 25.0 g of sucrose in 175.0 g of water. Characterize its concentration by the appropriate percent composition.

Answer: The appropriate percent composition is percent by mass since both measurements are given in grams.

Grams of solution = Grams solute + Grams solvent = 25.0 g + 175.0 g = 200.0 g

$$\text{Percent by mass concentration} = \frac{\text{Grams solute}}{\text{Grams solution}} \times 100 = \frac{25.0 \text{ g}}{200.0 \text{ g}} \times 100 = 12.5 \text{ wt \%}$$

10.20 Gasohol is a solution of gasoline and ethanol. Every liter of gasohol contains 90.0 mL of ethanol dissolved in gasoline. Characterize the solute concentration by the appropriate percent composition. [*Hint:* Assume you have 1 liter of gasohol solution.]

10.21 Suppose you have 65.0 g of the solution from Practice Problem 10.19. How many grams of sucrose do you have?

In this chapter we've looked at what solutions are, why solutes dissolve, and how to quantify the amount of solute dissolved. Since almost all the chemical reactions that are going on inside your body or in any living organism take place in aqueous solution, knowledge of solutions can be very important. Small changes in the concentrations of some of the gaseous and solid solutes dissolved in your blood can cause illness or even death. Modern medicine has determined the "normal" concentration levels for many of these solutes, while biochemists work to understand what determines and regulates their concentrations.

HAVE YOU LEARNED THIS?

Solution (p. 339)

Solvent (p. 339)

Solute (p. 339)

Aqueous solution (p. 341)

Ion–dipole force (p. 344)

Solvation (p. 345)

Hydration (p. 345)

Hydration energy (p. 345)

Coulomb's law (p. 345)

Forming a solution (p. 346)

Entropy (p. 351)

Second law of thermodynamics (p. 351)

Solubility (p. 353)

Henry's law (p. 355)

Saturated (pp. 356, 360)

Hydrophilic (p. 357)

Hydrophobic (p. 357)

Concentration (p. 359)

Molarity (p. 361)

Percent composition (wt %, vol %, wt/vol %) (p. 368)

WHAT IS A SOLUTION?

10.22 Define the terms solute, solvent, and solution.

10.23 Suppose you took a small amount of sugar, mixed it with a large amount of flour, and then spent hours grinding the mixture to a very fine powder. Is this mixture a solution? Explain.

10.24 Is it appropriate to call a soft drink an aqueous solution? Justify your answer.

10.25 Is it appropriate to call the atmosphere we breathe a solution? Justify your answer.

10.26 Vinegar is a common household solution that we consume. What is the solvent and what is the solute in vinegar?

10.27 What is a solid solution? Give some examples.

10.28 Classify the following as solutions or heterogeneous mixtures.
  (a) 14 karat gold (prepared from mixing 10 parts copper and 14 parts gold as the molten metals)
  (b) Filtered ocean water
  (c) A piece of wood
  (d) Your exhaled breath
  (e) A bottle of salad oil and vinegar shaken extremely well

10.29 Suppose you had a 50:50 homogeneous mixture of oxygen gas in helium gas. Which would you call the solvent and which would you call the solute?

10.30 A solution of salt and sugar in water is allowed to evaporate, leaving behind the two solid solutes. Is what remains a solution or a heterogeneous mixture? Explain your answer. [*Hint:* As solids, both solutes exist as lattices that would exclude each other.]

ENERGY AND THE FORMATION OF SOLUTIONS

10.31 Is there anything wrong with the following diagram? Explain your answer.

10.32 What evidence is there in your everyday experience to indicate that intermolecular attractive forces must exist between water molecules?

10.33 If there were no attractive forces between water molecules, what phase or phases of water would you expect to exist? Explain your answer.

10.34 What is the "sale at a crowded shopping mall" analogy for liquid water, and why is it a reasonable model for the solvent?

10.35 In what way is melting a piece of NaCl similar to dissolving it in water?

10.36 Why does it take such a high temperature to melt NaCl but a much lower temperature to dissolve it in water?

10.37 When an ionic substance such as NaCl dissolves, the crystal lattice has to break apart to release the individual ions into the solution. Does this part of the dissolving process require energy or release energy? Explain your answer.

10.38 In dissolving any solute, room must be made in the solvent to accommodate solute particles. Does making this room take energy or release energy? Explain why.

10.39 What do we mean by hydration, and why does it release energy?

10.40 List the three conceptual steps that must occur for NaCl to dissolve in water and illustrate each with a diagram.

10.41 Solid sucrose consists of individual sugar molecules fixed in a lattice. The sucrose molecules are hydrogen bonded to each other via OH groups in the molecules. Knowing this, comment on which takes more energy, breaking up an NaCl lattice or breaking up a sugar lattice. Justify your answer.

10.42 When NaCl dissolves, what helps keep the dissolved Na^+ and Cl^- ions from coming back together and reforming the lattice, precipitating the solid?

10.43 What is the name of the attractive force between dissolved Na^+ ions and water molecules? Diagram this force, showing how a water molecule would approach an Na^+ ion. Do the same for a Cl^- ion.

10.44 Some alcohols are quite soluble in water. For example, isopropyl alcohol, shown below, is sold as an aqueous solution we call rubbing alcohol. Show how a water molecule would be attracted to isopropyl alcohol, and name the most important (strongest) intermolecular force involved. Would you call this hydration?

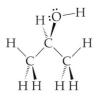

10.45 Why is the amount of energy associated with the third conceptual step of the dissolving process so critical to whether a solute will dissolve?

10.46 Consider a solid dissolving in a liquid. Suppose the energy released upon forming solute–solvent interactions is substantially less than the energy required to break up the lattice and to make room in the solvent. Would the solid be very soluble in this solvent? Explain your answer fully.

10.47 When some salts dissolve in water, the water heats up, even when no external heat is added. How is this possible?

10.48 When some salts dissolve in water, the water gets colder. How is this possible?

10.49 Suppose the hydration energy for an ionic compound is much less than the energy that it takes to pull apart the lattice. Would such a compound likely be soluble or insoluble? Explain your answer.

10.50 Sodium chloride is very soluble in water but it is insoluble in liquid hexane (C_6H_{14}). Why is this so?

10.51 Even though NaCl does not dissolve in hexane (C_6H_{14}), if we imagine the dissolving process for this system, there is one step that would take less energy than the corresponding step for NaCl dissolving in water. Which step is it? Why would it require less energy?

10.52 Which would have a greater hydration energy, $AlCl_3$ or NaCl? Explain your answer.

10.53 In theory, it is possible for one ionic compound to give rise to a greater hydration energy than another and still be the less soluble of the two. What would have to be true for this to be so?

10.54 What is meant by the rule of thumb, "like dissolves like"?

10.55 What do we mean by the total energy change (ΔE_{total}) for the dissolving process, and why is it important to know about it?

10.56 Consider the following diagram and answer the questions below:

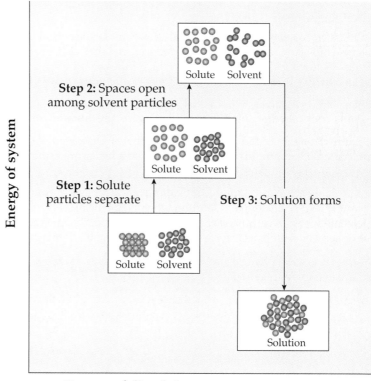

Step 2: Spaces open among solvent particles

Step 1: Solute particles separate

Step 3: Solution forms

Energy of system

Process of dissolving ⟶

(a) Why does step 1 require energy?
(b) Why does step 2 require energy?
(c) Why does step 3 release energy?
(d) On the basis of this diagram, would you expect the potential solute to be soluble or insoluble? Explain.
(e) Assuming the solute does dissolve, would you expect the solution to get warmer or colder as dissolving takes place? Explain.

10.57 When we write (*aq*) for a dissolved ion, such as Na^+(*aq*), exactly what does it mean? Answer both in words and with a diagram.

10.58 Suppose you have two ionic compounds, A and B. The only significant difference between them is that the ionic bonds that hold the lattice together in compound A are stronger than those in compound B. Which compound would you expect to be more soluble in water? Explain your answer.

10.59 When a gaseous solute dissolves in water, which step in the conceptual dissolving process is essentially skipped? Explain why.

10.60 When a liquid solute dissolves in water, there is still a conceptual "first step" that requires energy, but it doesn't require breaking up a crystal lattice as for a typical solid solute. What is it, and why does it take energy?

10.61 Consider the following three molecules, all liquids at room temperature:

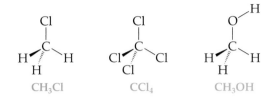

Which would be least soluble in water? Which would be most soluble in water? Explain your answers fully.

10.62 Why is oil insoluble in water? (Don't just say one is polar and one is nonpolar; give an answer based on energy considerations.)

ENTROPY AND FORMATION OF SOLUTIONS

10.63 What is entropy?

10.64 What does the second law of thermodynamics say?

10.65 What is the justification for the second law of thermodynamics?

10.66 When you release a drop of blue food dye into a beaker of water, the drop eventually dissolves to give a homogeneously light blue solution. No matter how long you wait, the dye molecules will never regroup to form the original drop. Why not?

10.67 In terms of energy balance, when is entropy an important factor for dissolving a solute in a solvent? When is it not?

10.68 True or false? Whenever something spontaneously dissolves in water, the entropy of the universe has increased. Justify your answer.

10.69 Certainly if NaCl dissolved in liquid hexane (C_6H_{14}), the entropy associated with the Na^+ and Cl^- ions would increase. Nevertheless, NaCl does not dissolve in hexane. Why not?

10.70 If the second law of thermodynamics is indeed true, then what is the ultimate fate of the universe?

SOLUBILITY, TEMPERATURE, AND PRESSURE

10.71 While increasing the temperature increases the solubility of most solids in water, the opposite is true for gaseous solutes. Explain.

10.72 What effect does increasing the pressure have on the solubility of gaseous solutes? Explain your answer.

10.73 Exactly how is the medical condition known as the bends related to solubility?

10.74 Examine the graphs of solubility in water versus temperature on page 354. What is the trend for most of the ionic substances shown?

10.75 Relating to Problem 10.74, one ionic compound in the graph shows almost no temperature-dependence, and one clearly violates the general trend. Identify these two ionic compounds.

10.76 Aquatic life is often damaged when hot water is discharged from power stations into rivers and lakes. What might this have to do with gas solubility in water?

10.77 Why does a can of soda pop go flat when left open to the atmosphere?

10.78 How many grams of sucrose are necessary to saturate 100.0 lb of water at 90°C? [See the bar chart on page 353; 1 lb = 453.6 g.]

10.79 Below are solubility versus temperature values for two ionic compounds in water:

Solubility (g/100 g water)

Temp (°C)	KCl	Li_2SO_4
0	27.6	35.3
10	31.0	35.0
20	34.0	34.2
30	37.0	33.5
40	40.0	32.7
50	42.6	32.5
100	56.7	29.9

(a) Make a plot similar to the graph on page 354 for these compounds.
(b) Using the plot, estimate the solubility of both compounds in water at 70°C.
(c) How much of each compound can be dissolved in a beaker containing 75 g of water at 70°C?

10.80 How many more grams of Li_2SO_4 can you dissolve in 250.0 g of water at 10.0°C than at 50.0°C? [See Problem 10.79.]

10.81 A saturated solution of KCl was prepared at 40.0°C using 5.00 lb of water. How many grams of KCl were required to prepare this solution? [See Problem 10.79; 1 lb = 453.6 g.]

10.82 A saturated solution of KCl was prepared at 50.0°C using 2.00 L of water. How many grams of KCl were required to prepare this solution? [See Problem 10.79; assume the density of water is 1.00 g/mL.]

GETTING UNLIKES TO DISSOLVE: SOAPS AND DETERGENTS

10.83 Soap molecules have two portions with very different properties. Draw a typical soap molecule and discuss the different natures of the two parts.

10.84 What do the terms hydrophobic and hydrophilic mean?

10.85 Micelles are spherical, although they are usually drawn as a flat cross-section (as shown on page 358). Why wouldn't micelles exist in water as flat, two-dimensional structures?

10.86 Soap molecules have a hydrophobic portion and yet they dissolve in water. Explain how they accomplish this.

10.87 Explain how soaps function to allow water to wash oily, nonpolar dirt off clothes and skin.

10.88 A very common molecule used in detergents is sodium lauryl sulfate, shown below:

$$CH_3CH_2CH_2CH_2CH_2CH_2CH_2CH_2CH_2CH_2CH_2CH_2O-\overset{\overset{\displaystyle O}{\|}}{\underset{\underset{\displaystyle O}{\|}}{S}}-O^- \quad Na^+$$

Identify the hydrophobic and hydrophilic portions of this molecule.

CONCENTRATION

10.89 Define molarity.

10.90 Give precise instructions to your laboratory assistant as to how to prepare 1.00 liter of a 1.00 molar solution of $CaCl_2$. Remember that your assistant will be weighing the $CaCl_2$ in grams. The assistant may use a 1 liter volumetric flask (a flask with a mark at 1 liter).

10.91 Give precise instructions to your laboratory assistant as to how to prepare 1.00 liter of a 1.00 molar aqueous solution of sucrose ($C_{12}H_{22}O_{11}$). Remember that your assistant will be weighing the sucrose in grams. The assistant may use a 1 liter volumetric flask.

10.92 Give precise instructions to your laboratory assistant as to how to prepare 1.00 liter of a 0.250 molar aqueous solution of sucrose ($C_{12}H_{22}O_{11}$). Remember that your assistant will be weighing the sucrose in grams. The assistant may use a 1 liter volumetric flask.

10.93 Give precise instructions to your laboratory assistant as to how to prepare 0.500 liter of a 1.50 molar solution of sucrose ($C_{12}H_{22}O_{11}$). Remember that your assistant will be weighing the sucrose in grams. The assistant may use a 0.50 liter volumetric flask.

10.94 Your assistant tells you that she weighed out 2.50 moles of NaCl and then added enough water to get 500.0 mL of solution to prepare a 5.00 M solution of NaCl.
(a) How much NaCl did she weigh out in grams?
(b) Did she successfully prepare a 2.5 M solution? Prove your answer.

10.95 Your assistant tells you he weighed out 116.886 g of NaCl and then added exactly 1.00 L of water to it to prepare a 2.00 M solution of NaCl. Do you fire him or give him a promotion? Explain completely.

10.96 How many grams of NaCl are in 2500.0 mL of a 0.250 M solution?

10.97 How many grams of sucrose ($C_{12}H_{22}O_{11}$) are in 45.0 mL of a 0.250 M solution?

10.98 How many milliliters of a 1.00 M solution of NaCl are required to obtain 5.00 g of NaCl?

10.99 How many milliliters of a 0.250 M solution of glucose ($C_6H_{12}O_6$) are required to obtain 100.0 g of glucose?

10.100 There is a bottle of 0.500 M sucrose stock solution in the laboratory. Give precise instructions to your assistant on how to use the stock solution to prepare 250.0 mL of a 0.348 M sucrose solution.

10.101 There is a bottle of 4.50 M NaCl solution in the laboratory. Give precise instructions to your assistant on how to use the stock solution to prepare 100.0 mL of a 4.00 M NaCl solution.

10.102 What do you always get when you multiply molarity times volume in liters?

Percent Composition

10.103 What are the three types of percent compositions? Give names and definitions.

10.104 In this chapter you learned that "proof" for an alcoholic drink equals twice the percentage of alcohol in the drink. The complete definition of proof is that it is twice the percentage by volume of alcohol. Knowing this, exactly what does it mean to have a 90 proof drink?

10.105 A solution of ethanol is prepared by combining 22.5 g of ethanol with 49.6 g of water. What is the percent by mass composition of alcohol in this solution?

10.106 A solution of a particular solid solute in water has a concentration of 25.0 wt %.
(a) Given 100.0 g of this solution, how many grams of solute do you have?
(b) Given 48.0 g of this solution, how many grams of solute do you have?
(c) How many grams of this solution do you need to obtain 56.5 g of solute?

10.107 How would you prepare 2.00 kilograms of an NaCl solution that is 30.0 wt % NaCl?

10.108 How would you prepare 1.00 liter of an alcohol–water solution that is 5.00 vol % alcohol?

10.109 An alcohol–water solution is 35.00 vol % alcohol. How much solution is required to obtain 200.0 mL of alcohol?

10.110 An aqueous solution is 25.0 wt % NaCl. The density of the solution is 1.05 g/mL. What is the molarity of the solution? [*Hint:* Assume you have 100.0 g of this solution.]

WorkPatch Solutions

10.1 To be a solution, a mixture must be homogeneous. This can occur only when the molecules are free to move about randomly (that is, in the liquid or gas phase), and thus completely mix the substances together.

10.2 Remember that the water molecule is polar, with its partially negative charge ($\delta-$) located on the oxygen atom and its partially positive charges ($\delta+$) located on the hydrogens. Because opposite charges attract, the $\delta-$ oxygen atoms are attracted to and closely approach the aqueous Na^+ ions, whereas the $\delta+$ hydrogen atoms are attracted to and closely approach the Cl^- ions.

10.3 **Step 1:** Requires lots of energy. Ionic bonds within a lattice are very strong.

Step 2: Requires moderate or little energy. These are nonpolar solvent molecules, so the only forces that need to be overcome here are the London forces.

Step 3: Releases little energy. This is a case of two very different particles trying to attract one another. CCl_4 is nonpolar whereas the Na^+ and Cl^- ions are charged. The only force that can exist between these two particles is a weak London force. Development of a weak attractive force will release only a small amount of energy.

10.4 Since ΔE_{total} is zero for this example, any mixing that occurs will be due solely to entropy. In the spontaneous direction (from left to right), we are going from a relatively ordered situation (pure A and pure B) to a more chaotic situation in which A and B are randomly mixed together. The latter is a very probable state, and it contributes to an entropy increase for the universe. The second law of thermodynamics would therefore predict that mixing should be spontaneous. The reverse direction goes from a disordered to an ordered situation. It contributes to an entropy decrease for the universe and is far less probable than the mixed state, because there is only one way to achieve the ordered arrangement and there are so many more ways to arrive at the mixed arrangement. This is in violation of the second law of thermodynamics, so it won't happen.

10.5 (a) The micelle would look like the diagram drawn below. It would exist this way because it would want to present its nonpolar tails to the nonpolar solvent, leading to a maximum attractive force. The polar heads would be buried inside so they could avoid the unlike solvent that surrounds the micelle.

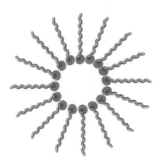

(b) Micelles in this form would be good at removing polar dirt from clothes that were being dry-cleaned.

10.6 Use the density as a conversion factor to convert the answer to grams:

$$79.6 \text{ mL ethanol} \times \frac{0.7893 \text{ g ethanol}}{1 \text{ mL ethanol}} = 63.1 \text{ g ethanol}$$

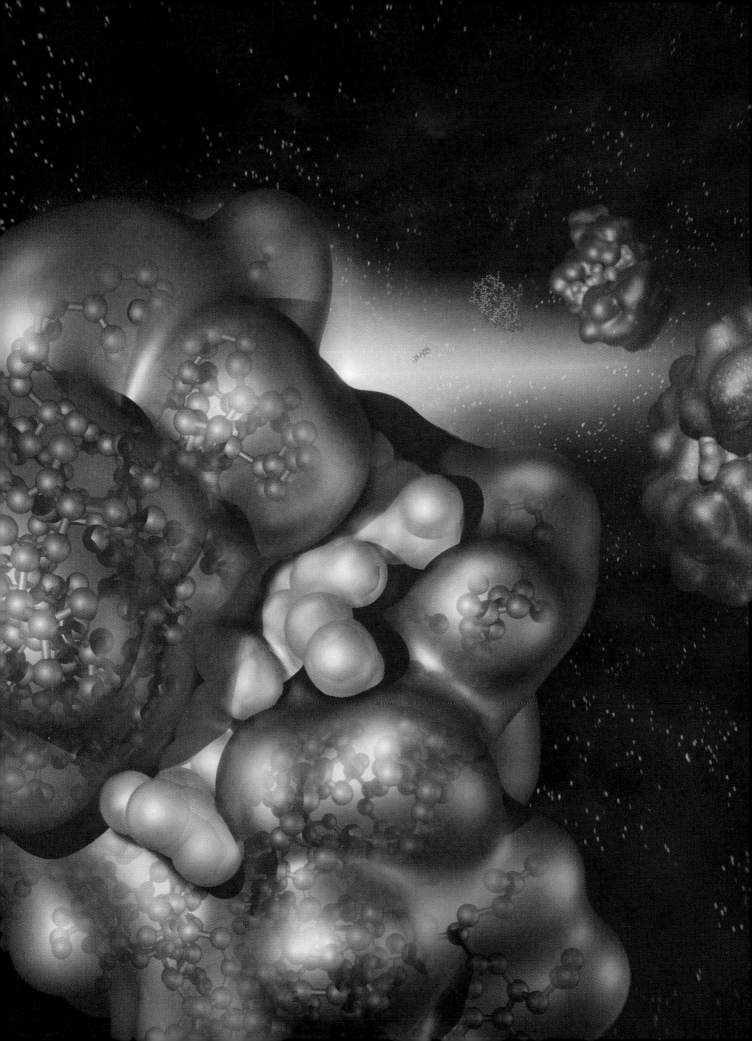

When Reactants Turn Into Products

CHEMICAL KINETICS 11.1

So far we have examined atoms, molecules, the forces within molecules (bonds), and the forces between molecules (intermolecular forces). All of these appetizers have prepared us to take a closer look at the main course of chemistry—chemical reactions.

Methane (CH_4) is one of the molecules that we have seen many times already. Now consider what happens in the combustion reaction between methane (CH_4) and oxygen (O_2):

$$CH_4 \quad + \quad 2\,O_2 \quad \longrightarrow \quad CO_2 \quad + \quad 2\,H_2O$$

One methane molecule reacts with two molecules of oxygen to produce one molecule of carbon dioxide (CO_2) and two water (H_2O) molecules. At the end of the reaction the carbon and hydrogen atoms that were bonded to each other in methane end up bonded to oxygen. For this to happen, all four carbon–hydrogen bonds must break and so must both oxygen–oxygen bonds. This is not a bad way to think about what happens. During most chemical reactions, atoms end up exchanging partners to form new compounds. But how do molecules do this? What process does the arrow that we draw between reactants and products actually represent? In this chapter we are going to examine step-by-step what happens to the reactants during the time they are being converted into products. It is usually not as simple as the reactants just banging into one another to form products instantaneously.

For the reaction of methane with oxygen, the reactant CH_4 and O_2 molecules have to collide with one another sufficiently hard to begin breaking the strong covalent bonds within them. But do they have to collide so hard that they are blasted completely apart into separate atoms before being reassembled into products?

Such a scenario is very unlikely. First of all, it requires a simultaneous collision of *three* molecules—one methane and two oxygen molecules. A collision of three molecules at the same place at the same time has a very low probability of occurrence (much lower than the likely probability of a collision between two molecules). Second, at normal reaction temperatures, most collisions would not supply nearly enough energy to tear the molecules completely to pieces. Instead, in most reactions, the reactants are converted into products via a series of one-molecule or two-molecule steps, each of which takes less energy than the crash scene depicted at left. The exact sequence of steps that molecules go through on changing from reactants to products is called a **reaction mechanism**.

11.1 WORKPATCH   Give another reason why the single-step mechanism shown above, in which reactant molecules are completely torn apart into atoms, is unlikely. [*Hint:* Remember, the reaction of methane and oxygen produces CO_2 and H_2O. Think about CH_2O (formaldehyde), CO (carbon monoxide), H_2O_2 (hydrogen peroxide), and CH_3OH (methanol).]

Could you answer the WorkPatch? Check your answer against ours.

To deduce a reaction's mechanism we must study the reaction while it is occurring. Unfortunately, this is often very difficult to do. Many reactions, like the combustion of methane, are essentially instantaneous, typically occurring within a few millionths of a second. Until recently, this was just too fast to allow a direct probe of what was going on. Other reactions, such as the rusting of steel, are inconvenient to study for the opposite reason.

A number of reactions, however, occur on a time scale suitable for study, and these examples have allowed chemists to determine what happens during a chemical reaction. But how exactly do we determine the mechanism of chemical transformation? Since we cannot see individual molecules or atoms, we cannot directly observe a reaction's mechanism. We have to examine it indirectly. One way to do this is to study the speed (rate) of the reaction and see how it changes as we vary the temperature and the reactant concentration. Chemists called *kineticists* devote their careers to these measurements, which provide our window on the mechanisms of reactions.

One reaction that has been studied to the point where most chemists feel confident they know exactly how the reactants turn into products is the substitution of OH for Br in methyl bromide (CH_3Br):

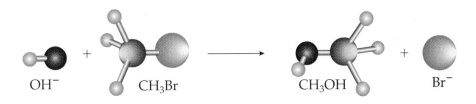

$$OH^- \quad + \quad CH_3Br \quad \longrightarrow \quad CH_3OH \quad + \quad Br^-$$

In this type of reaction, called a **substitution reaction**, one atom or group of atoms replaces another atom or group of atoms in a molecule. In this case, a hydroxyl (OH) group ends up replacing (substituting for) a bromine atom. We will be referring to this reaction throughout the chapter. Substitution reactions are a very important class of reactions. They are frequently used to modify molecules in the preparation of new pharmaceutical compounds. They are also thought to be the principal type of reaction used in living cells to turn on and off gene expression during cell development.

Why is it necessary to understand a reaction's mechanism? A major reason is that it gives us some practical advantages. If we understand exactly how reactants turn into products on a molecular level, then we can intelligently do things to influence the reaction. For example, rust (iron oxide, Fe_2O_3) is purposely produced in large quantities for use as the magnetic medium in audiotapes, videotapes, and computer disks. What if you decide to go into the business of producing rust? You know that you can't just set out a pile of iron nails and hope to have rust anytime soon. It would be helpful to understand the mechanism of rusting well enough so that you could do something to speed up the reaction. Or, consider

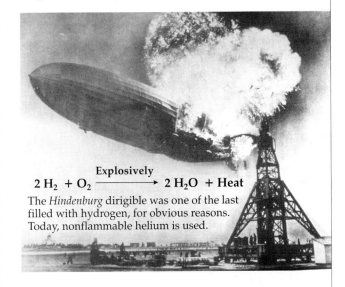

Explosively
$$2\,H_2 + O_2 \xrightarrow{\hspace{1.5cm}} 2\,H_2O + Heat$$

The *Hindenburg* dirigible was one of the last filled with hydrogen, for obvious reasons. Today, nonflammable helium is used.

the reaction of hydrogen and oxygen to produce water. This reaction is generally so fast and releases so much energy so quickly that it is explosive.

If you knew enough about the mechanism of the reaction, you might be able to slow it down to make it useful. This is precisely what is done inside a fuel cell, where hydrogen and oxygen are slowly combined to supply spacecraft not only with water but also with electricity.

So deducing a reaction's mechanism is important, and we get insight into the mechanism by studying how its rate depends on temperature and concentration. The next question is, "How do we get this insight?" Suppose we study a reaction and successfully determine how its rate depends on temperature and concentration. Now what? What is its mechanism? To answer this question, we must first understand the energy changes that take place during a reaction.

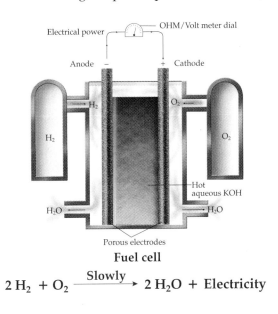

Fuel cell
$$2\,H_2 + O_2 \xrightarrow{\text{Slowly}} 2\,H_2O + Electricity$$

11.2 ENERGY CHANGES AND CHEMICAL REACTIONS

The key to understanding the mechanism of chemical change lies in knowing something about the changes in energy that take place during a reaction. Most chemical reactions occur with some **net change in energy**; that is, the products are at a different level of energy than the reactants. If the product molecules in a particular reaction contain more energy than the reactants, then the reaction must have absorbed energy from the surroundings as it occurred. Alternatively, a reaction can result in products that contain less energy than the reactants. In this case, the excess energy is released into the surroundings (typically in the form of light or heat) as the reaction takes place. The detonation of nitroglycerine is just such a reaction:

$$4\ C_3H_5N_3O_9 \longrightarrow 6\ N_2 + 10\ H_2O + 12\ CO_2 + O_2 +\ \boxed{5720\ kJ}$$
Nitroglycerine

Each nitroglycerine molecule contains a lot of energy, stored mostly in its chemical bonds. The products of the detonation, N_2, H_2O, CO_2, and O_2, are all lower in energy content. The difference in energy between nitroglycerine and all the products added together is the excess energy that is released when nitroglycerine explodes.

Reactions that release energy into the surroundings as they occur are called **exothermic** (from Greek roots meaning "heat out"). These are the reactions usually associated with the stereotype of the bumbling and careless "mad scientist" on television and in the movies.

The detonation of nitroglycerine is an extremely exothermic reaction. Combustion reactions, such as the burning of natural gas or methane in your stove or furnace, are also very exothermic, and we can make direct use of the energy released for cooking or heating.

Somewhat less familiar are reactions that absorb energy (or heat) from their surroundings. These reactions are called **endothermic** (from the Greek roots meaning "heat in"). If you hold a flask in which such a reaction is occurring, it will feel cold. This is because the reaction is drawing heat out of your hand to drive the reaction. If the reaction is very endothermic, such as the reaction shown in the photo, then you could even end up with the flask frozen to your skin.

To sum up:

- If a reaction releases energy in some form (usually as heat or light), then the reactants have more energy than the products and the reaction is called exothermic.

The reaction of hydrated barium hydroxide with ammonium chloride is strongly endothermic. The beaker containing the reactants was placed in a puddle of water on this block of wood; the reaction has frozen the water solid.

- If a reaction absorbs energy from its surroundings as it occurs, then the reactants contain less energy than the products and the reaction is referred to as endothermic.

- If a reaction is neutral—that is, it neither produces nor absorbs energy—then both the reactants and products must contain the same amount of energy. There is no special designation given to this type of reaction, which is much less common than the previous two.

Chemists often represent the energy changes that occur during a chemical reaction graphically with a **reaction energy profile**, which plots the relative energies of the reactants and the products. For example, let's consider the hypothetical exothermic reaction R $\longrightarrow$ P, where R represents the reactants and P the products. Since it is exothermic, the reaction releases energy, just like the reaction of nitroglycerine. Its reaction energy profile would look like this:

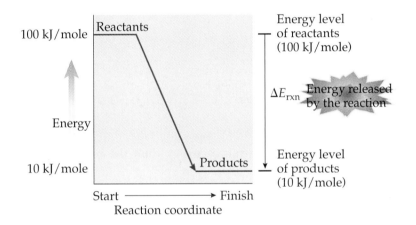

Energy is plotted on the vertical axis. The horizontal axis, called the *reaction coordinate*, traces the progress of the reaction from start to finish. Look closely at the graph and notice that the reactants R are higher up on the energy axis than the products P. The reactant molecules contain more energy than the product molecules. The difference in energy between the reactants and products is indicated on the diagram as ΔE_{rxn} (remember, the Greek letter delta, Δ, stands for difference or change; the subscript rxn is an abbreviation for reaction). It is this exact amount of energy, ΔE_{rxn}, that is released into the surroundings when the higher-energy reactants turn into the lower-energy products.

We can calculate the amount of energy released by the reaction, since the energy per mole of R and P is indicated on the reaction profile plot. To calculate ΔE_{rxn}, we subtract the energy of the reactants R from the energy of the products P. That is:

$$\Delta E_{rxn} = E_{products} - E_{reactants}$$

If we apply this formula to our reaction, we get

$$\Delta E_{rxn} = 10 \text{ kJ/mole} - 100 \text{ kJ/mole} = -90 \text{ kJ/mole}$$

Note the minus sign in our calculated value of ΔE_{rxn}. Because of the way ΔE_{rxn} is defined ($E_{\text{products}} - E_{\text{reactants}}$), exothermic (heat-releasing) reactions will always have a negative value for ΔE_{rxn}. In other words, in our reaction, 90 kJ of energy is released for every mole of reactant that becomes product. You should get used to this sign convention. Any reaction with a negative ΔE_{rxn} releases energy. Since exothermic reactions go from higher-energy reactants to lower-energy products, they are often said to go "downhill" in energy.

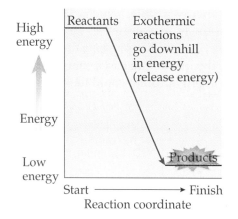

An endothermic reaction is just the opposite. It must absorb energy to proceed, since the products contain more energy than the reactants. The path from reactants to products is now "uphill" in energy.

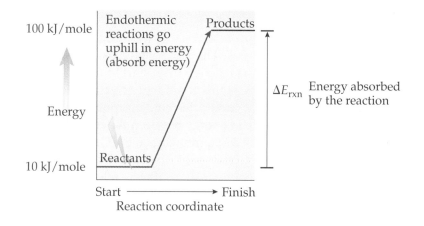

Once again we can calculate ΔE_{rxn}, but in this case the number we get is the amount of energy that must be *absorbed* from the surroundings on going from reactants to products. Using the energy values given on the plot for the endothermic reaction, we calculate

$$\Delta E_{\text{rxn}} = E_{\text{products}} - E_{\text{reactants}} = 100\,\text{kJ/mole} - 10\,\text{kJ/mole} = +90\,\text{kJ/mole}$$

This time the result is positive, which will always be the case for an endothermic reaction. Reactions that absorb energy always have a ΔE_{rxn} greater than zero.

It is important to keep straight the meaning of the sign of ΔE_{rxn}, so let's summarize what we've said:

For exothermic (energy-releasing) reactions, $\Delta E_{rxn} < 0$ (ΔE_{rxn} is negative).

For endothermic (energy-absorbing) reactions, $\Delta E_{rxn} > 0$ (ΔE_{rxn} is positive).

Finally, there are those reactions whose $\Delta E_{rxn} = 0$ kJ/mole. For these reactions, there is no net energy change. Overall, energy is neither released nor absorbed. That is because the reactants and products are at the same energy level.

PRACTICE PROBLEMS

11.1 Consider the following reaction profile for $A \longrightarrow B$ and answer the questions.

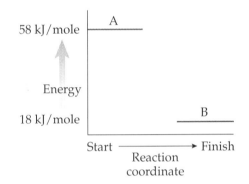

(a) Is this reaction exothermic or endothermic?
(b) Would the container in which this reaction was carried out feel hot or cold to the touch? Explain why.
(c) What is the value of ΔE_{rxn} (include the sign)?
(d) What does the sign of ΔE_{rxn} for this problem tell you?
(e) Which contains more energy, the reactants or the products?
(f) Does this reaction go uphill or downhill in energy?

Answer:
(a) Exothermic (energy is released)
(b) Hot, because it is exothermic
(c) $\Delta E_{rxn} = E_{products} - E_{reactants} = 18$ kJ/mole $- 58$ kJ/mole $= -40$ kJ/mole
(d) The fact that ΔE_{rxn} is negative tells you that the reaction releases energy.
(e) The reactants have more energy in them (they are higher on the plot).
(f) Downhill

11.2 A reaction is endothermic by 65 kJ/mole (that is, $\Delta E_{rxn} = +65$ kJ/mole).
(a) Draw a reaction energy profile for this reaction, assuming the energy of the reactants to be 100 kJ/mole.
(b) Does this reaction go uphill or downhill in energy?
(c) What is the value of ΔE_{rxn}?
(d) Would the container feel hot or cold to the touch? Explain your answer.

11.3   A reaction has a ΔE_{rxn} of -450 kJ/mole. The products have an energy of 20 kJ/mole.

(a) Are the reactants at a higher energy or lower energy than the products? Explain your answer.

(b) What is the starting energy of the reactants?

(c) Is the reaction exothermic or endothermic? How do you know?

(d) Does this reaction go uphill or downhill in energy?

(e) Draw a reaction energy profile for this reaction.

Now let's try a bit of a twist. So far we have been looking at reactions proceeding in one direction, from reactants to products (Reactants ⟶ Products), the so-called forward direction. This makes sense. After all, that's the way the arrow points, and that's the way the reaction goes. But we are always free to consider the reverse reaction in which the reaction proceeds from products to reactants (Products ⟶ Reactants). Even if a reaction doesn't proceed in the reverse direction, there is nothing to keep us from thinking about it. So let's return to Practice Problem 11.1, where we considered the exothermic reaction A ⟶ B ($\Delta E_{rxn} = -40$ kJ/mole). What about the reverse reaction where B turns into A, A ⟵ B? What is ΔE for the reverse reaction? Try the following WorkPatch.

11.2 WORKPATCH

Here is the reaction profile from Practice Problem 11.1 for the exothermic reaction A ⟶ B, for which $\Delta E_{rxn} = -40$ kJ/mole. What is ΔE for the reverse reaction A ⟵ B?

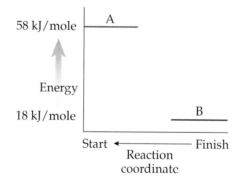

We've included this WorkPatch before explaining the energy change for a reverse reaction because we wanted you to answer it intuitively. Since the forward reaction A ⟶ B goes downhill in energy (it's exothermic; $\Delta E_{rxn} = -40$ kJ/mole), then the reverse reaction A ⟵ B goes uphill in energy by exactly the same amount (it's endothermic; $\Delta E_{reverse\ rxn} = +40$ kJ/mole). A reaction that releases energy in one direction must absorb exactly the same amount of energy in the reverse direction. Another way to say this is that the magnitude (size) of ΔE for the reverse reaction is exactly the same as it is for the forward reaction, but the sign has changed. We can express this with an equation:

$$\Delta E_{reverse\ rxn} = -\Delta E_{forward\ rxn}$$

This principle is always true. So remember that when you know ΔE for a reaction, you also know it for the reverse reaction. Just change the sign.

11.4 A reaction releases 400 kJ/mole of energy.
   (a) What is ΔE_{rxn}?   (b) What is $\Delta E_{reverse\ rxn}$?

Answers:
(a) $\Delta E_{rxn} = -400$ kJ/mole   *(b)* $\Delta E_{reverse\ rxn} = +400$ kJ/mole

11.5 As a reaction proceeds, its container feels cold to the touch. Measurements show a net energy change of 250 kJ/mole.
   (a) What is ΔE_{rxn}?   (b) What is $\Delta E_{reverse\ rxn}$?

11.6 As a reaction proceeds, its container feels hot to the touch. Measurements show a net energy change of 250 kJ/mole.
   (a) What is ΔE_{rxn}?   (b) What is $\Delta E_{reverse\ rxn}$?

Some questions logically arise at this point. Why do energy changes occur in a chemical reaction? If a reaction releases energy, where does the energy come from? If a reaction absorbs energy, where is the energy being stored? Molecules don't have pockets where they can store energy, but they do have bonds. The answer to these questions is that there are usually changes in the number and type of chemical bonds within the reacting molecules as reactants become products. For example, consider the balanced reaction that produces water from hydrogen molecules and oxygen molecules (the combustion of hydrogen). The large negative value for ΔE_{rxn} indicates that this reaction is highly exothermic:

$$2\ H_2(g) + O_2(g) \longrightarrow 2\ H_2O(g) \qquad \Delta E_{rxn} = -479\ kJ$$

Let's imagine that we can see this reaction occurring at the molecular level. At the start of the reaction we would see H_2 molecules with their single H–H covalent bonds and O_2 molecules with their double O=O covalent bonds. To understand the source of ΔE_{rxn}, we can postulate a simple reaction mechanism where the reactant molecules break apart into individual atoms. While this is not the actual mechanism for this reaction, it will take us from H_2 and O_2 to H_2O, so we can use this model to understand where ΔE_{rxn} comes from. In this case, we can break the H–H and O=O bonds in the reactant molecules to give individual H and O atoms. Of course, covalent bonds are quite strong, so it will take energy to break these covalent bonds.

It always takes energy to break a chemical bond—no exceptions, ever! So can we assume this reaction absorbs energy and is endothermic? Not necessarily (and we know this is not true in this case). Breaking the covalent bonds in the reactant molecules is only half the story. After the bonds have been broken, we can bring the atoms together in a new arrangement, forming the new bonds (in this case, O–H bonds) of the products. This releases energy. Energy is always released when a chemical bond is formed—no exceptions, ever!

Energy is absorbed in breaking old reactant bonds, and energy is released on forming new product bonds. *Whether the overall reaction releases or absorbs energy depends on the relative magnitudes of these bond-breaking and bond-making steps.* In principle, the following two schemes exist for our reaction:

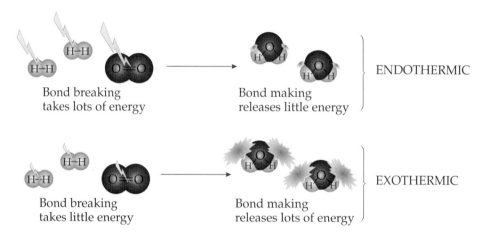

Simply compare the total energy it takes to break reactant bonds to the total energy released when forming product bonds. Whichever is larger in magnitude wins by the difference between them. That is, ΔE_{rxn} is the difference between the energy released and the energy absorbed; it has a + sign if the energy absorbed has a larger magnitude and a − sign if the energy released has a larger magnitude.

When we actually run the reaction above—that is, when we burn hydrogen—we observe that a tremendous amount of heat is released (remember the *Hindenburg!*). Therefore, more energy must be released upon forming the O–H bonds in the product than is required to break the H–H and O=O bonds in the reactants. The exothermic scheme shown above therefore holds true for this reaction.

PRACTICE PROBLEMS

11.7 Consider the hydrogen combustion reaction, where $\Delta E_{rxn} = -482$ kJ/mole. Suppose we also told you that the energy it takes to break the reactant bonds is 1370 kJ/mole. How much energy must be released on forming the bonds in the products?

Answer: To break the reactant bonds, the reactants must absorb 1370 kJ of energy. However, this reaction is exothermic (ΔE_{rxn} is negative), so more than this amount of energy (482 kJ more, to be exact) must be released. Therefore, the amount of energy released on forming the O–H bonds is 1370 kJ + 482 kJ = 1852 kJ.

11.8   A reaction occurs in which 2200 kJ/mole of energy is needed to break the reactant bonds and 2000 kJ/mole of energy is released on forming the product bonds.
(a) Is this reaction exothermic or endothermic? Explain your answer.
(b) What is the value of ΔE_{rxn}? (Get your sign right.)

11.9   A reaction with $\Delta E_{rxn} = -100$ kJ/mole releases 300 kJ/mole upon forming the bonds in the products. How much energy did it take to break the bonds in the reactants?

11.10   Draw reaction energy profiles for the reactions in Practice Problems 11.8 and 11.9. For both, label the gap that is ΔE_{rxn}.

Let's not lose sight of our original goal—gaining insight into the mechanism of a reaction by studying how temperature and concentration affect the rate. Remember, we said that to do this, we first had to know something about the energy changes that take place during a reaction. So far we have looked at the initial total energy of the reactants, the final total energy of the products, and the difference in energy between these two values (ΔE_{rxn}). To reach our goal, we also need to look at the energy changes that take place during the course of the reaction itself, as the reactants are transforming into products.

REACTION RATES AND ACTIVATION ENERGY—GETTING OVER THE HILL 11.3

Of the three reaction types, exothermic, endothermic, and energy-neutral, which do you think will be fastest? Which will be slowest? Since exothermic reactions are downhill in energy and endothermic reactions must proceed uphill, you might be tempted to declare that exothermic reactions are faster. In fact, this was a trick question, because there is no single answer. A particular endothermic, uphill reaction may be faster than many exothermic, downhill reactions. The total energy change, ΔE_{rxn}, does not determine the *rate* of the reaction. The rate of a chemical reaction is determined by the energy changes that occur between the reactants and products, during the process of conversion from reactants to products.

This region is where the steps in the reaction mechanism are taking place. In general, in order for reactants to convert to products, some bonds within the reacting molecules must break, and this always takes energy. In other words, almost all reactions must begin by going uphill and absorbing energy, even if the overall reaction from reactants to products is downhill (exothermic). This creates an energy barrier between the reactants and the products.

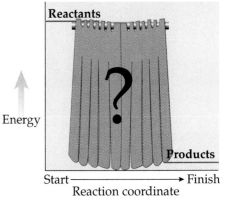

Reaction rate depends on what is going on behind the curtain

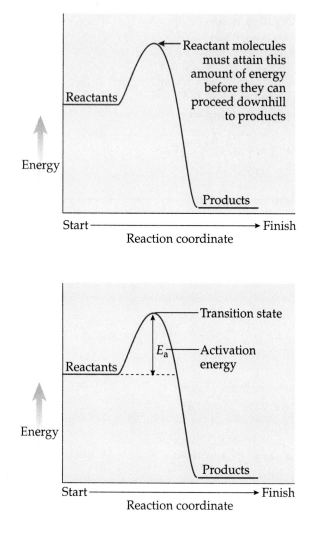

This energy barrier indicates how hard reactant molecules must collide (how much energy the collisions must have) in order for a reaction to occur (bonds begin to break). If reactant molecules collide with less energy than this, they just bounce off each other unchanged. This special amount of energy is called the **activation energy** of the reaction, and it is abbreviated as E_a. The activation energy is the amount of energy that must be absorbed by the reactants before they can proceed to products. It is indicated graphically by the difference in energy between the initial energy of the reactants and the top of the energy barrier.

Each reaction has its own unique value of activation energy. Some reactions have tremendously large barriers to overcome, while other reactions have smaller barriers. At the top of the barrier, the reactant molecules are in a special, high-energy and unstable condition called the *transition state*. We can't actually observe what the reactants look like at the transition state, but we can describe it for a typical one-step reaction. In the **transition state**, enough energy (E_a) has been added to the reactants to have stretched to the breaking point the bonds that are destined to break, but these bonds have not yet completely broken. In addition, the new bonds that will be present in the product have begun to form. This transition state condition occurs right at the top of the E_a barrier. It represents molecules in transition, or on the way from reactants to products.

We will talk a bit more about the transition state shortly, but for now let's keep focused on the activation energy. Try the following WorkPatch, and see if you can plot some reaction energy profiles.

11.3 WORKPATCH　Here are data for three hypothetical reactions:

Reaction	Energy of reactants (kJ/mole)	Energy of products (kJ/mole)	Energy of transition state (kJ/mole)
Case 1	50	60	70
Case 2	20	10	90
Case 3	10	20	90

For each of the three cases:
  (a) Sketch the reaction energy profile.
  (b) Indicate whether the reaction is endothermic or exothermic.
  (c) Write down the value of ΔE_{rxn}.
  (d) Write down the value of E_a.

WorkPatch 11.3 is extremely critical; everything we are about to discuss is based on your being able to answer it. Don't go on until you get it right.

Your WorkPatch plots will help to reveal what ultimately determines the rate of a reaction. In order for a chemical reaction to occur, the reacting molecules must absorb enough energy (E_a) to reach the top of the hill and form the transition state. It is the size of this hill that determines the speed or rate of the reaction.

The higher the activation energy barrier, the slower the reaction.

This is one of the most important fundamental concepts in chemistry. You should also note that it is an inverse relationship. As E_a gets big, the reaction rate gets small, and vice versa. Thus:

A large E_a gives rise to a slow reaction.

A small E_a gives rise to a fast reaction.

Given this inverse relationship, we can tell which of the three reactions plotted in WorkPatch 11.3 would be fastest. Case 1 would be fastest since it has the smallest E_a (20 kJ/mole). It makes intuitive sense that reactions that must pass over large barriers (large E_a) will be slower than reactions with smaller barriers. But we can use our model of reactant molecules colliding with one another to go even further.

PRACTICE PROBLEMS

11.11 Draw two reaction energy profiles, one for an endothermic reaction A and one for an exothermic reaction B. Make the profile for reaction A represent a faster reaction than the profile for reaction B by drawing the relative E_a barriers appropriately.

Answer:

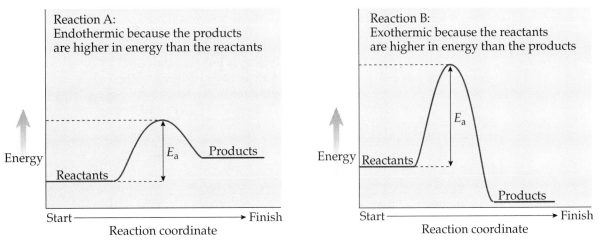

Reaction A is faster than reaction B because it has a smaller activation energy, E_a.

11.12 Draw two reaction energy profiles, one for an endothermic reaction (a) and one for an exothermic reaction (b). Make (b) represent a faster reaction than (a) by drawing the relative E_a barriers appropriately.

11.13 True or false? If reaction X is more exothermic than reaction Y, reaction X must be faster than reaction Y. Whatever your answer, back it up with reaction energy profiles.

TEMPERATURE AND THE COLLISION THEORY OF REACTION RATES

Every chemical reaction has its own characteristic activation energy. This energy depends to a large extent on the number and types of bonds that are broken and formed in the transition state. If many strong bonds are breaking but only a few weak bonds are forming in the transition state, then the activation energy for this reaction will be very high and it will inherently be a very slow reaction. Fortunately, with the understanding we now have as to why such a reaction is inherently slow, we can also suggest some things that can be done to speed it up.

One of the factors that can always be used to manipulate the speed of a chemical reaction is temperature. Temperature is a measure of the average kinetic energy of a collection of atoms or molecules. As such, it is also a relative measure of the average speed of the atoms or molecules in the collection.

We have said that in order to react with one another, reactant molecules must collide with a minimum amount of energy, the activation energy, E_a. For example, ozone (O_3) and nitrogen oxide (NO) molecules will not react unless they collide with sufficient energy, E_a for the reaction:

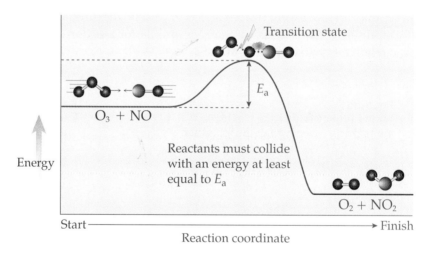

The force (or energy) of each collision depends on the speeds of the colliding molecules. Slow-moving molecules that result in collisions less energetic than E_a do not produce a reaction. They just bounce off each other and go their separate ways, chemically unchanged.

In a gas or in a liquid, molecules are traveling with a range of different speeds, some very fast and others more slowly. Thus, at any given temperature, some molecules are moving fast enough to lead to energetically sufficient collisions. If we now increase the temperature of the system, the average speed of the molecules also increases and the molecules are traveling faster than before. This means that more of the molecules are now capable of collisions with sufficient energy to react. At a higher temperature, more molecules collide (per unit time) with energy greater than E_a than at a lower temperature.

The fraction of fast-moving molecules increases with increasing temperature.

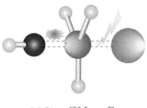

》◐ slow
》》◐ fast
》》》● very fast

Increasing temperature ⟶

Slow⟶Fast   Slow⟶Fast   Slow⟶Fast

This is why an increase in temperature always increases the rate of a chemical reaction (or, conversely, why cooling it slows it down). In fact, a general rule of thumb is that the rate of a reaction will double for every 10°C increase in temperature. Like all rules of thumb, this one is approximate; the actual factor by which a reaction rate increases for a fixed change in temperature depends on the actual magnitude of the activation energy for that reaction.

Another factor that helps determine a reaction's inherent rate has to do with the relative orientation of the colliding molecules. In many reactions, the reactants must collide not only with sufficient energy but also in an appropriate orientation with respect to each other. To see why, let's look again at the substitution reaction in which an OH^- ion substitutes for the Br atom in methyl bromide (CH_3Br):

$$OH^- + CH_3Br \longrightarrow CH_3OH + Br^-$$

In the transition state, the bond between the C and the Br atom is partially broken, and at the same time, the bond between the incoming OH^- and the C is partially formed.

$$HO \cdots CH_3 \cdots Br$$

The transition state

For the reaction to occur, it is not enough for the OH^- and CH_3Br to simply collide with sufficient energy; they must also come together with the correct geometry for the reaction to take place. Many different experiments have shown that the OH^- must approach the methyl bromide from the "back side," directly opposite the Br. Any other approach generally does not lead to this reaction, even if the energy of collision is sufficient to cause the C–Br bond to break.

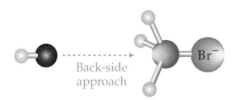

Back-side approach

Br⁻

So we now see that the rate of a reaction depends not only on the number of collisions (per unit time) that occur with sufficient energy, but also on the fraction of those collisions that are of the proper orientation. We can summarize this concept as a mathematical expression:

$$\text{Reaction rate} = \underbrace{\left(\begin{array}{c}\text{Number of collisions}\\ \text{with energy greater than } E_a\\ \text{that occur per unit time}\end{array}\right)}_{\text{Energy factor}} \times \underbrace{\left(\begin{array}{c}\text{Fraction of collisions with}\\ \text{proper orientation}\end{array}\right)}_{\text{Orientation factor}} \quad (11.1)$$

The *orientation factor* is a number between 0 and 1—that is, a fraction. An orientation factor of 1 means that orientation doesn't matter, and any sufficiently energetic collision between reactant molecules will do. For most reactions, the orientation factor is quite small (less than 0.1, meaning less than 10% of the collisions are oriented properly). Since molecules come in a variety of shapes and sizes, each reaction has its own orientation requirements for making collisions effective.

We now have enough information to understand the accepted mechanism for the substitution reaction we have been studying. It is a one-step mechanism in which an OH^- ion in solution collides with the CH_3Br molecule in a back-side approach to form the transition state shown below, which then goes on to form products:

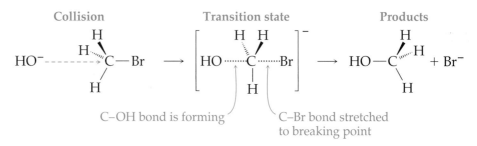

Examine the transition state. Notice the dotted lines that represent both bonds that are in the process of breaking and forming, and also notice how the hydrogen atoms have moved. This transition state lives at the very peak of the activation energy hill. Furthermore, this mechanism reflects the principle that collisions are rarely energetic enough to blast molecules entirely to pieces. The hydrogen atoms remain attached to carbon throughout the mechanism. Only one old bond breaks, and only one new bond forms.

PRACTICE PROBLEM

11.14 Consider our substitution reaction of OH^- with CH_3Br. Imagine it is occurring in a solution where there are 1000 collisions every second between OH^- ions and CH_3Br molecules. Suppose only 10% of these collisions are sufficiently energetic to lead to products. Also, assume that the orientation factor is equal to 0.2.

(a) What does an orientation factor of 0.2 mean?

(b) What is the rate of this reaction? Give your answer in units of number of CH_3OH molecules formed per second (CH_3OH molecules/sec).

Answer:
(a) An orientation factor of 0.2 means that only 20% of all the collisions are oriented properly
   to lead to products.
(b) The reaction rate expression (11.1) says that

$$\text{Reaction rate} = (\text{Number of effective collisions per second}) \times (\text{Orientation factor})$$
$$= (100 \text{ sufficiently energetic collisions/sec}) \times (0.2)$$
$$= 20 \text{ effective collisions/sec}$$

Since every effective collision of proper orientation gives product, then the rate of product
formation is 20 CH₃OH molecules/sec.

Notice from Practice Problem 11.14 that rate is expressed in units of molecules formed per second (molecules/second or, more typically, moles/second). Rate is always expressed as units of something over time; for example, the legal rate of travel on most highways is 55 miles/hour. So thinking about how fast a reaction goes (its rate) is no different from thinking about how fast anything else goes. Now try the remaining practice problems for this section.

PRACTICE PROBLEMS

11.15 Suppose the temperature is increased from that in Practice Problem 11.14 by 10°C.
   (a) What should happen to the number of effective collisions per second? (Give an estimate of the new value based on the rule of thumb given in the text.)
   (b) Why does the number of effective collisions per second change on increasing the temperature?
   (c) What is an approximate value for the new rate of this reaction?

11.16 Repeat Practice Problem 11.14, but this time assume that the orientation factor is only 0.1.

CATALYSTS

As we mentioned previously, a reaction with a large value for its activation energy will be inherently a very slow reaction. When confronted with such a reaction there are a number of things you might do to increase its rate. One is to raise the temperature. But there are limits to just how high a temperature you can use in many circumstances. Sometimes the reactants are sensitive to temperature, and if it gets too high, they will decompose before they can react. Often, a large increase in temperature will cause a number of additional reactions to take place, giving rise to undesired products (called *side reactions* and *by-products*). So temperature control is not a complete solution to inherently slow reactions. To examine what else we might do, let's continue with our substitution reaction. Here is the energy profile for this reaction:

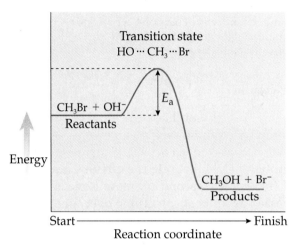

What if you were a reactant molecule and had to make the trip over the E_a hill to become a product molecule? Faced with having to climb the hill, you might consider instead an alternative, lower-energy path.

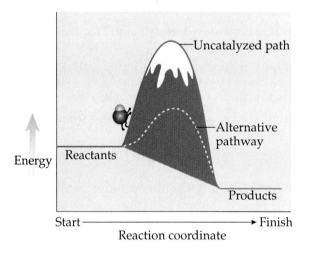

In fact, this is sometimes possible in a chemical reaction. Creating such a new path essentially involves changing the way in which the reaction occurs—that is, changing its mechanism. One way to do this is to use a *catalyst*. A **catalyst** is defined as something that, when added to a reaction, increases the rate of the reaction by lowering its activation energy. In addition, a catalyst does this without being used up in the process, and it usually needs to be present in only trace amounts. A catalyst works by providing a new mechanism—a different way for the reactants to combine and form products. Most importantly, this new pathway has a lower E_a than the one that occurs in the absence of the catalyst.

How can the addition of an extraneous substance (one not needed in the balanced equation for the reaction) change the mechanism? We can demonstrate this using another substitution reaction, the conversion of 2-propanol, $(CH_3)_2CHOH$, into 2-chloropropane, $(CH_3)_2CHCl$. If you try to do this by combining the 2-propanol with Cl^-, expecting a back-side attack and substitution like we saw earlier, you will see that the reaction barely proceeds at all; its rate is practically zero:

$$(CH_3)_2CHOH + Cl^- \xrightarrow{\ \ \oslash\ \ } (CH_3)_2CHCl + OH^-$$

That is because this substitution reaction has a different mechanism with a much larger E_a than the methyl bromide reaction. (Recall that each reaction has its own unique E_a.) The uncatalyzed version of this reaction takes place in two steps, as shown below:

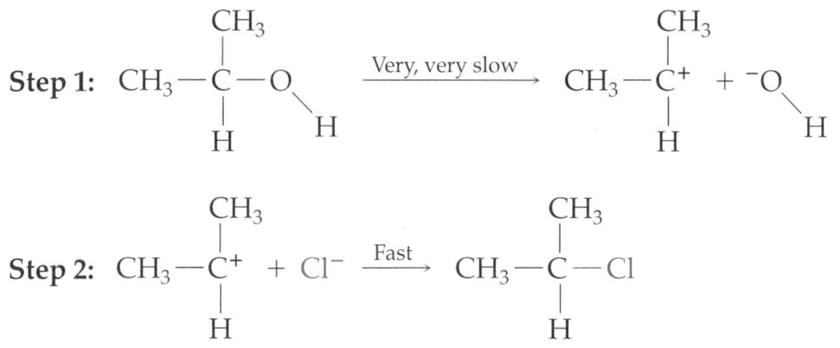

Step 1: $\displaystyle CH_3-\overset{\overset{\displaystyle CH_3}{|}}{\underset{\underset{\displaystyle H}{|}}{C}}-O{\diagdown}_H \xrightarrow{\text{Very, very slow}} CH_3-\overset{\overset{\displaystyle CH_3}{|}}{\underset{\underset{\displaystyle H}{|}}{C^+}} + {}^-O{\diagdown}_H$

Step 2: $\displaystyle CH_3-\overset{\overset{\displaystyle CH_3}{|}}{\underset{\underset{\displaystyle H}{|}}{C^+}} + Cl^- \xrightarrow{\text{Fast}} CH_3-\overset{\overset{\displaystyle CH_3}{|}}{\underset{\underset{\displaystyle H}{|}}{C}}-Cl$

The first step of this mechanism involves just the breaking of the bond between the C and the OH group (without simultaneous formation of a bond to Cl^-), and is the step responsible for the high activation energy of the reaction. However, adding a trace of Zn^{2+} ions and some acid (H^+) as a catalyst dramatically increases the rate of the reaction. Here is the alternative mechanism:

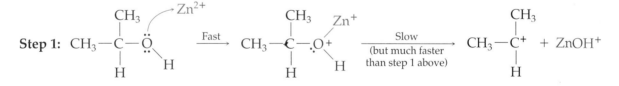

Step 1: $\displaystyle CH_3-\overset{\overset{\displaystyle CH_3}{|}}{\underset{\underset{\displaystyle H}{|}}{C}}-\overset{..}{\underset{..}{O}}{\diagdown}_H \xrightarrow[\ ]{\text{Fast}} \ \ \xrightarrow[\substack{\text{(but much faster}\\ \text{than step 1 above)}}]{\text{Slow}} \ \ CH_3-\overset{\overset{\displaystyle CH_3}{|}}{\underset{\underset{\displaystyle H}{|}}{C^+}} + ZnOH^+$

Step 2: $ZnOH^+ + H^+ \xrightarrow{\text{Fast}} Zn^{2+} + HOH$

Step 3: $\displaystyle CH_3-\overset{\overset{\displaystyle CH_3}{|}}{\underset{\underset{\displaystyle H}{|}}{C^+}} + Cl^- \xrightarrow{\text{Fast}} CH_3-\overset{\overset{\displaystyle CH_3}{|}}{\underset{\underset{\displaystyle H}{|}}{C}}-Cl$

In the first step, the Zn^{2+} ions form a bond with the −OH portion of the propanol molecule. The addition of the Zn^{2+} to the oxygen of the OH group makes it easier to break the C−OH bond. This makes it easier (faster) to form the **carbocation** (pronounced car-bo-cat-ion, the group with the C^+ in it). Once formed, the carbocation reacts rapidly with the Cl^- ions in solution to give the product.

The added trace amount of Zn^{2+} catalyst provides a new, lower E_a mechanism for turning reactants into products by making use of zinc's ability to remove the OH^- group from the propanol in step 1. Notice that in step 2, the Zn^{2+} is regenerated when an H^+ ion removes the OH^- group from $ZnOH^+$. This means that the Zn^{2+} is not used up in the reaction, and is thus available to be used again and again. This is why only a trace amount of Zn^{2+} is necessary to help promote the reaction. Finally, in the third step, the Cl^- forms a bond to the carbocation, giving us the final product.

Catalysts are tremendously valuable to the chemical industry. Reactions that would normally require great amounts of energy (and therefore dollars) due to their large E_a are run for less money and more profit. Even more important, though, is the role of catalysts in biological systems—you would not be alive without them. Many of the biochemical reactions responsible for keeping you alive would run too slowly at normal body temperature without catalysts. Very large molecules called **enzymes** are the major catalysts in living systems. These large biomolecules made of proteins speed up chemical reactions, but in a highly selective fashion—each is designed to interact with only certain specific reactants. Enzymes are selective because of their shape, as we saw in our discussion of sulfanilamide in Chapter 6. They contain specifically shaped pockets or depressions that fit only specific reactant molecules. A reactant molecule that fits into this pocket and undergoes the reaction is called a **substrate**. The pocket itself is known as the enzyme's **active site**. How an enzyme interacts with its substrate is often referred to as the "lock and key" mechanism, where the active site is the lock and the substrate molecule is the only key that will fit it. Once the "key" is inserted into the "lock," intermolecular interactions between the enzyme and the substrate weaken certain substrate bonds and bring the substrate into the proper orientation to react with a lower E_a.

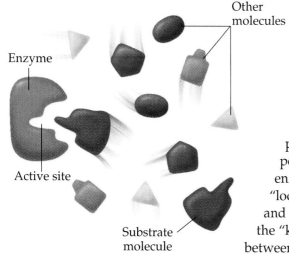

Enzyme

Active site

Other molecules

Substrate molecule

For example, the enzyme sucrase has an active site that conforms to the shape of sucrose, the sugar you sprinkle on cereal or put into coffee. Once a molecule of sucrose is in the active site, this enzyme catalyzes the conversion of sucrose into two simpler sugars, glucose and fructose.

Intermolecular interactions between the enzyme and the sucrose molecule weaken the bond, lowering the E_a for the reaction.

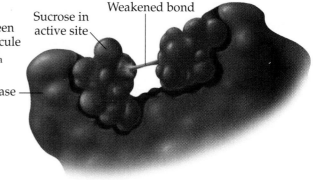

Weakened bond

Sucrose in active site

Sucrase

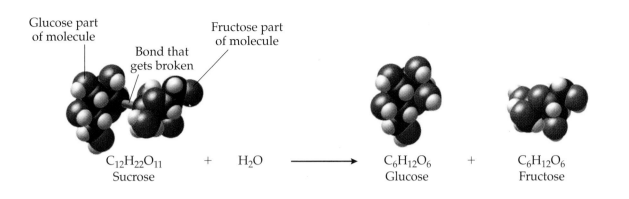

Glucose part of molecule

Bond that gets broken

Fructose part of molecule

$$C_{12}H_{22}O_{11} + H_2O \longrightarrow C_6H_{12}O_6 + C_6H_{12}O_6$$

Sucrose   Glucose   Fructose

This transformation is necessary to keep us alive, since it is glucose, not the sucrose we eat, that is used by our cells to produce the energy needed to carry out other life-sustaining reactions. Many diseases can be traced to hereditary defects that lead to the production of enzymes with malformed active sites, incapable of accommodating the substrate molecules for which they were meant.

The mathematical expression we gave for a reaction rate, equation (11.1), is repeated at the bottom of this page. Explain which factor or factors you think a catalyst can alter.

WORKPATCH **11.4**

Your answer to the WorkPatch should indicate how thorough nature can be when it has to solve a problem.

THE EFFECT OF CONCENTRATION ON RATE 11.4

Up to this point, if we wanted to discuss a reaction mechanism, we simply presented it to you. Now we want to show you one of the primary experimental methods used in determining a reaction's mechanism: varying the concentration of the reactants and observing how the rate changes. To show you how this works, we will begin by discussing why varying the reactant concentrations has an effect on a reaction's rate.

So far everything we have discussed has involved the inherent rate of a reaction. As we have seen, the inherent rate depends on factors that can't be changed (without help from a catalyst) because they are properties of the reacting molecules themselves, properties such as activation energy and orientation. We repeat the relationship we stated earlier below:

$$\text{Reaction rate} = \underbrace{\left(\begin{array}{c} \text{Number of collisions} \\ \text{with energy greater than } E_a \\ \text{that occur per unit time} \end{array} \right)}_{\text{Energy factor}} \times \underbrace{\left(\begin{array}{c} \text{Fraction of collisions with} \\ \text{proper orientation} \end{array} \right)}_{\text{Orientation factor}} \quad \textbf{(11.1)}$$

Now let's modify this expression a little bit by separating the first factor into two parts, as shown below:

$$\text{Reaction rate} = \underbrace{\left(\begin{array}{c}\text{Total number of}\\\text{collisions that occur}\\\text{per unit time}\end{array}\right) \times \left(\begin{array}{c}\text{Fraction of collisions}\\\text{with energy}\\\text{greater than } E_a\end{array}\right)}_{\text{Energy factor}} \times \underbrace{\left(\begin{array}{c}\text{Fraction of collisions}\\\text{with proper}\\\text{orientation}\end{array}\right)}_{\text{Orientation factor}} \quad (11.2)$$

This expression says exactly the same thing as (11.1), since the two new factors are equal to the one energy factor in that relationship. So why did we modify (11.1)? Because (11.2) highlights a very important fact: While the rate of a reaction depends on the inherent properties of E_a and orientation, it also depends on the total number of collisions that are occurring per unit time.

For example, suppose the inherent properties for a reaction at a given temperature are such that only 1% of all collisions are *effective collisions* (collisions that lead to product formation). Let's run a reaction in two equal-sized boxes, 1 and 2, both at the same temperature but with one important difference: In box 1 there will be 20 collisions between reactant molecules per second, while in box 2 there will be 20,000 collisions per second. Don't forget, in both boxes only 1% of the collisions are effective.

Box 1	Box 2
20 collisions/second occur in this box	20,000 collisions/second occur in this box

In which box would product be formed faster? Obviously, in box 2. Even though 1% of the collisions are effective for both boxes, 1000 times more collisions occur each second in box 2. The next question is, "How did we arrange to have 1000 times more collisions per second in box 2?" The answer may be seen by thinking about rush hour traffic. More cars crowding onto a highway will result in more collisions. To get 1000 times more collisions per second in box 2 without changing either the size of the box or the temperature, we simply put many more reactant molecules in box 2 than in box 1.

Both boxes at the same temperature;
the same reaction is occurring in both boxes;
in both boxes, only 1% of collisions are effective.

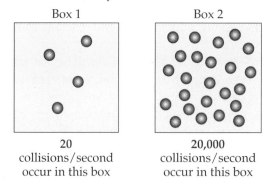

Box 1
20 collisions/second occur in this box

Box 2
20,000 collisions/second occur in this box

Since both boxes have the same size (volume), putting more reactant molecules in box 2 means that it has more reactant molecules per unit volume than box 1. Remember from our discussion of molarity in Chapter 10 that the number of molecules per unit volume is a unit of concentration. And so we arrive at the effect of reactant concentration on rate.

The greater the concentration of the reactants, the greater the total number of collisions occurring per second, and thus the greater the rate of the reaction.

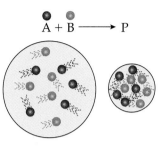

Another way to increase the concentration of molecules is by decreasing the volume of the container they are in. For example, imagine a gas-phase reaction in which A + B $\longrightarrow$ Products. There are 5 A molecules and 5 B molecules in each of two reaction vessels. In which reaction vessel shown would the reaction go faster?

The reaction on the right would have the greater rate, since the concentration of reactants is greater (there are more molecules per unit volume). Being crowded into a smaller volume, the molecules will travel less distance on average before bumping into each other. Since they are moving at the same average speed in both vessels (the temperature is the same), this means there will be more collisions per second in the vessel on the right.

However you do it, whether by injecting more molecules into the system or by squeezing the same number of molecules into a smaller volume, increasing the concentration of the reactants will speed up the reaction by increasing the total number of collisions per unit of time. The bottom line is that the magnitude of the first factor in our mathematical expression for the reaction rate depends on the concentration of the reactants.

$$\text{Reaction rate} = \left(\begin{array}{c} \text{Total number of} \\ \text{collisions that occur} \\ \text{per unit time} \end{array} \right) \times \left(\begin{array}{c} \text{Fraction of collisions} \\ \text{with energy} \\ \text{greater than } E_a \end{array} \right) \times \left(\begin{array}{c} \text{Fraction of collisions} \\ \text{with proper} \\ \text{orientation} \end{array} \right) \qquad (11.3)$$

This factor depends on concentration of reactants. When concentration increases, so does this factor.

These are the inherent rate factors. They depend on bond strengths and molecular shapes (*not* on concentration).

Notice the two kinds of factors: those that are inherent (factors that depend on unchanging characteristics of the reactant molecules, such as bond strengths and molecular shape) and the factor that depends on concentration. Chemists like to simplify equations. In this case, we replace the two unchanging, inherent factors with the symbol k:

$$k = \left(\begin{array}{c} \text{Fraction of collisions with} \\ \text{energy greater than } E_a \end{array} \right) \times \left(\begin{array}{c} \text{Fraction of collisions with} \\ \text{proper orientation} \end{array} \right) \qquad (11.4)$$

Inherent rate factors

k is called the **rate constant** for a reaction. Each different reaction has its own unique value for k. If k is large, then the reaction will be inherently fast. If k is small, then the reaction will be inherently slow. And, as the definition of k in (11.4) clearly shows, the value of the rate constant depends on the value of a reaction's E_a and the shapes (orientations) of the reactant molecules.

11.5 WORKPATCH   Suppose a reaction has a large value for k.

(a) What does this say about its E_a?

(b) What does this say about the orientation requirements?

(c) Is such a reaction inherently fast or slow?

Check your answer to the WorkPatch against ours. If you got it right, then you are catching on to how all the aspects of a reaction—rate, k, E_a, orientation, concentration—are related. So, to simplify our mathematical rate expression (11.3), we simply use the definition of k in equation (11.3) and let k replace the two inherent factors:

$$\text{Reaction rate} = k \left(\begin{array}{c} \text{Total number of collisions} \\ \text{that occur per unit time} \end{array} \right) \qquad (11.5)$$

A final word about the rate constant k. It is called a constant because the things it depends on—a reaction's E_a and the shape of the reacting molecules—do not change. But there are two things that can change the value of a reaction's rate constant. One is adding a catalyst. A catalyst provides a new mechanism with a lower E_a and a different set of orientation requirements, so it will yield a larger value of k than for the uncatalyzed reaction. The second thing that can change the value of k is temperature. As temperature increases, so does k (and vice versa). Why is this so? You already know the answer. Look at equation (11.4). Will either of the two factors of k get larger as the temperature increases? Yes, the first factor will. As temperature increases, molecules move faster, so a larger fraction of collisions will occur with energy greater than E_a, causing the reaction to speed up. Remember the rule of thumb that said a reaction's rate will approximately double for every 10°C increase in temperature? This is true because k nearly doubles.

Calculating the reaction rate is looking pretty simple now. If someone gave us the value of a reaction's rate constant (k), we could plug it into the rate expression (11.5) and calculate the rate. All we would need to do is multiply the rate constant k by the total number of collisions that are occurring per unit time. But now we have a problem. After all, you can't just look at a reaction and know how many collisions are occurring each second. You can't see molecules, and even if you could, you would be hard-pressed to count the billions of collisions that were occurring each second. But don't lose hope. We have seen that the total number of collisions occurring per unit time depends on the concentration of the reactants, and reactant concentration is something that we can measure. What comes next is extremely important: The total number of collisions occurring per unit time is proportional to the product of the concentrations of the reactants, each raised to some power (or exponent) called an **order**.

For the general reaction $A + B \longrightarrow P$,

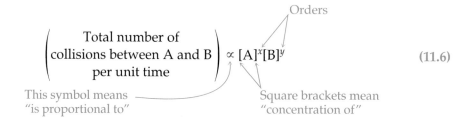

$$\left(\begin{array}{c} \text{Total number of} \\ \text{collisions between A and B} \\ \text{per unit time} \end{array} \right) \propto [A]^x [B]^y \qquad (11.6)$$

Orders

This symbol means "is proportional to"

Square brackets mean "concentration of"

Don't get lost in the details. Just keep in mind the basic concept that if you increase the concentration of the reactants, there will be more collisions between reactant molecules. The orders just fine-tune this idea, and we will discuss them more thoroughly soon. For the moment, the proportionality in expression (11.6) allows us to replace the factor for the total number of collisions that occur per unit time in the rate equation (11.5) with measurable concentrations. We do this below:

$$\text{Reaction rate} = k \underbrace{\left(\begin{array}{c} \text{Total number of} \\ \text{collisions that occur} \\ \text{per unit time} \end{array} \right)}_{} \tag{11.5}$$

We replace this factor with the equivalent expression from (11.6).

The result, for the general equation A + B ⟶ P, is the following:

$$\text{Reaction rate} = k[A]^x[B]^y \tag{11.7}$$

Equation (11.7) is called the **rate law** for the reaction. The rate constant k embodies the inherent rate factors (E_a, orientation requirements), while the concentrations $[A]^x[B]^y$ address the number of collisions that occur. Now, go back and take a quick look at expression (11.2), page 402, to see where we started. The rate law in (11.7) is just a symbolic expression of the same thing, where k represents the second and third terms of (11.2) and $[A]^x[B]^y$ represents the first term.

In principle, it is very easy to write the rate law for any chemical reaction. For example, consider the following general reaction (where the small letters a, b, and c represent the coefficients in the balanced equation):

$$a\,A + b\,B + c\,C \longrightarrow P$$

The rate law for the general chemical reaction is simply

$$\text{Reaction rate} = k[A]^x[B]^y[C]^z \tag{11.8}$$

Note that the concentrations of all the reactants are included, each raised to its respective order (x, y, and z). The general rate law for any reaction is the rate constant k times the concentration of each of the reactants, each raised to some power. This equation is the promised "window" on a reaction's mechanism, the reason we started looking at kinetics in the first place. It has the ability to reveal a reaction's mechanism. But there is one more thing: To look through this "window," you have to understand what the orders are all about and how we find them. We'll turn to them after you do the following practice problems.

PRACTICE PROBLEMS

11.17 Consider the reaction

$$2\,NO + O_2 \longrightarrow 2\,NO_2$$

Write a general rate law for this reaction, using x, y, etc., as orders.

Answer: Rate = $k[NO]^x[O_2]^y$
Note that only the reactants show up in the rate law, not the products. (Rate laws with products in them go beyond the scope of this book.)

11.18 Consider the reaction

$$H_2O_2(aq) + 3\,I^-(aq) + 2\,H^+(aq) \longrightarrow I_3^-(aq) + 2\,H_2O(l)$$

Write a general rate law for this reaction using x, y, etc., as orders.

11.19 Suppose that when the two reactions described in Practice Problems 11.17 and 11.18 are run at similar temperatures and concentrations, the reaction of Practice Problem 11.17 is much, much faster than the reaction of Practice Problem 11.18.
(a) What can you say about the relative sizes of k for each reaction?
(b) What can you say about the relative size of E_a for each reaction?
(c) What are three ways you might speed up the slow reaction to make it faster than the fast reaction?

11.20 Why should the number of collisions occurring per second between reactant molecules have anything to do with their concentration?

11.5 REACTION ORDER

There's still an important part of the rate law that we haven't explained—and that seems to have appeared out of thin air. What do the orders associated with the concentrations of the reactants mean? And where do they come from?

$$\text{Rate} = k[A]^x[B]^y[C]^z$$

Orders

We will start by telling you where we don't get them. In general, the orders do *not* come from the coefficients a, b, c of the balanced equation

$$a\,A + b\,B + c\,C \longrightarrow P$$

So, for example, for the reaction

$$H_2O_2(aq) + 3\,I^-(aq) + 2\,H^+(aq) \longrightarrow I_3^-(aq) + 2\,H_2O(l)$$

(from Practice Problem 11.18), you would be incorrect if you wrote Rate = $k[H_2O_2]^1[I^-]^3[H^+]^2$.

The correct way to find the orders in a rate law is by doing rate experiments, often called *kinetics experiments*. In **kinetics experiments**, we vary the reactant concentrations and observe what happens to the reaction rate. To see how this is done in principle, let's consider the hypothetical gas-phase reaction of A + B ⟶ Products. Suppose we start with a 1 liter vessel filled with 1.0 mole of A molecules and 1.0 mole of B molecules. This means that the starting concentrations of both A and B are 1.0 molar (1.0 mole/L, or 1.0 M). We can run the reaction for 1 minute and then measure the number of product molecules formed. Suppose after 1 minute, 0.1 mole of product is formed. Notice that as time proceeds, the reactants get used up (their concentrations decrease), while the product is formed.

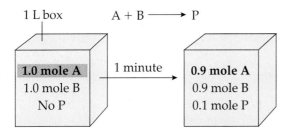

What is the rate of this reaction? The data from the kinetics experiment let us answer this question. Remember, the rate is always something over time. In this case we want to know how fast the concentration of product is changing during a certain time interval, so the "something" is the change in concentration of the product formed at the end of the time interval:

$$\text{Rate} = \frac{\text{Change in concentration of product}}{\text{Time}}$$

At the end of 1 minute, 0.1 mole of product has formed. And, since it is in a 1 liter vessel, the molar concentration of the product after 1 minute is (0.1 mole P)/(1 L) = 0.1 M. (The nice thing about working with a 1 liter vessel is that the moles of each substance present and its molarity are equal.) At the start of the reaction the concentration of the product was 0.0 M, so the change in concentration of product is just the difference between the two concentrations, or 0.1 M. Thus, for the first minute of this reaction, the rate is 0.1 M/minute.

Units of molarity/time may look strange to you, but they are the common units of reaction rate, just as miles/hour are common units for cars. Practice will help you get used to them. For example, suppose we repeat the kinetics experiment, this time doubling the starting concentration of A to 2.0 moles/L (2.0 M). We'll again measure the rate by determining how much product is produced within the first minute of the reaction.

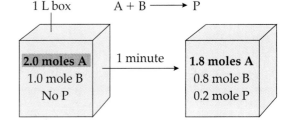

When we measure the amounts of everything in the vessel after 1 minute has passed, we find that 0.2 mole of P has been produced. Therefore, the change in the product concentration is 0.2 M, and the rate is 0.2 M/minute (twice its original value of 0.1 M/minute). So, when we doubled the concentration of A, the rate of product formation also doubled (the reaction was twice as fast). It thus appears that there is a one-to-one correspondence between the concentration of A and the reaction rate (as the concentration of A doubles, so does the rate). For any reactant where this is the case, we assign a value of 1 for its order. *An order of 1 means there is a one-to-one correspondence between that reactant's concentration and the rate.*

$$\text{Rate} = k[A]^1[B]^y$$

An order of 1 means that if you double the concentration of A, the rate also doubles.

We have found the value of order x for reactant A. To find this order, we ran two experiments and never changed the concentration of B; we changed only the concentration of A. This ensured that any change in rate was due to A only.

Now it's time to look at B. We run the reaction with 1.0 mole of A and 2.0 moles of B, doubling only the concentration of B and leaving A alone:

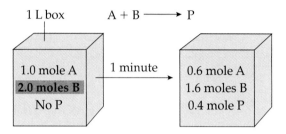

The kinetics data show that after 1 minute, 0.4 mole of product is formed. The rate of product formation is thus 0.4 M/minute, four times the rate we observed when we used only 1.0 mole each of A and B. Doubling the B concentration quadrupled the rate, quite a different result from doubling the concentration of A. How do we get this rate quadrupling effect of B into the rate law? The answer is by making the order of B equal 2. *An exponent of 2 tells you that the rate quadruples when the reactant's concentration is doubled.*

$$\text{Rate} = k[A]^1[B]^2$$

An order of 2 means that if you double the concentration of B, the rate quadruples.

To see the quadrupling effect of the exponent 2, just consider its effect on the concentration of B when it is doubled from 1 M to 2 M:

Concentration of B $[1]^2 = 1$ An exponent of 2 causes
Concentration of B doubled $[2]^2 = 4$ quadrupling upon doubling the concentration

When we double the concentration of B from 1 M to 2 M, the quantity $[B]^2$ increases by a factor of 4. Thus, including the term $[B]^2$ in the rate expression tells us that the rate will quadruple when the concentration of B is doubled.

Orders besides 1 and 2 are often observed in rate laws. For some reactions, a reactant may have an order of zero, which means that changing the concentration of that reactant has no effect on the rate; that is, the rate is not dependent on the concentration of that reactant. Such reactants usually are not included in the rate law for the reaction. It is even possible for reactants to have fractional or negative orders in some reactions, although we will not cover such situations in this book. The important thing to remember is that orders must be discovered by doing kinetics experiments, such as determining the effect of varying the concentration of one reactant on the rate while holding all other reactant concentrations constant.

Now let's switch over from the hypothetical A + B $\longrightarrow$ P to a real reaction and examine some real rate data from kinetics experiments. Consider the gas-phase reaction between nitrogen oxide and hydrogen to give nitrogen and water:

$$2\,NO(g) + 2\,H_2(g) \longrightarrow N_2(g) + 2\,H_2O(g) \tag{11.9}$$

The general rate law for this reaction is

$$\text{Rate} = k[NO]^x[H_2]^y \tag{11.10}$$

To find the orders, a kineticist decides to run this reaction three times, each time measuring the rate of formation of the product N_2 during the first second of the reaction. The data are presented in Table 11.1.

Table 11.1   Rate Data for Reaction (11.9) at 904°C

Experiment	Concentration of NO (mole/L)	Concentration of H₂ (mole/L)	Rate of N₂ formation (M/sec)
1	0.210	0.122	0.0339
2	0.210	0.244	0.0678
3	0.420	0.122	0.1356

First, look at the data for just experiments 1 and 2. For these two, the concentration of NO remains unchanged, but the concentration of H_2 is doubled. Notice that this also causes the rate to double. This means that the order (y) of H_2 is 1. Now let's find the order for the other reactant, NO. To do this, examine the data for just experiments 1 and 3. Why 1 and 3? Because we are looking for two experiments where the concentration of NO changes but the concentration of H_2 remains unchanged. For experiments 1 and 3, the concentration of NO doubles and the rate quadruples. This means that the order (x) for NO is 2. The complete rate law is thus

$$\text{Rate} = k[\text{NO}]^2[\text{H}_2]$$

Notice that when the order is equal to 1 we usually don't write the 1. Since the order for H_2 is 1, we say that the reaction is *first-order with respect to hydrogen*. Since the order for NO is 2, we say that the reaction is *second-order with respect to NO*. Chemists also often quote the **overall order of the reaction**, which is obtained by simply adding together all the individual reactant orders. For this example, the overall order of the reaction is 3 (*third-order*).

———————————————————————————————————————PRACTICE PROBLEMS

11.21 Suppose between experiments 1 and 3 in Table 11.1, the rate remained unchanged. Write the rate law. What is the overall order of the reaction?

Answer: If the rate remained unchanged upon doubling NO, then its order would be 0, meaning the rate does not depend on the NO concentration at all. The rate law would thus be Rate = k[H₂], and the overall reaction order would be first-order.

11.22 For the reaction

$$\text{BrO}_3^- + 5\,\text{Br}^- + 6\,\text{H}^+ \longrightarrow 3\,\text{Br}_2 + 3\,\text{H}_2\text{O}$$

the experimentally determined rate law is

$$\text{Rate} = k\,[\text{BrO}_3^-][\text{Br}^-][\text{H}^+]^2$$

(a) What is the order of this reaction with respect to Br^-?
(b) What is the order of this reaction with respect to H^+?
(c) What is the overall order of the reaction?
(d) What happens to the rate of this reaction when you double the H^+ concentration?

11.23 Consider the following reaction and kinetic data:

$$2\,NO + O_2 \longrightarrow 2\,NO_2$$

Experiment	[NO]	[O$_2$]	Rate of NO$_2$ formation (M/sec)
1	0.015 M	0.015 M	0.048
2	0.030 M	0.015 M	0.192
3	0.015 M	0.030 M	0.096
4	0.030 M	0.030 M	0.384

Write the rate law for this reaction.

There is one topic left to cover in this chapter. What determines the values of these orders? Why are orders sometimes 1, sometimes 2, sometimes something else? When we answer this question, you will understand how kinetics offers a window to a reaction's mechanism.

11.6 WHY REACTION ORDERS HAVE THE VALUES THEY DO—MECHANISMS

It would be wonderful if we could assign the orders in a rate law expression directly from the stoichiometric coefficients in front of the reactants in the balanced equation instead of having to do kinetics experiments. Unfortunately, this is generally not possible because the balanced chemical equation does not show you how the various reactants come together. For example, consider the balanced equation for the combustion of glucose:

$$C_6H_{12}O_6 + 6\,O_2 \longrightarrow 6\,CO_2 + 6\,H_2O$$
Glucose

Judging from the stoichiometric coefficients, the equation seems to imply that one glucose molecule and six oxygen molecules collide at once to form the combustion products CO_2 and H_2O. If this was how the reaction proceeded, then the reaction rate would be Rate = $k[C_6H_{12}O_6][O_2]^6$, but this is *not* correct because this is *not* what happens.

We said earlier that the chance of just three molecules colliding in the same spot was remote. The chance of seven molecules doing this is essentially zero. Most reactions occur in a series of simpler steps—called **elementary steps**— that involve collisions of just two molecules at a time. A reaction mechanism is a series of elementary steps that show exactly how the reactant molecules are converted into products. From the balanced equations for these elementary steps, we *can* assign orders directly from the stoichiometric coefficients— that is, *the balancing coefficients from the balanced elementary steps in a reaction mechanism are the orders!* Unfortunately, reactions don't come with their mechanism printed below them, and we can't observe the actual effective collisions responsible for the reaction, so how do we determine the elementary steps and the overall mechanism? The answer is, we work backward. That is, first we figure out the rate law (get the orders) by doing kinetics experiments, and then we postulate a mechanism that generates the same rate law. When postulating a mechanism, we always have to be sure to use only elementary steps that involve effective two-molecule collisions. If we postulate collisions involving three or more molecules, our postulated mechanism is not very likely to be correct. We are now going to take you through a few examples of this backward process.

We will start by considering the hypothetical reaction $A_2 + B_2 \longrightarrow 2\,AB$. Suppose we go into the lab, do the appropriate kinetics experiments, and find that the rate law is

$$\text{Rate} = k[A_2]^2$$

That is, the reaction is second-order with respect to A_2, and zero-order with respect to B (overall second-order for the entire reaction). At this point you should know what this means in terms of what happens to the rate when we double each of these concentrations. We will call this rate law the **experimental rate law**, because we found it by doing kinetics experiments. Now we are going to postulate a couple of different mechanisms and see if either of them generates this rate law.

Suppose this reaction occurs in one elementary step, involving a two-molecule collision of an A_2 molecule with a B_2 molecule to give 2 AB molecules. We'll call this mechanism I:

One-step mechanism I

$$A_2 \;+\; B_2 \;\longrightarrow\; 2\,AB$$

It is important to realize that this chemical equation represents an elementary step. It tells you exactly how the reaction occurs—that is, it reveals the actual mechanism of the reaction. If $A_2 + B_2 \longrightarrow 2\,AB$ is an elementary step, then the actual reaction must occur by the direct collision of an A_2 and a B_2 molecule with each other.

Since we are assuming that our reaction proceeds through mechanism I—a single, elementary step—we can write the rate law for the reaction directly from the balanced equation. In the balanced equation, the stoichiometric coefficients for the reactants are both 1, and therefore we can write the rate law predicted by this mechanism as

$$\text{Rate} = k[A_2]^1[B_2]^1 = k[A_2][B_2]$$

We will call this the **predicted rate law**, the one generated by the postulated reaction mechanism. Now we need to ask some questions: Is mechanism I correct? Does the reaction actually occur this way? To answer these questions, we compare the predicted rate law to the experimentally determined one:

Experimental rate law: $\text{Rate} = k[A_2]^2$
Predicted rate law: $\text{Rate} = k[A_2][B_2]$

The predicted rate law is different from the experimental one, so our postulated mechanism I cannot be correct. We must try again.

This time, we'll postulate a multistep mechanism, involving a sequence of three elementary two-molecule steps. We'll call this mechanism II:

Three-step mechanism II

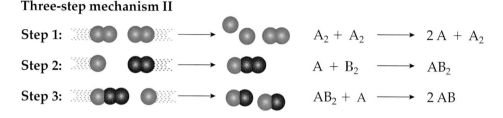

Step 1: $A_2 + A_2 \longrightarrow 2A + A_2$

Step 2: $A + B_2 \longrightarrow AB_2$

Step 3: $AB_2 + A \longrightarrow 2AB$

Multistep mechanisms must fulfill an important requirement: The individual elementary steps must add up to give the overall balanced reaction. If they don't, then the postulated mechanism can't be correct. Below we add the three elementary steps of mechanism II together. When we add elementary steps, we cross out (cancel) anything on the left side of one step that appears in exactly the same form on the right side of any other step.

Adding the three elementary steps together yields the overall net balanced reaction.

Step 1: $A_2 + \cancel{A_2} \longrightarrow 2\cancel{A} + \cancel{A_2}$ Be careful!

Step 2: $\cancel{A} + B_2 \longrightarrow A\cancel{B_2}$ There are 2 A's here! They cancel the 2 separate

Step 3: $\cancel{AB_2} + \cancel{A} \longrightarrow 2AB$ A's on the left.

Overall balanced reaction: $A_2 + B_2 \longrightarrow 2AB$

For any reaction with a mechanism that involves two or more steps, each step in the mechanism is an elementary step but the overall balanced reaction (the sum of the steps) is not. Most of the chemical equations you have seen in this book represent reaction mechanisms that involve more than one elementary step.

As you can see, when we add the three elementary steps, we do indeed get the correct balanced overall reaction. This does not mean, however, that this mechanism is necessarily the correct one. To be correct, it had to do at least this, but it must also do something else. It must generate the experimental rate law. So we now want to compare the rate law given by this mechanism to the experimental rate law and see if there is agreement between them. But mechanism II has three different steps from which we can write three different rate laws! Which one do we use? A reasonable choice would be to use the rate law that we get from the slowest of the three elementary steps. Why? Because the overall reaction can't go any faster than the slowest step in its mechanism. The slowest step in any reaction mechanism is the bottleneck in the whole process and is referred to as the **rate-determining step**. When we experimentally measure the rate of a reaction, it is really the rate of the slowest step that we are determining. Therefore, the experimental rate law should agree with the rate law derived from the slowest elementary step. Since we are postulating the mechanism, it's up to us to choose the rate-determining (slowest) step. We do know what the experimental rate law looks like (Rate = $k[A_2]^2$), however, and we can use that knowledge to help us make an intelligent choice. Look closely at the first step:

Step 1: $A_2 + A_2 \longrightarrow 2A + A_2$

It involves a collision between two A_2 molecules in which the bond in one molecule is broken to give 2 A atoms while the other A_2 molecule remains intact. The rate law for this elementary reaction step can be written directly from the balanced reaction as

$$\text{Rate} = k[A_2][A_2] = k[A_2]^2$$

The first step is second-order in A_2, and B_2 is not involved (it is zero-order in B_2). If we assume that step 1 is the slowest step of the three, then the speed of this step alone will govern the overall speed of the complete reaction, and the rate law for the first step becomes the predicted rate law for the entire reaction. Here are both the predicted rate law and the experimental rate law for comparison:

Experimental rate law: $\text{Rate} = k[A_2]^2$
Predicted rate law: $\text{Rate} = k[A_2]^2$

They match! Does this mean mechanism II is correct? Maybe. However, there are many other mechanisms that we could postulate that would predict a rate law that agrees with the experimentally determined one. Doing kinetics experiments to get the experimental rate law allows us to rule out incorrect mechanisms, but it can never alone prove that a particular mechanism is correct. It can only provide support for the validity of a possible mechanism.

This does not mean, however, that we can never truly know the exact mechanism for a chemical reaction. Kineticists can do additional experiments to further determine the validity of a particular mechanism. For example, in our mechanism II, the first step produces A atoms, which are not among the final products of the overall reaction (AB molecules are the only product of our hypothetical reaction). The second step produces an AB_2 molecule, which is also not a product of the overall reaction. These species, A and AB_2, are called *reaction intermediates*. A **reaction intermediate** is a species that is produced during one step of a reaction and then consumed in a subsequent step on the way toward making the final product. A chemist can try to observe or detect the presence of a particular intermediate as it comes and goes during the course of a reaction. This is usually a difficult task since these intermediates are often present in only very low concentrations and for very short periods of time. However, if intermediates can be detected, and if they match those predicted by the postulated mechanism, then this gives additional support for its validity.

Let's try this again using the substitution reaction we introduced at the beginning of the chapter:

$$OH^- + CH_3Br \longrightarrow CH_3OH + Br^-$$

There are at least two possible mechanisms for this overall reaction. The elementary steps and predicted rate laws are shown for both below:

Mechanism I

$$CH_3Br \xrightarrow{\text{Slow first step}} CH_3^+ + Br^-$$

$$CH_3^+ + OH^- \xrightarrow{\text{Fast second step}} CH_3OH$$

$$\text{Rate} = k[CH_3Br]^1 = k[CH_3Br]$$

Mechanism II

$$CH_3Br + OH^- \xrightarrow{\text{Only one step}} CH_3OH + Br^-$$

$$\text{Rate} = k[CH_3Br]^1[OH^-]^1 = k[CH_3Br][OH^-]$$

Both mechanisms have a rate-determining step. In mechanism I, the first step is clearly labeled as the slow step. It involves the dissociation (loss) of the Br^- ion to give CH_3^+, a reaction intermediate. This is followed by a second step in which a two-molecule collision between the intermediate and an OH^- group give the product. In mechanism II, there is only one step, so it must be rate-determining.

11.6 WORKPATCH

Why is the CH_3^+ cation in mechanism I properly considered to be a reaction intermediate?

Since mechanism I consists of more than one step, the steps must add up to give the proper balanced reaction before we can even consider it as the correct mechanism.

11.7 WORKPATCH

Prove that when adding together the steps in mechanism I, you arrive at the proper balanced equation:

$$OH^- + CH_3Br \longrightarrow CH_3OH + Br^-$$

From the WorkPatch we see that mechanism I is a possibility. Now come the important questions. Which of these two mechanisms is incorrect? Which might possibly be correct? We need to compare the predicted rate law from each mechanism to the experimental rate law. The experimental data from some kinetics experiments are presented in WorkPatch 11.8. See what you can find out.

11.8 WORKPATCH

Rate data for the reaction: $CH_3Br + OH^- \longrightarrow CH_3OH + Br^-$

Experiment	[CH$_3$Br]	[OH$^-$]	Rate of production of CH$_3$OH (M/minute)
1	0.200 M	0.200 M	0.015
2	0.400 M	0.200 M	0.030
3	0.400 M	0.400 M	0.060

Use the above rate data to find the experimental rate law; that is, determine the values of x and y in the rate equation

$$Rate = k[CH_3Br]^x[OH^-]^y$$

Then choose mechanism I or mechanism II as the possible correct mechanism of the reaction.

You should have found that mechanism I is wrong.

11.9 WORKPATCH

Why is mechanism I wrong?

You also should have found that mechanism II may be correct.

11.10 WORKPATCH

Why is mechanism II possibly correct?

Check your answers to these WorkPatches against ours. Once you understand the principles behind the answers, you will understand how the field of kinetics is indeed one of our best windows into a reaction's mechanism. The time spent doing kinetics experiments and determining how rate depends on concentration to arrive at the orders is time well spent. The hard part for even advanced students of chemistry is postulating the mechanisms, because it takes practice and experience to know what is sensible. We have given you a few guidelines. Let's review them here:

1. The elementary steps should never go beyond a two-molecule collision.
2. For a multistep mechanism, the elementary steps must add up to give the proper balanced equation.
3. The experimental rate law can help you get started, since it describes what is on the left side of the arrow of the rate-determining step. That is, whatever mechanism you postulate must generate the experimental rate law.
4. Remember that, after all is said and done, a match between a predicted rate law and the experimental rate law only supports the postulated mechanism. It does not prove it because other mechanisms may also generate the same experimental rate law. To prove any mechanism beyond a doubt, chemists must do many more experiments, which is still one of the most exciting challenges of modern chemistry.

PRACTICE PROBLEMS

11.24 Suppose the experimental rate law for the reaction

$$X_2 + Y_2 \longrightarrow 2\,XY$$

is

$$\text{Rate} = k[X_2][Y_2]$$

A student postulates a multistep mechanism in which the first step is the slow, rate-determining step. For this step, he postulates

$$2\,Y_2 \longrightarrow Y_3 + Y$$

Is such a step possible?

Answer: No. If this were the rate-determining step, the predicted rate law would be Rate = k[Y$_2$]^2, which does not match the experimental rate law.

11.25 Suppose the experimental rate law for the reaction in Practice Problem 11.24 was

$$\text{Rate} = k[Y_2]^2$$

Use the step postulated by the student in Practice Problem 11.24 to write a possibly correct complete mechanism for the reaction. Don't be afraid to be creative with your steps, but keep your collisions to two molecules. In addition:
(a) Show that the steps of the mechanism add to give the proper overall reaction.
(b) Circle all reaction intermediates.
(c) What might you do to prove this mechanism beyond a reasonable doubt?

11.26 Suppose the experimental rate law for the reaction in Practice Problem 11.24 was

$$Rate = k[Y_2]$$

Suppose also that the first step in the mechanism is the slow, rate-determining step. What might this first step look like? [*Hint:* It must generate the experimental rate law. Consider the first step in mechanism I on page 413.]

We began this chapter by saying we would take a close look at the arrow in Reactants $\longrightarrow$ Products, the mechanism of the reaction. We've seen that this simple arrow represents an often complex, multistep transformation process. The payoff for all the hard work of doing kinetics experiments is the insight we get into a reaction's mechanism. Today we have come to realize more than ever that such insight is important, for essentially all the diseases that afflict us, from viral to genetic in origin, operate at the molecular level. An inherited gene with a defect causes the production of a malformed enzyme that no longer interacts with the proper substrate. A virus produces a compound that allows it to attach to and penetrate our cell membranes. If we don't understand these diseases at the molecular level of their reaction mechanism, then we can never hope to cure them.

HAVE YOU LEARNED THIS?

Reaction mechanism (p. 382)

Substitution reaction (p. 383)

Net energy change (ΔE_{rxn}) (pp. 384, 385)

Exothermic reaction (p. 384)

Endothermic reaction (p. 384)

Reaction energy profile (p. 385)

Activation energy (E_a) (p. 392)

Transition state (p. 392)

Catalyst (p. 398)

Carbocation (p. 399)

Enzyme (p. 400)

Enzyme substrate (p. 400)

Active site (p. 400)

Effect of concentration on rate (p. 401)

Rate constant (k) (p. 403)

Orders (p. 404)

Rate law (p. 405)

Kinetics experiments (p. 406)

Overall order of the reaction (p. 409)

Elementary step (p. 410)

Experimental rate law (p. 411)

Predicted rate law (p. 411)

Rate-determining step (p. 412)

Reaction intermediate (p. 413)

CHEMICAL KINETICS

11.27 Describe the following reaction in terms of which bonds must be broken and which bonds must be formed.

$$H_2C{=}CH_2 + 3\,O_2 \longrightarrow 2\,CO_2 + 2\,H_2O$$

11.28 Is it likely that a single collision leads to the breaking of all the bonds that you indicated in Problem 11.27? Explain your answer.

11.29 Is the reaction

$$CH_3I + Cl^- \longrightarrow CH_3Cl + I^-$$

an example of a substitution reaction? Explain.

11.30 Consider the following reaction. Use labeled arrows to indicate which bonds must be broken and which bonds must be formed.

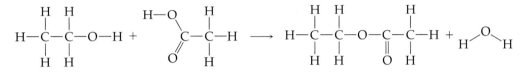

11.31 What is meant by "the mechanism of a chemical reaction"?

11.32 What branch of chemistry concerns itself with the study of rates of reactions and the factors that affect rates?

11.33 What do chemists typically do to indirectly "see" a reaction mechanism?

11.34 What are two benefits that can be gained from understanding a reaction's mechanism?

ENERGY CHANGES AND CHEMICAL REACTIONS

11.35 Compound A converts into compound B; ΔE_{rxn} is -100 kJ/mole. Is compound B at a higher or lower energy level than compound A? By how much?

11.36 Compound A has half as much energy in it as compound B. If compound A converts to B, will this reaction release energy into the surroundings or absorb it? Explain your answer.

11.37 In a chemical reaction, compound A is converted into compound B. In the process, energy is absorbed from the surroundings. Which compound is at a higher energy level? Explain your answer.

11.38 In a chemical reaction, compound C is converted into compound D. In the process, energy is released into the surroundings. Which compound is at a higher energy level? Explain your answer.

11.39 Referring to Problems 11.37 and 11.38, which reaction is exothermic and which is endothermic? Justify your answer, and go on to describe which reaction could be used to supply heat and which could be used to "supply cold" (actually, remove heat).

11.40 A reaction occurs in which a mole of A is converted into a mole of B. A mole of A has an energy content of 20 kJ. A mole of B has an energy content of 60 kJ. Is this reaction exothermic or endothermic? Calculate ΔE_{rxn}.

11.41 What does it mean when ΔE for a reaction is negative?

11.42 The reaction of Problem 11.40 is run in the reverse direction. Is it exothermic or endothermic? Calculate ΔE_{rxn} for the reverse reaction.

11.43 What do we mean by an energy-uphill reaction? What do we mean by an energy-downhill reaction? Include reaction energy profiles with your explanations.

11.44 Judging from the following reaction profile, is the reaction endothermic or exothermic? What is the value of ΔE_{rxn}?

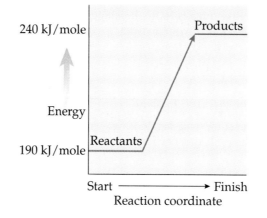

11.45 According to the energy profile in Problem 11.44, the products are higher in energy than the reactants. What must occur during the course of the reaction for this to occur? [*Hint:* The word "surroundings" must appear in your answer.]

11.46 The value of ΔE_{rxn} for exothermic, energy-releasing reactions is always negative. Why is this so?

11.47 For a particular chemical reaction, it takes 800 kJ/mole to break old bonds, and 400 kJ/mole is released on forming new bonds. Calculate ΔE_{rxn} and comment on whether this reaction is exothermic or endothermic. Explain why.

11.48 For a particular reaction, it takes 800 kJ/mole to break old bonds, and ΔE_{rxn} is equal to -800 kJ/mole. How much energy is released into the surroundings upon forming the new product bonds?

REACTION RATES AND ACTIVATION ENERGY

11.49 What do we mean by activation energy?

11.50 What is the relationship between the rate of a reaction and the size of its E_a?

11.51 True or false? An energy-downhill reaction can always be expected to be faster than an energy-uphill reaction. Explain your answer.

11.52 Use reaction energy profiles, complete with E_a's, to show that your answer to Problem 11.51 is correct.

11.53 A reaction is exothermic, with $\Delta E_{rxn} = -40$ kJ/mole, and the transition state is 20 kJ/mole higher in energy than the reactants. Sketch a reaction energy profile consistent with this information, complete with E_a.

11.54 Define transition state.

11.55 For a particular reaction, the reactants are at 30 kJ/mole, the products are at 60 kJ/mole, and the transition state is at 100 kJ/mole. Sketch a reaction energy profile showing both ΔE_{rxn} and E_a. Also, calculate the magnitude of ΔE_{rxn} and state whether this reaction is endothermic or exothermic.

11.56 Would decreasing the size of E_a increase or decrease the rate of a reaction? Explain your choice fully.

11.57 Is the reaction rate directly or inversely related to E_a?

11.58 Would increasing the temperature increase or decrease the rate of a reaction? Explain your choice fully.

11.59 At a given temperature, which reactant molecules can go on to become product molecules?

11.60 Using energy profiles, plot two exothermic reactions that have exactly the same ΔE_{rxn}, but make one reaction substantially faster than the other. Label the plots "fast" and "slow," and explain why you labeled them as you did.

11.61 What does changing the temperature do to the size of the E_a?

11.62 Why might one reaction have a much larger E_a than another reaction?

11.63 What is the rule of thumb for the rate of a reaction with regard to increasing the temperature?

11.64 How can orientation play a role in substitution reactions?

11.65 If there were no orientation requirement for collisions, would reactions be faster or slower than they are? Explain your answer.

11.66 The product

$$\left(\begin{array}{c}\text{Fraction of collisions with} \\ \text{energy greater than } E_a\end{array}\right) \times \left(\begin{array}{c}\text{Fraction of collisions with} \\ \text{proper orientation}\end{array}\right)$$

is often called the "inherent rate" of a reaction. Why?

11.67 What is a catalyst?

11.68 In general, how does a catalyst increase the rate of a chemical reaction?

11.69 In the substitution reaction of Cl^- for OH^- in 2-propanol, explain how Zn^{2+} acts as a catalyst to increase the reaction rate.

11.70 What is the general name given to biological catalysts?

11.71 What is meant by the term "substrate"?

11.72 Explain what is meant by the "lock and key" mechanism when discussing enzymes and their substrates.

11.73 The opening of Chapter 6 told of the development of the first antibiotics. Was this accomplished by making modified versions of the lock or the key? Explain your answer.

11.74 Why are catalysts important to industrial chemical processes? Why are they important to biological chemical processes?

THE EFFECT OF CONCENTRATION ON RATE; ORDERS

11.75 In general, increasing the concentration of a reactant will usually increase the rate of a reaction. Why is this true?

11.76 Why would decreasing the volume of a container in which a gas-phase reaction is taking place speed up the reaction?

11.77 Does the rate constant k increase, decrease, or stay the same when the temperature of a reaction is increased? Explain your choice fully.

11.78 Does the rate constant k increase, decrease, or stay the same when you add a catalyst? Explain your choice fully.

11.79 The rate of a reaction depends both on inherent factors and on concentration. The rate constant k is associated with the inherent factors. What are they?

11.80 A student says that an exothermic reaction will always have a larger rate constant k than an endothermic one, and will thus always be faster. What is wrong with her line of reasoning?

11.81 Given the general form of the rate law,

$$\text{Rate} = k[\text{Reactant 1}]^x[\text{Reactant 2}]^y$$

answer the following questions:
(a) Which part of the rate law reflects those factors that are thought of as the inherent rate of the reaction?
(b) What is the general name for the exponents x and y?
(c) How do we calculate the overall order of a chemical reaction?
(d) Suppose the reaction is second-order with respect to reactant 1 and first-order with respect to reactant 2. What are the values of x and y, and what is the overall order of a reaction with only these two reactants?
(e) Suppose reactant 1 does not appear in the rate law. What does this say about the value of its order? What is the meaning of the value of its order?

11.82 True or false? The orders x, y, and so on, in a rate law are written directly from the appropriate balancing coefficients in the overall balanced equation for the reaction.

11.83 How do we go about determining the orders in a rate law?

11.84 What does it mean when a reaction rate has a first-order dependence on a reactant concentration? Answer in terms of what will happen to the rate when the concentration of that reactant is doubled.

11.85 What does it mean when a reaction rate has a second-order dependence on a reactant concentration? Answer in terms of what will happen to the rate when the concentration of that reactant is doubled.

11.86 What does it mean when a reaction rate has a zero-order dependence on a reactant concentration in terms of how that reactant's concentration affects the rate?

11.87 A reaction A + B $\longrightarrow$ Product is run in a balloon. (Both A and B are gases.) The balloon has a volume of 1 liter and is initially loaded with 1 mole of A and 1 mole of B. The reaction has the rate law

$$\text{Rate} = k[A]$$

The reaction is run again using exactly the same amount of reactants, but this time in a smaller balloon with a volume of only 0.5 liter. How much faster will the reaction proceed in the smaller balloon? Explain your answer.

11.88 Repeat Problem 11.87 for a reaction that has the rate law

$$\text{Rate} = k[A]^2$$

11.89 Repeat Problem 11.87 for a reaction that has the rate law

$$\text{Rate} = k[A][B]^2$$

11.90 Repeat Problem 11.87 for a reaction that has the rate law

$$\text{Rate} = k[A][B]$$

11.91 What do we mean by a kinetics experiment, and how is it tied to the experimental rate law?

11.92 Given the rate data below from a series of kinetics experiments, determine the orders for the following reaction, and state the overall order of the reaction.

$$H_2O_2(aq) + 3\,I^-(aq) + 2\,H^+(aq) \longrightarrow I_3^-(aq) + 2\,H_2O(l)$$

Experiment	$[H_2O_2]$	$[I^-]$	$[H^+]$	Rate (M/sec)
1	0.010 M	0.010 M	0.000 50 M	1.15×10^{-6}
2	0.020 M	0.010 M	0.000 50 M	2.30×10^{-6}
3	0.010 M	0.020 M	0.000 50 M	2.30×10^{-6}
4	0.010 M	0.010 M	0.001 00 M	1.15×10^{-6}

11.93 In a kinetic study of the reaction

$$2\,A(g) + B(g) \longrightarrow P(g)$$

the following rate data were obtained. Write the rate law with proper orders. Give the overall order of the reaction. Finally, state what this problem confirms with respect to the orders and the overall balanced equation.

Experiment	[A]	[B]	Rate of disappearance of A (M/sec)
1	0.0125 M	0.0253 M	0.0281
2	0.0250 M	0.0253 M	0.0562
3	0.0125 M	0.0506 M	0.1124

11.94 In a kinetic study of the reaction

$$2\,ClO_2(aq) + 2\,OH^-(aq) \longrightarrow ClO_3^-(aq) + ClO_2^-(aq) + H_2O$$

the following rate data were obtained. Write a rate law complete with proper values for the orders. What is the overall order of the reaction?

Experiment	$[ClO_2]$	$[OH^-]$	Rate (M/sec)
1	0.060 M	0.030 M	0.024 84
2	0.020 M	0.030 M	0.002 76
3	0.020 M	0.090 M	0.008 28

REACTION MECHANISMS AND RATE LAWS

11.95 Why is it unlikely that the reaction $A + 2\,B + C \longrightarrow P$ occurs in one step?

11.96 True or false? The orders in a rate law are equal to the balancing coefficients in the slowest elementary step in a mechanism.

11.97 What is an elementary step?

11.98 Suppose the reaction in Problem 11.95 did occur in one step. What rate law would predict such a mechanism?

11.99 What is meant by the term "rate-determining step"?

11.100 Why can we ignore other steps and use only the rate-determining step in a mechanism to write the predicted rate law?

11.101 Is it wise to postulate a three-molecule collision as an elementary step in a reaction mechanism? Explain your answer.

11.102 Suppose a postulated reaction mechanism generates a rate law that does not agree with the experimentally determined rate law. What does this say about the postulated mechanism?

11.103 Suppose a postulated mechanism does generate the experimental rate law, but the elementary steps, when added together, do not generate the balanced overall reaction. What can you say about the postulated mechanism?

11.104 Suppose a postulated mechanism does generate the experimental rate law, and, when the elementary steps are added together, the proper overall reaction is generated. What can you say about the postulated mechanism?

11.105 Consider the reaction shown below:

$$H_3C-\underset{\underset{CH_3}{|}}{\overset{\overset{CH_3}{|}}{C}}-Br + H_2O \longrightarrow H_3C-\underset{\underset{CH_3}{|}}{\overset{\overset{CH_3}{|}}{C}}-OH + HBr$$

Kinetics studies reveal a first-order rate dependence on the concentration of the $(CH_3)_3C-Br$ and a zero-order dependence on the concentration of H_2O.
(a) What does this mean for the reaction rate in terms of the $(CH_3)_3C-Br$ concentration? In terms of the H_2O concentration?
(b) Two different mechanisms for this reaction are offered below. Can you rule out either of them? Is either mechanism plausible given the overall balanced equation and kinetic data? Explain your answer fully.

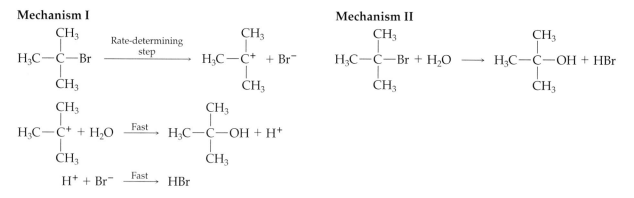

Mechanism I

Mechanism II

11.106 Write the overall balanced chemical equation that goes along with the following mechanism:

Elementary step 1: $Cl_2 \longrightarrow 2\,Cl$

Elementary step 2: $Cl + CHCl_3 \longrightarrow HCl + CCl_3$

Elementary step 3: $Cl + CCl_3 \longrightarrow CCl_4$

11.107 Suppose the first step in the reaction of Problem 11.106 was the rate-determining step. Would the rate law be $k[Cl_2][CHCl_3]$, $k[Cl_2]$, $k[CHCl_3]$, or $k[Cl_2]^2$? Explain.

WORKPATCH SOLUTIONS

11.1 The products of the reaction of methane with oxygen are carbon dioxide and water. If the mechanism completely blasted apart the reactant molecules into atoms, then there would be no particular reason for the atoms to recombine to form just CO_2 and H_2O. Instead, all the other molecules listed in the hint could also be produced. [Chemists describe this situation as a loss of product selectivity. Since the combustion of methane is very selective, producing pre-dominantly CO_2 and H_2O, there must be something about the mechanism of this reaction that steers toward the formation of these products.]

11.2 If ΔE_{rxn} for the forward reaction is downhill and equal to -40 kJ/mole (released), then ΔE_{rxn} for the reverse reaction is uphill and equal to $+40$ kJ/mole (absorbed).

11.3 (a)

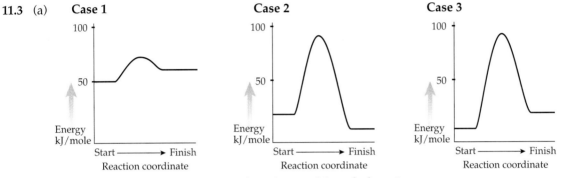

    (b) Case 1 is endothermic; case 2 is exothermic; case 3 is endothermic
    (c) Case 1: $\Delta E_{rxn} = +10$ kJ/mole; case 2: $\Delta E_{rxn} = -10$ kJ/mole;
       case 3: $\Delta E_{rxn} = +10$ kJ/mole
    (d) Case 1: $E_a = 20$ kJ/mole; case 2: $E_a = 70$ kJ/mole; case 3: $E_a = 80$ kJ/mole

11.4 A catalyst would likely have an effect on both the energy factor and the orientation factor. The number of collisions (per unit time) with energy greater than E_a would increase because a catalyst lowers E_a for the reaction. The fraction of collisions with proper orientation would also likely change because a catalyst changes the reaction mechanism by offering the reactants a new mechanism that involves the catalyst.

11.5 (a) A reaction with a large rate constant k has a small E_a (they are inversely related).
    (b) The orientation requirements are probably not very restrictive. If many orientations will do, then this will speed up a reaction, making k large.
    (c) A reaction with a large k is inherently fast. The larger the rate constant k, the faster the reaction.

11.6 The CH_3^+ cation is a reaction intermediate for two reasons. First, it is not one of the final products of the reaction (this reaction produces only CH_3OH and free Br^-). Second, it is produced in one step and then consumed in a subsequent step.

11.7 Adding together the two elementary steps, and crossing out identical entries on the left and right, we get the proper overall balanced reaction:

$$CH_3Br \longrightarrow \cancel{CH_3^+} + Br^-$$
$$+ \ \cancel{CH_3^+} + OH^- \longrightarrow CH_3OH$$
$$\overline{CH_3Br + OH^- \longrightarrow CH_3OH + Br^-}$$

11.8 Comparing experiments 1 and 2, where just the CH_3Br concentration doubles, so does the rate. Thus, $x = 1$. Comparing experiments 2 and 3, where just the OH^- concentration doubles, so does the rate. Thus, $y = 1$. This gives an experimental rate law of Rate $= k[CH_3Br][OH^-]$. Of the two mechanisms, only mechanism II predicts this rate law, so mechanism II is a valid choice.

11.9 Mechanism I is wrong because it predicts a rate law that doesn't match the experimental rate law.

11.10 Mechanism II is possibly correct because its predicted rate law matches the experimentally determined one. Of course, other mechanisms might also do this, so we are not sure that mechanism II is correct. We only know that it may be correct.

Dynamic Equilibrium

Chapter 12

Chemical Equilibrium

DYNAMIC EQUILIBRIUM—MY REACTION SEEMS TO HAVE STOPPED! 12.1

From an economic viewpoint, the reaction used for the production of sulfuric acid, H_2SO_4, is one of the most important chemical reactions. Millions of tons of sulfuric acid are produced by the chemical industry each year. In fact, no other chemical is produced in greater amounts. One major use of sulfuric acid is in the manufacture of many of the fertilizers on which worldwide food supply is so dependent.

Suppose you decided to go into the sulfuric acid production business. The first thing you would need to do is learn how to make it. A common method, called the *contact process*, involves three reactions. The first reaction involves the partial burning of elemental sulfur to make sulfur dioxide (SO_2) gas:

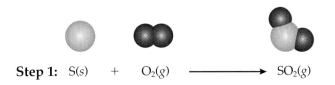

Step 1: $S(s)$ + $O_2(g)$ ⟶ $SO_2(g)$

This is called a partial burning because the sulfur dioxide can be made to combine with additional oxygen and form sulfur trioxide (SO_3) gas in the second step of the contact process:

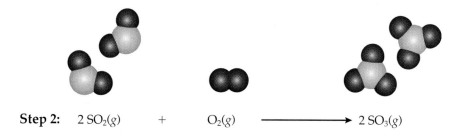

Step 2: $2\,SO_2(g)$ + $O_2(g)$ $\longrightarrow$ $2\,SO_3(g)$

The reaction is usually run over the surface of a metal catalyst, such as platinum, to speed it up. Finally, when the SO_3 gas is allowed to combine with water, sulfuric acid is produced (the third step):

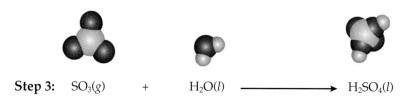

Step 3: $SO_3(g)$ + $H_2O(l)$ $\longrightarrow$ $H_2SO_4(l)$

Imagine getting into this business and attempting to turn your knowledge about chemistry into some personal profit. First you would need to find someone willing to invest money in a production facility. Having accomplished this, you would start with step 1, burning sulfur to produce SO_2 gas. Then it would be on to step 2, converting the SO_2 into SO_3. Following the instructions of the balanced reaction, $2\,SO_2(g) + O_2(g) \longrightarrow 2\,SO_3(g)$, you would combine 2 moles of the SO_2 gas with 1 mole of O_2 gas to produce an expected 2 moles of SO_3 gas. You even throw in some platinum to act as a catalyst to help the reaction run faster. Time is money, after all.

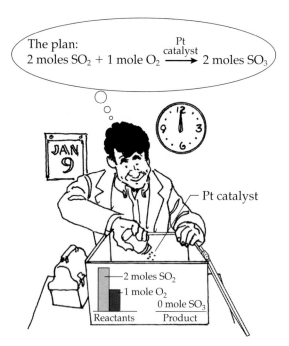

You start the reaction, step out for lunch, and return one hour later in eager anticipation of finding 2 moles of profit-yielding SO_3 gas. What you find instead is disturbing. After a full hour, only 1.8 moles of the SO_3 product has formed, while 0.2 mole of SO_2 and 0.1 mole of O_2 remain unreacted.

It appears that the reaction is not yet complete. Perhaps the reaction is just not as fast as you thought it would be. Maybe it needs more time to finish (to reach completion). The weekend is coming up, so you decide to check it again on Monday. Imagine your horror when you return on Monday, accompanied by your investor, to find absolutely no change. Your attempts to get it going again by shaking the reaction vessel do little to impress either the reaction or your investor. What went wrong?

"My reaction seems to be stuck!"

Your first mistake was going into business before reading this chapter. You see, your reaction hasn't really stopped. It has reached what chemists call *dynamic equilibrium*. In a chemical reaction, **dynamic equilibrium** occurs when the rate of the forward reaction becomes precisely equal to the rate of the reverse reaction.

We introduced the concept of a reaction running in the reverse direction briefly in Chapter 11. In principle, all chemical reactions can proceed in both the forward and reverse directions simultaneously. In chemical reactions where the reverse reaction makes a significant contribution, the reaction is often referred to as **reversible**. We indicate this property of a reaction by the use of a double arrow, one pointing to the right (forward) and the other pointing to the left (reverse):

$$2\,SO_2(g) + O_2(g) \underset{\text{Reverse}}{\overset{\text{Forward}}{\rightleftarrows}} 2\,SO_3(g)$$

When the forward rate (the rate at which reactants become products) exactly equals the reverse rate (the rate of converting products back into reactants), the reaction will appear to have stopped because the amounts of reactant and product present will no longer change with time. Hence the "equi" in the word equilibrium, for equal rates. We use the word "dynamic" to describe the equilibrium because it reminds us that although the reaction appears to have stopped, it is really still going on. It's just that the reverse reaction is now proceeding at exactly the same rate as the forward reaction.

$$2\,SO_2(g)\ +\ O_2(g)\ \underset{\text{Reverse}}{\overset{\text{Forward}}{\rightleftarrows}}\ 2\,SO_3(g)$$

Both rates are the same at equilibrium.

Dynamic equilibrium

Therefore, even though the reaction to make SO_3 seemed to go almost (90%) to completion and then "stop," we now know that it really didn't stop at all. At the 90% completion point it reached dynamic equilibrium—or equilibrium, for short. Since the forward and reverse reactions are now occurring at equal rates, the total number of reactant and product molecules present will no longer change. Every time two SO_2 molecules and an O_2 molecule react, producing two new SO_3 molecules, two SO_3 molecules react in the reverse direction to replace them.

Of course, not all reactions reach equilibrium at the 90% point. For example, consider the reaction that produces nitrogen monoxide from nitrogen and oxygen:

$$N_2(g)\ +\ O_2(g)\ \rightleftarrows\ 2\,NO(g)$$

If we load a reaction vessel with 1 mole each of N_2 and O_2 according to the balanced reaction and adjust the temperature to 2027°C, we find that the reaction appears to stop when only a small amount of product has formed:

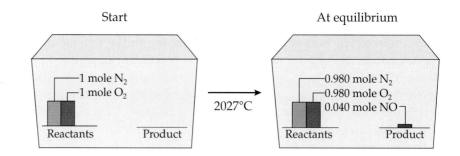

At equilibrium, the vessel contains mostly reactants. You could wait forever and this situation would not change. The reaction reaches equilibrium at only 2% completion.

As our examples demonstrate, some reactions reach equilibrium near completion (to the right side of the reaction, toward products) and some reach it close to the beginning (to the left side of the reaction, toward reactants). Some also reach it near the middle.

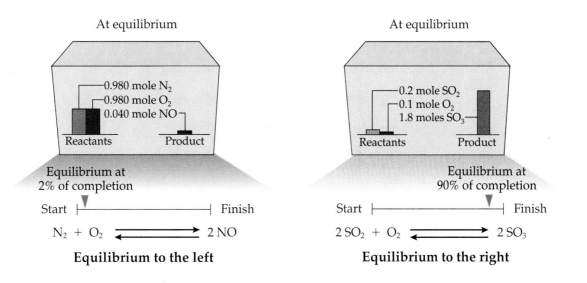

Equilibrium to the left

Equilibrium to the right

There are, however, two special cases. First, some reactions reach equilibrium so close to the beginning (left, reactant side) that practically no product is formed. These reactions appear to never even get started before they "stop." Since there is practically no product formed at equilibrium, we say that the reaction essentially does not occur:

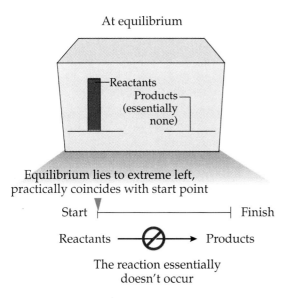

The other special case is just the opposite—reactions that reach equilibrium so close to the end (right, product side) that practically no reactant remains. For this special case we use only one arrow pointing from left to right instead of a double arrow and say that the reaction goes to completion.

It is important to realize that our treatment of these two special cases as "one-way" reactions is oversimplified. All reactions, given the correct conditions, reach equilibrium. It's just that for these special cases, the

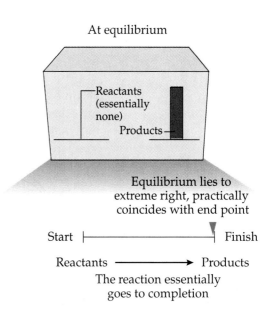

equilibrium lies so far to the left or to the right that we can ignore the concept of equilibrium since it is of no practical importance.

12.1 WORKPATCH

Below are five possibilities for the hypothetical reaction $A \longrightarrow B$.

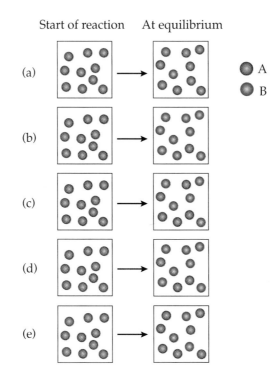

Which of the following best describes each of these possibilities?
1. The equilibrium lies to the left.
2. The equilibrium lies to the right.
3. The equilibrium lies in the middle.
4. The reaction essentially doesn't occur.
5. The reaction essentially goes to completion.

Now, let's return to our SO_3 production facility and our worried investor, who's concerned that the reaction has "stopped" before all the reactants were used up. You calmly explain that it has not stopped. It just appears to have stopped because the rates of the forward reaction and reverse reaction have become equal. It has reached equilibrium. "Great!" he exclaims, "What about the other reactions you plan to run? Will they reach equilibrium? Where will they reach equilibrium? Can't you do something about all this!!??" He's got some good questions that need answering (especially since he's holding the checkbook).

Questions about equilibrium

1. Why do reactions reach equilibrium?
2. What determines where a reaction reaches equilibrium?
3. Is there a way to shift the position of a reaction's equilibrium?

We will have to answer these questions if we are going to stay in business. We'll start with the first question, "Why do chemical reactions reach equilibrium?" But before we answer this, try these practice problems concerning the basic concept of dynamic equilibrium.

12.1 Consider a point during a chemical reaction at which the rate of the forward reaction is less than the rate of the reverse reaction.
   (a) Is the reaction at equilibrium at that point?
   (b) Which way does the overall reaction appear to be running?

Answer:
(a) No, the reaction is not at equilibrium at that point, since the forward and reverse rates are not equal.
(b) The reaction appears to be running to the left (in reverse, converting products into reactants).

12.2 From a practical point of view, why would you want a reaction's equilibrium to lie very far toward the right?

12.3 Why can we ignore equilibrium for reactions that go to completion?

WHY DO CHEMICAL REACTIONS REACH EQUILIBRIUM? 12.2

We have defined chemical equilibrium as the point along the reaction coordinate at which the forward and reverse rates become equal. Why does this happen? Why do the forward and reverse rates of a chemical reaction eventually become equal? To understand this we are going to have to look at what influences reaction rates in general. We discussed the effects of temperature and concentration in the previous chapter, where we saw that the rate of a reaction could be expressed mathematically by a rate law:

$$\text{Reactants} \xrightarrow{\text{Forward}} \text{Products}$$

$$\text{Rate}_{\text{forward}} = k_{\text{forward}}[\text{Reactants}]^{\text{orders}}$$

Assume that we are running a reaction at some constant temperature. Since the rate constant k_{forward} doesn't change at a given temperature, the only way for the rate of the reaction to change is for the concentration of one or more of the reactants to change. But this is exactly what happens during a chemical reaction. As a reaction proceeds in the forward direction, reactants are consumed, thus decreasing their concentration. Consequently, the rate of the forward reaction decreases as the reaction proceeds. The reaction "speedometers," or rate meters, shown below illustrate this behavior for most reactions:

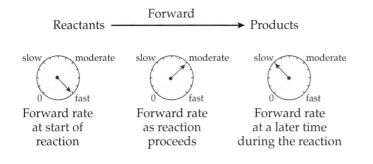

Reactants $\xrightarrow{\text{Forward}}$ Products

| Forward rate at start of reaction | Forward rate as reaction proceeds | Forward rate at a later time during the reaction |

If we now consider the above reaction to be reversible, then we must also take into account the reverse reaction. This reverse reaction would have its own rate law with its own rate constant $k_{reverse}$, which is different from $k_{forward}$:

$$\text{Reactants} \xleftarrow{\quad \text{Reverse} \quad} \text{Products}$$

$$\text{Rate}_{reverse} = k_{reverse}[\text{Products}]^{orders}$$

Now, looking closely at this rate law, we can see that the rate of the reverse reaction depends on the concentration of the products raised to some power. But at the very beginning of a reaction, there are no products, so the product concentration is zero. Therefore, the reverse reaction starts off with a zero rate. As the forward reaction progresses and product is formed, the product concentration increases, and the speed of the reverse reaction begins to increase. Whereas the forward reaction starts fast and slows down, the reverse reaction starts slow and speeds up. At some point the two rates will become equal. At that point, equilibrium is established and no further change in concentrations will occur.

Let's see what happens in our SO_3 reaction. Return to the start of the reaction when 2 moles of SO_2 and 1 mole of O_2 are combined. We will call the very beginning of the reaction *time zero*. At time zero, no SO_3 product has been produced; only reactants are present. Now, consider both possible reactions, the forward and the reverse:

Time 0

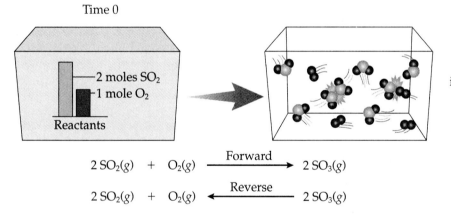

At time 0, no product is present, so collisions occur only among reactant molecules.

$$2\,SO_2(g) \;+\; O_2(g) \xrightarrow{\quad \text{Forward} \quad} 2\,SO_3(g)$$

$$2\,SO_2(g) \;+\; O_2(g) \xleftarrow{\quad \text{Reverse} \quad} 2\,SO_3(g)$$

At time zero, only the reactants SO_2 and O_2 are present, so only collisions between reactant molecules occur. This means that the forward reaction should be relatively fast at time zero. For the reverse reaction to occur, two SO_3 molecules must collide with each other, but there are no SO_3 molecules present at time zero, so the reverse reaction has a zero rate at time zero. As time passes, however, some of the SO_2 and O_2 reactants get used up, decreasing their concentration and thus decreasing the rate of the forward reaction. At the same time, SO_3 product is being produced, building up its concentration. The presence of SO_3 enables the reverse reaction to run, and as the concentration builds up, more and more collisions occur between SO_3 molecules, increasing the rate of the reverse reaction. We illustrate this with reaction rate meters:

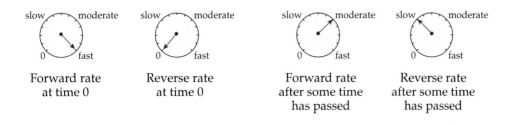

As time passes, the forward reaction keeps slowing down while the reverse reaction keeps speeding up. Eventually, a point is reached where their rates become equal. At that moment, reactants are being turned into product and product is being turned into reactants at exactly the same rate. The reaction appears to stop. Dynamic equilibrium has been reached.

At equilibrium

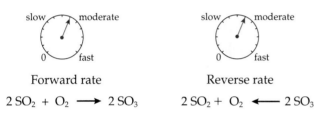

You should realize that we could just as easily have approached this reaction from the other direction. That is, we could have filled the reaction vessel with just product (2 moles of SO_3). At time zero, the reverse reaction would be quite fast while the forward reaction rate would be zero:

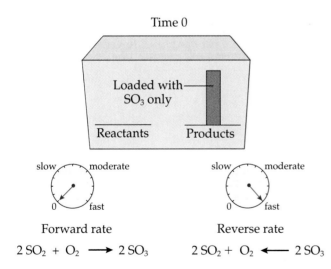

Given enough time, the reverse reaction would slow down and the forward reaction would speed up, until the two rates were once again equal. Thus, because the forward and reverse reaction rates change with time, we are guaranteed that at some point, the forward and reverse reactions will have the same rate.

See if you can answer the following WorkPatch.

12.2 WORKPATCH

Given a reaction vessel with the initial conditions shown below, which set of rate meters (a)–(e) describes the situation best? Explain your choice, and then explain why each of the other choices is incorrect.

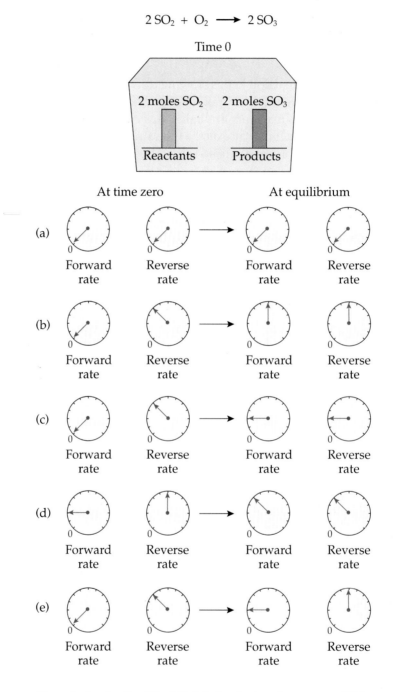

One choice in the WorkPatch is obviously wrong. Did you catch it? That choice was (e), the last one. At equilibrium, the forward and reverse rates must be equal—by definition.

Under the proper conditions, all chemical reactions will come to equilibrium. The question is where will equilibrium occur—toward the left? The right? The middle? Is there something that can tell us?

THE POSITION OF EQUILIBRIUM—THE EQUILIBRIUM CONSTANT K_{EQ} 12.3

Up to now we have described where equilibrium occurs in qualitative terms, saying things like a reaction's equilibrium lies to the left or to the right, or a reaction practically goes to completion. Such descriptions are fine until you need to worry about the exact amounts of reactants and products that will be present in your reaction vessel when equilibrium is reached. To quantify the position at which a particular reaction will come to equilibrium, chemists use a quantity known as an **equilibrium constant, K_{eq}**.

The equilibrium constant for a reaction can be derived from considering the reaction rates for both the forward and reverse reactions. This is easy to show only for the simplest of reactions, those that occur in one elementary step. We will derive K_{eq} for a hypothetical one-step reaction $A \rightleftharpoons B$, and then present the general result for all other, more complex reactions.

The general definition of dynamic chemical equilibrium is

Forward rate = Reverse rate

Therefore, when our simple one-step reaction is at equilibrium, we may write

$$k_f[A] = k_r[B]$$

where k_f represents $k_{forward}$, and k_r represents $k_{reverse}$.

Now try your hand at a little algebraic manipulation (Chapter 2) by answering the WorkPatch below.

Rearrange the equation $k_f[A] = k_r[B]$ algebraically, solving it for the ratio k_f/k_r.

WORKPATCH 12.3

Were you able to do this? (If not, see our solution at the end of the chapter.) The correct result is

$$\frac{k_f}{k_r} = \frac{[B]}{[A]}$$

— Concentration of the product at equilibrium
— Concentration of the reactant at equilibrium

During the course of the reaction, the concentrations of A and B will be changing until equilibrium is reached. But at a given temperature, the values of k_f and k_r will never change (they are rate *constants*). This rearrangement of the basic equilibrium equation collects all the constant terms (the rate constants) on one side of the equals sign and leaves the variable quantities (the concentrations) on the other.

Since k_f and k_r are constants, the ratio k_f/k_r is also a constant for the reaction. Chemists therefore replace the ratio k_f/k_r with a new symbol, K_{eq}, the equilibrium constant:

$$K_{eq} = \frac{k_f}{k_r} = \frac{[B]}{[A]}$$

— Concentration of the product at equilibrium
— Concentration of the reactant at equilibrium

We have arrived at a mathematical expression for the equilibrium constant for a one-step reaction. It is the ratio of the forward rate constant to the reverse rate constant. It is also the ratio of the concentrations of the product to reactant at equilibrium. These must be the equilibrium concentrations since we got to the above equation by insisting that the reaction was at equilibrium (remember, we started by saying $k_f[A] = k_r[B]$). Thus, in the above expression for K_{eq}, [B] is the concentration of product B at equilibrium, and [A] is the concentration of reactant A at equilibrium. This equation suggests two ways to determine the value of K_{eq} for a reaction. One is to determine the values of k_f and k_r for the one-step reaction in question; then directly calculate K_{eq} by dividing k_f by k_r. The second way is to go into the lab, let the reaction A $\rightleftharpoons$ B come to equilibrium, measure the concentrations of A and B, and then calculate [B]/[A]. You will get exactly the same result for K_{eq} either way.

Deriving the equilibrium expression for a multistep reaction is beyond the scope of this text. However, we'll give you the result of such a derivation because you will be needing it. For a multistep reaction that has the general form

$$a\,A + b\,B \;\rightleftharpoons\; c\,C + d\,D$$

where a, b, c, and d stand for the stoichiometric coefficients of the balanced reaction, and the double arrow shows that the reaction should be treated as an equilibrium reaction,

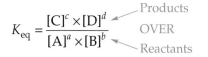

$$K_{eq} = \frac{[C]^c \times [D]^d}{[A]^a \times [B]^b} \quad\begin{array}{l}\nearrow\text{Products}\\ \text{OVER}\\ \searrow\text{Reactants}\end{array}$$

This general definition of the equilibrium constant, which applies to multistep as well as one-step reactions, is clearly related to the definition we derived for a one-step reaction. In both cases, terms for the concentration of products are divided by terms for the concentration of reactants. The product concentrations are always in the numerator and the reactant concentrations are always in the denominator. When there is more than one reactant or product, their concentrations are multiplied by each other. In addition, each concentration is raised to a power equal to its stoichiometric coefficient in the balanced equation for the reaction. For example, our balanced SO_3 reaction and its K_{eq} expression are shown below.

$$2\,SO_2 + O_2 \;\rightleftharpoons\; 2\,SO_3$$

$$K_{eq} = \frac{[SO_3]^2}{[SO_2]^2 \times [O_2]}$$

PRACTICE PROBLEMS

12.4 Write the equilibrium constant expression for the following reaction:

$$CH_4(g) + 2\,H_2S(g) \;\rightleftharpoons\; CS_2(g) + 4\,H_2(g)$$

Answer: $K_{eq} = \dfrac{[CS_2][H_2]^4}{[CH_4][H_2S]^2}$

12.5 Write the equilibrium constant expression for the following reaction:

$$H_2(g) + I_2(g) \rightleftharpoons 2\,HI(g)$$

12.6 Write the equilibrium constant expression for the following reaction:

$$Fe^{3+}(aq) + SCN^-(aq) \rightleftharpoons Fe(SCN)^{2+}(aq)$$

So in practice, it is easy to write the equilibrium constant expression for a reaction. And, as we said earlier, it is relatively simple to find the value of K_{eq} for a reaction. All you have to do is go into the lab, determine the concentrations of all the reactants and products at equilibrium, plug them into the K_{eq} expression, and evaluate it. For example, let's go back to our SO_3 reaction. We'll start this time with a 1.0 liter box containing 2.0 moles of the reactant SO_2 and 1.0 mole of the reactant O_2. When the reaction reaches equilibrium, the 1 liter box contains 0.2 mole of SO_2, 0.1 mole of O_2, and 1.8 moles of SO_3. Thus, at equilibrium, the concentrations of SO_2, O_2, and SO_3 are 0.2 M, 0.1 M, and 1.8 M, respectively.

To arrive at a value for K_{eq}, we write the equilibrium expression for the reaction, plug in the equilibrium concentrations, and do the required arithmetic:

−0.2 mole SO_2
−0.1 mole O_2
1.8 moles SO_3
Reactants   Product
At equilibrium

$2\,SO_2(g) + O_2(g) \rightleftharpoons 2\,SO_3(g)$

$$K_{eq} = \frac{[SO_3]^2}{[SO_2]^2 \times [O_2]} = \frac{[1.8\ M]^2}{[0.2\ M]^2 \times [0.1\ M]} = 810$$

We get a value of 810 for the equilibrium constant. Remember that this constant applies to the reaction only as long as the temperature doesn't change. Any time we come across this reaction at equilibrium at this temperature, then no matter what kind of vessel it is in, and no matter how much reactant was initially put into the vessel, if we measure the concentrations of all the reactants and products and plug them into the K_{eq} expression, we will get 810. This number is a fundamental constant for this reaction. For example, what if we changed the starting amounts of reactants to 5 moles of SO_2 and 3 moles of O_2? Remember, our box has a volume of 1 liter, so the starting concentrations at time zero are 5 M for SO_2 and 3 M for O_2. If you actually did this experiment, you would get the equilibrium situation shown below:

$$2\,SO_2 + O_2 \rightleftharpoons 2\,SO_3$$

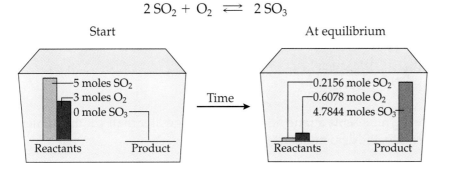

Start

—5 moles SO_2
—3 moles O_2
0 mole SO_3
Reactants   Product

Time

At equilibrium

—0.2156 mole SO_2
—0.6078 mole O_2
4.7844 moles SO_3
Reactants   Product

Changing the starting concentrations has given rise to a different set of equilibrium concentrations. Let's calculate the equilibrium constant for this second case, using our new equilibrium concentrations:

$$K_{eq} = \frac{[SO_3]^2}{[SO_2]^2 \times [O_2]} = \frac{[4.7844]^2}{[0.2156]^2 \times [0.6078]} = 810$$

So we see that K_{eq} is still equal to 810. The value of K_{eq} for a reaction is constant unless the temperature changes.

PRACTICE PROBLEMS

12.7 Consider the reaction

$$H_2(g) + I_2(g) \rightleftharpoons 2\,HI(g)$$

At the start of the reaction, the concentrations of the reactants and products are: $[H_2] = 0.100$ M; $[I_2] = 0.100$ M; $[HI] = 0.000$ M. At 427°C, the equilibrium concentrations of the reactants and products are: $[HI] = 0.158$ M; $[H_2] = 0.021$ M; $[I_2] = 0.021$ M. Calculate the value of K_{eq} for this reaction.

Answer: 57

12.8 Given the information in Practice Problem 12.7, does the equilibrium for the HI reaction lie to the left or to the right? Explain your choice.

12.9 Consider the reaction:

$$CH_4(g) + 2\,H_2S(g) \rightleftharpoons CS_2(g) + 4\,H_2(g)$$

The equilibrium concentrations of the reactants and products are: $CS_2 = 6.10 \times 10^{-3}$ M; $H_2 = 1.17 \times 10^{-3}$ M; $CH_4 = 2.35 \times 10^{-3}$ M; $H_2S = 2.93 \times 10^{-3}$ M. Calculate the value of K_{eq} for this reaction.

Now that we know how to determine the value of K_{eq} for a reaction, we can ask the question, "What exactly does the equilibrium constant tell us about the equilibrium?" For any reaction, if we neglect the stoichiometric balancing coefficients that are used as exponents, the equilibrium constant is essentially a ratio of product concentrations over reactant concentrations at equilibrium,

$$K_{eq} = \frac{[Products]}{[Reactants]}$$

Since K_{eq} represents a ratio, its value will be greater than 1 if the numerator is larger than the denominator—that is,

When the numerator is larger than the denominator

$$\frac{[Products]}{[Reactants]} > 1$$

Therefore, in general, any reaction with a K_{eq} value substantially greater than 1 will have significantly more products present at equilibrium than reactants (the equilibrium will lie toward the right). Looking back at our SO_3 example, you can see that this is so. We found that K_{eq} is large ($K_{eq} = 810$), and the reaction indeed reached equilibrium close to completion (90%). In general, we say that if K_{eq} is larger than 1, the equilibrium lies to the right. The larger K_{eq} is for a reaction, the farther the equilibrium lies to the right. A reaction with a K_{eq} that is very large (greater than 1×10^5, or 10,000) goes essentially to completion.

What if a reaction has a small value (less than 1) for its equilibrium constant? Mathematically, the way to make K_{eq} small is for the denominator (the concentration of the reactants at equilibrium) to be larger than the numerator (the concentration of the products at equilibrium):

When the denominator is larger than the numerator

$$\frac{[Products]}{[\mathbf{Reactants}]} < 1$$

This means, in general, that any reaction with a small K_{eq} (less than 1) will have significantly more reactants than products at equilibrium (the equilibrium will lie toward the left). The smaller K_{eq} is for a reaction, the farther the equilibrium lies to the left. A reaction with a K_{eq} that is very small (less than 1×10^{-6}, or 0.000 001) essentially does not proceed. The $N_2 + O_2 \rightleftharpoons 2\,NO$ reaction we looked at earlier has such a left-lying equilibrium point, as you'll see in the WorkPatch.

Calculate the value of the equilibrium constant K_{eq} for the NO reaction given the following data:

WORKPATCH 12.4

$$N_2(g) + O_2(g) \rightleftharpoons 2\,NO(g)$$

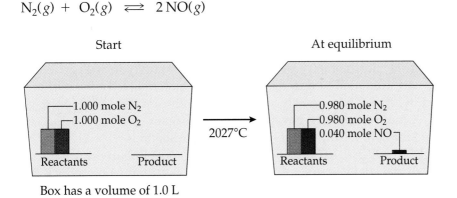

Box has a volume of 1.0 L

Looking at the equilibrium box in the WorkPatch leaves no question that this equilibrium lies to the left, since mostly reactants are present. What did you get for the value of K_{eq}? Was it less than 1? You may remember that we found earlier that this reaction goes only to 2% completion.

Finally, there are those reactions that reach equilibrium somewhere near the middle—around 50% completion. For these reactions, K_{eq} will have a value around 1, since the numerator and denominator of the K_{eq} expression will have roughly the same value.

Chemists have run many reactions, allowed them to reach equilibrium, measured the concentrations of products and reactants present, and then calculated their equilibrium constants. Tables of K_{eq} values for tens of thousands of reactions are recorded in many chemical reference books and scientific papers. Some of these reactions are presented in the following WorkPatch. See if you can match up the reactions with their measured equilibrium constants.

12.5 WORKPATCH   For each of the following reactions, write the K_{eq} expression:

(a) $HCl(aq) \longrightarrow H^+(aq) + Cl^-(aq)$ Reaction goes to completion

(b) $CH_3COOH(aq) \rightleftharpoons H^+(aq) + CH_3COO^-(aq)$ Equilibrium lies far to the left

(c) $PCl_5(g) \rightleftharpoons PCl_3(g) + Cl_2(g)$ Equilibrium lies to the left

(d) $H_2O(g) + CH_4(g) \rightleftharpoons CO(g) + 3 H_2(g)$ Equilibrium lies near the middle

Notice that next to each reaction we have written some qualitative information describing the position of equilibrium. Now decide which of the following equilibrium constants goes with each of the above reactions.

(1) $K_{eq} = 1.0 \times 10^{-5}$ (2) $K_{eq} = 2.2 \times 10^{-2}$
(3) $K_{eq} = 4.6$ (4) $K_{eq} > 10^6$

Are you getting a feel for the meaning of the size of an equilibrium constant? Check your WorkPatch answers against ours.

Finally, we want to emphasize an important point about reactions with very large or very small equilibrium constants. When K_{eq} is extremely large, then we know that the equilibrium lies so far to the right that the reaction will go essentially to completion. This usually means we can neglect the equilibrium altogether. One such reaction is $2 CO + O_2 \rightleftharpoons 2 CO_2$, for which $K_{eq} = 1.9 \times 10^{11}$. For this reaction, if you react 2 moles of CO with 1 mole of O_2, following the stoichiometry of the balanced equation, you will get 2 moles of CO_2 with essentially no unreacted starting materials left over. On the other hand, when K_{eq} is extremely small, the equilibrium lies so far to the left that the reaction essentially never starts. Once again, this means that we can ignore the equilibrium altogether. One such reaction is $N_2 + O_2 \rightleftharpoons 2 NO$, for which $K_{eq} = 2.3 \times 10^{-9}$ at 25°C. If you mix 1 mole of N_2 with 1 mole of O_2 at 25°C (typical daily temperature), you will get essentially no product. No reaction occurs. Even at 2027°C, $K_{eq} = 1.7 \times 10^{-3}$ (as you should have found out in WorkPatch 12.4). Be thankful that this is the case, since these gases make up most of the atmosphere you breathe. If K_{eq} for this reaction were large, there would be a substantial amount of NO in the atmosphere at equilibrium. The NO would sear your lungs, and you would die a rather painful death, as NO is a powerful oxidizing agent.

If we take the reaction to make NO and turn it around, then we have a reaction for the decomposition of NO into N_2 and O_2:

$$2\,NO \rightleftharpoons N_2 + O_2$$

WORKPATCH 12.6

(a) Write the K_{eq} expressions for the reaction above, and for the reaction to form NO ($N_2 + O_2 \rightleftharpoons 2\,NO$).

(b) How is the K_{eq} expression for the NO forming reaction related to the K_{eq} expression for the NO decomposition reaction?

(c) The text gives the value of K_{eq} for the NO forming reaction at 25°C as 2.3×10^{-9}. What is the value of K_{eq} for the decomposition reaction at 25°C? To which side does the equilibrium lie?

Did you discover something about how the equilibrium constants for the forward and reverse versions of a reaction are related? Be sure to check your answer at the end of the chapter before going on.

Finally, knowing the value of K_{eq} can do more than tell you where a reaction's equilibrium lies. As we said at the beginning of this section, knowing the value of K_{eq} quantifies the position of equilibrium. While this book does not show you how, the equilibrium constant for a reaction can be used to calculate the actual concentrations of the reactants and products that will be present when equilibrium is attained. For example, remember the sulfuric acid production business we tried to start at the beginning of this chapter? If our assistant had researched the SO_3 reaction adequately, he could have discovered that the K_{eq} for this reaction is 810. Then, using this information, he could have calculated that we would end up with 0.1 mole O_2 and 0.2 mole SO_2 left over. Instead, we found out the hard way.

DISTURBING A REACTION ALREADY AT EQUILIBRIUM—LE CHATELIER'S PRINCIPLE 12.4

Let's return to our SO_3 production reaction. When we last left our assistant, he was panicked that his reaction would not go to completion.

What can he do about it? Suddenly, our assistant has an idea! He'll simply suck all the leftover SO_2 and O_2 out of the box with a device designed to remove just these gases. Then we'll have pure SO_3 in the box—right?

—0.2 mole SO_2
—0.1 mole O_2
1.8 moles SO_3
Reactants   Product
At equilibrium
$2\,SO_2 + O_2 \rightleftharpoons 2\,SO_3$

On the road to disaster. . . . .

Only SO₃ left

He quickly does this and then runs down the hall, retrieves our investor, and proudly shows him the box. Imagine our assistant's panic when he sees the following results:

An equilibrium that is disturbed fights back. . . . .

0.186 mole SO$_2$
0.093 mole O$_2$
1.614 moles SO$_3$
Reactants   Product

What happened? A cruel joke? No! A lesson learned. Before the SO$_2$ and O$_2$ were sucked out, the reaction was at equilibrium. The rate of the forward reaction equaled the rate of the reverse reaction. When all the SO$_2$ and O$_2$ were removed, the rate of the forward reaction dropped down to zero. At that moment (time zero), only the reverse reaction was running with an appreciable rate, merrily turning SO$_3$ back into SO$_2$ and O$_2$:

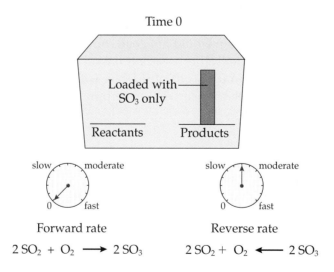

Time 0

Loaded with SO$_3$ only

Reactants   Products

slow   moderate   slow   moderate

0   fast   0   fast

Forward rate   Reverse rate

$2\,SO_2 + O_2 \longrightarrow 2\,SO_3$ $2\,SO_2 + O_2 \longleftarrow 2\,SO_3$

The rates of the forward and reverse reactions were no longer equal. Disturbing the equilibrium by removing SO$_2$ and O$_2$ from the mixture brought about a nonequilibrium situation. As SO$_3$ gets used up, the reverse reaction begins to slow down. As SO$_2$ and O$_2$ are produced, the forward reaction "turns back on" and begins to pick up speed. Eventually, the forward and reverse rates again become equal, and equilibrium is reestablished.

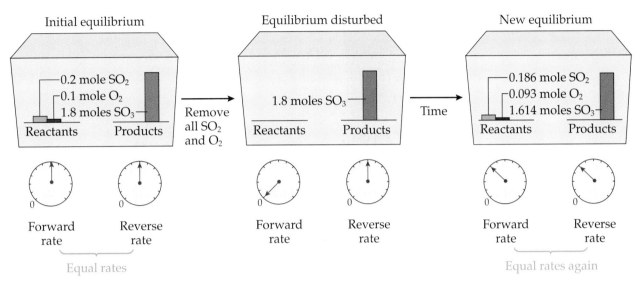

Notice that the amounts of reactants and products for the initial and reestablished equilibria are not exactly the same. When an equilibrium situation is disturbed by adding or removing reactant or product, the reestablished equilibrium concentrations will always be slightly different. However, there is something that does remain the same.

Calculate the values of K_{eq} for both equilibrium situations (before and after the disturbance).

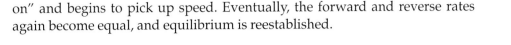

WORKPATCH **12.7**

As we stated previously, the value of the equilibrium constant K_{eq} does not change unless the temperature is changed. Both equilibrium situations described above still have the exact same value of K_{eq}, as you found out in the WorkPatch. This gives us a general result.

When you disturb a reaction at equilibrium by changing the concentration of one or more of the reactants or products, the reaction will return to equilibrium by driving the altered concentrations partially back toward what they were. This principle was summed up many years ago (1884) by the French chemist Henri Louis Le Chatelier:

Le Chatelier's Principle: If you disturb a reaction at equilibrium, the reaction will return to equilibrium by shifting in such a direction as to partially undo the disturbance.

Another way to say this is that any time you disturb an equilibrium, the reaction will partly undo what you did. If you remove a reactant or product, the reaction will shift in a direction to produce more of what was removed. If you add a reactant or product, the reaction will shift in a direction to remove it.

In our scenario, both SO_2 and O_2 were removed. The reaction responded by shifting to make more SO_2 and O_2, partly replacing what was removed.

Disturbing the equilibrium . . .

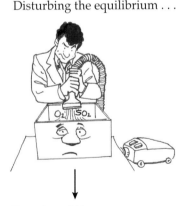

Reaction shifts to restore equilibrium . . .

$2\,SO_2 + O_2 \longleftarrow 2\,SO_3$

Since SO_2 and O_2 are on the left side of the chemical equation, we say that the reaction shifted to the left.

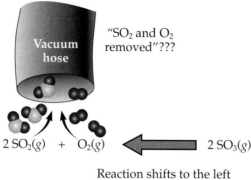

$2\,SO_2(g)$ + $O_2(g)$ $2\,SO_3(g)$

Reaction shifts to the left
to produce some more reactants

Try the following problem concerning the $2\,SO_2 + O_2 \rightleftharpoons 2\,SO_3$ reaction.

12.8 WORKPATCH

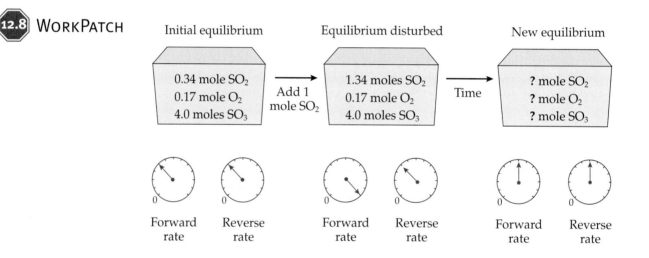

Which of the following is the best choice for the concentrations at the new equilibrium?

(a) 0.34 mole SO_2
    0.17 mole O_2
    4.0 moles SO_3

(b) 1.04 moles SO_2
    0.02 mole O_2
    4.3 moles SO_3

(c) 1.34 moles SO_2
    0.17 mole O_2
    4.0 moles SO_3

(d) 1.68 moles SO_2
    0.02 mole O_2
    4.3 moles SO_3

Your answer to the WorkPatch should be consistent with a shift to the right. Since we added SO_2, a reactant, the reaction should shift to consume the added SO_2 and produce more SO_3. In the process, some additional O_2 will also be consumed.

In other words, if you disturb an equilibrium by adding something to the left side, the reaction will respond by shifting to the right to partially remove it. The addition of SO_2 speeds up the forward reaction while having no immediate effect on the rate of the reverse reaction. The net result is consumption of some of the added SO_2 plus some O_2. As they are consumed, the forward reaction slows. As more SO_3 is produced, the reverse reaction speeds up. Eventually, the forward and reverse reaction rates once again become equal. Equilibrium is reestablished.

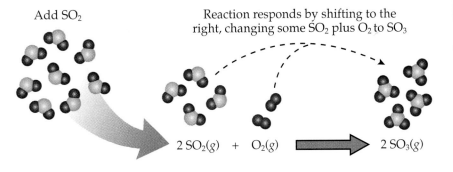

Add SO_2

Reaction responds by shifting to the right, changing some SO_2 plus O_2 to SO_3

$$2\,SO_2(g) \;+\; O_2(g) \;\longrightarrow\; 2\,SO_3(g)$$

PRACTICE PROBLEMS

12.10 In WorkPatch 12.8, why are the rates of the forward and reverse reactions greater at the new equilibrium than they were at the initial equilibrium?

Answer: In both cases, the forward rate equals the reverse rate. This must be true, since both situations are at equilibrium. However, when SO_2 is added, some of it is converted to SO_3, so both of these substances are present at higher concentration than before. Therefore, collisions between molecules are more frequent. In general, this will increase the rate of a reaction.

12.11 If you add some SO_3 to a vessel in which the reaction $2\,SO_2 + O_2 \rightleftarrows 2\,SO_3$ is at equilibrium, which way will the reaction shift?

12.12 Suppose you added N_2 rather than SO_3 to the vessel in Practice Problem 12.11. Which way would the reaction shift?

RESPONSE OF EQUILIBRIUM TO CHANGES IN TEMPERATURE 12.5

Another way to disturb a reaction at equilibrium is to change its temperature by adding or removing heat from it. We can use Le Chatelier's principle to predict how a reaction at equilibrium will respond to changes in temperature. To understand why temperature affects equilibrium, you must recall some things we discussed previously. In the last chapter you learned that some reactions give off heat (exothermic, ΔE_{rxn} is negative),

Reactants $\rightleftarrows$ Products + Heat

Exothermic reaction:
ΔE_{rxn} is negative

while other reactions absorb heat (endothermic, ΔE_{rxn} is positive):

Heat + Reactants $\rightleftarrows$ Products

Endothermic reaction:
ΔE_{rxn} is positive

Notice that we wrote the word "heat" directly into the reactions. For an exothermic reaction heat is a product (heat is released), so it is written on the right side. For an endothermic reaction heat is consumed (absorbed), just like a reactant, so it is written on the left side. We normally don't include heat in a chemical equation, since the sign of ΔE_{rxn} already tells us whether heat is released or absorbed by the reaction. Nevertheless, when deciding how a reaction at equilibrium will respond to changes in temperature, we recommend that you include heat on the appropriate side to make it easier to apply Le Chatelier's principle. As you are about to see, how a reaction at equilibrium responds to a temperature disturbance depends on whether the reaction is exothermic or endothermic.

A good example is our reaction to make SO_3. Since $\Delta E_{rxn} = -197$ kJ/mole for this reaction, it is exothermic, and we write heat on the right side of the reaction:

$$2\,SO_2(g) + O_2(g) \rightleftharpoons 2\,SO_3(g) + \text{Heat}$$

Exothermic reaction:
$\Delta E_{rxn} = -197$ kJ/mole

As we revisit our SO_3 production facility, we see that it is quite cold inside. That's because the investor has stopped paying the bills, including the heating bill. Shivering in the cold, our beleaguered assistant can hardly believe his eyes. The box is now almost completely filled with pure product (SO_3), and almost none of the starting materials (SO_2 and O_2). After all his efforts to purify the product by shaking the box and by removing the unreacted SO_2 and O_2 had failed, the reaction "fixed itself!" Does this reaction have a mind of its own?

Our assistant forgot about Le Chatelier's principle. As the room cooled, so did the reaction vessel. This removed heat from the reaction vessel. As far as the reaction at equilibrium was concerned, it was being disturbed. Something was removing heat. It "fought back." To partially undo the disturbance, the reaction shifted in a direction to produce heat. For an exothermic reaction like this one, this means shifting to the right to produce (release) more heat:

Approximately 2 moles SO_3

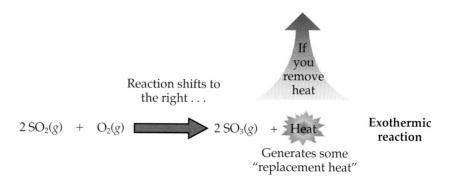

Reaction shifts to the right . . .

If you remove heat

$$2\,SO_2(g) + O_2(g) \longrightarrow 2\,SO_3(g) + \text{Heat}$$

Exothermic reaction

Generates some "replacement heat"

The opposite is also true. If you add heat to an exothermic reaction at equilibrium by raising the temperature, the equilibrium will shift in the direction that consumes some of the added heat. This means shifting to the left. (An exothermic reaction running in reverse is endothermic.)

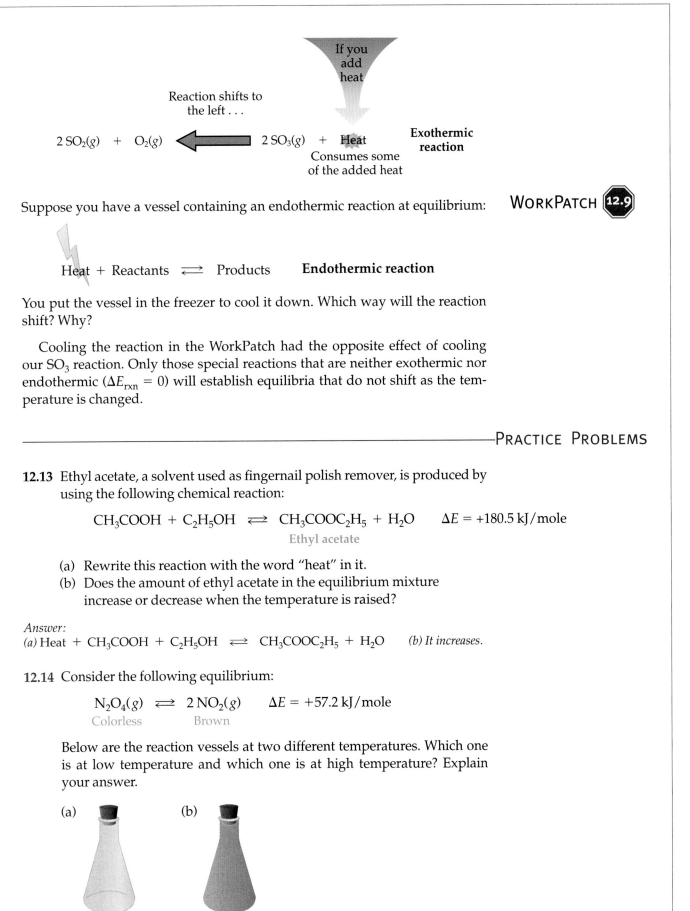

Suppose you have a vessel containing an endothermic reaction at equilibrium: WORKPATCH 12.9

Heat + Reactants ⇌ Products   **Endothermic reaction**

You put the vessel in the freezer to cool it down. Which way will the reaction shift? Why?

Cooling the reaction in the WorkPatch had the opposite effect of cooling our SO_3 reaction. Only those special reactions that are neither exothermic nor endothermic ($\Delta E_{rxn} = 0$) will establish equilibria that do not shift as the temperature is changed.

───────────────────────────────────PRACTICE PROBLEMS

12.13 Ethyl acetate, a solvent used as fingernail polish remover, is produced by using the following chemical reaction:

$$CH_3COOH + C_2H_5OH \rightleftharpoons CH_3COOC_2H_5 + H_2O \quad \Delta E = +180.5 \text{ kJ/mole}$$
Ethyl acetate

(a) Rewrite this reaction with the word "heat" in it.
(b) Does the amount of ethyl acetate in the equilibrium mixture increase or decrease when the temperature is raised?

Answer:
(a) Heat + CH_3COOH + C_2H_5OH ⇌ $CH_3COOC_2H_5$ + H_2O *(b) It increases.*

12.14 Consider the following equilibrium:

$$N_2O_4(g) \rightleftharpoons 2 NO_2(g) \quad \Delta E = +57.2 \text{ kJ/mole}$$
Colorless   Brown

Below are the reaction vessels at two different temperatures. Which one is at low temperature and which one is at high temperature? Explain your answer.

(a)   (b)

12.15 Consider the following reaction:

$$2\,CO(g) + O_2(g) \rightleftharpoons 2\,CO_2(g) \qquad \Delta E = -563.5 \text{ kJ}$$

(a) Rewrite this reaction with the "heat" in it.
(b) Which way will the reaction shift if the temperature is raised? Explain your answer.

Prior knowledge of Le Chatelier's principle would have benefited our relationship with our investor. The exothermic SO_3 reaction should have been run in the cold to drive it further toward products. Of course, there would have been a trade-off. In the previous chapter we saw that lowering the temperature will always slow down a chemical reaction. Thus, it would take longer to reach equilibrium, and time is money in the business world. Using a catalyst could help to get around this problem. In Chapter 11 we saw that a catalyst can be used to speed up a chemical reaction. The use of a catalyst does not change the position at which the equilibrium occurs, because a catalyst speeds up the forward and reverse reactions by exactly the same amount. There are other ways around a reaction with an undesirable equilibrium position, and chemical engineers spend much of their time applying these methods to industrial chemical processes. The moral is, that even in the business world, it pays to know some of the fundamental concepts of chemistry.

HAVE YOU LEARNED THIS?

Dynamic equilibrium (p. 427)   Equilibrium constant, K_{eq} (p. 435)
Reversible reaction (p. 427)   Le Chatelier's principle (p. 443)

DYNAMIC EQUILIBRIUM

12.16 Define equilibrium in terms of the forward and reverse reaction rates.

12.17 What do we mean when we say that a reaction is reversible?

12.18 When we said that our SO_3 reaction was "stuck," we meant that it had stopped. Did it really stop, and how does your answer help explain why we can also say "dynamic equilibrium"?

12.19 What do we mean by "the position of a reaction's equilibrium"?

12.20 In terms of the position of the equilibrium point, where is a reaction that goes to completion?

12.21 In terms of the position of the equilibrium point, where is a reaction that appears not to occur?

12.22 Sometimes reactions are drawn with a double arrow pointing in both directions instead of a single arrow going from reactants to products. What does the double arrow mean?

12.23 What surprises might be encountered if a chemist ran a reaction without ever worrying about equilibrium?

WHY DO CHEMICAL REACTIONS REACH EQUILIBRIUM?

12.24 If a reaction vessel is loaded with reactants, the forward reaction is fast at first but gradually slows down. Explain why this is so.

12.25 When a reaction vessel is first loaded with reactants, the reverse reaction has a rate of zero. Explain why this is so.

12.26 After a reaction vessel is loaded with reactants and the reaction begins, the rate of the reverse reaction gradually speeds up. Explain why this is so.

12.27 Suppose a reaction vessel is loaded only with the products of a reaction. Which would be fastest at the moment after loading, the forward or the reverse reaction? Explain your answer.

12.28 Is it possible for the reaction in Problem 12.27 to attain equilibrium? Explain your answer.

12.29 A beaker of water left in a room will slowly evaporate until the beaker is dry. However, place that same beaker in a sealed box and the water level in the beaker will drop a bit but then stop and remain filled to that level. Is the latter case an example of equilibrium? Explain your answer.

12.30 Write the definition of equilibrium in terms of the very general rate laws

$$\text{Rate}_{\text{forward}} = k_{\text{forward}}[\text{Reactants}] \quad \text{and} \quad \text{Rate}_{\text{reverse}} = k_{\text{reverse}}[\text{Products}]$$

12.31 Is the following behavior possible for a reaction run at constant temperature? Explain your answer.

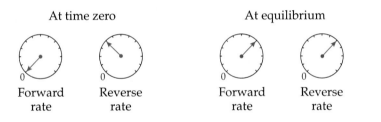

At time zero		At equilibrium	
Forward rate	Reverse rate	Forward rate	Reverse rate

12.32 Adjust the rate meters of Problem 12.31 to fix the problem and make it possible.

THE POSITION OF EQUILIBRIUM—THE EQUILIBRIUM CONSTANT K_{EQ}

12.33 Using the definition of equilibrium for the one-step reaction $R \rightleftharpoons P$, show how k_f/k_r is equal to a ratio of $[P]/[R]$.

12.34 At a given temperature, why is the ratio k_f/k_r a constant for a given reaction?

12.35 What symbol and name may be used to replace the ratio k_f/k_r for a reaction?

12.36 If $k_f > k_r$, will K_{eq} be less than 1 or greater than 1? Explain your answer.

12.37 Suppose you have a reaction with many different reactants. When you write the equilibrium expression for the reaction, do the corresponding terms all go in the numerator or the denominator? Do you add, subtract, multiply, or divide the concentrations of the different reactants?

12.38 Write the equilibrium expression for the reaction $2\,A + 3\,B \rightleftharpoons C + D$.

12.39 Write the equilibrium expression for the reaction $C + D \rightleftharpoons 2\,A + 3\,B$.

12.40 Compare your answers to Problems 12.38 and 12.39. What conclusion can you draw?

12.41 Write the expression for K_{eq} for the reaction $N_2O_4 \rightleftharpoons 2\,NO_2$.

12.42 Write the expression for K_{eq} for the reaction $4\,NH_3 + 5\,O_2 \rightleftharpoons 4\,NO + 6\,H_2O$.

12.43 Suppose you wanted the value of K_{eq} for the reaction of Problem 12.41. What would you do in the lab to obtain it?

12.44 Would the value you obtain for K_{eq} for a reaction depend on the initial concentrations of reactants and products you chose for the reaction? Explain your answer.

12.45 The equilibrium concentrations for the reaction $N_2 + O_2 \rightleftharpoons 2\,NO$ at 2000°C are: $[N_2] = 0.25$ M; $[O_2] = 1.2$ M; $[NO] = 0.011$ M. What is the value of K_{eq} for this reaction?

12.46 At 25°C, K_{eq} for the reaction in Problem 12.45 equals 2.3×10^{-9}. What can you say about the position of the equilibrium for this reaction at 25°C compared to its position at 2000°C?

12.47 What does a value of K_{eq} much greater than 1 imply? Prove that your answer is correct by using the general expression for the equilibrium constant, $K_{eq} = $ [Products]/[Reactants].

12.48 Suppose a reaction has a K_{eq} value of 2.05. When we write the reaction, can we use a single arrow to the right instead of a double set of equilibrium arrows? Explain your answer.

12.49 What does a value of K_{eq} much less than 1 imply? Prove that your answer is correct by using the general expression for the equilibrium constant, $K_{eq} = $ [Products]/[Reactants].

12.50 A certain reaction has a K_{eq} value of 1.5×10^{-6}. Would this be a practical reaction from which to isolate pure product? Explain your answer.

12.51 Consider the reaction

$$CO(g) + 3\,H_2(g) \rightleftharpoons CH_4(g) + H_2O(g)$$

It is run in a 10.0 liter vessel. The vessel is loaded with 1 mole of CO and 3 moles of H_2. At equilibrium, the composition of gases is 0.613 mole CO, 1.839 moles H_2, 0.387 mole CH_4, and 0.387 mole H_2O. What is the value of the equilibrium constant for this reaction? Describe the position of the equilibrium.

12.52 Consider the reaction in Problem 12.51. The reaction is run in the same vessel and at the same temperature, however, this time the vessel is loaded with 2 moles of CO and 3 moles of H_2. When the reaction reaches equilibrium, what will be the value of K_{eq}?

12.53 Why is the equilibrium constant called a constant?

12.54 Consider the following reaction:

$$CH_4(g) + 2\,H_2S(g) \rightleftharpoons CS_2(g) + 4\,H_2(g)$$

For this reaction at 900°C, $K_{eq} = 3.59$. The concentrations of the gases in the vessel at 900°C are $[CH_4] = 1.15$ M; $[H_2S] = 1.20$ M; $[CS_2] = 1.51$ M; $[H_2] = 1.08$ M. Is this reaction at equilibrium?

12.55 On the basis of the value of K_{eq}, which of the following reactions goes essentially to completion? How would you describe the other reaction?
(a) $2\,H_2(g) + O_2(g) \rightleftharpoons 2\,H_2O(g)$; $K_{eq} = 3 \times 10^{81}$
(b) $2\,HF(g) \rightleftharpoons H_2(g) + F_2(g)$; $K_{eq} = 1 \times 10^{-95}$

12.56 The equilibrium constant for a reaction is shown below. Write the balanced chemical equation that goes with it.

$$K_{eq} = \frac{[H_2O]^2[Cl_2]^2}{[HCl]^4[O_2]}$$

12.57 An 8.00 L reaction vessel at 491°C contains 0.650 mole H_2, 0.275 mole I_2, and 3.00 moles HI. Assuming that the reaction is at equilibrium, determine the value of K_{eq} and comment on where the equilibrium lies. The reaction is

$$H_2(g) + I_2(g) \rightleftharpoons 2\,HI(g)$$

12.58 How would the value of the equilibrium constant for a one-step reaction calculated as k_f/k_r compare to the value calculated from using the concentrations of all the substances present at equilibrium?

DISTURBING AN EQUILIBRIUM—LE CHATELIER'S PRINCIPLE

12.59 Suppose a reaction is at equilibrium and the equilibrium is then disturbed by adding reactants. What happens to the value of K_{eq}? Explain your answer.

12.60 State Le Chatelier's principle using the words "undo" and "partially."

12.61 Suppose we have an equilibrium mixture of reactants and products for the reaction

$$PCl_3(g) + Cl_2(g) \rightleftharpoons PCl_5(g)$$

Predict the direction that the reaction will shift when some:
(a) Chlorine (Cl_2) gas is added.
(b) Chlorine is removed.
(c) PCl_5 is added.
(d) PCl_3 is removed.
(e) H_2 gas is added. (Assume that H_2 does not react with any of the reactants or products in the reaction.)

12.62 Why can't you simply remove unreacted reactants from an equilibrium mixture to get pure product? Assume that K_{eq} is larger than 1 but not all that much larger.

12.63 When a reaction at equilibrium is disturbed by the addition of products:
(a) Which way will the reaction shift?
(b) After it is done shifting, will the product concentration be the same as before the disturbance, greater than before the disturbance, or less than before the disturbance? Explain your answer.
(c) Repeat part (b) but for the reactant concentration.

12.64 Consider the reaction

$$PCl_3(g) + Cl_2(g) \rightleftharpoons PCl_5(g)$$

Use forward and reverse rate meters to represent the forward and reverse reaction rates for:
(a) The reaction at equilibrium
(b) Moments after disturbing the equilibrium by adding PCl_5
(c) The restored equilibrium
(d) Which way did the reaction shift to get from the initial equilibrium, part (a), to the restored equilibrium, part (c)?

12.65 One way of preparing hydrogen is by the decomposition of water:

$$2\,H_2O(g) \rightleftharpoons 2\,H_2(g) + O_2(g) \qquad \Delta E_{rxn} = 484\ kJ/mole$$

Would you expect the decomposition to be more complete at high temperature or at low temperature? Explain.

12.66 Suppose you are making ammonia (NH_3) by the Haber reaction, shown below:

$$3\,H_2(g) \ + \ N_2(g) \ \rightleftharpoons \ 2\,NH_3(g) \qquad K_{eq} = 0.105 \text{ at } 472°C.$$

(a) Describe qualitatively where the equilibrium lies for this reaction.
(b) On the face of it, would this reaction be a good one for isolating pure ammonia?
(c) What would happen if you could keep feeding H_2 and N_2 into this reaction while at the same time selectively removing only NH_3?

RESPONSE OF EQUILIBRIUM TO CHANGES IN TEMPERATURE

12.67 The equilibrium constant for the synthesis of methanol, shown below, is 4.3 at 250°C and 1.8 at 275°C.

$$CO(g) \ + \ 2\,H_2(g) \ \rightleftharpoons \ CH_3OH(g)$$
Methanol

(a) Does this reaction shift to the left or to the right upon heating? Explain how you know.
(b) Is this reaction endothermic or exothermic? Explain how you know.
(c) Rewrite the reaction, including heat on the appropriate side.

12.68 The amount of nitrogen dioxide formed by dissociation of dinitrogen tetroxide, shown below, increases as the temperature rises.

$$N_2O_4(g) \ \rightleftharpoons \ 2\,NO_2(g)$$

(a) Is the reaction exothermic or endothermic? Explain how you know.
(b) Does K_{eq} increase or decrease as the temperature rises?
(c) Rewrite the reaction, including heat on the appropriate side.

12.69 Diamond and graphite are two forms of elemental carbon. Under the appropriate conditions they will be in equilibrium with each other:

$$C_{diamond} \ \rightleftharpoons \ C_{graphite}$$

If graphite is subjected to very high pressure and temperature, it will convert to the diamond form. Is the above equilibrium reaction exothermic or endothermic? Explain how you know.

12.70 Suppose you have an endothermic reaction with K_{eq} approximately equal to 1. How could you adjust the temperature of this reaction to drive it toward completion? Explain your answer.

12.71 Will K_{eq} for an exothermic reaction increase or decrease upon heating? Upon cooling? Explain your answer.

12.72 Will K_{eq} for an endothermic reaction increase or decrease upon heating? Upon cooling? Explain your answer.

12.73 Below you are given the value for ΔE_{rxn} for the production of methanol:

$$CO(g) \ + \ 2\,H_2(g) \ \rightleftharpoons \ CH_3OH(g) \qquad \Delta E_{rxn} = -90.8 \text{ kJ/mole}$$

Would the fraction of methanol present at equilibrium be increased or decreased by raising the temperature? Explain your answer.

12.74 Cooling an exothermic reaction can drive an "unfavorable" equilibrium toward products, but there is a trade-off. What is the downside of cooling such a reaction, as far as producing product is concerned?

12.75 What effect does a catalyst have on the position of equilibrium and on the value of the equilibrium constant?

12.76 What does a catalyst do to the time it takes for a reaction to reach equilibrium? Explain how it does it.

12.77 The value of K_{eq} for the $N_2 + O_2 \rightleftharpoons 2\ NO$ reaction was reported as 0.0017 at 2027°C and 2.3×10^{-9} at 25°C in this chapter.
(a) Judging from the values of K_{eq}, does this reaction shift to the left or to the right upon heating? Explain your answer.
(b) Is this reaction endothermic or exothermic? Explain your answer.

WORKPATCH SOLUTIONS

12.1 (a) The reaction essentially doesn't occur.
(b) The reaction essentially goes to completion.
(c) The equilibrium lies in the middle.
(d) The equilibrium lies to the right.
(e) The equilibrium lies to the left.

12.2 (c) is correct. Since at time zero, one of the reactants (O_2) is missing, the rate of the forward reaction must be zero. However, since there is a large concentration of product (SO_3), the reverse reaction will have a large rate. As time progresses, the forward reaction will speed up and the reverse reaction will slow down, so the rate meters at equilibrium also make sense for (c).
(a) is incorrect. Since the box is full of SO_3, the reverse reaction will have a large rate at time zero, not a zero rate.
(b) is incorrect. There is nothing wrong with the rate meters at time zero. However, as time progresses, the reverse reaction will slow down as SO_3 is consumed. The reverse reaction rate meter at equilibrium shows a greater rate than at time zero.
(d) is incorrect. The forward rate at time zero must be zero, since one of the reactants (O_2) is missing.
(e) is incorrect. There is nothing wrong with the meters at time zero. However, at equilibrium, the rate meters must be equal.

12.3 First, divide both sides by k_r and cross out anything identical appearing on top and bottom of a side:

$$\frac{k_f[A]}{k_r} = \frac{\cancel{k_r}[B]}{\cancel{k_r}}$$

$$\frac{k_f[A]}{k_r} = [B]$$

The left side now has k_f/k_r, but it is multiplied by [A]. To get rid of the [A], divide both sides by [A], crossing out anything identical appearing on top and bottom of a side:

$$\frac{k_f[\cancel{A}]}{k_r[\cancel{A}]} = \frac{[B]}{[A]}$$

$$\frac{k_f}{k_r} = \frac{[B]}{[A]}$$

12.4 $K_{eq} = \dfrac{[NO]^2}{[N_2][O_2]} = \dfrac{(0.040\ M)^2}{(0.980\ M)(0.980\ M)} = 0.0017$

12.5 (a) $\dfrac{[H^+] \times [Cl^-]}{[HCl]}$ (4) $K_{eq} > 10^6$

(b) $\dfrac{[H^+] \times [CH_3COO^-]}{[CH_3COOH]}$ (1) $K_{eq} > 1.0 \times 10^{-5}$

(c) $\dfrac{[PCl_3] \times [Cl_2]}{[PCl_5]}$ (2) $K_{eq} > 2.2 \times 10^{-2}$

(d) $\dfrac{[CO] \times [H_2]^3}{[H_2O] \times [CH_4]}$ (3) $K_{eq} = 4.6$

12.6 (a) $K_{eq(\text{decompose NO})} = \dfrac{[N_2] \times [O_2]}{[NO]^2}$ $K_{eq(\text{form NO})} = \dfrac{[NO]^2}{[N_2] \times [O_2]}$

(b) One is the inverse of the other. That is:

$$K_{eq(\text{decompose NO})} = \frac{1}{K_{eq(\text{form NO})}}$$

(c) $K_{eq(\text{decompose NO})} = \dfrac{1}{K_{eq(\text{form NO})}} = \dfrac{1}{2.3 \times 10^{-9}} = 4.8 \times 10^8$

This is a very large number, so the equilibrium lies far to the right (completion).

12.7 $2\,SO_2 + O_2 \;\rightleftharpoons\; 2\,SO_3$

$$K_{eq} = \frac{[SO_3]^2}{[SO_2]^2 \times [O_2]}$$

For the initial equilibrium:

$$K_{eq} = \frac{[1.8\ M]^2}{[0.2\ M]^2 \times [0.1\ M]} = 810$$

For the new equilibrium:

$$K_{eq} = \frac{[1.614\ M]^2}{[0.186\ M]^2 \times [0.093\ M]} = 810$$

They are the same. They should be! The value of K_{eq} is a constant at a given temperature.

12.8 (b) is the best choice. Remember, Le Chatelier's principle says that a reaction will shift to partially undo a disturbance to an equilibrium. We increased SO_2 from 0.34 mole to 1.34 mole. The reaction will shift so as to decrease SO_2 from 1.34 mole, but not all the way back to 0.34 mole.

12.9 If you cool an endothermic reaction, it will shift to the left, toward the reactants. In other words, the value of K_{eq} will decrease. Why it shifts this way is explained by Le Chatelier's principle. Since we removed heat by cooling, the reaction will shift in a direction that produces heat, attempting to partially undo our disturbance. The only way an endothermic reaction like this one can produce heat is to run in the reverse direction (toward reactants, including heat).

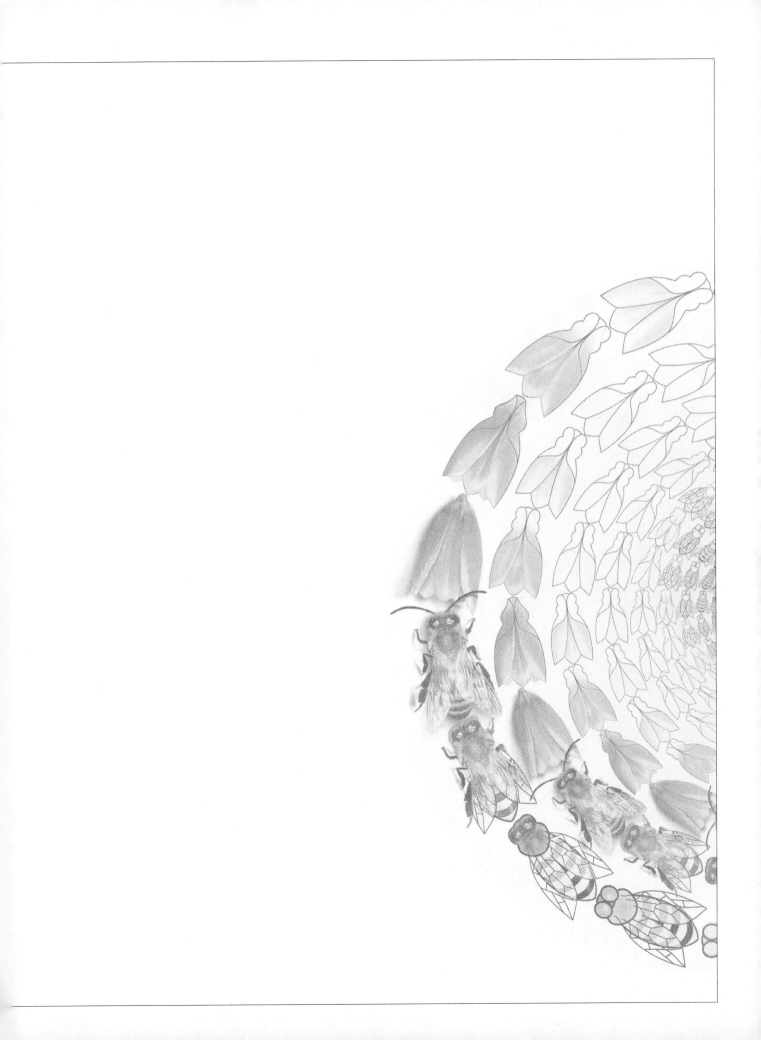

Chapter 13

Electrolytes, Acids, and Bases

ELECTROLYTES 13.1

We have seen that the science of chemistry attempts to understand matter in terms of the atoms and molecules from which all things are made. But chemistry had its beginnings long before the widespread acceptance of atoms and molecules. Many early chemists concerned themselves with finding, examining, and categorizing various "chemical" substances in the world around them. Much as the biologist uses the categories of phylum, genus, and species to group together living things with similar traits and behaviors, the pioneering early chemists did the same sort of thing by creating the categories of *acid* and *base*.

A chemical substance was considered to be an **acid** if it had the three following properties:

1. A sour taste. Early chemists used all of their senses to probe the nature of matter. Many substances were tasted to determine whether they had a sour taste and might therefore be classified as an acid. (The word "acid" comes from the Latin *acidus*, for sour.) Of course, such taste testing does not occur without some risk. We can only imagine the fate of the pioneering chemist who determined that hydrogen cyanide (HCN), a fast-acting and lethal poison, is an acid.

Tasting an unknown compound—not a good idea.

Not all acids are toxic. After all, vitamin C is ascorbic acid, and citric acid is found in all kinds of fruits, vegetables, sour candies, and so on. Clearly though, other tests for acids were needed if the science of chemistry, not to mention the chemists themselves, were to survive. The other two properties subsequently used to classify compounds as acids were:

2. Turns the plant dye called litmus (extracted from certain mosses) a bright red color.

3. Eats away at the more active metals such as zinc (Zn), reacting with them and producing hydrogen gas.

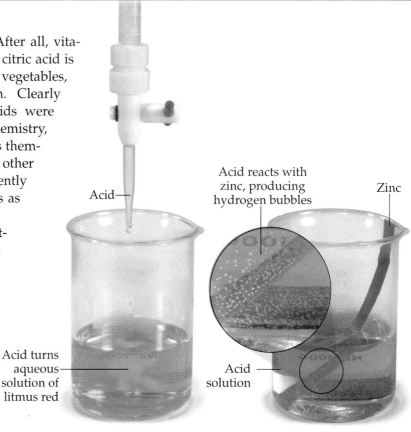

Acid

Acid reacts with zinc, producing hydrogen bubbles

Zinc

Acid turns aqueous solution of litmus red

Acid solution

Conversely, a chemical substance was classified as a **base** if it:

1. Has a bitter taste.
2. Turns an aqueous solution of the litmus plant dye dark blue.
3. Gives rise to aqueous solutions that feel slippery to the touch.

Compounds such as litmus are called *indicators* because they indicate by their color whether an acid or base is present. As early chemists tested the substances they found in nature, they discovered that some were acids, some were bases, and some were neither. Substances that gave no signs of being either acidic or basic were called **neutral**. Testing and categorizing numerous compounds in this way was an important first step in understanding the chemical behavior of many common substances.

The categories of acid and base proved to be so fundamental and useful that they remain with us even today. However, modern chemistry goes further, explaining at the molecular level why acids and bases have the properties they do. The key to understanding this is yet another classification scheme, that of *electrolyte* versus *nonelectrolyte*. This was one of the first classification schemes for compounds to be successfully explained on the molecular level, and it led directly to a clearer understanding of acids and bases. Therefore, we will spend a little time discussing electrolytes versus nonelectrolytes before we consider acids and bases.

Base

Base turns litmus blue

Very simply, a substance is considered to be an **electrolyte** if an aqueous solution of the substance can conduct an electric current. If the solution cannot conduct current, the substance is a **nonelectrolyte**. It is easy to determine whether a substance is an electrolyte or nonelectrolyte. All we need is a sample of the substance dissolved in pure water, a light bulb with a plug and cord, and an electrical outlet:

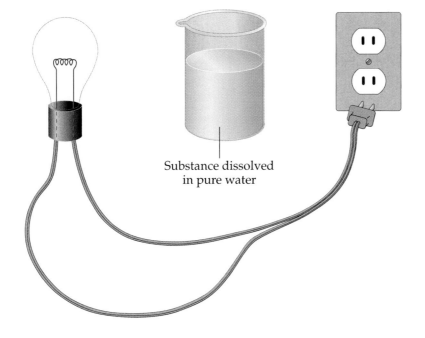

Substance dissolved
in pure water

If you plug the cord into the outlet, electrons flow through the circuit in the form of an electric current, and the bulb lights:

To convert the bulb into an electrolyte tester, all you need to do is unplug the bulb and then cut one of its wires. Our specially made electrolyte tester is exactly the same thing:

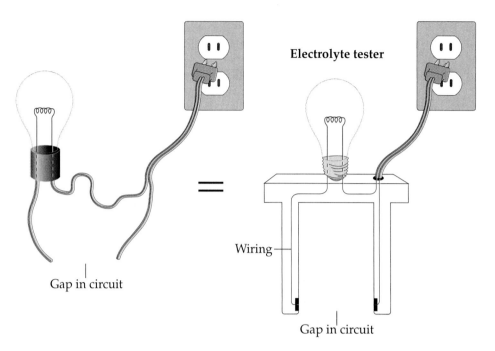

Gap in circuit

Electrolyte tester

Wiring

Gap in circuit

If we plug our tester into the wall outlet, the bulb won't light because electrons cannot flow across the gap in the circuit. Now, let's dip the prongs of the tester into a beaker of pure (distilled) water. If electrons could somehow "swim" through the water, they would be able to cross the gap between the prongs, and the bulb would light. But the bulb doesn't light. Therefore, pure water is a poor conductor of electricity.

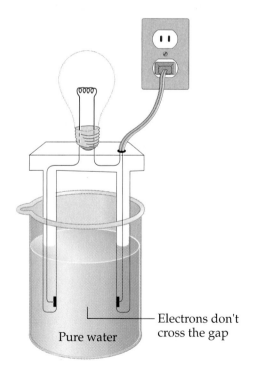

Pure water

Electrons don't cross the gap

Now let's test two other familiar substances: salt (NaCl) and table sugar (sucrose). We dissolve both substances in pure water and insert our tester:

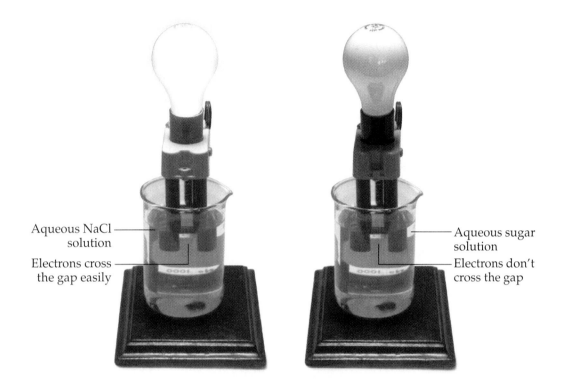

Aqueous NaCl solution
Electrons cross the gap easily

Aqueous sugar solution
Electrons don't cross the gap

Clearly, the aqueous solution of NaCl conducts electricity—the bulb lights up. Therefore, NaCl is an electrolyte. Sugar is a nonelectrolyte, since the bulb remains unlit. An electric current doesn't travel through sugar water any more than it does through pure water.

Using methods like these, chemists compiled lists of electrolytes and non-electrolytes. Table 13.1 lists some typical substances from each category.

Table 13.1 Some Well-Known Electrolytes and Nonelectrolytes

Electrolytes	Nonelectrolytes
Acetic acid (CH_3COOH, vinegar)	Acetone (C_3H_6O, nail polish remover)
Hydrogen chloride (HCl)	Carbon monoxide (CO)
Sodium bicarbonate ($NaHCO_3$, baking soda)	Ethanol (CH_3CH_2OH)
	Methane (CH_4)
Sodium chloride (NaCl)	Oxygen (O_2)
Sodium hydroxide (NaOH, lye)	Sucrose ($C_{12}H_{22}O_{11}$, sugar)
Sodium sulfate (Na_2SO_4)	Turpentine (hydrocarbons like $C_{10}H_{22}$)
Sulfuric acid (H_2SO_4)	

Until about 200 years ago, chemists had no idea *why* aqueous solutions of some substances conducted electricity whereas solutions of other substances did not. Michael Faraday, an English physicist who did much of the basic work on the nature of electricity during the early 1800s, was the first to shed

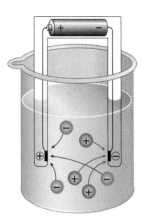

Faraday's mobile charged particles that carried electric current. He called them ions.

some light on what was going on. Faraday proposed that in order for an electric current to pass through an aqueous solution, some form of mobile charged particles must be present. He called these particles *ions*. Today we know that ions are charged atoms (like Na^+, Cl^-) or groups of atoms (like the polyatomic ions sulfate, SO_4^{2-}, and ammonium, NH_4^+). Faraday, however, had no real concept of what these ions were, and he incorrectly hypothesized that they were created in the water by the electricity. He was correct, though, about ions being mobile charged particles. He was also correct when he said that electricity can be carried through water only when these charged particles are present.

In 1884, after years of painstaking measurements on the electrical conductivity of hundreds of different aqueous solutions, the Swedish chemist Svante Arrhenius offered the first comprehensive (and correct) explanation for the behavior of electrolytes. In his Ph.D. thesis Arrhenius hypothesized that Faraday's ions came from the dissociation of the solute itself. In other words, when electrolytes were dissolved in water, the solute broke up into charged mobile particles (ions), and these ions were present in the solution whether or not an electric current was ever applied.

Dissolve solid NaCl in water

The response of Arrhenius' thesis advisor was essentially, "You have a new theory? That is very interesting. Please close the door on your way out." Arrhenius was awarded his degree grudgingly, in spite of what was considered a "radical and ridiculous" ion theory. Years later, the world caught up to Arrhenius, and in 1903 he received the Nobel Prize for his work.

Today we realize that a solution of saltwater conducts electricity because the Na^+ and Cl^- ions act as the mobile charge carriers that Faraday proposed, each migrating to the wire of opposite electrical charge. Electrons cannot flow through water, but ions can. These mobile charge carriers take the place of electrons, keeping the flow of charge (electricity) going.

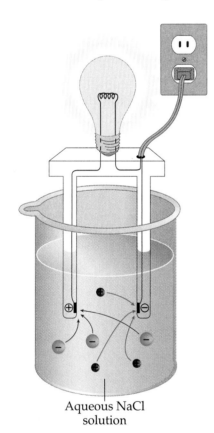

Aqueous NaCl solution

Positive cations (Na^+) are attracted to the negative wire, and negative anions (Cl^-) are attracted toward the positive wire.

The mobile charge carriers exist because the ionic compound NaCl not only dissolves in water, it also breaks up to give Na^+ and Cl^- ions. We call this breaking apart to produce ions **dissociation**. Soluble ionic compounds dissolve and dissociate. The alternative would be for a substance to dissolve without dissociation. For example, molecular, covalent substances such as sugar and ethanol dissolve without dissociation into ions, and therefore are nonelectrolytes. Their solutions consist of hydrated, intact, neutral molecules. Since no ions are present, no electricity can be passed through the solution.

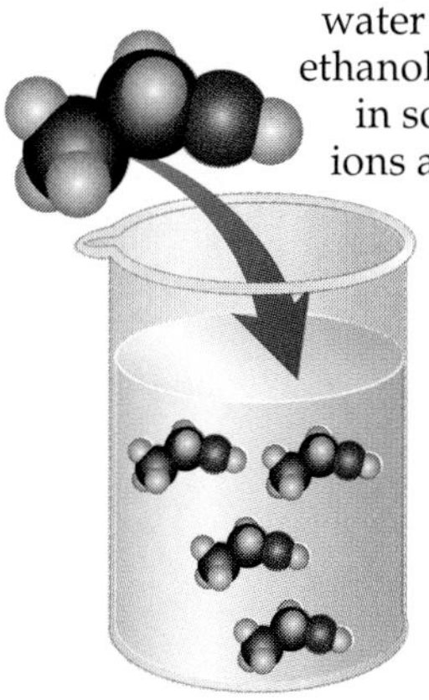

Dissolving ethanol (CH_3CH_2OH) in water puts intact ethanol molecules in solution. No ions are present.

What you want to remember here is this difference between how ionic and molecular solutes dissolve—the former with dissociation, the latter without. This makes sense when you stop to think about it. Molecular substances are made up of discrete, whole molecules, while ionic solids are made up of ions packed into a lattice. Dissolving an ionic solid simply puts into solution the particles that make up the lattice. We can now use the word "ion" to extend our definitions of electrolyte and nonelectrolyte:

An **electrolyte** is a solute that dissolves in water and dissociates into ions to yield a solution that conducts electricity.

A **nonelectrolyte** is a solute that dissolves in water without producing ions to yield a solution that does not conduct electricity.

You should be able to determine whether a compound is an electrolyte or not based on its formula. Water-soluble ionic substances are electrolytes. Most ionic compounds consist of a metal combined with a nonmetal or group of nonmetals. Therefore, when you see a formula like NaCl or $CaCO_3$, you can be reasonably confident that you are looking at an ionic compound and an electrolyte. Ionic compounds that do not contain a metal are exceptions to this rule. For example, NH_4^+ is a polyatomic ion. Compounds containing this ion, such as NH_4Cl (ammonium chloride) and $(NH_4)_2SO_4$ (ammonium sulfate), are water-soluble ionic compounds and thus are electrolytes.

Water-soluble molecular substances generally consist completely of nonmetals and are usually nonelectrolytes. Therefore, when you see a formula that includes no metals, such as CO_2 or $C_{12}H_{22}O_{11}$ (sucrose), you can be reasonably confident that you are looking at a covalent compound and a nonelectrolyte. Of course, there are exceptions to this rule as well. One that you should remember pertains to compounds of the formula HX, where X = Cl, Br, and I (the group VIIA halogens). The molecules hydrogen chloride (HCl), hydrogen bromide (HBr), and hydrogen iodide (HI) are polar covalent molecules. They are not ionic substances. Nevertheless, they dissolve in water and cause the light bulb to light, demonstrating that they are electrolytes.

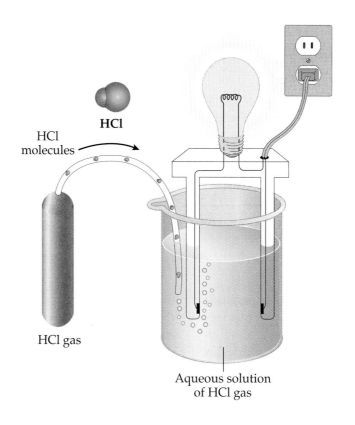

HCl

HCl molecules

HCl gas

Aqueous solution
of HCl gas

Unlike most molecular substances, which simply dissolve to give hydrated neutral molecules (such as O_2 and CH_3CH_2OH, which are nonelectrolytes), HX molecules go on to dissociate, producing H^+ and X^- (Cl^-, Br^-, I^-) ions. This makes them electrolytes.

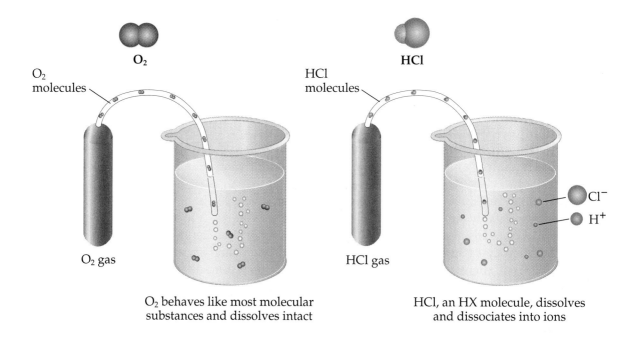

O_2

O_2 molecules

O_2 gas

O_2 behaves like most molecular
substances and dissolves intact

HCl

HCl molecules

Cl^-

H^+

HCl gas

HCl, an HX molecule, dissolves
and dissociates into ions

There are other molecular compounds that dissolve in water and also dissociate to give ions, making them electrolytes. We will show you some of these shortly. Most molecular substances, however, are nonelectrolytes.

13.1 Consider the compound H_2SO_4.
  (a) Is it molecular or ionic?
  (b) How did you decide on your answer to part (a)?
  (c) Based on your answer to part (a), would you call H_2SO_4 an electrolyte or a nonelectrolyte?
  (d) An aqueous solution of H_2SO_4 causes a light bulb to light. What does this tell you about H_2SO_4?

Answer:
(a) Molecular
(b) The formula has no metals in it, just nonmetals. This is a sign that the compound is probably molecular and not ionic.
(c) Most molecular compounds are not electrolytes, so the safe bet would be to call H_2SO_4 a nonelectrolyte.
(d) Since the light bulb lights, then H_2SO_4 is an electrolyte. This tells us two things: (1) Just like HCl, H_2SO_4 is one of those exceptions to the rule that says most molecular compounds are not electrolytes. (2) H_2SO_4 not only dissolves, it must also dissociate to produce ions.

13.2 Indicate whether each of the following compounds is an electrolyte or a nonelectrolyte when placed in water.
  (a) $Al(NO_3)_3$ (b) $(CH_3)_2O$ (c) $(NH_4)_2SO_4$
  (d) CH_3OH (e) $CuSO_4$ (f) KBr
  (g) HBr

13.3 (a) Draw a beaker full of water and show what ions, if any, are present when the compound ammonium bromide (NH_4Br) dissolves.
  (b) Does the NH_4Br dissociate? What does this mean?
  (c) Add a light bulb with positive and negative wires to your drawing from part (a). Indicate with arrows which way the dissolved species move.

————————————————————————————————WEAK AND STRONG ELECTROLYTES 13.2

Let's look a bit more closely at molecular solutes like HCl that are also electrolytes. Both HCl and HF dissociate in water to give H^+ ions and halide ions. However, an HCl solution would make a light bulb glow very brightly, while an HF solution of the same concentration would barely cause it to glow. Why the difference? We can find the answer by considering how efficient each molecule is at dissociating into ions once it dissolves.

Hydrogen chloride is particularly good at dissociating in water. In an HCl solution you would practically never come across an intact HCl molecule. All you would ever find are H^+ and Cl^- ions:

$$HCl(g) \xrightarrow[\text{in } H_2O]{\text{Dissolves}} H^+(aq) + Cl^-(aq) \quad \text{100\% dissociation into ions}$$

When we write $H^+(aq)$, we are really using an abbreviation. To be absolutely accurate we should write $H_3O^+(aq)$ to indicate a hydronium ion, since "naked" H^+ ions do not exist in water. Instead, they are solvated by the water to yield hydronium (H_3O^+) ions, because H^+, being a bare proton with no electrons, is too reactive to exist by itself in water.

$$HCl(g) + H_2O(l) \longrightarrow H_3O^+(aq) + Cl^-(aq)$$

The H^+ shares a lone pair of electrons from an oxygen atom in water, forming a new covalent bond to form H_3O^+. This fills the H^+ ion's valence electron shell and results in the more stable hydronium ion:

$$H^+ + H_2O \longrightarrow H_3O^+$$

Hydronium ion

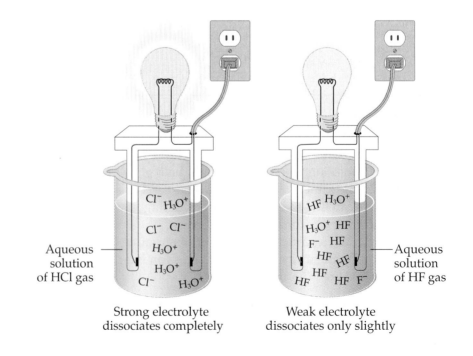

Here, brackets mean that the charge applies to the whole ion.

The situation is quite different for HF. If you examined an aqueous solution of HF, you would discover some H_3O^+ and F^- ions, but you would also come across many more intact, undissociated HF molecules. The dissociation reaction can be written as an equilibrium situation that lies toward the left, which we indicate by exaggerating the length of the left-pointing arrow:

$$HF(g) + H_2O(l) \overset{\longrightarrow}{\longleftarrow} H_3O^+(aq) + F^-(aq)$$

This puts fewer ions into solution, resulting in a decreased ability to conduct electricity and a light bulb that barely lights. Only a small percentage of the dissolved HF molecules dissociate (~3% in a 1.0 M solution) compared to the 100% dissociation of HCl. For this reason we call HCl a strong electrolyte and HF a weak electrolyte.

Aqueous solution of HCl gas

Aqueous solution of HF gas

Strong electrolyte dissociates completely

Weak electrolyte dissociates only slightly

A **strong electrolyte** is one that completely dissociates into ions upon dissolving.

A **weak electrolyte** is one that only partially dissociates into ions upon dissolving.

Molecular electrolytes are known that span the full range from very strong to very weak.

All of this relates back to the previous chapter on equilibrium. Both the HCl and the HF dissociation reactions, like all reactions in principle, are equilibrium reactions:

$$HCl(g) + H_2O(l) \rightleftharpoons H_3O^+(aq) + Cl^-(aq)$$

$$HF(g) + H_2O(l) \rightleftharpoons H_3O^+(aq) + F^-(aq)$$

In the case of HCl, however, the equilibrium lies so far to the right that we say the reaction goes to completion and we drop the double arrows:

$$HCl(g) + H_2O(l) \longrightarrow H_3O^+(aq) + Cl^-(aq) \qquad K_{eq} \text{ much greater than 1}$$

In the case of HF, by contrast, the equilibrium lies to the left:

$$HF(g) + H_2O(l) \xrightarrow{\longleftarrow} H_3O^+(aq) + F^-(aq) \qquad K_{eq} = 3.5 \times 10^{-4} \text{ (much less than 1)}$$

Almost all the dissolved HF remains undissociated.

Based on the diagrams below:

WORKPATCH **13.1**

  (a) Which is a weak electrolyte, which is a strong electrolyte, and which is a nonelectrolyte? How can you tell?
  (b) Which of these compounds is/are molecular? Which is/are ionic?

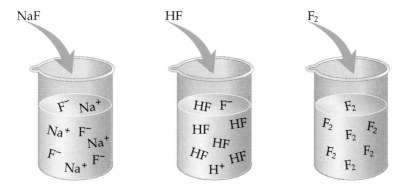

Were you able to tell which electrolyte was weak or strong just from examining the species in solution in the beakers? Once you understand what an electrolyte is, you are ready to go back to acids and bases, which, as you will see, are particular kinds of electrolytes.

ACIDS, WEAK AND STRONG 13.3

The molecular electrolytes that we have considered so far, HCl and HF, were known over 150 years ago. Indeed, quite a few molecular electrolytes were known, including HBr, HI, HNO_3, H_2SO_4, and HClO. At that time, little was understood about them, but some similarities among them had been discovered. For example, they were all known to dissolve in water to give aqueous solutions that taste sour and turn litmus red. They also react with active metals like zinc to liberate hydrogen (H_2) gas. As we saw at the beginning of the chapter, this behavior classified them as acids.

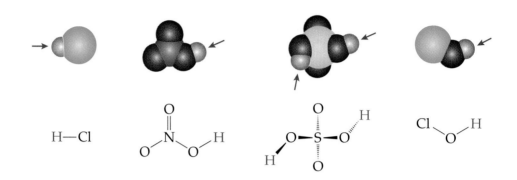

$$H—Cl$$

Why very different molecules like HCl, HNO_3, and H_2SO_4 should all give rise to aqueous solutions with similar acidic properties was a mystery, though one clue came from the fact that all these compounds are electrolytes. It was Arrhenius who in 1884 put it all together, postulating that all these compounds dissociate in water to yield H^+. That became the first modern definition of an acid. Chemists have modified this a bit to recognize that aqueous H^+ ions really exist as H_3O^+ (hydronium) ions, but, otherwise, we still use Arrhenius' definition of an acid.

Arrhenius definition of an acid: An acid is any electrolyte that contains hydrogen and produces H^+ (actually H_3O^+, hydronium) ions when dissolved in water.

Molecular substances like HCl, HBr, HI, HNO_3, and H_2SO_4 are all strong electrolytes, dissociating completely to give lots of H_3O^+ ions. This makes their aqueous solutions very acidic, and we refer to these compounds as strong acids.

A **strong acid** is a water-soluble compound that completely dissociates to give H_3O^+ ions.

On the other hand, HF is a weak electrolyte and therefore dissociates only partially, producing far fewer H_3O^+ ions. Since intact HF molecules do not make a solution acidic (only H_3O^+ ions do), an aqueous solution of HF is only weakly acidic. We therefore refer to HF as a weak acid.

A **weak acid** is a water-soluble compound that dissociates only partially, producing few H_3O^+ ions.

The figure at the top of the next page shows the difference in behavior between a strong acid and a weak acid. One mole of HCl produces 1 mole of H_3O^+ ions, making the solution strongly acidic. By contrast, 1 mole of HF produces far fewer H_3O^+ ions, making the solution only weakly acidic.

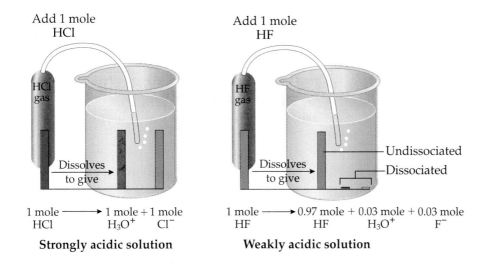

1 mole ⟶ 1 mole + 1 mole   1 mole ⟶ 0.97 mole + 0.03 mole + 0.03 mole
HCl H_3O^+ Cl^- HF   HF H_3O^+ F^-

Strongly acidic solution   **Weakly acidic solution**

Strong or weak, all these molecular electrolytes produce acidic solutions when dissolved in water. To acknowledge this fact, we change the names of these electrolytes slightly when they are dissolved in water. For example, the gaseous HCl molecule is called hydrogen chloride, but it would be incorrect to call HCl dissolved in water by that name, since it is completely dissociated into ions. Instead, an aqueous solution of HCl is called hydrochloric acid. Likewise, aqueous solutions of HF, HNO_3, and H_2SO_4 are called hydrofluoric acid, nitric acid, and sulfuric acid, respectively. They all produce H_3O^+ ions upon dissociation.

We can now see why all these electrolytes have similar acidic properties. The H_3O^+ ion is responsible for the sour taste. The H_3O^+ ion also reacts with litmus to produce the red color, and it reacts directly with active metals to yield hydrogen gas. Tables 13.2A and 13.2B list some common strong and weak acids.

Table 13.2A   Strong Acids

Formula	Name	Maximum number of H_3O^+ ions produced per molecule of acid
HCl	Hydrochloric acid	1
HBr	Hydrobromic acid	1
HI	Hydroiodic acid	1
HNO_3	Nitric acid	1
H_2SO_4	Sulfuric acid	2

Table 13.2B   Weak Acids

Formula	Name	Maximum number of H_3O^+ ions produced per molecule of acid
HF	Hydrofluoric acid	1
HClO	Hypochlorous acid	1
CH_3COOH	Acetic acid	1
H_2CO_3	Carbonic acid	2
H_3PO_4	Phosphoric acid	3

You may be familiar with some of these acids. Sulfuric acid, H_2SO_4, is the acid used in car batteries. Vinegar is a dilute solution of acetic acid, CH_3COOH, in water. Carbonic acid is present in carbonated soft drinks, as is phosphoric acid, which gives soda pop its ability to completely dissolve a tooth left soaking in it for just a few days.

Take a close look at the numbers in the last column in Tables 13.2A and 13.2B. Some acids are capable of producing more than one H_3O^+ ion per molecule. For example, sulfuric acid, H_2SO_4, can produce a maximum of two H_3O^+ ions. Such an acid is referred to as **diprotic**. Acids that give just one hydronium ion per molecule (HCl, HF) are called **monoprotic**. **Triprotic** acids, such as phosphoric acid, H_3PO_4, can produce a maximum of three H_3O^+ ions:

$$H_3PO_4 + 3\,H_2O \longrightarrow 3\,H_3O^+ + PO_4^{3-} \tag{13.1}$$

 **13.2** WORKPATCH

Sulfuric acid, H_2SO_4, is a diprotic acid. Write a balanced equation for sulfuric acid in water, showing the maximum number of H_3O^+ ions it can produce.

The reaction you just wrote for H_2SO_4 and the reaction for H_3PO_4 above show the *maximum* number of protons these acids can produce in water. In fact, though, these reactions do not go to completion when these acids are added to water. Read on.

A common misconception is that diprotic and triprotic acids are stronger than monoprotic acids because they can dissociate to give more hydronium ions. This is not necessarily true. For example, even though carbonic acid, H_2CO_3, is diprotic, it is a much weaker acid than monoprotic HCl. This is because carbonic acid barely dissociates in water, so even though it can technically produce two H_3O^+ ions per molecule of H_2CO_3, most of the dissolved H_2CO_3 molecules prefer to stay intact. So don't confuse the terms "strong" and "weak" with whether an acid is mono-, di-, or triprotic. They are not related. Only "strong" and "weak" can tell you to which side the dissociation equilibrium in water lies—to the right, toward completion, for strong acids, and to the left, toward nondissociation, for weak acids.

Strong acids have equilibrium constants for their dissociation that are very much greater than 1. So much greater than 1, in fact, that we say the dissociation goes to completion. For weak acids, however, knowing the value of K_{eq} for dissociation can help us gauge just how weak they are. For example, consider the three stepwise dissociations for triprotic phosphoric acid:

Loss of first H^+

$$H_3PO_4 + H_2O \rightleftharpoons H_3O^+ + H_2PO_4^- \qquad K_{eq} = 7.5 \times 10^{-3}$$

Loss of second H^+

$$H_2PO_4^- + H_2O \rightleftharpoons H_3O^+ + HPO_4^{2-} \qquad K_{eq} = 6.2 \times 10^{-8} \tag{13.2}$$

Loss of third H^+

$$HPO_4^{2-} + H_2O \rightleftharpoons H_3O^+ + PO_4^{3-} \qquad K_{eq} = 4.2 \times 10^{-13}$$

All these equilibria have a K_{eq} that is much smaller than 1, so all lie to the left. The acid H_3PO_4 gives a proton (H^+) to water to become $H_2PO_4^-$. The $H_2PO_4^-$ is also an acid, as it gives up a second proton to water in the second dissociation equilibrium, becoming HPO_4^{2-}. And HPO_4^{2-} is yet another acid, as it gives up a third and final proton to water in the third dissociation equilibrium. Of these three weak acids, which is the weakest? The values of K_{eq} tell us:

Weak acid	K_{eq} for dissociation
H_3PO_4	7.5×10^{-3}
$H_2PO_4^-$	6.2×10^{-8}
HPO_4^{2-}	4.2×10^{-13}

While all three acids are certainly weak, HPO_4^{2-} is clearly the weakest, since it has by far the smallest value of K_{eq} for dissociation.

Acetic acid, CH_3COOH, has a K_{eq} for dissociation of 1.8×10^{-5}. This means that it is a weaker acid than H_3PO_4, but a stronger acid than $H_2PO_4^-$. The value of K_{eq} can even be used to calculate exactly what percentage of the acid will dissociate. For example (without going into the details), based on the value of its K_{eq}, we can calculate that 99.58% of the acetic acid in a 1.0 M solution remains as intact CH_3COOH molecules. Only 0.42% dissociates to give acetate (CH_3COO^-) and hydronium ions.

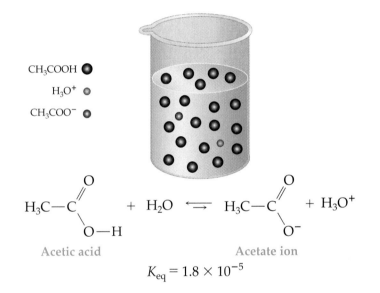

$K_{eq} = 1.8 \times 10^{-5}$

The K_{eq} for dissociation can reveal some interesting behaviors. For example, sulfuric acid is listed as a strong acid in Table 13.2A, but take a closer look at the dissociation equilibria for this diprotic acid:

Loss of first H^+

$$H_2SO_4 + H_2O \longrightarrow H_3O^+ + HSO_4^- \qquad K_{eq} = 1.0 \times 10^3$$

Loss of second H^+

$$HSO_4^- + H_2O \rightleftharpoons H_3O^+ + SO_4^{2-} \qquad K_{eq} = 1.2 \times 10^{-2}$$

The equilibrium constant for the first dissociation is quite large, so sulfuric acid is indeed a very strong acid, but only for the first proton dissociation; the second proton is only partially dissociated. So in a sense sulfuric acid is both a strong and a weak acid. In other words, even though sulfuric acid is diprotic, it does not lose both protons with equal ease. However, when one dissociation goes to completion, the acid is still considered to be strong.

PRACTICE PROBLEMS

13.4 True or false? For a compound to be a molecular substance and an acid in water, it must also be an electrolyte.

Answer: True. An acid must dissociate to produce H_3O^+ ions in water. Anything that dissociates into ions in water is an electrolyte.

13.5 Carbonic acid, H_2CO_3, is a diprotic acid. Write a balanced dissociation equation for carbonic acid in water that shows the maximum number of H_3O^+ ions that can be produced.

13.6 Your answer to Practice Problem 13.5 shows the dissociation yielding two moles of protons per mole of acid in water, but it actually produces fewer protons. Write the two balanced equilibrium equations for the dissociation of carbonic acid in water.

13.4 BASES—THE OPPOSITES OF ACIDS

Many different substances were discovered in the early days of chemistry that could be classified as acids. In addition, a large number of compounds were discovered that behaved similarly to one another but were not acidic. These compounds had a bitter taste instead of a sour taste, and they turned the natural pigment litmus a dark blue instead of the red caused by acids. These compounds were classified as bases. The most interesting property of bases, however, was that when they were added in the proper amount to a solution of an acid, they destroyed the acid properties of the solution and left the water neutral (neither acidic nor basic) and with a salty taste. The bases apparently "killed" or somehow neutralized the H_3O^+ ions responsible for the acidity. It was this observation, along with his definition of acids, that led Arrhenius to his definition of a base.

Arrhenius reasoned that the base took the H_3O^+ ions produced by acids and turned them into neutral water. The question of "What is a base?" then boiled down to the question, "What could react with H_3O^+ ions and turn them into neutral water?" Arrhenius' answer was the OH^- ion (the hydroxide ion). The reaction between an H_3O^+ ion supplied by an acid and an OH^- ion supplied by a base, shown below, is called the **acid–base neutralization reaction**. Hydronium plus hydroxide react to give two molecules of water:

$$H_3O^+ + OH^- \xrightarrow[\text{other to give}]{\text{Neutralize each}} 2\,H_2O \qquad (13.3)$$

$$\text{Acid} \quad \text{Base}$$

We can understand this reaction at the molecular level by examining dot diagrams. Basic hydroxide ions convert acidic hydronium ions into water by accepting a proton from them:

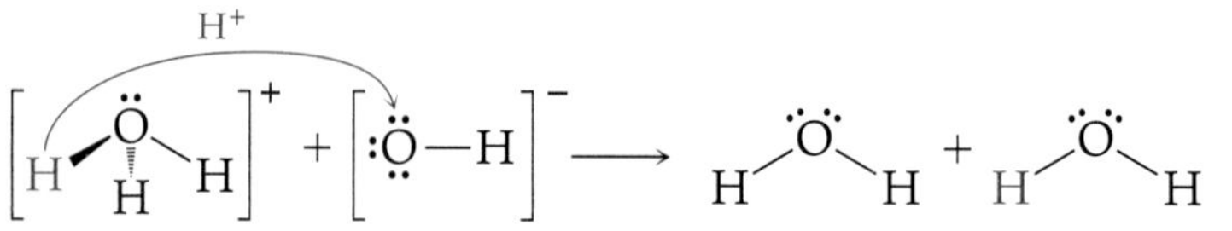

$$(13.4)$$

There was another hint that it was the hydroxide ion that gave basic compounds their properties. Most of the bases known at the time were ionic compounds with such formulas as NaOH, LiOH, CaO_2H_2, MgO_2H_2. No doubt, Arrhenius looked at those last two formulas and rewrote them in his mind as $Ca(OH)_2$ and $Mg(OH)_2$. Today we call these basic compounds metal hydroxides and represent them by the generic formula $M(OH)_n$, where M stands for a metal atom. Table 13.3 lists the common metal hydroxides.

Table 13.3   Some Common Bases

LiOH	Lithium hydroxide
NaOH	Sodium hydroxide
KOH	Potassium hydroxide
$Mg(OH)_2$	Magnesium hydroxide
$Ca(OH)_2$	Calcium hydroxide
$Ba(OH)_2$	Barium hydroxide

All have the formula $M(OH)_n$

All these bases are water-soluble to varying degrees. They are ionic compounds, and they dissociate to form OH^- ions in aqueous solution. Arrhenius put all of this together and defined a base as anything that dissolves in water to give hydroxide ions.

Arrhenius definition of a base: A base is any electrolyte that contains a metal and hydroxide (OH^-) group and produces OH^- (hydroxide) ions when dissolved in water.

Like acids, some bases are strong electrolytes, while others are weak electrolytes. And just like the acids, bases that are strong electrolytes are also strong bases, while those that are weak electrolytes are also weak bases. The group IA metal hydroxides in Table 13.3 are strong electrolytes, highly soluble and dissociating completely to produce lots of OH^- ions in aqueous solution. This makes them all **strong bases** because their dissociation reactions go essentially to completion:

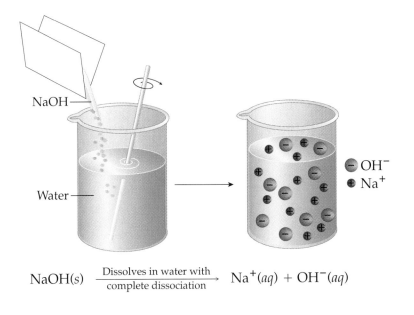

$$NaOH(s) \xrightarrow[\text{complete dissociation}]{\text{Dissolves in water with}} Na^+(aq) + OH^-(aq)$$

We will take a close look at weak bases a little later in the chapter.

Seeing bases as providers of hydroxide ions, Arrhenius could even explain the fact that after an acid–base neutralization, the resulting solution is salty. For example, consider what happens if we combine a solution of hydrochloric acid with a solution of sodium hydroxide. The solution of HCl has Cl^- ions as well as hydronium ions, and the solution of NaOH has Na^+ ions as well as hydroxide ions. After the OH^- ions have neutralized the H_3O^+ ions, the Na^+ and Cl^- ions are still present. We are left with an aqueous solution of NaCl, a salty solution.

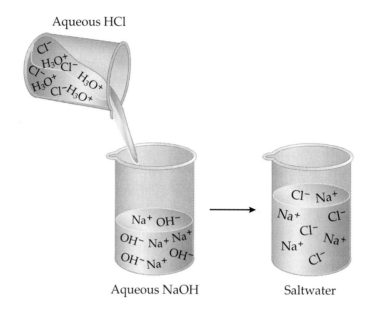

13.3 WORKPATCH

(a) What species are present in each of the beakers below?

(b) Ionic compounds are often called salts. What is the formula and name of the salt that gives rise to the salty solution in the beaker on the right?

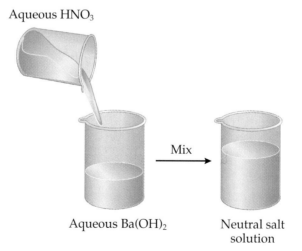

Did you get the correct salt for the WorkPatch? Now try the following practice problems.

13.7   What salt will form upon complete neutralization of sulfuric acid with sodium hydroxide? Give its name and formula.

Answer: Na_2SO_4, sodium sulfate

13.8   How many moles of NaOH would it take to neutralize 1 mole of hydrochloric acid? How many moles of NaOH would it take to neutralize 1 mole of sulfuric acid? Why aren't the answers to these two questions the same?

13.9   How many moles of $Ba(OH)_2$ would it take to neutralize 0.1 mole of hydrochloric acid?

13.10 Make a set of beaker drawings like the ones in WorkPatch 13.3 for Practice Problem 13.9. Label them with the species present in each beaker.

Arrhenius's definition of a base as anything that dissolves in water to give hydroxide ions is still used today. However, it is not the only definition.

HELP! I NEED ANOTHER DEFINITION OF ACIDS AND BASES **13.5**

Arrhenius's definition of acids and bases is not completely satisfactory. For example, a solution of ammonia dissolved in water turns litmus blue and can be used to neutralize a solution of hydrochloric acid:

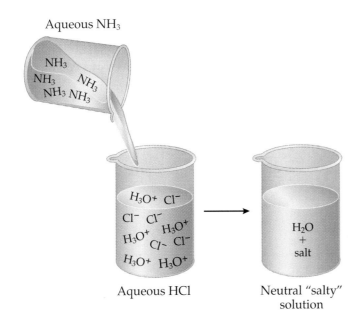

Aqueous NH_3

Aqueous HCl

Neutral "salty" solution

Since ammonia has the properties of a base, it must be a base. But we have a problem. By the Arrhenius definition, a base is a compound that dissociates in water to produce hydroxide ions. This is not a problem for ionic compounds like $NaOH$ or $Ba(OH)_2$, which are lattices packed with hydroxide ions. The Arrhenius definition of a base fits these compounds like a glove. But Arrhenius would have been hard-pressed to explain the basicity of NH_3. How can it produce OH^- ions in water if it doesn't even have an oxygen atom in it, let alone hydroxide ions? We need a broader definition.

In 1923, the Danish chemist J. N. Brønsted and the English chemist T. M. Lowry came up with the answer. Instead of thinking of a base as something that produces hydroxide ions, they defined a base more generally as any substance that removes H_3O^+ ions from solution. But they didn't stop there. They went on to detail exactly how this would occur at the molecular level. To remove H_3O^+ ions from solution, a molecule of base would have to accept a proton (H^+) from the hydronium ion. Remember, a hydronium ion is just a water molecule with a proton attached to it. If a base steals back this proton from an H_3O^+, then the H_3O^+ is converted back to water, and once again we have an acid–base neutralization reaction. Here is how it works:

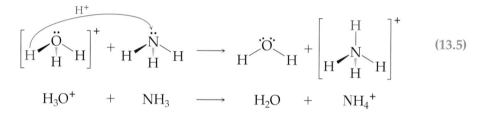

$$H_3O^+ \quad + \quad NH_3 \quad \longrightarrow \quad H_2O \quad + \quad NH_4^+ \qquad (13.5)$$

Take a very close look at this reaction. When the H moves from H_3O^+ to NH_3, it moves as an H^+ ion (a proton). This means that it leaves behind the two electrons in the O–H bond that initially held it to the oxygen atom in H_3O^+. These two electrons form the second lone pair on oxygen in the resulting H_2O molecule on the right. What comes next is very important. In order for the H^+ to be accepted by the NH_3 molecule, the NH_3 molecule must have a lone pair of electrons on the N atom. Remember, it takes two electrons to form a covalent bond between two nuclei. Since H^+ ions have no electrons, a molecule of base must have two electrons available with which to form the bond.

13·4 WORKPATCH

Below are dot diagrams for three different molecules, CH_4, PH_3, and BH_3. Only one could serve as a base according to Brønsted and Lowry's definition. Which one? Explain your answer.

(a)   H–C(–H)(–H)–H   (b)   H–P̈(–H)–H   (c)   H–B(–H)–H

The Brønsted and Lowry definition of a base then becomes "anything that can accept a proton." This is wonderfully general. It explains why species other than hydroxide ions can be bases, and it even explains why hydroxide ions act as base.

Go back and examine reaction (13.4), where a hydroxide ion reacts with a hydronium ion, and then answer the following questions:

(a) Use the Brønsted–Lowry definition of a base to explain why it is proper to call a hydroxide ion a base.

(b) Why is the hydroxide ion able to do what Brønsted and Lowry say a base should do?

Once Brønsted and Lowry defined a base as a proton acceptor, then the definition of an acid was fixed. Since an acid is the opposite of a base, an acid is a proton donor. Certainly the hydronium ion qualifies, as we've seen in reaction (13.5), where H_3O^+ donates a proton to ammonia, and reaction (13.4), where H_3O^+ donates a proton to hydroxide. So we now have a more general definition of acids and bases—the **Brønsted–Lowry definition**.

Anything that **donates** a proton is called an acid

Anything that **accepts** a proton is called a base.

Brønsted–Lowry definition of acids: Anything that *donates* a proton is called an acid.

Brønsted–Lowry definition of bases: Anything that *accepts* a proton is called a base.

Indeed, this definition is general enough that it can explain the concept of acid and base in the absence of water, OH^- ions, and H_3O^+ ions. For example, HCl molecules react with NH_3 molecules in the gas phase (in the absence of water) to give solid ammonium chloride, NH_4Cl:

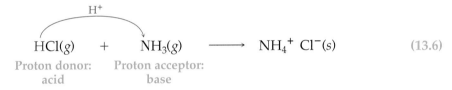

$$HCl(g) \quad + \quad NH_3(g) \quad \longrightarrow \quad NH_4^+ \, Cl^-(s) \qquad (13.6)$$

Proton donor: acid   Proton acceptor: base

Arrhenius would have had a tough time explaining this, but not Brønsted and Lowry. The HCl molecule is the acid because it donates a proton to the NH_3 molecule. The NH_3 molecule is the base because it accepts the proton. In fact, this definition allows us to classify substances as acids and bases that we normally don't think of as either. Water is an example. We usually think of water as being neutral, but when we put HCl into water, the HCl acts as an acid by donating a proton to water. By accepting the proton, water is acting as a base. In this situation, therefore, according to the Brønsted–Lowry definition, water is a base.

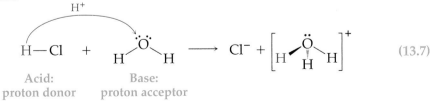

$$H—Cl \quad + \quad H\overset{..}{\underset{..}{O}}H \quad \longrightarrow \quad Cl^- + \left[H \overset{..}{\underset{|}{O}} \overset{H}{} H \right]^+ \qquad (13.7)$$

Acid: proton donor   Base: proton acceptor

Even chemical reactions that at first appearance seem totally strange to you can be understood in terms of the Brønsted–Lowry definition. For example, an important reaction in organic chemistry is that of acetylene (C_2H_2), the gas used in high-temperature oxyacetylene torches that can cut through steel, with the amide ion (NH_2^-):

$$NH_2^- + C_2H_2 \longrightarrow NH_3 + C_2H^-$$

Amide   Acetylene   Ammonia   Acetylide

Now we will show you this reaction using dot diagrams so we can emphasize exactly what is happening:

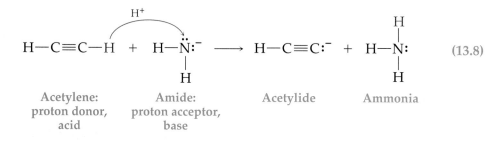

Acetylene: Amide: Acetylide   Ammonia

proton donor,   proton acceptor,

acid   base

(13.8)

By accepting a proton from acetylene, the amide ion is acting as a base. By donating the proton, acetylene is acting as an acid. The Brønsted–Lowry definition allows us to make sense of this unfamiliar reaction by recognizing it as a reaction between an acid and a base. Achieving such an understanding is always one of the most important goals of chemistry.

PRACTICE PROBLEMS

13.11 In the following reaction, which of the reactants is the acid, and which is the Brønsted–Lowry base?

$$NH_3 + PH_3 \longrightarrow NH_2^- + PH_4^+$$

Answer: PH_3 accepts a proton, so it is the base; NH_3 donates the proton, so it is the acid.

13.12 Rewrite the reaction of Practice Problem 13.11 using dot diagrams, and show with an arrow the proton being transferred.

13.13 In reaction (13.7), between HCl and water, water was shown acting as a base. How is water behaving in the following reaction? Explain your answer.

$$NH_2^- + H_2O \longrightarrow NH_3 + OH^-$$

13.6 WEAK BASES

One of the things that the Brønsted–Lowry definition does for us is to no longer require that acids and bases be electrolytes. Recall that Arrhenius said that an acid produces H_3O^+ ions in water, and a base produces OH^- ions in water. To produce ions in water, a substance must be an electrolyte, so as far as Arrhenius was concerned, acids and bases had to be electrolytes. Neither water nor electrolytes are necessary for a compound to be an acid or a base by the more general Brønsted–Lowry definition. Just look back at reaction (13.8), between acetylene and amide. This reaction does not occur in water, and even if it did, acetylene is a nonelectrolyte. Nevertheless, the Brønsted–Lowry definition has no problem assigning acetylene as an acid in the reaction.

Does this mean we should throw out the Arrhenius definition? Absolutely not. Much of the chemistry that is done on a daily basis—in a lab or in our own bodies—occurs in water, where the Arrhenius definition is quite adequate. So, let's go back to aqueous solutions where to be an acid, a compound has to produce H_3O^+ ions, and to be a base, a compound has to produce OH^- ions. In other words, a compound has to be an electrolyte.

We have discussed strong acids that are strong electrolytes, like HCl and H_2SO_4. We have discussed weak acids that are weak electrolytes, like HF and CH_3COOH (acetic acid). And we have discussed strong bases that are strong electrolytes, ionic compounds like NaOH and $Ba(OH)_2$. Are there compounds that are **weak bases** in water—that is, are there compounds that are weak electrolytes, producing only a low concentration of OH^- ions? The answer is yes. And just like weak acids, they tend to be molecular compounds that produce only a low concentration of ions in water.

One of the most commonly encountered weak bases is ammonia, NH_3. Its aqueous solutions turn litmus blue, a sign of basicity. In addition, ammonia is a weak electrolyte; its aqueous solutions cause a light bulb in electrolyte detection apparatus to glow dimly. By the Arrhenius definition, NH_3 must be producing OH^- ions in solution. Indeed, close examination of an aqueous solution of NH_3 reveals not only dissolved NH_3 molecules, but also some OH^- ions. But this brings us right back to our earlier problem. Ammonia has no OH^- ions in it, so where did they come from? We can help out the Arrhenius interpretation by invoking the Brønsted–Lowry definition. The ammonia molecule accepts a proton from water, converting it to a hydroxide ion:

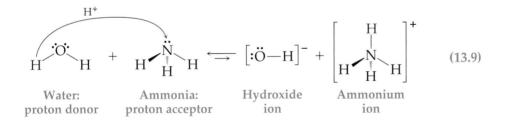

Water:	Ammonia:	Hydroxide	Ammonium
proton donor	proton acceptor	ion	ion

(13.9)

Because NH_3 produces ions in water, we are justified in calling it an electrolyte. However, the equilibrium lies far to the left (the equilibrium constant for the reaction is 1.8×10^{-5}). Thus, NH_3 is a weak electrolyte and a weak base. If you place 1 mole of NaOH in a liter of water, and 1 mole of NH_3 in a liter of water, the NaOH solution will be much more basic. That's because the beaker of aqueous NaOH will have 1 mole of OH^- ions in it, whereas the beaker of aqueous NH_3 will contain only 0.004 mole of OH^- (and NH_4^+) and almost 1 mole (actually $1.000 - 0.004 = 0.996$ mole) of intact NH_3 molecules.

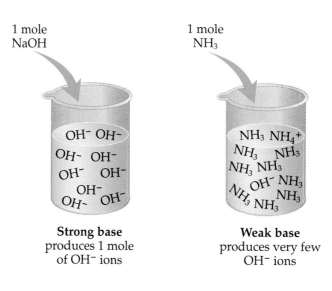

Strong base
produces 1 mole
of OH⁻ ions

Weak base
produces very few
OH⁻ ions

Since ammonia puts much less OH^- into the water than an equal molar amount of NaOH, it is referred to as a weak base. An aqueous solution of NH_3 consists mostly of intact, dissolved NH_3 molecules. Relatively speaking, there are just a few OH^- and ammonium (NH_4^+) ions in solution. However, because these ions are present, solutions of ammonia in water are sometimes called ammonium hydroxide solutions.

There are other weak bases. Two examples are the carbonate ion, CO_3^{2-}, and the acetate ion, CH_3COO^-. They usually occur as salts (that is, as the ionic compounds sodium carbonate, Na_2CO_3, and sodium acetate, $NaOOCCH_3$).

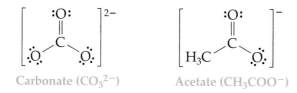

Carbonate (CO_3^{2-})   Acetate (CH_3COO^-)

Like ammonia, carbonate and acetate ions produce small amounts of OH^- in aqueous solution without actually having any OH^- ions themselves. Demonstrate that you understand why acetate and carbonate are bases in water by doing the following WorkPatch.

 WORKPATCH

Write equilibria similar to reaction (13.9), where ammonia produces hydroxide in water, that show how CO_3^{2-} and CH_3COO^- ions produce small amounts of hydroxide in aqueous solution.

Were you able to do WorkPatch 13.6? Check your equilibria against ours at the end of the chapter.

Here are the equilibrium constants for the three weak bases we have discussed:

Weak base	K_{eq}
NH_3 (ammonia)	1.8×10^{-5}
CH_3COO^- (acetate)	5.6×10^{-10}
CO_3^{2-} (carbonate)	2.1×10^{-4}

These K_{eq} values go along with the two equilibria you just wrote for Work-Patch 13.6 and with the one we wrote for ammonia in reaction (13.9). Just as we did for the weak acids, we can judge which of these weak bases is weakest and which is strongest based on their K_{eq}. Since carbonate has the largest equilibrium constant, it is the strongest of these three weak bases. Next comes ammonia, and the weakest—the one whose equilibrium lies farthest to the left—is acetate.

You may come across equilibrium constants for weak acids and weak bases that are given as K_a and K_b, respectively. For example, the K_b for NH_3 is 1.8×10^{-5}. Whenever you see such a symbol remember that K_b is just the equilibrium constant for the reaction with water in which a molecule accepts a proton from water and produces a small amount of OH^-. The same is true for weak acids, where K_a is the equilibrium constant for the reaction where the molecule donates a proton to water, converting a small amount of it to H_3O^+.

PRACTICE PROBLEMS

13.14 Aniline, C_6H_7N, is a molecular compound that is also a weak base. The molecule has a lone pair of electrons on the N atom.
  (a) Why can aniline serve as a base at all?
  (b) Write the equilibrium for aniline in water that shows how it makes the water basic.

Answer:
(a) *Aniline can serve as a base due to the lone pair of electrons on the N atom. This gives it the ability to accept a proton from an acid and bind to it.*

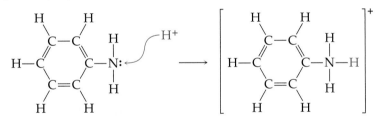

(b) $C_6H_7N + H_2O \rightleftharpoons C_6H_7NH^+ + OH^-$

13.15 The bicarbonate ion, HCO_3^-, can serve as both a Brønsted–Lowry acid and base in relation to water. Write both of these equilibrium equations, showing HCO_3^- acting as an acid and as a base.

13.16 The K_a for HCO_3^- is 4.8×10^{-11}; the K_b for HCO_3^- is 2.4×10^{-8}.
  (a) Which of the two reactions that you wrote in Practice Problem 13.15 has an equilibrium that lies farther to the right? Explain how you know.
  (b) Based on your answer to part (a), do you think an aqueous solution of bicarbonate would be acidic or basic? Explain your answer. [*Hint:* It will be acidic if the solution has more H_3O^+ in it than OH^-. It will be basic if the solution has more OH^- in it than H_3O^+.]

13.7 IS THIS SOLUTION ACIDIC OR BASIC? UNDERSTANDING WATER, AUTODISSOCIATION, AND K_w

We have spent a good deal of this chapter talking about what an acid is and what a base is, and how they make a solution acidic or basic. What we have yet to do is discuss a way to describe just how acidic or basic a particular solution is. An aqueous solution is described as being **acidic** when it contains more H_3O^+ ions than OH^- ions. Likewise, a solution is described as being **basic** when it contains more OH^- ions than H_3O^+ ions.

Acidic solutions have a higher concentration of hydronium ion than hydroxide ion.

$$[H_3O^+] > [OH^-]$$

Basic solutions have a higher concentration of hydroxide ion than hydronium ion.

$$[OH^-] > [H_3O^+]$$

When we ask exactly how acidic or basic an aqueous solution is, we are really asking about how many H_3O^+ or OH^- ions are in the solution. By now you know that when it comes to asking how much of something there is, chemists answer in moles. So when you ask how acidic a solution is, a chemist might tell you the molarity of hydronium ion, meaning how many moles of H_3O^+ ions there are per liter of solution. But that's not enough to let you know if the solution is acidic unless you first know something about the concept of a "neutral" solution.

When we say that a solution is **neutral**, we mean that it is neither acidic nor basic. Pure distilled water is neutral. But does neutral also mean that there are absolutely no H_3O^+ or OH^- ions in the water? The answer is no. In fact, it is impossible to have water that completely lacks these two ions, even if it is absolutely pure. The purest of water always has some H_3O^+ and OH^-, although not much. One liter of pure water at 25°C contains 1.0×10^{-7} (0.000 000 10) mole of H_3O^+ ions, and exactly the same amount of OH^- ions. This, in fact, is what it means to be neutral—not an absence of acidic H_3O^+ ions and basic OH^- ions, but an equal number of each. And not just any number, but 1.0×10^{-7} mole of each ion per liter, meaning the molar concentration of each is 1.0×10^{-7} M.

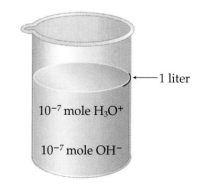

Neutral pure water has equal concentrations of hydronium and hydroxide ions, 10^{-7} M at 25°C.

Notice in the figure that we have written 10^{-7} mole of each ion instead of 1.0×10^{-7} mole. The number 1.0×10^{power} is the same as 10^{power}. We can use this shortened form only when the number that multiplies 10^{power} is 1.

Where do the H_3O^+ and OH^- ions in pure neutral water come from, and why is it impossible to have water without small amounts of these ions present? The answer is that water itself dissociates. That is, one water molecule donates a proton to another water molecule. The molecule that donates (loses) the proton turns into an OH^- ion, while the molecule that accepts the proton becomes an H_3O^+ ion. The equilibrium equation for this dissociation, often referred to as the **autodissociation** or **autoionization** of water, is shown below:

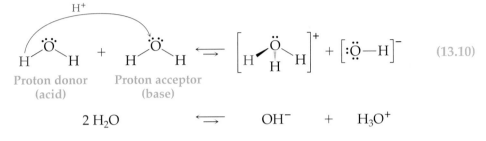

$$2\,H_2O \quad \rightleftharpoons \quad OH^- \quad + \quad H_3O^+$$

This equilibrium lies tremendously far to the left; only one out of every 555 million water molecules, on average, is dissociated at any given time. This is why the concentrations of H_3O^+ and OH^- in neutral pure water are so low. Notice also that water acts as both the acid (proton donor) and the base (proton acceptor) in this reaction. The result is the production of an equal number of H_3O^+ and OH^- ions (10^{-7} mole of each) in each liter of neutral pure water.

The fact that the autoionization equilibrium lies so far to the left means that the equilibrium constant has a value much lower than 1. We can determine the actual value of the equilibrium constant for the autoionization of water since we know that the equilibrium concentrations of H_3O^+ and OH^- in pure water are both 10^{-7} M. First, we write the K_{eq} equilibrium expression for the balanced equilibrium reaction in the usual way, as products over reactants, each raised to its own stoichiometric coefficient (note that there are two water molecules on the left side of the equilibrium):

$$\underbrace{H_2O + H_2O}_{\text{Reactants}} \rightleftharpoons \underbrace{OH^- + H_3O^+}_{\text{Products}} \qquad K_{eq} = \frac{[OH^-] \times [H_3O^+]}{[H_2O]^2}$$

Next, we simplify the equilibrium expression. Since this is an aqueous solution, there is a lot of water present compared to anything else. Pure water is 55.5 molar—that is, there are 55.5 moles of water in a liter of water. This is a number so large that it is not going to change much when we start dissolving other things in the water (at least for dilute solutions). In other words, the concentration of water in a dilute aqueous solution is essentially a constant 55.5 M. So while we can change the concentration of H_3O^+ ions and OH^- ions in water by adding acid or base, the concentration of H_2O is essentially unchanging no matter what we do. Therefore, chemists combine the constant $[H_2O]^2$ term with the equilibrium constant K_{eq} to give a new equilibrium constant called K_w (w stands for water):

$$\underbrace{H_2O + H_2O}_{\text{Reactants}} \rightleftharpoons \underbrace{OH^- + H_3O^+}_{\text{Products}} \qquad \begin{aligned} K_w &= K_{eq} \times [H_2O]^2 \\ &= [OH^-] \times [H_3O^+] \end{aligned}$$

This gives us the result that $K_w = [OH^-] \times [H_3O^+]$. This is an exceptionally important mathematical relationship for water. You should always remember it. In a moment we will show you why it is so useful, but first we need the numerical value of K_w. To find it, we simply plug in the equilibrium concentrations of H_3O^+ and OH^- in pure water. Remember, both are equal to 10^{-7} M:

$$K_w = [H_3O^+] \times [OH^-] = [10^{-7}\,M] \times [10^{-7}\,M]$$
$$= 10^{-14} \qquad \text{The value of } K_w \text{ for autoionization of water}$$

The value of K_w is tiny. Its value of 10^{-14} is much less than 1, which agrees with what we said before, that the autodissociation equilibrium lies very far to the left.

Now we are finally ready to answer the questions, "How acidic?" or "How basic?" because we have something to compare our particular solution to—neutral water. Since the OH^- and the H_3O^+ concentrations are both 10^{-7} M in neutral water, a solution whose H_3O^+ concentration is greater than 10^{-7} M is acidic. A solution whose OH^- concentration is greater than 10^{-7} M is basic.

Acidic solutions are solutions that have $[H_3O^+] > 10^{-7}$ M.

Basic solutions are solutions that have $[OH^-] > 10^{-7}$ M.

At this point you might be wondering about our earlier definitions of when a solution is acidic or basic. (Acidic solutions have $[H_3O^+] > [OH^-]$; basic solutions have $[OH^-] > [H_3O^+]$.) Which definitions of acidic and basic are right? They both are, because the OH^- and the H_3O^+ concentrations in an aqueous solution are not independent of one another. When one goes up in a solution, the other must go down. They can never be equal except for the special case of neutrality, when both are equal to 10^{-7} M. Why? Because of the autodissociation equilibrium and its K_w expression. This equilibrium, which is always occurring, tells us that for any aqueous solution, the product of the H_3O^+ concentration times the OH^- concentration must always equal 10^{-14}. The relationship $K_w = [H_3O^+] \times [OH^-] = 10^{-14}$ is an equilibrium constant for water. It always holds true.

For example, suppose we add 1 mole of the strong acid HCl to enough water to make a liter of aqueous solution. It will completely dissociate, producing 1 mole of H_3O^+ ions and making the H_3O^+ concentration 1 M:

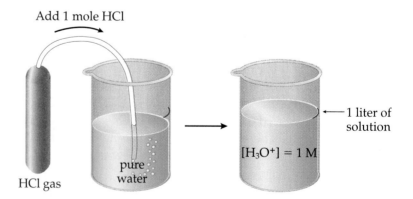

Add 1 mole HCl

HCl gas

pure water

$[H_3O^+] = 1\,M$

1 liter of solution

This solution is very acidic since the H_3O^+ concentration is much larger than 10^{-7} M. But is there any OH^- (base) present? Absolutely. There always is, no matter how acidic we make the solution, due to the autodissociation equilibrium. How much OH^- is present? What is its concentration? The K_w

expression will tell us. Since $[H_3O^+] = 1$ M, and $[H_3O^+] \times [OH^-]$ must always equal 10^{-14}, then the $[OH^-]$ concentration must be 10^{-14} M in our beaker of acid solution, as you can see below. Follow the steps to see how it works.

Solving for $[OH^-]$ when you know $[H_3O^+]$

Step 1: Plug the known concentration for H_3O^+ into the K_w expression:

$$K_w = 10^{-14} = [H_3O^+] \times [OH^-]$$

1 M

$$K_w = 10^{-14} = [1] \times [OH^-]$$

Step 2: Divide both sides by the concentration of H_3O^+ to isolate $[OH^-]$:

$$\frac{10^{-14}}{[1]} = [OH^-]$$

Step 3: Solve for $[OH^-]$:

$$10^{-14} = [OH^-] \qquad \frac{10^{-14}}{[1]} = 10^{-14}$$

The K_w expression is extremely useful for an aqueous solution. As soon as you know either the H_3O^+ or OH^- concentration, you can immediately use it to figure out the other one. The solution will be acidic or basic depending on which one is greater than 10^{-7} M. You try it now.

The following list gives the concentration of H_3O^+ in 15 solutions. For each solution, determine the concentration of OH^- and decide whether the solution is acidic, basic, or neutral. Arrange your results in tabular form.

WORKPATCH 13.7

Solution	$[H_3O^+]$, M	Solution	$[H_3O^+]$, M	Solution	$[H_3O^+]$, M
1.	10^0 Same as 1.0×10^0 M or 1.0 M	6.	10^{-5}	11.	10^{-10}
2.	10^{-1}	7.	10^{-6}	12.	10^{-11}
3.	10^{-2}	8.	10^{-7}	13.	10^{-12}
4.	10^{-3}	9.	10^{-8}	14.	10^{-13}
5.	10^{-4}	10.	10^{-9}	15.	10^{-14}

When you did the WorkPatch, were you able to see how as one concentration goes up, the other goes down? The autodissociation equilibrium and the K_w expression are so important for water that we present it to you one more time along with its important aspects:

$$K_w = [H_3O^+] \times [OH^-] = 10^{-14}$$

- *Both* H_3O^+ and OH^- ions are *always* present in aqueous solutions.
- The product of their molar concentrations *always* equals 10^{-14} at 25°C.

Of course, you should be able to solve the equation for concentrations that are not so simple. For example, suppose the H_3O^+ concentration in your solution was 2.56 M. What would the OH^- concentration be? Is the solution acidic? The answer to the second question is easy. Yes, it's acidic, because the H_3O^+ concentration is greater than 10^{-7} M. To find the OH^- concentration, we just run through the same procedure as above:

Step 1: Plug the known concentration for H_3O^+ into the K_w expression:

$$K_w = 10^{-14} = [H_3O^+] \times [OH^-] \quad \overset{2.56\ M}{\frown}$$

$$K_w = 10^{-14} = [2.56] \times [OH^-]$$

Step 2: Divide both sides by the concentration of H_3O^+ to isolate $[OH^-]$:

$$\frac{10^{-14}}{[2.56]} = [OH^-]$$

Step 3: Solve for $[OH^-]$:

$$3.91 \times 10^{-15} = [OH^-] \qquad \frac{10^{-14}}{[2.56]} = 3.91 \times 10^{-15}$$

WorkPatch 13.8 requires you to use the K_w expression to solve for H_3O^+ and OH^-. Make sure you can do both.

(13.8) WORKPATCH   For each of the solutions shown below, determine the unknown concentration, and then decide whether the solution is acidic or basic. Why did you decide as you did?

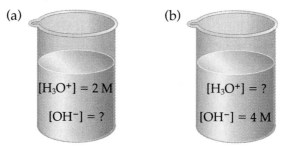

(a) $[H_3O^+] = 2\ M$
      $[OH^-] = ?$

(b) $[H_3O^+] = ?$
      $[OH^-] = 4\ M$

PRACTICE PROBLEMS

13.17 In the example we worked above, the concentration of H_3O^+ was 2.56 M and that of OH^- was 3.91×10^{-15} M. Suppose you are told to multiply these values together, and you are forbidden to use a calculator or a pencil. What is the answer?

Answer: It must be 10^{-14}. The product of $[H_3O^+] \times [OH^-]$ for any aqueous solution is always 10^{-14}.

13.18 The OH^- concentration of an aqueous solution is 0.000 155 M.
   (a) Is the solution acidic or basic? Explain how you know.
   (b) What is the molar H_3O^+ concentration in this solution?

13.19 Suppose that 1.00 mole of HCl is dissolved in enough water to give 500.0 mL of solution.
   (a) What is the H_3O^+ concentration? [*Hint:* When 1 mole of HCl dissociates, we get 1 mole of H_3O^+ ions, since the acid is strong and monoprotic. Your answer should be in moles H_3O^+/liter of solution.]
   (b) What is the OH^- concentration? [*Hint:* Use the K_w relationship.]

13.20 Suppose that 0.0100 mole of $Ba(OH)_2$ is dissolved in enough water to give 500.0 mL of solution.
  (a) What is the OH^- concentration? [*Hint:* Remember that every time 1 mole of $Ba(OH)_2$ dissociates, we get 2 moles of OH^- ions.]
  (b) What is the H_3O^+ concentration? [*Hint:* Use the K_w relationship.]

THE pH SCALE 13.8

Let's take quick stock of where we are. Arrhenius has taught us that in water, H_3O^+ is the acid species and OH^- is the basic species. The fact that water undergoes an autodissociation equilibrium has taught us that both species are always present in water, their concentrations tied to one another via the K_w expression. Only when the water is neutral will the concentrations of both be the same, 10^{-7} M. However, if you add acid to neutral water, then the H_3O^+ concentration will become larger than 10^{-7} M and the solution will be acidic. If you add base to neutral water, then the OH^- concentration will become larger than 10^{-7} M and the solution will be basic. The concentration of 10^{-7} M therefore takes on a special significance. It is the neutral value to which we must compare the H_3O^+ and OH^- molar concentrations to determine if a solution is acidic or basic.

As much as chemists are used to molar concentrations, they have devised a special way to indicate how acidic or basic a solution is. Rather than report the molar concentration of the H_3O^+ ion, they report the **pH** of a solution:

$$pH = -\log [H_3O^+]$$

All scientific calculators have a [log] key, so it is easy to simply punch in a solution's molar H_3O^+ concentration, hit the log key, and then multiply the result by -1. When you do this, you arrive at the solution's pH. Thus, to get a pH you do not even have to understand what a log is. But you should. A brief explanation can make it clear.

The "log" stands for the *logarithm to the base 10*, and we speak of "taking the log of a number." Given any number x, you can enter it into your calculator and press the log key. Your calculator will return some new number y, which is the log of the number x that you originally entered. We write this as $\log x = y$. Try it for yourself. Enter the number 100 into your calculator, then hit the log key and see what you get. Your calculator should report that the log of 100 equals 2; that is, $\log 100 = 2$.

What does it mean? A log is an exponent, or power, of 10. In other words, when you take the log of 100, you are really asking your calculator the following question, "To what power must I raise 10 to get the number 100?" The answer is 2, because $10^2 = 100$. If we ask you for the log of 1000, we are really asking, "To what power must you raise 10 to get the number 1000?" You don't even need a calculator for this. You know that $10^3 = 1000$, so the log of 1000 is 3. Any time you ask for the log of a number, you are really asking the question, "To what power must I raise 10 to get that number?" That power is the log of that number.

PRACTICE PROBLEMS

13.21 Without using a calculator, what is the log of 0.0010?

Answer: We are really asking, "To what power must you raise 10 to get 0.0010?" The number 0.0010 can be written as 1.0×10^{-3}. Thus, the log of 0.0010 is -3.

13.22 Without using a calculator, what is the log of 10,000?

13.23 Without using a calculator, what is the log of 10? The log of 10^1 is the same. Why?

13.24 Without using a calculator, what is the log of 0.010? The log of 10^{-2} is the same. Why?

Are you getting a feel for what a log is? It's easy to find the log of numbers that are whole powers of 10, like the numbers in Practice Problems 13.21–13.24. If such numbers are written in scientific notation, it's even easier. For example, the number 0.000 000 10 can be written as 1.0×10^{-7}, or simply as 10^{-7}. The log of 10^{-7} is just -7. After all, when you ask for the log of 10^{-7}, you are really asking, "To what power must you raise 10 to get 10^{-7}?"

PRACTICE PROBLEMS

Do not use a calculator for the following problems.

13.25 What is the log of 10^3?

Answer: 3

13.26 What is the log of 1.0×10^{-11} and of 10^{-11}?

13.27 What is $-\log$ of 10^{-7}? [*Hint:* Don't forget the minus sign in front of the log.]

You will need to use a calculator when asked for the log of a number that is not a whole power of 10. For example, what is the log of 558? This number is not a whole power of 10, and is between 100 and 1000. The answer will be a number between 2 and 3, since the log of 100 is 2 ($10^2 = 100$), and the log of 1000 is 3 ($10^3 = 1000$). A calculator reveals that the log of 558 is 2.75 (that is, $10^{2.75} = 558$).

Now let's go back to pH. Table 13.4 lists the results you obtained in Work-Patch 13.7 plus a new column, the pH of the solution. Each row represents a solution with a different H_3O^+ concentration, going from highly acidic (1 M H_3O^+) on top, through neutral, to highly basic (1 M OH^-) on the bottom. The pH column was filled out by simply taking the negative of the log of the number in the [H$_3$O$^+$] column. For example, in the first row, $-\log 10^0 = -(0) = 0$. In the second row, $-\log 10^{-1} = -(-1) = 1$; and so on.

Table 13.4   The pH Scale

[H$_3$O$^+$], M	[OH$^-$], M		pH
10^0	10^{-14}	Very acidic	0
10^{-1}	10^{-13}		1
10^{-2}	10^{-12}		2
10^{-3}	10^{-11}		3
10^{-4}	10^{-10}	Acidic	4
10^{-5}	10^{-9}		5
10^{-6}	10^{-8}		6
10^{-7}	10^{-7}	◀ Neutral	7
10^{-8}	10^{-6}		8
10^{-9}	10^{-5}		9
10^{-10}	10^{-4}		10
10^{-11}	10^{-3}	Basic	11
10^{-12}	10^{-2}		12
10^{-13}	10^{-1}		13
10^{-14}	10^0	Very basic	14

*Remember that $10^0 = 1.0 \times 10^0$ M $= 1.0$ M, and so forth, for the other entries.

Examine the pH column in Table 13.4 and you will learn how a solution's pH varies as it goes from very acidic, through neutral, to very basic. A neutral solution has pH = 7, because its H$_3$O$^+$ concentration is 10^{-7} M. A lower pH means increased acidity, and the pH has to be lower than 7 for a solution to be considered acidic ([H$_3$O$^+$] > [OH$^-$]). Likewise, as the solution becomes increasingly basic, the pH goes up, and the pH has to be higher than 7 for a solution to be considered basic ([OH$^-$] > [H$_3$O$^+$]). In addition, because it is a logarithmic scale, each unit decrease in pH represents a tenfold increase in acidity, and each unit increase in pH represents a tenfold decrease in acidity. For example, when the pH decreases from 1 to 0, the H$_3$O$^+$ concentration increases from 0.1 M to 1.0 M, a tenfold increase. A solution whose pH = 5 is 100 times more acidic than a solution whose pH = 7 (2 pH units = $10 \times 10 = 100$). A solution whose pH = 10 is 1000 times less acidic than a solution whose pH = 7 (3 pH units = $10 \times 10 \times 10 = 1000$).

Practice Problems

13.28 Which solution is less acidic, solution A with pH 2, or solution B with pH 6, and by how much?

Answer: Solution B, with the higher pH, is less acidic by 10,000 times. The difference between pH = 6 and pH = 2 is 4, and 4 pH units = 10^4, or 10,000.

13.29 Basic solution A has pH = 9. Basic solution B is 10 times more basic than A. What is the pH of solution B?

13.30 Consider a solution 10 times more acidic than the most acidic solution in Table 13.4.
(a) What is the pH of this solution?
(b) What is the concentration of H_3O^+ in this solution?

The pH of a solution can be quite important. For example, the pH of your blood is approximately 7.4, just to the basic side of neutrality. If the pH should fall below 7.4, a condition called acidosis (too much acid) exists. The blood pH of severely diabetic patients can drop to a level as low as 6.8. Such severe acidosis can cause coma and death. You wouldn't think that a drop of just 0.6 of a pH unit could kill you, until you remember that the scale is logarithmic and that a drop of a full pH unit represents a tenfold increase in acidity.

A weakened forest due to acid rain

Another area where pH is extremely important is our environment. Natural rainwater has a pH close to 5.6, acidic due to the formation of the weak carbonic acid (H_2CO_3) from atmospheric CO_2. However, the atmospheric pollutants SO_2 gas and SO_3 gas, which come from cars and power plants burning the sulfur impurities in gasoline and coal, respectively, form sulfurous acid (H_2SO_3) and sulfuric acid (H_2SO_4). When rain falls in areas with such pollutants in the air, the result is what we call *acid rain*. In many areas of the industrialized world, rain has been measured to have a pH as low as 3! A drop from pH 5.6 to 3 is almost 3 full pH units, meaning that acid rain is almost 1000 times as acidic as normal rain. The result is often the devastation of forests and the eradication of fish from lakes and streams—sometimes many miles from the source of the pollutants. Environmental agencies devote much time to monitoring the pH of rain, soil, and lakes with a simple device known as a pH meter.

Once the pH is known, it is a simple matter to convert it back into the molar concentration of H_3O^+. After all, a pH is a log (actually, the negative of a log), and a log is an exponent of 10. Therefore, to convert pH back to molar concentration, we simply put a minus sign in front of it and make it an exponent of 10:

The formula for converting pH back into a molar H_3O^+ concentration

$$[H_3O^+] = 10^{-(pH)}$$

You can do this with a scientific calculator. Simply enter the pH, multiply it by

−1, and then press the key labeled $\boxed{10^x}$. For those calculators that do not have this key, depress the $\boxed{\text{INV}}$ (inverse) key followed by the $\boxed{\log}$ key. For example, the molar H_3O^+ concentration of your pH 7.4 blood is $10^{-7.4}$, which will give you 3.98×10^{-8} M when entered correctly into your calculator.

PRACTICE PROBLEMS

13.31 A basic solution has a pH of 9.8. What is its molar H_3O^+ concentration?

Answer: $[H_3O^+] = 10^{-9.8} = 1.58 \times 10^{-10}$ M

13.32 What is the hydronium ion concentration in a solution that is 100 times less acidic than one having a pH of 2.56?

13.33 What is the OH^- concentration in a solution having a pH of 5.55? [*Hint:* You will need to use the K_w expression.]

The chart shows the pH of various common substances:

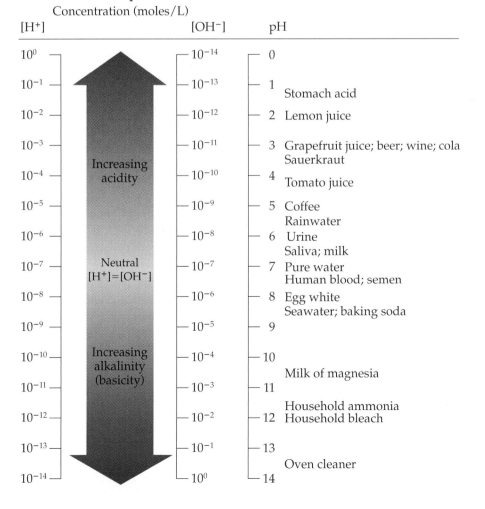

The pH of some common substances

Concentration (moles/L)

$[H^+]$		$[OH^-]$	pH	
10^0		10^{-14}	0	
10^{-1}		10^{-13}	1	Stomach acid
10^{-2}		10^{-12}	2	Lemon juice
10^{-3}		10^{-11}	3	Grapefruit juice; beer; wine; cola
	Increasing acidity			Sauerkraut
10^{-4}		10^{-10}	4	Tomato juice
10^{-5}		10^{-9}	5	Coffee
				Rainwater
10^{-6}		10^{-8}	6	Urine
				Saliva; milk
10^{-7}	Neutral $[H^+]=[OH^-]$	10^{-7}	7	Pure water
				Human blood; semen
10^{-8}		10^{-6}	8	Egg white
				Seawater; baking soda
10^{-9}		10^{-5}	9	
10^{-10}	Increasing alkalinity (basicity)	10^{-4}	10	
				Milk of magnesia
10^{-11}		10^{-3}	11	
				Household ammonia
10^{-12}		10^{-2}	12	Household bleach
10^{-13}		10^{-1}	13	
				Oven cleaner
10^{-14}		10^0	14	

You can see in the chart on the previous page that human blood, at 7.4, is slightly basic. Pay particular attention to stomach acid, all the way down at about pH 1.3 and thus very strongly acidic. We'll have more to say about stomach acid in the next section.

13.9 RESISTING pH CHANGES—BUFFERS

In the last section we saw that a decrease in the pH of your blood from its normal value of 7.4 down to 6.8 could be fatal. How much acid would it take to do this, and where might it come from? Diabetes was mentioned as one source. Another source is your stomach, which produces strong gastric acids with pH as low as 1.3. Your stomach is lined to prevent this acid from entering your bloodstream. But suppose you should develop a bleeding ulcer, or suppose you take too much aspirin (acetylsalicylic acid) for a headache, or catch a stomach flu, all of which could cause some bleeding in your stomach lining. Now your blood is in contact with your stomach acid.

How much of this acid would have to get into your bloodstream to lower your blood pH to a fatal level? Assuming that you have 5 L of blood (a typical adult blood volume), some quick calculations can show that it should take only 0.01 mL of stomach acid to do the job. A milliliter is about 10 drops from an eyedropper, so 0.01 mL is only a fraction of a drop. Only a fraction of a drop of your own stomach acid getting into your bloodstream could lower your blood pH from 7.4 to 6.8. So what has been keeping you alive?

The answer is that your blood has the extremely important ability to resist changes to its pH, even upon addition of reasonably large amounts of acids or bases. The ability of a solution to resist changes to its pH is called its **buffering ability**, and the solution is called a **buffered solution**, or just a **buffer** for short. The buffering ability of your blood makes it quite unlike pure water, which has no buffering ability. For example, while that fraction of a drop of stomach acid in 5 L of your blood actually causes almost no change in pH, the same amount added to 5 L of neutral water would decrease its pH a full unit, from 7 to 6, as we predicted above. If your blood behaved like water, then you wouldn't be reading this now. Chemists have discovered where blood gets its buffering ability from, and in fact, we can buffer any aqueous solution to resist pH changes. In order to understand how a buffer works, you must learn one more thing about acids and bases—the concept of *conjugates*.

If you look up the word "conjugate" in a dictionary, you will see synonyms such as "couple" or "marry." Every acid and every base has a conjugate, a "mate" to which it is forever tied. Every acid has a **conjugate base**, and every base has a **conjugate acid**. An acid and its conjugate base make up what is called a **conjugate pair**, and they are called conjugates of each other. A base and its conjugate acid also make up a conjugate pair and are called conjugates of each other.

Conjugate pair

Acid   and its   Conjugate base

Conjugate pair

Base   and its   Conjugate acid

Notice that a conjugate pair always consists of an acid and a base. The only difference between the conjugates in a conjugate pair is a proton (H^+). The acid member of the pair has the proton, the basic member does not. This means that an acid turns into its conjugate base by losing a proton, and a base turns into its conjugate acid by gaining a proton. The examples listed in Table 13.5 should make this clear. Notice that the only difference between each acid in the left column and its conjugate base in the right column is a proton. The acid has it, whereas the conjugate base does not.

Table 13.5   Some Weak Acids and Their Conjugate Bases

Weak acid	Conjugate base
Acetic acid, CH_3COOH	Acetate ion, CH_3COO^-

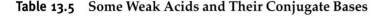

Carbonic acid, H_2CO_3	Bicarbonate ion, HCO_3^-

Hydrofluoric acid, HF	Fluoride ion, F^-
Ammonium ion, NH_4^+	Ammonia, NH_3
Phosphoric acid, H_3PO_4	Dihydrogen phosphate ion, $H_2PO_4^-$
Hypochlorous acid, HClO	Hypochlorite ion, ClO^-

What is the conjugate base of citric acid, formula $C_5H_7O_6OH$? [*Hint:* The last H written is the one donated by this acid.]   **WORKPATCH**

Did you get both the formula and the charge of the conjugate base correct? Remember, if a molecule or ion loses something that has a +1 charge (like an H^+), then it must become more negative by −1. This conserves charge. Try the next WorkPatch to make sure you understand the difference between conjugates.

Carbonic acid (H_2CO_3) is a weak acid. Carbonate ion (CO_3^{2-}) is a weak base. A student claims that these are conjugates of each other. Is she right or wrong? Explain your answer.   **WORKPATCH** 13.10

We will give you a hint. She is absolutely wrong. Make sure that you know why. Check your answer against ours at the end of the chapter.
In Table 13.5, we listed some weak acids and their conjugate bases. Table 13.6 lists some weak bases and their conjugate acids.

Table 13.6   Some Weak Bases and Their Conjugate Acids

Conjugate acid	Weak base
Acetic acid, CH_3COOH	Acetate ion, CH_3COO^-
Carbonic acid, H_2CO_3	Bicarbonate ion, HCO_3^-
Hydrofluoric acid, HF	Fluoride ion, F^-
Ammonium ion, NH_4^+	Ammonia, NH_3
Phosphoric acid, H_3PO_4	Dihydrogen phosphate ion, $H_2PO_4^-$
Hypochlorous acid, HClO	Hypochlorite ion, ClO^-

Take a moment to compare Tables 13.5 and 13.6. Do you notice anything similar about them? We hope so. Except for the column headings, the entries are identical. The point is that in any conjugate pair, it is arbitrary which we call the conjugate. Either the acid or the base can be considered the conjugate because they are conjugates of each other. It's up to you.

PRACTICE PROBLEMS

13.34 The bicarbonate ion, HCO_3^-, is interesting because it can act as both a weak base and a weak acid.
   (a) Show how HCO_3^- can act as an acid.
   (b) Show how HCO_3^- can act as a base.
   (c) Since HCO_3^- can act as either an acid or a base, can it be its own conjugate?

Answer:
(a) $HCO_3^- + H_2O \rightleftarrows CO_3^{2-} + H_3O^+$ *(donates a proton to water)*
(b) $HCO_3^- + H_2O \rightleftarrows H_2CO_3 + OH^-$ *(accepts a proton from water)*
(c) No, HCO_3^- cannot be its own conjugate. Conjugates must differ by a proton.

13.35 If HCO_3^- is considered to be a weak acid, then what is its conjugate base?

13.36 If HCO_3^- is considered to be a weak base, then what is its conjugate acid?

There is one more thing we need to tell you about conjugates before we go on to the workings of a buffer. If you compare Table 13.5 or 13.6 to Tables 13.2A and 13.2B on page 469, you will notice that all the acids listed in Tables 13.5 and 13.6 are weak acids and all the bases are weak bases. Why are there no strong acids, like HCl, or strong bases, like NaOH, in these tables? Doesn't hydrochloric acid have a conjugate base? Doesn't sodium hydroxide have a conjugate acid? On paper, yes, they do, but in practice, no, they don't. Consider HCl, a very strong acid. Its conjugate base is just HCl minus a proton (H^+), which is the chloride (Cl^-) ion:

Strong acid:	HCl	⎫
		⎬ They differ by a proton
Its conjugate base:	Cl^-	⎭

So on paper, HCl has a conjugate base. The problem is that chloride ions don't act like a base in water. Solutions of chloride ions are not bitter, they are not slippery to the touch, they do not turn litmus blue, and they possess no excess of hydroxide ion. This is very different behavior from the conjugate base of a weak acid, like acetic acid:

Weak acid: CH$_3$COOH
                          Acetic acid
                                        — They differ by a proton
Its conjugate base: CH$_3$COO$^-$
                          Acetate

If you dissolve acetate ions in water, the water gets moderately basic. Acetate is in fact a weak base.

The rule then is as follows: Weak acids give rise to conjugates that are weak bases (and vice versa). However, very strong acids and bases give rise to conjugates that are so weak they are not acids or bases at all in an Arrhenius sense. Chloride ion, for example, refuses to accept a proton from water. Therefore, no OH$^-$ ions are produced, and the solution is not basic.

$$Cl^- \ + \ H_2O \ \nrightarrow \ HCl \ + \ OH^-$$
Conjugate   Strong
base of HCl   acid

This should not surprise you. After all, HCl is a very strong acid, which means that it has a tremendous urge to donate its proton. Therefore, the opposite must also be true: Cl$^-$ must have no urge at all to accept it and get it back. The same can be said of H$_2$O, the conjugate acid of OH$^-$. Hydroxide ion is an extremely strong base. It has a tremendous urge to accept a proton. Therefore its conjugate acid, H$_2$O, must have no real tendency to donate it. Thus, while on paper, H$_2$O is the conjugate acid of OH$^-$, in practice pure water is not acidic at all. So, when it comes to applying the concept of conjugates, we usually limit it to weak acids and bases. Weak acetic acid has a conjugate base that behaves like a base. Hydrochloric acid does not. The weak base acetate has a conjugate acid that behaves like an acid. The strong base NaOH does not.

WORKPATCH 13.11

Suppose you have a bottle of acetic acid and a bottle of sodium acetate (NaOOCCH$_3$), and you place some of both into a beaker of water. Consult Tables 13.5 and 13.6 to help you answer the following questions.
   (a) Is there a weak acid in this solution? If so, what is it?
   (b) Is there a weak base in this solution? If so, what is it?
   (c) Is there a conjugate pair in this solution? If so, what is it?

If you answered the WorkPatch correctly, then you are closer than you think to understanding a buffer solution and its ability to resist changes to its pH. That's because the solution described in WorkPatch 13.11 is a buffered solution.

When we dissolve a conjugate weak acid/base pair in water, the resulting aqueous solution is a buffer. In WorkPatch 13.11, the acetic acid is a weak acid, and the acetate ion is a weak base. Of course, they are not just any acid and base; they are conjugates of each other. The source of the acetate ions was sodium acetate (NaOOCCH$_3$), an ionic compound that completely dissociates

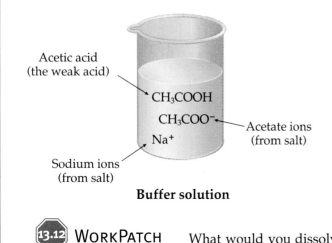

Acetic acid
(the weak acid)

CH$_3$COOH

CH$_3$COO$^-$ — Acetate ions
(from salt)

Na$^+$

Sodium ions
(from salt)

Buffer solution

into ions upon dissolving. Since ionic compounds are often referred to as salts, we can say that this buffer solution was prepared by dissolving a weak acid and its salt. This is a very common way to prepare a buffer solution.

The Na$^+$ ions that are present in the buffer solution shown at left play no role in the solution's buffering ability, and are thus often referred to as *spectator ions*. It is the presence of the weak acid and its conjugate weak base that give the solution its buffering ability. Without having a good amount of both, you don't have a buffer.

13.12 WORKPATCH What would you dissolve in water along with the weak acid HF in order to prepare a buffer solution?

Now that you know how to prepare a buffer, the question is, when you add a strong acid like HCl or a strong base like NaOH to a buffered solution, why doesn't the pH of the solution change much? In other words, how does a buffer work? The answer is actually quite simple. A buffer in effect replaces the added strong acid or strong base by converting it into a weak acid or weak base, respectively. As a result, the added acid or base has much less effect on the pH of the solution.

How does a buffer replace the added strong acid with weak acid and the added strong base with weak base? It sends its conjugate weak acid/base pair to do battle with them. Remember, acids and bases react with each other. Lurking in a buffer solution like sharks are a weak acid and its conjugate weak base, waiting to react with any added strong base (OH$^-$) or strong acid (H$_3$O$^+$), respectively.

In a buffer, the weak acid (acetic acid, CH$_3$COOH)
goes after any added strong base (OH$^-$),
and the weak base (acetate ion, CH$_3$COO$^-$)
goes after any added strong acid (H$_3$O$^+$).

The acid–base reactions that occur are shown on this page. To get rid of added strong base, the buffer sends in the weak acid to donate a proton to it. This "kills" the OH⁻ (turns it into water) and converts the weak acid into weak base. The net result is that the OH⁻ gets replaced by the much weaker base acetate. Because OH⁻ is a strong base, this reaction goes to completion, even though acetic acid (CH_3COOH) is a weak acid. As long as one of the reactants in an acid–base reaction is strong, the reaction will go to completion.

To get rid of added strong acid, the buffer sends in the weak base to accept a proton from it. This "kills" the H_3O^+ (turns it into water) and converts the weak base into weak acid. The net result is that the H_3O^+ gets replaced by the much weaker acetic acid. This reaction goes to completion, even though acetate (CH_3COO^-) is a weak base, because H_3O^+ is a strong acid.

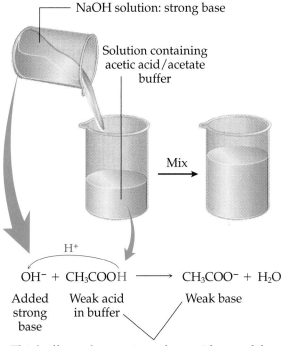

$$\overset{H^+}{\overbrace{OH^- + CH_3COOH}} \longrightarrow CH_3COO^- + H_2O$$

Added strong base   Weak acid in buffer   Weak base

This buffer replaces a strong base with a weak base at the expense of some buffering weak acid.

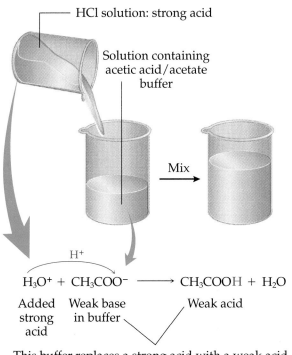

$$\overset{H^+}{\overbrace{H_3O^+ + CH_3COO^-}} \longrightarrow CH_3COOH + H_2O$$

Added strong acid   Weak base in buffer   Weak acid

This buffer replaces a strong acid with a weak acid at the expense of some buffering weak base.

WORKPATCH 13.13

(a) How does a weak acid in a buffer eliminate added strong OH⁻ base? What does it turn the base into? What happens to the weak acid?
(b) How does a weak base in a buffer eliminate added strong H_3O^+ acid? What does it turn the acid into? What happens to the weak base?

It is essential that you be able to answer WorkPatch 13.13. Examine the shark buffer solution on the facing page to help you. Each buffer species (shark) always attacks its opposite. Buffer acid goes after added strong base. Buffer base goes after added strong acid. Let's see if you are catching on.

13.14 WORKPATCH   Imagine a buffer made by combining an aqueous solution of the weak acid HF with its sodium salt, NaF.

(a) Write the reaction responsible for removing added strong base (OH⁻) and converting it into weak base.

(b) Write the reaction responsible for removing added strong acid (H_3O^+) and converting it into weak acid.

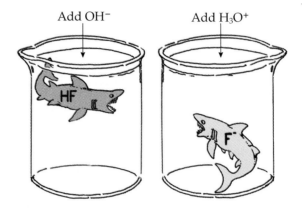

Compare your answers to the WorkPatch to ours, and make sure you get it right.

The main buffer in our own blood is a bicarbonate/carbonate (HCO_3^-/CO_3^{2-}) system that regulates the blood pH at around 7.4. Buffers like the one in our blood or the ones discussed above ($CH_3COOH/NaOOCCH_3$ and HF/NaF) will keep working until their ingredients, the weak acid or the conjugate weak base, get used up. If you add enough strong acid, you will consume all the weak base. Or, if you add enough strong base, you will consume all the weak acid. Once either occurs, we say that the buffer is exhausted, or that its capacity has been reached. At that point, the buffer starts acting like plain old water. Therefore, when you prepare a buffer solution, you need to make sure that it has good amounts of both the weak acid and its conjugate base or salt.

Finally, make sure that you understand that even the best of buffers only resists changes to its pH upon addition of strong acids and bases; no buffer entirely cancels out their effect. The pH still goes up when strong base is added, and it still goes down when strong acid is added. It just changes a lot less than it would in pure water. That the pH still changes somewhat should not surprise you. After all, if all the added H_3O^+ is replaced by acetic acid in an acetic acid/acetate buffer, then the solution still has all that newly produced acetic acid. Since acetic acid is a weak acid, some of it will dissociate to produce some more H_3O^+ ions, decreasing the pH. Likewise, adding OH⁻ forms more acetate ion, a weak base, which will slightly raise the pH.

PRACTICE PROBLEMS

13.37 Ammonium ion (NH_4^+) is a weak acid. Write its reaction with water.

Answer: $NH_4^+ + H_2O \rightleftharpoons NH_3 + H_3O^+$

Being weak, the equilibrium lies far to the left.

13.38 Ammonia (NH_3) is a weak base. Write its reaction with water.

13.39 What do you call an aqueous solution that has a large amount of NH_4Cl and NH_3 dissolved in it? Explain your answer.

13.40 Write a reaction to show how the solution in Practice Problem 13.39 resists pH change upon addition of strong base (OH^-).

13.41 Write a reaction to show how the solution in Practice Problem 13.39 resists pH change upon the addition of strong acid (H_3O^+).

HAVE YOU LEARNED THIS?

Acid (p. 457)

Base (p. 458)

Neutral (p. 458)

Electrolyte (pp. 459, 463)

Nonelectrolyte (pp. 459, 463)

Dissociation (p. 462)

Strong electrolyte (p. 467)

Weak electrolyte (p. 467)

Arrhenius definition of an acid (p. 468)

Strong acid (p. 468)

Weak acid (p. 468)

Monoprotic, diprotic, and triprotic acids (p. 470)

Acid–base neutralization reaction (p. 472)

Arrhenius definition of a base (p. 473)

Strong base (p. 473)

Brønsted–Lowry definition of an acid and a base (p. 477)

Weak base (p. 479)

K_a and K_b (p. 481)

Acidic solution (p. 482)

Basic solution (p. 482)

Neutral solution (p. 482)

Autodissociation or autoionization (p. 483)

K_w (pp. 483, 485)

pH scale (pp. 487, 489)

Buffer (p. 492)

Conjugate acid/base pairs (p. 492)

ELECTROLYTES

13.42 List the three criteria early chemists used to classify a compound as an acid.

13.43 List the three criteria early chemists used to classify a compound as a base.

13.44 What is an indicator? Give an example of one.

13.45 Describe an experimental setup to determine whether a compound is an electrolyte or a nonelectrolyte. What would you look for?

13.46 What is an electrolyte? What is a nonelectrolyte? Give some examples of each.

13.47 Ethanol (C_2H_5OH) dissolves in water. So does magnesium chloride ($MgCl_2$). Yet there is a fundamental difference in the way these two substances dissolve. What is this fundamental difference?

13.48 What is meant by the term "dissociation"? Give an example.

13.49 If an ionic compound is water-soluble, then it is an electrolyte. Explain why this is so.

13.50 True or false? Since a molecular compound has no ions, it cannot dissociate upon dissolving. Back up your answer with an explanation and an example.

13.51 What must be present in an aqueous solution in order for it to conduct electricity?

13.52 The molecular compound HCl is also an electrolyte.
(a) What do we mean when we say HCl is a molecular compound?
(b) Is it incorrect to call HCl an ionic compound?
(c) What must a molecular compound like HCl do when dissolved in aqueous solution in order to function as an electrolyte?

13.53 The compound NaCl is also an electrolyte.
(a) Would it be incorrect to call NaCl a molecular compound? Explain your answer.
(b) Why is NaCl expected to be an electrolyte?

13.54 Explain why the following statement is true: "Water-soluble compounds that consist of a metal combined with nonmetal atoms are electrolytes."

13.55 Which of the following are electrolytes? Which are nonelectrolytes?
(a) Calcium bromide ($CaBr_2$)
(b) Iodine (I_2)
(c) Hydrogen (H_2)
(d) Hydrogen bromide (HBr) [*Hint:* Look at the other elements in group VIIA in the periodic table.]
(e) Ammonium fluoride (NH_4F)
(f) Potassium chlorate ($KClO_3$)
(g) Sucrose ($C_{12}H_{22}O_{11}$)

13.56 Write a balanced equation to show what happens to Na_2SO_4 (sodium sulfate) when it dissolves in water. Use the (*aq*) symbol when necessary.

13.57 Write the formula, complete with charge, for the ammonium ion.

13.58 Write a balanced equation to show what happens when $CaBr_2$ (calcium bromide) dissolves in water. Use the (*aq*) symbol when necessary.

13.59 What is wrong with the following request? "Write a balanced equation for the dissociation in water of the sugar glucose, $C_6H_{12}O_6$."

13.60 What do we mean by saying that some molecular compounds dissolve in water and also dissociate? Give an example, and draw a picture showing exactly what is in solution.

WEAK VERSUS STRONG ELECTROLYTES

13.61 Two beakers each contain 1 liter of water. Into beaker A is dissolved 0.1 mole of HCl gas. Into beaker B is dissolved 0.1 mole of HF gas. The solution in beaker A causes a light bulb to glow intensely. The solution in beaker B causes the same light bulb to glow dimly. Explain these observations.

13.62 Regarding Problem 13.61, in which solution is the halide ion concentration equal to 0.1 M, and in which solution is the halide ion concentration much less than 0.1 M? Explain how you know.

13.63 To which side does the dissociation equilibrium lie for a strong electrolyte? For a weak electrolyte?

13.64 What do we mean by the term "partially dissociates"? Give an example of a compound that partially dissociates in water, and draw a picture of the solution showing what is in it and the relative amounts.

13.65 Most molecular electrolytes produce a particular ion in water. What is that ion?

13.66 Is it more correct to write $H^+(aq)$ or $H_3O^+(aq)$? Explain your answer.

13.67 A particular molecular electrolyte has an equilibrium constant for dissociation of 8.2×10^{-6}. Is it strong or weak? Explain your answer.

13.68 A particular molecular electrolyte has an equilibrium constant for dissociation of approximately 10^8. Is it appropriate to forget about the equilibrium for this electrolyte? Explain your answer.

13.69 Write the formulas and names of two strong molecular electrolytes and two weak molecular electrolytes. For all four, write the appropriate balanced dissociation reaction.

ACIDS, WEAK AND STRONG

13.70 Give the names of the following acids:
(a) HCl   (b) HNO_3 (c) H_2SO_4 (d) HF   (e) CH_3COOH

13.71 Which of the compounds in Problem 13.70 are molecular and which are ionic? Explain how you know.

13.72 Which of the acids in Problem 13.70 are molecular electrolytes, and what common ion do they produce in water?

13.73 According to the Arrhenius definition, why are all the compounds in Problem 13.70 called acids?

13.74 True or false? The acidity of a beaker of 1.0 M hydrofluoric acid is the same as that of a beaker of 1.0 M hydrochloric acid. Fully explain your answer.

13.75 Which of the acids in Problem 13.70 are weak? Which are strong? What do these terms mean with respect to equal concentrations of all these acids?

13.76 What does "diprotic" mean when applied to an acid? Give an example and show both dissociation equilibria.

13.77 Give an example of a triprotic acid, and write down all the dissociation equilibria.

13.78 Other than water, what would you expect to find in highest concentration in an aqueous solution of CH_3COOH? Explain your answer.

13.79 Other than water, what would you expect to find in highest concentration in an aqueous solution of HNO_3? Explain your answer.

13.80 Imagine that the eight identical molecules shown below are all molecules of a weak molecular acid. The beaker is full of water. After the eight molecules dissolve in the water and equilibrium is reached, draw a picture showing all the expected ions or molecules in the water, taking into account the expected equilibrium concentrations.

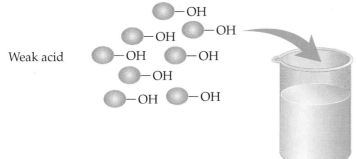

13.81 Answer the same question as in Problem 13.80, but now consider the acid to be very strong.

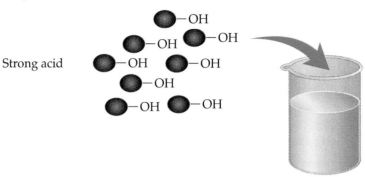

Strong acid

13.82 Draw the molecular structure (a dot diagram) of acetic acid. Indicate which proton is the one that dissociates.

13.83 What is the value of the equilibrium constant for the dissociation of acetic acid's proton? What does it tell us?

13.84 What would you estimate the value of the equilibrium constant for the dissociation of the first proton from sulfuric acid to be? Explain your estimate.

13.85 True or false? Since carbonic acid, H_2CO_3, is diprotic, it is a stronger acid than HCl, which is only monoprotic. Explain your answer.

BASES—THE OPPOSITE OF ACIDS

13.86 What is the Arrhenius definition of a base?

13.87 What is meant by acid–base neutralization? Write a balanced reaction to go along with your answer.

13.88 What kinds of compounds are typically strong bases?

13.89 When calcium hydroxide reacts with HCl in a neutralization reaction, the resulting solution is found to conduct electricity very well. Explain this.

13.90 Write a balanced neutralization equation for the reaction of potassium hydroxide with aqueous hydrochloric acid.

13.91 How many moles of $Ba(OH)_2$ would it take to neutralize 0.50 mole of H_2SO_4?

13.92 How many moles of $Ba(OH)_2$ would it take to neutralize 2.5 moles of HNO_3?

13.93 What salts will be formed from the neutralization reactions of Problems 13.91 and 13.92?

13.94 Draw a set of beakers similar to those in WorkPatch 13.3 for Problems 13.91 and 13.92. Fill in the species present in each beaker. For H_2SO_4, imagine that it dissociates completely to give two H_3O^+ ions (although it actually doesn't do this).

ANOTHER DEFINITION OF ACIDS AND BASES

13.95 Ammonia is a weak base. Write a balanced equation for what happens when gaseous NH_3 is dissolved in water.

13.96 The equilibrium constant for the balanced equation in Problem 13.95 is equal to 1.8×10^{-5}. What does this indicate about which way the equilibrium lies? What are the predominant species in solution?

13.97 Why would Arrhenius have a problem explaining why NH_3 is a base?

13.98 Use the Brønsted–Lowry definition to explain why NH_3 is a base in water.

13.99 According to the Brønsted–Lowry definition, why is it proper to say that water is acting as an acid upon the addition of ammonia?

13.100 Solid ammonium chloride (NH_4Cl) will react with solid sodium hydroxide to make water, sodium chloride (NaCl), and ammonia gas.
(a) Write a balanced equation for this reaction.
(b) According to the Brønsted–Lowry definition, which is the acid and which is the base? Explain your answer.

13.101 Consider the following equilibrium:

$$NH_4^+(aq) + H_2O(l) \rightleftharpoons NH_3(aq) + H_3O^+(aq)$$

The K_{eq} for this reaction is 5.6×10^{-10}.
(a) Is a solution of ammonium ion weakly acidic or strongly acidic?
(b) Is water acting as an acid or a base according to the Brønsted–Lowry definition? Explain.
(c) Should we call the K_{eq} an acidity constant, K_a, or a basicity constant, K_b? Explain your answer.

13.102 The carbonate ion (CO_3^{2-}) is a weak base. An aqueous solution of it turns litmus blue.
(a) Write the equilibrium equation that shows how carbonate makes water basic.
(b) What is the acid and what is the base according to the Brønsted–Lowry definition?
(c) The basicity constant K_b for carbonate is 2.1×10^{-4}. What does this tell you about the strength of carbonate as a base?

13.103 Hydride ion (H^-) is an exceptionally strong base, reacting with water to produce lots of hydroxide ion and H_2 gas. The K_{eq} for this reaction is huge.
(a) Write the balanced reaction for hydride with water.
(b) Explain why H_2 gas forms. [*Hint:* Use the Brønsted–Lowry definition.]

13.104 Rewrite the equation of Problem 13.103 using dot diagrams, and show with an arrow the proton being transferred. Note that the H^- is an H with a lone pair of electrons on it.

13.105 What is it about the hydride ion that allows it to accept a proton (that allows it to act as a base)?

13.106 If NH_3 can act as a base, then why can't CH_4? [*Hint:* Draw dot diagrams.]

13.107 The reaction between ethanol, C_2H_5OH, and hydride ion, H^-, to produce H_2 gas and the $C_2H_5O^-$ anion looks strange. Arrhenius could not tell you what is going on, but Brønsted and Lowry would have no trouble explaining it. How would they explain this reaction?

WEAK BASES

13.108 How would Arrhenius define a weak base?

13.109 Why is it proper to call ammonia a weak base?

13.110 Why aren't ionic compounds like NaOH and $Mg(OH)_2$ weak bases?

13.111 Consider a 1 M solution of NH_3 and a 1 M solution of LiOH. Are these solutions equally basic, or is one more basic than the other? Explain your answer.

13.112 Draw a set of beakers similar to those in WorkPatch 13.3 for the solutions in Problem 13.111. Fill in the species present in each beaker.

13.113 The bisulfate ion, HSO_4^-, has the ability to act as both a weak base and a weak acid in water.
(a) Write an equilibrium reaction that shows bisulfate acting as a weak acid.
(b) Write an equilibrium reaction that shows bisulfate acting as a weak base.
(c) What information would you need to help you determine whether a solution of bisulfate was going to be slightly acidic or slightly basic? How would you use that information to find out?

13.114 Amines are organic compounds that contain an NH_2 group. If they are water-soluble, they are weak bases in water. For example, the compound methylamine, H_3C-NH_2, is a weak base.
(a) Draw a dot diagram for methylamine.
(b) Using dot diagrams, show the equilibrium reaction between methylamine and water.
(c) To which side does the equilibrium in part (b) lie? What did we tell you that allowed you to figure out the answer?
(d) The similar compound ethane, H_3C-CH_3, does not act as a weak base. Why can methylamine act as a weak base, while H_3C-CH_3 can't? [*Hint:* Draw a dot diagram for H_3C-CH_3.]
(e) Is it appropriate to call methylamine an electrolyte? If so, is it weak or strong? Explain your answer.

13.115 To be a weak base in water, a molecular compound must also be a weak electrolyte. What must be one of the ions that it produces in water?

13.116 Consider phosphoric acid, H_3PO_4.
(a) List all of the weak bases that it can produce via successive losses of its protons.
(b) Of the bases you listed in part (a), which has no ability to also serve as a weak acid?

ACIDIC OR BASIC? WATER, AUTODISSOCIATION, AND K_w

13.117 What do we mean by the autoionization of water? Write a balanced equation to go along with your explanation.

13.118 We can also call the autoionization of water the autodissociation of water. Exactly how is water dissociating?

13.119 Does the equilibrium for the autoionization of water lie to the left or to the right? What constant verifies your answer?

13.120 Write the mathematical expression that allows you to solve for the OH^- concentration in water when you know only the H_3O^+ concentration.

13.121 True or false? "A liter of pure water contains no ions in it whatsoever." Explain your answer.

13.122 What is the concentration of hydronium ion and hydroxide ion in pure water at 25°C?

13.123 If pure water has both hydronium (acid) and hydroxide (base) in it, then how can pure water be neutral?

13.124 Based solely on concentrations, when is an aqueous solution judged to be acidic? Give two answers to this question.

13.125 Based solely on concentrations, when is an aqueous solution judged to be basic? Give two answers to this question.

13.126 True or false? "Even in a strongly basic aqueous solution there is some H_3O^+ (acid) present." Explain your answer.

13.127 True or false? "In an aqueous solution, as the H_3O^+ concentration increases, the OH^- concentration must decrease." Explain your answer.

13.128 True or false? "In an aqueous solution at 25°C, you will always get the same number when you multiply the equilibrium H_3O^+ concentration by the equilibrium OH^- concentration." Explain your answer.

13.129 An aqueous solution has an H_3O^+ concentration of 1.0 M. What is the OH^- concentration? Is this solution acidic or basic? Justify your answer.

13.130 An aqueous solution has an OH^- concentration of 1.0×10^{-11} M. What is the H_3O^+ concentration? Is this solution acidic or basic? Justify your answer.

13.131 A solution is prepared by dissolving 2.50 moles of LiOH in enough water to get 4.00 L of solution.
(a) What is the OH^- concentration?
(b) What is the H_3O^+ concentration?

13.132 A solution is prepared by dissolving 0.250 mole of $Ba(OH)_2$ in enough water to get 4.00 L of solution.
(a) What is the OH^- concentration?
(b) What is the H_3O^+ concentration?

13.133 A solution is prepared by dissolving 2.40 g of $Mg(OH)_2$ in enough water to get 4.00 L of solution. [*Hint:* You will need to calculate the molar mass of $Mg(OH)_2$.]
(a) What is the OH^- concentration?
(b) What is the H_3O^+ concentration?

13.134 A solution is prepared by dissolving 2.00 moles of HNO_3 in enough water to get 800.0 mL of solution.
(a) What is the H_3O^+ concentration?
(b) What is the OH^- concentration?

13.135 A solution has an OH^- concentration of 10^{-11} M.
(a) Is this solution basic or acidic? Explain how you know.
(b) Without using a calculator, explain how you could quickly determine the H_3O^+ concentration of this solution. What is the concentration?

THE pH SCALE

13.136 What is a log?

13.137 Without using a calculator, what is the log of 10^{-34}?

13.138 Without using a calculator, what is the log of 10^{13}?

13.139 Without using a calculator, what is the log of 10^0? Of 1?

13.140 Multiple choice: The log of 60 is a number:
(a) Between −2 and −1   (b) Between −1 and 0
(c) Between 0 and 1   (d) Between 1 and 2
(e) Between 2 and 3

13.141 Explain how you arrived at your answer to Problem 13.140.

13.142 Multiple choice: The log of 0.73 is a number:
(a) Between −2 and −1   (b) Between −1 and 0
(c) Between 0 and 1   (d) Between 1 and 2
(e) Between 2 and 3

13.143 Explain how you arrived at your answer to Problem 13.142.

13.144 True or false? "As a solution's acidity increases, its pH decreases."

13.145 Write the mathematical definition of pH.

13.146 Why is a pH of 7 equal to neutrality?

13.147 What is the pH of a solution whose hydronium ion concentration is 0.0010 M? Is the solution acidic or basic?

13.148 What is the pH of a solution whose H_3O^+ concentration is 10^{-6} M? Is the solution acidic or basic?

13.149 What is the pH of a solution whose H_3O^+ concentration is 6.40×10^{-9} M? Is the solution acidic or basic?

13.150 What is the pH of a solution whose OH^- concentration is 10^{-14} M? Is the solution acidic or basic?

13.151 What is the pH of a solution whose hydroxide ion concentration is 2.00×10^{-3} M? Is the solution acidic or basic?

13.152 Solution A has a pH of 3. Solution B has a pH of 6. Which solution is more acidic, and by how much?

13.153 What is the pH of a solution whose H_3O^+ concentration is 10.0 M? Is this solution acidic or basic?

13.154 The pH of a solution is 4. What is the H_3O^+ concentration? Is the solution acidic or basic?

13.155 The pH of a solution is 8. What is the OH^- concentration? Is the solution acidic or basic?

13.156 The pH of a solution is -1. What are the H_3O^+ and OH^- concentrations? Is the solution acidic or basic?

13.157 Suppose two students dissolved 1 mole of an acid in enough water to get 1 liter of solution. Student A does this with acetic acid. Student B does this with hydrochloric acid. Will their solutions have the same pH, or will each have a different pH? If they are different, then how will they be different? Why should they be different?

RESISTING pH CHANGES—BUFFERS

13.158 Aniline, formula $C_6H_5NH_2$, is a weak base, possessing a lone pair of electrons on the nitrogen atom.
(a) According to Brønsted and Lowry, what must aniline do to serve as a base?
(b) Why can aniline serve as a base?
(c) What would be the formula and charge of the conjugate acid of aniline?

13.159 Can a weak acid and its conjugate base ever have the same charge? Explain your answer.

13.160 If aniline (Problem 13.158) is a weak base, then what can you say about the strength of its conjugate acid?

13.161 Nitric acid, HNO_3, is a very strong acid. Solutions of sodium nitrate, $NaNO_3$, have lots of nitrate ions in solution. Would you expect such a solution to be acidic, basic, or neutral? Fully explain your answer. [*Hint:* Think about the conjugate base of HNO_3.]

13.162 Why is the conjugate base of a weak acid like acetic acid often referred to as its salt?

13.163 If Cl^- is the conjugate base of HCl, then why isn't an aqueous solution of NaCl acidic?

13.164 What do we mean when we say that a solution has buffering ability?

13.165 Can pure liquid water demonstrate buffering ability? Explain your answer.

13.166 What is the general recipe for making a buffer? Explain the functions of the ingredients.

13.167 Name one biological system in which control of pH (buffering ability) is important, and name the buffer that is in control.

13.168 Can a buffer resist changes to its pH for any added amount of strong acid or base? Explain your answer.

13.169 Would the mixture of carbonic acid (H_2CO_3, a weak acid) and the compound $NaHCO_3$ (sodium bicarbonate) in water constitute a buffer? If no, explain why. If yes, explain why and use chemical equations to show what would happen upon addition of OH^- or H_3O^+ to the solution.

13.170 When acid is added to a buffer, the pH changes a little bit.
(a) Will the pH increase or decrease?
(b) Why does the pH change at all? Why doesn't the buffer hold the pH completely constant?

13.171 One way to make an acetic acid buffer is to mix substantial amounts of both acetic acid and its salt, sodium acetate, in water. Another way to make the same buffer is to add a substantial amount of acetic acid to water and then half as much NaOH to the water. Explain how and why this makes a buffer.

13.172 How would you make a buffer based on ammonia? [*Hint:* Refer to Table 13.5.]

13.173 How would you make a buffer based on hypochlorous acid? [*Hint:* Refer to Table 13.5.]

13.174 It is possible to make two completely different buffers using the dihydrogen phosphate ion, $H_2PO_4^-$.
(a) In one buffer, dihydrogen phosphate serves as the weak acid. What would be the conjugate weak base?
(b) Write the equations that show how the buffer in part (a) works upon addition of H_3O^+ and OH^-.
(c) In the other buffer, dihydrogen phosphate serves as the weak base. What would be the conjugate weak acid?
(d) Write the equations that show how the buffer in part (c) works upon addition of H_3O^+ and OH^-.

13.175 Suppose 2.0 moles of sodium acetate are dissolved in some water, and then 1.0 L of a 1.0 M HCl solution is added to it.
(a) Write the chemical reaction that occurs.
(b) After the reaction, what are the predominant species in solution? How many moles of each are present?
(c) Is the resulting solution a buffer? If it is, explain why.

13.176 A buffer works by replacing added strong acid with weak acid. Explain how it does this.

13.177 A buffer works by replacing added strong base with weak base. Explain how it does this.

13.178 How does a buffer "kill" added strong base?

13.179 How does a buffer "kill" added strong acid?

13.180 Write the equations that show how a hypochlorous acid buffer defends against added strong acid and base.

WorkPatch Solutions

13.1 (a) NaF is a strong electrolyte because it completely dissociates to give only ions. HF is a weak electrolyte because very little of it dissociates into ions. F_2 is a nonelectrolyte because it produces no ions whatsoever upon dissolving.
(b) HF and F_2 are molecular compounds. HF is a polar covalent molecule, and F_2 is a nonpolar covalent molecule. Only NaF is ionic (metal plus nonmetal).

13.2 Since it is a diprotic acid, a molecule of sulfuric acid can produce a maximum of two H_3O^+ ions:

$$H_2SO_4 + 2\,H_2O \longrightarrow 2\,H_3O^+ + SO_4^{2-}$$

13.3 (a) Aqueous HNO_3

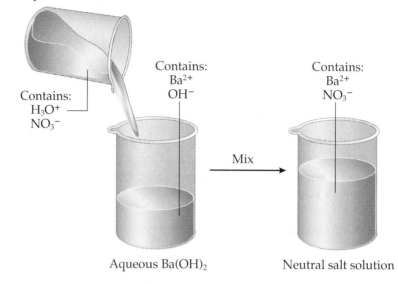

Contains:
H_3O^+
NO_3^-

Contains:
Ba^{2+}
OH^-

Contains:
Ba^{2+}
NO_3^-

Mix

Aqueous $Ba(OH)_2$ Neutral salt solution

(b) $Ba(NO_3)_2$, barium nitrate. Note that it takes two NO_3^- ions to counterbalance the charge of one Ba^{2+} ion; hence the formula has two nitrates per barium ion.

13.4 The only molecule that can serve as a Brønsted–Lowry base is PH_3 (phosphine). This is because it is the only molecule of the three that possesses an atom with a lone pair of electrons.

13.5 (a) It is proper to call the hydroxide ion (OH^-) a Brønsted–Lowry base because it accepts a proton (H^+) from hydronium (H_3O^+).
   (b) A hydroxide ion can accept a proton and act as a base because it has lone pairs of electrons on the oxygen atom.

13.6 Both weak bases will accept a proton from water:

$$CO_3^{2-} + H_2O \rightleftharpoons HCO_3^- + OH^-$$

$$CH_3COO^- + H_2O \rightleftharpoons CH_3COOH + OH^-$$

Because both are weak bases, these equilibria lie far to the left.

13.7

Solution	$[H_3O^+]$, M	$[OH^-]$, M	Acidic, basic, or neutral
1.	10^0	10^{-14}	The most acidic solution
2.	10^{-1}	10^{-13}	Acidic
3.	10^{-2}	10^{-12}	Acidic
4.	10^{-3}	10^{-11}	Acidic
5.	10^{-4}	10^{-10}	Acidic
6.	10^{-5}	10^{-9}	Acidic
7.	10^{-6}	10^{-8}	Acidic
8.	10^{-7}	10^{-7}	Neutral
9.	10^{-8}	10^{-6}	Basic
10.	10^{-9}	10^{-5}	Basic
11.	10^{-10}	10^{-4}	Basic
12.	10^{-11}	10^{-3}	Basic
13.	10^{-12}	10^{-2}	Basic
14.	10^{-13}	10^{-1}	Basic
15.	10^{-14}	10^0 (1.0 M)	The most basic solution

13.8 (a) $[OH^-] = \dfrac{K_w}{[H_3O^+]} = \dfrac{10^{-14}}{2} = 5 \times 10^{-15}$ M

This solution is acidic because $[H_3O^+] > 10^{-7}$ M.

(b) $[H_3O^+] = \dfrac{K_w}{[OH^-]} = \dfrac{10^{-14}}{4} = 2.5 \times 10^{-15}$ M

This solution is basic because $[OH^-] > 10^{-7}$ M.

13.9 The conjugate base of citric acid has the formula $C_5H_7O_6O^-$. When citric acid, $C_5H_7O_6OH$, loses an H^+, then it becomes $C_5H_7O_6O^-$.

13.10 She is wrong. A weak acid and its conjugate base differ by a single proton (H^+). The difference between weak carbonic acid (H_2CO_3) and carbonate (CO_3^{2-}) is two protons. That is: $H_2CO_3 \longrightarrow 2\,H^+ + CO_3^{2-}$. Thus they are not conjugates of each other. The conjugate base of H_2CO_3 is HCO_3^-.

13.11 (a) Yes; the weak acid is acetic acid (CH_3COOH).
(b) Yes; the weak base is acetate ion (CH_3COO^-)
(c) Yes; the conjugate pair is acetic acid and its conjugate base, the acetate ion.

13.12 To make a buffer, you would also dissolve a salt of this weak acid, such as sodium fluoride (NaF). Other ionic fluoride salts would do, such as LiF, or MgF_2. Anything that puts F^- ions into solution would do.

13.13 (a) A weak acid in a buffer eliminates added strong OH^- base by donating a proton to it and turning it into water. In turn, the weak acid is converted into its own conjugate base.
(b) A weak base in a buffer eliminates added strong H_3O^+ acid by accepting a proton from it and turning it into water. In turn, the weak base is converted into its own conjugate acid.

13.14 (a) $HF + OH^- \longrightarrow F^- + H_2O$
(b) $F^- + H_3O^+ \longrightarrow HF + H_2O$

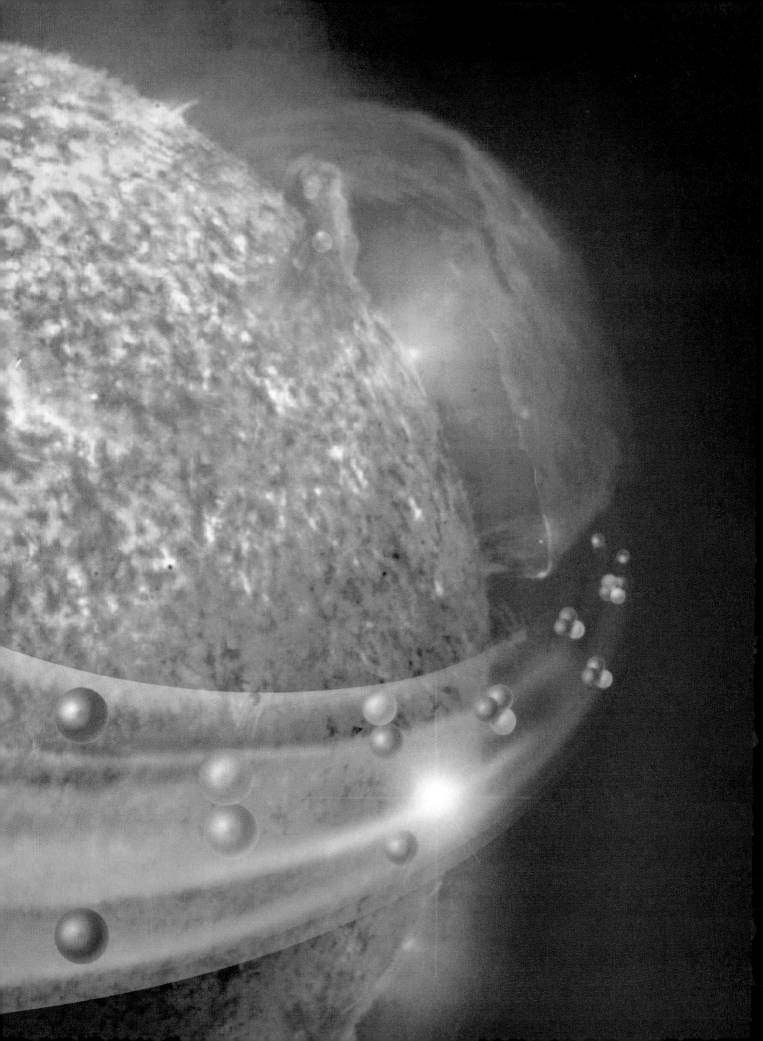

Nuclear Chemistry

THE CASE OF THE MISSING MASS: 14.1
MASS DEFECT AND THE STABILITY OF THE NUCLEUS

My name is Otto Bismarck Wahn, private investigator. The name is a mouthful, so I go by O. B. on the street. This case was unusual from the start. A woman, she claimed to be a nuclear chemist, showed up at my office late one night. Funny, I would have pegged her for a molecular biologist. She went on about some theft that had apparently occurred in her lab. Seems that she and her lab partners had noticed some missing mass regarding the carbon isotope $^{12}_{6}C$. "One mole of $^{12}_{6}C$ has a mass of exactly 12 grams," she said. "Sure, everybody knows that," I said, "so what's your point?" She was starting to get on my nerves. "I calculate that a mole of $^{12}_{6}C$ should have a mass greater than 12 grams," she said, pointing out that 1 mole of $^{12}_{6}C$ atoms was made from 6 moles of protons, 6 moles of neutrons, and 6 moles of electrons. I was reaching the limit of my patience. "Cut to the chase lady!" She got huffy. "Don't you get it?" she said. "If we add up the masses of 6 moles of protons, 6 moles of neutrons, and 6 moles of electrons, we get more than 12 grams!" She proceeded to show me:

$$6 \text{ moles protons} \times \frac{1.007\,30 \text{ g}}{\text{mole protons}} = 6.043\,80 \text{ g}$$

$$6 \text{ moles neutrons} \times \frac{1.008\,70 \text{ g}}{\text{mole neutrons}} = 6.052\,20 \text{ g}$$

$$6 \text{ moles electrons} \times \frac{0.000\,55 \text{ g}}{\text{mole electrons}} = \underline{0.003\,3 \ \text{ g}}$$
$$12.099\,3 \ \text{ g}$$

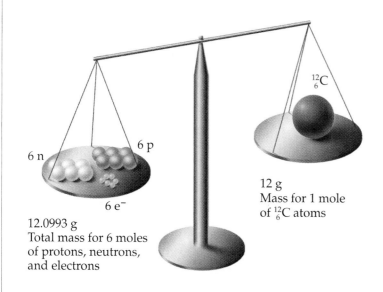

6 n

6 p

6 e⁻

12.0993 g
Total mass for 6 moles
of protons, neutrons,
and electrons

$^{12}_{6}$C

12 g
Mass for 1 mole
of $^{12}_{6}$C atoms

"Do you understand now? I calculate that 1 mole of $^{12}_{6}$C atoms should have a mass of 12.0993 g, but in actuality its mass is only 12 g. So $^{12}_{6}$C is lighter than it's supposed to be! Someone has stolen 0.0993 g of mass from each mole of $^{12}_{6}$C." She was stumped, and desperate. "Help me O. B. Wahn, you're my only hope!" Ouch! How could I turn her down? I went to my Rolodex and looked up the number for Professor Albert Einstein. I had used him as a source before. It was no surprise to me that he had already figured this out. The answer involved his famous equation $E = mc^2$, where m is mass and c^2 is the speed of light squared. He said that the missing mass was called the **mass defect**, and all atoms have it to some degree. All atoms are a little bit lighter than they are supposed to be. According to Einstein, $^{12}_{6}$C's mass defect of 0.0993 g per mole is converted into energy and released when the atoms are formed from their subatomic pieces. He even calculated it for me by plugging the mass defect and the speed of light into his equation and solving for the energy released:

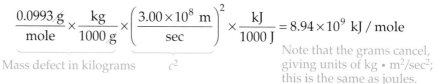

$$\underbrace{\frac{0.0993\ \cancel{g}}{\text{mole}} \times \frac{\text{kg}}{1000\ \cancel{g}}}_{\text{Mass defect in kilograms}} \times \underbrace{\left(\frac{3.00 \times 10^8\ \text{m}}{\text{sec}}\right)^2}_{c^2} \times \frac{\text{kJ}}{1000\ \text{J}} = 8.94 \times 10^9\ \text{kJ / mole}$$

Note that the grams cancel, giving units of kg · m²/sec²; this is the same as joules.

"Al, that's incredible!" I said. I realized that 10^9 kJ is a billion kilojoules, so he was talking almost 9 billion kJ of energy released when 1 mole of $^{12}_{6}$C is formed from its subatomic particles. Now, I know something about chemistry. Your most highly exothermic chemical reaction might release a few thousand kilojoules per mole of reactant, but he was talking billions of kilojoules! "Al, do you have a name for all this energy?" Of course he did. "The energy released when an atom forms from its subatomic particles is called its **binding energy**," he said. "Think of it as the energy that holds the nucleus together. You would have to put this much energy back into a mole of $^{12}_{6}$C atoms to break their nuclei apart into separate protons and neutrons. The more binding energy a nucleus has, the more stable it is." A light went on in my head. "Hold on Al," I said. I quickly did a similar calculation for $^{13}_{6}$C, the heavier carbon isotope with 6 protons, 6 electrons, and 7 neutrons:

What the mass of 1 mole of $^{13}_{6}$C should be:

$$6\ \text{moles protons} \times \frac{1.007\ 30\ \text{g}}{\text{mole protons}} = 6.043\ 80\ \text{g}$$

$$7\ \text{moles neutrons} \times \frac{1.008\ 70\ \text{g}}{\text{mole neutrons}} = 7.060\ 90\ \text{g}$$

$$6\ \text{moles electrons} \times \frac{0.000\ 55\ \text{g}}{\text{mole electrons}} = \frac{0.003\ 3\ \ \text{g}}{13.108\ 0\ \ \text{g}}$$

What the mass of 1 mole of $^{13}_6C$ actually is:

13.003 35 g

Mass defect:

$$13.108\ 0 \text{ g} \longleftarrow \text{What the mass of 1 mole } ^{13}_6C \text{ should be}$$
$$-13.003\ 35 \text{ g} \longleftarrow \text{What the mass of 1 mole } ^{13}_6C \text{ actually is}$$
$$\overline{0.104\ 65 \text{ g}} \longleftarrow \text{Missing mass per mole}$$

Binding energy

$$\frac{0.104\ 65\ \cancel{g}}{\text{mole}} \times \frac{\text{kg}}{1000\ \cancel{g}} \times \left(\frac{3.00 \times 10^8 \text{ m}}{\text{sec}}\right)^2 \times \frac{\text{kJ}}{1000\ \text{J}} = 9.42 \times 10^9 \text{ kJ / mole}$$

"Al, I just calculated the mass defect and binding energy for $^{13}_6C$ and got bigger numbers than you did for $^{12}_6C$. Does that mean the $^{13}_6C$ nucleus is more stable than the $^{12}_6C$ nucleus?" Einstein gave me one of those low chuckles that always got my goat. "O. B. Wahn, my friend, you must not consider the total binding energy. You must instead trust the binding energy per nucleon." He went on to explain that a **nucleon** is the term used for a proton or a neutron. The lighter $^{12}_6C$ has 12 nucleons in its nucleus (6 protons and 6 neutrons), whereas $^{13}_6C$ has 13 nucleons (6 protons and 7 neutrons). He instructed me to take the total binding energy for each isotope and divide it by the total number of nucleons in the nucleus:

For $^{12}_6C$:

$$\frac{8.94 \times 10^9 \text{ kJ / mole}}{12 \text{ nucleons}} = 7.45 \times 10^8 \frac{\text{kJ / mole}}{\text{nucleon}}$$

For $^{13}_6C$:

$$\frac{9.42 \times 10^9 \text{ kJ / mole}}{13 \text{ nucleons}} = 7.25 \times 10^8 \frac{\text{kJ / mole}}{\text{nucleon}}$$

"Well I'll be! It looks like the $^{12}_6C$ nucleus is more stable than the $^{13}_6C$ nucleus after all, since it has more binding energy per nucleon. Thanks professor!" I hung up before I had to listen to another one of his chuckles, and turned to my new client. I was about to explain that no one had stolen mass from her $^{12}_6C$, but she was out the door. She left without saying a word, but she did leave behind a graph on which she scrawled the message "Thanks! This makes sense now." It was a plot of the average binding energy versus the number of nucleons in the nucleus.

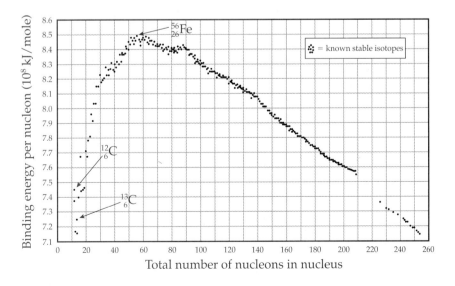

WORKPATCH

(a) What does the plot say about the stability of the nucleus of the iron isotope $^{56}_{26}$Fe?
(b) What is the total binding energy for the $^{56}_{26}$Fe nucleus in kilojoules per mole?

The plot made it clear that while all nuclei have binding energy, one nucleus is the most stable of all, and now my client, you, and I know which one. The case is closed.

The investigation into mass defect and binding energy also solves another mystery, the failure of medieval alchemists to transform base metals like lead into gold via chemical means. For example, an alchemist might have tried to react lead carbonate ($PbCO_3$) with acid to form carbon dioxide, water, and aqueous gold ions.

$$PbCO_3(s) + 2\,H_3O^+ \;\not\longrightarrow\; 3\,H_2O(l) + CO_2(g) + Au^{2+}(aq)$$

Forming gold ions from lead in this manner would be quite a trick. Unfortunately, no chemical reaction has ever accomplished this. If you start with lead, it stays lead. That goes for the hydrogen, oxygen, and carbon as well.

$$PbCO_3(s) + 2\,H_3O^+ \longrightarrow 3\,H_2O(l) + CO_2(g) + Pb^{2+}(aq)$$

In a chemical reaction, the atoms may change partners, but they don't change elemental identity. A chemical reaction simply cannot transform one element into another. The reason for this has to do with energy. A chemical reaction involves the valence electrons of atoms, redistributing them between atoms, breaking old bonds and making new ones. The energies associated with redistributing valence electrons are large, but not tremendous. That is, the bonds within lead carbonate are strong, but they can be broken with the addition of a reasonable amount of heat. On the other hand, nuclear binding energies are huge—billions of kilojoules per mole. There is just no way that a chemical reaction can produce enough energy to overcome this binding energy and affect the nucleus. Lead has an atomic number of 82, whereas gold has an atomic number of 79. To turn lead into gold would require removing 3 protons from lead's nucleus.

WORKPATCH

Consider $^{207}_{82}$Pb, which has a mass number of 207.

(a) How many nucleons are in the $^{207}_{82}$Pb nucleus?
(b) Read the plot of binding energy per nucleon as best you can, and calculate how many kilojoules per mole it would take to remove 3 protons from lead's nucleus.

An Exothermic Reaction

Nucleus

The lead carbonate plus acid reaction releases 1.1 kJ of heat energy per mole of $PbCO_3$. What is the difference between this number and the number you just calculated in the WorkPatch? That is how far short the reaction falls from being able to turn lead into gold. No chemical reaction produces remotely enough energy to cause change to the nucleus. You might as well try to blow a brick wall down with your breath.

Having said all this, it still turns out that some nuclei undergo spontaneous change—without the input of any energy at all. These atoms eject pieces of their nuclei, transforming themselves into new elements. We call these atoms *radioactive*. A **radioactive** atom, therefore, is an atom that possesses a nucleus that undergoes spontaneous change. The reasons for spontaneous nuclear change are well beyond the scope of this text—it is not simply a matter of insufficient binding energy. However, scientists have made some empirical observations that can help us to predict what kind of nucleus will be radioactive and how its nucleus will change. Try the following WorkPatch to get ready for what comes next. It reviews some key concepts from Chapter 3.

Three isotopes of the element carbon have mass numbers 12, 13, and 14.
  (a) Write the full atomic symbols for these three isotopes of carbon, including mass number and atomic number.
  (b) For each isotope, indicate the number of neutrons and protons in the nucleus.
  (c) For each isotope, indicate the total number of nucleons in the nucleus.
  (d) How can all three isotopes be the element carbon if they have different mass numbers?

Of the three carbon isotopes mentioned in the WorkPatch, only the heaviest one, $^{14}_{6}C$, is radioactive. What kind of change will its nucleus undergo? We will show you how to predict this, but don't go on until you feel confident about this WorkPatch, or what comes next will not make much sense.

HALF-LIFE AND THE BAND OF STABILITY 14.2

Nuclei of atoms are collections of positively charged protons (p) and uncharged neutrons (n). Nuclei are also tiny. This brings up an interesting question: "Why don't nuclei blow apart?" After all, they have all those positively charged protons jammed into a tiny space, and like charges repel each other. The answer is that the nucleons in a nucleus are glued together by the **strong force**, the strongest force known in nature. This force exists only between nucleons, and it develops fully only when they are touching. Any time nucleons touch, the strong force kicks in and gives rise to an energy of attraction between the nucleons. There also exists an electrostatic repulsive energy between the positively charged protons, which is extremely large when the protons touch one another. The presence of neutrons spreads out the positive charge of the protons and decreases the electrical repulsion between them. For a nucleus to exist at all, the attractive energy must be greater than the repulsive energy. The difference is the binding energy, which holds the nucleus together.

   However, it is possible to have too many or too few neutrons. Here is where empirical observations can help us to predict whether a nucleus will be radioactive and how it will change. While studying the composition of the nuclei of all known isotopes of the elements, nuclear physicists noticed something about the number of neutrons present relative to the number of protons. The known isotopes of the lighter atoms, those with an atomic number between 1 and 20, have nuclei in which the number of neutrons tends to equal the number of protons. For these nuclei, the **neutron-to-proton (n/p) ratio** is close to 1.

Things change for the heavier elements with atomic numbers greater than 20. For these atoms, the known isotopes have more neutrons than protons. For example, the most abundant isotope of iron, $^{56}_{26}$Fe, has 26 protons and 30 neutrons, giving it a neutron-to-proton ratio of 1.15. By the time we get to the $^{209}_{83}$Bi isotope of bismuth, this ratio has increased to 1.52 (126 neutrons/83 protons). In other words, as atoms get heavier, the n/p ratio increases. Evidently, by the time a nucleus has more than 20 protons, it has become so positive that additional neutrons are required to help keep it together. By the time we get to bismuth, with its 83 protons, 1.5 times as many neutrons as protons are required to do the job.

Scientists have plotted the n/p ratios for all the known isotopes. On this plot, as we'll see shortly, the known isotopes form a compact band called the **band of stability**. This plot can be used to predict whether and how a nucleus will change. To show you how this plot is constructed, we will use the three carbon isotopes mentioned in WorkPatch 14.3. They are shown below along with their calculated n/p ratios:

	$^{12}_{6}$C	$^{13}_{6}$C	$^{14}_{6}$C
Number of protons	6	6	6
Number of neutrons	6	7	8
n/p ratio	1.0	1.17	1.33

To construct the plot of band of stability, we simply put a point at the intersection of the number of neutrons (vertical axis) and the number of protons (horizontal axis) for each isotope.

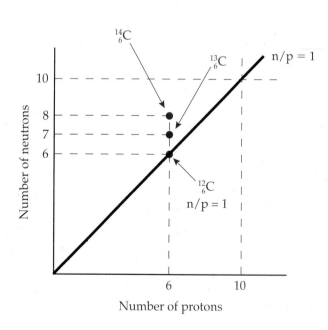

Any point that lies on the thick red diagonal line in the graph represents a nucleus with an n/p ratio of 1. The farther a point is above this line, the greater the n/p ratio of the isotope it represents. In other words, points above this line represent nuclei with more neutrons than protons (they are neutron-rich). Note that we included $^{14}_{6}$C in this plot even though its nucleus is radioactive. For the purposes of the band of stability plot, *stable* is interpreted to include both nonradioactive isotopes and radioactive isotopes with measurable *half-lives*. **Half-life** is defined as the time it takes for the amount of radioactive nuclei in a sample of a given isotope to drop to half the initial amount due to spontaneous nuclear change. Exactly what the radioactive isotope changes into will be covered in the next section. The radioactive iodine isotope $^{123}_{53}$I has a half-life of 13.1 hours. This means that if you have a 10 g sample of $^{123}_{53}$I, 5 g ($\frac{1}{2} \times$ 10 g) will be left after 13.1 hours. After another 13.1 hours (two half-lives), 2.5 g ($\frac{1}{2} \times \frac{1}{2} \times$ 10 g) will be left, and so on. Some radioactive isotopes and their half-lives are listed in Table 14.1.

Table 14.1   Some Radioactive Isotopes

Element	Radioactive isotope	Half-life
Iodine	$^{123}_{53}I$	13.1 hours
Phosphorus	$^{32}_{15}P$	14.28 days
Carbon	$^{14}_{6}C$	5715 years
Uranium	$^{235}_{92}U$	7.04×10^8 years

All the isotopes listed in Table 14.1 are included in the band of stability plot shown below.

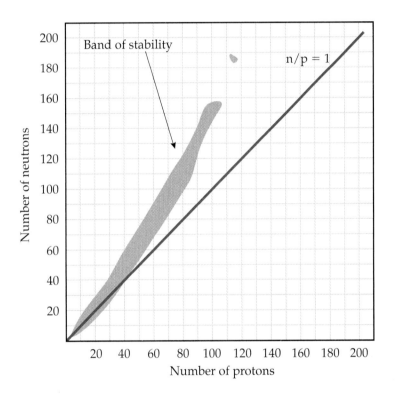

Remember that all the isotopes within the band are either nonradioactive (their nuclei are stable indefinitely) or radioactive with measurable half-lives (their nuclei undergo spontaneous change). The little island at the top corresponds to superheavy elements that have been predicted to be stable. Amazingly, as this chapter was being written, scientists produced the first elements within this island, possessing 114, 116, and 118 protons and half-lives of less than a millisecond. The regions outside the band can be thought of as an "ocean of instability." A nucleus composed of a number of neutrons and protons represented by a point in this ocean would be so unstable that its half-life would be immeasurably short.

PRACTICE PROBLEMS

14.1 The radioactive iodine isotope $^{123}_{53}I$ is used to treat thyroid disease. Suppose a patient is given a 30 microgram (μg) dose. How much will be left in the patient after 29.3 hours? [See Table 14.1.]

Answer: 29.3 hours is three half-lives, so there will be 3.75 μg left.

$$30 \ \mu g \times \frac{1}{2} \times \frac{1}{2} \times \frac{1}{2} = 3.75 \ \mu g$$

14.2 What do we call atoms possessing nuclei that undergo spontaneous change?

14.3 Calculate the binding energy (in kJ/mole) for $^{4}_{2}\text{He}$.

14.4 The band of stability curves up above the diagonal line in the plot. Why does this occur?

You are now ready to learn about the types of spontaneous changes that radioactive nuclei undergo, and how the band of stability can help us to predict which type of change will occur.

14.3 SPONTANEOUS NUCLEAR CHANGES: RADIOACTIVITY

Let's take a moment to summarize what we have covered so far. All nuclei have a mass defect and considerable binding energy. Nevertheless, some atoms have nuclei that are unstable. We call these atoms radioactive, and their nuclei undergo some sort of spontaneous change at a rate characterized by each atom's half-life. While the reason for nuclear instability is beyond the scope of this book, there are some empirical observations involving the neutron-to-proton ratio that can help us predict whether a nucleus will be unstable and how it will change. Light atoms have a neutron-to-proton ratio of $n/p = 1$, but as they get heavier, the n/p ratio increases to 1.5 and above, as extra neutrons are needed to help keep the nucleus together.

All existing isotopes of every atom, both stable (unchanging) and radioactive, have nuclei with compositions inside the band of stability. Through the interior of the band of stability lies a path that contains all the indefinitely stable (nonradioactive) nuclei. The story is different above and below this path, and beyond atomic number 83 (past the bismuth buoy). All nuclei within the band of stability that lie outside the central path of indefinite stability are radioactive. Nuclei above this path have an n/p ratio that is too large (they have too many neutrons). One way they could move toward the fully stable, nonradioactive path would be to convert one or more of their neutrons into protons. Nuclei below the path of full stability have an n/p ratio that is too small (they have too few neutrons). One way they could move toward the nonradioactive path would be to convert one or more of their protons into neutrons. The far-lying nuclei beyond bismuth are just too big. Evidently, a nucleus with more than 83 protons, if it exists, will be radioactive no matter how many neutrons it has.

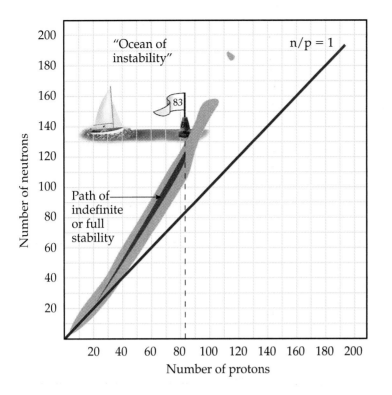

Neutrons changing into protons? Protons changing into neutrons? Nuclei losing protons? Believe it or not, these are exactly the processes that spontaneously occur for radioactive nuclei. We call these and other spontaneous nuclear changes **radioactive decay**. The best way to understand these decay processes is to start by reviewing the subatomic particles.

KNOWING YOUR SUBATOMIC PARTICLES

To completely understand the radioactive decay processes, you need to be familiar with the subatomic nuclear particles involved and how they are related to each other. We begin with the neutron (n) and the proton (p). From this point on, we will use the full symbols for the neutron and proton. In other words, we will include the mass number and atomic number with each symbol, just as we do for atoms.

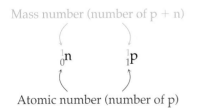

Mass number (number of p + n)

$$_0^1n \qquad _1^1p$$

Atomic number (number of p)

Do you understand the full symbols for the neutron and proton?
  (a) Explain why they both have the same mass number.
  (b) Explain why the atomic number for the neutron is 0.
  (c) Explain why the atomic number for the proton is 1.

WORKPATCH 14.4

Make sure you understand the superscripts and subscripts for these sub-atomic particles. Understanding them will be absolutely crucial for understanding the nuclear decay processes that follow.

We need to consider two additional subatomic particles—the *electron* and the *positron*—to understand how a neutron can turn into a proton or vice versa. You are already quite familiar with the electron (e^-). Its charge is -1, and it is almost 2000 times lighter than a proton or neutron. A **positron** is, in essence, an electron that has a positive ($+1$) charge. Their full symbols are given below:

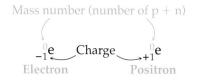

The superscripts (mass numbers) of 0 for both particles should make sense to you, since both particles contain no protons or neutrons. The subscripts, however, take on a new meaning for these particles. Instead of being the number of protons, the subscripts now represent the charge of the particle.

Positrons are examples of what physicists refer to as **antimatter**, the oppositely charged version of matter. A positron is the antimatter version of an electron—an antielectron—identical in size but opposite in charge. If an electron and positron meet, they instantaneously annihilate each other. This means that their combined mass disappears from the universe, to be replaced by the equivalent in energy ($E = mc^2$). Positrons and electrons are the key to understanding how radioactive nuclei convert neutrons into protons or protons into neutrons.

 **WORKPATCH**

Each of the following symbols has one mistake in it. Identify the mistake, explain why it is incorrect, and then fix it.

$$^{0}_{0}n \qquad ^{1}_{0}p \qquad ^{1}_{1}e$$

Was the last symbol in the WorkPatch meant to be an electron or a positron? How did you know? Check our answer to find out.

CONVERSION OF A NEUTRON INTO A PROTON: RADIOACTIVE DECAY VIA β^- EMISSION

A neutron changes into a proton when a nucleus ejects an electron. Now, you might be asking how a nucleus can eject an electron when it has no electron in the first place. The electron comes into being when the neutron-to-proton conversion occurs. In other words, a neutron turns into a proton *and* an electron, and the latter is ejected from the nucleus. This conversion is summarized below:

$$^{1}_{0}n \longrightarrow ^{1}_{1}p + ^{0}_{-1}e$$

We call the ejected electron a β^- **particle** (β is the Greek letter beta), and we call the ejection β^- **emission**. The β^- symbol and the $^{0}_{-1}e$ symbol can be used interchangeably. Note that charge is conserved in the above conversion. That is, the total charge on the right side of the arrow (sum of the subscripts) is equal to the charge (subscript) on the left side. You can use this fact to help you remember that when a neutron converts into a proton, an electron must also be formed.

β^- emission

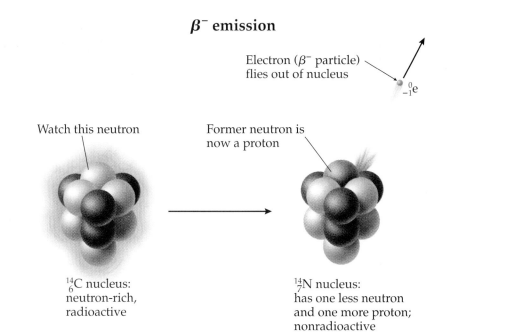

Electron (β^- particle)
flies out of nucleus $\quad _{-1}^{0}e$

Watch this neutron

Former neutron is
now a proton

$_{6}^{14}C$ nucleus:
neutron-rich,
radioactive

$_{7}^{14}N$ nucleus:
has one less neutron
and one more proton;
nonradioactive

Since a neutron is turning into a proton, the net effect of β^- particle emission is to increase the atomic number by 1 but leave the mass number unchanged. An example of β^- emission is the radioactive decay of $_{6}^{14}C$. Look at the above figure and you will see a $_{6}^{14}C$ nucleus undergoing β^- decay. We can summarize this event with the following nuclear reaction:

$$_{6}^{14}C \longrightarrow {}_{7}^{14}N + {}_{-1}^{0}e$$

This is called a **nuclear reaction** because it represents a change in the atom's nucleus. The $_{7}^{14}N$ is called a *daughter isotope*. A **daughter isotope** (or nucleus) is one that results from a nuclear decay process. The $_{6}^{14}C$ is referred to as the **parent** because it "gave birth" to the daughter. In this case, the $_{6}^{14}C$ parent lies on the upper (radioactive) part of the band of stability and the $_{7}^{14}N$ daughter is on the fully stable (nonradioactive) path.

Before going on, we want to show you an easy way to use the superscripts and subscripts to check that a nuclear reaction has been written correctly and in a balanced fashion. The rule is as follows: The sum of the superscripts on the left side of the reaction must equal the sum of the superscripts on the right side of the reaction. The same is true for the subscripts. Notice how this works for our $_{6}^{14}C$, β^- emission example.

$$14 + 0 = 14$$
$$_{6}^{14}C \longrightarrow {}_{7}^{14}N + {}_{-1}^{0}e$$
$$7 + (-1) = 6$$

By using this method of balancing, you can actually predict the product of a nuclear reaction. Try it now in the next WorkPatch. First, determine the value of the superscript and subscript question marks. Once you have determined the subscript (the atomic number), you'll know which elemental symbol to use.

14.6 **WORKPATCH**   Balance the following nuclear reaction and predict the product:

$$^{214}_{82}\text{Pb} \longrightarrow \, ^{?}_{?}? + \, ^{0}_{-1}\text{e}$$
$$\beta^- \text{ particle}$$

Do you understand why the mass number didn't change but the elemental identity did? Remember that β^- emission tends to occur for nuclei with too many neutrons (the radioactive upper part of the band of stability).

PRACTICE PROBLEMS

14.5 What would the answer from WorkPatch 14.6 become if there were another β^- emission?

Answer: $^{214}_{83}\text{Bi} \longrightarrow \, ^{214}_{84}\text{Po} + \, ^{0}_{-1}\text{e}$

14.6 Suppose $^{35}_{14}\text{Si}$ undergoes β^- emission. Write a nuclear reaction for this spontaneous change.

14.7 Suppose $^{40}_{20}\text{Ca}$ is the product of a radioactive isotope that underwent β^- emission. What was the radioactive isotope?

RADIOACTIVE DECAY VIA POSITRON EMISSION OR ELECTRON CAPTURE

Now let's consider radioactive isotopes that need some more neutrons (the radioactive lower part of the band of stability). These isotopes tend to decay by converting a proton into a neutron. They can do that in two ways: *positron emission* and *electron capture*.

A proton changes into a neutron when a nucleus ejects a positron via a process known as **positron emission**.

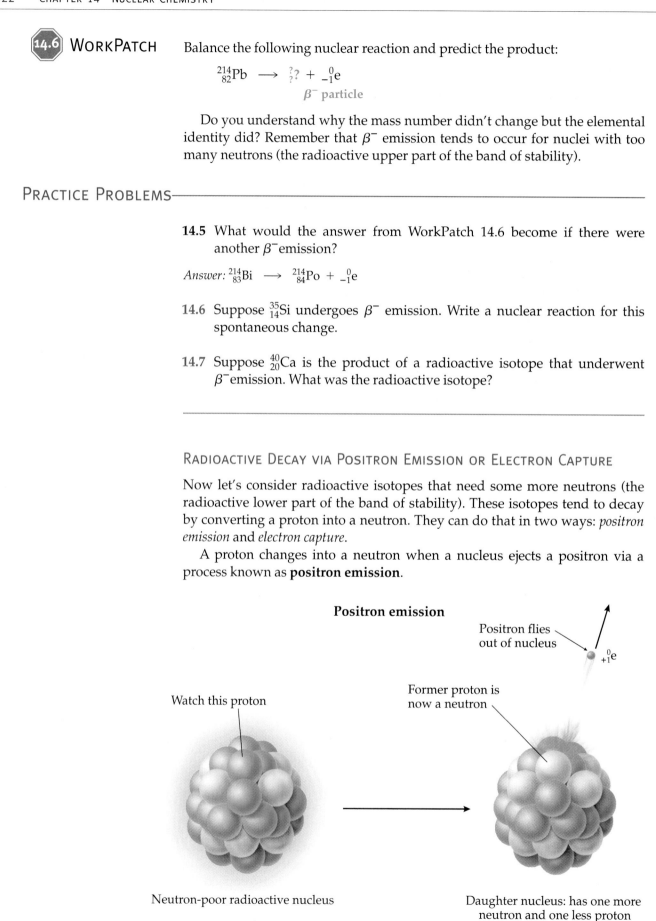

Positron emission

Positron flies out of nucleus

$^{0}_{+1}\text{e}$

Watch this proton

Former proton is now a neutron

Neutron-poor radioactive nucleus

Daughter nucleus: has one more neutron and one less proton

The positron comes into being when the proton-to-neutron conversion occurs. In other words, a proton turns into a neutron *and* a positron, and the latter is ejected from the nucleus. This conversion is summarized below:

$$\,^{1}_{1}p \longrightarrow \,^{1}_{0}n + \,^{0}_{+1}e$$

Once again, conservation of charge can help you remember this. If a proton turns into a neutron, then a particle with positive charge, the positron, must also be formed.

Another way to represent a positron is as β^+, although we will use the full $\,^{0}_{+1}e$ symbol shown above when writing out nuclear reactions. An example of positron emission is shown below. Notice how the reaction is properly balanced.

$$\overbrace{\,^{40}_{19}K \longrightarrow \underbrace{\,^{40}_{18}Ar + \,^{0}_{+1}e}_{18 + 1 = 19}}^{40 + 0 = 40}$$

Besides positron emission, a proton can convert into a neutron via the radioactive decay process known as *electron capture*. In **electron capture**, the nucleus absorbs, or captures, an inner-shell (core) electron. In doing so, the proton is converted into a neutron.

Electron capture

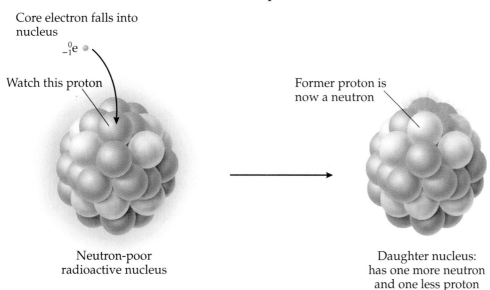

Core electron falls into nucleus
$\,^{0}_{-1}e$

Watch this proton

Neutron-poor radioactive nucleus

Former proton is now a neutron

Daughter nucleus: has one more neutron and one less proton

Unlike the proton-to-neutron conversion that occurs with positron emission, this time a particle is captured by the nucleus instead of ejected from it. This conversion is summarized below:

$$\,^{1}_{1}p + \,^{0}_{-1}e \longrightarrow \,^{1}_{0}n$$

Once again, check the superscripts and subscripts to see that the above nuclear reaction is balanced. Electron capture has the effect of decreasing the atomic number by 1 while leaving the mass number unchanged. Note that this is exactly the same result as in positron emission. An example of electron capture is shown below:

$$197 + 0 = 197$$
$$^{197}_{80}\text{Hg} + ^{0}_{-1}\text{e} \longrightarrow ^{197}_{79}\text{Au}$$
$$80 + (-1) = 79$$

Both positron emission and electron capture tend to occur for radioactive isotopes that need to convert a proton into a neutron (on the radioactive lower part of the band of stability). Which decay mode actually occurs depends on the particular radioactive isotope.

PRACTICE PROBLEMS

14.8 The fluorine isotope $^{17}_{9}\text{F}$ undergoes positron emission.
  (a) Write a nuclear reaction for this emission.
  (b) Where on the band of stability would you expect to find $^{17}_{9}\text{F}$?

Answer:
(a) $^{17}_{9}\text{F} \longrightarrow ^{0}_{+1}\text{e} + ^{17}_{8}\text{O}$
(b) On the radioactive lower part of the band of stability.

14.9 The argon isotope $^{37}_{18}\text{Ar}$ undergoes electron capture.
  (a) Write a nuclear reaction for this process.
  (b) Where on the band of stability would you expect to find $^{37}_{18}\text{Ar}$?
  (c) What is the daughter nucleus produced?

14.10 The magnesium isotope $^{25}_{12}\text{Mg}$ is the daughter that occurs when a radioactive parent undergoes electron capture. What is the full symbol for the parent?

14.11 The magnesium isotope $^{25}_{12}\text{Mg}$ is the daughter that occurs when a radioactive parent undergoes positron emission. What is the full symbol for the parent?

14.12 Write full nuclear reactions for Practice Problems 14.10 and 14.11.

RADIOACTIVE DECAY VIA α PARTICLE EMISSION

Go back and take a look at the band of stability plot where we indicated the fully stable, nonradioactive path. Notice how the nonradioactive path ends at the bismuth buoy. That's because all isotopes within the band of stability that have more than 83 protons are radioactive. The nuclei in these isotopes are just too big to be nonradioactive. They need to be smaller if they are to produce daughters nearer to or on the nonradioactive path. They tend to eject a sizable chunk of their nuclei via a process called *alpha (α) decay.* **Alpha decay** is the emission of an α *particle* from the nucleus. An **α particle** is a small piece of the nucleus that consists of 2 protons, 2 neutrons, and no electrons. This gives it an overall +2 charge. An α particle is therefore equivalent to the nucleus of a helium atom, so the full symbol for an α particle makes use of the helium symbol, He:

$^4_2$He

An α particle

Even though the α particle has a +2 charge, we usually don't include it in the full symbol.

By emitting an α particle, a large nucleus can get rid of a reasonably large chunk of itself all at once. In doing so, the atomic number decreases by 2, while the mass number decreases by 4.

α particle emission

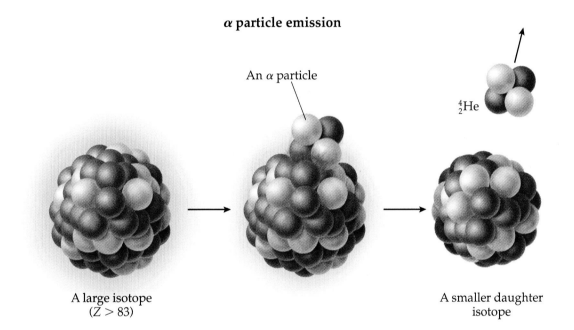

An α particle

$^4_2$He

A large isotope
$(Z > 83)$

A smaller daughter
isotope

An example of α emission is the conversion of uranium into thorium (note that the nuclear reaction is balanced):

$$234 + 4 = 238$$
$$^{238}_{92}\text{U} \longrightarrow ^{234}_{90}\text{Th} + ^4_2\text{He}$$
$$90 + 2 = 92$$

The thorium daughter is also radioactive, but at atomic number 90 it is closer to the fully stable nonradioactive path than its uranium parent.

Test yourself now to see if you are getting the hang of α decay.

Balance this nuclear reaction and predict the product:

WORKPATCH **14·7**

$$^{209}_{84}\text{Po} \longrightarrow ^{?}_{?}? + ^4_2\text{He}$$
α particle

Have the last few WorkPatches made you realize something? We hope so. These different decay modes are, in effect, moving us around on the periodic table in predictable ways. For example, suppose you start with some radioactive isotope of an element somewhere on the periodic table. Let's represent this element with the symbol $_Z^m E$, where m is the mass number and Z is the atomic number. When this parent isotope undergoes β^- emission, the daughter will be one element to the right. When this parent isotope undergoes either positron (β^+) emission or electron capture, the daughter will be one element to the left. And when this parent isotope undergoes α emission, the daughter will be two elements to the left:

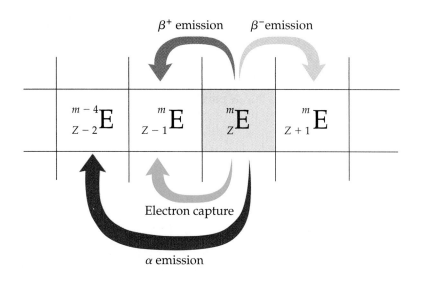

For example, here is the answer to WorkPatch 14.7, brought to you by simply looking at the periodic table and knowing in advance which way and how much to move for α decay. Just follow the curved arrow:

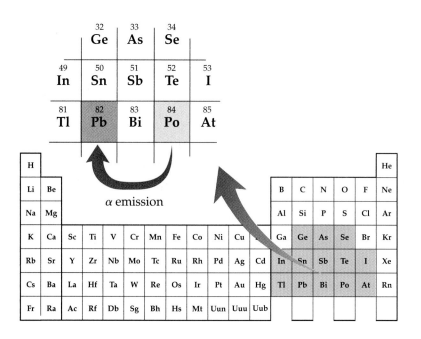

This summary can also be done in table form. We have started the table for you in the WorkPatch.

Fill in the rest of this table.

WORKPATCH **14.8**

	Particle	Change in atomic number	Change in mass number	Change in neutron number
α emission	^4_2He	-2	-4	-2
β^- emission	$^0_{-1}\text{e}$	?	?	?
Positron (β^+) emission	$^0_{+1}\text{e}$	?	?	?
Electron capture	$^0_{-1}\text{e}$	?	?	?

According to your table, which two decay processes give identical results? Does this agree with the figure on the facing page, which shows how the various decay modes move us around the periodic table? It should.

Uranium We'll end this discussion of α particle emission by highlighting one of the most famous (or infamous, depending on your point of view) elements of the twentieth century. Uranium is used to power nuclear reactors for generation of electricity. It was also used in the form of a bomb to flatten the Japanese city of Hiroshima during World War II. Uranium has an atomic number of 92. This is well beyond our bismuth buoy, so all isotopes of uranium are radioactive. Most of the uranium in uranium ore is the isotope $^{238}_{92}\text{U}$, which spontaneously decays to become, ultimately, the nonradioactive lead isotope $^{206}_{82}\text{Pb}$:

$$^{238}_{92}\text{U} \xrightarrow{\text{Many steps}} {}^{206}_{82}\text{Pb}$$

Radioactive   Nonradioactive
                     (fully stable)

Mass number decreases by 32
Atomic number decreases by 10

As you can see, the atomic number must decrease by 10 and the mass number by 32. None of the simple decay processes that we have discussed can accomplish this in one step. The decay of $^{238}_{92}\text{U}$ to $^{206}_{82}\text{Pb}$ is a multistep process, involving eight α emission steps and six β^- emission steps, as shown at right. All the isotopes in the process are radioactive and decay spontaneously over time, except for the lead isotope $^{206}_{82}\text{Pb}$, which is on the nonradioactive path.

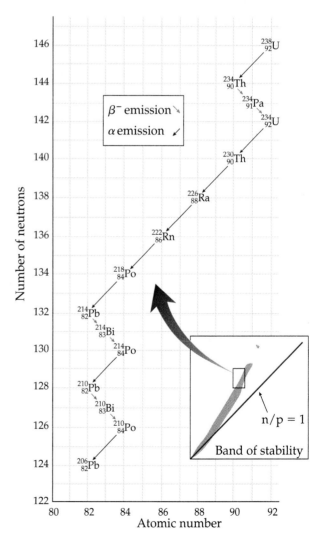

PRACTICE PROBLEMS

14.13 Positron (β^+) emission moves us one step to the left in the periodic table, and α emission moves us two steps to the left. Does this mean that two successive β^+ emissions give exactly the same result as one α emission with respect to the isotope obtained?

Answer: No. Starting from the same radioactive species, two β^+ emissions will give the same element for a product as one α emission (they will have the same atomic number, Z), but they will be different isotopes. The product of two successive β^+ emissions will have a mass number that is 4 larger than the mass number obtained from one α emission.

14.14 Starting from $^{238}_{92}U$, demonstrate that the answer to Practice Problem 14.13 is correct.

14.15 If a radioactive element undergoes a single decay process and transforms into a different element two spaces away on the periodic table, then which decay process must have taken place? Explain why it could not be the other processes.

14.16 If a radioactive element undergoes a single decay process and transforms into a different element one step to the right in the periodic table, then did a proton turn into a neutron or did a neutron turn into a proton? What do we call this type of decay?

GAMMA (γ) RADIATION

When a nucleus undergoes radioactive decay, it often releases a great deal of energy. For nuclei that decay via particle emission (α, β^-, and β^+ emission), the particle itself can carry away some of this energy in the form of kinetic energy. That is, the ejected particles generally move very fast. However, many radioactive nuclei also release energy in the form of electromagnetic radiation, most usually as **gamma rays (γ rays)**. Gamma rays are more energetic than X rays (see Chapter 3), and thus can be very harmful to living organisms.

Unlike the α, β^-, and β^+, and electron-capture decay modes, emission of a gamma ray from the nucleus causes no change to either the mass number or the atomic number. That's because electromagnetic radiation has no mass. Therefore, if the nucleus of a radioactive isotope does nothing more than emit a gamma ray, the elemental identity of the isotope remains unchanged. However, γ ray emission usually accompanies the other decay modes we have discussed. For example, the equation we gave earlier for $^{238}_{92}U$ undergoing α decay should have included the accompanying release of a gamma ray. Both the α particle and the γ ray carry energy away from the nucleus.

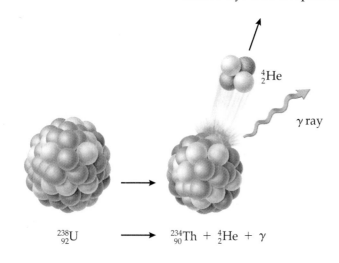

$^{238}_{92}U \quad \longrightarrow \quad ^{234}_{90}Th + ^{4}_{2}He + \gamma$

$^{4}_{2}He$

γ ray

USING RADIOACTIVE ISOTOPES TO DATE OLD OBJECTS 14.4

We defined half-life earlier in the chapter as the time it takes for the amount of a particular radioactive isotope in a sample to drop to half of its initial amount due to spontaneous nuclear change. We want to show you how the measured half-lives of various radioactive isotopes can be used to date old objects. The theory behind this process is actually quite simple. All we have to do is determine the percent of the initial amount of radioactive isotope remaining in some object. Radioactive isotope in an object? "There's no radioactive isotopes in me!" you say. Actually, there are. One in particular is the radioactive carbon isotope $^{14}_{6}C$, which we saw earlier is a β^- emitter. Everything alive has a roughly constant amount of $^{14}_{6}C$ in it, because $^{14}_{6}C$ is constantly being produced in the upper atmosphere as neutrons produced by cosmic radiation bombard nitrogen nuclei:

$$^{14}_{7}N + {}^{1}_{0}n \longrightarrow {}^{14}_{6}C + {}^{1}_{1}H$$

This radioactive carbon isotope enters the food chain and behaves chemically like the nonradioactive carbon isotopes. This means that we metabolize it and excrete it as CO_2 at the same rate as the nonradioactive isotopes, maintaining a constant amount of $^{14}_{6}C$ in our bodies. However, once we die, we no longer take in any $^{14}_{6}C$. The $^{14}_{6}C$ that is present begins to decay at a predictable rate characterized by its half-life of 5715 years. Now consider Table 14.2, which was generated by taking 100% (the original amount present) and repeatedly multiplying it by $\frac{1}{2}$.

Table 14.2

Number of half-lives that have passed	Percentage of isotope remaining
0	100%
1	50%
2	25%
3	12.5%
4	6.25%
5	3.13%
6	1.56%
7	0.78%
8	0.39%
9	0.19%
10	0.098%

Suppose someone digs up your bones a long time from now, does an analysis, and finds that your bones contain 75% of the $^{14}_{6}C$ they contained while you were alive (this would be the amount present in anyone living at the time of the analysis). According to Table 14.2, that puts your bones between 0 and 1 half-lives old. Your bones are between 0 and 5715 years old. You might be thinking that since 75% is exactly halfway between 100% and

50% of the isotope remaining, your bones are 5715/2 or 2858 years old. But it's not that simple. To find the actual age, we must use the following formula:

$$\text{Age} = \frac{-2.303 \times \log\left(\dfrac{\text{Percent } {}^{14}_{6}\text{C remaining}}{100}\right) \times \text{Half-life}}{0.693}$$

$$= \frac{-2.303 \times \log\left(\dfrac{75\%}{100}\right) \times 5715 \text{ years}}{0.693}$$

$$= 2.3 \times 10^3 \text{ years (2300 years old)}$$

Of course, there is some uncertainty in this number, depending on the accuracy to which the percent ${}^{14}_{6}$C remaining was measured. This calculation also assumes that there is a constant amount of ${}^{14}_{6}$C in the food chain over the years—that is, there hasn't been any change in the amount of ${}^{14}_{6}$C found in living humans from the time you died until your bones were analyzed. This may not be true. Still, this method works pretty well within its limitations. Now, suppose your bones were dug up 1 million years after you died.

 **WORKPATCH**   Consult Table 14.2 and then answer the following question: What would be wrong with trying to determine the age of fossils that are 1 million years old using the ${}^{14}_{6}$C isotope?

Do you see the problem? One million years is almost 175 half-lives for ${}^{14}_{6}$C. Check your answer against ours to be sure you understand. The solution to this problem would be to choose a different radioactive isotope present in the fossils—one that has a longer half-life, closer to the age of the fossils. The Earth itself has been dated by determining the amount of the radioactive potassium isotope ${}^{40}_{19}$K in rocks, because ${}^{40}_{19}$K has a half-life of 1.26×10^9 years. Measurements indicate that our planet is approximately 4.5 billion years old.

PRACTICE PROBLEMS

14.17 Analysis of a rock from an asteroid shows that it contains 2.57 g of ${}^{238}_{92}$U and 3.83 g of ${}^{206}_{82}$Pb. The molar mass of ${}^{206}_{82}$Pb is 205.974 46 g/mole; the molar mass of ${}^{238}_{92}$U is 238.029 g/mole. Assume that all the ${}^{206}_{82}$Pb came from the radioactive decay of the ${}^{238}_{92}$U.
   (a) How many atoms of each isotope are present in the rock?
   (b) How many atoms of ${}^{238}_{92}$U were in the rock when it was first formed?
   (c) What is the percent of ${}^{238}_{92}$U atoms remaining in the rock compared to when it was first formed?
   (d) How old is the asteroid?

Answer:

(a) $2.57 \text{ g } {}^{238}_{92}\text{U} \times \dfrac{1 \text{ mole } {}^{238}_{92}\text{U}}{238.029 \text{ g } {}^{238}_{92}\text{U}} \times \dfrac{6.02 \times 10^{23} \text{ atoms } {}^{238}_{92}\text{U}}{1 \text{ mole } {}^{238}_{92}\text{U}} = 6.50 \times 10^{21} \text{ atoms } {}^{238}_{92}\text{U}$

$3.83 \text{ g } {}^{206}_{82}\text{Pb} \times \dfrac{1 \text{ mole } {}^{206}_{82}\text{Pb}}{205.974 \, 46 \text{ g } {}^{206}_{82}\text{Pb}} \times \dfrac{6.02 \times 10^{23} \text{ atoms } {}^{206}_{82}\text{Pb}}{1 \text{ mole } {}^{206}_{82}\text{Pb}} = 1.12 \times 10^{22} \text{ atoms } {}^{206}_{82}\text{Pb}$

(b) Just add the number of $^{238}_{92}U$ atoms and $^{206}_{82}Pb$ atoms present in the old rock:

$$6.50 \times 10^{21} + 1.12 \times 10^{22} = 1.77 \times 10^{22} \text{ atoms } ^{238}_{92}U \text{ initially present}$$

(c) % $^{238}_{92}U$ remaining in rock $= \dfrac{\text{Number of } ^{238}_{92}U \text{ atoms present now}}{\text{Number of } ^{238}_{92}U \text{ atoms intially present}} \times 100$

$$= \dfrac{6.50 \times 10^{21} \text{atoms } ^{238}_{92}U}{1.77 \times 10^{22} \text{atoms } ^{238}_{92}U \text{ initially present}} \times 100$$

$$= 36.7\% \text{ remaining}$$

(d) Age $= \dfrac{-2.303 \times \log\left(\dfrac{\text{Percent } ^{238}_{92}U \text{ remaining}}{100}\right) \times \text{Half-life} \overset{\displaystyle 4.46 \times 10^9 \text{ years}}{}}{0.693}$

$$= \dfrac{-2.303 \times \log\left(\dfrac{36.7\%}{100}\right) \times (4.46 \times 10^9 \text{ years})}{0.693}$$

$= 6.45 \times 10^9$ years (6.45 billion years old)   *This is a very old rock, indeed, predating our solar system!*

14.18 In part (b) of Practice Problem 14.17, why is the number of $^{238}_{92}U$ atoms initially present in the rock equal to the sum of the number of $^{238}_{92}U$ atoms and $^{206}_{82}Pb$ atoms present in the old rock?

14.19 An even older asteroid is found. A rock from it yields 1.82 g of $^{238}_{92}U$ and 4.02 g of $^{206}_{82}Pb$. How old is this asteroid?

NUCLEAR ENERGY (FISSION AND FUSION) 14.5

Humanity has come a long way since the early alchemists. Indeed, now we can convert one element into another by inducing nuclear reactions, but quite often the energy cost is tremendous. No one is getting rich turning lead into gold. There are two types of induced nuclear reactions, however, that can release tremendous amounts of energy—*nuclear fission* and *nuclear fusion*. **Nuclear fission** is the breaking up of a large nucleus into smaller ones. **Nuclear fusion** is the combining of small nuclei into larger ones. Both processes release energy so long as the combined mass defects of the products is greater than the combined mass defects of the reactants (in other words, as long as mass is lost on going from reactants to products). Not much mass needs to be lost to generate a tremendous amount of energy. As we saw earlier, this is because the energy equivalent of mass is $E = mc^2$, where $c = 3.0 \times 10^8$ m/sec. The value of c^2 is such a large number that when it multiplies the mass m, even a tiny m, a lot of energy is released. We have the ability to carry out these nuclear reactions quickly in an uncontrolled manner or slowly in a controlled manner. When the energy is released all at once, we have a nuclear bomb. When it is released slowly in a nuclear reactor, the energy can be used to generate steam for turning turbines in an electrical power plant. Let's look a bit more closely now at some examples of induced nuclear fission and fusion.

Fission

In nuclear reactors, fission is induced by firing neutrons at a heavy radioactive isotope such as $^{235}_{92}U$, causing the uranium atoms to split into smaller atoms. One of the fission reactions that occurs is shown below:

$$^{235}_{92}U + ^1_0n \longrightarrow ^{142}_{56}Ba + ^{91}_{36}Kr + 3\,^1_0n$$

Loss of mass for fission reaction is 0.1868 g/mole.

Energy released upon fission of $^{235}_{92}U$ = 16,800,000, 000 kJ/mole.

There are two important things to notice here. First, a tremendous amount of energy is released in the fission reaction of 1 mole (235.04 g) of $^{235}_{92}U$. You would have to burn approximately 131,885 gallons of gasoline to get the same amount of energy. Second, this fission reaction produces more neutrons than it consumes. The 3 neutrons that are produced can cause fission of 3 more $^{235}_{92}U$ atoms, which each produce 3 neutrons, which can go on to cause fission of 9 more $^{235}_{92}U$ atoms, which each produce 3 neutrons, which can go on to cause fission of 27 more $^{235}_{92}U$ atoms, and so on. Things can get out of hand pretty quickly.

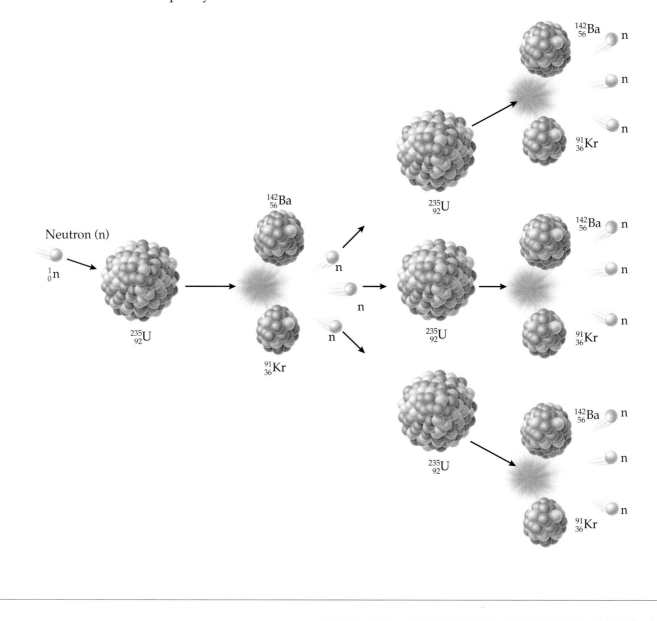

The third generation of fission of $^{235}_{92}U$ produces 27 neutrons. How many neutrons would the tenth generation of fission produce?

WORKPATCH 14.10

This scenario is called a *chain reaction*. In a **chain reaction**, each reaction event gives rise to more than one subsequent reaction event. In a nuclear reactor, the chain reaction is regulated with control rods containing neutron-absorbing material such as boron or cadmium.

Even without the control rods, there is no danger of a nuclear power plant exploding like an atom bomb. A nuclear explosion can occur only if the $^{235}_{92}U$ is of the proper mass and shape so that most of the neutrons produced by the chain reaction stay within the uranium sample rather than escaping before they initiate more fission. This mass, 56 kg for pure $^{235}_{92}U$, is called the **critical mass**. For $^{235}_{92}U$, however, the critical mass can be reduced to 15 kg by reflecting escaped neutrons back into the uranium mass.

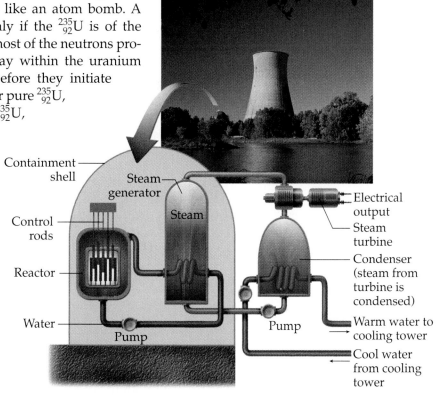

No nuclear reactor has ever been designed that would allow for 56 kg of pure $^{235}_{92}U$ to come together. Still, nuclear reactors account for only 22% of the total electricity generated in the United States. While there is no chance that a fission reactor could ever explode like an atom bomb, there are other problems. There is the potential for accidents that release radioactive materials into the environment. And there is the problem of what to do with the nuclear waste. The spent uranium fuel consists of highly radioactive fission products, some with half-lives of thousands of years. During the writing of this book, the United States opened its first nuclear waste storage facility located deep under the New Mexico desert, where the government plans to bury nuclear waste for thousands of years. Not everyone is pleased with this solution. Nuclear power plants are also tremendously expensive to build and maintain. Of course, there are also benefits to nuclear power generation. It lessens our dependence on fossil fuels such as petroleum and coal. Currently, coal and petroleum are the main fuels used in the United States to produce steam for turning electrical turbines. As a by-product, these fuels produce millions of tons of atmospheric pollution each year, contributing to our acid rain and smog. In addition, mining coal and transporting petroleum have their own drawbacks. Nevertheless, the general fear of anything nuclear will probably prevent nuclear power generation via fission from becoming the predominant source of electrical energy in the United States. This is not true everywhere, however. For example, France produces 73% of its electricity via nuclear fission.

FUSION

Nuclear fusion is the process responsible for the production of energy in stars like our Sun. The Sun is essentially a giant ball of hydrogen (approximately 80%) and helium (approximately 20%). Inside its core, hydrogen atoms are fusing together to form helium atoms. In the process, mass is lost and converted into energy. Scientists calculate that the Sun converts 4×10^{12} g of mass into energy every second. That's 4 trillion grams per second! One sequence of fusion reactions thought to be occurring in our Sun is shown below:

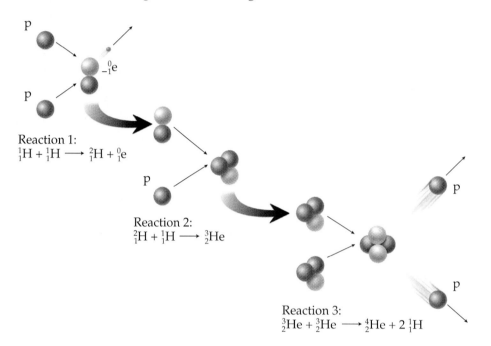

Reaction 1:
$$^1_1H + ^1_1H \longrightarrow ^2_1H + ^0_1e$$

Reaction 2:
$$^2_1H + ^1_1H \longrightarrow ^3_2He$$

Reaction 3:
$$^3_2He + ^3_2He \longrightarrow ^4_2He + 2\,^1_1H$$

There are no fusion-based nuclear reactors producing electricity here on Earth. The problem with fusion is to get two nuclei to combine. Nuclei repel each other as they get close, because all nuclei are positive and like charges repel. In the core of the Sun, hydrogen is heated to incredibly high temperatures (estimated to be 15,000,000 K). At these high temperatures, nuclei are moving so fast that collisions between them can overcome their mutual repulsion, and fusion can occur. Here on Earth, generating the temperatures required to sustain fusion is a major problem. This is not a problem for the Sun. When the Sun formed, gravitational collapse provided the energy to heat the gas until fusion started. Now the reaction is self-sustaining; the energy to overcome the repulsion between nuclei comes from the fusion itself. The Sun's immense gravity keeps the solar core dense, so that fusion can continue.

On Earth we have yet to figure out a way to heat hydrogen up to 15,000,000 K and keep it contained in a dense enough form to sustain fusion. Such heat would instantly vaporize the thickest concrete or steel vessel that we could build to contain the reaction, so most of the experimental fusion reactors use magnetic fields to confine the reaction. While we do have the ability to produce fusion in reactors, we have yet to achieve the *break-even point*, where we would get as much energy back from the reaction as we put in to make it happen. Beyond the break-even point, a fusion reactor would actually become a power source—that is, it would give us back more energy than we feed into it.

For the past 45 years, nuclear physicists have been predicting that practical nuclear fusion was just 20 years away. Today, while we are closer than ever, we are still probably at least 20 years away, and the cost of getting there will be astronomical. However, the benefits of generating electricity via fusion instead of fission would be tremendous. Heavy isotopes of hydrogen, the likely fuel, are present in ocean water and are plentiful. In addition, there would be no long-lived radioactive waste products. Fusion reactors may one day help us meet our energy requirements.

Meanwhile, scientists have perfected ways to perform uncontrolled nuclear fusion in the form of the hydrogen bomb. In a hydrogen bomb, isotopes of hydrogen surround a nuclear fission bomb. When the fission bomb goes off, sufficient temperatures are reached to induce fusion in the surrounding hydrogen. With no attempt being made to contain the energy released, the result is rather powerful, to say the least. While the Hiroshima $^{235}_{92}U$ fission bomb destroyed almost everything within a 1 mile radius of the explosion, hydrogen bombs have been produced that will obliterate everything within a 15–20 mile radius.

PRACTICE PROBLEMS

14.20 Complete the following nuclear reaction, and state whether it is fission or fusion.

$$^{239}_{94}Pu + {}^{1}_{0}n \longrightarrow {}^{90}_{38}Sr + \text{?} + 3\,{}^{1}_{0}n$$

Answer: It is fission—you figure out the missing element.

14.21 Could the nuclear reaction of Practice Problem 14.20 be used to produce a chain reaction? Fully explain your answer.

14.22 Explain how you would determine whether the nuclear reaction of Practice Problem 14.20 is exothermic or endothermic.

BIOLOGICAL EFFECTS AND MEDICAL APPLICATIONS OF RADIOACTIVITY 14.6

Many people have a fear bordering on the irrational of anything nuclear. This is somewhat understandable, given the media's preoccupation with the lethal effects of nuclear radiation that would be unleashed by an atomic war or from the meltdown of a reactor core. Much less front-page coverage is given to the useful applications of nuclear radiation. Consider, for example, a technique called *magnetic resonance imaging (MRI)*, used in modern hospitals for the last decade or so. In this technique, a person is placed in a magnetic field, and radio waves are used to probe certain nuclei in the body. The results are spectacular images of the insides of the body, allowing doctors to locate tumors and other internal problems without dangerous exploratory surgery. This

"I'm not climbing into that nuclear contraption!"

technique is completely safe and involves no nuclear radiation whatsoever. Nevertheless, the original name for this technique, *nuclear magnetic resonance (NMR)*, had to be scrapped and replaced by MRI simply because people so feared the word nuclear that they would often refuse to be examined by the instrument.

The term **nuclear radiation** refers to the high-energy particles and electromagnetic radiation emitted by a nucleus during nuclear change. Among these are the α particles, β particles, and γ rays that we have discussed. Certainly it is true that sufficient exposure to nuclear radiation can cause harm ranging from simple burns to death. The high energy associated with the radiation can cook living tissue as well as any stove. Another way that radiation's energy can do harm is by *ionizing* the biological molecules in your body. By **ionizing** we mean that the radiation deposits sufficient energy into a biological molecule to knock an electron out of it, converting it into a positive cation:

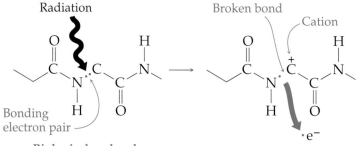

Biological molecule

Since biological molecules such as proteins, carbohydrates, and enzymes are held together by covalent bonds, ionizing them can cause irreparable damage in the form of weakened and broken bonds. Damage enough biological molecules within an organism, and it will die. However, the different types of radiation have different biological effects.

For a given amount of energy, α particles are less dangerous than β particles when the radioactive source is outside the body. This is because α particles are much more massive than β particles, causing them to move more slowly and decreasing their ability to penetrate matter. Normal clothing is enough to protect the skin from α radiation. Lighter β particles of similar energy move much more rapidly, but they can be stopped by a block of wood or heavy protective clothing. With proper precautions to protect the skin, α and β particles are relatively safe, but we must qualify this. If an α or β emitter is ingested, the radiation is then in intimate contact with your tissues and is extremely hazardous. Indeed, α particles are more dangerous than β particles when the emitter is inside the body. Gamma rays are the most dangerous type of external radiation. They have 1000 times the penetrating power of α particles, and can thus penetrate deep into the body and damage internal organs. It takes a lead block several inches thick to stop γ rays.

We don't want to leave you with the impression that radiation can only cause harm. Quite the opposite is true, and the medical field has become dependent on radioactive substances for a variety of applications. Radiation is part of many therapeutic procedures. For example, the γ rays given off by radioactive cobalt, $_{27}^{60}$Co, are directed toward tumors in a procedure called *external radiation therapy*. Because γ rays can penetrate into the deepest cells in the body and kill them by breaking covalent bonds in proteins and DNA, carefully controlled and focused doses of γ radiation can be used to destroy cancer cells. The trick is to expose the cancer cells to the γ rays while not exposing too much of the surrounding healthy tissue.

Another example, called *radiation imaging*, is very useful in diagnosis. For instance, while a healthy thyroid gland will absorb iodine, a hyperthyroid gland will absorb too much, and a hypothyroid gland will absorb too little. A patient is injected with radioactive $_{53}^{131}$I, a β^- emitter, and then either a piece of film or a radiation detector is placed close to the patient's neck. The thyroid absorbs the $_{53}^{131}$I and becomes a "hot" organ, emitting both β^- and γ radiation. This will develop the film or cause the detector to produce a signal. An estimate of thyroid function can be made from the amount of radiation detected. The dose of radioactive iodine is not sufficient to cause harm, and the half-life of $_{53}^{131}$I is only 8 days, so it is mostly gone after a month.

Still other procedures called *radioimmunoassays* are done outside the body using blood samples. In these procedures, radioactive substances are bound to substances in the blood such as drugs or hormones in an effort to determine their concentration. Still, the irrational fear of anything nuclear can rear its ugly head. Each year, thousands are sickened and many die from food poisoning. In 1999, over 20 people died in the midwest from exposure to *E. coli* bacteria in meat contaminated at one plant. Salmonella in poultry is also a large problem. A proven way to kill harmful bacteria in meats, vegetables, and fruits is to briefly irradiate them with ionizing radiation. In 1986 the U.S. Food and Drug Administration approved the irradiation of fruits, vegetables, herbs, spices, and pork. In 1990 they approved irradiation for poultry. Indeed, astronauts and cosmonauts eat irradiated foods while in space to protect them from food poisoning. However, public fear regarding anything nuclear has limited its commercial application. More education on this subject may eventually overcome this fear.

HAVE YOU LEARNED THIS?

Mass defect (p. 512)

Binding energy (p. 512)

Nucleon (p. 513)

Radioactive (p. 515)

Strong force (p. 515)

Neutron-to-proton (n/p) ratio (p. 516)

Band of stability (p. 516)

Half-life (p. 516)

Radioactive decay (p. 519)

Positron (p. 520)

Antimatter (p. 520)

β^- particle (p. 520)

β^- emission (p. 520)

Nuclear reaction (p. 521)

Daughter and parent (p. 521)

Positron emission (p. 522)

Electron capture (p. 523)

α particle (p. 524)

α decay (p. 524)

γ ray (p. 528)

Nuclear fission (p. 531)

Nuclear fusion (pp. 531, 534)

Chain reaction (p. 533)

Critical mass (p. 533)

Nuclear radiation (p. 536)

Ionizing radiation (p. 536)

MASS DEFECT AND THE STABILITY OF THE NUCLEUS

14.23 What do we mean when we say that an atom has a mass defect?

14.24 What do we mean by the binding energy for an atom's nucleus?

14.25 True or false? The number of nucleons for a nucleus is equal to its mass number.

14.26 Consider the following facts regarding two hypothetical nuclei, one heavy and one light: (1) The heavy nucleus has a greater total binding energy than the light nucleus. (2) The lighter nucleus is more stable than the heavy nucleus. Explain how both facts can be true.

14.27 Explain why a chemical reaction can never cause changes to an atom's nucleus.

14.28 Of all the isotopes of the elements, which has the greatest mass defect? What does this mean for that isotope?

14.29 What in Einstein's famous energy equation ensures that a tiny mass defect will result in a tremendous amount of energy? Explain your answer.

14.30 The mass of 1 mole of radioactive $^{14}_{6}C$ is 14.003 24 g. Calculate its binding energy (in kJ/mole).

14.31 Examine the plot of binding energy per nucleon versus the number of nucleons (page 513). Where would your answer to Problem 14.30 put the $^{14}_{6}C$ point relative to $^{12}_{6}C$ and $^{13}_{6}C$? What does this imply about the stability of the $^{14}_{6}C$ nucleus relative to the lighter isotopes?

14.32 What is a radioactive atom?

14.33 True or false? Radioactive atoms have nuclei with no mass defect or binding energy. Explain your answer.

HALF-LIFE AND THE BAND OF STABILITY

14.34 Why are neutrons thought to be important for making a nucleus stable?

14.35 As we go from light atoms to heavier ones:
(a) What happens to the value of the neutron-to-proton ratio?
(b) Why does the answer to part (a) make sense?

14.36 Write the full symbols for the isotopes of oxygen possessing 8, 9, and 11 neutrons.

14.37 Calculate the n/p ratios for the isotopes in Problem 14.36.

14.38 What is the band of stability?

14.39 Define half-life.

14.40 How long would it take for a 10 g sample of $^{123}_{53}I$ to drop to 0.039 g? [Half-life of $^{123}_{53}I$ is 13.1 hours.]

14.41 According to the band of stability, when would an atom with 60 protons in its nucleus be unstable? (Read the plot as best you can.)

14.42 What do we mean by the "ocean of instability"?

14.43 Why do radioactive isotopes appear on the band of stability?

SPONTANEOUS NUCLEAR CHANGES: RADIOACTIVITY

14.44 What can we say about the n/p ratios for nuclei on the part of the band of stability above the path of full stability?

14.45 What can we say about the n/p ratios for nuclei on the part of the band of stability below the path of full stability?

14.46 What is true of all atoms with atomic numbers greater than 83?

14.47 What is meant by the term "radioactive decay"?

14.48 Write the full symbols for a neutron, proton, electron, and positron.

14.49 Why is a positron referred to as antimatter?

14.50 How do we interpret the subscripts for the full symbols of an electron and a positron?

14.51 What kind of nucleus would ever want to convert a proton into a neutron, and why would it want to do it?

14.52 What kind of nucleus would ever want to convert a neutron into a proton, and why would it want to do it?

14.53 True or false? Radioactive decay ends up changing the elemental identity of the isotope that is undergoing the decay. Explain your answer.

14.54 What happens to an atom's nucleus when it undergoes β^- emission?

14.55 How is it possible for a nucleus to eject an electron when it has no electrons within it?

14.56 The tantalum $^{186}_{73}$Ta isotope is radioactive. Its radioactive decay process involves conversion of a neutron into a proton.
(a) Where is this atom likely to be in the band of stability?
(b) Write a nuclear reaction for this process.
(c) Name this decay process.

14.57 What happens to an atom's nucleus when it undergoes positron emission?

14.58 What happens to an atom's nucleus when it undergoes electron capture?

14.59 The tungsten $^{162}_{74}$W isotope is radioactive. Its radioactive decay process involves conversion of a proton into a neutron.
(a) Where is this atom likely to be in the band of stability?
(b) Write two nuclear reactions that could account for this process.
(c) Name these decay processes.

14.60 What kind of nucleus would ever want to eject 2 neutrons and 2 protons?

14.61 What happens to an atom's nucleus when it undergoes α emission?

14.62 The thorium $^{232}_{90}$Th isotope is radioactive. Its radioactive decay process involves the ejection of 2 protons and 2 neutrons from its nucleus.
(a) Where is this atom likely to be in the band of stability?
(b) Write a nuclear reaction for this process.
(c) Name this decay process.

14.63 Name two general forms in which energy can be carried away from a nucleus undergoing radioactive decay.

14.64 In principle, which of Problems 14.56, 14.59, and 14.62 could occur along with γ radiation?

14.65 How do you check to see if a nuclear reaction is balanced? Give an example.

14.66 Near some deposits of radioactive ores are also found pockets of trapped helium gas. How can you explain this?

14.67 What happens to the mass number of a nucleus when it:
(a) Ejects a β^- particle?   (b) Ejects a β^+ particle?
(c) Undergoes electron capture?   (d) Ejects an α particle?

14.68 What happens to the atomic number of a nucleus when it:
(a) Ejects a β^- particle?   (b) Ejects a β^+ particle?
(c) Undergoes electron capture?   (d) Ejects an α particle?

14.69 Explain how it is possible to quickly predict the daughter that will result for each of the following by simply examining the periodic table.
(a) α decay   (b) β^- decay   (c) β^+ decay   (d) Electron capture

14.70 Why doesn't γ emission change the elemental identity of a nucleus?

14.71 Starting with the lead isotope $^{207}_{82}Pb$, postulate a sequence of radioactive decays that would convert it to an isotope of gold.

14.72 Complete the following nuclear reaction, and name the decay process that is occurring.

$$^8_4Be + ? \longrightarrow {}^8_3Li$$

14.73 Complete the following nuclear reaction, and name the decay process that is occurring.

$$^{47}_{20}Ca \longrightarrow ? + {}^{47}_{21}Sc$$

14.74 Complete the following nuclear reaction, and name the decay process that is occurring.

$$^{235}_{92}U \longrightarrow {}^4_2He + ?$$

14.75 Complete the following nuclear reaction, and name the decay process that is occurring.

$$? \longrightarrow {}^{11}_5B + {}^0_{-1}e$$

14.76 Complete the following nuclear reaction, and name the decay process that is occurring.

$$^0_{-1}e + ? \longrightarrow {}^{40}_{18}Ar$$

USING RADIOACTIVE ISOTOPES TO DATE OLD OBJECTS

14.77 Suppose you have 100 g of $^{123}_{53}I$. How much of it will be left after 26.2 hours? How much of it will be left after 39.3 hours? [Half-life of $^{123}_{53}I$ is 13.1 hours.]

14.78 Would $^{14}_6C$ be a useful isotope to date a sample of a fossil that is 120 million years old? Explain.

14.79 Given the half-life of $^{14}_6C$, why is any of it present in the environment?

14.80 Measurements show that the percentage of $^{14}_6C$ remaining in a particular artifact is 22.8%. What is the age of the object in years?

14.81 Measurements show that a sample of rock contains 14.90 g of $^{238}_{92}U$ and 26.50 g of $^{206}_{82}Pb$. [See Practice Problem 14.17, page 530, for molar masses of isotopes.]
(a) To date this rock, what assumption must be made as to where the $^{206}_{82}Pb$ came from?
(b) How many atoms of each isotope are present in the rock?
(c) How many atoms of $^{238}_{92}U$ were in the rock when it was new?
(d) What is the percentage of $^{238}_{92}U$ remaining in the rock compared to when it was new?
(e) How old is the rock (in years)? [Half-life of $^{238}_{92}U$ is 4.46×10^9 years.]

NUCLEAR ENERGY (FISSION AND FUSION)

14.82 For a fission or fusion reaction to be exothermic, what must be true regarding the mass defect? Explain your answer fully.

14.83 What is nuclear fission?

14.84 What is nuclear fusion?

14.85 Why does nuclear fission often proceed as a chain reaction?

14.86 Complete the following fission reaction:

$$^{239}_{94}Pu + ^{1}_{0}n \longrightarrow ? + ^{140}_{54}Xe$$

14.87 Complete the following fusion reaction:

$$^{8}_{4}Be + ^{4}_{2}He \longrightarrow ? + \gamma$$

14.88 The hydrogen in our Sun is undergoing fusion and turning into helium. As the hydrogen begins to run out billions of years from now, the helium will begin to fuse, forming even heavier atoms. Eventually, these heavier atoms will also begin to undergo fusion. Interestingly, when astronomers examine the remnants of burnt-out stars, they find them to be extremely rich in iron. Explain why this would be so.

14.89 What do we mean by critical mass for a sample of pure $^{235}_{92}U$?

14.90 What is done with the heat produced by a fission reactor to produce electricity?

14.91 Discuss the benefits and problems associated with using nuclear fission to produce electricity.

14.92 Why can't a nuclear reactor explode like a nuclear bomb?

14.93 Why does it take so much input of energy to initiate nuclear fusion?

14.94 Why are there as yet no fusion reactors?

14.95 What would be the advantages of fusion reactors over fission reactors?

BIOLOGICAL EFFECTS AND MEDICAL APPLICATIONS OF RADIOACTIVITY

14.96 Of all the emissions from an unstable nucleus, which is least likely to damage you upon external exposure? Which is most likely? Explain fully.

14.97 How does radiation do its damage to living organisms?

14.98 Radioactivity is often called ionizing radiation. Explain why.

14.99 Briefly discuss the different medical uses of radioactivity presented in this chapter.

WORKPATCH SOLUTIONS

14.1 (a) Being at the top of the plot, the $^{56}_{26}Fe$ nucleus has the most binding energy per nucleon, making it the most stable isotope of all the elements.
(b) Reading the binding energy off the plot, we get 8.48×10^8 kJ/mole per nucleon. Multiplying this by the 56 nucleons that are in the $^{56}_{26}Fe$ nucleus, the total binding energy is 4.75×10^{10} kJ/mole.

14.2 (a) There are 207 nucleons (the mass number).
(b) Going to 207 nucleons on the plot of binding energy, we read 7.57×10^8 kJ/mole per nucleon. To remove 3 nucleons would take $3 \times 7.57 \times 10^8$ kJ/mole $= 2.27 \times 10^9$ kJ/mole.

14.3 (a) $^{12}_{6}C$, $^{13}_{6}C$, $^{14}_{6}C$
(b) 6 p and 6 n in $^{12}_{6}C$, 6 p and 7 n in $^{13}_{6}C$, 6 p and 8 n in $^{14}_{6}C$
(c) 12 nucleons in $^{12}_{6}C$, 13 nucleons in $^{13}_{6}C$, 14 nucleons in $^{14}_{6}C$
(d) They are all carbon because they all have the same atomic number, 6, which means they all have 6 protons.

14.4 (a) They both have the same mass number (1) because they both consist of 1 nucleon.
(b) The atomic number of a neutron is 0 because it has no protons in it.
(c) The atomic number for a proton is 1 because it has one proton (it *is* a proton).

14.5 $_{0}^{0}n$ This should be 1. The superscript is the mass number, the sum of protons (0) + neutrons (1).

$_{0}^{1}p$ This should be 1. This is the number of protons.

This should be 0. The superscript is the mass number, the sum of protons (0) + neutrons (0).

$_{1}^{1}e$ Since you are told there is only one mistake, then this must be a positron. The bottom subscript can be +1 = 1 for a positron or −1 for an electron.

14.6 $_{82}^{214}Pb \longrightarrow {}_{83}^{214}Bi + {}_{-1}^{0}e$

14.7 $_{84}^{209}Po \longrightarrow {}_{82}^{205}Pb + {}_{2}^{4}He$

14.8

	Particle	Change in atomic number	Change in mass number	Change in neutron number
α emission	$_{2}^{4}He$	−2	−4	−2
β^{-} emission	$_{-1}^{0}e$	+1	0	−1
Positron (β^{+}) emission	$_{+1}^{0}e$	−1	0	+1
Electron capture	$_{-1}^{0}e$	−1	0	+1

14.9   After 1 million years, or 175 half-lives, the percentage of $_{6}^{14}C$ isotope remaining in the fossil would be too tiny to detect, let alone measure.

14.10 The tenth generation of fission would produce $3^{10} = 59,049$ neutrons.

The Chemistry of Carbon

CARBON—A UNIQUE ELEMENT 15.1

By applying the fundamental concepts of chemistry covered in the previous chapters, you can begin to understand some of the mysteries of nature. So, what should we look at? The glow of a firefly? The production of acid rain? The elasticity of rubber? Why limit ourselves? Let's examine the biggest subject of them all—life. Knowledge of the fundamental concepts of chemistry has brought humanity to the brink of some astounding possibilities regarding life, from the possibility of conquering viruses and cancer to the manipulation of our own genetic code and cloning. If we are to explore the chemistry of life using what we have learned, then we must start with the one element upon which all life as we know it is based—carbon.

Carbon is the element in the periodic table with atomic number 6. Its most abundant isotope, $^{12}_{6}C$, has 6 protons and 6 neutrons in the nucleus, with 6 electrons outside the nucleus. Of those 6 electrons, 4 are valence electrons. When a carbon atom bonds to another atom, it is these 4 valence electrons that are involved:

Carbon bonds via its $\longrightarrow \cdot \overset{\displaystyle \cdot}{\underset{\displaystyle \cdot}{C}} \cdot$
4 valence electrons.

One of the things that carbon atoms are very good at is bonding to other carbon atoms. In fact, no other element in the periodic table is able to covalently bond with other atoms of its own kind as well as carbon. Carbon atoms bond to each other by sharing their valence electrons, forming long chains, rings, and a large variety of other carbon frameworks, as shown on the next page.

Some chains, rings, and other structures formed by carbon atoms (attached hydrogen atoms not shown)

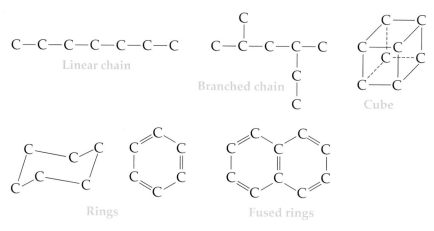

Linear chain

Branched chain

Cube

Rings

Fused rings

This ability to connect and link together is called **catenation.** No other atoms have quite the ability to do this. For example, oxygen atoms can bond to each other to form the diatomic molecule O_2. Also known but much more reactive is the ozone molecule O_3, a bent chain of 3 oxygen atoms covalently bound to one another. But that is it; longer chains of oxygen are unknown. Nitrogen, which falls between carbon and oxygen in the periodic table, can bond to other atoms of itself, but cannot manage to form long chains. For example, you are currently breathing N_2 molecules, but the N_3^- azide ion (with 3 nitrogen atoms in a chain) is so unstable that it explodes when heated. Thus, carbon is unique.

Elemental carbon comes in three different structural forms (called *allotropes*): diamond; graphite; and the more recently discovered fullerene form, found in soot. The structures of these allotropes of carbon are shown below:

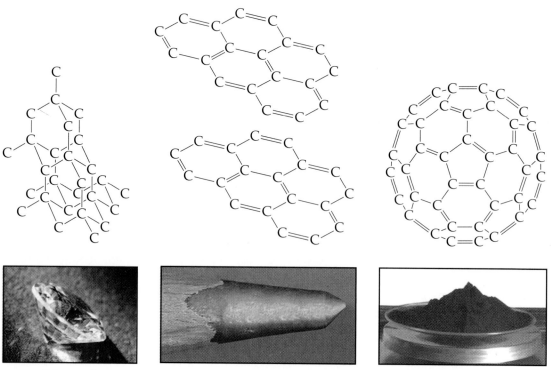

Diamond   Graphite   Soot containing fullerene

In each allotrope, the carbon atoms are covalently bound to other carbon atoms. In diamond, each carbon atom is bound by single bonds to 4 other carbon atoms in a tetrahedron, as we would expect from VSEPR theory (see Chapter 6). In graphite and fullerene, each carbon atom is bound to 3 other carbon atoms in a triangular manner, also expected from VSEPR theory. In other words, the geometry about a carbon atom depends on how many other atoms it is bound to.

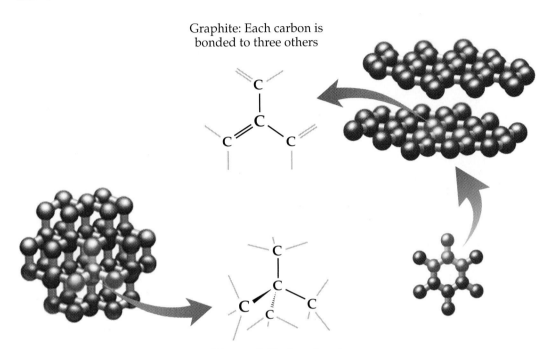

Graphite: Each carbon is
bonded to three others

Diamond: Each carbon is
bonded to four others

These seemingly small differences give rise to substances with drastically different properties. Diamond has all its carbons locked in place by strong covalent bonds. The result is the hardest known natural substance, which is used to coat saw blades that can cut through virtually any other material. Graphite, on the other hand, consists of sheets of carbon atoms attracted to one another by weak London forces, allowing the layers to slide over one another. The result is a soft, crumbly material that can be used as a lubricant.

Take another look at the structures of the diamond, graphite, and fullerene forms of carbon shown on the facing page. Can you see what they have in common? It is not the number of attached carbon atoms. Every carbon atom in diamond is bound to 4 other carbon atoms, but in fullerene and graphite, each carbon atom is bound to only 3 other carbon atoms. The common feature is that in all three allotropes, every carbon atom forms 4 covalent bonds:

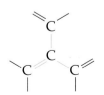

Four bonds
(diamond)

Four bonds
(graphite and fullerene)

Because carbon has 4 valence electrons, it will always "want" to form 4 covalent bonds to arrive at an octet of electrons. This can be accomplished with just single bonds or with a combination of single and double bonds:

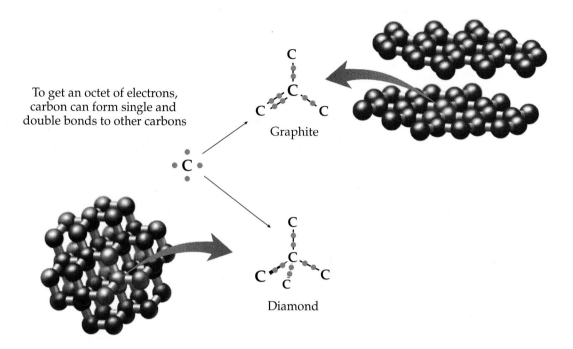

To get an octet of electrons, carbon can form single and double bonds to other carbons

Graphite

Diamond

In one case, carbon is attached to 4 other carbon atoms. In the other case, carbon is attached to 3 other carbon atoms. But in both cases, every carbon atom has 4 covalent bonds. This is always true; carbon always forms 4 covalent bonds.

Another way that carbon can form 4 bonds is to make use of triple bonds:

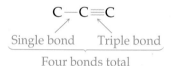

Here, 1 single and 1 triple bond result in a total of 4 bonds and a complete octet for the central carbon. While triple bonds do not occur in the three allotropic forms of elemental carbon, they do occur in compounds of carbon.

15.1 WORKPATCH

There is a problem with at least one carbon atom in each of the three molecules shown below. Identify the problem carbon(s) for each molecule, and then fix the problem by either removing or adding one hydrogen atom, or removing or adding a bond.

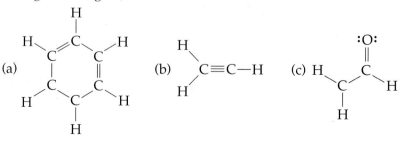

Because of its ability to catenate and form long chains of variable lengths, carbon gives rise to a larger number of compounds than any other element. Indeed, a whole branch of chemistry called **organic chemistry** is devoted to the study of carbon-based molecules. The name "organic" comes from the fact that carbon-based molecules make up the majority of molecules present in living organisms. No other branch of chemistry is based on just one element, an indication of how important carbon is. This chapter and the next will introduce you to the rich variety of molecules that carbon forms.

NATURALLY OCCURRING COMPOUNDS OF CARBON 15.2
—WITH HYDROGEN—HYDROCARBONS

Everyone in industrialized societies has heard of crude oil, also known as petroleum. The majority of the chemical compounds in petroleum are **hydrocarbons**, molecules made up entirely of carbon and hydrogen. The carbon atoms in a hydrocarbon molecule are bound to one another to form the chains described in the previous section, and the hydrogen atoms are bound to the carbon atoms. While the carbon atoms may be bound to each other with single, double, or even triple bonds, the hydrogen atoms are always bound to the carbon atoms with single bonds. A typical hydrocarbon molecule is octane, which has an eight-carbon chain:

$$H-\overset{\displaystyle\overset{H}{|}}{\underset{\displaystyle\underset{H}{|}}{C}}-\overset{\displaystyle\overset{H}{|}}{\underset{\displaystyle\underset{H}{|}}{C}}-\overset{\displaystyle\overset{H}{|}}{\underset{\displaystyle\underset{H}{|}}{C}}-\overset{\displaystyle\overset{H}{|}}{\underset{\displaystyle\underset{H}{|}}{C}}-\overset{\displaystyle\overset{H}{|}}{\underset{\displaystyle\underset{H}{|}}{C}}-\overset{\displaystyle\overset{H}{|}}{\underset{\displaystyle\underset{H}{|}}{C}}-\overset{\displaystyle\overset{H}{|}}{\underset{\displaystyle\underset{H}{|}}{C}}-\overset{\displaystyle\overset{H}{|}}{\underset{\displaystyle\underset{H}{|}}{C}}-H$$

The chief use of petroleum is as a fuel source. Hydrocarbons burn well, and the heat released can be used to warm your house, drive the wheels of your automobile, push a jet aircraft through the sky, or spin the turbines in an electric power generating plant. Petroleum is also used as a source of chemicals for the production of fertilizers, insecticides, plastics (polyethylene, nylon, Dacron, Rayon, Teflon, polyester, etc.), food preservatives, paints, inks, lubricants, detergents, solvents, modern medicines (such as antibiotics, steroids, pain relievers, anesthetics, cancer and AIDS medications), and so on. In other words, petroleum is the raw material for a large fraction of the things we use.

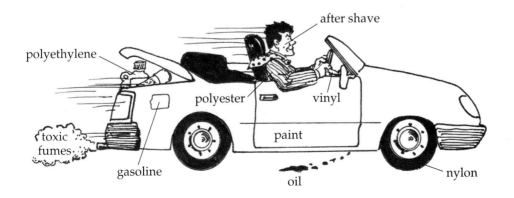

This leads to an ethical dilemma, for as we burn cheap petroleum to support our energy-intensive lifestyle, we are also polluting the environment and using up a limited resource. The most optimistic estimates are that at our present rate of consumption, we will have burned up most of the planet's petroleum reserves within 200 years. Research to develop alternative energy sources (such as conversion of sunlight, wind, and oceanic wave motion into electricity; and advanced forms of nuclear power) will help extend our petroleum reserves. The pace of research and deployment will increase as petroleum becomes increasingly scarce and the price of fuels, plastics, detergents, and so on, climb dramatically.

CHAINS OF CARBON

Since hydrocarbon molecules consist of chains of carbon atoms, chemists often describe them in terms of their *chain length*. The **chain length** is simply the number of carbon atoms that are bound together to form the longest continuous chain in the molecule. Our earlier hydrocarbon example, octane, has a chain length of 8:

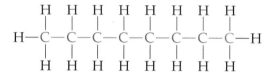

Octane has a *linear* chain of carbon atoms. A **linear hydrocarbon** is one in which all but the end (or *terminal*) carbons are bound to two other carbon atoms. We have to be careful with the term "linear," because the actual shape of the octane molecule is anything but a straight line. Recall from Chapter 6 that VSEPR theory predicts tetrahedral bond angles of 109° around each carbon atom. Therefore, the real shape of octane is like this:

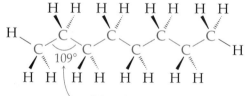

All bond angles are tetrahedral.

"Linear" therefore refers to how the carbon atoms are connected, not to the actual shape of the molecule.

Hydrocarbons can be *branched* as well as linear:

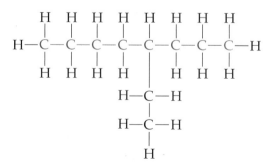

A **branched hydrocarbon** has at least one carbon atom that is attached to more than two other carbon atoms. You can also think about a branched hydrocarbon molecule as having a backbone with a smaller branch growing off it. The backbone (called the **main chain**) is the longest continuous chain in the molecule (not necessarily written in a straight line). The shorter chain or chains are the branches. In our example, the main chain is 8 carbons long, and the branch has a chain length of 2.

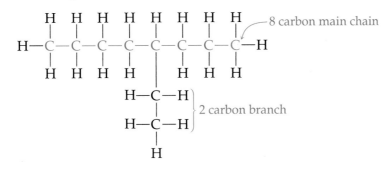

Consider the following branched hydrocarbon molecule:

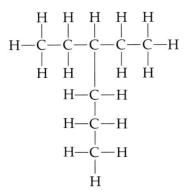

(a) What is the length of the main chain?
(b) What is the length of the branch?

Did you get a main chain length of 5 and a branch chain length of 3? If you did, you were fooled. Remember, the main chain is the longest continuous chain of carbon atoms in the molecule, but it does not have to be written in a straight line. Check your answer against ours.

One of the most useful rules for understanding the structure of hydrocarbon molecules is the "four bonds to carbon" rule we discussed earlier. Carbon always forms four covalent bonds in a hydrocarbon molecule. This four-bond rule is so reliable that chemists have taken advantage of it to develop a shorthand way to represent hydrocarbon molecules, called the **line-drawing** method. In this method, we show just the C–C bonds, which stand for the carbon "skeleton" of the molecule. The line drawing for octane is drawn like this:

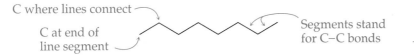

Since each line segment in this zigzag represents a C–C bond, there is a carbon atom at each point where the lines connect and at each end of the zigzag. The hydrogen atoms and C–H bonds are not shown. Here is how the line drawing of octane corresponds to the complete structure:

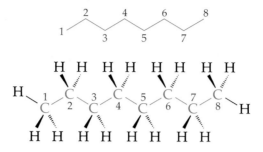

The C–H bonds are not shown in the line-drawing method because you can fill them in automatically by using the four-bond rule. For instance, look at the carbon at position 1 (called C1). The four-bond rule tells us that this carbon must have four bonds. The zigzag line drawing shows only one bond attaching to this carbon, so the other three must be C–H bonds. The complete structure confirms this. The line drawing shows two bonds for each of the carbons at positions 2 through 7 (C2 through C7), so the additional two bonds required by the four-bond rule must be C–H bonds. The carbon at position 8 (C8) is like C1, so it must have three attached hydrogens. We can also figure out the formula from the line drawing. The formula for octane is C_8H_{18}.

Double or triple bonds between carbons reduce the number of hydrogens in the molecule. For example, consider this eight-carbon molecule:

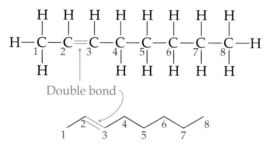

The only changes are at C2 and C3. Each of these carbon atoms now has a double bond and needs only one hydrogen instead of two to fulfill the four-bond rule. The formula is therefore C_8H_{16}, two hydrogens less than before.

A chemist must be able to deduce the formula for a hydrocarbon molecule from its line drawing, as well as be able to turn a line drawing into a full drawing and vice versa. The following practice problems will give you a chance to do this.

PRACTICE PROBLEMS————————————————————————————

15.1 Give the formula for each of the following hydrocarbon molecules:

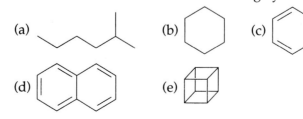

Answer: (a) C_7H_{16} *(b)* C_6H_{12} *(c)* C_6H_6 *(d)* $C_{10}H_8$ *(e)* C_8H_8

15.2 Draw full drawings for the hydrocarbons in Practice Problem 15.1, showing all carbons, hydrogens, and bonds.

15.3 Turn the following full drawing into a line diagram:

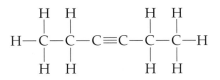

15.4 Consider the following hydrocarbon molecule:

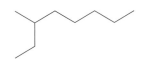

   (a) Is it linear or branched? Explain how you can tell.
   (b) What is the length of the main chain?
   (c) What is the length of the branch chain?
   (d) What is the formula of this hydrocarbon?

SATURATED VERSUS UNSATURATED

You are probably familiar with some of the hydrocarbon molecules fo[...] petroleum:

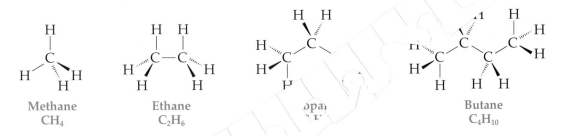

Methane and ethane are the s[...]st [...] hydrocarbon molecules. They are gases at room temperatur[...]o'[...] principal components of the natural gas burned in h[...]e st[...] and [...]aces. Natural gas is relatively inexpensive, burns cleanly, [...]ransport through pipes. Propane, also a gas at room temp[...]rat[...] often used in torches. Butane, at chain length 4, is also a gas at r[...]rature, but it can be compressed to a liquid. This is the fuel i[...]nos[...]nters.

[...]ple hydrocarbons all have single bonds between the carbon [...]ut as we have seen, hydrocarbons can also employ double and triple [...]nds. For example, there are three ways to make a hydrocarbon with a chain length of 2:

Ethane, C_2H_6 Ethene, C_2H_4 Ethyne, C_2H_2

In all three molecules, each carbon atom has four covalent bonds. However, the molecules have different numbers of hydrogen atoms. The hydrocarbon molecule with all single bonds has the most hydrogens (C_2H_6). The hydrocarbon molecule with the triple bond has the fewest hydrogens (C_2H_2). To talk about these differences, we use the terms *saturated* and *unsaturated*. A hydrocarbon molecule that contains the most possible hydrogen atoms is said to be **saturated**. A molecule with the same number of carbons but fewer hydrogens is said to be **unsaturated**. Unsaturated molecules have fewer hydrogens because they have double or triple bonds. Among the above three structures, ethane is saturated, whereas ethene and ethyne are unsaturated.

You have probably heard the terms "saturated" and "unsaturated" used in relation to fats and oils. Unsaturated oils contain double bonds in their linear hydrocarbon portions. They may be healthier to eat than saturated fats, since the body doesn't metabolize them into artery-clogging cholesterol (although recent evidence suggests it is not this simple).

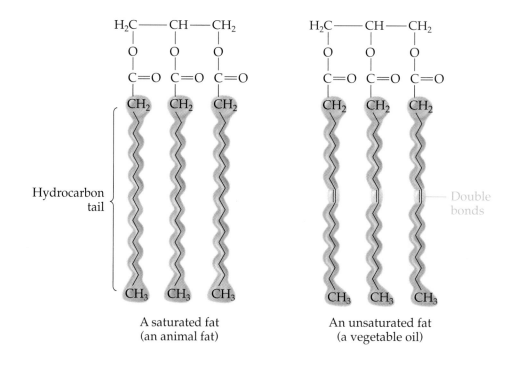

A saturated fat
(an animal fat)

An unsaturated fat
(a vegetable oil)

It is important to remember that the multiple (double and triple) bonds in hydrocarbon molecules that are responsible for unsaturation occur only between carbon atoms. The carbon-to-hydrogen bonds are always single bonds.

PRACTICE PROBLEMS

15.5 A particular linear hydrocarbon molecule has 6 carbons and 10 hydrogens. Is it unsaturated or saturated?

Answer: It is unsaturated. The best way to discover this is to make a line drawing for a fully saturated 6 carbon hydrocarbon (give it all C–C single bonds) and then count the number of hydrogens it has:

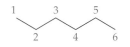

Carbons 1 and 6 have 3 hydrogens each. Carbons 2–5 have 2 hydrogens each. The total number of hydrogens is 14. This is the maximum number of hydrogens a 6 carbon chain can have. The hydrocarbon in this question has 4 fewer than the maximum, so it is unsaturated.

15.6 Draw possible line drawings for the unsaturated hydrocarbon molecule in Practice Problem 15.5.

15.7 Of the three hydrocarbon molecules shown below, which is least unsaturated? Explain your answer.

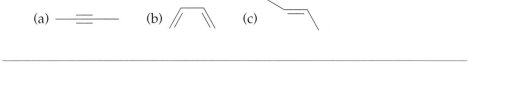

NAMING HYDROCARBONS 15.3

As chemists discovered more and more hydrocarbon molecules they began to run into a problem. Since carbon atoms are so good at forming chains of varying length, along with single, double, and triple bonds, the possible number of different hydrocarbon molecules is astronomical. There can even be many different hydrocarbon molecules with the same formula. Consider, for example, the formula C_5H_{12}. Three different hydrocarbon molecules have this formula:

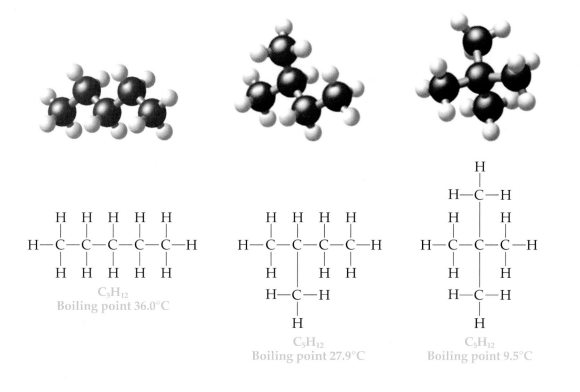

These three hydrocarbons are all completely saturated and all have the same formula, but they have different structures, as you can see. Because they are put together differently, they are different compounds, possessing different physical and chemical properties. For example, their boiling points are different, as indicated below the structures. We call these molecules *isomers* of each other. **Isomers** are molecules that have the same formula but different structures.

How shall we name all these molecules? We could use common names, but these wouldn't give us any clues as to the formulas and structures of the compounds. For example, water is the common name for the molecule H_2O, but this name tells us nothing about the molecule; you must simply memorize that water means H_2O. Even if we wanted to make up a different common name for every different hydrocarbon molecule, that would get out of hand rather quickly. While there are only 3 isomers with the formula C_5H_{12}, there are 5 with the formula C_6H_{14}, 75 with the formula $C_{10}H_{22}$, and over 4 billion with the formula $C_{30}H_{62}$. What we need is a naming system that never runs out of names and somehow instantly paints a picture of the structure of the molecule in your mind.

In Chapter 5 you learned about a better naming system. For example, H_2O is called dihydrogen oxide, a name that clearly indicates the formula. This naming system is known as the **IUPAC (International Union of Pure and Applied Chemistry) nomenclature system**. The IUPAC nomenclature system does what we want, although the rules for naming organic compounds are a bit more involved than those for naming simple compounds like water. We will introduce you to the basics of this system. The best way to start is with linear hydrocarbons. We will tackle branched hydrocarbons afterward.

The IUPAC nomenclature system for organic molecules begins by developing a general name for all simple hydrocarbons. Years ago, organic chemists were busy studying the structures of fats and oils. The Greek word for fat is *aleiphas*. Since, as we have seen, fats and oils contain long chains of carbon atoms, organic chemists began referring to long-chain hydrocarbons as *aliphatic hydrocarbons*. The root *alk-* later came into general use in the names of these molecules. Simple hydrocarbons that have only C–C single bonds are called **alkanes.** Hydrocarbons that have at least one C=C double bond are called **alkenes**, and those that have at least one C≡C triple bond are called **alkynes** (pronounced alk-eye-ns).

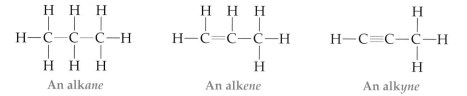

An alkane   An alkene   An alkyne

Thus, the suffixes *-ane*, *-ene*, and *-yne* have special significance. If you see *-ene* in the name of a hydrocarbon, you automatically know it has at least one C=C double bond. However, the terms alkane, alkene, and alkyne are general, and fit a chain of any length. We want a name that also tells us the length of the main chain for a given molecule. To do this, we replace the *alk-* in alkane, alkene, and alkyne with a root name that indicates the length of the main chain. Table 15.1 lists the names for main chains with 1–10 carbon atoms.

Table 15.1 Hydrocarbons of Increasing Length

Chain length (n)	Alkanes	Alkenes	Alkynes
1	Methane	—	—
2	Ethane	Ethene	Ethyne
3	Propane	Propene	Propyne
4	Butane	Butene	Butyne
5	Pentane	Pentene	Pentyne
6	Hexane	Hexene	Hexyne
7	Heptane	Heptene	Heptyne
8	Octane	Octene	Octyne
9	Nonane	Nonene	Nonyne
10	Decane	Decene	Decyne
n	C_nH_{2n+2}	C_nH_{2n}	C_nH_{2n-2}

From Table 15.1 you can see that the root *meth-*, as in *meth*ane, means that the hydrocarbon chain consists of only 1 carbon. For this reason, there can be no methene or methyne. The root *eth-*, as in *eth*ane, *eth*ene, and *eth*yne, means that the hydrocarbon chain consists of 2 carbons; *prop-* (pronounced pr-oh-p) means 3 carbons; and *but-* (pronounced b-you-t) means 4 carbons. For 5 carbon chains and beyond, we use Greek number prefixes *pent-* (5), *hex-* (6), *hept-* (7), *oct-* (8), *non-* (9, pronounced n-oh-n), and *dec-* (10), as indicated in Table 15.2.

Table 15.2   Method for Indicating the Length of the Carbon Chain

Chain length	Root name	Alkane example	Chain length	Root name	Alkane example
1	meth	*meth*ane	8	oct	*oct*ane
2	eth	*eth*ane	9	non	*non*ane
3	prop	*prop*ane	10	dec	*dec*ane
4	but	*but*ane	11	undec	*undec*ane
5	pent	*pent*ane	12	dodec	*dodec*ane
6	hex	*hex*ane	20	eicos	*eicos*ane
7	hept	*hept*ane	30	triacont	*triacont*ane

As Table 15.1 illustrates, hexane is a linear 6 carbon molecule with all single bonds. Hexene and hexyne are the same except that they have one C=C double bond and one C≡C triple bond, respectively. While we are not done showing you the basics of this nomenclature system, you can begin to see how the name tells you what the molecule looks like.

The compounds listed in the three columns of Table 15.1 represent three *homologous series*. **Homologous** simply means that all the compounds in the series are similar. For example, the compounds from methane (CH_4) to decane ($C_{10}H_{22}$) and beyond make up one homologous series of compounds. They are all alkanes, having only C–C single bonds. In addition, they all have the same general formula C_nH_{2n+2}, where n is an integer that goes from 1 to very large numbers. Table 15.3 shows the beginning of this homologous series.

Table 15.3   The Alkanes: A Homologous Series of Hydrocarbons

Chain length (n)

All the compounds listed under alkenes in Table 15.1 make up another homologous series. All these alkenes have one double bond. Because they have only one double bond, we can also refer to them as **monoalkenes**, and they have the same general formula, C_nH_{2n}.

Table 15.4   The Monoalkenes: A Homologous Series of Hydrocarbons

Chain length (n)		
2	$\begin{array}{c} H\ \ H \\ |\ \ \ | \\ C=C \\ |\ \ \ | \\ H\ \ H \end{array}$ Ethene, C_2H_4	
3	$\begin{array}{c} H\ \ H\ \ H \\ |\ \ \ |\ \ \ | \\ C=C-C-H \\ |\ \ \ \ \ \ \ | \\ H\ \ \ \ \ \ H \end{array}$ Propene, C_3H_6	All these compounds have the same general formula, C_nH_{2n}
4	$\begin{array}{c} H\ \ H\ \ H\ \ H \\ |\ \ \ |\ \ \ |\ \ \ | \\ C=C-C-C-H \\ |\ \ \ \ \ \ |\ \ \ | \\ H\ \ \ \ \ \ H\ \ H \end{array}$ Butene, C_4H_8	
5	$\begin{array}{c} H\ \ H\ \ H\ \ H\ \ H \\ |\ \ \ |\ \ \ |\ \ \ |\ \ \ | \\ C=C-C-C-C-H \\ |\ \ \ \ \ \ |\ \ \ |\ \ \ | \\ H\ \ \ \ \ \ H\ \ H\ \ H \end{array}$ Pentene, C_5H_{10}	

Note that the general formula for a monoalkene, C_nH_{2n}, has 2 fewer hydrogen atoms in it than the general formula for an alkane, C_nH_{2n+2}. As we have seen, every double bond in a carbon chain reduces the number of hydrogen atoms in the molecule by 2 compared to the corresponding hydrocarbon having only C–C single bonds. Alkenes are unsaturated.

Finally, all the compounds listed under alkynes in Table 15.1 make up yet another homologous series of hydrocarbon molecules. All these alkynes have a C≡C triple bond. Because they have only one triple bond, we can also call them **monoalkynes**, and they all have the same general formula, C_nH_{2n-2}. With 2 fewer hydrogens than an alkene, alkynes are even more unsaturated.

Table 15.5   The Monoalkynes: A Homologous Series of Hydrocarbons

Chain length (n)		
2	$H-C\equiv C-H$ Ethyne, C_2H_2	
3	$\begin{array}{c} H \\ | \\ H-C\equiv C-C-H \\ | \\ H \end{array}$ Propyne, C_3H_4	
4	$\begin{array}{c} H\ \ H \\ |\ \ \ | \\ H-C\equiv C-C-C-H \\ |\ \ \ | \\ H\ \ H \end{array}$ Butyne, C_4H_6	All these compounds have the same general formula, C_nH_{2n-2}
5	$\begin{array}{c} H\ \ H\ \ H \\ |\ \ \ |\ \ \ | \\ H-C\equiv C-C-C-C-H \\ |\ \ \ |\ \ \ | \\ H\ \ H\ \ H \end{array}$ Pentyne, C_5H_8	

The IUPAC nomenclature system as we have described it so far works fine for alkanes, but we have to do more for alkenes and alkynes. For example, consider the molecule butene. The *but-* tells you that there are 4 carbon atoms in the main chain. The *-ene* tells you that there is a C=C double bond. Put it all together and the formula must be C_4H_8 (C_nH_{2n}, with $n = 4$). But there are three ways to draw this formula:

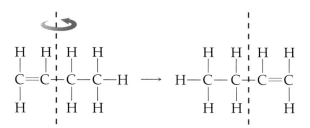

Where should we put the C=C double bond? Does it matter where we put it? Let's consider the left and right versions first. These are really the same, because if we spin the left version about the axis, as shown below, we generate the right version:

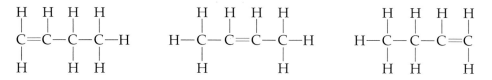

Any time you can convert one molecule into another by simply reorienting it, then the molecules are identical.

However, the butene with the double bond in the middle is truly unique. No amount of rotating or flipping will make this molecule the same as the others. In fact, these two butenes have different physical properties:

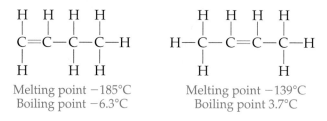

Melting point −185°C   Melting point −139°C
Boiling point −6.3°C   Boiling point 3.7°C

What we have here are two different compounds with the same formula but different structures. They are isomers, just as the linear and branched versions of C_5H_{12} we saw earlier were isomers. Since they are truly different compounds, they need different names. The IUPAC nomenclature system takes care of this by numbering the carbon atoms sequentially, starting with the number 1. This results in two different names:

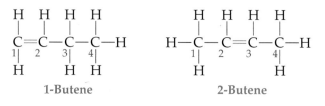

1-Butene   2-Butene

The isomer with the double bond at the end is called 1-butene. The number indicates the lowest-numbered carbon that has the double bond. The isomer with the double bond in the middle is called 2-butene.

You have to be careful with this numbering system when assigning numbers to the carbon atoms. For example, we began numbering the carbons from the left, but why not number them starting from the right? That would not change the name of the butene isomer with the double bond in the middle. In this particular case, both ways of numbering the carbon atoms would give the name 2-butene:

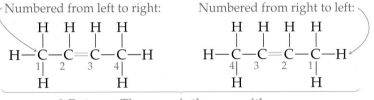

2-Butene—The name is the same either way

However, the direction of numbering does affect the name of the other butene isomer:

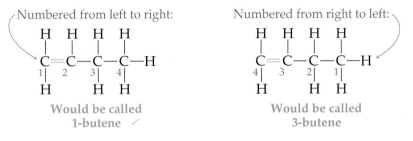

The IUPAC system solves this problem by insisting that you number the carbon atoms starting at the end closest to the double bond:

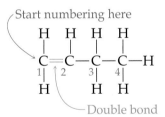

Therefore, 1-butene is the only correct name for this compound. The same rule applies to a chain containing a triple bond: You number the chain starting at the end closest to the triple bond. Try your hand at this now.

What is the IUPAC name of the hydrocarbon shown below?

WORKPATCH 15.3

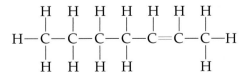

Did you consult Table 15.1 for the proper root name? Did you number the chain starting at the end nearest the double bond? Check your answer against ours.

All the hydrocarbons that we have named so far have been linear molecules. But more numerous are the branched hydrocarbons, which have one or more shorter branches attached to a main chain. Let's see how the branched alkane shown below is named.

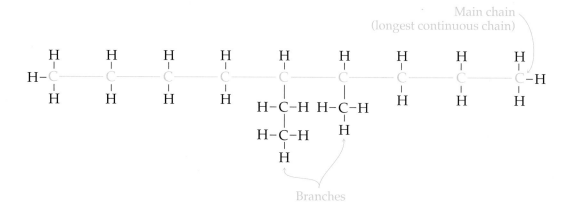

Naming a branched hydrocarbon molecule requires three steps:

Step 1: Find and name the main chain—the longest continuous chain in the molecule.

Step 2: Name the branches off the main chain.

Step 3: Give the location of the branches on the main chain.

We follow these steps to name the above alkane:

Step 1: The longest continuous chain in our example is 9 carbons long. There are no carbon–carbon multiple bonds, so you can tell from Table 15.2 that this main chain is called nonane.

Step 2: Off the main chain are a 1 carbon branch and a 2 carbon branch. The 1 carbon branch is CH_3, which looks a lot like methane, CH_4, except that one hydrogen is missing. So we call the CH_3 branch a *methyl* branch. The 2 carbon branch is CH_3CH_2, which looks a lot like ethane, CH_3CH_3, except that, once again, a single hydrogen is missing. So we call the CH_3CH_2 branch an *ethyl* branch.

Indeed, any alkane can serve as a branch by losing a single hydrogen atom so that it can bond to the main chain. The branch is then referred to as an *alkyl* branch, with the *-yl* suffix replacing the *-ane* in alkane.

Chain (alk*ane*)	Branch (alk*yl*)
Methane, CH_4	Methyl, $-CH_3$
Ethane, CH_3CH_3	Ethyl, $-CH_2CH_3$
Propane, $CH_3CH_2CH_3$	Propyl, $-CH_2CH_2CH_3$

So our hydrocarbon molecule is a nonane with a methyl branch and an ethyl branch. In the IUPAC system, the branch names precede the name of the main chain and are listed in alphabetical order. Therefore, we could call this molecule an ethylmethylnonane.

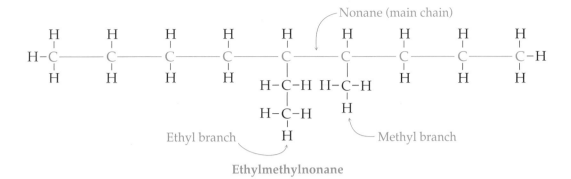

Ethylmethylnonane

Step 3: All that is left is to give the locations of the branches on the main chain. To do this, we number the main chain, but once again, we can number the main chain from left to right or from right to left. Which way is correct? Both ways are shown below.

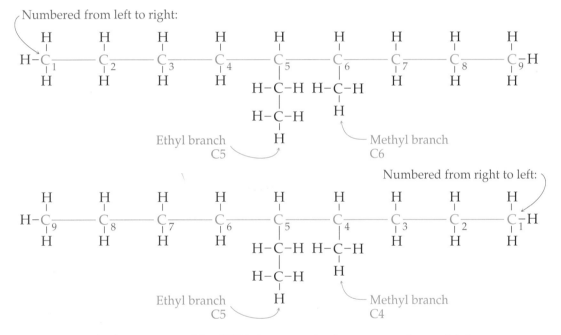

Which is correct? The IUPAC answer is the same as for multiple bonds: You start numbering the chain at the end nearest a branch. Look again at our example, and you will see that the main chain should be numbered starting from the right—that is the end closest to a branch. The correct name of our compound is 5-ethyl-4-methylnonane. There are two things to notice about this name. First, the IUPAC rules say that the position number for each branch should precede the name of the branch and be set off by hyphens. Second, in accordance with the alphabetical rule, the branches are listed in alphabetical order.

There is one more possibility to deal with. What do you do if a molecule has two or more of the same branch? For example, suppose our nonane had two methyl branches:

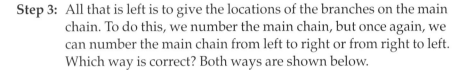

This molecule is *not* named 4-methyl-5-methylnonane. Instead, it is named 4,5-dimethylnonane. Branches of the same type get lumped together in the name. We then use the Greek prefixes *di, tri, tetra*, and so on, to indicate how many of that type of branch are in the molecule.

It's time for you to try your hand at this. Perhaps the most challenging part of naming a branched hydrocarbon is finding the main chain, so your first task is always to identify the longest continuous chain in the molecule. Also, remember to number the hydrocarbon from the end nearest a branch.

15·4 WORKPATCH   Name the following branched hydrocarbons:

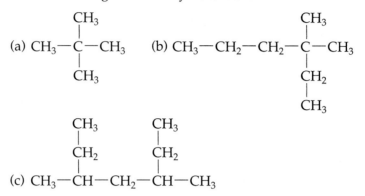

Make sure you can do WorkPatch 15.4, because we are going to make things a little more complicated now.

The final level of naming complexity for this book will be to name a branched hydrocarbon that is also unsaturated. For example, consider the following branched, unsaturated molecule:

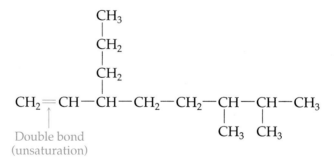

As always, we start by finding the main chain, but now there is a catch. If the hydrocarbon is unsaturated, the main chain must contain the multiple bond. That means the main chain in the molecule may not be the longest chain. Such is the case in our example:

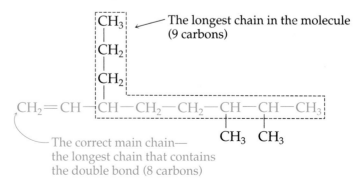

By the IUPAC system, the longest chain that contains the double bond becomes the main chain, not the absolute longest chain in the molecule. Everything else off the main chain is a branch. This molecule is thus an octene with one propyl and two methyl branches. Now all we need to do is number the main chain to locate the positions of the branches and the double bond. We're supposed to number the main chain from the end nearest the double bond—but also from the end nearest a branch point. Which gets priority?

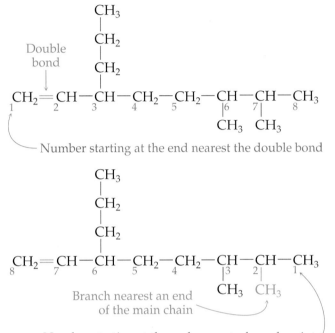

The IUPAC naming system gives priority to multiple bonds. The multiple bond must get the lowest number possible. Thus, our example is correctly numbered from left to right, and named 6,7-dimethyl-3-propyl-1-octene. Notice also in the name that propyl comes after methyl. We are still listing branches in alphabetical order, but we ignore the Greek prefixes di, tri, and so on, when considering the alphabetical order of the branches.

The beauty of the IUPAC naming system is not only that it comes up with a unique name for each different hydrocarbon, but also that the name can paint a picture of that molecule in your mind. Try it now.

WORKPATCH 15.5

Draw the molecule 2-methyl-1-butene. [*Hint:* Start by drawing the main chain, number the carbons, then position the branches and the multiple bond.]

When you can do this WorkPatch correctly, you will know that you understand the basics of the IUPAC nomenclature system. It can also tackle molecules with more than one double or triple bond, or combinations of both of them, although we will not consider such cases in this book. The next time you put on some deodorant or shampoo your hair, stop and read the bottle label. You will see IUPAC names of hydrocarbon molecules—names that will now have some meaning for you.

PRACTICE PROBLEMS

15.8 Give the IUPAC names for the following hydrocarbons:

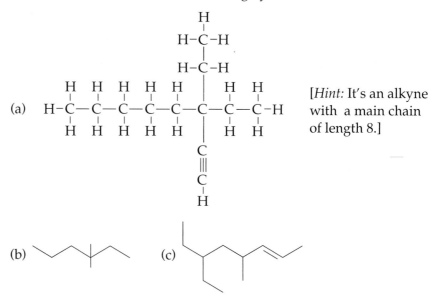

(a)

[*Hint:* It's an alkyne with a main chain of length 8.]

(b)   (c)

Answer:
(a) 3,3-Diethyl-1-octyne
(b) 3,3-Dimethylhexane
(c) 6-Ethyl-4-methyl-2-octene

15.9 Do a full drawing, showing all C and H atoms, for 6-methyl-4-propyl-2-octene.

15.10 The following line drawing for a hydrocarbon molecule is numbered and named incorrectly.

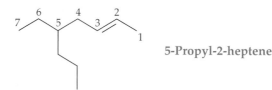

5-Propyl-2-heptene

(a) Why is the name incorrect?
(b) Number the atoms and name the molecule correctly.

15.4 PROPERTIES OF HYDROCARBONS

As a class, hydrocarbons are essentially nonpolar and are thus quite insoluble in water (which, of course, is polar). However, since like dissolves like, those hydrocarbons that are liquids are good solvents for dissolving fats, oils, and many other nonpolar organic substances. The four smallest linear alkanes are all gases at room temperature: methane, CH_4; ethane, CH_3CH_3; propane, $CH_3CH_2CH_3$; and butane, $CH_3CH_2CH_2CH_3$. At a chain length of 5 (pentane, $CH_3CH_2CH_2CH_2CH_3$), hydrocarbons become liquids. As they increase in length, their boiling points increase. At chain lengths of 20 carbons and above, hydrocarbons become solids. You know some of these hydrocarbons as waxes and asphalt. Look at Table 15.6, and you will see how the boiling points of linear alkanes steadily increase as their chain length increases.

Table 15.6   Boiling Points for Some Linear Alkanes (at 1 atm pressure)

Chain length (n)	Alkane	Boiling point (°C)	
1	Methane, CH_4	−164	Gases at 25°C
2	Ethane, CH_3CH_3	−88.6	
3	Propane, $CH_3CH_2CH_3$	−42.1	
4	Butane, $CH_3CH_2CH_2CH_3$	−0.5	
5	Pentane, $CH_3CH_2CH_2CH_2CH_3$	36.1	Liquids at 25°C
6	Hexane, $CH_3CH_2CH_2CH_2CH_2CH_3$	69.0	
7	Heptane, $CH_3CH_2CH_2CH_2CH_2CH_2CH_3$	98.4	
8	Octane, $CH_3CH_2CH_2CH_2CH_2CH_2CH_2CH_3$	124.7	
9	Nonane, $CH_3CH_2CH_2CH_2CH_2CH_2CH_2CH_2CH_3$	150.8	
10	Decane, $CH_3CH_2CH_2CH_2CH_2CH_2CH_2CH_2CH_2CH_3$	174.1	Solid at 25°C
20	Eicosane, $CH_3CH_2CH_2CH_2CH_2CH_2CH_2CH_2CH_2CH_2CH_2CH_2CH_2CH_2CH_2CH_2CH_2CH_2CH_2CH_3$	343	

By now you should be able to explain why the boiling points increase as the molecules get bigger. A higher boiling point means stronger intermolecular forces between molecules. Therefore, since the boiling point increases with size, the strength of the intermolecular forces must also be increasing with size. Why should that be so? Remember that hydrocarbons are essentially nonpolar, so the intermolecular forces operating between hydrocarbon molecules can't be dipolar or hydrogen bonding in nature. This leaves just London forces (see Chapter 9), which are due to instantaneous imbalances in the electron distributions within the molecules. The larger the hydrocarbon molecule, the more electrons there are to be unbalanced, and the greater the molecular surface area, allowing for greater contact between molecules. These factors lead to stronger London attractive forces between molecules as hydrocarbon size increases, leading to the steady increase in boiling point observed in Table 15.6.

Functional Groups

While alkanes are important as fuels, they are otherwise unreactive and not very exciting chemically. However, if we add other atoms besides hydrogen onto the carbon chains—atoms such as the halogens (F, Cl, Br, I), oxygen, or nitrogen—we obtain what are called **functionalized hydrocarbons**. These additional atoms often give functionalized hydrocarbons new and exciting properties and increased reactivity. The number of functionalized hydrocarbons that chemists have prepared and/or characterized is absolutely mind-boggling; hundreds of thousands of them are known. A few familiar ones are shown here.

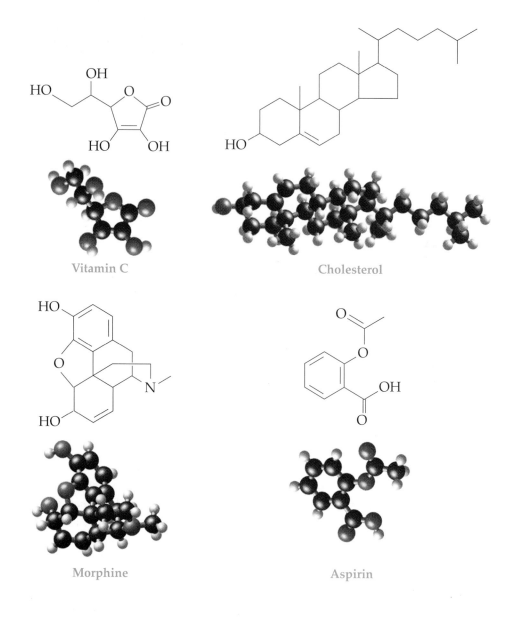

Vitamin C

Cholesterol

Morphine

Aspirin

Because there are so many different kinds of functionalized hydrocarbons, each with its own unique properties and reactivities, organizing them in a way that makes sense and allows us to understand their behavior as well as name them is a difficult task. The IUPAC nomenclature system can do all of this for us. The portion of a functionalized hydrocarbon molecule that gives it unique properties is called a **functional group**. There are many different types of functional groups, each with its own name. For example, let's start with the simplest hydrocarbon, methane, and remove one hydrogen atom. In its place we will put something other than a carbon or hydrogen atom:

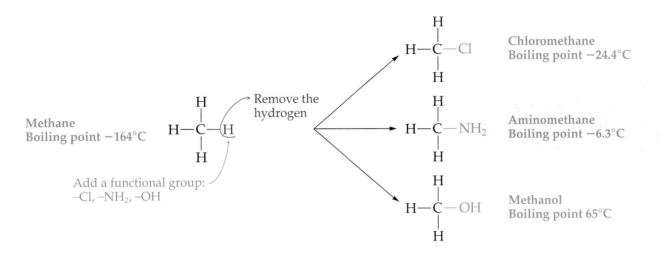

The molecules with different functional groups have drastically different properties than the parent hydrocarbon, methane. They boil at different temperatures, they have different odors, they have different solubilities in water, and they have different chemical reactivities.

To name a hydrocarbon with a halogen on it (fluorine, chlorine, bromine, or iodine), we replace the *-ine* ending of the halogen's elemental name with *-o* and then put it in front of the name of the parent hydrocarbon. For example, the molecule CH_3Cl above is called *chloro*methane. For hydrocarbons with chains of more than 2 carbon atoms, we report the numerical position of the halogen just as we did for branches and multiple bonds. We number the main chain starting at the end nearest the halogen:

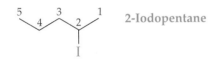   2-Iodopentane

The molecule CH_3NH_2 that we showed above is called aminomethane. To name a hydrocarbon with an NH_2 group on it, we put the prefix *amino-* before the name of the parent hydrocarbon, and (when the chain has more than 2 carbons) number the carbon chain from the end nearest the amino group. The NH_2 group in aminomethane gives this functionalized hydrocarbon properties similar to those of NH_3 (ammonia).

Finally, the molecule CH_3OH shown above has an OH group and is therefore a member of the general class of functionalized hydrocarbons known as *alcohols*. Alcohols are named differently from halogen or amino functional hydrocarbons. Instead of putting a prefix in front of the parent alkane's name, we just use the parent name and replace the last letter *-e* with *-ol*. Thus, the name methan*e* becomes methan*ol* for CH_3OH. Once again, we number the carbon chain from the end nearest the OH group (when necessary) as in the following example:

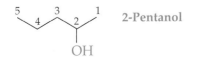   2-Pentanol

PRACTICE PROBLEMS

Name each of the following functionalized hydrocarbons:

15.11

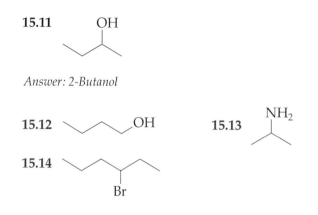

Answer: 2-Butanol

15.12   OH   **15.13** NH_2

15.14

                    Br

15.15 Draw a line drawing for the molecule 3-methyl-2-pentanol.

We have given you just a very brief look at the IUPAC nomenclature system for functionalized hydrocarbons. The functional groups halogen, OH, and NH_2, are just three among an entire zoo of functional groups. If you take a course in organic chemistry, you will spend a lot of time learning the details of this system.

We would rather discuss the properties of different types of functionalized hydrocarbons than their names, for it is their properties that make them so interesting and useful. We will therefore spend the rest of this chapter introducing you to some of the more interesting functional groups and the properties they give to molecules. As we proceed, keep in mind that a functional group has a large effect on an organic molecule, giving it unique properties that are quite different from those of the parent hydrocarbon. Indeed, the effect of a functional group is generally so large that all organic molecules with the same functional group tend to have similar chemical and physical properties. We will start with a functional group that frequently appears in the environmental headlines—halogens.

HALOGEN FUNCTIONALIZED HYDROCARBONS

Depending on who you talk to, halogens are either evil or wonderful functional groups. Halogen functionalized hydrocarbons do not occur naturally. They are prepared from hydrocarbons, often by reacting them with halogens:

$$
\begin{array}{c}
\underset{\displaystyle \text{H}}{\overset{\displaystyle \text{H}}{\text{H}-\text{C}-\text{H}}} + \text{Cl}_2
\xrightarrow{\text{UV}}
\underset{\displaystyle \text{H}}{\overset{\displaystyle \text{H}}{\text{H}-\text{C}-\text{Cl}}} +
\underset{\displaystyle \text{H}}{\overset{\displaystyle \text{Cl}}{\text{H}-\text{C}-\text{Cl}}} +
\underset{\displaystyle \text{H}}{\overset{\displaystyle \text{Cl}}{\text{Cl}-\text{C}-\text{Cl}}} +
\underset{\displaystyle \text{Cl}}{\overset{\displaystyle \text{Cl}}{\text{Cl}-\text{C}-\text{Cl}}}
\end{array}
$$

The above reaction is not balanced—it is just meant to show that many products may be obtained. Industrial chemical producers often add chlorine (and to lesser extents fluorine, bromine, and iodine) to hydrocarbon chains in order to produce compounds with specific properties. Some examples are shown below:

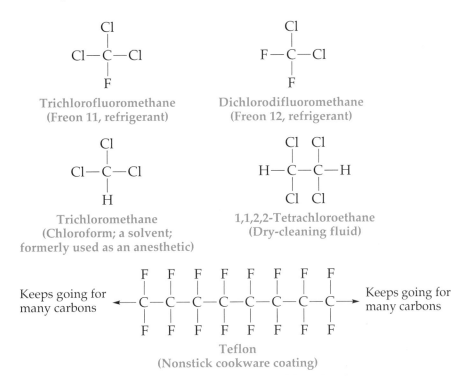

Trichlorofluoromethane
(Freon 11, refrigerant)

Dichlorodifluoromethane
(Freon 12, refrigerant)

Trichloromethane
(Chloroform; a solvent;
formerly used as an anesthetic)

1,1,2,2-Tetrachloroethane
(Dry-cleaning fluid)

Keeps going for
many carbons

Keeps going for
many carbons

Teflon
(Nonstick cookware coating)

These are interesting compounds. Trichlorofluoromethane (CCl_3F) and dichlorodifluoromethane (CCl_2F_2), known as Freon 11 and Freon 12, are often referred to as chlorofluorocarbons. They were used as the refrigerant fluids inside refrigerators and air-conditioners and, until recently, as propellants in cans of hair spray, deodorant, and paint. They are relatively unreactive, nonflammable, and nontoxic—just what's needed if you are going to spray them toward your body or use them in refrigerators to cool your food. Unfortunately, once released into the atmosphere, their unreactivity allows them to survive and reach the high-altitude ozone layer that protects us from the Sun's harmful ultraviolet radiation. Once there, they are activated by sunlight and begin to destroy the ozone layer. One chlorofluorocarbon molecule can destroy tens of thousands of ozone molecules by acting as a catalyst for its decomposition into oxygen. For this reason, chlorofluorocarbons are now banned in most of the industrialized world for use as refrigerants and propellants.

Halogen functionalized hydrocarbons can be made to undergo a variety of reactions, allowing them to serve as starting materials for putting other functional groups onto hydrocarbons. For example, while there is no efficient way to turn methane (CH_4) into methanol (CH_3OH) directly, chemists can convert methane into methanol by reacting chloromethane (CH_3Cl) with a strong base such as hydroxide (OH^-):

Before we proceed any further with our tour of functional groups, we want to introduce a shorthand notation that chemists commonly use when drawing functionalized hydrocarbons. Instead of writing down the entire hydrocarbon chain that is attached to a functional group, chemists often simply write the capital letter R. In other words, R represents the hydrocarbon portion of the molecule. Thus, for example, R–Cl stands for all monochlorinated hydrocarbons:

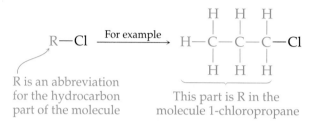

R is an abbreviation for the hydrocarbon part of the molecule

This part is R in the molecule 1-chloropropane

This notation allows chemists to quickly draw an abbreviated version of a molecule that emphasizes the functional group.

HYDROXY (OH) FUNCTIONALIZED HYDROCARBONS — ALCOHOLS (R–OH)

As we noted earlier, a hydrocarbon with an OH group is called an **alcohol**. By our shorthand notation, all alcohols have the general formula R–OH. The alcohol CH_3CH_2OH, called ethanol, is consumed in the form of wine, beer, whiskey, or champagne, depending on the occasion. All these drinks are solutions of ethanol in water, with other substances present for flavor. Ethanol acts as a depressant on the central nervous system. High concentrations in the blood lead to impaired physical ability and mental judgment. Pure ethanol (100%, or 200 proof), called absolute ethanol, is available only to licensed institutions for research and industrial applications. The general public can purchase only the denatured form to which poisons have been added to make it undrinkable. Still, pure 200 proof ethanol is nothing you would want to drink. It is extremely hygroscopic, which means that it readily absorbs water. A drink of it would severely dehydrate the tissues of your tongue and throat, damaging them as it went down.

Although ethanol can be produced from ethane, its main source in beer, wine, and champagne is the fermentation of various grains and fruits. For this reason, ethanol is sometimes called grain alcohol. Enzymes found in naturally occurring yeasts change the sugars and carbohydrates from these sources into ethanol and carbon dioxide, as shown below:

$$C_6H_{12}O_6 \xrightarrow{\text{Enzymes}} 2\,CH_3CH_2OH(aq) \;+\; 2\,CO_2(g)$$

Glucose (sugar)   Ethanol   Carbon dioxide

The carbon dioxide is what supplies the fizz and bubbles in champagne and beer.

Other alcohols are very toxic to humans. Only a few milliliters of methanol, CH_3OH, can cause nausea and blindness. Larger quantities can be fatal. Why is there such a large difference between methanol and ethanol, two seemingly similar molecules? The answer is that your body metabolizes methanol and ethanol differently. Enzymes in the human liver convert methanol to formaldehyde, while they convert ethanol to acetaldehyde. Acetaldehyde is produced in the body by other metabolic cycles, and your body knows how to deal with it. Formaldehyde is not normally found in your body. Until recently, formaldehyde was used to pickle and preserve dead animals. In your body, formaldehyde quickly poisons and pickles your cells, resulting in blindness and death. Most methanol is produced from methane, but it can also be isolated from the fermentation of wood. For this reason, methanol is often called wood alcohol. Certain animals have livers that lack the enzyme needed to metabolize wood alcohol into formaldehyde. Horses, for example, can ingest large quantities of wood alcohol with no apparent ill effects. (This is why extending animal toxicity tests to humans must be done very carefully and with an understanding of the biochemistry involved.) It might surprise you to know that the treatment for wood alcohol poisoning in humans is, believe it or not, large doses of ethanol! The ethanol keeps the liver enzymes busy, and the methanol, instead of being metabolized, just gets flushed out.

2-Propanol, also known as rubbing alcohol, is very poisonous. The dialcohol (or *diol*) 1,2-ethanediol, commonly called ethylene glycol, is the major component of automobile antifreeze and is also highly toxic.

2-Propanol
(Rubbing alcohol)

1,2-Ethanediol
(Ethylene glycol; antifreeze)

Interestingly, large alcohols containing many carbons tend to be much less toxic. Small alcohols like the ones we have discussed so far tend to be soluble in water, because the OH group is polar and capable of hydrogen bonding. Since blood is largely water, small alcohols of 1–5 carbons have no trouble getting into the bloodstream. Larger alcohols are much less water-soluble because the larger nonpolar R portion "outweighs" the smaller polar OH part. This makes the larger alcohols less toxic as they become less soluble in aqueous body fluids. (Another factor that makes larger alcohols less toxic to humans is an automatic bodily response to regurgitate them.)

With all this talk of toxicity, we don't want you to get the impression that alcohols are not important and useful. Quite the contrary. Alcohols are used as solvents for a variety of substances such as paints and glues. Also, the high reactivity of the OH functional group makes alcohols useful as starting materials, much as the halogenated hydrocarbons can be used to make alcohols.

ETHERS (R—O—R)

An **ether** has the general formula R—O—R. Here are some examples:

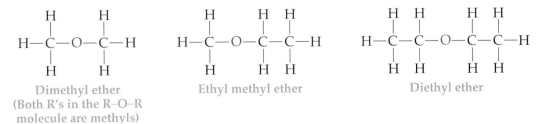

Dimethyl ether
(Both R's in the R—O—R
molecule are methyls)

Ethyl methyl ether

Diethyl ether

Ethers are commonly used as solvents for organic solutes. Diethyl ether is a very volatile liquid, readily producing vapors at room temperature. This is the "ether" that once was used as an anesthetic. (One of the authors of this book remembers having a mask put over his face so that he could breathe diethyl ether vapors when he was seven years old. This put him right to sleep, only to awaken later without tonsils.) Today, diethyl ether is no longer used as an anesthetic because the vapors are highly flammable. An errant spark from nearby electrical equipment could easily cause an explosion.

ORGANIC ACIDS—CARBOXYLIC ACIDS (R–COOH)

A hydrocarbon with a COOH group on it is called a **carboxylic acid**:

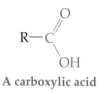

A carboxylic acid

As you can see, this functional group has both a C–O single bond and a C=O double bond in it. All hydrocarbons that have this group act as weak acids, due to the following dissociation equilibrium:

The equilibrium constants K_a for these carboxylic acid dissociations tend to be much smaller than 1. This means that the equilibria lie to the left, and carboxylic acids are weak acids. In a carboxylic acid, only the H on the oxygen is acidic. Acetic acid, which we've seen in previous chapters, is a carboxylic acid with R = CH_3, a methyl group:

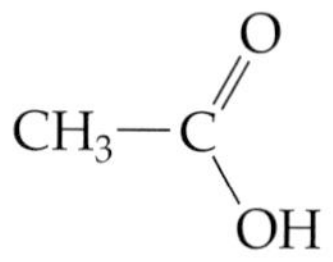

Acetic acid

Vinegar is an approximately 5% by mass solution of acetic acid in water. It is naturally produced by the bacteria *Acetobacter*, which, in the presence of oxygen, oxidizes ethanol to acetic acid. Wine spoils when *Acetobacter* oxidizes some of the ethanol into acetic acid.

Acetic acid consists of a 2 carbon chain. Its systematic name is thus ethanoic acid. The ending *-oic* is used to indicate a carboxylic acid, just as *-ol* indicates an alcohol. Acetic acid is the common, and more frequently used, name for ethanoic acid.

The smallest carboxylic acid is methanoic acid, commonly called formic acid. Methanoic acid is excreted by some ants as a defense against threats. Try drawing this compound now.

Draw a dot diagram for methanoic acid, explicitly showing all hydrogen, carbon, and oxygen atoms.

Your drawing for the WorkPatch should have just one carbon atom in it.

ORGANIC BASES—AMINES (H_2NR, HNR_2, NR_3)

Hydrocarbons containing a nitrogen atom bound by a single bond to carbon are called **amines**. They are related to ammonia (NH_3), with one, two, or three of the H atoms replaced by hydrocarbon (R) chains. Like ammonia, they all have a single lone pair of electrons on the nitrogen atom.

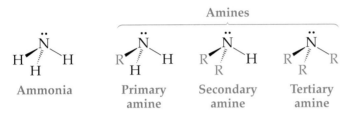

All amines are Brønsted–Lowry bases because they can accept a proton (H^+). They form a bond to a proton using the lone pair of electrons on the nitrogen. The result is a positively charged ammonium ion.

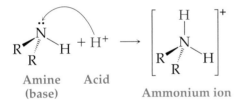

The smaller amines are soluble in water and act as weak bases, producing small amounts of OH^- (hydroxide ions) in water, just as ammonia does:

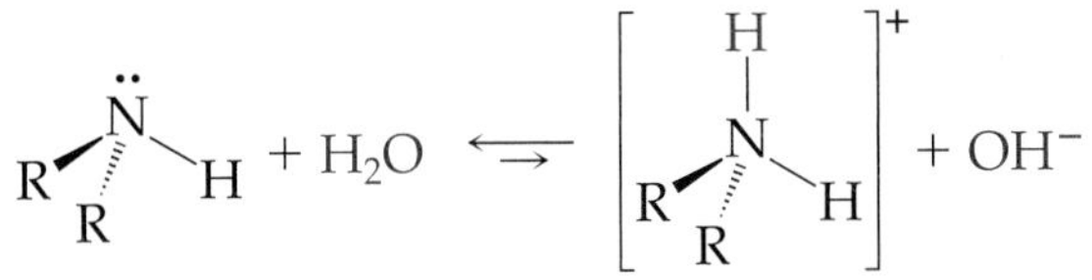

The equilibrium constants K_b for water-soluble amines are less than 1. This means the reaction equilbria lie to the left, producing only small amounts of hydroxide ion.

Ammonia itself has a very characteristic, sharp odor that jolts your sense of smell. Spirits of ammonia, also called smelling salts, are sometimes used to revive someone who has fainted. Small amines like methyl amine (CH_3NH_2) have a similar odor. Larger amines simply stink. The common names of the two amines shown below give you some idea as to how bad they smell:

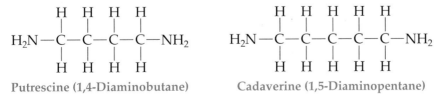

Putrescine (1,4-Diaminobutane)   Cadaverine (1,5-Diaminopentane)

You can guess where cadaverine is likely to be formed. Putrescine is formed in rotting fish. Knowing something about chemistry, you can deal effectively with putrescine if you are served some bad fish. Simply rub the fish with lemon. The lemon is a source of carboxylic acids such as citric acid.

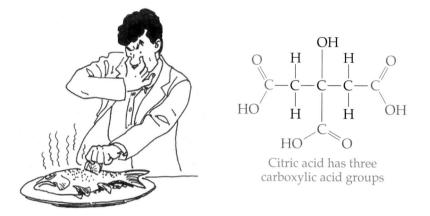

Citric acid has three
carboxylic acid groups

You know that when an acid and a base meet, they neutralize each other. When citric acid meets putrescine (an amine base), an acidic COOH proton is transferred, and an ammonium salt is formed:

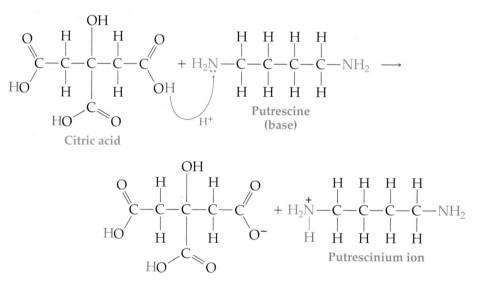

The ammonium salt doesn't have a bad smell, but of course, the fish is still bad, so don't eat it.

KETONES ($R_2C{=}O$) AND ALDEHYDES ($RHC{=}O$)

Organic chemists refer to atoms other than C and H as *heteroatoms*. The prefix *hetero-* means different; in this case, the atoms are different from C and H. Typical heteroatoms that occur in organic compounds are the halogens, oxygen, nitrogen, sulfur, and phosphorus. However, the heteroatom oxygen leads to the largest variety of functional groups. So far we have seen oxygen in alcohols (R–OH), ethers (R–O–R), and carboxylic acids (R–COOH). Each has very different properties from the others. In alcohols and ethers, the bonds between C and O are single bonds. In a carboxylic acid, there is one C–O single bond and one C=O double bond.

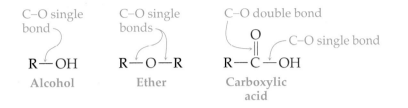

The C=O unit is referred to as a **carbonyl**. Of the three functional groups above, only the carboxylic acid includes a carbonyl. The functional groups known as **aldehydes** and **ketones** also have a carbonyl. They differ from carboxylic acids because they lack the OH portion, with its weakly acidic hydrogen. The shorthand representations for aldehydes and ketones are shown below:

A ketone can be written as R(CO)R or R_2CO, and an aldehyde as R(CO)H or RHCO. When we put the CO in parentheses, we are emphasizing that it is a carbonyl unit and the atoms to the right or left of the parentheses are bound to the carbonyl carbon.

Ketones and aldehydes are closely related. For a molecule to be a ketone, two hydrocarbon chains (R) must be attached to the carbonyl. In an aldehyde, one R chain and one hydrogen must be attached to the carbonyl. The systematic names for ketones end with the suffix *-one* (pronounced own) and for aldehydes with the suffix *-al*. For example, a 3 carbon ketone is called 2-propanone, and a 3 carbon aldehyde is called propanal:

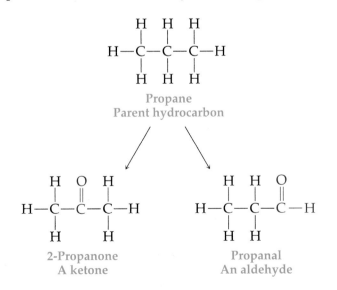

Some ketones and aldehydes are already familiar to you. The 2-propanone shown above has the common name of acetone. Acetone is a widely used industrial solvent, capable of dissolving many glues and paints. It is sold as a solution in water as nail polish remover. The simplest aldehyde, including just 1 carbon, is called methanal. Its common name is formaldehyde.

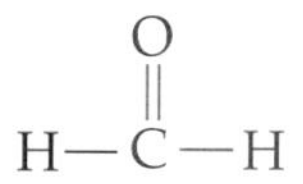

Methanal
(Formaldehyde)

PRACTICE PROBLEMS

15.16 Consider the molecule ethanal.
  (a) What functional group does the molecule contain? How can you tell?
  (b) Draw a dot diagram for the molecule.

Answer:
(a) The molecule is an aldehyde. The ending -al tells you this.

(b)

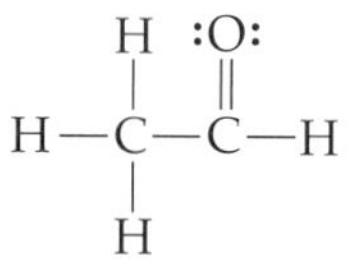

$$H-\overset{\displaystyle H}{\underset{\displaystyle H}{\overset{|}{\underset{|}{C}}}}-\overset{\displaystyle :O:}{\overset{\|}{C}}-H$$

15.17 There is no molecule called ethanone. Explain why not.

15.18 Circle and name the type of functional group present in each of the following molecules:

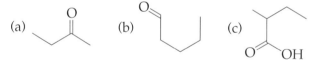

15.19 Give the systematic names for the molecules in Practice Problem 15.18.

We have only scratched the surface of organic chemistry. A full course in the subject would go on to examine additional kinds of functional groups, the properties they bestow upon hydrocarbons, and the reactions they undergo. It is a fascinating area of chemistry to study because when we examine functionalized hydrocarbons we examine ourselves. We are made of carbon in the form of functionalized hydrocarbons. Functional groups like the NH_2 of amines and the COOH of carboxylic acids are key to building the proteins from which our skin, hair, blood, cells, and enzymes are made. This realization has lead to chemistry's current and perhaps greatest challenge—the manipulation of the molecules of life in living organisms. We will take a glimpse of this subject in the final chapter of this book.

HAVE YOU LEARNED THIS?

CARBON—A UNIQUE ELEMENT

15.20 Explain what is meant by catenation with regard to the element carbon.

15.21 What are the three known allotropes of elemental carbon? How do their structures differ from one another, and how are they similar?

15.22 Explain why a carbon atom will always form four bonds in a covalent substance.

15.23 Consider the following incomplete drawing of a 3 carbon hydrocarbon:

$$H-C-C-\overset{\overset{\displaystyle H}{|}}{\underset{\underset{\displaystyle H}{|}}{C}}-H$$

(a) Complete the drawing employing only C–C single bonds and adding hydrogen atoms as needed.
(b) Complete the drawing employing one C=C double bond and adding hydrogen atoms as needed.
(c) Complete the drawing employing one C≡C triple bond and adding hydrogen atoms as needed.

15.24 A hydrocarbon molecule consists of a chain of 4 carbon atoms, one C=C double bond, and two C–C single bonds. What is the formula of this hydrocarbon? Draw a possible structure.

15.25 The geometry about each carbon in a particular hydrocarbon molecule is tetrahedral. What does this say about the possibility of there being C=C double or C≡C triple bonds in the molecule? Explain your answer fully.

15.26 The following molecule has four bonds to every carbon, yet it doesn't exist. Explain why.

15.27 What is the name of the branch of chemistry that is devoted to carbon-based molecules? Why is it called this?

15.28 Why are there more different compounds of carbon than of any other element?

HYDROCARBONS (CHAINS OF CARBON)

15.29 Define the term "hydrocarbon."

15.30 Make a drawing of a linear 6 carbon hydrocarbon that has only C–C single bonds. Explicitly show all H and C atoms.

15.31 Make as many different drawings as you can of a branched 6 carbon hydrocarbon that has only C–C single bonds. Explicitly show all H and C atoms.

15.32 How can you tell if a hydrocarbon is branched or linear?

15.33 Why is it more correct to draw a linear hydrocarbon that has only C–C single bonds as a zigzag rather than a straight line?

15.34 Is it more proper to draw the following hydrocarbon as a straight line or as a zigzag? Explain your answer fully.

$$H-C\equiv C-C\equiv C-H$$

15.35 What is the source of most hydrocarbons?

15.36 What ethical dilemmas are involved with our society's love for burning hydrocarbons as fuel?

15.37 Consider the following hydrocarbon:

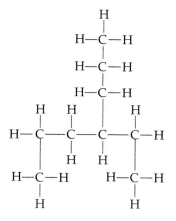

(a) What is the length of the main chain?
(b) What is the chain length of the branch?

15.38 Hydrocarbons are often drawn in an abbreviated form as line drawings. How is it possible to look at a line drawing and deduce the formula?

15.39 Give a full drawing for each line drawing shown below, as well as the formula for the molecule. Also, describe the hydrocarbon as linear or branched.

15.40 Produce a line drawing for the hydrocarbon in Problem 15.37.

15.41 Produce a line drawing for each of the following hydrocarbons:

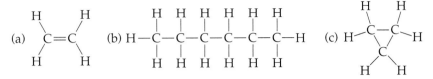

15.42 Define the terms "saturated" and "unsaturated" as applied to hydrocarbons.

15.43 Which molecule in Problem 15.39 is unsaturated?

15.44 Which of the following molecules is more unsaturated? Explain your answer fully.

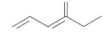

15.45 Suppose you were asked to remove 2 H atoms from hydrocarbon molecule (a) in Problem 15.44 to make it even more unsaturated.
(a) What is wrong with the following answer?

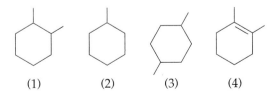

(b) What would be a correct answer?

15.46 A molecule has 5 carbons atoms and 12 hydrogen atoms. Is it saturated or unsaturated? Explain how you knew.

15.47 Consider the molecule with formula C_7H_{14}. Is it unsaturated or saturated? How can you tell?

NAMING HYDROCARBONS

15.48 Which of the following hydrocarbons are isomers of each other?

15.49 Draw line drawings for all the isomers of C_6H_{14}.

15.50 What must be true for two molecules to be isomers of one another with respect to their formula, structure, and properties?

15.51 Are H_2O and H_2O_2 isomers of one another? Explain your answer fully.

15.52 What is wrong with using common names for hydrocarbon molecules?

15.53 *Cyclic hydrocarbons* contain a ring of carbon atoms. Consider the following cyclic hydrocarbons:

(1)   (2)   (3)   (4)

(a) Which are isomers of each other?
(b) Give the reasons why the molecules you chose in part (a) are isomers.

15.54 Relate the words "saturated" and "unsaturated" to the general formulas C_nH_{2n+2}, C_nH_{2n}, and C_nH_{2n-2}.

15.55 What kind of hydrocarbons are called aliphatic hydrocarbons, and where does the word "aliphatic" come form?

15.56 How is the word "aliphatic" modified to name hydrocarbons with all C–C single bonds, hydrocarbons with one C=C double bond, and hydrocarbons with one C≡C triple bond?

15.57 What do the letters in IUPAC stand for?

15.58 What is meant by the expression "homologous series"? Give an example of one using a few molecules.

15.59 How is the general formula of the alkenes different from the alkanes? Explain this difference.

15.60 Name and draw the alkanes that do not use Greek prefixes in their names.

15.61 The common name for ethyne is acetylene. Draw a dot diagram and a line drawing for acetylene, and indicate whether acetylene is more or less unsaturated than ethene.

15.62 Define the prefixes *meth-*, *eth-*, *prop-*, and *but-*.

15.63 Consider the molecule with formula $C_{22}H_{46}$. Is it an alkane, an alkene, or an alkyne? How can you tell?

15.64 The IUPAC name for C_3H_6 needs no number in it, but the IUPAC name for C_4H_8 does. Explain why.

15.65 Do propane, propene, and propyne belong to a homologous series? Explain.

15.66 Draw a full drawing and a line drawing for 2-hexene.

15.67 What is the relationship between the molecules 2-hexene and 3-hexene? Justify your answer.

15.68 Explain why the molecules 1-hexene, 2-hexene, and 3-hexene exist, but not 4-hexene and 5-hexene.

15.69 True or false? We always number hydrocarbons from left to right because that is how we read. Explain your answer, and include an example to back up your explanation.

15.70 What should be your first task in naming any hydrocarbon molecule?

15.71 Name the following linear hydrocarbon molecules:
(a)   (b)   (c)   (d)

15.72 Give formulas and make line drawings for the following hydrocarbons:
(a) Propene   (b) 1-Octene   (c) Hexane
(d) 2-Butyne   (e) 1-Butyne

15.73 Referring to Problem 15.72, which molecules represent a pair of isomers?

15.74 Where do we start numbering carbons on the main chain when both branches and C=C or C≡C bonds are present?

15.75 There is something wrong with the name 2-methyl-4-hexene. What is wrong? What is the correct name?

15.76 The molecule 2,2-dimethyl-1-butene cannot exist. Explain why.

15.77 Give line drawings and formulas for the following hydrocarbons:
(a) 3-Methylpentane   (b) 2,3-Dimethyl-2-butene
(c) 3-Methyl-1-pentene   (d) 5-Methyl-1-hexyne

15.78 Give IUPAC names for the following molecules:

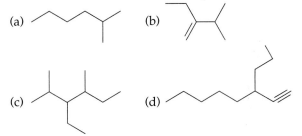

(a)   (b)

(c)   (d)

15.79 How many branches does the following hydrocarbon have? What are they?

15.80 Draw a line drawing of 5,5-dimethyl-1-hexene. Then explain what would be wrong with calling it 2,2-dimethyl-5-hexene.

15.81 True or false? The longest continuous chain of carbons in a hydrocarbon is always the chain to use in naming a hydrocarbon. Explain your answer.

15.82 Although it was not discussed in the chapter, it is possible for a hydrocarbon molecule to have more than one site of unsaturation. One such molecule is 1,3-butadiene; the *di* tells us that there are two double bonds in the molecule. Make a line drawing of this molecule, and give its formula.

PROPERTIES OF HYDROCARBONS; FUNCTIONAL GROUPS

15.83 The simple alkanes up through 4 carbons are gases. However, as alkanes get larger they become liquids and then solids. Explain why this is so.

15.84 What do we mean by the expression "functional group"?

15.85 In general, explain how halogenated hydrocarbons are named.

15.86 In general, explain how hydrocarbons that have an NH_2 group are named.

15.87 In general, explain how hydrocarbons that have an OH group are named.

15.88 Name the following molecules:

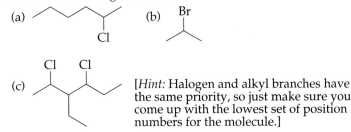

(a)   (b) Br

Cl

(c) Cl   Cl   [*Hint:* Halogen and alkyl branches have the same priority, so just make sure you come up with the lowest set of position numbers for the molecule.]

15.89 Draw the Teflon molecule.

15.90 Name the following molecules:

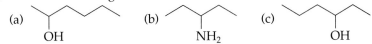

(a) OH   (b) NH₂   (c) OH

15.91 Why is chlorine an important functional group?

15.92 Why have Freons been banned?

15.93 What class of compounds does R–OH represent? What does the R represent?

15.94 Given an R–Cl, how can you convert it to an alcohol?

15.95 A student claims there are three isomers of propanol: 1-propanol, 2-propanol, and 3-propanol. Is he correct? Explain your answer.

15.96 Draw all the isomers of the alcohols with the formula $C_4H_{10}O$.

15.97 Why is wood alcohol toxic to humans but not to horses?

15.98 Why is wood alcohol poisoning treated by having the patient consume ethanol?

15.99 2-Propanol is quite soluble in water, but 2-decanol is insoluble in water. Draw both molecules and explain why one is soluble but the other isn't.

15.100 The line drawing below may look like a flying bat, but it is really a functionalized hydrocarbon. Give its formula and name.

15.101 Name the following ethers:

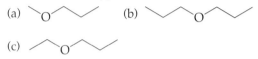

15.102 Draw the molecule butyl propyl ether and give its formula.

15.103 Produce a line drawing for ethanoic acid and also give its common name.

15.104 Write the equilibrium reaction that is associated with molecules of the type R–COOH. To which side does the equilibrium lie? What does this imply?

15.105 Show how a general carboxylic acid (R–COOH) will react with a tertiary amine (NR_3).

15.106 Write an equilibrium to show how a primary amine (H_2NR) makes water basic. To which side does the equilibrium lie? What does this imply?

15.107 The compound diethylamine is a secondary amine (HNR_2) with the formula $C_4H_{11}N$. Produce a dot diagram for diethylamine.

15.108 Produce a line drawing for 2-aminohexane and indicate whether it is a primary, secondary, or tertiary amine.

15.109 The molecule methanal has the common name formaldehyde. Produce a complete dot diagram for formaldehyde.

15.110 Draw the molecules 1-chloropropane, 1-propanol, 1-aminopropane, 2-propanone, propanoic acid, and propanal. Then assign each to its proper functional type (for example, alcohol, ketone, aldehyde, and so on).

15.111 Draw a line drawing for 2-propanone and give its common name.

15.112 Where are you likely to run across acetic acid outside a chemistry lab?

15.113 Circle the portion of the molecule that is the functional group and name the molecule:

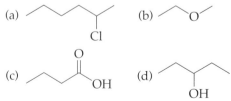

15.114 What classes of molecules have a carbonyl (C=O) in them?

15.115 The following compound is sold as Prozac, used for the treatment of depression.

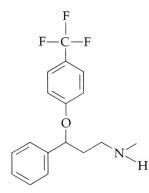

(a) Circle all the functional groups present and name them.
(b) What is the formula of Prozac?

WorkPatch Solutions

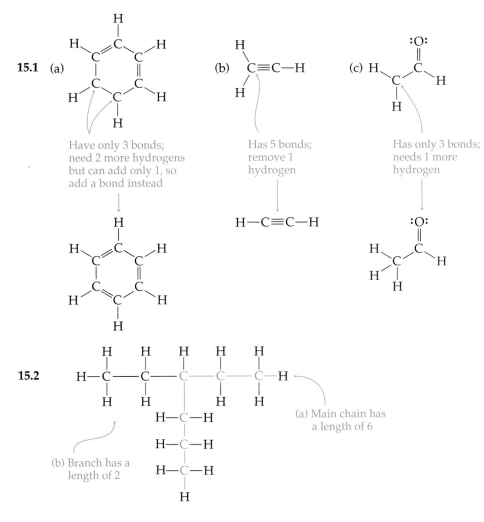

15.1 (a)

Have only 3 bonds;
need 2 more hydrogens
but can add only 1, so
add a bond instead

(b) Has 5 bonds;
remove 1
hydrogen

(c) Has only 3 bonds;
needs 1 more
hydrogen

15.2

(a) Main chain has
a length of 6

(b) Branch has a
length of 2

15.3 2-Heptene

15.4 (a) 2,2-Dimethylpropane
(b) 3,3-Dimethylhexane
(c) 3,5-Dimethylheptane

15.5

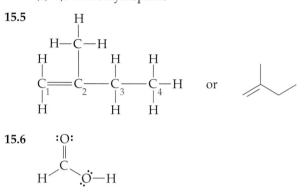

or

15.6

-8.763 1.00 59.53   1HBB1319
87 -8.194 1.00 57.58   1HBB1320
98 -8.451 1.00 55.51   1HBB1321
7.611 1.00 55.94   1HBB1322
-8.912 1.00 58.92   1HBB1323
-10.186 1.00 59.11   1HBB1324
-9.707 1.00 59.45   1HBB1325
-9.640 1.00 53.55   1HBB1326
-19.934 -10.030 1.00 52.74   1HBB1327
ATOM   1   O GLU B   -18.651 -9.234 1.00 49.33   1HBB1328
ATOM   1159 B GLU B 6   -17.862 -9.106 1.00 49.2   1HBB1329
ATOM   115   G GLU B 6   19.018 -19.719 -11.522 1.00 7   1330
ATOM   11   GLU B 6   19.705 -20.803 -12.358 1.00   331
ATOM   OE1 GLU B 6   20.918 -21.496 -11.825 1.00   1HBB1332
ATOM   1161 OE2 GLU B 6   21.968 -20.811 -11.9   1HBB1333
ATOM   162 N GLU B 7   20.953 -22.631 -11.355 1.00   1HBB1334
ATOM   1163 CA GLU B 7   17.594 -18.464 -8.653 1.00 44.   1HBB1335
ATOM   1164 C GLU B 7   17.334 -17.291 -7.808 1.00 40.32   1HBB1336
ATOM   1165 O GLU B 7   17.870 -17.589 -6.409 1.00 40.33   1HBB1337
ATOM   1166 CB GLU B 7   18.461 -   59 1.00 40.21   1HBB1338
ATOM   1167 CG GLU B 7   15.267 -16   8 -7.708 1.00 30.42   1HBB1339
ATOM   1168 CD GLU B 7   -15.817 -8.924 1.00 32.35   1HBB1340
ATOM   116   GLU B 7   13.118 -16.654 -8.818 1.00 33.93   1HBB1341
ATOM   OE2 GLU B 7   13.443 -14.754 -8.233 1.00 34.14   1HBB1342
ATOM   171 N GLU B 7   832 -9.275 1.00 39.84   1HBB1343
ATOM   172 C   19.383 -4.727 1.00 38.38   1HBB1344
19.594 -19.449 -4.727 1.00 37.24   1HBB1345
B L   20.195 -19.278 -4.627 1.00 35.54   1HBB1346
CG LY   17.453 -20.749 -3.547 1.00 37.44   1HBB1347
CD LYS   17.638 -21.461 -4.522 1.00 44.41   1HBB1348
LY   16.376 -21.990 -3.206 1.00 60.83   1HBB1349
LY   15.450 -22.824 -2.563 1.00 72.76   1HBB1
14.142 -23.091 -3.404 1.00 75.03   1HBB
20.24 19.692 -2.735 1.00 66.43   1HBB
5.751 1.00 34.32   1HB

Polymers: Synthetic and Biological

You are similar to a collection of plastic sandwich bags filled with water. This may not be a very flattering view, but it is essentially correct. Your body is almost 75% by weight water and is enclosed by your skin, the biological equivalent of a plastic bag. If we are ever visited by some nonaqueous life form from space, we will appear to them to be living bags of water.

The analogy between your skin and a plastic lunch bag goes even further. Both are made of large molecules known as *polymers*. The word "polymer" means many parts (*poly* meaning many, and *mer* from the Greek *meros* meaning part). A **polymer** is a very large molecule made from small units, called **monomers**, that are linked together:

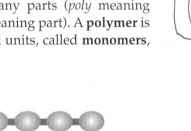

Monomer units

Polymer

You can think of monomers as the building blocks from which polymers are made. Some polymers are made of only one kind of monomer; others contain more than one kind.

In Chapter 15, we mentioned the molecules eicosane and triacontane, which are linear hydrocarbon chains containing 20 and 30 carbon atoms, respectively. While these may be considered polymers of the monomer (repeating) unit $-CH_2-$, the term polymer is typically used to describe molecules much larger than these, with chains that are hundreds or thousands of atoms long and with molecular masses ranging from the thousands to hundreds of thousands of grams per mole. Because of their enormous size, polymer molecules are often referred to as **macromolecules** (*macro* meaning large). The best way to start examining the subject of polymers is to look at the simplest ones first. This means that we'll start with a real sandwich bag and work our way up to your skin.

16.2 POLYETHYLENE AND ITS RELATIVES: TODAY'S MIRACLE PLASTICS

The common name for ethene is ethylene. Ethylene is the simplest of the alkenes and is very important to the chemical industry.

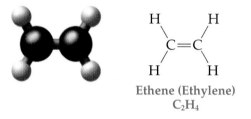

Ethene (Ethylene)
C_2H_4

Thirteen billion kilograms of ethylene are produced annually by the U.S. chemical industry. Ethane from natural gas is heated to a high temperature in the absence of air over a metal catalyst such as palladium. The saturated ethane loses hydrogen and becomes the unsaturated ethylene:

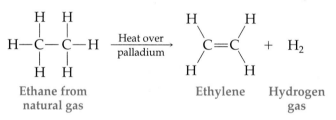

Ethane from   Ethylene   Hydrogen
natural gas   gas

When ethylene gas is heated under pressure in the presence of oxygen, or exposed to high-energy ultraviolet light, huge polymer molecules form, with molecular weights of up to 50,000 g/mole. These macromolecules consist of thousands of ethylene molecules linked together:

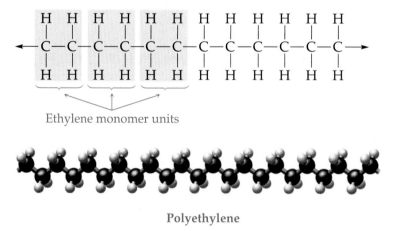

Ethylene monomer units

Polyethylene

The polymer is called polyethylene because it is made from many ethylene molecules. You might have noticed that there are no C=C double bonds in polyethylene. Nevertheless, it is still called polyethylene and not poly-ethane. We keep the *-ene* ending to indicate that it was made from ethylene monomers. Sandwich bags, plastic bottles, and bottle caps are made from polyethylene.

The double bond is missing in the polymer because one of the electron pairs from the C=C double bond is used to bond the monomer units to each other:

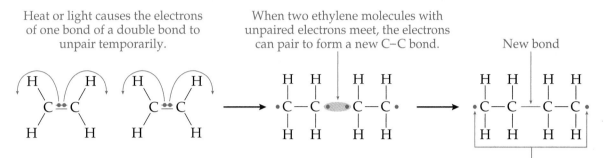

Heat or light causes the electrons of one bond of a double bond to unpair temporarily.

When two ethylene molecules with unpaired electrons meet, the electrons can pair to form a new C–C bond.

New bond

Monomers can continue to hook onto the ends of the new molecule, allowing the polymer to grow.

Many other polymers with different properties can be produced by using monomers that are close relatives of ethylene. For example, consider chloroethene, commonly called vinyl chloride. The polymer made from vinyl chloride is called poly(vinyl chloride), often abbreviated as PVC. PVC is most familiar to you as plastic sandwich wrap, and it is also used to make floor tiles, raincoats, and pipe for carrying water and sewage.

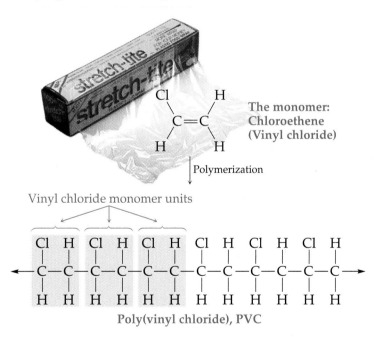

The monomer:
Chloroethene
(Vinyl chloride)

Polymerization

Vinyl chloride monomer units

Poly(vinyl chloride), PVC

Notice the distinction between the terms "monomer" and "monomer unit." Monomer refers to the actual small molecule (or molecules) used as the building blocks in the synthesis of the polymer. Monomer unit refers to the repeating pattern in the polymer—derived from the individual monomer molecules—that produces the structure of the polymer. There will always be a difference between the monomer and the monomer unit due to the rearrange-ment of electrons and/or atoms required to bond the monomers together.

Teflon, the familiar nonstick coating used on pots and pans, is made by the polymerization of tetrafluoroethene:

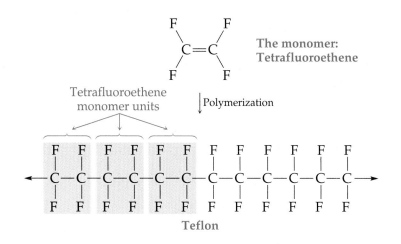

PRACTICE PROBLEMS

16.1 Consider the monomer 1,2-dibromoethene.
(a) Draw the monomer.
(b) Draw the polymer, making it at least three monomer units long.

Answer:

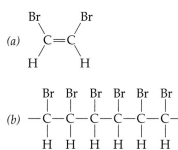

16.2 Consider the monomer 1-butene.
(a) Draw the monomer.
(b) Draw the polymer, making it at least three monomer units long.
[*Hint:* There are no double bonds in the polymer.]

16.3 Consider the monomer ethyne (acetylene).
(a) Draw the monomer.
(b) Draw the polymer, making it at least three monomer units long.
[*Hint:* The polymer contains double bonds, not triple bonds.]

So far we have been showing you a monomer first and then the polymer it forms. In order to better understand the relationship between a polymer and its monomer, it is useful to do a reverse analysis and attempt to identify the monomer unit by examining the polymer. For example, consider the polymer poly(propylene oxide):

Poly(propylene oxide)

What is the monomer unit? Remember that the monomer unit is the small piece that gets repeated over and over again in the polymer. Here are two candidates, (a) and (b):

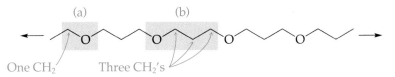

Both candidates contain one oxygen atom. Choice (a) also contains a CH_2, so the formula for this candidate is CH_2O. The formula for candidate (b) is $CH_2CH_2CH_2O$. So which is the monomer unit? To find out, all we have to do is repeat them and see what we get:

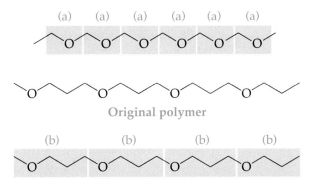

Original polymer

You can see that repeating (a) doesn't reproduce the original polymer, so it can't be the monomer unit. The correct choice is (b).

For the remainder of this chapter we will be examining some very interesting polymers. Being able to look at them and identify the monomer unit will help you understand them. Try your skill at this now.

The following polymer is an example of a *polyester* and is often used to make clothing. The monomer unit is called an *ester*. Which unit—(a), (b), or (c)—correctly shows the monomer unit? Explain why your choice is correct.

WORKPATCH 16.1

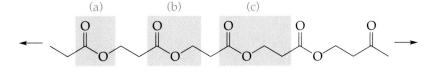

NYLON: A POLYMER YOU CAN WEAR 16.3

Unless you insist on wearing clothes made only of cotton or wool, chances are that your shirt, blouse, or slacks are made from some petroleum-based synthetic fabric. Polyester, nylon, and Dacron are woven from fibers made from synthetic polymers. **Nylon polymers** have the following structure:

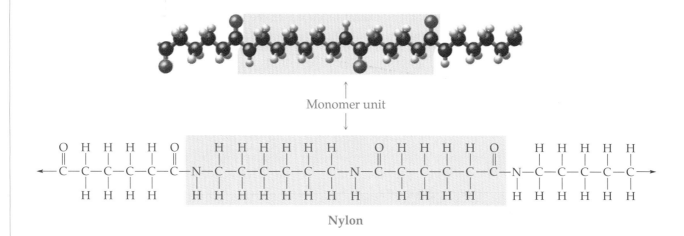

Monomer unit

Nylon

Nylon fibers wound about one another are so strong they can be used to make ropes for tying up ships. Much of this strength is due to hydrogen bonds that form between adjacent polymer chains:

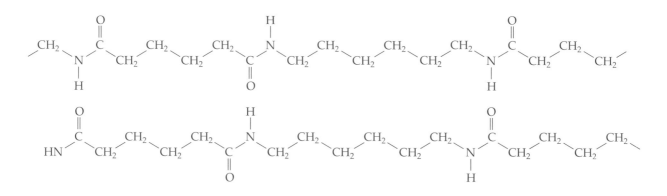

Nylon polymers are actually made by starting with two different monomers. We'll show how this works using the specific nylon shown above, which is called nylon 66. To make this nylon, we start with the following two monomers, each of which is based on a hexane with functional groups at each end:

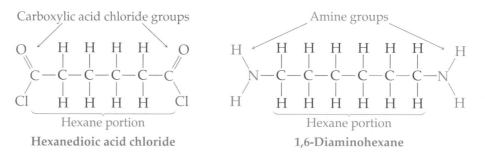

Carboxylic acid chloride groups

Amine groups

Hexane portion

Hexane portion

Hexanedioic acid chloride

1,6-Diaminohexane

In the case of hexanedioic acid chloride, the functional groups are carboxylic acid chloride groups—essentially, carboxylic acid groups (COOH) in which the −OH has been replaced by a −Cl. In 1,6-diaminohexane, the functional groups are amines.

The functional groups on these two monomers can react with each other, forming a bond that links the two monomers together. Specifically, the Cl from the carboxylic acid chloride combines with an H from an amine group to form HCl. The HCl departs, leaving the two monomers linked together by a bond called an **amide bond**:

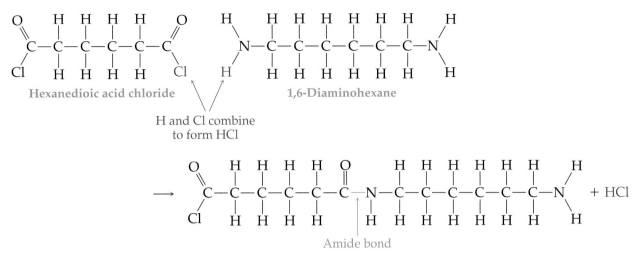

Hexanedioic acid chloride

1,6-Diaminohexane

H and Cl combine
to form HCl

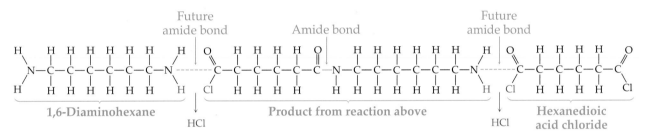

Amide bond

Because *both* ends of each monomer carry a functional group, monomer molecules can add onto both ends of this unit. The reactions are exactly the same as shown above, and the result is a nylon chain held together by amide bonds:

Future amide bond

Amide bond

Future amide bond

1,6-Diaminohexane

HCl

Product from reaction above

HCl

Hexanedioic acid chloride

The chains can be grown long enough to make fibers that can be woven into stockings, shirts, and ropes.

We have chosen to highlight nylon for a reason. The same amide bonds that are in nylon are also part of your skin, muscles, enzymes, hormones, blood cells, and so on, in the form of *proteins*. **Proteins** are biological polymers in which amide bonds link monomers together, just as they do in nylon. This makes recognizing amide bonds and understanding amide bond formation very important. Try the next WorkPatch to make sure you understand the amide bond.

Show how the amine and acid chloride below react to form an amide bond. Draw the product, and indicate the new amide bond that forms.

WORKPATCH 16.2

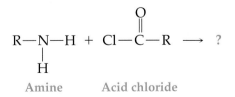

Amine   Acid chloride

Compare your drawing to ours (at the end of the chapter) to make sure you understand the amide bond.

Nature uses a variety of polymers—called **biopolymers**—in constructing living organisms. To give you an introduction to this topic, we will consider two of the most important classes of biopolymers, polysaccharides and proteins.

16.4 POLYSACCHARIDES AND CARBOHYDRATES

Billions of years ago when the first single-cell form of plant life was developing, something was needed to form a boundary that would separate the contents of the cell from the surroundings and protect the life processes occurring inside. This boundary, called the *cell wall*, is made from a type of biological polymer known as a *polysaccharide*. The particular polysaccharide that makes up cell walls is *cellulose*. You are probably more familiar with other polysaccharides such as *starch*, an important food source. The exoskeletons of insects also consist largely of a type of polysaccharide. The word "saccharide" comes from *saccharine*, which means sugar or sweet. Sugar molecules are the monomers in polysaccharides. **Polysaccharides** are therefore polymers in which tens, hundreds, or even thousands of sugar molecules are linked together.

Let's begin our examination of polysaccharides by looking at the monomer, a sugar molecule. Sugars belong to a class of molecules known as *carbohydrates*. **Carbohydrates** are generally aldehydes or ketones that also have many OH groups in them. The most common sugar molecule is glucose, $C_6H_{12}O_6$:

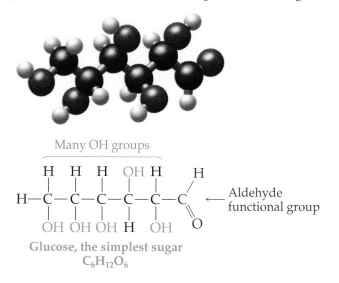

Glucose, the simplest sugar
$C_6H_{12}O_6$

Notice the aldehyde group and the five OH groups. Glucose is the principal sugar in your blood. It is consumed by your cells to obtain energy necessary for sustaining life processes.

The linear structure shown above for glucose is not its most common form. Usually, glucose exists in a ring or cyclic structure:

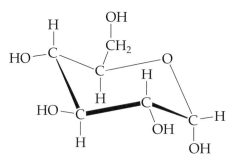

The formula is still the same, but the structure is different. The cyclic form exists in equilibrium with the linear form. In the cyclic form, glucose no longer has an aldehyde functional group, but it is still called a carbohydrate. One reason we can call this cyclic sugar a carbohydrate comes from the meaning of the name itself. The *carbo* portion comes from carbon, while *hydrate* means water. Thus, carbohydrate literally means hydrated carbon, or carbon with water attached. Clearly, the cyclic glucose molecule shown above is not carbon with water molecules attached. However, most of the carbons do have the makings of water, an H atom and an OH group, bound to them.

Glucose is a **monosaccharide** because it consists of only one sugar ring. The sugar that you eat on your cereal and in sweets is sucrose, a **disaccharide**. Sucrose consists of two rings: a glucose ring joined to the ring of another monosaccharide called fructose. Sucrose has the formula $C_{12}H_{22}O_{11}$:

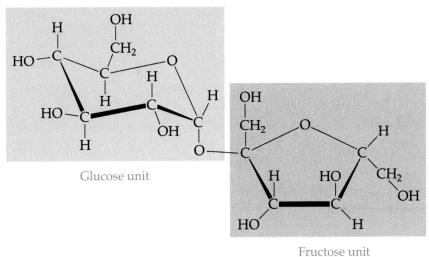

Glucose unit

Fructose unit

Sucrose (table sugar), a disaccharide

Polysaccharides are essentially huge polymers of carbohydrate molecules. Two polysaccharides made from glucose monomers are starch and cellulose:

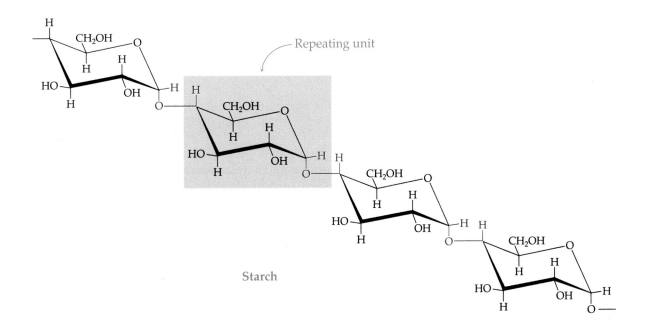

Starch

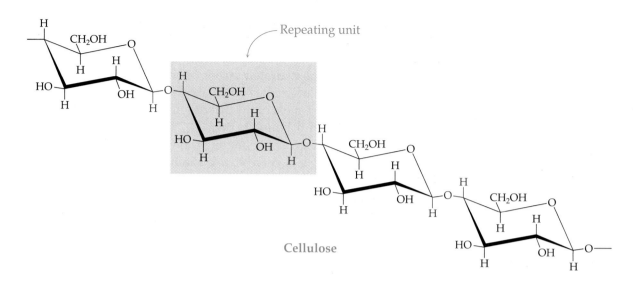

Repeating unit

Cellulose

The difference between starch and cellulose is slight. Look at the structures carefully, focusing on the oxygen atoms that connect the glucose rings, to see the difference. This slight change renders cellulose indigestible to humans. While we can eat a potato or rice and gain nutritional value from its starch, we get nothing but an upset stomach if we try to eat wood, which is largely made of cellulose. Termites, on the other hand, have microorganisms in their gut that can digest wood and get the same nutritional value from it that we do from eating starch.

When you consider that wood is almost 50% cellulose—a polysaccharide biopolymer—then you start to get an appreciation for how strong these polymers can be. Nature uses cellulose and similar polymers much as we use steel girders—as structural supports for building things. While cellulose works well for plants, you won't find any in yourself. Most forms of animal life use a different, somewhat more flexible biopolymer.

16.5 PROTEINS

The main biopolymers used for both structure and function in all living things are the proteins. As we mentioned earlier, our skin is made of protein. So are our hair, nails, and muscles. Proteins are also used as enzymes and hormones, to store iron, to transport oxygen, to act as disease-fighting antibodies, and to clot blood. Proteins are polymers of a group of small molecules called *amino acids*. An **amino acid** has (at the least) two functional groups, an amine (H_2NR or HNR_2) and a carboxylic acid (COOH). Here is the structure of the simplest common amino acid (called glycine) found in proteins:

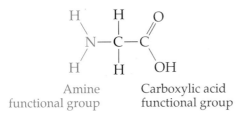

Amine
functional group   Carboxylic acid
functional group

Any hydrocarbon molecule containing both an amine and a carboxylic acid functional group could be called an amino acid. This means that there are essentially an unlimited number of different amino acid molecules possible. However, investigation has shown that only 20 amino acids are necessary for synthesizing the proteins in our bodies. All 20 of these amino acids have an important similarity. They have just one carbon atom between the amine and the acid functional groups:

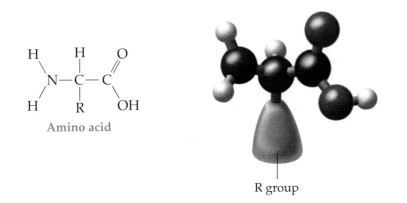

Amino acid

R group

The R group in the diagram varies, from an H atom in the simplest amino acid to functionalized hydrocarbon chains in the more complex amino acids. Table 16.1 (on the next page) shows the 20 amino acids we need, along with their names and three-letter abbreviations. Your body can synthesize half the amino acids from simple molecules. The other 10 amino acids, called **essential amino acids**, cannot be synthesized by your body (or, in the case of arginine, cannot be synthesized in sufficient amounts). These must be present in the food you eat to provide the monomers needed to manufacture many proteins. Failure to take in enough essential amino acids in your diet can result in a wide variety of nutritional deficiency diseases. (The essential amino acids are indicated in Table 16.1 by blue shading.)

To build protein polymers, living organisms string the amino acids listed in Table 16.1 together using the same type of amide bonds we observed in nylon. If we wanted to join glycine and alanine in the lab, for example, we could heat a mixture of them together. One possible result is that the acid end of glycine would approach the basic end of alanine, as shown at the top of page 601.

Table 16.1   **The 20 Amino Acids Used to Build Proteins**
(Essential Amino Acids Are Indicated by Blue Shading)

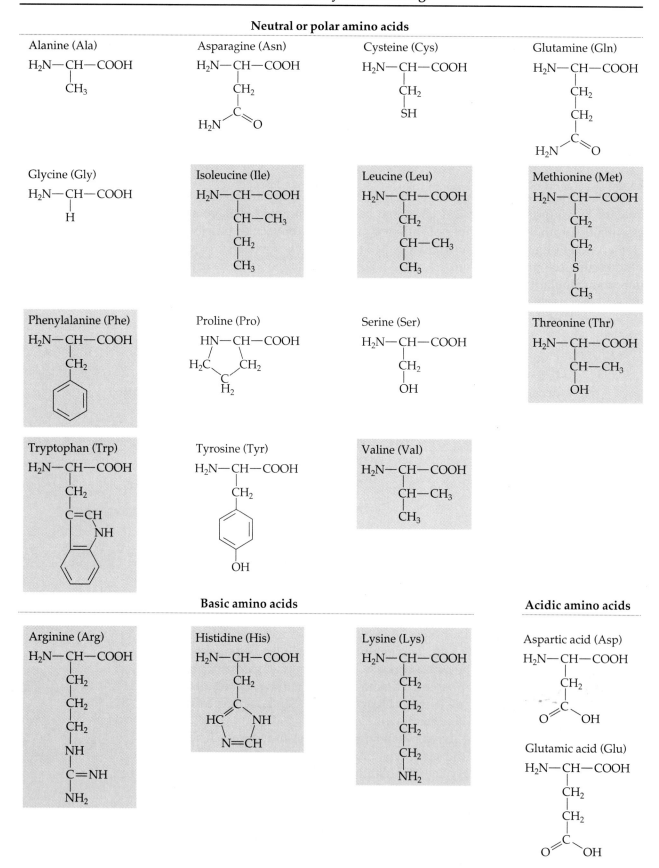

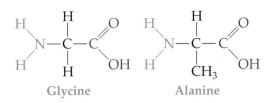

Glycine   Alanine

In a process similar to the formation of nylon, the OH from the carboxylic acid (COOH) and an H from the amine group combine to make water. In the process, a C–N bond forms between the carbonyl carbon and the amine nitrogen. In nylon we called this C–N bond an amide bond. In proteins this same type of bond is often called a **peptide linkage**.

Glycine   Alanine   Gly-Ala dipeptide

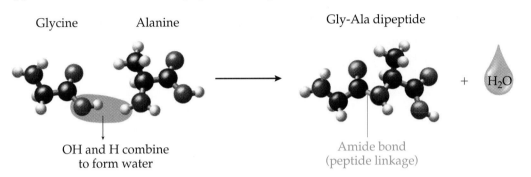

OH and H combine
to form water

Amide bond
(peptide linkage)

The resulting molecule is called a **dipeptide** and is named in abbreviated form as Gly-Ala. A close look shows that this dipeptide still has a free (unreacted) carboxylic acid group on one end and a free amine group on the other. This means that other amino acids can approach these ends and join by forming additional peptide linkages. Chains with fewer than about 50 amino acids are called **polypeptides**. Chains larger than that are called proteins.

Test yourself now and see if you understand the peptide linkage.

Draw the two dipeptides Ala-Ser and Ser-Ala that can result from the combination of the amino acids alanine and serine:

WORKPATCH 16.3

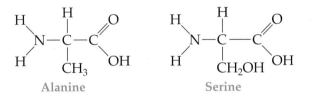

Alanine   Serine

Do you understand why it is possible to make two different dipeptides from alanine and serine? Check your answer with ours and make sure you do.

How a protein functions in the body depends on its sequence of amino acids and its three-dimensional shape. For example, hemoglobin (shown on the first page of this chapter) is the protein in red blood cells that is responsible for carrying oxygen from the lungs to other cells in the body. It consists of 574 amino acids. For hemoglobin to function correctly, each of the 574 amino acids must be in its proper place in the protein. Any change in the position or identity of even one of these amino acids can change the shape of the hemoglobin protein and affect its function. That is what happens in sickle-cell anemia, an inherited disease of the blood. People with sickle-cell anemia have hemoglobin proteins in which a valine has replaced a glutamic acid. This single substitution causes

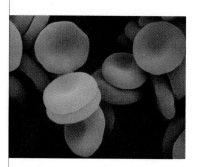

Normal red blood cells

Sickled red blood cells

the three-dimensional shape of the hemoglobin molecule to change whenever the amount of oxygen in the blood is reduced. This, in turn, changes the shape of the red blood cells from round to crescent (sickle-shaped), causing them to clog up small blood vessels and restrict the flow of blood. The result is extremely painful and debilitating.

Given that one misplaced amino acid can wreak such havoc, it is amazing that any of us can function at all. For example, consider the much smaller polypeptide called angiotensin II, a hormone that regulates blood pressure in humans. Angiotensin II contains eight different amino acids:

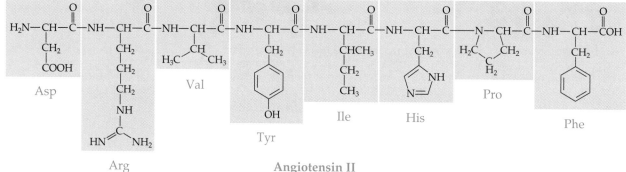

Angiotensin II

From a strictly mathematical point of view, there are over 40,000 different ways to arrange eight unique amino acids in a chain. Only one of those 40,000 combinations will work to regulate your blood pressure. These are tremendous odds for having something go wrong, and yet mistakes are very rare. Something is serving as a blueprint for correctly putting your proteins together, something you inherited.

PRACTICE PROBLEMS

16.4 Draw the dipeptide Asp-Arg. Point out the peptide linkage.

Answer:

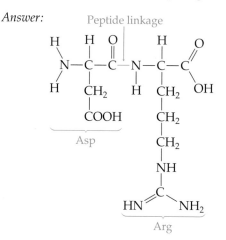

16.5 Draw the dipeptide Arg-Asp. Point out the peptide linkage.

16.6 Consider the following polypeptide. Give its amino acid sequence using the three-letter abbreviations separated by hyphens. [Refer to Table 16.1.]

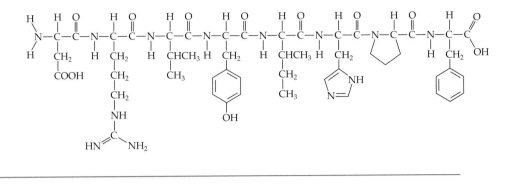

DNA: THE MASTER BIOPOLYMER 16.6

The master blueprint for all your proteins is called **DNA**, short for **deoxyribonucleic acid**. You received half of your DNA from your mother and half from your father. This master blueprint for building protein polymers is a polymer itself. The chromosomes and genes present in the nuclei of most cells are giant DNA polymers ranging in molecular mass from a few million grams per mole for small bacteria to billions of grams per mole in higher animals. The DNA polymer consists of a chain, or "backbone," made of alternating units: sugar molecules called *deoxyribose* and phosphates, as shown in the illustration.

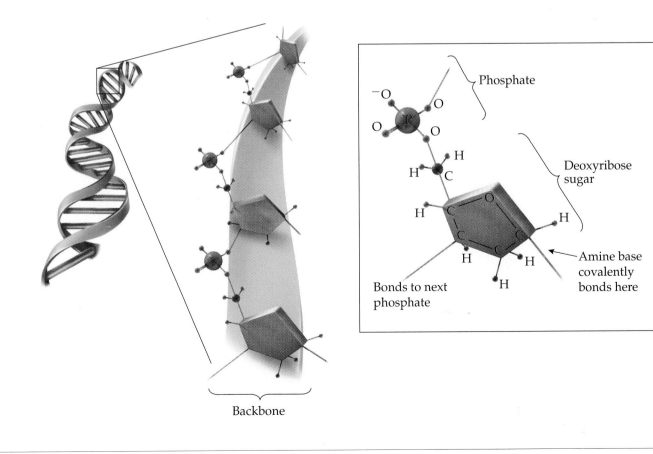

Backbone

This is just the backbone of the DNA polymer. Notice the arrow pointing to the carbon atom in the deoxyribose sugar. Covalently bonded to each of these carbon atoms is one of the four amine bases shown below. Here, the arrows indicate the point where each amine base is attached to the carbon atom of the deoxyribose sugar.

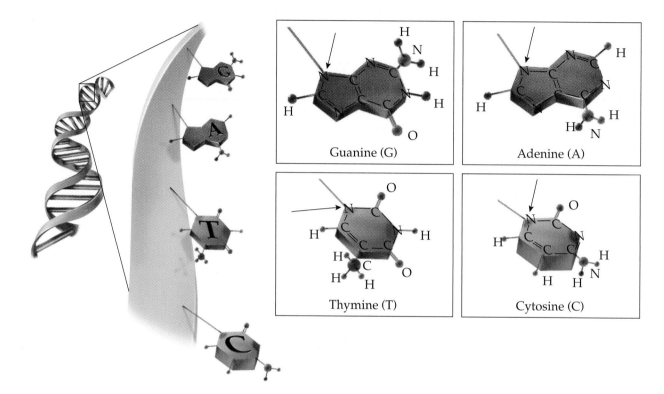

Guanine (G)

Adenine (A)

Thymine (T)

Cytosine (C)

Part of a single strand of DNA polymer would look something like this:

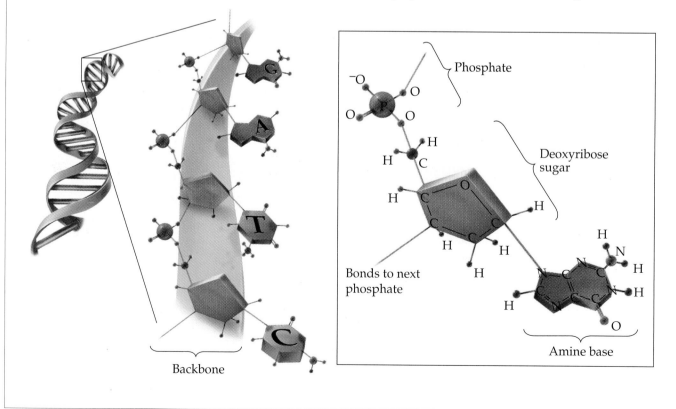

Phosphate

Deoxyribose sugar

Bonds to next phosphate

Amine base

Backbone

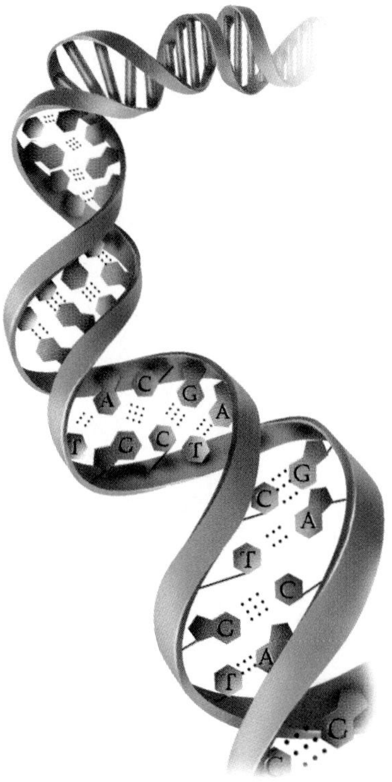

We emphasize *single strand* because the DNA that makes up the genes in your cells is actually double-stranded. Two of the single strands shown above wind around each other to form a **double helix**, somewhat analogous to a spiral staircase.

These two strands of DNA are held together in the double-helix form by hydrogen bonds that form between the amine bases of both strands. These hydrogen bonds are shown as dotted lines in the diagram at left. An enlarged detail is shown below so that you can see these hydrogen bonds more clearly:

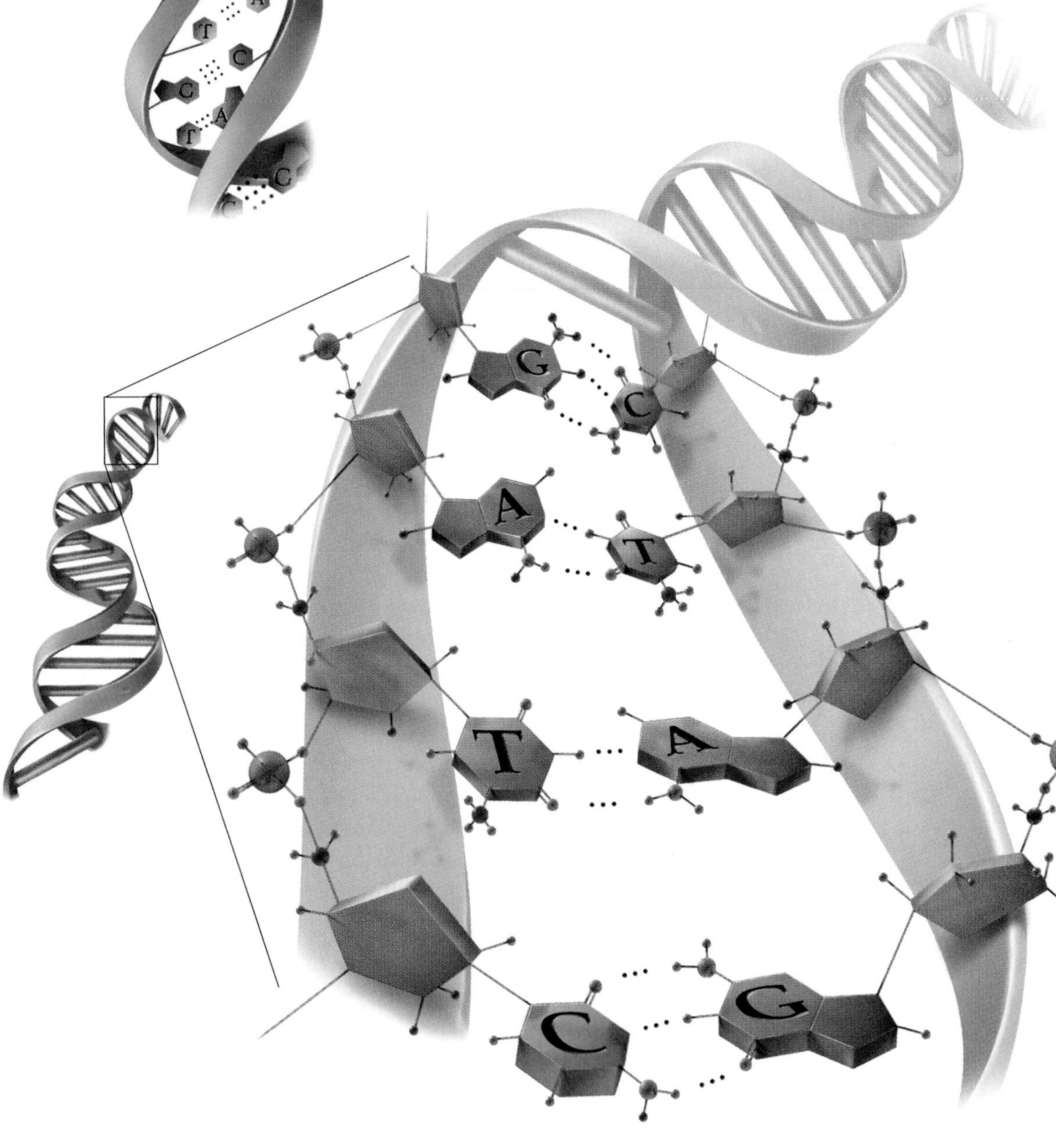

The structures of the four amine bases are complementary to each other. A thymine is always paired up with an adenine, and a guanine is always hydrogen bonded to a cytosine. The helical form is adopted because it maximizes the amount of hydrogen bonding that can occur.

As we discussed in Chapter 9, the hydrogen bonding between the bases is crucial to how DNA works. When DNA serves as a template for protein synthesis, the two strands have to separate. If they were held together by covalent bonds, it would take too much energy to pull them apart. If they were held together by London forces, they would be too likely to fall apart on their own. Hydrogen bonds are just right for the job.

DNA serves as a blueprint for protein synthesis because the sequence of the bases adenine, guanine, cytosine, and thymine on a strand determine how the protein is built. It is the exact sequence of these bases that determines all your inherited traits. Each triplet of adjacent DNA bases is a "word" or *codon* that codes for the incorporation of a particular amino acid into a growing protein chain. For example, the triplet sequence guanine-cytosine-adenine codes for the incorporation of the amino acid alanine. Within the first few years of the twenty-first century, biochemists will have completed a project begun in the 1980s to map the sequence of DNA bases in all our genes. In essence, they will have mapped out the entire human blueprint. At the same time, they are also making tremendous progress at determining what the different genes do.

This brings us to the subject of **viruses**. The word "virus" comes from Latin and means poison. Viruses are not "alive," since they cannot reproduce without a host. They are lifeless parasites, far smaller than bacteria. A basic virus consists of a protein shell surrounding strands of genetic material, often DNA. Transmitted through the air, in food or water, or by body fluids, viruses enter your bloodstream and penetrate your cell membranes. What happens next could come from the script of a low-budget horror movie. Once inside one of your cells, your own cell's enzymes remove the protein coat from the virus, releasing the viral DNA. The viral DNA then takes over your cell's protein-building apparatus, forcing it to replicate more viruses. Eventually, there are so many virus particles that they may actually burst from your cell, killing it in the process. The newly formed viruses go on to infect more of your cells. Up to 300 viruses can be produced by a single cell in an hour. They can go on to infect 300 other cells, which can each produce 300 more viruses. At this rate, one virus could generate billions upon billions of virus particles within 24 hours. If it were not for your immune system's quick response, every virus infection—such as a simple cold or flu—would quickly convert you into a giant virus-producing factory, and you would die. One virus that has found a way around our immune system is HIV, human immunodeficiency virus. HIV attacks the very cells of the human immune system itself and is responsible for the fatal disease known as AIDS, acquired immune deficiency syndrome.

Viral DNA

Protein coat

A typical virus

Strangely enough, viruses may turn out to be our salvation. We are just now beginning to learn how to incorporate snippets of foreign DNA into the DNA of living cells, instructing them to do our bidding. For example, the portion of human DNA that codes for the production of the protein insulin has been spliced into the DNA of the bacteria *E. coli*. These genetically engineered bacteria go on to produce human insulin, which can be isolated and given to people with diabetes. Previously, diabetics were dependent on slightly different and less effective forms of insulin from pigs or cattle. A better solution would be to splice the DNA that codes for insulin production into the cells of human diabetics. This would forever cure their diabetes. Viruses may allow us to do this. Correct genes for insulin production could be loaded into a host virus and used to "infect" a diabetic patient. The virus could then penetrate the patient's cells, release its genetic load, and replace the defective genes. While this is still science fiction, procedures like this will be science fact in the near future.

We may soon be able to cure almost every illness known, from cancer, to inherited abnormalities, to the common cold, but there is a flip side to this. With the same techniques, it will be possible to select or influence almost every human trait, from eye color to personality, from height to sex, from intelligence to life span. We will be able to preselect these traits for our unborn children, change them in ourselves, or order them up in our clones. Do not doubt it. We will have this knowledge, and soon.

In about 2000 years we have gone from Democritus and Aristotle philosophizing about the nature of matter to the brink of taking over our own evolution. The implications and effects of the biochemical revolution that will occur in the twenty-first century will dwarf those of the industrial revolution and the nuclear age combined. For better or worse, what is about to come will truly astound us all.

HAVE YOU LEARNED THIS?

Polymer (p. 589)

Monomer (p. 589)

Macromolecule (p. 590)

Nylon (p. 593)

Amide bond (p. 595)

Protein (pp. 595, 598)

Biopolymer (p. 595)

Polysaccharides (p. 596)

Carbohydrates (p. 596)

Monosaccharide (p. 597)

Disaccharide (p. 597)

Amino acid (p. 599)

Essential amino acids (p. 599)

Peptide linkage (p. 601)

Dipeptide (p. 601)

Polypeptide (p. 601)

DNA, deoxyribonucleic acid (p. 603)

Double helix (p. 605)

Virus (p. 606)

POLYMERS

16.7 What is a polymer?

16.8 What is meant by the term "macromolecule"? Are polymers macromolecules? Explain your answers.

16.9 What is a monomer? How does a monomer unit differ from a monomer? Give an example of each.

16.10 In the polymer polyethylene there are no double bonds. Why, then, is it called polyethylene?

16.11 Use a drawing to show how one ethylene monomer reacts with another to form a bond between ethylene monomer units in the polymer polyethylene. Show how the chain can keep growing.

16.12 Consider the monomer propene. What would a polymer made from propene look like? Draw it, at least a few repeat units long. What would the polymer be called?

16.13 Consider the following polymer, shown with both a line drawing and a structural drawing:

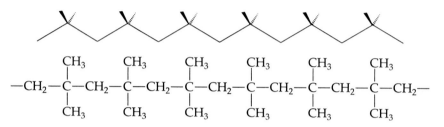

(a) Put a set of parentheses around one of the monomer units, and give its formula.
(b) Draw an isolated monomer molecule. [*Hint:* It has a double bond in it.]
(c) Name the monomer molecule, and name the polymer.
(d) Show how the electrons in the double bonds of the monomer must be moved to generate the polymer.

16.14 What is plastic lunch wrap made of?

16.15 What is Teflon? Draw the polymer and an isolated monomer molecule.

16.16 The following polymer is called a silicone and has a noncarbon, "inorganic" backbone of alternating silicon (Si) and oxygen (O) atoms:

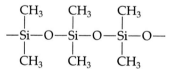

(a) Put a set of parentheses around one of the monomer units.
(b) Another name for this polymer is polydimethylsiloxane. Explain why it has this name.

16.17 Ethylene is a gas, but polyethylene is a solid. How do you explain this?

16.18 What is it about the monomers used to make nylon that allows them to grow into large macromolecules?

16.19 Define an amide bond. Then show the amide bond that would form between dimethylamine and acetic acid chloride.

16.20 Draw a nylon 66 polymer that is two monomer units long.

16.21 Where do you think the monomers for making synthetic polymers ultimately come from?

BIOLOGICAL POLYMERS

16.22 Compare the linear and cyclic forms of glucose shown on page 596. Indicate which oxygen atom in the cyclic form was the aldehyde oxygen in the linear form. Is there any similarity to an ether in the cyclic form?

16.23 What is meant by the terms "monosaccharide," "disaccharide," and "polysaccharide"? Give examples of each.

16.24 How does glucose differ structurally from sucrose?

16.25 What does the word "carbohydrate" mean literally?

16.26 Are sugar molecules examples of carbohydrates? Explain.

16.27 How are cellulose and starch similar? How are they different?

16.28 When you burn a wood log in a fire, what is it that you are actually burning? What are the combustion products?

16.29 What is an amino acid?

16.30 What is the difference between the amino acids glycine and alanine?

16.31 What does the expression "essential amino acid" mean?

16.32 What is a peptide linkage?

16.33 What is a polypeptide? How does it differ from a protein?

16.34 Draw the polypeptide Gly-Ala-Ser.

16.35 What two peptides are possible from the combination of the amino acids alanine and glycine? Draw both of them.

16.36 What are some of the functions of proteins in living organisms?

16.37 How does the sequence of amino acids influence a protein?

16.38 How does DNA serve as the master blueprint for building proteins?

16.39 What is the backbone of the DNA polymer made of? Name the parts, and draw two repeat units of the backbone.

16.40 DNA is double-stranded. What holds the strands together? What would be wrong with holding the strands together with covalent bonds?

16.41 How many DNA bases are there? Name them and give their abbreviations.

16.42 The DNA bases always hydrogen bond to each other in specific pairs. What are those pairs?

16.43 What is meant by the term "codon"? What is the function of a codon?

16.44 What do viruses do (on the cellular level) to make you sick?

16.45 How might viruses be used for good purposes?

WORKPATCH SOLUTIONS

16.1 Only choice (c) can be repeated to reproduce the polymer:

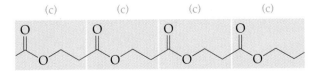

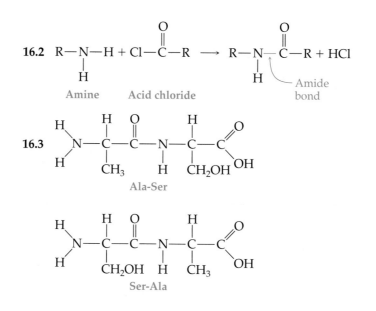

Glossary

absolute temperature scale Another term for Kelvin temperature scale.

accuracy The degree to which a measured value is close to the true value. Compare precision.

acid A substance that increases the concentration of H^+ in an aqueous solution.

acidic solution An aqueous solution in which the concentration of hydronium ions (H_3O^+) is less than 10^{-7} M, giving the solution a pH value below 7. Compare basic solution, neutral solution.

activation energy (E_a) The minimum amount of energy, specific to each chemical reaction, that the reactants must absorb from their surroundings in order for the reaction to occur.

active metal In any pair of metals, the one that more easily gives up its electrons.

active site A precisely shaped location (on an enzyme molecule, for example) at which substrate molecules having a complementary shape may bind.

activity series See electromotive force series.

actual yield The actual or experimental amount of product obtained in a chemical reaction. Compare theoretical yield.

air pressure The force exerted on a surface by air molecules colliding with the surface.

alchemist A practitioner of science as it was known in the Middle Ages, before the scientific method came into universal use. One of the main goals of alchemists was to change base (nonprecious) metals into gold.

alcohol A functionalized hydrocarbon containing one or more −OH functional groups. General formula: R−OH.

aldehyde A functionalized hydrocarbon containing a carbonyl group (C=O) bonded to at least one hydrogen. General formula: R−COH.

algebraic manipulation Rearrangement of an equation in order to solve for a desired quantity.

aliphatic hydrocarbon An early general name for long-chain hydrocarbons, stemming from the Greek *aleiphas* meaning "fat"; applied to these compounds because many fats and oils contain long-chain hydrocarbon molecules. A general term for all alkanes, alkenes, and alkynes.

alkali metal A general name for any element in group IA (except H).

alkaline earth metal A general name for any element in group IIA.

alkane A hydrocarbon containing only carbon–carbon single bonds; a saturated hydrocarbon. General formula: C_nH_{2n+2}.

alkene A hydrocarbon containing at least one carbon–carbon double bond; an unsaturated hydrocarbon. General formula: C_nH_{2n}.

alkyne A hydrocarbon containing at least one carbon–carbon triple bond; an unsaturated hydrocarbon. General formula: C_nH_{2n-2}.

allotrope One of a set of different structural forms of an element, such as the graphite, fullerene, and diamond forms of carbon.

alloy A solid solution in which both solute (or solutes in some cases) and solvent are metals.

α-amino acid Any amino acid whose amine functional group is separated from the acid (carboxyl) functional group by only a single carbon atom.

alpha decay Another term for alpha emission.

alpha (α) emission A nuclear reaction in which a radioactive nucleus ejects an alpha particle (^4_2He) and thereby becomes a daughter isotope containing two fewer protons and two fewer neutrons.

alpha (α) particle A helium nucleus (^4_2He) ejected from some radioactive nuclei during a process known as alpha emission.

amide A functionalized hydrocarbon that contains a derivative of the carboxyl group (COOH) in which the −OH has been replaced by an amine. General formulas: primary, $R−CONH_2$; secondary, $R−CONHR$; tertiary, $R−CONR_2$.

amide bond The covalent bond formed between the carbon atom of the carbonyl group and the nitrogen atom in an amide.

amine An organic derivative of ammonia (NH_3) in which one, two, or all three of its hydrogen atoms is replaced by a hydrocarbon chain. See primary amine; secondary amine; tertiary amine.

amino acid An organic molecule that contains both a carboxylic acid group and an amino group.

amino group See amine.

angstrom (Å) A non-SI unit of length, equal to 1×10^{-10} meter. Often used to express bond lengths in molecules.

anion An ion that has a negative charge. Compare cation.

anode The negative electrode in a galvanic cell; the electrode at which oxidation takes place. Compare cathode.

aqueous solution A solution in which the solvent is water.

Arrhenius acid The earliest identification of acids as a particular class of compounds, with "acid" defined as any electrolyte that ionizes to give H^+ ions when dissolved in water.

Arrhenius base The earliest identification of bases as a particular class of compounds, with "base" defined as any electrolyte that ionizes to give OH^- ions when dissolved in water.

atmosphere The collection of gases that surround the Earth. Also, a non-SI unit of pressure; 1 atm = 760 mm Hg = 29.9 in. Hg at sea level.

atmospheric pressure See air pressure.

atom The smallest unit of an element that has the properties of that element. An atom consists of a central nucleus plus one or more electrons outside the nucleus; the number of electrons is equal to the number of protons, making an atom electrically neutral. Compare ion.

atomic number (Z) The number of protons in the nucleus of an atom. All atoms of a particular element have the same atomic number, which is indicated by a subscript to the left of the element symbol.

atomic radius The distance from the center of the nucleus to the outermost electron shell.

atomic weight (mass) A unitless number expressing the mass of a particular atom relative to the mass of an atom of carbon-12, which is arbitrarily assigned a mass of exactly 12.

atomic weight scale The weighted average of the masses of all the naturally occurring isotopes of an element relative to carbon-12.

autodissociation The process by which a small number of the water molecules in any given volume of water dissociate into hydronium and hydroxide ions. See K_w.

autoionization A synonym for autodissociation.

Avogadro's number 6.02×10^{23}, the number of units in 1 mole.

balanced equation A chemical equation written such that matter is conserved, which means the total number of atoms of each element on the right side of the equation is exactly equal to the number of atoms of those elements on the left side.

balancing coefficient A number placed in front of a chemical formula in a chemical equation to indicate the number of moles (or molecules) of the substance that take part in the chemical reaction represented by the equation.

band of stability The area on a band of stability plot representing all existing nuclei, both nonradioactive isotopes and all radioactive isotopes having a measurable half-life.

band of stability plot A plot of number of neutrons in a nucleus (represented on the vertical axis of the graph) as a function of number of protons in the nucleus (represented on the horizontal axis), with some indication as to whether an atom exists with that combination.

barometer A device used to measure atmospheric pressure.

barometric pressure Another term for air pressure or atmospheric pressure.

base A substance that increases the concentration of OH^- in an aqueous solution.

base unit Any one of seven fundamental SI units: meter, kilogram, second, kelvin, mole, ampere, candela. Compare derived unit.

basic solution An aqueous solution in which the concentration of hydroxide ions (OH^-) is greater than 10^{-7} M, giving the solution a pH value above 7. Compare acidic solution; neutral solution.

battery Common term for galvanic cell.

beta decay Another term for beta emission.

beta (β^-) emission A nuclear reaction in which a neutron changes to a proton plus a beta particle (electron); the proton remains in the nucleus, and the beta particle is ejected from the atom.

beta (β^-) particle An electron created during a nuclear reaction when a neutron changes to a proton; the particle is ejected from the nucleus.

bimolecular Involving two molecules.

binary compound A compound containing atoms of only two different elements. The compound may contain more than two atoms, but only two different elements.

binary covalent compound A covalent compound containing atoms of only two elements, both usually nonmetals.

binary ionic compound An ionic compound containing only one type of cation and one type of anion. The compound may contain more than two ions, but all the cations (usually a metal) are identical and all the anions (usually a nonmetal) are identical.

binding energy The energy released when a nucleus is created from protons and neutrons. Also, the energy required to break up a nucleus into its constituent protons and neutrons.

binding energy per nucleon The binding energy divided by the number of nucleons (protons and neutrons) in the nucleus.

biopolymer A polymer made by and used in living organisms.

Bohr diagram A diagram showing the electron shells of an atom as a series of concentric circles centered on the nucleus.

Bohr model of the atom A model proposed by Niels Bohr, in which the atom is pictured as a central nucleus surrounded by orbiting electrons. Because the energy of the electrons is quantized, the electrons can orbit the nucleus only at certain allowed distances. Compare Rutherford model; Thomson model.

boiling point (bp) The temperature at which a pure substance changes from a liquid to a gas at standard atmospheric pressure. Because condensation is the reverse of evaporation (boiling), the boiling point is numerically equal to the condensation point.

bond See chemical bond.

bonding pair Two electrons forming a covalent bond between two atoms, usually with one electron of the pair coming from one atom and the other electron coming from the other atom.

branch In a branched hydrocarbon, any side chain that extends from, but is not a part of, the main chain.

branched hydrocarbon A hydrocarbon in which at least one carbon atom is bonded to more than two other carbon atoms. Compare linear hydrocarbon.

break-even point For a nuclear reaction, the point at which the amount of energy recovered from the reaction equals the amount of energy input to start the reaction.

Brønsted–Lowry acid A substance that is a proton (H^+ ion) donor in aqueous solution.

Brønsted–Lowry base A substance capable of accepting a proton (H^+ ion) from a proton donor.

buffer A shorthand name for buffered solution.

buffered solution A solution that is able to maintain an essentially constant pH when acids or bases are added to it. The solution is made up of either a weak acid and a salt of its conjugate base or a weak base and a salt of its conjugate acid.

c The symbol representing the speed of light, 3.00×10^8 meters per second $= 186,000$ miles per second.

calorie A non-SI unit of energy; 1 cal $= 4.184$ J. The amount of heat necessary to raise 1 g of water 1°C.

Calorie (Note capital C.) The energy unit used to measure the amount of energy contained in food; equal to 1000 calories.

carbocation A cation containing a C^+ ion.

carbohydrate A general term for molecules made up of carbon, hydrogen, and oxygen, with many of the hydrogen and oxygen atoms forming alcohol (–OH) groups.

carbonyl group A functional group consisting of a carbon atom and an oxygen atom joined by a double bond: $\rangle C{=}O$.

carboxyl group A functional group consisting of a carbon atom bonded to an –OH group by a single bond and to an oxygen atom by a double bond: –COOH.

carboxylic acid A functionalized hydrocarbon containing one or more carboxyl groups. General formula: R–COOH.

catalyst A substance that, by lowering the activation energy of a chemical reaction, changes the rate at which the reaction proceeds but is not itself consumed in the reaction.

cathode The positive electrode in a galvanic cell; the electrode at which reduction takes place. Compare anode.

cation An ion with a positive charge. Compare anion.

Celsius temperature scale A metric (but non-SI) temperature scale used for everyday purposes in all countries except the United States, and often used for scientific work in the United States. On this scale, the freezing point of water is set at 0°C and the boiling point of water is set at 100°C. The unit of temperature on this scale is the degree Celsius.

centi- The SI prefix meaning one-hundredth (10^{-2}).

chain length The number of carbon atoms in the main chain of an organic molecule.

chain reaction (nuclear) A sequence of nuclear reactions in which the first reaction produces sufficient energy and daughter particles to enable multiple repeat reactions in the next step, and so on.

chalcogen A general name for any group VIA element.

chemical bond Any attractive electrostatic force between two atoms that binds them together. Ionic bonds, covalent bonds, double bonds, and so forth, are all types of chemical bonds.

chemical change Another term for chemical reaction.

chemical equation A notation for showing what takes place in a chemical reaction. The chemical formulas for all reactants are written on the left side of the arrow, and the chemical formulas of the products are written on the right side.

chemical equilibrium The stage in a chemical reaction at which the rate of the forward reaction equals the rate of the reverse reaction.

chemical formula A combination of element symbols and numbers that give the exact composition and makeup of a chemical compound.

chemical kinetics The study of the rates at which chemical reactions take place.

chemical periodicity The fact that, if you arrange the elements in order of increasing atomic number, elements spaced at periodic (regular) intervals have similar chemical properties. For the first 18 elements, an element's properties are similar to those of the elements 8 ahead of it and 8 behind it. Beyond element 18, the interval (periodicity) increases to 18 or 32 elements. This periodicity is the basis for the periodic table.

chemical property A property of a substance that can be studied only by having the substance undergo a chemical change; a property that determines the chemical reactions the substance can undergo. Compare physical property.

chemical reaction A process in which one or more reactants are transformed into one or more products through the breaking of chemical bonds in the reactant molecules and the formation of new chemical bonds in the product molecules.

chemical transformation Another term for chemical reaction.

chemistry The branch of science that studies matter and the changes it undergoes.

classical physics Physics as described by Newton, and as it was understood until the end of the nineteenth century, before the discovery that energy can be quantized. The laws of classical physics work well when applied to everyday objects because at that level the fact that energy can be quantized can be ignored, but the laws fail when applied to objects that are the size of atoms and electrons. Compare quantum physics.

combustion A synonym for burning. In any combustion reaction, the substance being burned combines with oxygen from the air.

combustion analysis A procedure for determining the chemical formula of an unknown substance. The substance is burned in a closed container, and the products of the combustion are then isolated and weighed. The weights are then used to calculate the empirical formula of the unknown.

common name A traditional name for a substance, usually one that gives no information about its composition.

compound A pure substance composed of two or more atoms of different elements. Compare elemental substance.

concentration A measure of the amount of solute dissolved in a given amount of solvent or solution.

condensation point The temperature at which a substance changes from a gas to a liquid at standard atmospheric pressure. Because evaporation (boiling) is the reverse of condensation, the condensation point is numerically equal to the boiling point.

condensed phase Either the liquid state or the solid state of matter.

conjugate acid For a given base, the substance that contains all the atoms of the base plus one additional H^+ per molecule. The part of a basic molecule that exists after chemical reaction with an acid. Example: base NaOH; conjugate acid Na^+. Compare conjugate base.

conjugate base For a given acid, the substance that contains all the atoms of the acid minus one H^+ per molecule. The part of an acid molecule that exists after chemical reaction with a base. Example: acid HCl; conjugate base Cl^-. Compare conjugate acid.

conjugate pair An acid and its conjugate base, or a base and its conjugate acid.

contact process An industrial process for producing sulfuric acid from elemental sulfur and oxygen gas.

continuous spectrum An electromagnetic spectrum in which the wavelengths of visible light blend smoothly from one color to the next. When you pass sunlight through a prism, the rainbow you see is a continuous spectrum. Compare line spectrum.

conversion factor A ratio used in unit analysis to change given units to the units wanted in the answer.

core electron An electron in an atom that occupies any shell other than the valence shell.

corrosion The oxidation of metal at a very slow pace, with the oxidizing agent being O_2 from the air. Rust (iron oxide) is the most common evidence of this type of reaction.

Coulomb's law A statement of the fact that the attractive force between two objects of opposite electric charge is directly proportional to the magnitude of the charges and inversely proportional to the distance between them.

covalent bond An attractive force between two atoms in a molecule, resulting from the sharing of valence electrons. Compare ionic bond; metallic bond.

covalent compound A compound containing all covalent or polar covalent bonds.

critical mass In a nuclear fission reaction, the amount of radioactive material needed to sustain a chain reaction.

crystal lattice The regular, three-dimensional, geometric arrangement of atoms in a crystalline solid.

cubic meter The SI unit of volume.

D The element symbol for deuterium.

Δ The Greek capital letter delta; used to represent either a change in a quantity or a difference between two values.

ΔE The symbol used to represent a change in the energy of a system.

ΔE_{rxn} The symbol representing the net energy change for a chemical reaction, defined to be the difference between the energy of the products and the energy of the reactants. This amount of energy is released during an exothermic reaction and is absorbed during an endothermic reaction. Exothermic reactions have a negative value of ΔE_{rxn}, and endothermic reactions have a positive value of ΔE_{rxn}.

δ+, δ− The Greek lowercase letter delta, used with a plus or minus sign to indicate fractional or partial charge.

Dalton's atomic theory The theory of matter formulated by John Dalton: *(1)* All matter is made of atoms; *(2)* atoms can neither be created nor destroyed; *(3)* all atoms of a given element are identical; *(4)* atoms of different elements are different from one another; *(5)* all chemical reactions involve the combining or separating of atoms in the reactant compounds.

daughter isotope An isotope created in a nuclear reaction.

daughter nucleus Another name for daughter isotope.

deci- The SI prefix meaning one-tenth (10^{-1}).

density A measure of the compactness of matter. For any substance, it is the amount of mass per unit volume; $D = Mass/Volume$.

deoxyribonucleic acid (DNA) The biopolymer present in every living cell that carries the hereditary information the cell uses to synthesize all the protein molecules it needs to function and to reproduce itself.

derived unit Any SI unit of measurement obtained from a combination of two or more SI base units.

deuterium (D) The isotope of hydrogen containing one proton and one neutron; $_1^2H$.

dialcohol An alcohol containing two $-OH$ groups. Also known as a glycol or diol.

diamond The allotrope of carbon in which every carbon atom is bonded to four other carbon atoms, all arranged in a three-dimensional crystal lattice. Compare fullerene; graphite.

diatomic Consisting of two atoms. Compare monatomic; polyatomic.

diol A synonym for dialcohol.

dipeptide A small protein molecule containing only two amino acids joined by a peptide linkage. Compare polypeptide; protein.

dipolar force A synonym for dipole–dipole force.

dipolar molecule A synonym for polar molecule.

dipole–dipole attractive force The intermediate-strength intermolecular attractive force exerted by polar molecules on each other. Compare hydrogen bond; London force.

dipole moment An experimental measure of the polarity of a molecule. A measure of both the strength of the dipole charges and the distance between them.

dipole vector A synonym for dipole moment; a form of notation used to represent the size and direction of the dipole moment; $\longmapsto$.

diprotic acid An acid that, when it dissociates, can yield a maximum of two H_3O^+ ions per molecule of acid.

disaccharide A sugar molecule consisting of two monosaccharide (simple sugar) units joined together.

dissociation The process by which either the molecules of a covalent compound or the cations and anions bound in the lattice of an ionic compound break up into individual cations and anions.

DNA The acronym for deoxyribonucleic acid.

dot diagram A notation used to show valence electrons and bonding in covalent compounds.

double bond A multiple covalent bond between two atoms that is formed by two bonding electron pairs.

dry cell A galvanic cell in which both the species being reduced and the salt bridge are present not as aqueous solutions but rather as semisolid ("dry") pastes.

dynamic equilibrium Another name for chemical equilibrium.

E_a The symbol for the activation energy of a chemical reaction.

effective collision In a chemical reaction, a collision between reactant molecules that leads to formation of a product molecule.

electric charge The property of any particle that determines how it behaves in an electric (or magnetic) field. Particles carrying like electric charges repel each other, those carrying opposite electric charges attract each other, and those carrying no electric charge are neutral.

electricity The flow of electrons through a conducting medium, such as a wire.

electrochemical cell See galvanic cell.

electrode The anode or cathode of a galvanic cell.

electrolyte A water-soluble substance that dissociates into anions and cations when dissolved in water, creating a solution that can conduct an electric current. Compare nonelectrolyte.

electromagnetic radiation All wavelengths of radiant energy.

electromagnetic spectrum The continuum comprising all wavelengths of electromagnetic radiation, from low-energy forms (radio waves and microwaves) to high-energy forms (X rays and gamma rays), with visible light midway between these extremes.

electromotive force (EMF) series A ranking scheme indicating the relative (oxidizing) activities (tendency to give up valence electrons) of metals.

electron A subatomic particle that has a mass $1/1836$ that of a hydrogen atom, carries an electric charge of -1, and moves about at high velocity around the nucleus of an atom.

electron capture A nuclear reaction in which a proton in a nucleus changes to a neutron by absorbing (capturing) an electron.

electron configuration The way the electrons of an atom are distributed among the atom's shells and subshells.

electron configuration notation A way of indicating the distribution of electrons in an atom, consisting of the principal quantum number (indicating the main shell) followed by the subshell letter followed by a superscript indicating the number of electrons in that subshell.

electron shell See shell.

electron tunneling The ability of electrons to pass through barriers that supposedly should stop them.

electronegativity An atom's ability to attract shared electrons in a chemical bond to itself.

electrostatic force Forces of attraction between unlike charges and repulsion between like charges.

element A pure substance that cannot be broken down into simpler substances. Atoms of an element get their identity from the unique number of protons in the nucleus.

elemental substance A pure substance composed of atoms of a single element. Compare compound.

elementary step In a multistep chemical reaction, any one of the steps in the mechanism that occurs exactly as written in the equation.

EMF series A shortened name for electromotive force series.

empirical Based solely on observation or on data collected from experiments.

empirical formula A chemical formula in which the numeric subscripts indicate the ratio of elements in a substance expressed as the lowest possible numbers. One molecule of the substance may contain the number of atoms shown in the empirical formula or a multiple of those numbers. Compare molecular formula.

endothermic reaction A chemical reaction that absorbs heat from its surroundings; ΔE_{rxn} is positive. Compare exothermic reaction.

energy (E) A measure of the capacity to do work.

energy barrier A graphical representation of the activation energy of a chemical reaction.

energy level diagram A diagram that shows the energies of the shells and subshells that electrons can occupy in an atom.

entropy A measure of disorder in a system. The more disordered the system, the higher its entropy value.

enzyme A class of large protein molecules that act as catalysts in biochemical reactions.

equilibrium A shorthand name for chemical equilibrium.

equilibrium constant (K_{eq}) A number that expresses the position of equilibrium for a particular chemical reaction at a particular temperature. See also K_a; K_b; K_w.

essential amino acids The ten amino acids that are required in the human diet because they cannot be synthesized in the body.

ester A functionalized hydrocarbon containing a carboxyl group in which the hydrogen atom of the –OH group has been replaced by another hydrocarbon group. General formula: R–COOR.

ether A functionalized hydrocarbon consisting of a central oxygen atom attached to two hydrocarbon groups. General formula: R–O–R.

evaporation A change from the liquid state to the gaseous state, usually occurring below the boiling point.

exact number A number that has no uncertainty. Compare measured number.

excited state For an atom, a state in which one or more electrons have absorbed energy and jumped from their ground-state position to a higher-energy shell or subshell, leaving behind an electron vacancy. Compare ground state.

exothermic reaction A chemical reaction that, as it proceeds, releases energy into the surroundings; ΔE_{rxn} is negative. Compare endothermic reaction.

experiment A procedure carried out to study a phenomenon or to test a theory.

extensive property Any property of a substance that depends on the amount present. Compare intensive property.

Fahrenheit temperature scale The (non-SI) temperature scale used for nonscientific work in the United States, in which the freezing point of water is set at 32°F and the boiling point of water is set at 212°F. The unit of temperature on this scale is the degree Fahrenheit.

first ionization energy The amount of energy needed to remove an electron from the outermost shell of a neutral (uncharged) atom.

fission A shorthand name for nuclear fission.

formula Another term for chemical formula.

forward rate In a reversible chemical reaction, the rate at which the forward reaction takes place. Compare reverse rate.

forward reaction In a reversible chemical equation, the reaction reading from left to right (by convention).

freezing point The temperature at which a substance changes from a liquid to a solid at standard atmospheric pressure. Because melting is the reverse of freezing, the freezing point is numerically equal to the melting point.

fuel cell A galvanic cell in which the chemical reaction is the direct combination of oxygen and hydrogen to form water and create electricity, with both the water and the electricity being useable products.

fullerene An allotrope of carbon in which every carbon atom is bonded to three other carbon atoms, creating a series of hexagons and pentagons forming a sphere. Compare diamond; graphite.

functional group A commonly occurring atom or group of atoms (other than carbon or hydrogen), attached to a hydrocarbon. The properties of the funtionalized hydrocarbon are different from the properties of the parent hydrocarbon.

functionalized hydrocarbon A hydrocarbon to which one or more functional groups have been added.

fused ring A cyclic hydrocarbon consisting of two or more rings of carbon atoms that share one or more carbon atoms.

fusion A shorthand name for nuclear fusion.

galvanic cell The name used by scientists for what is commonly called a battery, consisting of two compartments connected by a salt bridge. One compartment contains a cathode (where reduction takes place) and the other an anode (where oxidation takes place). The oxidation–reduction reactions create electricity.

gamma decay Another name for gamma emission.

gamma emission A nuclear reaction in which a nucleus emits energy in the form of gamma radiation; no particles are emitted, so the identity of the decaying nucleus is not changed.

gamma (γ) radiation The form of very high-energy electromagnetic radiation most frequently emitted by radioactive nuclei during decay.

gamma ray Another name for gamma radiation.

gas A state of matter in which fast-moving atoms or molecules behave relatively independent of one another, only interacting upon collision. A gas expands or contracts as necessary to assume the shape and volume of any vessel used to contain it. Compare liquid; solid.

gas-phase reaction A chemical reaction in which all reactants are gases.

gaseous solution A solution in which solute and solvent are gases. Example: air.

giga- The SI prefix meaning 1 billion (10^9).

glycol A synonym for dialcohol.

graphite An allotrope of carbon in which every carbon atom is bonded to three other carbon atoms, creating a series of hexagons arranged in two-dimensional sheets stacked one on top of another. Compare diamond; fullerene.

ground state For an atom, the state in which all of its electrons are arranged so as to have the lowest possible total energy. The arrangement in which each electron occupies the lowest-energy subshell available. Compare excited state.

group A column of the periodic table. All the elements in each column have similar valence electron configurations.

h The symbol representing Planck's constant, 6.626×10^{-34} J · S.

half-life The time needed for one-half of the radioactive nuclei in a sample to decay.

halide rule A statement of the fact that any halogen atom in a covalent compound most typically has an oxidation state of -1.

halogen A general name for any group VIIA element.

heavy hydrogen Another name for deuterium.

heavy water Water in which each molecule contains two deuterium atoms instead of two hydrogen-1 atoms; D_2O.

Henry's law A statement of the fact that the solubility of a gas in a liquid increases with increasing pressure.

heteroatom In the language of organic chemistry, any atom other than a carbon or hydrogen.

heterogeneous mixture A mixture whose composition is not uniform throughout. Compare homogeneous mixture.

homogeneous mixture A mixture whose composition is uniform throughout; a solution. Compare heterogeneous mixture.

homologous series A group of similar compounds in which one member differs from the next by a constant number of atoms. Example: alkanes.

hydration Solvation with water as the solvent.

hydration energy As a substance dissolves, the energy given off as solute ions are solvated by water molecules.

hydride A binary covalent compound consisting of one or more hydrogen atoms bonded to atoms that are less electronegative than hydrogen. This gives the H atom a -1 oxidation number. Example: LiH.

hydrocarbon A compound containing only carbon and hydrogen atoms.

hydrocarbon tail The nonpolar section of a soap or detergent molecule, consisting of nonpolar C and H atoms.

hydrogen bond An intermolecular dipole–dipole attraction between a $\delta+$ H atom covalently bonded to either an O, N, or F atom in one molecule and an O, N, or F atom in another molecule. Compare dipole–dipole attractive force; London force.

hydronium ion (H_3O^+) The cation created when an H^+ ion is hydrated with one water molecule.

hydrophilic Referring to the section of a soap or detergent molecule (the polar head) that is soluble in water ("water-loving"). Compare hydrophobic.

hydrophobic Referring to the section of a soap or detergent molecule (the nonpolar hydrocarbon tail) that is insoluble in water ("water-fearing"). Compare hydrophilic.

hydroxide ion An anion with formula OH^-.

hyperbaric High-pressure.

ideal gas A gas in which absolutely no intermolecular forces are present between molecules, and whose molecules occupy no volume whatsoever. Such a gas, which is only an abstraction and does not exist in the real world, remains a gas at all temperatures and all pressures, and obeys the ideal gas law.

ideal gas constant (R) The constant of proportionality in the ideal gas law; $R = 0.0821$ L · atm/K · mole.

ideal gas law A mathematical expression relating the pressure P, volume V, temperature T (in kelvins), and number (n) of moles of a sample of an ideal gas: $PV = nRT$, where R is the ideal gas constant.

indicator Any substance that changes color with the strength (or concentration) of acid (or base) present.

inert gas Outdated name for a group VIIIA element. A synonym for noble gas.

inherent rate factor In a reaction rate equation, any term that does not depend on the concentrations of the reactants. All inherent factors are collected in the rate constant k for a given reaction.

inner-shell electron Another term for core electron.

intensive property Any property of a substance that does not depend on the amount present. Examples: density; boiling point. Compare extensive property.

intermolecular force A force exerted by one molecule on another molecule, arising from the separation of positive and negative electric charges in the molecules. The positive part of one molecule is attracted to the negative part of the other molecule. The three types of intermolecular forces are London forces (momentary charge separation), dipole–dipole forces (permanent charge separation), and hydrogen bonds (permanent charge separation). Compare intramolecular force.

intramolecular force A force exerted by one atom in a molecule on another atom in the same molecule. Ionic bonds and covalent bonds are two types of intramolecular forces. Compare intermolecular force.

ion Any atom or group of atoms carrying either a net negative or a net positive electric charge. See anion; cation.

ion–dipole attraction Another name for ion–dipole force.

ion–dipole force The force of attraction exerted by an ion and a polar molecule on each other.

ionic bond A bond between two ions, formed when the atoms of one give up electrons (form cations) and the atoms of the other accept electrons (form anions). Compare covalent bond; metallic bond.

ionic compound A compound containing all ionic bonds.

ionic lattice The three-dimensional ordered arrangement of the ions making up an ionic solid, consisting of alternating cations and anions.

ionic solid A solid made up of individual ions arranged in an ionic lattice held together by ionic bonds. Compare molecular solid; network solid.

ionization energy The amount of energy needed to remove an electron from an atom or ion. Each electron in any atom or ion has a specific ionization energy. Compare first ionization energy.

ionizing radiation Radiation capable of ionizing atoms, especially those in the cells of biological organisms.

isomer Any one of a set of molecules that have the same chemical formula but different structures.

isotopes Different forms of an element having the same number of protons but different numbers of neutrons (and therefore different atomic weights).

IUPAC The acronym for International Union of Pure and Applied Chemistry.

IUPAC nomenclature system A set of rules for naming chemical compounds uniquely and unambiguously.

joule (J) The SI unit of energy.

K_a The equilibrium constant for the reaction in which a weak acid partially dissociates in aqueous solution.

K_b The equilibrium constant for the reaction in which a weak base partially dissociates in aqueous solution.

K_{eq} The equilibrium constant of a chemical reaction.

K_w A modified equilibrium constant for the autodissociation of water; $K_w = [OH^-] \times [H_3O^+] = 10^{-14}$.

kelvin (K) The SI base unit of temperature.

Kelvin temperature scale The SI temperature scale, where the freezing point of water is set at 273.15 K and the boiling point of water is set at 373.15 K. The unit of temperature on this scale is the kelvin. Sometimes called the absolute temperature scale.

ketone A functionalized hydrocarbon in which the functional group is a carbonyl group attached to two hydrocarbon groups. General formula: R–CO–R.

kilo- The SI prefix meaning 1000 (10^3).

kilogram The SI base unit of mass; equal to about 2.2 pounds.

kinetic energy The energy an object has whenever it is moving; the faster the object is moving, the higher its kinetic energy.

kineticist A chemist who studies reaction rates.

kinetics experiment An experiment run to determine the rate of a chemical reaction.

law A statement that describes the way things are consistently observed to behave under a given set of circumstances. Compare theory.

law of conservation of matter A statement of the fact that no matter is ever created or destroyed in any chemical reaction.

law of constant composition Another name for law of definite proportions.

law of definite proportions A statement of the fact that the amounts of the elements in a compound are determined by the identity of the compound.

law of Mendeleev Another name for chemical periodicity. See also law of octaves.

law of multiple proportions A statement of the fact that, for two elements A and B capable of forming more than one compound with each other, a given amount of A yields compounds having the general chemical formula A_xB_y in which the ratio

$$\frac{\text{Amount of B in compound 1}}{\text{Amount of B in compound 2}}$$

is always a small whole number. This law provides compelling evidence for the existence of atoms.

law of octaves The pattern discovered by Mendeleev in which each element in his version of the periodic table has properties similar to those of the elements eight positions to the left and eight positions to the right.

Le Chatelier's principle A statement of the fact that any time the equilibrium of a chemical reaction is disturbed, the forward and reverse reaction rates shift in the direction that acts to partially remove the disturbance.

leading zero Any zero to the left of the first nonzero digit in a number.

Lewis dot diagram Another name for dot diagram.

limiting reactant In a chemical reaction, a reactant that, because there is less of it than is needed to react with the total amount of the other reactants, determines how much product is produced. All of the limiting reactant is used up in the reaction.

line-drawing method A shorthand way of representing hydrocarbon molecules using line segments.

line spectrum An electromagnetic spectrum consisting of discrete lines separated by dark spaces of various widths. The lines indicate radiation (energy) emitted by an atom when excited electrons in the atom lose energy and thereby fall to a lower energy state. Compare continuous spectrum.

linear hydrocarbon A hydrocarbon in which each carbon atom, except the two end carbons, is bonded to two other carbon atoms. Compare branched hydrocarbon.

liquid A state of matter in which molecules or atoms are in contact (touch) and are also in motion, jostling by one another. A liquid changes shape to match the shape of the vessel used to contain it but maintains a constant volume. Compare gas; solid.

liter A non-SI unit of volume, equal to 1.057 quarts.

litmus A plant-derived dye that is red in the presence of an acid and blue in the presence of a base.

lock-and-key model A model describing enzyme action as the binding of a substrate molecule (the key) at an active site (the lock) on an enzyme molecule.

log A shorthand term for logarithm.

logarithm For any numeric value, the power to which 10 must be raised to get that value.

London force The relatively weak intermolecular attractive forces exerted by molecules on each other, caused by brief, constantly shifting electron imbalances in the molecules that cause them to act as momentary dipoles. Compare dipole–dipole attractive force; hydrogen bond.

lone pair A valence-shell electron pair on an atom that is not part of a chemical bond.

macromolecule A general term for any very large organic or biochemical molecule, where molecular weights can be as high as 50,000 g/mole and more.

magnetic resonance imaging (MRI) A technique for creating images of the interior of a body.

main chain In a branched hydrocarbon, the longest continuous chain of carbon atoms in the molecule. When a multiple bond occurs in the molecule, the longest continuous chain that includes the multiple bond.

main group element Another name for representative element.

mass A measure of the quantity of matter in an object. Mass is independent of the location of the object, with a given object having the same mass no matter where in the Universe it is located. Compare weight.

mass defect The difference between the actual mass of an atom and the theoretical mass calculated by adding up the masses of all the protons, neutrons, and electrons in the atom.

mass number The number of protons plus neutrons in the nucleus of an atom.

mass percent The mass of any component of a sample divided by the total mass of the sample, with this quotient multiplied by 100.

matter Anything that has mass and takes up space.

measured number A number obtained by using a measuring device. Measured numbers always have some uncertainty because no measuring device is perfect. Compare exact number.

mega- The SI prefix meaning 1 million (10^6).

melting point (mp) The temperature at which a substance changes from a solid to a liquid at standard atmospheric pressure. Because freezing is the reverse of melting, the melting point is numerically equal to the freezing point.

mercury barometer A device that uses mercury as the liquid to measure the pressure exerted by the atmosphere.

metal An element that tends to lose one or more valence electrons in chemical reactions. In their elemental form, metals typically are shiny and bendable, and conduct heat and electricity well.

metal hydroxide A compound consisting of a metal cation and one or more hydroxide anions, having the general formula $M(OH)_n$. Those that are soluble produce basic solutions when dissolved in water.

metallic bond Bonds between metal atoms that are the result of electrons being shared by all the atoms. Compare covalent bond; ionic bond.

metalloid An element whose properties are intermediate between those of metals and those of nonmetals.

meter The SI base unit of length, equal to 39.37 inches.

metric system A system of units in which the basic length unit is the meter.

micelle The spherical structure formed by a group of soap or detergent molecules in water. The polar heads orient relative to each other so that they create the surface of the sphere, and the hydrocarbon tails all point to the interior. The polarity of the surface causes the micelle to dissolve in water, while (nonpolar) grease and oil molecules dissolve in the nonpolar interior.

micro- The SI prefix meaning one-millionth (10^{-6}).

micron A synonym for micrometer.

milli- The SI prefix meaning one-thousandth (10^{-3}).

millimeters of mercury A non-SI unit of pressure; 760 mm Hg = 1 atm.

mixture Matter that consists of two or more substances in variable amounts.

model A verbal description or physical construction used to visualize something a theory describes.

molar (M) A unit of concentration for solutions, indicating the number of moles of solute contained in each liter of a solution; M = moles/liter.

molar mass The mass of 1 mole of any substance, expressed in grams.

molarity A measure of solution concentration, defined as the number of moles of solute in 1 liter of solution.

mole The SI base unit of amount of substance, defined as the amount of any substance that contains 6.02×10^{23} atoms or molecules of the substance. For elements, this amount is equal to the atomic weight expressed in grams; for molecules, this amount is equal to the molar mass expressed in grams.

molecular electrolyte A substance that, despite containing only covalent bonds, either partially or completely dissociates into ions when dissolved in water.

molecular formula A chemical formula in which the numeric subscripts indicate the actual composition of one molecule of a substance. Compare empirical formula.

molecular solid A solid made up of discrete molecules attracted to each other by intermolecular forces. Compare ionic solid; network solid.

molecular weight A synonym for molar mass.

molecule A stable collection of two or more atoms bound together by covalent bonds.

molten In the liquid state.

monatomic Consisting of one atom. Compare diatomic; polyatomic.

monoalkene An alkene containing only one carbon–carbon double bond.

monoalkyne An alkyne containing only one carbon–carbon triple bond.

monomer A hydrocarbon molecule that forms chains to create a polymer.

monomer unit A monomer once it has been incorporated into a polymer.

monoprotic acid An acid that, when it dissociates in water, can yield only one H_3O^+ ion per molecule of acid.

monosaccharide A sugar molecule consisting of one simple sugar.

MRI An acronym for magnetic resonance imaging.

multiple bond More than one bond between two atoms, formed by the sharing of more than one electron from each atom. Can be either a double bond (two bonding pairs) or a triple bond (three bonding pairs).

n/p ratio Shorthand for neutron-to-proton ratio.

nano- The SI prefix meaning one-billionth (10^{-9}).

net energy change (ΔE_{rxn}) The difference between the energy of the products of a chemical reaction and the energy of the reactants. This amount of energy is released during an exothermic reaction (ΔE_{rxn} is negative) and is absorbed during an endothermic reaction (ΔE_{rxn} is positive).

net molecular dipole moment The overall dipole moment of a molecule, obtained by adding all the dipole moments.

network solid A nonmolecular solid made up of atoms held together by covalent bonds. Compare ionic solid; molecular solid.

neutral solution A solution in which the concentration of hydronium ions is the same as the concentration of hydroxide ions, giving the solution a pH value of exactly 7. Compare acidic solution; basic solution.

neutralization reaction The reaction of an acid with a base to form a salt and water. The combining of one hydronium ion (H_3O^+) and one hydroxide ion (OH^-) to form two molecules of water.

neutron A subatomic particle that has a mass approximately equal to that of the hydrogen atom, carries zero electric charge, and is located in the nucleus of an atom.

neutron-to-proton ratio The ratio of neutrons to protons in an atomic nucleus.

Newton's laws of motion Three laws formulated by Isaac Newton to describe how applied forces cause changes in the motion of any object.

Newtonian physics Another name for classical physics.

NMR An acronym for nuclear magnetic resonance.

noble gas Any element in group VIIIA of the periodic table.

noble gas notation A shorthand form of electron configuration notation in which the element symbol of the noble gas immediately preceding the atom under consideration is used to represent the electron configuration up to that point.

nonelectrolyte A water-soluble substance that does not dissociate into ions when dissolved in water, with the result that the solution cannot conduct an electric current. Compare electrolyte.

nonmetal An element that tends to gain one or more valence electrons in a chemical reaction. In their elemental form, nonmetals do not conduct heat or electricity well, meaning they are thermal and electrical insulators.

nonpolar molecule A covalent molecule in which there is no net separation of electric charge in the molecule. Compare polar molecule.

normal atmospheric pressure The air pressure that raises the liquid column in a mercury barometer to a height of 760 mm = 29.9 in. = 1 atm.

normal boiling point The boiling point of a substance at normal atmospheric pressure (1 atm).

normal melting point The melting point of a substance at normal atmospheric pressure (1 atm).

NR An abbreviation for no reaction.

nuclear decay Another name for radioactive decay.

nuclear fission The splitting apart of a large parent nucleus to two smaller daughter nuclei.

nuclear fusion The fusing (joining together) of two small nuclei into one larger nucleus.

nuclear magnetic resonance (NMR) An alternative name for magnetic resonance imaging.

nuclear radiation A general term for the particles and electromagnetic radiation emitted by a nucleus undergoing radioactive decay.

nuclear reaction A process in which change occurs to the nuclei of the atoms involved. Compare chemical reaction.

nuclear transformation Another term for nuclear reaction.

nucleon A particle found in the nucleus of an atom; a general term for protons and neutrons.

nucleus The tiny, positively charged core of an atom, made up of protons plus neutrons (except in the case of the hydrogen nucleus, which contains no neutrons). The electric charge of the nucleus binds the electrons to the atom, and the number of protons determines which element the atom is.

ocean of instability All areas on a band of stability plot except the band of stability and the small "island" representing predicted stable isotopes having atomic numbers 114 and above.

octet rule A statement of the fact that when two or more elements react to form a compound, the atoms usually combine in such a way that, once the compound is formed, each atom has eight electrons in its valence shell.

orbital A synonym for subshell.

order An experimentally determined number, shown as an exponent in the rate law for a chemical reaction, that expresses how the concentration of a reactant affects the reaction rate. Example: Rate = $k[A]^1[B]^2$ says that the reaction is first-order with respect to A and second-order with respect to B.

organic chemistry The branch of chemistry that deals with carbon and the compounds it forms.

orientation factor A number between 0 and 1 indicating how important proper orientation between reactants is in a chemical reaction, with a value of 1 indicating orientation does not matter.

outer-shell electron Another name for valence electron.

oxidation The loss of electrons by an atom, resulting in an increase in oxidation state. Compare reduction.

oxidation number Another term for oxidation state.

oxidation–reduction reaction An electron transfer reaction in which one reactant loses electron(s) (is oxidized) and another reactant gains electron(s) (is reduced).

oxidation state A numeric value indicating the change in the number of valence electrons in a covalently bonded atom, more (−) or less (+) than in the atom in its free (unbonded) state.

oxidation state bookkeeping method A technique for determining the number of valence electrons "owned by" a covalently bonded atom.

oxidizing agent The reactant in an electron transfer reaction that accepts electrons. The oxidizing agent is reduced during the reaction. Compare reducing agent.

parent isotope In a nuclear reaction, the radioactive isotope that undergoes decay.

pascal The SI unit of pressure.

peptide linkage An amide bond when it occurs in a protein molecule.

percent by mass (wt %) A unit of concentration for solutions, used most frequently for solutions made up of a solid solute dissolved in a liquid solvent and defined as

$$\frac{\text{Grams of solute}}{\text{Grams of solution}} \times 100$$

percent by volume (vol %) A unit of concentration for solutions, used most frequently for solutions made up of a liquid solute dissolved in a liquid solvent and defined as

$$\frac{\text{Volume of solute}}{\text{Volume of solution}} \times 100$$

percent composition A measure of solution concentration defined in terms of mass (percent by mass), volume (percent by volume), or mass and volume (percent mass/volume).

percent mass/volume (wt/vol %) A unit of concentration for solutions, defined as

$$\frac{\text{Grams of solute}}{\text{Volume of solution}} \times 100$$

percent yield The actual yield of product from a chemical reaction divided by the theoretical yield, with this quotient multiplied by 100 to give a value in percent.

period A row of the periodic table.

periodic behavior Another name for chemical periodicity.

periodic table A table of the elements arranged in order of increasing atomic number. The elements are arranged in rows (periods) in such a way that all the elements in a given column (group) have similar chemical properties.

pH value The negative of the logarithm of the hydronium ion concentration in an aqueous solution.

phase In describing matter, a synonym for state.

photon A packet of energy considered to be the basic particle of electromagnetic radiation.

physical change A change that alters a physical property of a substance without changing its identity.

physical property A property of a substance that can change without the substance undergoing a chemical reaction; examples are density, boiling point, and color. Compare chemical property.

physics See classical physics; quantum physics.

pico- The SI prefix meaning one-trillionth (10^{-12}).

Planck's constant (*h*) An unchanging numeric value used in calculating the energy of electromagnetic radiation; 6.626×10^{-34} J · S.

polar covalent bond A covalent bond in which the electrons are shared unequally.

polar force A synonym for dipole–dipole attractive force.

polar molecule A covalent molecule containing atoms of different electronegativities arranged in such a way that there is a net (unbalanced) separation of electric charge in the molecule. Compare nonpolar molecule.

polyatomic Containing three or more atoms. Compare diatomic; monatomic.

polymer A very large molecule made from repeating units called monomers.

polypeptide A protein molecule containing between three and fifty amino acids joined by peptide linkages. Compare dipeptide; protein.

polysaccharide A sugar molecule consisting of many monosaccharide (simple sugar) units.

positron (*β⁺*) A subatomic particle having the same mass as an electron but carrying an electric charge of +1; the antimatter of the electron.

positron (*β⁺*) emission A nuclear reaction in which a proton changes to a neutron plus a positron, with the neutron staying in the nucleus and the positron being ejected from the atom.

precision The degree to which a set of measured values of the same quantity agree with each other. Compare accuracy.

pressure The force per unit area exerted on a surface.

primary amine An amine in which the amine functional group is bonded to one hydrocarbon chain and two hydrogen atoms. General formula: $R-NH_2$.

principal quantum number (*n*) The number that indicates which shell an electron occupies in the atom. Electrons with $n = 1$ occupy the first (lowest-energy) shell; electrons with $n = 2$ occupy the second shell counting outward from the nucleus, and so forth.

prism A piece of colorless, clear glass or plastic that has a triangular cross-section and is used to separate white light into its various colors.

product A substance produced in the course of a chemical reaction by the transformation of one or more reactants. Compare reactant.

proof A unit used to indicate the concentration of ethanol in alcoholic beverages; equal to twice the percent by volume of ethanol in the beverage.

protein A biopolymer containing fifty or more amino acid monomer units linked by peptide linkages. Compare dipeptide; polypeptide.

proton A subatomic particle that has a mass approximately equal to that of the hydrogen atom, carries an electric charge of +1, and is located in the nucleus of an atom.

pure substance Matter with constant composition that consists of just one element or compound in just one chemical form.

quantum jump The movement of an electron in an atom from one allowed energy level to another.

quantum mechanics See quantum physics.

quantum physics Physics that takes into account the fact that energy is quantized and therefore can be used to describe the behavior of atoms and subatomic particles. Compare classical physics.

R group notation A shorthand method for drawing functionalized hydrocarbon molecules, in which the hydrocarbon chain is represented by the symbol R.

radioactive dating The process of determining an object's age by measuring the amount of some radioactive isotope contained in the object and then using that information plus the half-life of the isotope to calculate age.

radioactive decay The spontaneous change that an unstable nucleus undergoes.

radioactive isotope An isotope whose nucleus is unstable and therefore tends to undergo a nuclear reaction in which it changes spontaneously (without the addition of energy from an external source) to the nucleus of an isotope of some other element.

rare earth element The elements with atomic numbers 58 through 71.

rare gas A synonym for noble gas.

rate constant (*k*) A numeric constant, specific to each chemical reaction, that represents the product of all the inherent rate factors for the reaction.

rate-determining step The slowest step in any chemical reaction.

rate law An equation that expresses the rate of a chemical reaction as the product of the rate constant for the reaction times the concentration of each reactant raised to an exponent called an order.

rate of reaction Another term for reaction rate.

reactant A starting material for a chemical reaction; a substance that enters into the reaction and is transformed into one or more product substances. Compare product.

reaction See chemical reaction.

reaction coordinate The horizontal axis of a reaction energy profile; represents the progress of the reaction.

reaction energy profile A graph showing the relative energies of the reactants and products in a chemical reaction, with the horizontal axis representing the progress of the reaction and the vertical axis representing the energy of the reaction.

reaction intermediate A product of one step of a multistep chemical reaction that serves as a reactant in a subsequent step.

reaction mechanism The sequence of elementary steps in a chemical reaction.

reaction rate The rate (change in concentration of a reactant or product per unit time) at which a chemical reaction proceeds.

reagent See reactant.

redox reaction A shorthand name for an oxidation–reduction (electron transfer) reaction.

reducing agent The reactant in an electron transfer reaction that loses electrons. The reducing agent is oxidized during the reaction. Compare oxidizing agent.

reduction The gain of electrons by an atom, resulting in a decrease in oxidation state. Compare oxidation.

relative atomic weight Another name for atomic weight.

relaxation The process by which an excited electron falls to a lower energy level. The difference in energy between the two levels is often emitted as electromagnetic radiation and creates a line spectrum.

representative element Any element in groups IA, IIA, IIIA, IVA, VA, VIA, VIIA, and VIIIA of the periodic table.

resonance forms Two or more equally correct representations of the same molecule; the average of these representations is the actual structure of the molecule.

reverse rate In a reversible chemical reaction, the rate at which the reverse reaction takes place. Compare forward rate.

reverse reaction In a reversible chemical equation, the reaction reading from right to left (by convention). See forward reaction.

reversible reaction A chemical reaction in which the rate at which products are converted back to reactants is significant.

Rutherford model of the atom A model proposed by Ernest Rutherford, in which the atom is pictured as being mostly empty space surrounding a tiny, but massive, positively charged nucleus. The atom's electrons take up a small portion of the "empty" space, and fly around the nucleus. Compare Bohr model; Thomson model.

sacrificial metal A metal used in a structure to protect other metals in the structure from corrosion.

salt Another name for ionic compound.

salt bridge A connector between the two compartments of a galvanic cell, containing an electrolyte that provides anions to the compartment containing the metal being oxidized and cations to the compartment containing the metal ion being reduced.

saturated hydrocarbon A hydrocarbon in which all the carbon–carbon bonds are single bonds, which means that the maximum number of hydrogen atoms are present.

saturated solution A solution containing the maximum amount of a given solute.

scientific method The method of making observations, proposing theories, and testing those theories through experimentation.

scientific notation A convention in which a numeric quantity is written as a number between 1 and 10 multiplied by 10 raised to the appropriate power.

second law of thermodynamics A statement of the fact that the entropy of the universe increases whenever a system undergoes a spontaneous change.

secondary amine An amine in which the amine functional group is bonded to two hydrocarbon chains (which may or may not be identical) and one hydrogen atom. General formula: R_2-NH.

semi-metal A synonym for metalloid.

shell One of the distances at which electrons orbit the nucleus (corresponding to allowed electron energy levels). Each shell is assigned a principal quantum number $n = 1, 2, 3, \ldots$, with $n = 1$ being the shell of lowest energy.

SI See Système Internationale d'Unités.

side product An unwanted product of a chemical reaction, resulting from a side reaction.

side reaction A chemical reaction that takes place along with a main reaction, usually yielding unwanted products.

significant digits A synonym for significant figures.

significant figures The digits in a measured number (or a number calculated from measured numbers) that are known with certainty.

single bond A covalent bond between two atoms that is formed by one bonding pair of electrons.

solid A state of matter in which molecules or atoms are in contact with one another and remain in fixed positions. A solid maintains a constant shape and volume. Compare gas; liquid.

solid solution A solution in which both solute and solvent are solids at room temperature. The solution is prepared from molten solute and molten solvent.

solubility The maximum amount of a solute that dissolves in a given amount of solvent at a given temperature, and for a gaseous solute, at a given pressure.

solute In a solution, the substance dissolved in the solvent. Compare solvent.

solution A homogeneous mixture of a solute or solutes dissolved in a solvent.

solvation The process whereby solvent molecules in a solution completely surround a solute ion.

solvent In a solution, the substance present in the greatest amount. Compare solute.

spectator ion An ion present in the vessel in which a chemical reaction takes place that does not take part in the reaction.

stable isotope An isotope that is not radioactive, or, for the band of stability, a radioactive isotope that has a measurable half-life.

states of matter The physical states in which matter can occur: solid, liquid, and gas.

stock solution A solution of a known concentration that can be used to make more dilute solutions of the same solute.

stoichiometry The branch of chemistry that deals with measuring the quantities of reactants and products involved in chemical reactions.

strong acid An acid that dissociates completely when dissolved in water.

strong base A base that dissociates completely when dissolved in water.

strong electrolyte An electrolyte that dissociates completely when dissolved in water. Compare weak electrolyte.

strong force The very strong attractive force that holds an atomic nucleus together.

structural formula A drawing showing all the atoms and bonds in a compound arranged in their correct relative positions, as they actually appear in the molecule. Compare chemical formula.

subatomic particle Any particle that is more fundamental than the atom, such as protons, neutrons, and electrons.

sublimation A change from the solid state directly to the gaseous state.

subshell A subdivision of an electron shell, designated by letters $s, p, d, f, \ldots$, with the s subshell being the one of lowest energy

substitution reaction A chemical reaction in which one atom or group of atoms on a reactant molecule is replaced by an atom or group of atoms from another reactant molecule.

substrate For a given enzyme, a molecule that reacts at the active site of the enzyme.

superconductor Any material that offers zero resistance to the passage of electricity.

Système Internationale d'Unités (SI) The system of units used in science, including seven base units, with the five most important to chemistry being the meter for length, the kilogram for mass, the second for time, the kelvin for temperature, and the mole for amount of substance.

technology The application to practical problems of knowledge gained by doing science.

termolecular Involving three molecules.

tertiary amine An amine in which the amine functional group is bonded to three hydrocarbon chains (which may or may not be identical). General formula: R_3–N.

tetrahedron A four-sided polygon in which all sides are identical equilateral triangles.

theoretical yield For a chemical reaction, the maximum amount of product possible for a given amount of reactants. Compare actual yield.

theory A tentative explanation for a set of observations. A theory must be consistent with every one of the observations and is tested through experiments. Compare law.

Thomson model of the atom A model proposed by J. J. Thomson, in which the atom is pictured as a cloud of positive electricity with electrons embedded in the cloud. Compare Bohr model; Rutherford model.

trailing zero Any zero to the right of the final non-zero digit in a number.

transition metal Any element in groups IB through VIIIB of the periodic table.

transition state A transient state of the reactants in a chemical reaction; the reactants have absorbed the required activation energy and are therefore in a high-energy, unstable condition.

triple bond A multiple covalent bond between two atoms that is formed by three bonding pairs of electrons.

triprotic acid An acid that, when it dissociates in water, can yield a maximum of three H_3O^+ ions per molecule of acid.

tritium The isotope of hydrogen containing one proton and two neutrons; ^3_1H.

tunneling See electron tunneling.

uncertainty The amount by which a measured number may vary from the true value of the quantity being measured.

uncertainty principle A statement of the fact that it is impossible to know, at a given instant, both the velocity and the position of any particle.

unimolecular Involving one molecule.

unit analysis A problem-solving strategy in which data given in the problem statement are multiplied by appropriate conversion factors in order to obtain the correct units for the answer.

unsaturated hydrocarbon A hydrocarbon containing at least one carbon–carbon double or triple bond, which means that the carbon atoms in the multiple bond(s) have fewer than their full complement of hydrogen atoms.

valence electron An electron in the outermost shell of an atom.

valence shell The occupied shell of an atom that has the highest principal quantum number.

valence shell electron pair repulsion (VSEPR) theory A model used to predict molecular shape, the basic tenet being that the valence electrons in a molecule repel each other because they carry negative charges.

van der Waals attraction Intermolecular forces as a whole, including London forces.

vapor A synonym for gas.

vaporization A change from the liquid state to the gaseous state.

variable Any quantity in an algebraic equation that can change.

vector A quantity having both magnitude (size) and direction.

visible light The part of the electromagnetic spectrum between approximately 380 nm and 750 nm.

voltaic cell See galvanic cell.

VSEPR An acronym for valence shell electron pair repulsion.

wavelength The distance between any two adjacent comparable points on a wave, such as the distance from one crest to the next or the distance from one trough to the next.

weak acid An acid that dissociates only partially when dissolved in water.

weak base A base that yields a very low concentration of OH^- ions when dissolved in water.

weak electrolyte An electrolyte that dissociates only partially when dissolved in water. Compare strong electrolyte.

weight A measure of how strongly the Earth's gravitational force (or the gravitational force exerted by any other body) pulls on an object. As the gravitational force changes, the object's mass stays constant but its weight changes. Compare mass.

weighted average atomic weight The atomic weight of an element, obtained by multiplying the atomic weight of each isotope by the fraction of its naturally occurring abundance and then summing all the products.

white light Light containing all the wavelengths of the visible part of the electromagnetic spectrum, from 380 to 750 nm.

Z The symbol used to represent atomic number.

Selected Answers

Note: Blue numbers in the text indicate practice problems and end-of-chapter problems for which answers are provided here. For complete solutions to these problems, see the *Study Guide and Solutions Manual* that accompanies this text.

Chapter 1

1.2 Heterogeneous mixture
1.5 (d) Sulfur dioxide, SO_2, and (e) ammonia, NH_3, are compounds; (a) sulfur, S_8, is an element; (b) and (c) are mixtures.
1.8 (c) Freezing
1.10 Heat the metal until it begins to melt; then use a thermometer to measure the temperature, and compare it to the melting point of gold.
1.12 Methane (CH_4) and oxygen (O_2) are the reactants; water (H_2O) and carbon dioxide (CO_2) are the products.
1.16 If the theory is used to predict the results of proposed experiments and the experimental data agree with what the theory predicts, this is good evidence that a theory is correct.
1.17 Science is the experimental investigation and explanation of natural phenomena. Technology is the application of scientific knowledge.
1.18 Chemistry is the study of matter and the transformations that matter undergoes.
1.22 Matter is anything that has mass and occupies space (or volume).
1.23 Yes. For example, salt water is a mixture of NaCl and H_2O.
1.25 Heterogeneous mixture
1.26 (a) Heterogeneous mixture (b) Solution (c) Solution
1.28 An element is the fundamental form of matter that consists of atoms. There are 115 known elements.
1.30 The smallest piece of most elements is an atom. However, there are a few elements, such as oxygen (O_2), nitrogen (N_2), and sulfur (S_8), whose smallest unit contains 2 or more atoms of the element.
1.31 Lead, Pb; molybdenum, Mo; tungsten, W; chromium, Cr; mercury, Hg
1.33 Ti, titanium; Zn, zinc; Sn, tin; He, helium; Xe, xenon; Li, lithium
1.35 An elemental substance contains only one type of atom, whereas a compound contains two or more different types of atoms.
1.36 The formula tells how many of each type of atom are present in the smallest possible piece of the substance.
1.38 F_2, P_4, Ar, and Al are elemental substances; $BrCl_3$, C_2H_2, HCl, and Al_2O_3 are compounds.
1.40 H_2O_2

1.42 $C_6H_{12}O_6$
1.44 Solid, liquid, and gas
1.46 Condensation
1.48 $-117.3°C$
1.50 Melting point of water (ice): 0°C or 32°F; boiling point of water: 100°C or 212°F
1.53 Physical change
1.55 Chemical reactions represent the chemical transformations of one set of substances into different substances.
1.57 Chemical change
1.58 A chemical change has taken place, and the product of the reaction (NH_3) has different properties from the reactants (N_2 and H_2).
1.59 A law is a statement that summarizes data. A theory is a model that attempts to explain why a law is true.

Chapter 2

2.2 Ike is more accurate. Mike is more precise.
2.3 Jack
2.5 ± 0.05 gal
2.6 (b) ± 0.005 V
2.8 600.
2.9

	Number of significant figures	Uncertainty
10.0	3	0.05
0.004 60	3	0.000 005
123	3	0.5

2.11 0.473
2.12 47,325
2.14 0.002 35
2.15 6000
2.17 $4.710\,000\,0 \times 10^{13}$
2.18 $4.710\,000 \times 10^{13}$
2.20 44
2.21 $220. \times 3 = 660.$, but this must be rounded to 700 or 7×10^2.
2.23 (a) 6.1×10^2 lb (b) 6.11×10^2 or 611 lb
2.24 8×10^1
2.26 1556, rounded from 1555.801
2.27 142
2.29 4.736 km
2.30 25 mm
2.32 2500 mL
2.33 246.7 mL

2.34 K = °C + 273.15; therefore, °C = K − 273.15;
263.5 K − 273.15 = −9.6°C. To convert to °F:

$$°F = \frac{9}{5}°C + 32 = 14.6°F$$

2.36 $\dfrac{4.70\ g}{1000\ mm^3} = \dfrac{470\ g}{1.00\ cm^3} = 4.70\ g/cm^3 = 4.70\ g/mL$

2.37 $\dfrac{500.0\ g}{150.5\ mL} = 3.322\ g/mL$

2.39 $\dfrac{1\ day}{24\ hours}$ and $\dfrac{24\ hours}{1\ day}$

2.40 $50.0\ miles \times \dfrac{1\ hour}{600\ miles} = 0.0833\ hour$

2.41 $\dfrac{600\ miles}{1\ hour} \times 50.0\ hours = 3.00 \times 10^4\ miles$

2.43 $500.0\ L \times \dfrac{1000\ mL}{1\ L} \times \dfrac{0.001\ 30\ g}{1\ mL} = 650.\ g$ or

$6.50 \times 10^2\ g;\ 6.50 \times 10^2\ g \times \dfrac{1\ kg}{1000\ g} = 0.650\ kg$

2.44 $1.50\ lb \times \dfrac{453.6\ g}{1\ lb} \times \dfrac{1\ mL}{11.4\ g} = 59.7\ mL$

2.45 $6955\ g\ flour \times \dfrac{1\ cup\ flour}{120.0\ g} \times \dfrac{1\ cake}{6\ cups\ flour}$

= 9.660 cakes;
it will be possible to bake 9 cakes (you can't bake a partial cake).

2.49 The 3 in "3 feet in a yard" is an exact number. The 3 in "a piece of wood is 3 feet long" comes from a measurement and therefore has some uncertainty associated with it.

2.51 You should choose the accurate result.

2.53 The person with the tape measure.

2.55 The last digit written in a number is assumed to be where the uncertainty lies.

2.57 (a) 12.60 ±0.005 cm   (b) 12.6 ±0.05 cm
(c) 0.000 000 03 ±0.000 000 005 inch
(d) 125 ±0.5

2.58 (a) Four   (b) Three   (c) One   (d) Three

2.61 (a) 12.20̲2   (c) 20̲5   (d) 0.01̲0

2.63 It is not clear whether the number 30 has one or two significant figures. The number 30. has two significant figures.

2.65 (a) 56.0 (three significant figures)
(b) 0.000 25 (two significant figures)
(c) 5,600,000 (four significant figures)
(d) 2 (one significant figure)

2.67 (a) 3×10^1 ft   (b) 3.0×10^1 ft   (c) 3.00×10^1 ft

2.69 (a) 2.26×10^2 (b) 2.260×10^2 (c) 5.0×10^{-10}
(d) 3×10^{-1} (e) 3.0×10^{-1} (f) 9.00×10^8
(g) $9.000\ 006 \times 10^8$

2.71 102 inches; ±0.5 inch

2.73 3.873 14 miles (six significant figures because 5280 is an exact number)

2.74 (a) 2.55×10^5 km
(b) 1.000×10^{18} J

(c) 2.11×10^2 m
(d) 4.00×10^4 L

2.76 L; mL; they are volumes more commonly encountered in everyday situations and in the laboratory

2.78 To eliminate confusion

2.79 (a) 2.31×10^9 m   (b) 5.00×10^{-6} m   (c) 1.004 m
(d) 5.00×10^{-12} m   (e) 2.5×10^2 m

2.81 Celsius and Fahrenheit can have negative values; Kelvin cannot have negative values because the zero point on the Kelvin temperature scale is absolute zero.

2.82 (a) 72.5°F; 295.6 K   (b) −19.4°C; 253.7 K
(c) −273.2°C; −459.7°F   (d) 149.0°F; 338.2 K

2.85 (a) Left cylinder: ±0.05 mL;
right cylinder: ±0.5 mL
(b) 98 + 1.2 = 99 ±0.5

2.87 The student who reports 1.5 is scolded.

2.90 997 g

2.92 9×10^2 g

2.94 Both students will measure the same density.

2.96 1.08×10^5 sec

2.98 $0.007 083 per second

2.100 0.264 gal

2.103 (a) 101.4 cm   (b) 1.042×10^6 cm^3
(c) 0.106 g/mL

2.104 In order for both sides to remain equal, whatever is done to one side of an equation must be done to the other.

2.106 $x = z - y$

2.108 57.5 g

CHAPTER 3

3.5 The law of conservation of matter requires that the total mass of the substances produced equal the mass of coal plus the mass of oxygen reacting. The coal seems to disappear because it is being converted to carbon dioxide, a colorless, odorless gas. If the carbon dioxide were captured and weighed, the mass would equal the mass of the coal and oxygen reacting to produce them.

3.6

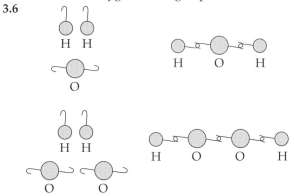

3.8 $^{79}_{35}Br,\ ^{81}_{35}Br$

3.9 The atoms of both isotopes have 35 electrons.

3.10

	$^{14}_{7}N$	$^{24}_{12}Mg$	$^{23}_{11}Na$	$^{59}_{26}Fe$
Mass number	14	24	23	59
Atomic number	7	12	11	26
Number of protons	7	12	11	26
Number of neutrons	7	12	12	33
Number of electrons	7	12	11	26

3.12 $\dfrac{48.0}{16} = 3.$

3.13 (a) $100\% - 75.77\% = 24.23\%$

(b) $\left(34.969 \times \dfrac{75.77}{100}\right) + \left(36.966 \times \dfrac{24.23}{100}\right) = 35.45$

(c) $^{37}_{17}Cl$ is $\dfrac{36.966}{34.969} = 1.057$ times heavier than $^{35}_{17}Cl$.

3.15 No

3.16 Similar physical and chemical properties

3.18 Iodine (I) and astatine (At); periodicity for I is 18 and for At is 32

3.19 The periodicity would be 2, 8, 8, 8, etc., instead of 2, 8, 8, 18, 18, 32, 32.

3.21 $O < S < Mg < Sr < Rb$

3.22 $^{56}_{26}Fe^{3+}$; group VIIIB

3.23 $^{16}_{8}O^{2-}$; group VIA

3.27 Radio waves; this is very low-energy electromagnetic radiation

3.29 (a) %S: $\dfrac{94.08 \text{ g S}}{100.0 \text{ g sample}} \times 100\% = 94.08\%$ S

%H: $\dfrac{5.92 \text{ g H}}{100.0 \text{ g sample}} \times 100\% = 5.92\%$ H

(b) The law of constant composition

3.31 (a) %I: $\dfrac{126.9 \text{ g I}}{162.4 \text{ g compound A}} \times 100\% = 78.1\%$ I

%Cl: $100.0\% - 78.1\%$ I $= 21.9\%$ Cl

(b) %I: $\dfrac{126.9 \text{ g I}}{233.3 \text{ g compound B}} \times 100\% = 54.4\%$ I

%Cl: $100.0\% - 54.4\%$ I $= 45.6\%$ Cl

(c) For the same amount of I (126.9 g):

$\dfrac{106.4 \text{ g Cl in compound B}}{35.45 \text{ g Cl in compound A}} = \dfrac{3}{1}$

(d) ICl_3, since compound B has 3 times as much Cl as compound A for a given amount of I

3.33 The law of multiple proportions states that when two elements combine to form more than one compound, then for a fixed mass of one, the masses of the other elements always form a ratio that is expressible in small, whole numbers. Since the numbers are always small whole numbers, this means that the atoms must be whole atoms.

3.35 (a)

Compound	Mass of Na present
A	11.50 g
B	45.98 g

(b)

Compound	%Na	%O
A	59.0	41.0
B	74.2	25.8

(c) Divide the masses of compound B by 2 to obtain: mass of sample = 30.99 g, mass of O = 8.00 g, mass of Na = 22.99 g. The ratio of the mass of Na in compound A to the mass of Na in compound B is $\dfrac{22.99}{11.50} = \dfrac{2}{1}$.

Since 1 and 2 are small whole numbers, the law of multiple proportions is satisfied.

3.37 The positively charged α particles were repelled by the nucleus.

3.39 The positive charges would have been evenly spread throughout the atom and not present in a tiny area in the center. This would mean that the structure of the atom would be more consistent with Thomson's "plum pudding" model.

3.41 Yes. In a neutral atom the number of electrons is equal to the number of protons.

3.43

	$^{15}_{8}O$	$^{16}_{8}O$	$^{37}_{17}Cl$	$^{23}_{11}Na$
Mass number	15	16	37	23
Atomic number	8	8	17	11
Number of protons	8	8	17	11
Number of neutrons	7	8	20	12
Number of electrons	8	8	17	11

3.45 The atomic number and the symbol do not agree. Carbon has an atomic number of 6, not 7. The element whose atomic number is 7 is nitrogen.

3.47 $^{1}_{1}H$

3.49 The atomic weight of an element is the weighted average of the masses of all its naturally occurring isotopes. The mass number of an istotope is the sum of the protons and neutrons in an atom.

3.51 $^{12}_{6}C$ is the only isotope whose atomic mass and mass number are equal. This is true because chemists have assigned $^{12}_{6}C$ a weight of exactly 12.

3.53 An "average" titanium atom is 3.99 times heavier than the $^{12}_{6}C$ atom: $\dfrac{47.88}{12} = 3.99$.

3.56 (a) 235

(b) The lighter elements would be a very small fraction of 235.

3.57 (a) 49.31%

(b) Atomic weight of Br

$= \left(78.918\,336 \times \dfrac{50.69}{100}\right) +$

$\left(80.916\,289 \times \dfrac{49.31}{100}\right) = 79.90$

3.58 The naturally occurring hydrogen on the other planet must contain more of the heaver $^{2}_{1}H$ isotope than on Earth.

3.60 (a) According to atomic weights
(b) He discovered repeating behavior every 8 elements.
(c) Mendeleev's ordering was by atomic weight; the modern ordering is by atomic number.

3.62 Periodicity refers to the fact that the chemical properties repeat themselves every 8, 18, and 32 elements.

3.63 8

3.65 IA, alkali metals; IIA, alkaline earth metals; VIA, chalcogens; VIIA, halogens; VIIIA, noble gases

3.67 A group is a vertical column of elements with similar chemical properties. A period is a horizontal row of elements.

3.69 (a) $BeCl_2$, $CaCl_2$, $SrCl_2$, $BaCl_2$, and $RaCl_2$
(b) $BeBr_2$, $MgBr_2$, $CaBr_2$, $SrBr_2$, $BaBr_2$, and $RaBr_2$
(c) The chemical behavior of elements in the same group will be similar.

3.70 They are among the most unreactive substances known, and they are the only group in the periodic table that contains only gases.

3.72 The transition metal portion is 10 elements wide; the rare earths portion is 14 elements wide.

3.73 The change from 8 to 18 is due to the transition metals; the change from 18 to 32 is due to the lanthanides and actinides.

3.75 As you go across the periodic table from left to right, one more proton is being added to the nucleus and one more electron is being added outside the nucleus. The added positive charge is pulling the elements in closer to the nucleus, causing the size of the atoms to decrease.

3.77 $F < Ar < S < Br$

3.79 The $^{15}_{7}N$ anion is negatively charged. Therefore there are 3 more electrons than protons. The number of electrons is 10. The full atomic symbol is: $^{15}_{7}N^{3-}$

3.81

	$^{15}_{8}O^{+}$	$^{27}_{13}Al^{3+}$	$^{31}_{15}P^{3-}$	$^{58}_{28}Ni^{+}$
Mass number	15	27	31	58
Atomic number	8	13	15	28
Number of protons	8	13	15	28
Number of neutrons	7	14	16	30
Number of electrons	7	10	18	27
Charge on ion	+1	+3	−3	+1

3.83 True

3.85 As you go down a group, the size of the atom increases and the electron is farther away from the nucleus. It thus feels less pull from the nucleus, is easier to remove, and requires less energy. As you go across a period, the atomic size decreases, and the electron is closer to the positively charged nucleus. More energy is thus needed to remove the electron.

3.87 Mg is the most difficult to ionize; K has the smallest ionization energy.

3.89 (a) Sodium easily loses an electron to form the cation Na^{+}, and Cl easily gains an electron to form the anion Cl^{-}.
(b) Less reactive
(c) More reactive

3.92 Since the energy is inversely related to wavelength, the shorter the wavelength the higher the energy. Therefore X rays, with shorter wavelength, have higher energy.

3.94 They are at the high-energy end of the electromagnetic spectrum and can cause damage to cells and tissues.

3.96 8.3 minutes for both

3.98 1.00 nm $= 1.00 \times 10^{-9}$ meters; 3.9×10^{-8} inch

3.100 $E = \dfrac{(6.63 \times 10^{-34} \text{ J} \cdot \text{sec})(3.00 \times 10^{8} \text{ m/sec})}{10 \text{ m}}$
$= 1.99 \times 10^{-26}$ J

3.102 A line spectrum consists of individual lines of certain colors separated by regions of blackness. A continuous spectrum consists of all colors blended into each other with no regions of blackness.

3.104 It tells us that something inside the atom is allowed to possess only certain energies, as opposed to any energy.

CHAPTER 4

4.2 2 more electrons could fit in the $n = 3$ shell

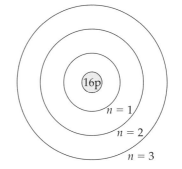

4.3 An electron in a shell with a low value of n is closer to the nucleus and feels a stronger attraction than one that is farther from the nucleus. Thus, the closer electron is more stabilized than one located in a shell with a higher value of n that is farther away from the nucleus.

4.4 $2 \times 5^2 = 50$ electrons

4.6

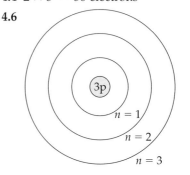

4.7

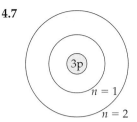

4.8 (a) Carbon   (b) Anion with −2 charge
(d)

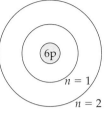

4.9 The $n = 1$ shell has 3 electrons in this diagram; the maximum is 2 electrons for this shell.
The $n = 2$ shell has only 6 electrons; the maximum is 8 for this shell, and it should be filled before an electron is placed in the $n = 3$ shell.

4.11 As has a total of 33 electrons:
$1s^2 2s^2 2p^6 3s^2 3p^6 3d^{10} 4s^2 4p^3$; yes, there are 5 valence electrons, and the group number is VA.

4.12 Sc has 21 electrons: $1s^2 2s^2 2p^6 3s^2 3p^6 3d^1 4s^2$

4.14 $1s^2 2s^2 2p^6 3s^2 3p^6 4s^2 3d^{10} 4p^6$; it is proper for Kr to be in group VIIIA because it has 8 valence electrons.

4.15 Pd has 46 electrons:
$1s^2 2s^2 2p^6 3s^2 3p^6 4s^2 3d^{10} 4p^6 5s^2 4d^8$

4.17 $1s^2 2s^2 2p^6 3s^2 3p^6 4s^2 3d^{10} 4p^6 5s^2 4d^{10} 5p^6 6s^2 4f^{14} 5d^{10} 6p^6 7s^2$;   [Rn]$7s^2$

4.18 $1s^2 2s^2 2p^6 3s^2 3p^6 4s^2 3d^{10} 4p^6 5s^2 4d^{10} 5p^6 6s^2 4f^{14} 5d^{10} 6p^6 7s^2 6d^1 5f^3$;   [Rn]$7s^2 6d^1 5f^3$

4.20 The highest value of n is 4, so it is in the fourth period; the element is vanadium (V).

4.21 Period 6; group IA; the element is cesium (Cs).

4.23

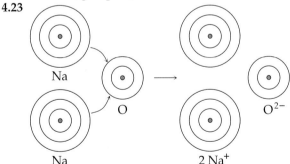

4.24 Ba is in group IIA and will lose 2 electrons to become Ba^{2+}.
F is in group VIIA and will gain 1 electron to become F^-.
The formula of the compound is BaF_2.

4.25 Al is in group IIIA and will lose 3 electrons to become Al^{3+}.
O is in group VIA and will gain 2 electrons to become O^{2-}.
The formula of the compound is Al_2O_3.

4.26 You would appear at the starting line, then a certain time later you would instantaneously appear some distance ahead, and later still some farther distance ahead, until finally you would appear instantaneously at the finish line.

4.28 "Quantized energy" means that the energy can have only certain allowable values and no values in-between.

4.30 Energy and stability are inversely related to each other. The higher the energy of an object, the less stable it is; the lower the energy, the more stable.

4.31 (a) is the most stable situation; (b) and (c) are least stable.

4.33 As n increases, an electron's energy and distance from the nucleus increase.

4.35 Electron shell

4.36 Energy is required

4.39 The lower shells are of lower energy; the electrons will go into lower-energy positions before beginning to go into higher-energy (less stable) positions.

4.40 Excited state, because the $n = 2$ energy shell is not filled, yet there is an electron in the $n = 3$ shell.

4.43 The energy of an electron depends on its distance from the nucleus. If an electron can have only certain energies, then it can be only certain specific distances from the nucleus.

4.45 Atoms in the same group have identical valence shell configurations.

4.46 Atomic line spectra are produced when energy is given off as a result of an excited electron falling back into a lower-energy shell.

4.49 (a) 12.1 eV   (b) 12.1 eV   (c) Ultraviolet

4.51 The energy of the shells would have to be continuous, not quantized.

4.53

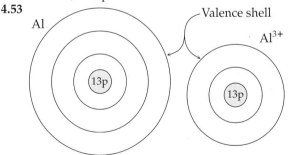

4.54 (a) There is nothing wrong with diagram (1). The fact that it is lacking 1 electron just makes it a cation.
Diagram (2) is not a ground-state electron configuration, but it is an allowable excited-state configuration. The $n = 2$ shell would contain 5 electrons in the ground-state configuration.
Diagram (3) has 3 electrons in the $n = 1$ shell, which is impossible. This level holds a maximum of 2 electrons.

4.54 (b) The full atomic symbol for (1) is $^{14}_{7}N^{+}$;
for (2) is $^{14}_{7}N$.
(c) Diagram (1) is a ground state; diagram (2) is an excited state.

4.55 The number of valence electrons is equal to the group number for the representative elements.

4.58 (a) Since the electrons are being placed in the same valence shell, one might expect the size to stay the same.
(b) The atomic size decreases because 1 electron is added to the valence shell and 1 proton is added to the nucleus. The added proton makes the nucleus more positive, which causes it to more strongly attract the surrounding electrons. The increased pull on the electrons shrinks the atom.

4.59 The simple Bohr model predicts that there should be 9 electrons in the $n = 3$ shell. However, the chemical behavior of K suggests that it has 1 valence electron in the $n = 4$ shell.

4.61 (a) $n = 1, 2, 3, 4$, etc.
(b) s, p, d, f, etc.
(c) The number of subshells in a given shell is equal to n; for example, $n = 1$ has one subshell (s); $n = 2$ has two subshells (s and p); $n = 3$ has three subshells (s, p, and d); etc.

4.63 In the $4s$ subshell instead of in the $3d$ subshell

4.65 (a) $1s^2 2s^2 2p^1$
(b) $1s^2 2s^2 2p^6 3s^2 3p^6 4s^2 3d^1$
(c) $1s^2 2s^2 2p^6 3s^2 3p^6 4s^2 3d^7$
(d) $1s^2 2s^2 2p^6 3s^2 3p^6 4s^2 3d^{10} 4p^4$
(e) $1s^2 2s^2 2p^6 3s^2 3p^6 4s^2 3d^{10} 4p^6 5s^2 4d^6$

4.68 (a) $[Xe]6s^2$
(b) $[Xe]6s^2 4f^{14} 5d^4$
(c) $[Xe]6s^2 4f^{14} 5d^{10} 6p^2$
(d) $[Xe]6s^2 5d^1 4f^2$
(e) $[Rn]7s^2 6d^1 5f^2$

4.69 (a) Group VA; period 2
(b) Group IA; period 3
(c) Group VIIA; period 4

4.70 (a), (b), (c), (d), and (e) are all incorrect; (f) is correct.
(a) is incorrect because $2s$ should come after $1s$.
(b) is incorrect because the $2s$ subshell should be followed by the $2p$ subshell.
(c) is incorrect because the $1s$ subshell must come first.
(d) is incorrect because the $2p$ subshell holds a maximum of 6 electrons.
(e) is incorrect because the $2p$ subshell would accommodate 6 electrons before the $3s$ subshell would be used.

4.71 (a) 5   (b) 1   (c) 2   (d) 8   (e) 2

4.73 O: $1s^2 2s^2 2p^4$; O^{2+}: $1s^2 2s^2 2p^2$; O^{2-}: $1s^2 2s^2 2p^6$;
O^{2-} should be found in most compounds of oxygen because the octet rule is satisfied.

4.75 The s block is 2 elements wide because each s subshell holds 2 electrons.

The p block is 6 elements wide because each p subshell holds 6 electrons.
The d block is 10 elements wide because each d subshell holds 10 electrons.
The f block is 14 elements wide because each f subshell holds 14 electrons.

4.76 The octet rule states that elements will react in such a way as to put 8 electrons in the valence shell. Having 8 valence electrons gives them a noble gas valence electron configuration, and makes them exceptionally stable.

4.78 By gaining enough electrons to have 8 in their outer shell

4.80 Each lithium atom has 1 valence electron, which it will lose to obtain an octet. Each nitrogen atom has 5 valence electrons, and will gain 3 to have an octet. Thus, each nitrogen atom will need three lithium atoms, and the formula of the compound will be Li_3N.

4.82 (a) $^{23}_{11}Na$ and $^{31}_{15}P$
(b) Na is the metal, and P is the nonmetal.
(c) Na_3P
(d)

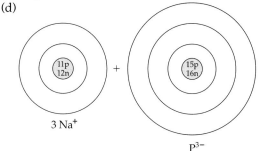

4.84 Hydrogen will sometimes lose its valence electron, similar to what the other elements in group IA do to obtain an octet. However, when hydrogen loses its electron, it does not obtain an octet; it is left with no electrons.

4.85 Aluminum (Al) has 3 valence electrons, which it will lose to form a +3 cation. Both oxygen and sulfur have 6 valence electrons and will gain 2 electrons to form −2 anions. Since Al needs to lose 3 electrons and oxygen and sulfur both need to gain 2 electrons, each compound will contain 2 Al^{3+} cations and 3 anions. The formulas are similar because both oxygen and sulfur have the same number of valence electrons.

4.87 True; the H^+ cation has no electrons.

4.89 It is true that Mg^{2+} and Na^+ have identical configurations, but they do not have identical properties. The ions have different charges, different sizes, and different chemical reactivities.

4.90 The group number is equal to the number of valence electrons the metal will lose. The cation will have a positive charge equal to the group number.

4.92 The charges on the cation and anion tell how many electrons are lost and gained, respectively, to form the ions. The formula of the compound is simply the minimum number of each ion needed to have an equal number of electrons transferred.

4.93 Be; the greater nuclear charge for Be (+4) compared to Li (+3) pulls the electrons in closer to the nucleus, shrinking the subshells.

4.96 Na; atoms get larger going down a group.

4.99 Si is the smallest, next is Mg, then Na, and then Rb is the largest.

4.100 We cannot know precisely the position and the direction in which an electron is traveling.

4.102 An orbital is a region of space around the nucleus of an atom in which it is most probable that the electron will be found.

4.103

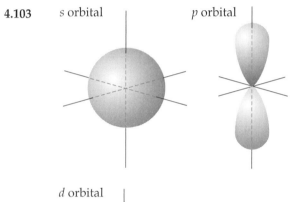

s orbital   p orbital

d orbital

4.104 The picture would be very fuzzy, indicating the probability that you would be in a certain region of the photograph.

CHAPTER 5

5.2 PH_3

5.3 $SiBr_4$

5.5 H:Ḟ:

5.6 Oxygen has 2 unpaired electrons. To achieve an octet, it will form two bonds with hydrogen to give H:Ö:H

5.8 H—Ö—Ċl:

5.9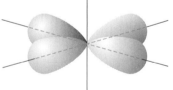

5.11
$$H-\underset{\underset{H}{|}}{\overset{\overset{H}{|}}{C}}-C\equiv C-H$$

5.12 H—C≡N:

5.13
$$H-\underset{\underset{H}{|}}{\overset{\overset{H}{|}}{C}}-\overset{\overset{:O:}{\|}}{C}-\underset{\underset{H}{|}}{\overset{\overset{H}{|}}{C}}-H$$

5.15
$$:\ddot{O}-\overset{\overset{:O:}{\|}}{S}-\ddot{O}:$$

5.16
$$\left[:\ddot{O}-\overset{\overset{:\ddot{O}:}{|}}{S}-\ddot{O}:\right]^{2-}$$

5.17 $[:N\equiv O:]^{+}$

5.18 No, since the atoms are arranged differently

5.20 EN C = 2.5, EN H = 2.1; ΔEN = 0.4; this puts us right on our somewhat arbitrary dividing line—since C–H bonds are considered nonpolar, we call them nonpolar covalent.

5.21 EN Si = 1.8, EN O = 3.5; ΔEN = 1.7; a polar covalent

5.22 Since Be is at the top of group IIA, she knows that Be is more electronegative than Ba.

5.24

Formula	Cation	Anion	Name
CaF_2	Ca^{2+}	F^-	Calcium fluoride
CsBr	Cs^+	Br^-	Cesium bromide
Al_2S_3	Al^{3+}	S^{2-}	Aluminum sulfide
K_2O	K^+	O^{2-}	Potassium oxide

5.25 All the group IIA bromides contain two bromide ions.

5.27 Titanium(III) chloride

5.28 SnF_2

5.30 Na_2CO_3

5.31 Calcium hypochlorite

5.32 $Mg(HCO_3)_2$

5.34 Nitrogen monoxide

5.35 Phosphorus pentachloride

5.36 P_4O_{10}

5.38 No; molecules that contain atoms of the same element are elemental substances (for example, O_2).

5.45 Because the electrons are attracted to both nuclei; this extra positive–negative attractive force is what bonds the atoms to each other.

5.47 Due to the extra attractive force of the covalent bond between the H atoms

5.50 Only the valence electrons are used for bonding because any inner electrons are too close to and too strongly attracted to the nucleus to be shared with other nuclei.

5.53 We mean that the electrons shared by two atoms are counted as part of the valence electrons for each atom that is sharing them. Hence, these bonding electrons are counted twice.

5.55 1 electron

5.56 The nitrogen atom is correct. The oxygen atom should be ·Ö:. Fluorine should have 7 valence electrons and should be ·F̈:.

5.61 Oxygen has 6 valence electrons; hence, it combines with two hydrogens to form H_2O. HO would leave the oxygen short of 1 electron.

5.63 ·P̈· and ·B̈r: combine to form

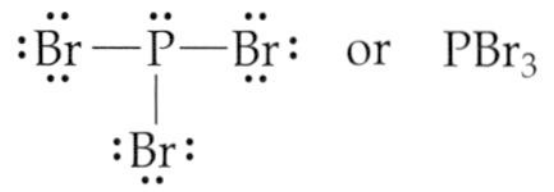

:B̈r—P̈—B̈r: or PBr_3
     |
    :B̈r:

5.65

H   H
| |
H—C—Ö—C—H   There are 8 bonding pairs
| | and 2 lone pairs.
H   H

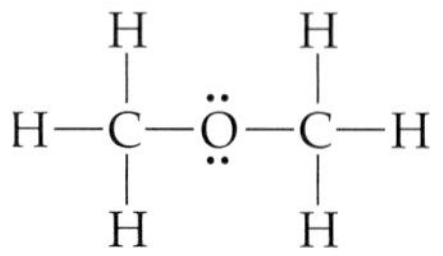

5.70 Three bonds; the atom has 5 valence electrons and needs 3 to achieve an octet.

5.71

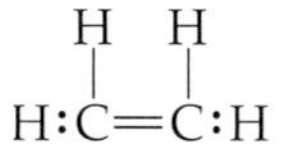

H   H
| |
H:C=C:H

5.75 There are two resonance forms:

:Ö—Ö=Ö: and :Ö=Ö—Ö:

5.77 There are two resonance forms for ozone. The bond on the left is single in one resonance form and double in the other. This averages out to 1.5 bonds.

5.79 Yes, the structure shown has too many electrons. The correct structure is :C≡O:

5.83 :O≡O:

5.85

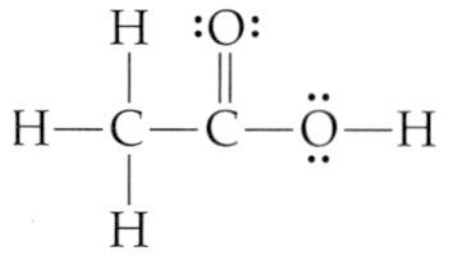

H :O:
| ‖
H—C—C—Ö—H
|
H

5.90 8 valence electrons on each carbon atom and 2 valence electrons on each hydrogen atom

5.92 An ionic bond is the attractive force between oppositely charged ions that were formed by a transfer of electrons. It is similar to a covalent bond in that both have approximately the same strength and both are due to the attraction between relatively large positive and negative charges. An ionic bond differs from a covalent bond in that an ionic bond is due to a transfer of electrons, whereas a covalent bond is due to a sharing of electrons.

5.94 Between a metal and a nonmetal

5.96 (a) $MgBr_2$ (b) BeO   (c) NaI

5.101 Fluorine (F) is the most electronegative element; francium (Fr) is the least electronegative element; they would form the ionic compound FrF.

5.106 The electronegativities are C, 2.5 and H, 2.1. The ΔEN is 0.4. This is a borderline case that is usually interpreted as nonpolar covalent.

5.108 H_2O

5.111 X is more electronegative because it carries the partially negative charge, indicated by δ−.

5.114 If the two atoms are the same

5.117 The suffix "ide" indicates the negative part in a binary ionic compound; the negative ion gets this suffix.

5.121 (a) Calcium nitride   (b) Aluminum fluoride
(c) Sodium oxide   (d) Calcium sulfide

5.124 (a) CuCl is copper(I) chloride or cuprous chloride; $CuCl_2$ is copper(II) chloride or cupric chloride.
(b) $Fe(OH)_2$ is iron(II) hydroxide or ferrous hydroxide; $Fe(OH)_3$ is iron(III) hydroxide or ferric hydroxide.

CHAPTER 6

6.2 Tetrahedral

6.3 Linear

6.4 Trigonal planar

6.6 Tetrahedral

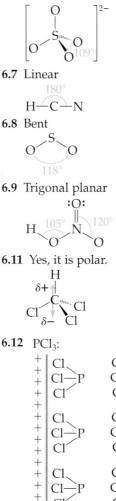

6.7 Linear

180°
H—C—N

6.8 Bent

S
O   O
118°

6.9 Trigonal planar

:O:
‖
H   120°
105° N
O   O

6.11 Yes, it is polar.

H
δ+ |
C····Cl
Cl   Cl
δ− Cl

6.12 PCl_3:

Cl   Cl
Cl—P   Cl—P
Cl   Cl

Cl   Cl
Cl—P   Cl—P
Cl   Cl

Cl   Cl
Cl—P   Cl—P
Cl   Cl

CHCl₃:

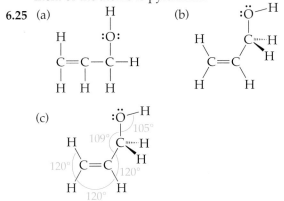

6.13 Because the bonding pair electrons can be farther apart

6.16

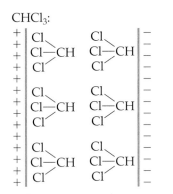

6.18 (a) (1)   (2)

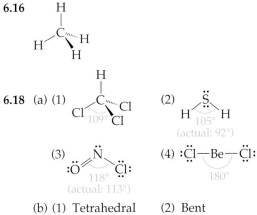

(3)   (4) :Cl̈—Be—C̈l:

(b) (1) Tetrahedral   (2) Bent
      (3) Bent   (4) Linear

6.20 All the electrons in a multiple bond occupy roughly the same region of space.

6.21 In the case where the bond angles are approximately 120°, the central atom has two bonding pairs and one lone pair. In the case where the bond angles are approximately 109°, the central atom has two bonding pairs and two lone pairs.

6.23 The shape of a molecule is determined by the arrangement of the *atoms* in the molecule, ignoring the position of the lone pairs. The arrangement of the atoms is pyramidal.

6.25 (a)   (b)

(c)

6.26 The shape of a molecule depends on the number of electron groups around the central atom.

6.28

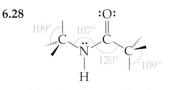

6.30 Yes, if we know what bonds in the molecule are polar (based on the relative electronegativities of the connected atoms) and the shape of the molecule.

6.32 (a) HCl, because the electronegativity difference is greater in HCl than in HBr.
(b) HCl, because the dipole moment is greater.

6.34 Because a dipole moment has a direction associated with it

6.37 The linear shape of the molecule results in the two bond dipoles canceling each other.

6.38 The NO₂⁻ ion is bent, so the bond dipoles do not cancel each other.

6.41 (a)   (b)

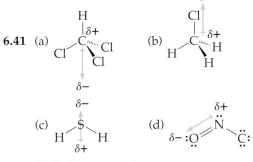

(c)   (d)

(e) BeCl₂ is nonpolar

6.43 No, dipole–dipole interactions are only a fraction of the interactions between ions because ions carry full charges whereas the dipole charges are partial charges.

6.45

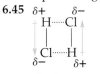

6.47 The lone pairs are important in determining the total number of electron groups around an atom and thus determining the shape of the molecule.

6.49 (a) :C̈l—Si—C̈l: (b)

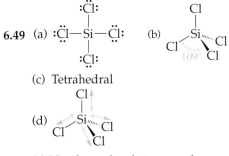

(c) Tetrahedral

(d)

(e) No, the molecule is nonpolar.

6.51 (a)   :O: (b)   :O:
         :Ö:S:Ö: ‖
         ̈ ̈ ̈   S
                        :Ö   Ö:
                           120°

6.51 (c) Trigonal planar   (d)

$$O = \overset{+}{S} \underset{O \quad O}{}$$

(e) No, the molecule is nonpolar.

6.53 (a)

$$H—\overset{..}{N}—\overset{..}{N}—H$$
$$\quad |\quad |$$
$$\quad H\ \ H$$

(b) All angles 105° (predicted by simple 2° compression rule)

(c)   (d) Yes

6.55 (a) $:N≡N—\overset{..}{\underset{..}{O}}:$ (b) $:N≡N—\overset{..}{\underset{..}{O}}:$ (c) Linear

180°

(d) $:N≡N—\overset{..}{\underset{..}{O}}:$ (e) Yes $:N≡N—\overset{..}{\underset{..}{O}}:$

Net

6.56 (a) NO_2 has an odd number of valence electrons: $5 + 6(2) = 17$. This makes it impossible to show only electron pairs in the dot diagram.

(b) $:\overset{..}{\underset{..}{O}}—\overset{.}{N}=\overset{..}{\underset{..}{O}}:$

(c) Bent; it would be polar.

6.58 As the dipole moment is increased, the intermolecular attractive forces would become stronger, and the boiling point would increase.

CHAPTER 7

7.2 $4\ Fe + 3\ O_2 \longrightarrow 2\ Fe_2O_3$

7.3 $C_6H_{12}O_6 + 6\ O_2 \longrightarrow 6\ CO_2 + 6\ H_2O$

7.4 $2\ HCl + Na_2CO_3 \longrightarrow 2\ NaCl + CO_2 + H_2O$

7.6 1 cup of sugar requires 5 eggs; therefore 3 cups of sugar require 15 eggs.

7.7 10 eggs will make two cheesecakes. Since the eggs will run out after two cheesecakes are made, we can make only two cakes.

7.9 (a) $0.1\ \text{mole U} \times \dfrac{6.02 \times 10^{23}\ \text{atoms U}}{1\ \text{mole U}}$

$= 6.02 \times 10^{22}\ \text{atoms U}$

(b) $0.1\ \text{mole U} \times \dfrac{238.029\ \text{g U}}{1\ \text{mole U}} = 23.803\ \text{g U}$

7.10 (a) $120.11\ \text{g C} \times \dfrac{1\ \text{mole}}{12.011\ \text{g}} = 10.00\ \text{moles}$

(b) $10.00\ \text{moles} \times \dfrac{6.02 \times 10^{23}\ \text{atoms C}}{1\ \text{mole C}}$

$= 6.02 \times 10^{24}\ \text{atoms C}$

7.12 From Practice Problem 7.11, 1 mole of propane has a mass of 44.096 g; 2 moles of propane have a mass of 88.192 g.

7.13 6 moles carbon atoms

7.14 1 mole of propane and 5 moles of oxygen react to give 3 moles of carbon dioxide and 4 moles of water.

7.16 (a) $C_6H_6 + 3\ H_2 \longrightarrow C_6H_{12}$

(b) 1 mole of benzene and 3 moles of hydrogen react to give 1 mole of cyclohexane.

(c) $(6 \times 12.011\ \text{g C}) + (6 \times 1.0079\ \text{g H}) = 78.019\ \text{g C}_6\text{H}_6$; $(2 \times 1.0079\ \text{g H}) = 2.0158\ \text{g H}_2$ (mass of 1 mole), since the balanced equation calls for 3 moles of hydrogen, we need 6.0474 g H_2

(d) $(6 \times 12.011\ \text{g C}) + (12 \times 1.0079\ \text{g H}) = 84.161\ \text{g C}_6\text{H}_{12}$

(e) $\dfrac{24.0\ \text{g}}{84.161\ \text{g}} \times 100 = 28.5\%$

7.17 (a) $C_6H_{12}O_6 + 6\ O_2 \longrightarrow 6\ CO_2 + 6\ H_2O$

(b) 1 mole of glucose and 6 moles of oxygen react to give 6 moles of carbon dioxide and 6 moles of water.

(c) $(6 \times 12.011\ \text{g C}) + (12 \times 1.0079\ \text{g H}) + (6 \times 15.999\ \text{g O}) = 180.155\ \text{g glucose}$; $(2 \times 15.999\ \text{g O}) = 31.998\ \text{g O}$ (mass of 1 mole), since the balanced equation calls for 6 moles of oxygen, we need 191.988 g O_2

(d) $(12.011\ \text{g C}) + (2 \times 15.999) = 44.009\ \text{g}$ (mass of 1 mole); the mass of 6 moles is 264.054 g.

(e) $\dfrac{196.0\ \text{g}}{264.054\ \text{g}} \times 100 = 74.23\%$

7.19 $10.0\ \text{g glucose} \times \dfrac{1\ \text{mole glucose}}{180.155\ \text{g glucose}}$

$\times \dfrac{6\ \text{moles H}_2\text{O}}{1\ \text{mole glucose}} \times \dfrac{18.015\ \text{g H}_2\text{O}}{1\ \text{mole H}_2\text{O}}$

$= 6.00\ \text{g H}_2\text{O}$

7.20 $10.0\ \text{g glucose} \times \dfrac{1\ \text{mole glucose}}{180.155\ \text{g glucose}}$

$\times \dfrac{6\ \text{moles CO}_2}{1\ \text{mole glucose}} \times \dfrac{44.009\ \text{g CO}_2}{1\ \text{mole CO}_2}$

$= 14.7\ \text{g CO}_2$

7.21 $10.0\ \text{g CO}_2 \times \dfrac{1\ \text{mole CO}_2}{44.009\ \text{g CO}_2} \times \dfrac{1\ \text{mole glucose}}{6\ \text{moles CO}_2}$

$\times \dfrac{180.155\ \text{g glucose}}{1\ \text{mole glucose}} = 6.82\ \text{g glucose}$

7.23 $10.0 \text{ g Al}_2\text{O}_3 \times \dfrac{1 \text{ mole Al}_2\text{O}_3}{101.96 \text{ g Al}_2\text{O}_3} \times \dfrac{2 \text{ moles Al}}{1 \text{ mole Al}_2\text{O}_3}$

$\times \dfrac{6.02 \times 10^{23} \text{ Al atoms}}{1 \text{ mole Al}} = 1.18 \times 10^{23} \text{ Al atoms}$

7.24 $10.0 \text{ g H}_2\text{O} \times \dfrac{1 \text{ mole H}_2\text{O}}{18.015 \text{ g H}_2\text{O}}$

$\times \dfrac{6.02 \times 10^{23} \text{ H}_2\text{O molecules}}{1 \text{ mole H}_2\text{O}}$

$= 3.34 \times 10^{23} \text{ H}_2\text{O molecules}$

7.26 (a) C_2H_6O
(b) 52.0% C; 13.1% H; 34.9% O
(c) $C_2H_6O + 3 O_2 \longrightarrow 2 CO_2 + 3 H_2O$

7.28 (a) CH_2
(b) C_2H_4
(c) $C_2H_4 + 3 O_2 \longrightarrow 2 CO_2 + 2 H_2O$

7.30 37.02% C; 2.22% H; 18.50% N; 42.26% O

7.31 The empirical and molecular formula is $CHCl_3$.

7.32 The empirical formula is $C_2H_4O_1$; the molecular formula is $C_4H_8O_2$.

7.34 (a) $100.0 \text{ g C}_3\text{H}_8 \times \dfrac{1 \text{ mole C}_3\text{H}_8}{44.096 \text{ g C}_3\text{H}_8}$

$= 2.268 \text{ moles C}_3\text{H}_8$

$2.268 \text{ moles C}_3\text{H}_8 \times \dfrac{5 \text{ moles O}_2}{1 \text{ mole C}_3\text{H}_8}$

$= 11.34 \text{ moles O}_2$

$11.34 \text{ moles O}_2 \times \dfrac{31.998 \text{ g O}_2}{1 \text{ mole O}_2} = 362.87 \text{ g O}_2$

(b) $2.268 \text{ moles C}_3\text{H}_8 \times \dfrac{4 \text{ moles H}_2\text{O}}{1 \text{ mole C}_3\text{H}_8}$

$\times \dfrac{18.015 \text{ g H}_2\text{O}}{1 \text{ mole H}_2\text{O}} = 163.4 \text{ g H}_2\text{O}$

This is the theoretical yield of water.

7.35 (a) $2 ZnS + 3 O_2 \longrightarrow 2 ZnO + 2 SO_2$
(b) 10.0 g Zns = 0.103 mole ZnS and 10.0 g O_2 = 0.312 mole O_2;

$ZnS: \dfrac{0.103}{2} = 0.0515; \quad O_2: \dfrac{0.312}{3} = 0.104$

Since 0.0515 is the smaller number, the theoretical yield must be based on ZnS. Theoretical yield of ZnO:

$\text{moles ZnO} = 0.103 \text{ mole ZnS} \times \dfrac{2 \text{ moles ZnO}}{2 \text{ moles ZnS}}$

$= 0.103 \text{ mole ZnO}$

$\text{grams ZnO} = 0.103 \text{ mole ZnO} \times \dfrac{81.389 \text{ g ZnO}}{1 \text{ mole ZnO}}$

$= 8.38 \text{ g ZnO}$

Theoretical yield of SO_2:

$\text{moles SO}_2 = 0.103 \text{ mole ZnS} \times \dfrac{2 \text{ moles SO}_2}{2 \text{ moles ZnS}}$

$= 0.103 \text{ mole SO}_2$

$\text{grams SO}_2 = 0.103 \text{ mole SO}_2 \times \dfrac{64.064 \text{ g SO}_2}{1 \text{ mole SO}_2}$

$= 6.60 \text{ g SO}_2$

(c) 5.06 g O_2 left over
(d) 88.2%

7.36 Only (a) represents a chemical reaction, since the substances on the right (water and oxygen) are different from the substance on the left (hydrogen peroxide, H_2O_2). In (b), all substances remain chemically unchanged.

7.39 None, it always takes energy to break a bond.

7.40 Two Cl–Cl bonds and two C–H bonds must be broken. Two H–Cl bonds and two C–Cl bonds are formed.

7.42 Cooling the reaction mixture slows down the molecules.

7.44 There will be more frequent collisions.

7.46 Changing the subscripts changes the chemical identity of a substance.

7.47 $C_2H_4 + 3 O_2 \longrightarrow 2 CO_2 + 2 H_2O$

7.48 1 mole of methane and 2 moles of oxygen react to give 1 mole of carbon dioxide and 2 moles of water. Or 1 molecule of methane and 2 molecules of oxygen react to give 1 molecule of carbon dioxide and 2 molecules of water.

7.50 $Al_2Cl_6 + 6 H_2O \longrightarrow 2 Al(OH)_3 + 6 HCl$

7.52 $2 C_2H_2 + 5 O_2 \longrightarrow 2 H_2O + 4 CO_2$

7.55 A reactant that is present in short supply and therefore determines (limits) the amount of product that will be made.

7.56 The reaction was run with B in short supply (the limiting reactant).

7.57 (a) $\dfrac{1 \text{ egg}}{2 \text{ cups flour}}, \dfrac{1 \text{ egg}}{3 \text{ cups sugar}}, \dfrac{2 \text{ cups flour}}{3 \text{ cups sugar}},$

$\dfrac{\frac{1}{4} \text{ pound butter}}{3 \text{ cups sugar}}, \dfrac{1 \text{ cup milk}}{1 \text{ dozen sugar cookies}},$

$\dfrac{1 \text{ dozen sugar cookies}}{\frac{1}{4} \text{ pound butter}}, \text{ etc.}$

(b) $30 \text{ cookies} \times \dfrac{1 \text{ dozen cookies}}{12 \text{ cookies}}$

$\times \dfrac{2 \text{ cups flour}}{1 \text{ dozen cookies}} = 5 \text{ cups flour}$

(c) You don't know how many cups of milk are in the container.

(d) $1 \text{ container of milk} \times \dfrac{4 \text{ cups milk}}{1 \text{ container of milk}}$

$\times \dfrac{1 \text{ egg}}{1 \text{ cup milk}} = 4 \text{ eggs}$

(e) 1 egg + 2 cups flour + 3 cups sugar + $\frac{1}{4}$ pound butter + 1 cup milk $\longrightarrow$ 1 dozen sugar cookies

(f) (ii) 1 dozen cookies
(iii) 1 cup of flour will be left over.

7.59 (a) $3 \text{ cakes} \times \dfrac{5 \text{ eggs}}{1 \text{ cake}} = 15 \text{ eggs}$

7.59 (b) To see which is limiting, divide the amounts by the coefficients in the recipe:

$$\frac{25 \text{ eggs}}{5} = 5, \quad \frac{9 \text{ blocks of cream cheese}}{3} = 3,$$

$$\frac{4 \text{ cups sugar}}{1} = 4$$

Since 3 is the smallest number, the cream cheese is limiting, and the number of cheesecakes will be determined by the amount of cream cheese:

9 blocks of cream cheese

$$\times \frac{1 \text{ cheesecake}}{3 \text{ blocks of cream cheese}} = 3 \text{ cheesecakes}$$

(c) 63 blocks of cream cheese

$$\times \frac{5 \text{ eggs}}{3 \text{ blocks of cream cheese}} = 105 \text{ eggs}$$

7.61 Always

7.63 There are 6.02×10^{23} bicycles in 1 mole of bicycles. Since there are two tires on every bicycle, the number of bicycle tires in 1 mole of bicycles is 2 times 6.02×10^{23}, or 1.20×10^{24}.

7.65 2.5 moles of pennies $\times \dfrac{6.02 \times 10^{23} \text{ pennies}}{1 \text{ mole of pennies}}$

$$= 1.5 \times 10^{24} \text{ pennies}$$

$$1.5 \times 10^{24} \text{ pennies} \times \frac{1 \text{ dollar}}{100 \text{ pennies}}$$

$$= 1.5 \times 10^{22} \text{ dollars}$$

7.67 2 molecules of sufur dioxide with 1 molecule of oxygen to give 2 molecules of sulfur trioxide. 2 moles of sufur dioxide with 1 mole of oxygen to give 2 moles of sulfur trioxide.

7.68 There is no such thing as a fractional molecule, but we *can* have a fraction of a mole.

7.70 In general, 1 mole of any molecule has a mass equal to its molar mass in grams.

7.72 6.02×10^{23}

7.73 6.02×10^{23}

7.75 (a) 3 moles
(b) 6 moles
(c) 2 times $6.02 \times 10^{23} = 1.20 \times 10^{24}$

7.77 The actual yield is the amount of product that is actually recovered after a chemical reaction.

7.79 % yield $= \dfrac{\text{Actual yield}}{\text{Theoretical yield}} \times 100$

7.81 Molar mass allows us to form conversion factors between moles and grams.

7.83 (a) $2 \text{ HCl} + \text{Zn} \longrightarrow \text{H}_2 + \text{ZnCl}_2$
(b) 2 moles of HCl and 1 mole of zinc react to give 1 mole of hydrogen and 1 mole of ZnCl_2.
(c) Molar mass of HCl is $(1.0079 \text{ g H}) + (35.453 \text{ g Cl}) = 36.461 \text{ g HCl}$. Since the reaction calls for 2 moles of HCl, the mass of HCl needed is 72.922 g HCl.

Molar mass of Zn is 65.39 g. This is the mass of 1 mole of Zn, which is what the equation calls for.
(d) Molar mass of H_2 is 2.0158 g, which is the theoretical yield.
(e) $\dfrac{2.00 \text{ g}}{2.0158 \text{ g}} \times 100 = 99.22\%$

7.85 $\dfrac{6 \text{ moles C atoms}}{1 \text{ mole C}_6\text{H}_{12}\text{O}_6}, \quad \dfrac{12 \text{ moles H atoms}}{1 \text{ mole C}_6\text{H}_{12}\text{O}_6},$

$\dfrac{1 \text{ mole C}_6\text{H}_{12}\text{O}_6}{6 \text{ moles O atoms}},$

7.87 (a) $\text{I}_2 + 3 \text{ Cl}_2 \longrightarrow 2 \text{ ICl}_3$
(b) From the equation:

$\dfrac{3 \text{ moles Cl}_2}{1 \text{ mole I}_2}, \quad \dfrac{2 \text{ moles ICl}_3}{1 \text{ mole I}_2}, \quad \dfrac{2 \text{ moles ICl}_3}{3 \text{ moles Cl}_2},$

$\dfrac{1 \text{ mole I}_2}{3 \text{ moles Cl}_2}, \quad \dfrac{1 \text{ mole I}_2}{2 \text{ moles ICl}_3}, \quad \dfrac{3 \text{ moles Cl}_2}{2 \text{ moles ICl}_3}$

From the formulas:

$\dfrac{2 \text{ moles I atoms}}{1 \text{ mole I}_2}, \quad \dfrac{2 \text{ moles Cl atoms}}{1 \text{ mole Cl}_2},$

$\dfrac{1 \text{ mole I atoms}}{1 \text{ mole ICl}_3}, \quad \dfrac{3 \text{ moles Cl atoms}}{1 \text{ mole ICl}_3}$

7.89 $24.0 \text{ g O}_2 \times \dfrac{1 \text{ mole O}_2}{31.998 \text{ g O}_2} = 0.750 \text{ mole O}_2$

7.90 $24.0 \text{ g O}_2 \times \dfrac{1 \text{ mole O}_2}{31.998 \text{ g O}_2} \times \dfrac{2 \text{ O atoms}}{1 \text{ mole O}_2}$

$= 1.50 \text{ O atoms}$

7.93 $1.00 \times 10^9 \text{ molecules H}_2\text{O}$

$\times \dfrac{1 \text{ mole H}_2\text{O}}{6.02 \times 10^{23} \text{ molecules H}_2\text{O}}$

$\times \dfrac{18.015 \text{ g H}_2\text{O}}{1 \text{ mole H}_2\text{O}} = 2.99 \times 10^{-14} \text{ g H}_2\text{O}$

7.95 (a) $20.0 \text{ g H}_2\text{O}_2 \times \dfrac{1 \text{ mole H}_2\text{O}_2}{34.014 \text{ g H}_2\text{O}_2}$

$\times \dfrac{2 \text{ moles H}_2\text{O}}{2 \text{ moles H}_2\text{O}_2} \times \dfrac{18.015 \text{ g H}_2\text{O}}{1 \text{ mole H}_2\text{O}}$

$= 10.6 \text{ g H}_2\text{O}$

(b) $20.0 \text{ g H}_2\text{O} \times \dfrac{1 \text{ mole H}_2\text{O}}{18.015 \text{ g H}_2\text{O}} \times \dfrac{2 \text{ moles H}_2\text{O}_2}{2 \text{ moles H}_2\text{O}}$

$\times \dfrac{34.014 \text{ g H}_2\text{O}_2}{1 \text{ mole H}_2\text{O}_2} = 37.8 \text{ g H}_2\text{O}_2$

(c) $20.0 \text{ g O}_2 \times \dfrac{1 \text{ mole O}_2}{31.998 \text{ g O}_2} \times \dfrac{2 \text{ moles H}_2\text{O}_2}{1 \text{ mole O}_2}$

$\times \dfrac{34.014 \text{ g H}_2\text{O}_2}{1 \text{ mole H}_2\text{O}_2} = 42.5 \text{ g H}_2\text{O}_2$

7.97 (a) $5.00 \text{ g H}_2 \times \dfrac{1 \text{ mole H}_2}{2.0158 \text{ g H}_2} \times \dfrac{2 \text{ moles H}_2\text{O}}{2 \text{ moles H}_2}$

$\times \dfrac{18.015 \text{ g H}_2\text{O}}{1 \text{ mole H}_2\text{O}} = 44.7 \text{ g H}_2\text{O}$

(b) $5.00 \text{ g H}_2\text{O} \times \dfrac{1 \text{ mole H}_2\text{O}}{18.015 \text{ g H}_2\text{O}} \times \dfrac{1 \text{ mole O}_2}{2 \text{ moles H}_2\text{O}}$

$\times \dfrac{31.998 \text{ g O}_2}{1 \text{ mole O}_2} = 4.44 \text{ g O}_2$

(c) $100.0 \text{ g H}_2 \times \dfrac{1 \text{ mole H}_2}{2.0158 \text{ g H}_2} \times \dfrac{1 \text{ mole O}_2}{2 \text{ moles H}_2}$

$\times \dfrac{31.998 \text{ g O}_2}{1 \text{ mole O}_2} = 793.7 \text{ g O}_2$

(d) $50.0 \text{ g O}_2 \times \dfrac{1 \text{ mole O}_2}{31.998 \text{ g O}_2} \times \dfrac{2 \text{ moles H}_2\text{O}}{1 \text{ mole O}_2}$

$\times \dfrac{18.015 \text{ g H}_2\text{O}}{1 \text{ mole H}_2\text{O}} = 56.3 \text{ g H}_2\text{O}$

(e) $56.3 \text{ g H}_2\text{O} \times \dfrac{1 \text{ mole H}_2\text{O}}{18.015 \text{ g H}_2\text{O}}$

$\times \dfrac{6.02 \times 10^{23} \text{ molecules H}_2\text{O}}{1 \text{ mole H}_2\text{O}}$

$= 1.88 \times 10^{24} \text{ molecules H}_2\text{O}$

7.99 The balanced equation is: $3 \text{ H}_2 + \text{N}_2 \longrightarrow 2 \text{ NH}_3$

(a) $10.0 \text{ g H}_2 \times \dfrac{1 \text{ mole H}_2}{2.0158 \text{ g H}_2} \times \dfrac{1 \text{ mole N}_2}{3 \text{ moles H}_2}$

$\times \dfrac{28.014 \text{ g N}_2}{1 \text{ mole N}_2} = 46.3 \text{ g N}_2$

(b) $10.0 \text{ g H}_2 \times \dfrac{1 \text{ mole H}_2}{2.0158 \text{ g H}_2} \times \dfrac{2 \text{ moles NH}_3}{3 \text{ moles H}_2}$

$\times \dfrac{17.031 \text{ g NH}_3}{1 \text{ mole NH}_3} = 56.3 \text{ g NH}_3$

(c) $56.3 \text{ g NH}_3 \times \dfrac{1 \text{ mole NH}_3}{17.031 \text{ g NH}_3}$

$\times \dfrac{6.02 \times 10^{23} \text{ molecules NH}_3}{1 \text{ mole NH}_3}$

$= 1.99 \times 10^{24} \text{ molecules NH}_3$

7.101 (a) C_2H_2
(b) C_4H_4
(c) C_6H_6

7.103 (a) 76.57% C; 6.42% H; 17.00% O
(b) $\text{C}_6\text{H}_6\text{O}$
(c) $\text{C}_6\text{H}_6\text{O}$

7.105 (a) 85.63% C; 14.37% H
(b) CH_2
(c) C_5H_{10}

7.106 Yes; an example is CHCl_3.

7.108 (a) 24.8%C; 2.08% H; 73.1% Cl
(b) CHCl

7.110 $\text{C}_4\text{H}_4\text{ON}$

7.112 KClO

7.114 38.7% C; 9.74% H; 51.6% O

7.116 23.19% C; 1.43% H; 1.80% N; 8.24% O; 65.34% I

7.118 (a) $\text{Cl}_2 + 3 \text{ F}_2 \longrightarrow 2 \text{ ClF}_3$
(b) F_2
(c) 4.10 moles ClF_3
(d) 379.0 g ClF_3
(e) 0.45 mole Cl_2
(f) 31.91 g Cl_2

7.119 (a) $2 \text{ Na}(s) + \text{Br}_2(l) \longrightarrow 2 \text{ NaBr}(s)$
(b) Na
(c) 22.4 g NaBr
(d) 12.6 g Br_2
(e) 65.6%

7.121 (a) $2 \text{ Na} + \text{H}_2 \longrightarrow 2 \text{ NaH}$
(b) H_2
(c) 0.5595 g NaH
(d) 9.46 g Na
(e) 76.4%

CHAPTER 8

8.2 (a) $\ddot{\text{O}}_1{=}\text{C}{=}\ddot{\text{O}}_2$

C oxidation state is +4; O oxidation state is −2
(b) None
(c) 4 valence electrons fewer
(d) It is correct to say that the C atom in CO_2 has a complete octet, because the valence electrons are double counted for purposes of determining satisfaction of the octet rule.

8.3 (a) $\text{H}:\overset{\displaystyle \text{H}}{\underset{\displaystyle \text{H}}{\ddot{\text{C}}}}:\text{H}$

C oxidation state is −4; H oxidation state is +1
(b) 8 electrons
(c) 4 valence electrons more

8.4 (a) $:\ddot{\text{C}}\text{l}:\overset{}{\underset{\displaystyle :\ddot{\text{C}}\text{l}:}{\ddot{\text{C}}}}:\ddot{\text{C}}\text{l}:$

C oxidation state is +2; H oxidation state is +1; Cl oxidation state is −1
(b) 2 electrons
(c) 2 valence electrons fewer
(d) It is correct to say that the C atom in CHCl_3 has a complete octet because the valence electrons are double counted for purposes of determining satisfaction of the octet rule.

8.6 Mg is +2 (rule 5); Br is −1 (halide rule applies since Br is more electronegative than Mg)

8.7 O is −2 (rule 2); Fe is +3 (rule 7)

8.8 H is +1 (rule 3); O is −2 (rule 2); Cl is +1 (rule 7)

8.9 O is −2 (rule 2); C is +4 (rule 7)

8.10 Cu is +2 (rule 6)

8.12 This is a redox reaction. The oxidation number of Fe changes from 0 to +3 (each Fe loses 3 electrons), and the oxidation number of O changes from 0 to −2 (each O gains 2 electrons). The Fe is oxidized and the O is reduced.

8.13 This is a redox reaction. The oxidation number of Ca changes from 0 to +2 (Ca loses 2 electrons), and the oxidation number of H changes from +1 to 0 (each H gains 1 electron). The Ca is oxidized and the H is reduced.

8.14 This is not a redox reaction.

8.16 Fe is the reducing agent, and O_2 is the oxidizing agent.

8.17 Ca is the reducing agent, and H^+ is the oxidizing agent.

8.19 (a) Pb^{2+} (b) Ni
(c) Pb cathode   (d) Ni anode

8.20 (a) Fe   (b) Ag^+

(c)
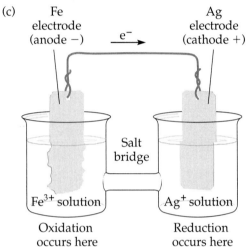

Fe electrode (anode −) →e^- Ag electrode (cathode +)

Salt bridge

Fe^{3+} solution Ag^+ solution

Oxidation occurs here   Reduction occurs here

8.22 (a) Will occur spontaneously

8.23
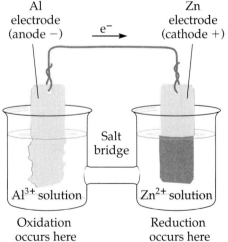

Al electrode (anode −) →e^- Zn electrode (cathode +)

Salt bridge

Al^{3+} solution Zn^{2+} solution

Oxidation occurs here   Reduction occurs here

8.24 Lead (Pb) is higher than copper (Cu) in the EMF series. The oxidizing agent is Cu^{2+}, and the reducing agent is Pb.

8.26 By moving copper wire through a magnetic field. Coils of copper wire are wound around a rotor and spun in a magnetic field by a turbine powered by steam or moving water (hydroelectric).

8.28 To keep track of how many electrons an atom owns, so we can determine whether electron transfer has occurred.

8.30 Both methods are correct.

8.32 The more electronegative atom owns the shared electrons.

8.34 H:Ö:Ö:H
Each O owns 7 electrons (both of its lone pairs, both electrons from the O–H bonds, and 1 electron from the O–O bond). Each H owns 0 electrons, since H is less electronegative than O.

8.35 $\left[:\ddot{O}:N::\ddot{O}: \atop :\ddot{O}: \right]^-$

Each O owns 8 electrons (all of its lone pairs and the electrons from the N–O bond). The N owns 0 electrons since N is less electronegative than O.

8.37 (a)–(c) All have oxidation state 0.

8.39 The sum must always be the overall charge on the molecule.

8.41 (a) More
(b) 2 more electrons

8.42 Add up all the other oxidation numbers of the atoms in the compound. Subtract this sum from the total charge on the species. The difference is the oxidation number of the unknown atom.

8.44 (a) O is −2; C is +2
(b) H is +1; Cl is −1; C is 0
(c) H is +1; O is −2; C is +2
(d) Cl is −1; Pt is +4

8.46 (a) The halide rule works for $AlCl_3$ because Cl is more electronegative than Al, and will own 8 electrons in the dot structure.
(b) The halide rule doesn't work for ClO^- because Cl is less electronegative than O. The O will therefore own the electrons in the bond, and the Cl will have a positive oxidation state.
(c) ClO^-: O is −2; Cl is +1
$AlCl_3$: Cl is −1; Al is +3

8.48 (a) Two oxygens at −2 each and six hydrogens at +1 each (using shortcut rules) sums to +2. This means the sum of the oxidation numbers of the three carbons must be −2 to get neutrality (the molecule is neutral). This is not enough information to assign the carbons a unique set of oxidation numbers.

(b)
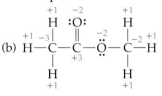

8.51 If the oxidation number of any atom changes during a reaction, a redox reaction has occurred.

8.53 The oxidation state increases.

8.55 Whenever electron transfer occurs, there must be both reduction and oxidation occurring simultaneously; hence the name redox.

8.58 (b), (c), and (d) are electron transfer reactions.

8.59 (a) For reaction (b): Fe is oxidized; N is reduced
For reaction (c): C is oxidized; O (in O_2) is reduced
For reaction (d): Br is oxidized; Ag is reduced
(b) For reaction (b): NO_3^- is the oxidizing agent
For reaction (c): O_2 is the oxidizing agent
For reaction (d): AgBr is the oxidizing agent
(c) For reaction (b): Fe is the reducing agent
For reaction (c): C_2H_6 is the reducing agent
For reaction (d): AgBr is the reducing agent

8.61 Yes, because it is an electron transfer (redox) reaction.

8.63 (a) In the hydroquinone the two carbons bonded to the oxygens are oxidized from +1 to +2.
(b) Ag is reduced.
(c) The oxidizing agent is AgBr.
(d) The reducing agent is hydroquinone.

8.65

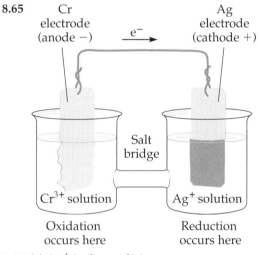

8.66 (a) Ag^+ is the oxidizing agent.
(b) Cr is the reducing agent.
(c) Ag electrode
(d) Cr electrode

8.69 (a) The tin (Sn) is getting oxidized (from 0 to +2).
(b) The Cu^{2+} is getting reduced (from +2 to 0).
(c)

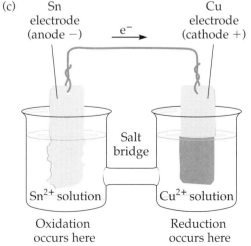

8.71 Electrons are negative and flow from the negative electrode (like charges repel) to the positive electrode (unlike charges attract). Since electrons always flow from the anode to the cathode, the anode must be negative and the cathode positive. Or, you can remember that a *cat*ion is positive (so is a *cat*hode) and an *an*ion is negative (so is an *an*ode).

8.72 (a) $B + A^+ \longrightarrow B^+ + A$
(b) The oxidizing agent is A^+.
(c) The reducing agent is B.
(d) Electrons flow from B to A^+.

8.74 You would have to do competitive redox experiments, two metals at a time. For example, you could first test metals X and Y, and see which electrode gets eaten away (and which one gains mass). The metal that gets eaten away (let's say it's X) would be the more active metal. Next, you would repeat the experiment for Y and Z; then again for X and Z. By comparing the results, you could properly order X, Y, and Z according to their relative activities.

8.77 Nothing happens.

8.79 Reaction (a) is spontaneous because Ag is more active than Au.

8.80 (a)

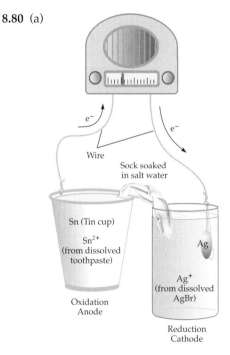

(b) The tin would be eaten away as it oxidized.
(c) Ag^+ is reduced, so it is the oxidizing agent.
(d) Sn is oxidized, so it is the reducing agent.

8.82 (a) $Pb + Cu^{2+} \longrightarrow Pb^{2+} + Cu$
(b) $Pb^{2+} + Cu \longrightarrow Pb + Cu^{2+}$

8.86 The next best metal would be Pt (platinum), since it is the next least active metal.

8.87 The Statue of Liberty is made of sculpted sheets of copper metal placed over a steel (iron) skeleton. Iron is higher on the EMF series than copper. When the copper sheets began to become oxidized, the iron skeleton gave up its electrons to reduce the copper ions. Therefore, the iron skeleton was literally being eaten away and converted into ions. (The asbestos that had originally been placed in the Statue to separate the copper and the iron had rotted away.) The solution to the problem was to replace the dissolving steel skeleton with a reinforced, corrosion-resistant, stainless steel skeleton incorporating high-tech plastics to insulate it from copper.

8.88 If the casing is not made thick enough, the conversion of Zn to Zn^{2+} could result in weak spots along the surface of the battery.

CHAPTER 9

9.2 HF

9.3

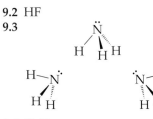

9.4 H_2O

9.6 Both ethylene and polyethylene are nonpolar molecules. The only intermolecular forces involved are London forces. Polyethylene is a very large molecule that has many more electrons than ethylene. Whereas the London forces in the small ethylene molecule are relatively weak, the London forces in polyethylene are quite substantial. Hence, ethylene is a gas at room temperature, and polyethylene is a solid.

9.7 (a) H:N̈:H   H:P̈:H
            Ḧ   Ḧ

(b) NH_3 is polar; PH_3 is nonpolar.

(c) PH_3 has the stronger London forces because it has more electrons.

(d) PH_3 because it has stronger London forces. However, this is not the case. See Practice Problem 9.8 for more information.

9.9

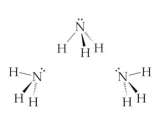

9.10 (a) Because it meets the criterion for hydrogen bonds, which is a very strong intermolecular attractive force.

(b) Because it has more electrons and therefore stronger London forces.

9.12 Both iron and sodium chloride are nonmolecular solids. The forces holding iron and sodium chloride together are true bonds—metallic bonds in iron and ionic bonds in NaCl.

9.13 Intermolecular London dispersion forces must be overcome to melt the solid.

9.14 (a) Metallic   (b) Metallic   (c) Ionic
(d) Network covalent   (e) Molecular

9.16 $10.50 \text{ g } H_2 \times \dfrac{1 \text{ mole } H_2}{2.0158 \text{ g } H_2} = 5.209 \text{ moles } H_2$

9.17 (a) $5 \times 760 \text{ mm Hg} = 3800 \text{ mm Hg}$

(b) $3800 \text{ mm Hg} \times \dfrac{1 \text{ cm}}{10 \text{ mm}} \times \dfrac{1 \text{ inch}}{2.54 \text{ cm}}$
$= 149.6 \text{ inches Hg}$

9.18 Too low, since the air in the tube will push down on the mercury

9.20 Use $P = \dfrac{nRT}{V}$ and solve for T: $T = \dfrac{PV}{nR}$

Moles $N_2 = 3.00 \text{ g } N_2 \times \dfrac{1 \text{ mole } N_2}{28.014 \text{ g } N_2}$
$= 0.107 \text{ mole } N_2$

Pressure in atm $= 450.5 \text{ mm Hg} \times \dfrac{1 \text{ atm}}{760 \text{ mm Hg}}$
$= 0.5928 \text{ atm}$

$T = \dfrac{0.5928 \text{ atm} \times 2.00 \text{ L}}{0.107 \text{ mole } N_2 \times \dfrac{0.0821 \text{ L} \cdot \text{atm}}{\text{K} \cdot \text{mole}}} = 135 \text{ K}$

$°C = 135 \text{ K} - 273.15 = -138°C$

9.21 Use $P = \dfrac{nRT}{V}$ and solve for n: $n = \dfrac{PV}{RT}$

Pressure in atm $= 40 \text{ lb/in.}^2 \times \dfrac{760 \text{ mm Hg}}{14.696 \text{ lb/in.}^2}$

$\times \dfrac{1 \text{ atm}}{760 \text{ mm Hg}} = 2.72 \text{ atm}$

Volume in liters $= 10.5 \text{ gal} \times \dfrac{3.785 \text{ L}}{1 \text{ gal}} = 39.7 \text{ L}$

$T = 22.5°C + 273.15 = 295.65 \text{ K}$

$n = \dfrac{2.72 \text{ atm} \times 39.7 \text{ L}}{\dfrac{0.0821 \text{ L} \cdot \text{atm}}{\text{K} \cdot \text{mole}} \times 295.65 \text{ K}} = 4.45 \text{ moles } O_2$

grams $O_2 = 4.45 \text{ moles } O_2 \times \dfrac{31.998 \text{ g } O_2}{1 \text{ mole } O_2}$
$= 142 \text{ g } O_2$

9.23 Cooling a gas slows the molecules down to a speed at which they are not moving fast enough to overcome the attractive forces that will make them condense into a liquid.

9.25 Yes. Molecules in the liquid phase are still very much in motion. They are jostling past each other constantly, but remain in close proximity to each other.

9.27 Water has stronger intermolecular attractive forces because it will condense into the liquid phase with less cooling than nitrogen requires.

9.29 Gas molecules are not bound to each other, as are the molecules of solids and liquids. Gas molecules are essentially independent of each other, and can travel the entire volume of their container.

9.31 (a) C–D; the greater the intermolecular attractive forces (the greater the diople moment), the less energy needs to be removed from the liquid in order to form a solid.
(b) C–D; the greater the intermolecular attractive forces, the more energy is needed for the molecules to overcome the attractive forces and escape into the gas phase.

9.32 H—Br

H—Br   H—Br

Dipole–dipole interactions and London forces

9.34 CBr_4; since it is the larger molecule, has more electrons, and therefore has stronger London forces.

9.35 (1) Chloromethane is a polar molecule and is subject to dipolar interactions.
(2) Chloromethane is a larger molecule and has stronger London forces.

9.37 The strength of London forces depends on the size of the molecule.

9.38 (a) HF, because the electronegativity difference is greater.
(b) HF, because the partial charges on the H and F atoms is greater, leading to a larger dipole moment in HF. Only HF displays hydrogen bonding; HCl does not.
(c) HF; the greater the intermolecular attractive forces (the greater the dipole moment), the less energy needs to be removed from the gas in order to form a liquid. Since HF has stronger intermolecular attractive forces, it will liquefy at a higher temperature; that is, HF has the higher boiling point.
(d) Yes, both molecules experience London forces. HCl has greater London forces because it is a larger molecule with more electrons.

9.41 Since long-chain hydrocarbons are very large molecules, they have many electrons, and therefore quite strong London forces. The London forces, in these cases, are strong enough to make the hydrocarbons solid at room temperature.

9.43 Two water molecules feel the greatest attraction when the partially positive H on one molecule is oriented toward the partially negative O on the other (see top of next column).

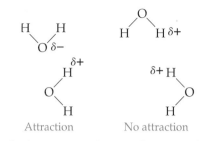

Attraction   No attraction

9.45 The hydrogen atom has no electrons other than those it is sharing and resembles a naked proton when bonded to a very electronegative atom such as O, N, or F.

9.47 (a) Because the size of the molecules is increasing and London forces are stronger
(b) H_2O has a much higher boiling point than expected because it experiences hydrogen bonding and is therefore subject to stronger intermolecular attractive forces than the other molecules in the group.

9.49 Because hydrogen bonds are not covalent bonds within a molecule (which we represent by solid lines). Hydrogen bonds are unusually strong intermolecular attractions.

9.51 DNA molecules must be very stable, but still able to break into two strands when the DNA is involved in protein building. Covalent bonds would be too strong (the strands would not be able to break apart), and London forces would be too weak (the DNA could not be kept intact).

9.52 If a nonpolar molecule is very large with a large number of electrons, its London forces will be quite substantial, and can be stronger than the dipolar interactions present in small polar molecules.

9.54 Sometimes

9.56 Because the nonmolecular solids are held together by much stronger forces, covalent, ionic, or metallic bonds

9.58 The high melting points of some metals, such as gold and iron, are evidence that metallic bonds can be as strong as ionic or covalent bonds.

9.60 (a) Metallic solid   (b) Ionic solid
(c) Network covalent solid   (d) Molecular solid

9.62 The assumption is that the molecules of an ideal gas are moving independently of each other and feel no attractive forces from each other. We are justified in making this assumption because the molecules are moving very fast in straight line paths and are relatively far from each other. If they attracted each other, they would not move in straight line paths.

9.64 By colliding with the walls of its container

9.67 760 mm Hg or 1 atm

9.69 (a) $29.7 \text{ inches Hg} \times \dfrac{2.54 \text{ cm}}{1 \text{ inch}} \times \dfrac{10 \text{ mm}}{1 \text{ cm}}$

$= 754 \text{ mm Hg}$

9.69 (b) $754 \text{ mm Hg} \times \dfrac{1 \text{ atm}}{760 \text{ mm Hg}} = 0.992 \text{ atm}$

9.71 $\text{K} = {}^\circ\text{C} + 273.15$
$\text{K} = 22.0{}^\circ\text{C} + 273.15 = 295.15 \text{ K}$

9.74 $600.2 \text{ lb/in.}^2 \times \dfrac{760 \text{ mm Hg}}{14.696 \text{ lb/in.}^2} \times \dfrac{1 \text{ atm}}{760 \text{ mm Hg}}$
$= 40.84 \text{ atm}$

9.76 The pressure decreases.

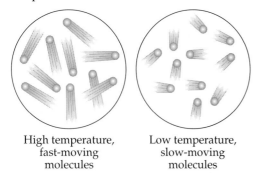

High temperature,   Low temperature,
fast-moving   slow-moving
molecules   molecules

9.78 (a) 1: It has the fewest molecules in the available space, and will therefore experience fewer collisions with the walls of the container.
(b) 3: It has the most molecules in the available space, and will therefore have the most collisions with the walls of the container.

9.79 2: In fact, the temperature has been lowered to the point where the gas has condensed into the liquid phase, which is why the molecules are clumped together at the bottom of the container.

9.80 The pressure of a gas is inversely proportional to its volume.

9.81 The pressure of a gas is directly proportional to its temperature.

9.83 Inverse proportionality means that as one value *increases* the other one *decreases*, and vice versa. Direct proportionality means that as one value *increases* the other one *also increases.*

9.85 Inverse proportion; as his age increases, his hair decreases.

9.87 $\text{L} \cdot \text{atm}/\text{K} \cdot \text{mole}$; the units indicate the units in which the volume, pressure, temperature, and quantity of gas must be expressed when using the ideal gas law.

9.89 $T = PV/nR$

9.91 You should get the value of R, which is $0.0821 \text{ L} \cdot \text{atm}/\text{K} \cdot \text{mole}$.

9.93 The gas pressure should increase sixfold.

9.94 The temperatures must be expressed in kelvins when using the ideal gas law. To get the correct answer, we must convert 25.0°C to kelvins, double that number, then reconvert to °C:

$$\text{K} = 25.0{}^\circ\text{C} + 273.15 = 298 \text{ K}$$

The pressure would double at 596 K, which is equivalent to 323°C.

9.96 $\text{K} = 0{}^\circ\text{C} + 273.15 = 273.15 \text{ K}$

$$V = \frac{nRT}{P} = \frac{(2.0 \text{ moles})\dfrac{0.0821 \text{ L} \cdot \text{atm}}{\text{K} \cdot \text{mole}}(273.15 \text{ K})}{2.5 \text{ atm}}$$
$$= 18 \text{ L}$$

9.98 Since $V = nRT/P$, if $T = 0$, V will equal 0. The volume of the gas would be 0 L. That does not make sense. What really happens is that the gas becomes a liquid as it approaches this temperature, and the ideal gas law no longer governs its behavior.

9.101 (a) $22.5{}^\circ\text{C} + 273.15 = 296 \text{ K}$

$$V = \frac{nRT}{P} = \frac{(1 \text{ mole})\dfrac{0.0821 \text{ L} \cdot \text{atm}}{\text{K} \cdot \text{mole}}(296 \text{ K})}{1.00 \text{ atm}}$$
$$= 24.3 \text{ L H}_2$$

(b) $1 \text{ mole H}_2 \times \dfrac{2 \text{ moles HCl}}{1 \text{ mole H}_2} = 2 \text{ moles HCl}$

$2 \times 24.3 \text{ L} = 48.6 \text{ L HCl}$

CHAPTER 10

10.2 (a) Water is the solvent; acetone is the solute.
(b) Nitrogen is the solvent; oxygen, water vapor, and other gases are the solutes.
(c) Iron is the solvent; chromium, nickel, and carbon are the solutes.
(d) Water is the solvent; aspirin is the solute.

10.3 Since 135 proof vodka would be 67.5% alcohol and 32.5% water, alcohol would be the major component, so it would be the solvent.

10.5 The energy released in dissolving NaCl will be less in carbon tetrachloride than in water. Since carbon tetrachloride is a nonpolar molecule, it has no dipoles that will attract ions. With little or no attraction between the ions and carbon tetrachloride, there will be little or no solvation energy released.

10.6 The energy required for pulling apart the lattice would be greater for $MgCl_2$. The attraction of the $+2$ magnesium ions for the negative chloride ions would be stronger than the attraction of the $+1$ sodium ions because the attractive force between oppositely charged particles increases as the magnitude of the charge increases.

10.8 $\Delta E_{\text{step 1}}$ would be larger for dissolving the solid ionic solute. Gaseous molecules have very little attraction for each other and are already separated. $\Delta E_{\text{step 1}}$ is negligible for a gaseous solute.

10.9 The more negative ΔE_{total}, the more negative $\Delta E_{\text{step 3}}$ is, since this is the only negative step in the process. Since $\Delta E_{\text{step 3}}$ is a measure of how strongly the solute and solvent are attracted to each other, the more negative

ΔE_{total}, the stronger the solute–solvent attraction and the more likely the solute will dissolve. Also, a very negative ΔE_{total} means that step 3 releases more than enough energy to accomplish steps 1 and 2.

10.10 (a) Because Mg is in group IIA of the periodic table and has 2 valence electrons that are lost when Mg forms an octet.

(b) $\Delta E_{step\ 1}$ is a measure of the energy needed to break up the lattice of the solute into individual ions. Since the Mg^{2+} ions have a stronger attraction for the Cl^- ions than do the Na^+ ions, more energy is needed to break the $MgCl_2$ lattice, and $\Delta E_{step\ 1}$ is greater for $MgCl_2$.

(c) $MgCl_2$ is more soluble in water than NaCl because even though $\Delta E_{step\ 1}$ is greater for $MgCl_2$ than for NaCl, $\Delta E_{step\ 3}$ is much larger for Mg^{2+} than for Na^+. The $\Delta E_{step\ 3}$ (hydration energy) is higher for Mg^{2+} because the strength of the ion–dipole attractions with water increases as the magnitude of the charge increases.

10.12 $500.0 \text{ mL water} \times \dfrac{1.00 \text{ g water}}{1 \text{ mL water}} = 500.0 \text{ g H}_2\text{O}$

$500.0 \text{ g water} \times \dfrac{56.7 \text{ g KCl}}{100 \text{ g water}} = 2.84 \times 10^2 \text{ g KCl}$

10.13 The biggest difference will be in step 1. Gases do not exist in a lattice, and the particles are already separated. Hence, step 1 will be essentially zero for gases.

10.15 $25.0 \text{ g glucose} \times \dfrac{1 \text{ mole glucose}}{180.155 \text{ g glucose}}$

$\times \dfrac{1 \text{ L solution}}{2.55 \text{ moles glucose}} \times \dfrac{1000 \text{ mL solution}}{1 \text{ L solution}}$

$= 54.4 \text{ mL solution}$

10.16 $200.0 \text{ mL solution} \times \dfrac{1 \text{ L solution}}{1000 \text{ mL solution}}$

$\times \dfrac{2.00 \text{ moles ethanol}}{1 \text{ L solution}} \times \dfrac{46.068 \text{ g ethanol}}{1 \text{ mole ethanol}}$

$= 18.4 \text{ g ethanol}$

10.17 $400.0 \text{ mL solution} \times \dfrac{1 \text{ L solution}}{1000 \text{ mL solution}}$

$\times \dfrac{2.00 \text{ moles NaCl}}{1 \text{ L solution}} \times \dfrac{58.443 \text{ g NaCl}}{1 \text{ mole NaCl}}$

$= 46.7 \text{ g NaCl required};$

add water until a volume of 400.0 mL is reached.

10.18 First get moles of NaCl needed:

$400.0 \text{ mL solution} \times \dfrac{1 \text{ L solution}}{1000 \text{ mL solution}}$

$\times \dfrac{2.00 \text{ moles NaCl}}{1 \text{ L solution}} = 0.800 \text{ mole NaCl}$

Next get volume of stock solution:

$0.800 \text{ mole NaCl} \times \dfrac{1 \text{ L solution}}{3.00 \text{ moles}}$

$\times \dfrac{1000 \text{ mL solution}}{1 \text{ L solution}}$

$= 2.67 \times 10^2 \text{ mL stock solution}$

267 mL of stock solution is needed; add water until a volume of 400.0 mL is reached.

10.20 Percent by volume $= \dfrac{90.0 \text{ mL ethanol}}{1000 \text{ mL solution}} \times 100$

$= 9.00 \text{ vol } \%$

10.21 $65.0 \text{ g solution} \times \dfrac{12.5 \text{ g sucrose}}{100 \text{ g solution}}$

$= 8.12 \text{ g sucrose}$

10.23 No, the two solids would have to be dissolved in a liquid in order to form a solution.

10.25 Yes, it is a homogeneous mixture of N_2, O_2, water vapor, and other gases.

10.26 The solute is acetic acid; the solvent is water.

10.28 (a) Solution
(b) Solution
(c) Heterogeneous mixture
(d) Solution
(e) Heterogeneous mixture

10.30 A heterogeneous mixture remains, since the mixture would not be homogeneous.

10.31 The diagram is incorrectly drawn since the + end of the dipole vector arrow is near the more electronegative oxygen atom. The oxygen should be near the head of the arrow, and the + end should be near the hydrogens.

10.33 Water would always exist in the gaseous state if there were no attractive forces between water molecules.

10.34 The analogy likens water to a crowded shopping mall because it models moving, jostling molecules that are touching one another.

10.36 In melting there are no solute–solvent interactions to provide energy for the lattice of ions to break apart, whereas in dissolving in water, the attractions between the ions and the water molecules aid in the solution process, therefore requiring less external energy.

10.38 Making room for the solute particles requires energy because the attractive forces in the solvent must be overcome.

10.40 The three steps that must occur to dissolve NaCl in water are:
(1) Overcoming the solute–solute attractive forces.
(2) Overcoming the solvent–solvent attractive forces.
(3) Forming solute–solvent attractive forces.

10.42 The associations formed between the ions and water molecules (hydration) help keep the ions from coming back together.

10.44 A water molecule could form hydrogen bonds with the alcohol, as shown below. This is an example of hydration.

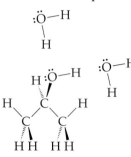

10.46 The solid solute will probably not be soluble in the solvent.

10.48 If the energy released in step 3 is less than the energy required in steps 1 and 2, the solution components will remove heat from the water and the water will get colder.

10.50 Since hexane is a nonpolar compound, there will be little interaction between the ions and the solvent. There will be little or no energy released in step 3, and the NaCl will not form a solution with C_6H_{14}.

10.52 $AlCl_3$, because the Al^{3+} ions have a much stronger interaction with water than the Na^+ ions.

10.54 "Like dissolves like" means that polar solutes dissolve in polar solvents, and nonpolar solutes dissolve in nonpolar solvents.

10.56 (a) Step 1 requires energy because attractive solute forces are being broken.
(b) Step 2 requires energy because attractive solvent forces are being broken.
(c) Step 3 requires energy because attractive forces between the solute and the solvent are being formed.
(d) Soluble, since the total energy balance is negative.
(e) Warmer, since the third step releases more energy than is absorbed by steps 1 and 2 combined.

10.57 The (aq) means that the ion is surrounded by and attracted to water molecules.

10.59 Step 1, because there is no lattice to be broken, and the molecules are already separated.

10.61 CCl_4 would be least soluble in water because it is nonpolar and will not form attractive forces with the polar water solvent molecules. CH_3OH would be most soluble because it can form very strong hydrogen bonds with water molecules.

10.64 The second law of thermodynamics says that for anything to happen, the entropy of the universe must increase.

10.66 The greater entropy state is the one in which the dye is dispersed.

10.67 Entropy is important when the net energy balance is either slightly positive or slightly negative. If the net energy balance is either a large positive or a large negative value, the entropy will not be the deciding factor.

10.69 Since NaCl is a polar compound and hexane is a nonpolar compound, they form no significant solute–solvent interactions. The net energy balance is too positive for the entropy to be the deciding factor.

10.71 Increasing the temperature adds energy to break down the attractive forces in the solid solutes (step 1). However, since there are very little attractive forces in gaseous solutes, the higher temperature merely serves to destabilize the attractive solute–solvent interactions of step 3.

10.73 The bends occurs when dissolved nitrogen in the blood is released too rapidly when a diver ascends too quickly. The decrease in pressure when the diver rises to the surface causes a decrease in the solubility of nitrogen, and the gas escapes rapidly.

10.75 NaCl shows very little increase in solubility as the temperature increases. Cerium sulfate becomes less soluble as the temperature increases.

10.76 Aquatic life suffers from oxygen depletion due to a decreased concentration of dissolved oxygen upon warming.

10.78 $100.0 \text{ lb sugar water} \times \dfrac{453.6 \text{ g sugar water}}{1 \text{ lb sugar water}}$

$\times \dfrac{420.0 \text{ g sucrose}}{100 \text{ g sugar water}} = 1.90 \times 10^5 \text{ g sucrose}$

10.79 (a)

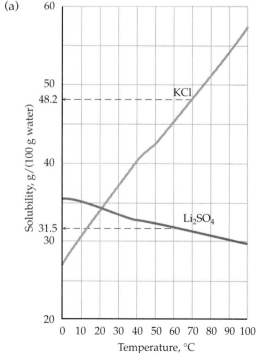

(b) KCl: 48.2 g/100 g water;
Li$_2$SO$_4$: 31.5 g/100 g water

(c) $75 \text{ g water} \times \dfrac{48.2 \text{ g KCl}}{100 \text{ g water}} = 36 \text{ g KCl}$

$75 \text{ g water} \times \dfrac{31.5 \text{ g Li}_2\text{SO}_4}{100 \text{ g water}} = 24 \text{ g Li}_2\text{SO}_4$

10.81 $5.00 \text{ lb water} \times \dfrac{453.6 \text{ g water}}{1 \text{ lb water}} \times \dfrac{40.0 \text{ g KCl}}{100 \text{ g water}}$

$= 907 \text{ g KCl}$

10.83 CH$_3$–CH$_2$–CH$_2$–CH$_2$–CH$_2$–CH$_2$–C

$\begin{array}{c} \text{O} \\ \| \\ \diagdown \\ \text{O}^- \text{ Na}^+ \end{array}$

Nonpolar hydrocarbon tail   Polar head

The nonpolar chain has an affinity for nonpolar molecules like grease and oils. The polar head has an affinity for polar molecules like water.

10.85 Since the nonpolar hydrophobic chain needs to be isolated from the water, the soap molecule forms a sphere to allow the chains to reside inside and away from the water. If the micelle were flat and two-dimensional, the water would be in direct contact with the hydrophobic chains.

10.88 The hydrophobic portion is the hydrocarbon chain. The hydrophilic portion includes the

$\begin{array}{c} \text{O} \\ \| \\ \text{O}-\text{S}-\text{O}^- \text{ Na}^+ \\ \| \\ \text{O} \end{array}$ head group.

10.89 Molarity is the number of moles of solute present in 1 liter of solution; Molarity = Moles of solute/Liter of solution.

10.91 $1.00 \text{ L solution} \times \dfrac{1.00 \text{ mole C}_{12}\text{H}_{22}\text{O}_{11}}{1 \text{ L solution}}$

$\times \dfrac{342.295 \text{ g C}_{12}\text{H}_{22}\text{O}_{11}}{1 \text{ mole C}_{12}\text{H}_{22}\text{O}_{11}}$

$= 342 \text{ g C}_{12}\text{H}_{22}\text{O}_{11}$

Place 342 g C$_{12}$H$_{22}$O$_{11}$ in a volumetric flask, dissolve in water, and then add enough water to bring the total volume to 1.00 L.

10.93 $0.500 \text{ L solution} \times \dfrac{1.50 \text{ moles C}_{12}\text{H}_{22}\text{O}_{11}}{1 \text{ L solution}}$

$\times \dfrac{342.295 \text{ g C}_{12}\text{H}_{22}\text{O}_{11}}{1 \text{ mole C}_{12}\text{H}_{22}\text{O}_{11}}$

$= 257 \text{ g C}_{12}\text{H}_{22}\text{O}_{11}$

10.95 You would fire him. To prepare a 2.00 M solution of NaCl enough water should be added to 116.886 g of NaCl to bring the total volume of solution to 1.00 L. Adding exactly

1.00 L of water will not result in a solution volume of 1.00 L.

10.97 $45.0 \text{ mL} \times \dfrac{1 \text{ L solution}}{1000 \text{ mL}} \times \dfrac{0.250 \text{ mole sucrose}}{1 \text{ L solution}}$

$\times \dfrac{342.295 \text{ g sucrose}}{1 \text{ mole sucrose}} = 3.85 \text{ g sucrose}$

10.99 $100.0 \text{ g glucose} \times \dfrac{1 \text{ mole glucose}}{180.1548 \text{ g glucose}}$

$\times \dfrac{1 \text{ L solution}}{0.250 \text{ mole glucose}} \times \dfrac{1000 \text{ mL solution}}{1 \text{ L solution}}$

$= 2.22 \times 10^3 \text{ mL solution}$

10.101 To calculate how much NaCl is needed:

$100.0 \text{ mL} \times \dfrac{1 \text{ L solution}}{1000 \text{ mL}} \times \dfrac{4.00 \text{ moles NaCl}}{1 \text{ L solution}}$

$= 0.400 \text{ mole NaCl needed}$

To find the volume of 4.50 M NaCl needed to obtain 0.400 mole NaCl:

$0.400 \text{ mole NaCl} \times \dfrac{1 \text{ L stock solution}}{4.50 \text{ moles NaCl}}$

$= 0.0889 \text{ L stock solution}$

Instructions: Put 0.0889 L (88.9 mL) of stock solution into a flask, and add enough water to bring the total volume to 100.0 mL of solution.

10.103 (1) Percent by mass $= \dfrac{\text{Grams of solute}}{\text{Grams of solution}} \times 100$

(2) Percent by volume =

$\dfrac{\text{Volume of solute}}{\text{Volume of solution}} \times 100$

(3) Percent by mass/volume =

$\dfrac{\text{Grams of solute}}{\text{Volume of solution}} \times 100$

10.105 $\dfrac{22.5}{22.5 + 49.6} \times 100 = 31.2 \text{ wt }\%$

10.107 $2.00 \text{ kg solution} \times \dfrac{1000 \text{ g solution}}{1 \text{ kg solution}}$

$\times \dfrac{30.0 \text{ g NaCl}}{100 \text{ g solution}} = 600 \text{ g NaCl}$

Place 600 g (0.600 kg) of NaCl into a flask. Add 1.40 kg of water to give a total of 2.00 kg of solution.

10.109 $200.0 \text{ mL alcohol} \times \dfrac{100 \text{ mL solution}}{35.0 \text{ mL alcohol}}$

$= 571.4 \text{ mL solution}$

CHAPTER 11

11.2 (a)

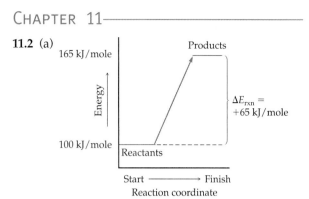

(b) Uphill   (c) $\Delta E_{rxn} = +65$ kJ/mole
(d) Cold, because it is endothermic

11.3 (a) Higher energy; since ΔE_{rxn} is negative, the reactants are higher on the energy profile.
(b) 470 kJ/mole (450 kJ up from the 20 kJ/mole of the products)
(c) Exothermic; energy is being released during the reaction because the sign of ΔE_{rxn} is negative.
(d) Downhill
(e)

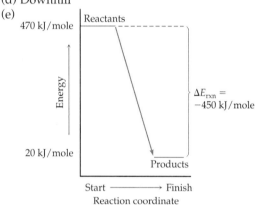

11.5 (a) $\Delta E_{rxn} = +250$ kJ/mole
(b) $\Delta E_{reverse\ rxn} = -250$ kJ/mole

11.6 (a) $\Delta E_{rxn} = -250$ kJ/mole
(b) $\Delta E_{reverse\ rxn} = +250$ kJ/mole

11.8 (a) Endothermic; the overall reaction absorbs energy because it takes more energy to break the reactant bonds than is released on forming the bonds in the product.
(b) $\Delta E_{rxn} = +200$ kJ/mole

11.9 200 kJ/mole

11.10 Energy profile for the reaction in Practice Problem 11.8:

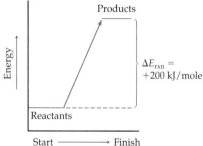

Energy profile for the reaction Practice Problem 11.9:

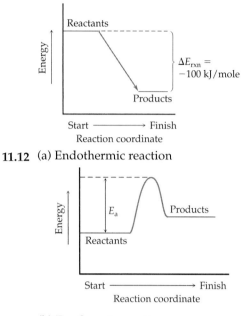

11.12 (a) Endothermic reaction

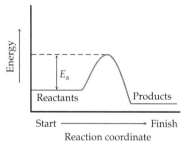

(b) Exothermic reaction

11.13 False; it is the E_a barrier that determines the speed of the reaction.

11.15 (a) The general rule is that the rate of a reaction will double with every 10°C increase in temperature. Thus, the number of collisions with sufficient energy will double from 100 collisions/sec to 200 collisions/sec.
(b) Increasing the temperature causes the molecules to possess more energy, move faster, and have more collisions. If the molecules have greater energy, more of them will have the minimum energy required for a reaction to occur.
(c) Approximately 40 CH_3OH molecules/sec

11.16 (a) An orientation factor of 0.1 means that only 10% of all the collisions are oriented properly to lead to products.
(b) Reaction rate = (100 sufficiently energetic collisions/sec) × (0.1)
= 10 effective collisions/sec
The rate of product formation is therefore 10 CH_3OH molecules/sec.

11.18 Rate = $k[H_2O_2]^x[I^-]^y[H^+]^z$

11.19 (a) k for the reaction in Practice Problem 11.17 is much larger than k for the reaction in Practice Problem 11.18.

(b) E_a for the reaction in Practice Problem 11.17 is much smaller than E_a for the reaction in Practice Problem 11.18. (Recall that k and E_a are inversely related.)

(c) (1) Increase the concentrations of the reactants; (2) increase the temperature of the reaction; (3) add a catalyst.

11.20 Concentration is proportional to molecules per unit volume. The more molecules there are in a certain volume, the more collisions will occur.

11.22 (a) 1   (b) 2   (c) 4

(d) The rate of reaction is quadrupled.

11.23 Rate $= k[NO]^2[O_2]$

11.25 Answers will vary, but here is one.

(a)–(b)

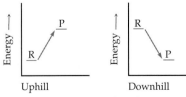

$$X_2 + Y_2 \longrightarrow 2\,XY$$

(c) See if you can detect intermediates y, y_3, or x during the course of the reaction.

11.26 $Y_2 \longrightarrow Y + Y$

11.27 Bonds broken: 1 C=C bond, 4 C–H bonds, 3 O=O bonds; bonds formed: 4 C=O bonds, 4 O–H bonds

11.29 Yes; the Cl is substituted for the I during the reaction.

11.31 The mechanism of a chemical reaction is the exact set of steps that molecules go through on changing from reactants to products.

11.33 Chemists study the rate of a reaction and how the rate changes as the temperature and the reactant concentration are varied.

11.36 This reaction will absorb energy from the surroundings since the product has twice as much energy in it as the reactant.

11.38 Compound C is at a higher energy level; D has less energy than C because some of the energy was released into the surroundings when D was formed from C.

11.40 This reaction is endothermic; $\Delta E_{rxn} = +40$ kJ/mole

11.41 The reaction is exothermic.

11.43 An energy-uphill reaction is one in which the products have more energy than the reactants. An energy-downhill reaction is one in which the products have less energy than the reactants.

11.45 The reactants must absorb energy from the surroundings.

11.46 In an energy-releasing reaction, the products are at a lower energy level than the reactants. For all reactions, $\Delta E_{rxn} = E_{products} - E_{reactants}$. If the energy of the reactants is larger than the energy of the products, the ΔE_{rxn} will always be a negative number.

11.48 The overall reaction releases 800 kJ/mole ($\Delta E_{rxn} = -800$ kJ/mole). You are told it takes 800 kJ/mole to break old bonds, so forming new ones must release 1600 kJ/mole.

11.50 The larger the E_a of a reaction, the slower the reaction.

11.52

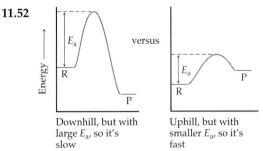

Downhill, but with large E_a, so it's slow   Uphill, but with smaller E_a, so it's fast

11.54 The transition state is the form that exists when enough energy has been added to the reaction mixture to cause the reactant bonds to be almost broken and the product bonds to be almost formed.

11.55 $\Delta E_{rxn} = +30$ kJ/mole; the reaction is endothermic; $E_a = 70$ kJ/mole.

11.58 Increasing the temperature would increase the rate of a reaction because it increases the energy of the reactants. This means that more reactant molecules have enough energy to climb the energy barrier (E_a).

11.59 The reactant molecules that collide with enough energy (greater than or equal to E_a) and the proper orientation can become product molecules.

11.61 No effect

11.62 The size of the E_a depends on the nature of the reactant molecules. One reaction might have a much larger E_a than another because its reactants have very strong bonds. It would therefore take a large amount of energy to break the reactant bonds, resulting in a large value for the E_a.

11.64 In some substitution reactions the new substituting group must approach the molecule from the back-side in order for the reaction to occur.

11.66 Because it is the predicted rate based on energy and orientation of the particular reactants, which are inherent properties.

11.68 By lowering the activation energy, E_a.

11.69 The Zn^{2+} forms a bond to the OH group, weakening the bond between the OH group and the C to which H is bound. This makes it much easier to break than the bond between OH and C, lowering the activation energy for the reaction and speeding it up.

11.71 The substrate is the reactant molecule that fits into an enzyme's active site.

11.73 Scientists made modified versions of the PABA key to fit into the enzyme lock and produce faulty vitamin B.

11.75 Increasing the concentration of a reactant will increase the number of collisions between reactant molecules. The more collisions, the faster the rate of reaction.

11.77 The rate constant k increases with temperature because k partially depends on the fraction of collisions with energy greater than E_a. As the temperature increases, more collisions will occur with energy greater than E_a, and k will increase.

11.79 The inherent factors are the fraction of collisions with energy greater than E_a and the fraction of collisions with the proper orientation.

11.81 (a) k

(b) Orders

(c) Find the sum of the individual reactant orders.

(d) $x = 2$ and $y = 1$; the overall order of the reaction is 3

(e) The value of its order is 0; that means the rate of the reaction does not depend on the concentration of this reactant.

11.83 The orders are obtained by running experiments that determine how the rate of a reaction changes when the concentrations of the individual reactants are changed.

11.85 When the concentration of that reactant is doubled, the rate will quadruple.

11.87 The rate of the reaction will double. When the volume of the balloon is halved, this has the effect of doubling the concentration of the reactants.

11.89 The rate of the reaction will increase eightfold.

11.91 A kinetics experiment is one in which we vary the reactant concentrations and observe what happens to the rate of the reaction. The exponents in the experimental rate law are determined directly from kinetics experiments.

11.92 The order of the reaction with respect to H_2O_2 can be determined from experiments 1 and 2. We see that doubling the concentration of H_2O_2 causes the reaction rate to double. This means that the reaction is first-order with respect to H_2O_2.

The order of the reaction with respect to I^- can be determined from experiments 1 and 3. We see that doubling the concentration of I^- causes the reaction rate to double. This means that the reaction is first-order with respect to I^-.

The order of the reaction with respect to H^+ can be determined from experiments 1 and 4. We see that doubling the concentration of H^+ has no effect on the rate. This means that the reaction is zero-order with respect to H^+.

The rate law for the reaction is
Rate = $k[H_2O_2]^1[I^-]^1 = k[H_2O_2][I^-]$
The overall order of the reaction is 2.

11.93 The order of the reaction with respect to A can be determined from experiments 1 and 2. We see that doubling the concentration of A causes the reaction rate to double. This means that the reaction is first-order with respect to A.

The order of the reaction with respect to B can be determined from experiments 1 and 3. We see that doubling the concentration of B causes the reaction rate to quadruple. This means that the reaction is second-order with respect to B.

The rate law for the reaction is
Rate = $k[A]^1[B]^2 = k[A][B]^2$
The overall order of the reaction is 3.
The reaction confirms that the exponents in the rate law cannot be taken from the overall balanced equation.

11.95 Four molecules would have to come together with the proper orientation. This is very unlikely.

11.97 An elementary step is a step in a mechanism that shows the actual effective collisions that take place between molecules during a reaction.

11.98 Rate = $k[A][B]^2[C]$

11.99 The slowest step in a mechanism.

11.102 The postulated mechanism cannot be correct if it does not agree with the experimentally determined rate law.

11.104 The postulated mechanism may possibly be the correct mechanism, but there may be others that will also agree with the experimental data. Incorrect mechanisms can be ruled out by experimental evidence, but correct mechanisms cannot be easily proved.

11.105 (a) The reaction rate will double when the concentration of CH_3C-Br is doubled, triple when it is tripled, etc. The rate is not affected by the concentration of H_2O.

(b) Mechanism II can be ruled out because it does not agree with the experimental evidence. Its one-step mechanism would mean that the reaction would be first-order in both CH_3C-Br and H_2O.

Mechanism I is plausible. The mechanism agrees with the experimental data, and the sum of the individual steps yields the overall balanced equation.

11.107 Rate = $k[Cl_2]$; only the reactants in the rate-determining step (raised to the power of their coefficients) appear in the rate law.

CHAPTER 12

12.2 You want almost all of the reactants to be converted to products.

12.3 For reactions that go to completion the equilibrium lies so far to the right (products side) that there are essentially no reactants present at equilibrium. The reverse reaction can be ignored, and the reaction is considered a one-way reaction in the forward direction.

12.5 $K_{eq} = \dfrac{[HI]^2}{[H_2][I_2]}$

12.6 $K_{eq} = \dfrac{[Fe(SCN)^{2+}]}{[Fe^{3+}][SCN^-]}$

12.8 To the right. If the reaction went to completion (all the way to the right), the concentrations would be [HI] = 0.200 M; [H$_2$] = 0.000 M; [I$_2$] = 0.000 M. The actual equilibrium concentrations are close to these.

12.9 $K_{eq} = \dfrac{[CS_2][H_2]^4}{[CH_4][H_2S]^2}$

$= \dfrac{(6.10\times10^{-3})(1.17\times10^{-3})^4}{(2.35\times10^{-3})(2.93\times10^{-3})^2} = 5.71\times10^{-7}$

12.11 The reaction will shift to the left.

12.12 No shift in the reaction since N$_2$ is neither a reactant nor a product in this reaction.

12.14 (a) is at low temperature; (b) is at high temperature. At low temperature the equilibrium will lie toward the left, the colorless N$_2$O$_4$ gas. At higher temperature the equilibrium will lie toward the right, the brown NO$_2$ gas.

12.15 (a) $2\,CO(g) + O_2(g) \rightleftharpoons 2\,CO_2(g) + \text{Heat}$
A negative ΔE value means that the reaction is exothermic.
(b) The equilibrium will shift to the left. The reaction will shift in such a way as to use up the additional heat.

12.16 Equilibrium is the state that occurs when the rate of the forward reaction equals the rate of the reverse reaction.

12.18 No; it only appears to have stopped because no more product is being formed since the rate of the reverse reaction equals the rate of the forward reaction and product is disappearing at the same rate it is being formed. Since reactions are still occurring, the state is called "dynamic equilibrium."

12.19 The position of a reaction's equilibrium refers to whether there are more products or more reactants present at equilibrium. If there are more reactants, the position of equilibrium is to the left; if there are more products, the position of equilibrium is to the right.

12.21 Far to the left.

12.23 The reaction might produce very little product if equilibrium lies to the left.

12.25 Initially no products of the forward reaction are present. Since these products are the reactants for the reverse reaction, the reverse reaction rate is initially zero.

12.27 The reverse reaction would be fastest since only reactants for the reverse reaction would be present.

12.28 Yes. The reverse reaction will produce products that are the reactants for the forward reaction. As the reaction proceeds the reverse reaction will slow down and the forward reaction will speed up until the rates become equal. At this point equilibrium is attained.

12.30 $k_{forward}[\text{Reactants}]^{order} = k_{reverse}[\text{Products}]^{order}$

12.31 No; the reverse rate should decrease when equilibrium is reached.

12.34 k values change only with temperature, not with concentration.

12.36 K_{eq} will be greater than 1 because the concentration of products is greater than the concentration of reactants if $k_f > k_r$, and so $k_f/k_r > 1$.

12.38 $K_{eq} = \dfrac{[C][D]}{[A]^2[B]^3}$

12.40 The two K_{eq} values are inversely related.

12.41 $K_{eq} = \dfrac{[NO_2]^2}{[N_2O_4]}$

12.43 You could place some N$_2$O$_4$ in a container, and allow the reaction to proceed until it reaches equilibrium. Then determine the concentration of N$_2$O$_4$ and NO$_2$ at equilibrium, and substitute those values into the K_{eq} expression.

12.44 No. The value of K_{eq} depends only on the temperature.

12.46 The position of equilibrium for this reaction at 25°C lies farther to the left than at 2000°C since the K_{eq} is smaller at 25°C (2.3×10^{-9}) than at 2000°C (4.0×10^{-4}).

12.48 No; an equilibrium constant of 2.05 means that the equilibrium lies near the middle. Only reactions whose equilibria lie extremely far to the right may use a single arrow to the right, indicating completion.

12.50 No; $K_{eq} = 1.5\times10^{-6}$ means that the equilibrium lies very far to the left.

12.51 $K_{eq} = \dfrac{[CH_4][H_2O]}{[CO][H_2]^3}$

$[CO] = \dfrac{0.613 \text{ mole}}{10.0 \text{ liters}} = 0.0613 \text{ M},$

$[H_2] = \dfrac{1.839 \text{ moles}}{10.0 \text{ liters}} = 0.1839 \text{ M},$

$[CH_4] = \dfrac{0.387 \text{ mole}}{10.0 \text{ liters}} = 0.0387 \text{ M},$

$[H_2O] = \dfrac{0.387 \text{ mole}}{10.0 \text{ liters}} = 0.0387 \text{ M}$

$K_{eq} = \dfrac{(0.0387)(0.0387)}{(0.0613)(0.1839)^3} = 3.93$

This is close in magnitude to 1, so the position of equilibrium is roughly near the middle.

12.53 Because the value always remains the same if the temperature is not changed.

12.55 Reaction (a) goes to completion.
Reaction (b) would not occur.

12.56 $4\,HCl + O_2 \rightleftharpoons 2\,H_2O + 2\,Cl_2$

12.58 They would be the same.

12.59 The value of K_{eq} would remain the same.

12.61 (a) The reaction will shift to the right.
(b) The reaction will shift to the left.
(c) The reaction will shift to the left.
(d) The reaction will shift to the left.
(e) The reaction will not shift.

12.62 When you remove reactants the reaction will shift to the left, converting some of the products into reactants.

12.64 (a) Initially at equilibrium

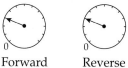

Forward   Reverse

(b) PCl_5 added

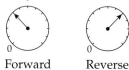

Forward   Reverse

(c) Restored equilibrium

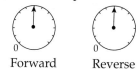

Forward   Reverse

(d) Reaction shifts left

12.66 (a) The equilibrium lies to the left ($K_{eq} < 1$).
(b) No; since the equilibrium lies to the left, there are significant amounts of reactants present at equilibrium.
(c) You could "drive" the reaction to completion by continually causing the reaction to shift to the right.

12.68 (a) Endothermic, because increasing the temperature causes the reaction to shift to the right. Heat must therefore be a reactant in the reaction.
(b) K_{eq} increases
(c) Heat + $N_2O_4(g) \rightleftharpoons 2\,NO_2(g)$

12.69 Exothermic, because the equilibrium shifts to the left when the reaction is heated.

12.71 K_{eq} for an exothermic reaction will decrease upon heating and increase upon cooling. For an exothermic reaction heat is a product. Adding heat will cause the equilibrium to shift to the left (decreasing K_{eq}), and removing heat will cause the equilibrium to shift to the right (increasing K_{eq}).

12.73 Decreased; adding heat will cause the equilibrium to shift to the left, decreasing the amount of methanol.

12.74 Cooling the reaction mixture will slow down the rate of the reaction. It will take longer for the reaction to reach the new equilibrium state.

12.75 The catalyst has no effect on the position of equilibrium or on the value of K_{eq}. A catalyst speeds up the forward and reverse reactions by the same amount.

12.77 (a) K_{eq} gets larger as the reaction is heated, so the reaction shifts to the right upon heating.
(b) Endothermic; shifting to the right upon heating means the reaction consumes heat on going from left to right.

CHAPTER 13

13.2 (a) Electrolyte   (b) Nonelectrolyte
(c) Electrolyte   (d) Nonelectrolyte
(e) Electrolyte   (f) Electrolyte
(g) Electrolyte

13.3 (a)

NH_4^+ Br^-
Br^- NH_4^+
NH_4^+ Br^-
Br^- NH_4^+

(b) Yes; this means that the NH_4Br breaks up into ions.
(c)

Br^- NH_4^+

13.5 $H_2CO_3 + 2\,H_2O \rightleftharpoons 2\,H_3O^+ + CO_3^{2-}$

13.6 $H_2CO_3 + H_2O \rightleftharpoons H_3O^+ + HCO_3^-$
$HCO_3^- + H_2O \rightleftharpoons H_3O^+ + CO_3^{2-}$

13.8 1 mole of $NaOH$ will neutralize 1 mole of HCl; 2 moles of $NaOH$ will neutralize 1 mole of H_2SO_4.

13.9 0.05 mole

13.10

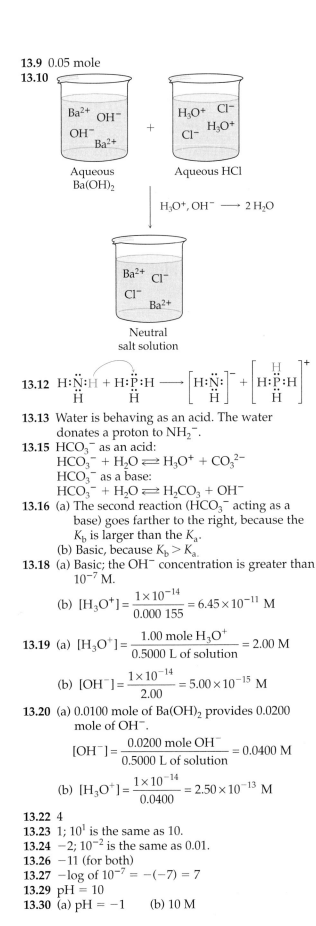

Aqueous
Ba(OH)$_2$ +   Aqueous HCl

H$_3$O$^+$, OH$^-$ ⟶ 2 H$_2$O

Neutral
salt solution

13.12 H:N̈:H + H:P̈:H ⟶ [H:N̈:H]$^-$ + [H:P̈:H (with H on top)]$^+$

13.13 Water is behaving as an acid. The water donates a proton to NH$_2^-$.

13.15 HCO$_3^-$ as an acid:
HCO$_3^-$ + H$_2$O ⇌ H$_3$O$^+$ + CO$_3^{2-}$
HCO$_3^-$ as a base:
HCO$_3^-$ + H$_2$O ⇌ H$_2$CO$_3$ + OH$^-$

13.16 (a) The second reaction (HCO$_3^-$ acting as a base) goes farther to the right, because the K_b is larger than the K_a.
(b) Basic, because $K_b > K_a$.

13.18 (a) Basic; the OH$^-$ concentration is greater than 10^{-7} M.

(b) $[H_3O^+] = \dfrac{1 \times 10^{-14}}{0.000\ 155} = 6.45 \times 10^{-11}$ M

13.19 (a) $[H_3O^+] = \dfrac{1.00\ \text{mole H}_3\text{O}^+}{0.5000\ \text{L of solution}} = 2.00$ M

(b) $[OH^-] = \dfrac{1 \times 10^{-14}}{2.00} = 5.00 \times 10^{-15}$ M

13.20 (a) 0.0100 mole of Ba(OH)$_2$ provides 0.0200 mole of OH$^-$.

$[OH^-] = \dfrac{0.0200\ \text{mole OH}^-}{0.5000\ \text{L of solution}} = 0.0400$ M

(b) $[H_3O^+] = \dfrac{1 \times 10^{-14}}{0.0400} = 2.50 \times 10^{-13}$ M

13.22 4
13.23 1; 10^1 is the same as 10.
13.24 −2; 10^{-2} is the same as 0.01.
13.26 −11 (for both)
13.27 −log of 10^{-7} = −(−7) = 7
13.29 pH = 10
13.30 (a) pH = −1   (b) 10 M

13.32 A solution 100 times less acidic than a solution with pH = 2.56 has pH = 4.56;
$[H_3O^+] = 10^{-4.56} = 2.75 \times 10^{-5}$
13.33 $[H_3O^+] = 10^{-5.55} = 2.82 \times 10^{-6}$;

$[OH^-] = \dfrac{1 \times 10^{-14}}{2.82 \times 10^{-6}} = 3.55 \times 10^{-9}$

13.35 CO$_3^{2-}$ is the conjugate base of HCO$_3^-$.
13.36 H$_2$CO$_3$ is the conjugate acid of HCO$_3^-$.
13.38 NH$_3$ + H$_2$O ⇌ NH$_4^+$ + OH$^-$
13.39 A buffer solution; any solution containing a large amount of weak acid and its conjugate base, or a weak base and its conjugate acid is a buffer solution.
13.40 NH$_4^+$ + OH$^-$ ⟶ NH$_3$ + H$_2$O
13.41 NH$_3$ + H$_3$O$^+$ ⟶ NH$_4^+$ + H$_2$O
13.42 Sour taste, turned litmus dye bright red, and reacted with active metals such as zinc, producing hydrogen gas
13.44 An indicator is a compound that is a different color in the presence of an acid than it is in the presence of a base. Litmus dye is an example of an indicator.
13.46 An electrolyte is a substance whose aqueous solution is capable of conducting electricity. A nonelectrolyte is a substance whose aqueous solution is not capable of conducting electricity. See Table 13.1.
13.47 Ethanol does not break up into ions when dissolved in water, whereas magnesium chloride does.
13.50 False. Some molecular compounds, such as HX acids, dissociate into H$^+$ (or H$_3$O$^+$) ions when placed in water.
13.51 Ions
13.53 (a) Yes; because NaCl does not exist as discrete molecules whose atoms are held together by covalent bonds.
(b) Because it is a soluble ionic compound
13.55 (a) Electrolyte   (b) Nonelectrolyte
(c) Nonelectrolyte   (d) Electrolyte
(e) Electrolyte   (f) Electrolyte
(g) Nonelectrolyte
13.56 Na$_2$SO$_4$(aq) ⟶ 2 Na$^+$(aq) + SO$_4^{2-}$(aq)
13.59 Glucose is a molecular compound that does not dissociate when dissolved in water.
13.61 HCl is a strong electrolyte that completely dissociates into ions in solution; 0.1 mole of HCl gas will produce 0.1 mole of H$_3$O$^+$ and 0.1 mole of Cl$^-$. HF, on the other hand, is a weak electrolyte that only partially dissociates into ions in solution; 0.1 mole of HF gas will produce fewer ions (and conduct less electricity) than 0.1 mole of HCl.
13.63 The dissociation equilibrium lies very far to the right for a strong electrolyte. (We say the reaction goes to completion.) The equilibrium lies to the left for a weak electrolyte.

13.65 H_3O^+

13.67 It is a weak electrolyte. An equilibrium constant value of 8.2×10^{-6} means that the equilibrium lies to the left, and most of the molecules are undissociated.

13.69 Answers will vary. Two possible strong molecular electrolytes are HBr (hydrobromic acid) and HCl (hydrochloric acid):
(1) $HBr + H_2O \longrightarrow H_3O^+ + Br^-$
(2) $HCl + H_2O \longrightarrow H_3O^+ + Cl^-$
Two possible weak molecular electrolytes are HF (hydrofluoric acid) and NH_3 (ammonia):
(1) $HF + H_2O \rightleftharpoons H_3O^+ + F^-$
(2) $NH_3 + H_2O \rightleftharpoons NH_4^+ + OH^-$

13.71 All the compounds are molecular. None of them contain a metal ion or two polyatomic ions.

13.73 Because they all produce H_3O^+ ions in solution.

13.74 False; HF is a weak acid and HCl is a strong acid. Therefore, a beaker of 1.0 M HCl is more acidic than a beaker of 1.0 M HF.

13.76 Diprotic means that there are two dissociable hydrogens in the acid. H_2SO_4 is a diprotic acid.
$$H_2SO_4 + H_2O \longrightarrow H_3O^+ + HSO_4^-$$
$$HSO_4^- + H_2O \rightleftharpoons H_3O^+ + SO_4^{2-}$$

13.78 CH_3COOH (acetic acid) would be in highest concentration after water, because it is a weak acid that dissociates to a very small extent in water. The equilibrium lies to the left, meaning that there is mostly undissociated acid present in solution.

13.79 H_3O^+ and NO_3^- would be in highest concentration, because HNO_3 (nitric acid) is a strong acid that dissociates completely in water, converting all the HNO_3 into H_3O^+ and NO_3^-.

13.81

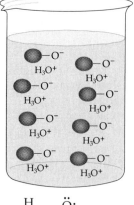

13.82

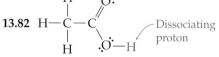

13.85 False; the degree of dissociation in water, not the number of dissociable hydrogens, determines whether an acid is weak or strong.

13.87 Acid–base neutralization is the production of water from the reaction between the H_3O^+ supplied by an acid with the OH^- supplied by a base.
$$H_3O^+ + OH^- \longrightarrow 2\,H_2O$$

13.89 When calcium hydroxide, $Ca(OH)_2$, reacts with HCl, a water-soluble salt, $CaCl_2$, is formed. All water-soluble salts are strong electrolytes, completely dissociating into ions in solutions.

13.91 $0.50 \text{ mole } H_2SO_4 \times \dfrac{2 \text{ moles } H_3O^+}{1 \text{ mole } H_2SO_4}$
Diprotic
$\times \dfrac{1 \text{ mole } OH^-}{1 \text{ mole } H_3O^+} \times \dfrac{1 \text{ mole } Ba(OH)_2}{2 \text{ moles } OH^-}$
$H_3O^+ + OH^- \longrightarrow 2\,H_2O$ From formula
Neutralization $Ba(OH)_2$
$= 0.50 \text{ mole } Ba(OH)_2$

13.94

Aqueous $Ba(OH)_2$ [OH^-, OH^-, Ba^{2+}] + Aqueous H_2SO_4 [H_3O^+, H_3O^+, SO_4^{2-}] → Salt solution ($BaSO_4$) [Ba^{2+}, SO_4^{2-}]

Aqueous $Ba(OH)_2$ [OH^-, OH^-, Ba^{2+}] + Aqueous HNO_3 [H_3O^+, NO_3^-, H_3O^+, NO_3^-] → Salt solution $Ba(NO_3)_2$ [NO_3^-, Ba^{2+}, NO_3^-]

13.95 $NH_3 + H_2O \rightleftharpoons NH_4^+ + OH^-$

13.97 NH_3 contains no OH^- and would not be able to dissociate to produce OH^- ions in solutions, which is required for it to be classified as a base by the Arrhenius definition of bases.

13.99 When ammonia reacts with water, the water can be thought of as an acid because water donates a proton to ammonia:
$$NH_3 + H_2O \rightleftharpoons NH_4^+ + OH^-$$

13.101 (a) Weakly acidic
(b) Base, because it is accepting a proton from NH_4^+
(c) K_a

13.103 (a) $H^- + H_2O \longrightarrow H_2 + OH^-$
(b) The water donates a proton to H^-, forming H_2 gas.

13.105 The lone pair of electrons

13.107 The C_2H_5OH donates a proton to the H^-.

13.109 Ammonia produces relatively few OH^- in aqueous solution.

13.111 LiOH is a stronger base than NH_3. LiOH is a hydroxide of a group IA metal, and is therefore a strong base. NH_3 is a weak base.

13.113 (a) $HSO_4^- + H_2O \rightleftharpoons H_3O^+ + SO_4^{2-}$
(b) $HSO_4^- + H_2O \rightleftharpoons H_2SO_4 + OH^-$
(c) You need the K_a and the K_b for the reactions in parts (a) and (b). If the K_a is larger, the solution is weakly acidic. If the K_b is larger, the solution is weakly basic.

13.115 OH^-

13.116 (a) $H_2PO_4^-$, HPO_4^{2-}, PO_4^{3-}
(b) PO_4^{3-}

13.118 It is breaking up into its component ions, H^+ (or H_3O^+) and OH^-.

13.120 $K_w = 1 \times 10^{-14} = [H_3O^+] \times [OH^-]$

13.122 $[H_3O^+]$ and $[OH^-]$ are both equal to 1.0×10^{-7} mole/L at 25°C

13.124 An aqueous solution is acidic if $[H_3O^+] > [OH^-]$. An aqueous solution is acidic if $[H_3O^+] > 1.0 \times 10^{-7}$.

13.126 True; there is always some H_3O^+ present in any aqueous solution due to the autoionization of water.

13.127 True; the product of $[H_3O^+]$ and $[OH^-]$ must always equal 1×10^{-14} in an aqueous solution. If one concentration increases, the other must decrease.

13.129 $K_w = 1 \times 10^{-14} = [H_3O^+] \times [OH^-]$
$= [1.0\ M] \times [OH^-] = [OH^-]$
So $[OH^-] = 1.0 \times 10^{-14}$ M. The solution is acidic because $[H_3O^+] > [OH^-]$.

13.131 (a) LiOH dissociates to produce 1 mole of OH^- for every mole of LiOH.

$$[OH^-] = \frac{2.50 \text{ moles } OH^-}{4.00 \text{ L of solution}} = 0.625 \text{ M}$$

(b) $[H_3O^+] = \dfrac{1 \times 10^{-14}}{0.625} = 1.6 \times 10^{-14}$ M

13.133 (a) Molar mass of $Mg(OH)_2 = 58.319$ g/mole.

$$\text{Moles } Mg(OH)_2 = 2.40 \text{ g} \times \frac{1 \text{ mole}}{58.319 \text{ g}}$$
$$= 0.0412 \text{ mole } Mg(OH)_2$$

$Mg(OH)_2$ dissociates to produce 2 moles of OH^- for every mole of $Mg(OH)_2$; 0.0412 mole $Mg(OH)_2$ produces 0.0824 mole OH^-.

$$[OH^-] = \frac{0.0824 \text{ mole } OH^-}{4.00 \text{ L of solution}} = 0.0206 \text{ M}$$

(b) $[H_3O^+] = \dfrac{1 \times 10^{-14}}{0.0206} = 4.85 \times 10^{-13}$ M

13.135 (a) Acidic; the $[OH^-] < 10^{-7}$, so the $[H_3O^+] > 10^{-7}$.
(b) If the $[OH^-]$ is 1×10^{-11} (same as 10^{-11}), then by the K_w relationship, the H_3O^+ concentration must be 1×10^{-3} (same as 10^{-3}), since $(10^{-11}) \times (10^{-3}) = 10^{-14}$.

13.137 -34

13.139 The log of both 10^0 and 1 is 0 ($10^0 = 1$).

13.140 (d) Between 1 and 2

13.141 The log of 10 is 1, and the log of 100 is 2; 60 is between 10 and 100, so the log of 60 must be between 1 and 2.

13.144 True

13.146 At pH = 7, $[H_3O^+] = [OH^-] = 10^{-7}$ M.

13.148 pH $= -\log [H_3O^+] = -\log 10^{-6} = 6$; acidic

13.149 pH $= -\log [H_3O^+] = -\log(6.40 \times 10^{-9}) = 8.19$; basic

13.151 If $[OH^-] = 2.00 \times 10^{-3}$, $[H_3O^+] = 5.00 \times 10^{-12}$ (from the K_w equation).
pH $= -\log [H_3O^+] = -\log 5.00 \times 10^{-12} = 11.3$; basic

13.152 Solution A is 1000 times more acidic.

13.154 $[H_3O^+] = 10^{-pH} = 10^{-4}$ M; acidic

13.156 $[H_3O^+] = 10^{-pH} = 10^{-(-1)} = 10$ M; using the K_w, $[OH^-] = 10^{-15}$ M; acidic

13.157 The solutions will have different pH's. The pH of the HCl solution will be lower than that of the acetic acid solution, because acetic acid is a weak acid, and HCl is a strong acid.

13.159 No; the weak acid will have one more proton than its conjugate base. Since protons have a positive charge, the acid will have one more positive charge than its conjugate base.

13.160 The conjugate acid of aniline is a weak acid.

13.161 The solution would be neutral. NO_3^- has no tendency to accept protons since it is from the strong acid HNO_3. Strong acids give rise to conjugates that are so weak that they are not acids or bases.

13.164 A solution with buffering ability resists a change in pH.

13.165 No; pure water does not contain significant amounts of a weak acid and its conjugate base or a weak base and its conjugate acid. This is required for a solution to demonstrate buffering ability.

13.168 No; if you add enough strong acid or strong base, the buffer will be exhausted.

13.170 (a) Decrease
(b) The pH changes because the added strong acid is converted to a weak acid (the conjugate acid in the buffer system).

13.171 The NaOH will react with half of the acetic acid to produce sodium acetate:

$$CH_3COOH + NaOH \longrightarrow NaOOCCH_3 + H_2O$$

The solution will then contain significant amounts of acetic acid and sodium acetate, components of a buffer solution.

13.173 Dissolve HClO and NaClO (to provide the conjugate base ClO^-) in water.

13.174 (a) HPO_4^{2-}
(b) $HPO_4^{2-} + H_3O^+ \longrightarrow H_2PO_4^- + H_2O$
$H_2PO_4^- + OH^- \longrightarrow HPO_4^{2-} + H_2O$
(c) H_3PO_4

13.174 (d) $H_3PO_4 + OH^- \longrightarrow H_2PO_4^- + H_2O$
$H_2PO_4^- + H_3O^+ \longrightarrow H_3PO_4 + H_2O$

13.176 The conjugate base of the weak acid reacts with added strong acid to convert it to the weak acid.

13.178 The weak acid in the buffer converts added strong base to weak base.

13.180 The components of a hypochlorous acid buffer system are $HClO$ and ClO^-.

$$ClO^- + H_3O^+ \longrightarrow HClO + H_2O$$
$$HClO + OH^- \longrightarrow ClO^- + H_2O$$

CHAPTER 14

14.2 Radioactive

14.3 $2 \text{ moles protons} \times \dfrac{1.007\ 30\ g}{\text{mole protons}} = 2.014\ 60\ g$

$2 \text{ moles neutrons} \times \dfrac{1.008\ 70\ g}{\text{mole neutrons}} = 2.017\ 40\ g$

$2 \text{ moles electrons} \times \dfrac{0.000\ 55\ g}{\text{mole electrons}} = 0.001\ 10\ g$

What the mass of 1 mole $\longrightarrow$ 4.033 10 g
of ^4_2He should be
What the mass of 1 mole $\longrightarrow$ -4.003 g
of ^4_2He actually is
Mass defect $\longrightarrow$ 0.030 10 g

$\text{Binding energy} = \dfrac{0.030\ g}{\text{mole}} \times \dfrac{kg}{1000\ g} \times$

$\left(\dfrac{3.00 \times 10^8\ m}{sec}\right)^2 \times \dfrac{kJ}{1000\ J}$

$= 2.74 \times 10^9\ kJ/\text{mole}$

14.4 The band of stability curves upward because the greater the number of protons, the more neutrons necessary to prevent proton–proton repulsions and the greater the n/p ratio.

14.6 $^{35}_{14}Si \longrightarrow ^{35}_{15}P + ^{0}_{-1}e$

14.7 $^{40}_{19}K$

14.9 (a) $^{37}_{18}Ar + ^{0}_{-1}e \longrightarrow ^{37}_{17}Cl$
(b) Electron capture increases the n/p ratio. This means that the n/p ratio for $^{37}_{18}Ar$ must have been too low, placing it below the path of fully stable nuclei.
(c) $^{37}_{17}Cl$

14.10 $^{25}_{13}Al$

14.11 $^{25}_{13}Al$

14.12 Reaction from Practice Problem 14.10:
$^{25}_{13}Al + ^{0}_{-1}e \longrightarrow ^{25}_{12}Mg$
Reaction from Practice Problem 14.11:
$^{25}_{13}Al \longrightarrow ^{0}_{+1}e + ^{25}_{12}Mg$

14.14 Two B^+ emissions: $^{238}_{92}U \longrightarrow ^{0}_{+1}e + ^{238}_{91}Pa$;
$^{238}_{91}Pa \longrightarrow ^{0}_{+1}e + ^{238}_{90}Th$
One α emission: $^{238}_{92}U \longrightarrow ^4_2He + ^{234}_{90}Th$

14.15 The decay process must have been an α emission. All the other decay processes would have transformed the element into a different element that is only one space away on the periodic table.

14.16 A neutron turned into a proton; it is a β^- emission.

14.18 The source of the $^{206}_{82}Pb$ in the old rock is the decay of some of the $^{238}_{92}U$ initially present in the rock.

14.19   Percent $^{238}_{92}U$ remaining $= \dfrac{1.82\ g}{5.84\ g} \times 100\%$

$= 31.2\%$

Age of the asteroid

$= \dfrac{-2.303 \log\left(\dfrac{31.2\%}{100}\right) \times 4.46 \times 10^9\ \text{years}}{0.693}$

$= 7.50 \times 10^9\ \text{years}$

14.20 $^{239}_{94}Pu + ^1_0n \longrightarrow ^{90}_{38}Sr + ^{147}_{56}Ba + 3\ ^1_0n$

14.21 Yes; because the reaction produces more neutrons (3) than it uses up (1).

14.22 The reaction would be exothermic because the $^{90}_{38}Sr$ and $^{147}_{56}Ba$ are more stable (larger binding energy and mass defect per nucleon) than the $^{239}_{94}Pu$.

14.24 Binding energy is the energy released when an atom forms its subatomic particles. You can think of it as the energy that holds the nucleus together (the energy it would require to break the nucleus apart into its individual subatomic particles).

14.26 In general, the more nucleons a nucleus has, the greater will be its overall binding energy. However, a lighter nucleus can still be more stable than a heavier one because nuclear stability is judged on the basis of binding energy per nucleon. The example of $^{13}_6C$ and $^{12}_6C$ in the text demonstrated this.

14.28 $^{56}_{26}Fe$ has the greatest mass defect. It is the most stable of all the isotopes that exist.

14.30 $6 \text{ moles protons} \times \dfrac{1.007\ 30\ g}{\text{mole protons}} = 6.043\ 80\ g$

$8 \text{ moles neutrons} \times \dfrac{1.008\ 70\ g}{\text{mole neutrons}} = 8.069\ 60\ g$

$6 \text{ moles electrons} \times \dfrac{1.000\ 55\ g}{\text{mole electrons}} = 0.003\ 30\ g$

What the mass of 1 mole $\longrightarrow$ 14.116 70 g
of $^{14}_6C$ should be
What the mass of 1 mole $\longrightarrow$ $-14.003\ 24$ g
of $^{14}_6C$ actually is
Mass defect $\longrightarrow$ 0.113 46 g

$$\text{Binding energy} = \frac{0.113\,46\,\cancel{g}}{\text{mole}} \times \frac{\cancel{kg}}{1000\,\cancel{g}} \times$$

$$\left(\frac{3.00 \times 10^8\,\cancel{m}}{\cancel{sec}}\right)^2 \times \frac{kJ}{1000\,\cancel{J}}$$

$$= 1.02 \times 10^{10}\ \text{kJ/mole}$$

14.31 On a per nucleon basis, $^{14}_{6}\text{C}$ has a binding energy of $(1.02 \times 10^{10}$ kJ/mole$)/14$, or 7.28×10^8 kJ/mole. This would put it lower down on the binding energy axis than either $^{12}_{6}\text{C}$ or $^{13}_{6}\text{C}$.

14.33 False; all nuclei have a mass defect, even radioactive ones.

14.35 (a) The n/p ratio increases.
(b) The more positively charged protons there are in the nucleus, the more neutrons are needed to reduce the proton–proton repulsions.

14.37 $^{16}_{8}\text{O}$: n/p ratio $= 8/8 = 1.0$
$^{17}_{8}\text{O}$: n/p ratio $= 9/8 = 1.12$
$^{19}_{8}\text{O}$: n/p ratio $= 11/8 = 1.38$

14.38 The band of stability is the band of points, on a plot of number of neutrons (vertical axis) versus number of protons (horizontal axis), that represents all known stable isotopes.

14.40 It would take 8 half-lives, since 10 g $\times$ (0.5 $\times$ 0.5 $\times$ 0.5 $\times$ 0.5 $\times$ 0.5 $\times$ 0.5 $\times$ 0.5 $\times$ 0.5) $=$ 0.039 g; 8 $\times$ 13.1 hours $=$ 104 hours.

14.41 When it has less than 65 or more than 92 neutrons.

14.43 Because of how we define the word "stable"; for this plot, stable means any nonradioactive nucleus as well as any radioactive nucleus with a measureable half-life.

14.44 The n/p ratios for these nuclei are larger than for those on the path of full stability, meaning these nuclei are neutron-rich compared to the nonradioactive, fully stable nuclei.

14.46 All of their isotopes are radioactive.

14.48 $^{1}_{0}\text{n}$; $^{1}_{1}\text{p}$, $^{0}_{-1}\text{e}$, $^{0}_{+1}\text{e}$

14.50 As charges (-1 for the electron and $+1$ for the positron).

14.52 A nucleus above the path of full stability would want to convert a neutron into a proton because its n/p ratio is too high.

14.54 Upon β^- emission, a neutron converts into a proton; the atomic number increases by 1, while the mass number remains unchanged.

14.55 When a neutron converts into a proton, an electron is created and emitted as a β^- particle.

14.57 Upon positron emission, a proton converts into a neutron. This decreases the atomic number by 1 and leaves the mass number unchanged.

14.59 (a) It will be below the path of fully stable nuclei. Conversion of a proton into a neutron will raise the n/p ratio.
(b)–(c) Positron emission: $^{162}_{74}\text{W} \longrightarrow {}^{0}_{+1}\text{e} + {}^{162}_{73}\text{Ta}$
Electron capture: $^{162}_{74}\text{W} + {}^{0}_{-1}\text{e} \longrightarrow {}^{162}_{73}\text{Ta}$

14.61 Its atomic number decreases by 2, and its mass number decreases by 4.

14.63 Electromagnetic radiation (γ rays) and kinetic energy of motion associated with the ejected particles.

14.65 You would add the superscripts and subscripts on each side of the equation. For a nuclear reaction to be balanced, the sum of the superscripts and subscripts on both sides of the equation must be equal.

14.67 (a) The mass number remains the same.
(b) The mass number remains the same.
(c) The mass number remains the same.
(d) The mass number decreases by 4.

14.69 (a) α decay will result in an isotope 2 elements to the left of the parent, since the atomic number decreases by 2.
(b) β^- decay will result in an isotope 1 element to the right of the parent, since the atomic number increases by 1.
(c)–(d) β^+ decay and electron capture will result in an isotope 1 element to the left of the parent, since the atomic number decreases by 1.

14.72 $^{8}_{4}\text{Be} + {}^{0}_{-1}\text{e} \longrightarrow {}^{8}_{3}\text{Li}$; electron capture is occurring.

14.74 $^{235}_{92}\text{U} \longrightarrow {}^{4}_{2}\text{He} + {}^{231}_{90}\text{Th}$; α particle emission is occurring.

14.76 $^{0}_{-1}\text{e} + {}^{40}_{19}\text{K} \longrightarrow {}^{40}_{18}\text{Ar}$; electron capture is occurring.

14.78 No; its half-life is only 5715 years, so essentially all of it would have already decayed in 120 million years.

14.80 $\text{Age} = \dfrac{-2.303 \log\left(\dfrac{22.8\%}{100}\right) \times 5715 \text{ years}}{0.693}$

$= 1.22 \times 10^4$ years

14.81 (a) The assumption is that all the $^{206}_{82}\text{Pb}$ came from the decay of $^{238}_{92}\text{U}$.

(b) $14.90\ \cancel{g\ ^{238}_{92}U} \times \dfrac{1\ \cancel{\text{mole}\ ^{238}_{92}U}}{238.029\ \cancel{g\ ^{238}_{92}U}}$

$\times \dfrac{6.02 \times 10^{23}\ \text{atoms}\ ^{238}_{92}\text{U}}{1\ \cancel{\text{mole}\ ^{238}_{92}U}}$

$= 3.768 \times 10^{22}$ atoms $^{238}_{92}\text{U}$

$26.50\ \cancel{g\ ^{206}_{82}Pb} \times \dfrac{1\ \cancel{\text{mole}\ ^{206}_{82}Pb}}{205.974\,46\ \cancel{g\ ^{206}_{82}Pb}}$

$\times \dfrac{6.02 \times 10^{23}\ \text{atoms}\ ^{206}_{82}\text{Pb}}{1\ \cancel{\text{mole}\ ^{206}_{82}Pb}}$

$= 7.745 \times 10^{22}$ atoms of $^{206}_{82}\text{Pb}$

(c) $3.768 \times 10^{22} + 7.745 \times 10^{22} = 1.1513 \times 10^{23}$ atoms of $^{238}_{92}\text{U}$ originally present in the rock

14.81 (d) % $^{238}_{92}$U remaining

$$= \frac{3.768 \times 10^{22} \text{ atoms}}{1.1513 \times 10^{23} \text{ atoms}} \times 100\%$$

$$= 32.73\%$$

(e) Age

$$= \frac{-2.303 \log\left(\dfrac{32.73\%}{100}\right) \times (4.46 \times 10^9 \text{ years})}{0.693}$$

$$= 4.98 \times 10^9 \text{ years}$$

14.82 The mass defect on going from reactant to product nuclei must increase. This ensures that mass will be lost for the nuclear reaction, which by Einstein's equation $E = mc^2$ is converted into energy and released.

14.85 Because each fission event produces more neutrons than it consumes.

14.87 $^8_4$Be + $^4_2$He $\longrightarrow$ $^{12}_6$C + γ

14.89 For a sample of pure $^{235}_{92}$U, the critical mass is the amount necessary to produce a runaway (explosive) chain reaction.

14.90 The heat produced is used to produce steam. The steam is used to power the turbines of electric generators.

14.93 Nuclei repel each other, and large amounts of energy are needed to force the nuclei close enough together for fusion to occur.

14.95 Fusion fuel (heavy isotopes of hydrogen) is cheap and readily available; reactors would produce no nuclear waste; and there is no danger that an accident would release toxic substances into the environment.

14.96 Alpha radiation is least likely to hurt you because it is not very penetrating and can be stopped by paper, clothing, or the top layer of skin. Gamma rays are most harmful.

14.98 Radiation can deposit enough energy into a molecule to knock an electron out.

CHAPTER 15

15.2 (a)

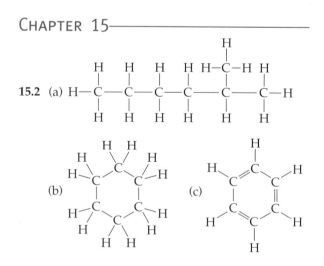

(d)

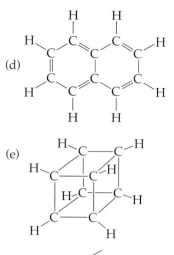

(e)

15.3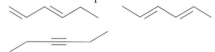

15.4 (a) Branched   (b) 8 carbons long
(c) 1 carbon long   (d) C_9H_{20}

15.6 The unsaturated hydrocarbon in Practice Problem 15.5 is missing 4 hydrogens. This could be due to the molecule having two double bonds or one triple bond.

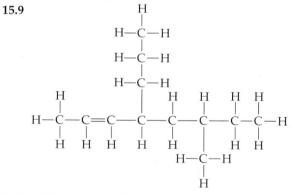

15.7 (c) is the least unsaturated, with only one double bond; it has 8 hydrogens, while the others have 6 hydrogens

15.9

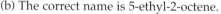

15.10 (a) It is incorrect because the main chain is the chain containing 8 carbons, not 7.
(b) The correct name is 5-ethyl-2-octene.

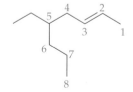

15.12 1-Butanol
15.13 2-Aminopropane
15.14 3-Bromohexane

15.15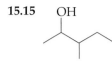

15.17 Ethanone does not exist because the carbonyl in a ketone must be connected to two other carbons. Hence, the smallest number of carbons that can exist in a ketone is 3.

15.18 (a) Ketone group   (b) Aldehyde group

(c) Carboxylic acid group

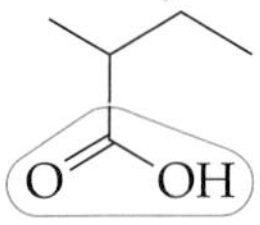

15.19 (a) 2-Butanone   (b) Pentanal
(c) 2-Methyl-butanoic acid

15.20 Catenation refers to the ability of carbon to connect and link together chains of carbon atoms.

15.22 Because carbon has four valence electrons and needs four more to form an octet; it acquires the additional four electrons by forming four covalent bonds.

15.23 (a) 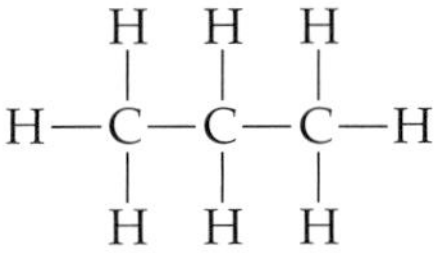 H—C—C—C—H (with H substituents)

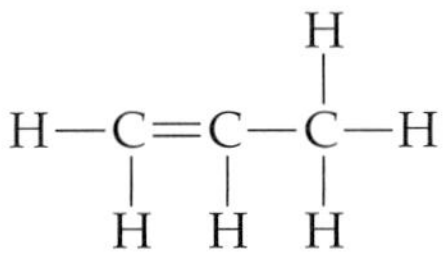

(b) H—C=C—C—H

(c) H—C≡C—C—H

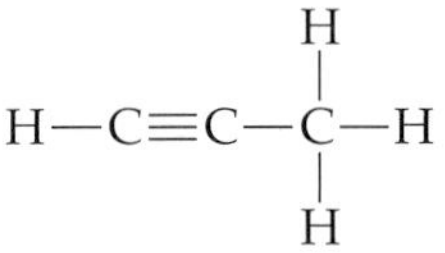

15.25 There can be no double or triple bonds in the molecule. Only carbon with all single bonds to it can be tetrahedral.

15.26 The C=H double bond is impossible; H can only form single bonds.

15.30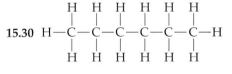

15.31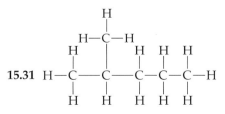

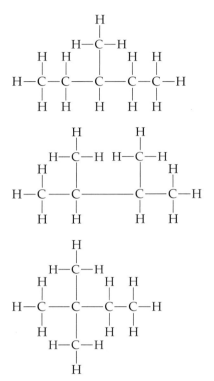

15.33 The geometry about C–C single bonds is tetrahedral, with bond angles of 109°. The zigzag line drawings more accurately represent the true shape of the molecule than straight line drawings, in which the angles appear to be 180°.

15.34 The molecule should be drawn as a straight line.

15.37 (a) 7 carbons   (b) 2 carbons

15.39 (a) C_3H_8, linear

H—C—C—C—H

(b) C_4H_{10}, linear

H—C—C—C—C—H

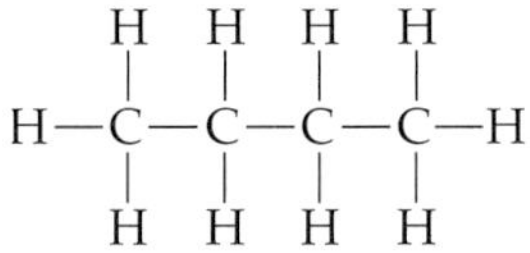

(c) C_5H_{10}, branched

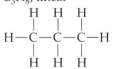

15.41 (a)  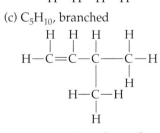   (b) 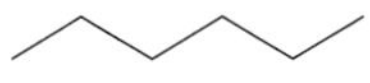   (c)

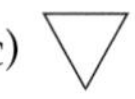

15.43 (c) is unsaturated

15.45 (a)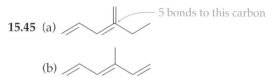
5 bonds to this carbon

(b)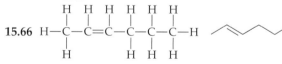

15.47 Unsaturated; the maximum number of hydrogens for a 7 carbon chain is 16. This compound has 2 fewer, and is therefore unsaturated.

15.48 (a) and (c) are isomers of each other. Each has 4 carbons and 10 hydrogens. (b) and (d) are not isomers of (a) and (c) because (b) has 5 carbons, and (d) has only 8 hydrogens.

15.51 No, because they don't have the same chemical formula.

15.53 (a) (1) and (3) are isomers
(b) They have the same formula but different structures.

15.56 Hydrocarbons with all C–C single bonds are called alkanes, hydrocarbons with C=C double bonds are called alkenes, and hydrocarbons with C≡C triple bonds are called alkynes.

15.58 A homologous series is a group of compounds that are similar to each other. For example, the alkanes are a homologous series, where each member differs from the next by a CH₂ group.

15.60

H—C—H   H—C—C—H

Methane   Ethane

H—C—C—C—H   H—C—C—C—C—H

Propane   Butane

15.63 An alkane because it fits the general formula C_nH_{2n+2}

15.66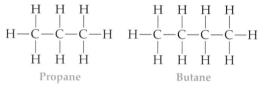

15.68 1-Hexene, 2-hexene, and 3-hexene all represent 6 carbon alkenes with the double bond in different positions. However, 4-hexene is really 2-hexene, and 5-hexene is really 1-hexene when the molecule is numbered from the correct end of the chain.

15.70 To identify and name the main chain in the molecule

15.72 (a) C_3H_6;

(b) C_8H_{16};

(c) C_6H_{14};

(d) C_4H_6;

(e) C_4H_6;

15.74 The multiple bond must be a part of the main chain, and the numbering starts at the end closer to the multiple bond. Multiple bonds have priority over branches.

15.76 If there are two methyl groups on the second carbon, as the name 2,2-dimethyl indicates, there can't be a double bond between the first and second carbons. The second carbon would have to form more than four bonds, which is impossible.

15.78 (a) 2-Methylhexane
(b) 2-Ethyl-3-methyl-1-butene
(c) 3-Ethyl-2,4-dimethylhexane
(d) 3-Propyl-1-octyne

15.80

The compound can't be named 2,2-dimethyl-5-hexene because the main chain should be numbered starting near the multiple bond. This gives the double bond the lower number.

15.82 C_4H_6;

15.83 As the alkanes get larger, the intermolecular attractive forces get progressively stronger, leading to a steady increase in the boiling point.

15.85 Replace the -ine ending of the halogen's elemental name with o, and then put it in front of the name of the parent hydrocarbon. For example, CH_3CH_2Br is called bromoethane.

15.87 Replace the last letter -e in the parent alkane with -ol. For example, CH_3CH_2OH is called ethanol.

15.88 (a) 2-Chlorohexane   (b) 2-Bromopropane
(c) 2,4-Dichloro-3-ethylhexane

15.90 (a) 2-Hexanol   (b) 3-Aminopentane
(c) 3-Hexanol

15.93 R–OH represents the alcohols. The R represents a hydrocarbon chain.

15.95 No; when the numbering of the chain is done correctly, 3-propanol is really 1-propanol.

15.96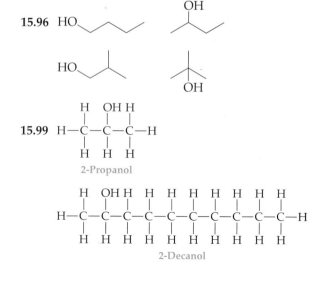

15.99

H—C—C—C—H

2-Propanol

H—C—C—C—C—C—C—C—C—C—C—H

2-Decanol

The 2-propanol is soluble in water because its nonpolar portion is small, and the polar OH group, which forms hydrogen bonds, dominates the solubility properties. However, 2-decanol has a larger nonpolar R portion that "outweighs" the smaller polar OH group, so the molecule is not soluble in water.

15.101 (a) Methyl propyl ether   (b) Dipropyl ether
(c) Ethyl propyl ether

15.103 Acetic acid;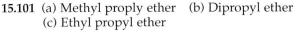

15.105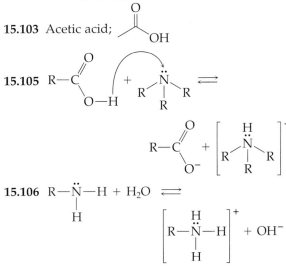

15.106

This equilibrium lies to the left, which implies that a primary amine is a weak base in water.

15.108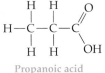

2-Aminohexane is a primary amine.

15.110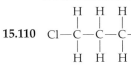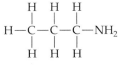

1-Chloropropane
(Halogen functionalized hydrocarbon)

1-Propanol
(Alcohol)

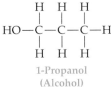 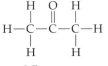

1-Aminopropane
(Amine)

2-Propanone
(Ketone)

Propanoic acid
(Carboxylic acid)

Propanal
(Aldehyde)

15.112 In vinegar

15.115 (a)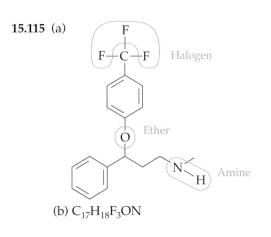

(b) $C_{17}H_{18}F_3ON$

CHAPTER 16

16.2 (a), (b)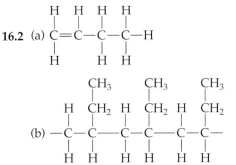

16.3 (a), (b)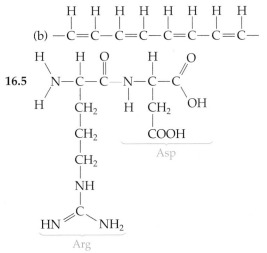

16.5

16.6 Asp-Arg-Val-Tyr-Ile-His-Pro-Phe

16.8 A macromolecule is a large molecule. Polymers are macromolecules because they are large molecules.

16.10 Because ethylene is the monomer from which the polymer is made

16.12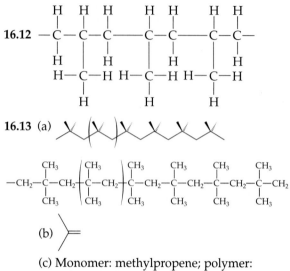

16.13 (a)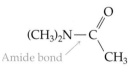

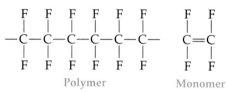

(b)

(c) Monomer: methylpropene; polymer: polymethylpropylene

(d)

16.15 Teflon is the nonstick coating used on pots and pans. It is made by the polymerization of tetra-fluoroethene.

Polymer   Monomer

16.17 Ethylene is a small molecule with very weak intermolecular attractive forces. Polyethylene is a huge macromolecular polymer, with very strong intermolecular attractive forces due to its size.

16.19 An amide bond is a bond between a carbonyl carbon and the nitrogen of an amine.

$(CH_3)_2N-C$
Amide bond CH_3
O

16.20

16.22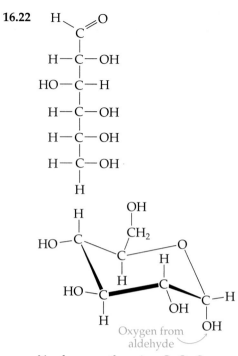

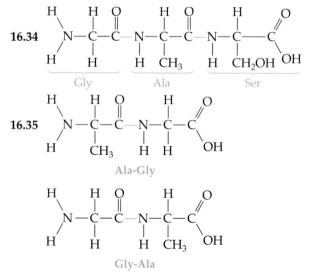

Oxygen from aldehyde

Yes, because there is a C–O–C group.

16.24 Glucose is a monosaccharide consisting of one glucose ring. Sucrose is a disaccharide consisting of one glucose ring and one fructose ring.

16.26 Yes, because most of the carbons in sugar molecules have both an H atom and an OH group attached.

16.28 You are actually burning cellulose. The combustion products are carbon dioxide and water.

16.30 Alanine has a methyl group in place of one of the hydrogens in glycine.

16.32 A peptide linkage is the name give to the

amide $\left(\begin{array}{c} O \\ \| \\ C-N \end{array} \right)$ linkage in proteins and polypeptides.

16.34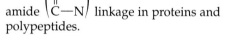

Gly   Ala   Ser

16.35

Ala-Gly

Gly-Ala

16.37 The sequence of amino acids determines the three-dimensional structure of a protein, which determines how the protein will function in the body.

16.39 The backbone of the DNA polymer is made of alternating sugar molecules (deoxyribose) and phosphates.

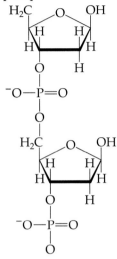

16.40 Hydrogen bonds hold the strands together. Covalent bonds would be too strong. The strands need to be able to separate for the DNA to serve as a template for protein synthesis.

16.42 Guanine to cytosine, and adenine to thymine

16.43 A codon is a sequence of 3 adjacent DNA bases that codes for the incorporation of a specific amino acid into a growing protein chain.

16.45 Viruses may allow physicians to incorporate portions of foreign DNA into a cell for the good of the cell. For example, correct genes for making insulin might be incorporated into the cells of a diabetic patient.

Index

USEFUL CONVERSION FACTORS

Length

SI unit: meter (m)

$1 \text{ km} = 0.621\,37 \text{ mi}$

$1 \text{ mi} = 5280 \text{ ft}$

$= 1.6093 \text{ km}$

$1 \text{ m} = 1.0936 \text{ yd}$

$1 \text{ in.} = 2.54 \text{ cm (exactly)}$

$1 \text{ cm} = 0.393\,70 \text{ in.}$

$1 \text{ Å} = 10^{-10} \text{ m}$

$1 \text{ nm} = 10^{-9} \text{ m}$

Mass

SI unit: kilogram (kg)

$1 \text{ kg} = 10^3 \text{ g} = 2.2046 \text{ lb}$

$1 \text{ lb} = 16 \text{ oz} = 453.6 \text{ g}$

Temperature

SI unit: Kelvin (K)

$0 \text{ K} = -273.15°\text{C}$

$= -459.67°\text{F}$

$\text{K} = °\text{C} + 273.15$

$°\text{C} = \frac{5}{9}(°\text{F} - 32)$

$°\text{F} = \frac{9}{5}(°\text{C}) + 32$

Energy

SI unit: Joule (J)

$1 \text{ J} = 0.239\,01 \text{ cal}$

$= 1 \text{ C} \times 1 \text{ V}$

$1 \text{ cal} = 4.184 \text{ J}$

$1 \text{ eV} = 1.602 \times 10^{-19} \text{ J}$

Pressure

SI unit: Pascal (Pa)

$1 \text{ Pa} = 1 \text{ kg}/(\text{m} \cdot \text{s}^2)$

$1 \text{ atm} = 101,325 \text{ Pa}$

$= 760 \text{ mm Hg (torr)}$

$= 29.9 \text{ in. Hg}$

$= 14.696 \text{ lb}/\text{in.}^2$

Volume

SI unit: cubic meter (m^3)

$1 \text{ L} = 10^{-3} \text{ m}^3$

$= 1 \text{ dm}^3$

$= 10^3 \text{ cm}^3$

$= 1.057 \text{ qt}$

$1 \text{ gal} = 4 \text{ qt}$

$= 3.7854 \text{ L}$

$1 \text{ cm}^3 = 1 \text{ mL}$

$= 10^{-6} \text{ m}^3$

$1 \text{ in.}^3 = 16.4 \text{ cm}^3$

FUNDAMENTAL CONSTANTS

Avogadro's number $= 6.02 \times 10^{23}/\text{mole}$

Gas constant, $R = 0.0821 \text{ L} \cdot \text{atm}/\text{K} \cdot \text{mole}$

Mass of electron, $m_e = 9.109\,390 \times 10^{-31} \text{ kg}$
$= 1/1836 \text{ of mass of H}$

Mass of neutron, $m_n = 1.674\,929 \times 10^{-27} \text{ kg}$
$\approx \text{mass of H}$

Mass of proton, $m_p = 1.672\,623 \times 10^{-27} \text{ kg}$
$\approx \text{mass of H}$

Planck's constant, $h = 6.626 \times 10^{-34} \text{ J} \cdot \text{s}$

Speed of light, $c = 3.00 \times 10^8 \text{ m/s}$